Nutrition and Bone Health

NUTRITION ◊ AND ◊ HEALTH

Adrianne Bendich, Series Editor

Nutrition and Bone Health, edited by **Michael F. Holick and Bess Dawson-Hughes,** 2004

Nutrition and Oral Medicine, edited by **Riva Touger-Decker, David A. Sirois, and Connie C. Mobley,** 2004

IGF, Nutrition, and Health, edited by **M. Sue Houston, Jeffrey M. P. Holly, and Eva L. Feldman,** 2004

Epilepsy and the Ketogenic Diet, edited by **Carl E. Stafstrom and Jong M. Rho,** 2004

Handbook of Drug–Nutrient Interactions, edited by **Joseph Boullata and Vincent T. Armenti,** 2004

Beverages in Nutrition and Health, edited by **Ted Wilson and Norman J. Temple,** 2004

Diet and Human Immune Function, edited by **David A. Hughes, L. Gail Darlington, and Adrianne Bendich,** 2004

Handbook of Clinical Nutrition and Aging, edited by **Connie W. Bales and Christine S. Ritchie,** 2004

Fatty Acids: Physiological and Behavioral Functions, edited by **David I. Mostofsky, Shlomo Yehuda, and Norman Salem, Jr.,** 2001

Nutrition and Health in Developing Countries, edited by **Richard D. Semba and Martin W. Bloem,** 2001

Preventive Nutrition: The Comprehensive Guide for Health Professionals, Second Edition, edited by **Adrianne Bendich and Richard J. Deckelbaum,** 2001

Nutritional Health: Strategies for Disease Prevention, edited by **Ted Wilson and Norman J. Temple,** 2001

Clinical Nutrition of the Essential Trace Elements and Minerals: The Guide for Health Professionals, edited by **John D. Bogden and Leslie M. Klevey,** 2000

Primary and Secondary Preventive Nutrition, edited by **Adrianne Bendich and Richard J. Deckelbaum,** 2000

The Management of Eating Disorders and Obesity, edited by **David J. Goldstein,** 1999

Vitamin D: Physiology, Molecular Biology, and Clinical Applications, edited by **Michael F. Holick,** 1999

Preventive Nutrition: The Comprehensive Guide for Health Professionals, edited by **Adrianne Bendich and Richard J. Deckelbaum,** 1997

NUTRITION AND BONE HEALTH

Edited by

MICHAEL F. HOLICK, PhD, MD

Department of Medicine, Dermatology, Physiology, and Biophysics, Boston University School of Medicine, Boston, MA

and

BESS DAWSON-HUGHES, MD

Human Nutrition Research Center on Aging, Tufts University, Boston, MA

Foreword by

Robert Lindsay, MBChB, PhD, FRCP

Regional Bone Center, Helen Hayes Hospital, West Haverstraw, NY

999 Riverview Drive, Suite 208
Totowa, New Jersey 07512

humanapress.com

For additional copies, pricing for bulk purchases, and/or information about other Humana titles, contact Humana at the above address or at any of the following numbers: Tel.: 973-256-1699; Fax: 973-256-8341; E-mail: humana@humanapr.com or visit our website at www.humanapress.com

Production Editor: Robin B. Weisberg.
Cover Illustration: From Fig. 1 in Chapter 13, "Nutrition and Bone Health in the Elderly," by Clifford J. Rosen.
Cover design by Patricia F. Cleary.

This publication is printed on acid-free paper. ∞
ANSI Z39.48-1984 (American National Standards Institute)
Permanence of Paper for Printed Library Materials.

Printed in the United States of America. 10 9 8 7 6 5 4 3 2 1

E-ISBN 1-59259-740-8

Library of Congress Cataloging-in-Publication Data

Nutrition and bone health / edited by Michael F. Holick and Bess Dawson-Hughes.
p. ; cm. -- (Nutrition and health)
Includes bibliographical references and index.
ISBN 1-58829-248-7 (alk. paper)
1. Bones. 2. Nutrition
[DNLM: 1. Bone and Bones--physiology. 2. Bone Development--physiology. 3. Minerals--metabolism. 4. Nutritional Requirements. 5. Osteoporosis--prevention & control. 6. Vitamins--metabolism. WE 200 N9762 2004] I. Holick, M. F. (Michael F.) II. Dawson-Hughes, Bess. III. Series: Nutrition and health (Totowa, N.J.)
>> QP88.2.N876 2004
>> 612.7'5--dc22
>> 2004005045

Series Introduction

The *Nutrition and Health* series of books have, as an overriding mission, to provide health professionals with texts that are considered essential because each includes (1) a synthesis of the state of the science; (2) timely, in-depth reviews by the leading researchers in their respective fields; (3) extensive, up-to-date, fully annotated reference lists; (4) a detailed index; (5) relevant tables and figures; (6) identification of paradigm shifts and the consequences; (7) virtually no overlap of information between chapters, but targeted, interchapter referrals; (8) suggestions of areas for future research; and (9) balanced, data-driven answers to patient /health professionals' questions that are based on the totality of evidence rather than the findings of any single study.

The series volumes are not the outcome of a symposium. Rather, each editor has been asked to examine a chosen area with a broad perspective, both in subject matter as well as in the choice of chapter authors. The international perspective, especially with regard to public health initiatives, is emphasized where appropriate. The editors, whose trainings are both research- and practice-oriented, have the opportunity to develop a primary objective for their book, define the scope and focus, and then invite the leading authorities from around the world to be part of their initiative. The authors are encouraged to provide an overview of the field, discuss their own research, and relate the research findings to potential human health consequences. Because each book is developed *de novo*, the chapters are coordinated so that the resulting volume imparts greater knowledge than the sum of the information contained in the individual chapters.

Nutrition and Bone Health, edited by Michael Holick and Bess Dawson-Hughes clearly exemplifies the goals of the *Nutrition and Health* series. Both editors are internationally recognized leaders in the field of bone biology and nutrition. Both are excellent communicators and have worked tirelessly to develop a book that is destined to be the benchmark in the field because of its extensive, in-depth chapters covering the most important aspects of the role of the diet and its nutrient components on the development, growth, maintenance, and disease prevention in bone. The editors have chosen 66 of the most well-recognized and respected authors from around the world to contribute the 36 informative chapters in the volume.

The book chapters are logically organized to provide the reader with all of the basics in both bone biology and nutrition in the first section, Basics of Nutrition and Bone Biology. Unique chapters in this section include the evolution of humans and the role of diet in the development of bone, bone physiology and the genetics of bone development. Assessment tools for both bone (with a detailed chapter that provides guidelines for use of dual energy X-ray absorptiometry equipment for accurate assessment of bone mineral density) and diet (including evaluation of food frequency, diary, and recall methodologies), as well as an in-depth analysis of the methods to measure calcium status including the newest stable isotope opportunities are provided in unique chapters. Specialized topics such as fracture healing, as well as the critical role of diet in the maintenance of healthy teeth and bones of the oral cavity, are reviewed in separate, comprehensive chapters.

Cutting edge discussions of the roles of growth factors, hormones, cellular and nuclear receptors and their ligands, gene promoters, prostaglandins, lymphocytes, muscle, adipose tissue, and all of the cells directly involved in bone biology are included in well-organized chapters that put the molecular aspects into clinical perspective. Of great importance, the editors and authors have provided chapters that balance the most technical information with discussions of their importance for clients and patients, as well as graduate and medical students, health professionals, and academicians.

Separate sections include Nutrition and Bone: Effect of Life Stages and Race, Effects of Dietary Macronutrients, Minerals and Fat-Soluble Vitamins/Micronutrients. Areas covered in these sections include separate chapters for each sex and age group (including detailed discussions of the effects of puberty, pregnancy, lactation, menopause, and male osteoporosis) and a chapter on the skeletal health of Blacks. Detailed tables and figures assist the reader in comprehending the complexities of the absorption, metabolism, and excretion of essential nutrients including, but not limited to calcium, vitamin D, potassium, phosphorus, fluoride, zinc, magnesium, copper, and boron. Novel chapters that are of great interest to clients and patients include chapters on vegetarian diets, fluoride, and lead toxicity as well as the pluses and minuses of vitamin K and vitamin A for bone health.

The last two sections of the book include chapters dealing with lifestyle effects, supplements, and diseases/disorders that result in secondary osteoporosis. Areas such as smoking, alcohol consumption, exercise, and obesity are critically reviewed. There is a separate chapter that examines the importance of evaluating the diet based on the effects of food groups, such as fruits and vegetables. The role of protein and acid–base balance in maintaining bone integrity is analyzed in separate, yet complementary chapters. Supplemental intakes of omega-3 fatty acids, soy, and other sources of phytoestrogens are reviewed with regard to the potential to reduce the risk of osteoporosis. Separate, extensively referenced chapters are provided in the areas of eating disorders, cystic fibrosis, epilepsy, and glucocorticoid-induced osteoporosis.

Hallmarks of the chapters include complete definitions of terms, with the abbreviation fully defined for the reader, and consistent use of terms between chapters. There are numerous relevant tables, graphs, and figures as well as up-to-date references; all chapters include a conclusion section that provides the highlights of major findings. The volume contains a highly annotated index and, within chapters, readers are referred to relevant information in other chapters.

This important text provides practical, data-driven resources based on the totality of the evidence to help the reader evaluate the critical role of nutrition, especially in at-risk populations, in optimizing health, and in preventing bone diseases and fractures. The overarching goal of the editors is to provide fully referenced information to health professionals so they may have a balanced perspective on the value of foods and nutrients that are routinely consumed and how these help to maintain bone health.

In conclusion, *Nutrition and Bone Health*, edited by Michael Holick and Bess Dawson-Hughes provides health professionals in many areas of research and practice with the most up-to-date, well-referenced, and easy-to-understand volume on the importance of nutrition in optimizing bone health. This volume will serve the reader as the most authoritative resource in the field to date and is a very welcome addition to the *Nutrition and Health* series.

Adrianne Bendich, PhD, FACN
Series Editor

Foreword

A rational approach to understanding the skeleton, its physiology, and pathology requires an integrated approach. Often in textbooks on metabolic bone disease, nutrition is given short shrift. In 36 chapters, *Nutrition and Bone Health* is a comprehensive review of all aspects of nutrition and the skeleton, and the interrelationships between nutrition and skeletal homeostasis. From a teleological perspective, prevention of phosphate deficiency in our saltwater ancestors represented an early environmental adaptation. The move to terrestrial hunter–gatherers changed the requirements, imposing the need to find large quantities of food of low caloric density to satisfy energy requirements. Today's challenges are different. Now a surfeit of high-calorie foods and a corollary obesity has become epidemic. In this environment are we achieving a diet that is adequate enough to allow us to build and maintain a healthy skeleton? To answer this question we need to understand the nutritional requirements of the skeleton, and how these requirements interact with, for example, genetic control of bone growth and remodeling. It is most propitious that this volume, which addresses those very issues, should be published at a time when there is much discussion about the various fad diets that potentially could modify skeletal behavior.

It has become commonplace to think of the skeleton only in terms of calcium nutrition. Indeed, calcium is stressed to be the "building block" of the skeleton and the backbone (so to speak) of all pharmacological interventions. But it is much more complex than that simple view. The skeleton is required to be strong, but flexible. It must support the everyday stresses placed on it, but must also resist sudden traumatic forces. The skeleton is also the depository for potentially harmful metals, and of course it is the major source for ions required in carefully controlled concentrations in serum. That it can achieve these three functions is one of nature's miracles, requiring a carefully controlled remodeling process to maintain its health and vigor, to provide one means for ion storage and release, and to repair stress-related damage. Because we are told that we are what we eat, it is not surprising that the skeleton requires a variety of nutrients for its own health. This remarkable volume tries to place into context the role of nutrition, both good and bad, in the overall health of the skeleton, and consequently of the organism. In putting together a unique group of internationally respected authors, the editors, themselves international experts in vitamin D and calcium homeostasis, have synthesized for the reader the wide variety of impacts that nutrition can have on the skeleton.

The achievement of adequate skeletal mass and strength requires complex interactions among genetics, health, nutrition, and physical stress during growth, and toward the end of growth, a normal transition through puberty. We often forget the many interactions required for this process. From early fetal development, nutrients supplied by the mother provide the basis for bone growth. Maternal nutrition is key here. Recent data suggest that a maternal diet inadequate in protein may result in a deficient stem cell population in the developing fetal skeleton, implying that we may have much more to learn about the role of maternal nutrition and its interactions with genetics in determining the mass of the skeleton at birth. Why might that be important? It is suggested, but by no means proven, that even early in life, a skeleton that has failed to develop adequately may set the stage for osteoporosis in later life.

The growth of the skeleton in the child is no less important. The interplay between genetics and environment creates an adult skeleton sufficient to withstand the stresses placed on it in everyday life. Big people grow big bones (simplistically put), and by the tests that we use to measure bones, have "denser" bones as an artifact of the test rather than a biological fact. Nevertheless, in the absence of proper nutrition, clearly the skeleton cannot respond to the variety of endocrine factors stimulating its growth and expansion, nor to the stress of childhood play and sport. Here again, nutrition means much more than assuring optimal calcium intake. It is not commonly recognized that the period of transition through puberty is a period of adaptation by the growing organism to increased needs to sustain the accelerated phase of growth. It is only then that the recognized gender differences in the skeleton become evident. A not infrequent disaster at this point is the appearance of an eating disorder, which can have catastrophic effects on the final maturation of the skeleton. Anorexia, coupled as it often is with failure of the hypothalamic–pituitary–ovarian axis, can lead to significant fracture risk at a young age, as the organism steals from the skeleton the essential nutrients it is failing to get otherwise.

Maintaining an adequate skeleton during adult life is equally complex. Here the effects of poor nutrition more often result in obesity, a highly prevalent feature of our adult society (and increasingly of our pediatric population). Although there is some suggestion that bone density, at least at some sites, may be increased in obese individuals, this by no means offsets the other multiple health problems that besiege the obese. Consequently, efforts at controlling weight abound, largely because of the relative ineffectiveness (usual among individuals not fully committed to the concept, but perhaps not always). Several of these will have detrimental effects on the skeleton. For example, gastric surgery and ketogenic diets will induce nutritional effects that result in excess bone loss. It is transition through menopause that alters the relationship between skeletal homeostasis and nutrition, by rendering the organism less efficient at absorbing and retaining calcium. The consequence is increased bone turnover and loss, with osteoporosis and fractures being the outcome.

At an even later stage in life, the efficiency of calcium absorption across the intestine declines and may be accompanied by vitamin D insufficiency. Secondary hyperparathyroidism ensues, with further loss of bone mass. Nutritional requirements thus change with age, with a need for higher vitamin D intakes, since skin synthesis declines at the same time. Finally, in old age, when hip fractures are common, protein nutrition assumes an important role, and it is clear that recovery from hip fracture, perhaps repair of the bone, and reduction in risk of a second hip fracture can be mediated by improved protein nutrition.

Nutrition and Bone Health, crafted by two international experts on calcium and vitamin D, brings together in one place the nutritional aspects of skeletal health and integrates them with other aspects of the control of mineral homeostasis. It begins with teleology, traverses genetics and the control of bone growth and metabolism, and includes discussion of other factors (such as medications, etc.) that might alter the nutritional requirements of bone. As the chapters unwind, they interweave the fundamental importance of good nutrition in maintaining the health of the skeleton. The editors have chosen their authors with care and have created a volume that should be read by all interested in bone health and nutrition.

Robert Lindsay, *MBChB, PhD, FRCP*
Regional Bone Center, Helen Hayes Hospital, West Haverstraw, NY

Preface

The skeleton is often perceived as an inert structure that simply acts as the scaffolding for the musculature and to house the brain and other essential organs. Thus, the skeleton is taken for granted. However, just as the intricate scaffolding of a suspension bridge requires constant maintenance, so too does the skeleton require nutritional maintenance. It has a voracious appetite for calcium and other macro- and micronutrients in order for it to maximize its size and to maintain its maximum structural strength.

The consequences of not providing the skeleton with its nutritional requirements can be quite severe. Infants and young children who do not get an adequate amount of calcium and vitamin D in their diet suffer from growth retardation and bony deformities of their skull, rib cage, arms, and legs. For adolescents and young adults, inadequate nutrition results in not being able to attain their genetically prescribed maximum peak bone mineral density. For middle-aged and older adults, inadequate calcium, vitamin D, protein, and macro- and micronutrient nutrition leads to a more rapid loss of bone that can precipitate and exacerbate osteoporosis. Twenty-five million Americans, and an equal number of Europeans and an untold hundreds of millions of adults worldwide are at risk for osteoporosis and its unfortunate consequences. In the United States, 1.5 million fractures will occur in women this year. Approximately 300,000 of these fractures will be of the hip. Twenty-five percent of women and 15% of men will suffer a hip fracture by the age of 80. It is estimated that $10 billion a year is expended for the acute and chronic care of patients suffering hip fracture. However, the most serious consequences of a hip fracture is that 50% of patients will never have the quality of life they once had and often become infirm, and 20% die within the first year after the fracture owing to complications.

The first objective of *Nutrition and Bone Health* is to provide practicing health professionals, including physicians, dietitians, nutritionists, dentists, pharmacists, health educators, policymakers, research investigators, graduate students, and medical students with comprehensive, well-balanced reviews of the newest clinical findings as well as up-to-date research discoveries regarding the role of nutrition in maintaining a healthy skeleton. It is a given that adequate calcium and vitamin D are important for skeletal health. However, the skeleton craves other nutrients that are equally essential for bone health. *Nutrition and Bone Health* includes three sections devoted to examining the effects of specific dietary components on bone health: macronutrients, minerals, and micronutrients. Additionally, dietary components such as soy and other bioactive factors from the diet are discussed in separate chapters. As examples, the importance of proper acid-based balance and the effect of minerals such as calcium, sodium, potassium, phosphorus, and magnesium as well as micronutrients including fat-soluble vitamins, zinc, and selenium are reviewed. This volume not only provides important information about which nutrients are important for bone health, but it also describes how excessive amounts of certain nutrients or ingestion of toxic substances such as lead and cadmium can negatively affect the health of the skeleton.

Nutrition and Bone Health explores how our earliest ancestors evolved in a relatively calcium-rich environment that served them well in providing a structurally sound skeleton in a hostile environment. The role of nutritional assessment, genetics, and molecular biology as they relate to nutritional requirements and bone health sets the stage for the chapters that provide up-to-date reviews of nutritional requirements during pregnancy, for fetal, neonatal, childhood, adolescent, young, middle-aged, and older adult's skeletal health presented in extensively referenced individual chapters.

Another goal of *Nutrition and Bone Health* is to put into perspective the impact of eating disorders, body weight, and body weight change on bone health. There are a multitude of drugs and other environmental and behavioral factors that negatively affect bone health. Among these are oral anticoagulants, glucocorticoid therapy, antiepileptic drugs, lead, smoking, and alcohol abuse. Another important section of the book includes detailed discussions of the consequences of diseases that either directly result in increased risk of osteoporosis or the therapies that cause secondary osteoporosis.

The heightened awareness of both the medical community and the public to the adverse effects of hormone replacement therapy has resulted in finding alternatives for not only preventing hot flashes resulting from menopause, but also the consequences of estrogen loss on the skeleton. The role of phytoestrogens as well as other dietary factors, including omega-3 fatty acids and soy protein, and exercise and sun exposure on bone cell function and bone mineral density are reviewed in detail to assure that the totality of the evidence presented to the reader provides up-to-date information on these topical, controversial subjects.

As editors, we are very excited about the contents of *Nutrition and Bone Health*. Chapters are written by experts who provide not only an overview of the subject, but also specific recommendations for how this information can be effectively utilized for practical application by health care professionals. The volume includes numerous tables and figures to help the reader quickly glean the essentials of each chapter. There is an extensive index that also helps provide a road map to easily cross-reference how particular nutrients, drugs, environmental factors, and age affect bone health.

Metabolic bone diseases, such as rickets and osteomalacia, as well as osteoporosis are diseases of neglect. Vigilance for satisfying the nutrient requirements of the skeleton is a small price to pay for remaining erect and fracture free throughout life. *Nutrition and Bone Health* should serve as a critical resource for health care professionals interested in utilizing nutrition, exercise, and other positive lifestyle factors to enhance the overall health and well-being for skeletal health throughout life.

Acknowledgments

Michael F. Holick thanks his wife Sally and children Michael and Emily for their continued support. Drs. Holick and Dawson-Hughes thank the authors of the chapters in the book for taking time out of their busy lives to write comprehensive, yet very readable reviews on subjects that affect bone health.

We also would like to acknowledge the technical assistance of Catherine St. Clair, Nancy Palermo, and Michele Wright-Nealand. In addition, the authors express their sincerest appreciation to Paul Dolgert, Editorial Director, Humana Press and Adrianne Bendich who is our Series Editor.

Michael F. Holick
Bess Dawson-Hughes

Contents

Series Introduction v
Foreword vii
Preface ix
Contributors xv

I Basics of Nutrition and Bone Biology

1 Evolutionary Aspects of Bone Health: *Development in Early Human Populations* 3
Dorothy A. Nelson, Norman J. Sauer, and Sabrina C. Agarwal

2 Genetics, Nutrition, and Bone Health 19
Serge Ferrari

3 Bone Physiology: *Bone Cells, Modeling, and Remodeling* 43
Lawrence G. Raisz

4 Interpretation of Bone Mineral Density As It Relates to Bone Health and Fracture Risk 63
Leon Lenchik, Sridhar Vatti, and Thomas C. Register

5 Importance of Nutrition in Fracture Healing 85
Sanjeev Kakar and Thomas A. Einhorn

6 Nutritional Assessment of Nutrients for Bone Health 105
Edith M. C. Lau and Winny W. Y. Lau

7 Nutritional Assessment: *Analysis of Relations Between Nutrient Factors and Bone Health* 113
John J. B. Anderson, Boyd R. Switzer, Paul Stewart, and Michael Symons

8 Nutrition and Oral Bone Status 129
Elizabeth A. Krall

II Nutrition and Bone: Effects of Life Stages and Race

9 Nutrition in Pregnancy and Lactation 139
Bonny L. Specker

10 Nutritional Requirements for Fetal and Neonatal Bone Health and Development 157
Stephanie A. Atkinson

11 Nutrition and Bone Health in Children and Adolescents 173
Velimir Matkovic, Nancy Badenhop-Stevens, Eun-Jeong Ha, Zeljka Crncevic-Orlic, and Albert Clairmont

12 Calcium and Vitamin D for Bone Health in Adults 197
Bess Dawson-Hughes

13 Nutrition and Bone Health in the Elderly 211
Clifford J. Rosen

14 Nutrition and Skeletal Health in Blacks 227
Susan S. Harris

III Effects of Dietary Macronutrients

15 Food Groups and Bone Health 235
Susan A. New

16 Vegetarianism and Bone Health in Women.................................. 249
Susan I. Barr

17 Protein Intake and Bone Health 261
Jean-Philippe Bonjour, Patrick Ammann, Thierry Chevalley, and René Rizzoli

18 Acid–Base Balance and Bone Health 279
David A. Bushinsky

IV Minerals

19 Quantitative Clinical Nutrition Approaches to the Study of Calcium and Bone Metabolism 307
Connie M. Weaver, Meryl Wastney, and Lisa A. Spence

20 Sodium, Potassium, Phosphorus, and Magnesium 327
Robert P. Heaney

21 Fluoride and Bone Health... 345
Johann D. Ringe

22 Lead Toxicity in the Skeleton and Its Role in Osteoporosis 363
J. Edward Puzas, James Campbell, Regis J. O'Keefe, and Randy N. Rosier

23 Microminerals and Bone Health 377
Steven A. Abrams and Ian J. Griffin

V Fat-Soluble Vitamins/Micronutrients

24 Vitamin A and Bone Health 391
Peter Burckhardt

25 Vitamin D 403
Michael F. Holick

26 Vitamin D Utilization in Subhuman Primates: Lessons Learned at the Los Angeles Zoo 441
John S. Adams, Rene F. Chun, Shaoxing Wu, Songyang Ren, Mercedes A. Gacad, and Hong Chen

27 Vitamin K, Oral Anticoagulants, and Bone Health 457
Sarah L. Booth and Anne M. Charette

VI Lifestyle Effects/Supplements

28 Smoking, Alcohol, and Bone Health 481
Douglas P. Kiel

29 Exercise and Bone Health 515
Maria A. Fiatarone Singh

30 Body Weight/Composition and Weight Change: *Effects on Bone Health* 549
Sue A. Shapses and Mariana Cifuentes

31 Attenuation of Osteoporosis by *n-3* Lipids and Soy Protein 575
Gabriel Fernandes

32 Phytoestrogens: *Effects on Osteoblasts, Osteoclasts, Bone Markers and Bone Mineral Density* 593
Lorraine A. Fitzpatrick

VII Secondary Osteoporosis/Diseases

33 Eating Disorders and Their Effects on Bone Health 617
Madhusmita Misra and Anne Klibanski

34 The Role of Nutrition for Bone Health in Cystic Fibrosis 635
Kimberly O. O'Brien and Michael F. Holick

35 Antiepileptic Drugs and Bone Health 647
Marielle Gascon-Barré

36 Glucocorticoid-Induced Osteoporosis 667
Barbara P. Lukert

Index 687

Contributors

STEVEN A. ABRAMS, MD • *Department of Pediatrics, Baylor College of Medicine, USDA/ARS Children's Nutrition Research Center, Texas Children's Hospital, Houston, TX*

JOHN S. ADAMS, MD • *Burns and Allen Research Institute, Division of Endocrinology, Diabetes and Metabolism Cedars-Sinai Medical Center, UCLA School of Medicine, Los Angeles, CA*

SABRINA C. AGARWAL, PhD • *McMaster University, Hamilton, ON, Canada*

PATRICK AMMANN, MD • *Service of Bone Diseases, WHO Collaborating Center for Osteoporosis Prevention, Department of Rehabilitation and Geriatrics, University Hospital, Geneva, Switzerland*

JOHN J. B. ANDERSON, PhD • *Department of Nutrition, Schools of Public Health and Medicine, University of North Carolina, Chapel Hill, NC*

STEPHANIE A. ATKINSON, PhD, RD • *Department of Pediatrics, McMaster University, Hamilton, ON, Canada*

NANCY BADENHOP-STEVENS, MS, RD • *Osteoporosis Prevention and Treatment Center & Bone and Mineral Metabolism Laboratory, Davis Medical Research Center, Ohio State University, Columbus, OH*

SUSAN I. BARR, PhD, RDN • *Department of Nutrition, University of British Columbia, Vancouver, Canada*

JEAN-PHILIPPE BONJOUR, MD • *Service of Bone Diseases, WHO Collaborating Center for Osteoporosis Prevention, Department of Rehabilitation and Geriatrics, University Hospital, Geneva, Switzerland*

SARAH L. BOOTH, PhD, *Jean Mayer USDA Human Nutrition Research Center on Aging, Tufts University, Boston, MA*

PETER BURCKHARDT, MD • *Department of Internal Medicine, University Hospital, Lausanne, Switzerland*

DAVID A. BUSHINSKY, MD • *Nephrology Unit, Departments of Medicine and of Pharmacology and Physiology, University of Rochester School of Medicine and Dentistry, Rochester, NY*

JAMES CAMPBELL, MD • *Department of Pediatrics, University of Rochester School of Medicine and Dentistry, Rochester, NY*

ANNE M. CHARETTE, MSN, *Jean Mayer USDA Human Nutrition Research Center on Aging, Tufts University, Boston, MA*

HONG CHEN, MD • *Burns and Allen Research Institute, Division of Endocrinology, Diabetes and Metabolism, Cedars-Sinai Medical Center, UCLA School of Medicine, Los Angeles, CA*

THIERRY CHEVALLEY, MD • *Service of Bone Diseases, WHO Collaborating Center for Osteoporosis Prevention, Department of Rehabilitation and Geriatrics, University Hospital, Geneva, Switzerland*

RENE F. CHUN, PhD • *Burns and Allen Research Institute, Division of Endocrinology, Diabetes and Metabolism, Cedars-Sinai Medical Center, UCLA School of Medicine, Los Angeles, CA*

MARIANA CIFUENTES, PhD • *Department of Nutritional Sciences, Rutgers University, New Brunswick, NJ and Institute of Nutrition and Technology, University of Chile, Santiago, Chile*

ALBERT CLAIRMONT, MD • *Osteoporosis Prevention and Treatment Center & Bone and Mineral Metabolism Laboratory, Davis Medical Research Center, Ohio State University, Columbus, OH*

ZELJKA CRNCEVIC-ORLIC, MD, PhD • *Department of Endocrinology, Medical Faculty University of Rijeka, Croatia*

BESS DAWSON-HUGHES, MD • *Calcium and Bone Metabolism Laboratory, Jean Mayer USDA Human Nutrition Research Center on Aging, Tufts University, Boston, MA*

THOMAS A. EINHORN, MD • *Department of Orthopaedic Surgery, Boston University Medical Center, Boston, MA*

SERGE FERRARI, MD • *Division of Bone Diseases, WHO Collaborating Center for Osteoporosis, Departments of Geriatrics and Internal Medicine, Geneva University Hospital, Geneva, Switzerland*

GABRIEL FERNANDES, PhD • *Division of Clinical Immunology, Department of Medicine, University of Texas Health Science Center at San Antonio, San Antonio, TX*

MARIA A. FIATARONE SINGH, MD, FRACP • *School of Exercise and Sport Science, University of Sydney, New South Wales, Australia and Hebrew Rehabilitation Center for Aged Research and Training Institute, Boston, MA*

LORRAINE A. FITZPATRICK, MD • *Departments of Endocrinology, Diabetes, Metabolism, Nutrition, and Internal Medicine Mayo Clinic College of Medicine, Rochester, MN*

MERCEDES A. GACAD, MS • *Burns and Allen Research Institute Division of Endocrinology, Diabetes and Metabolism, Cedars-Sinai Medical Center, UCLA School of Medicine, Los Angeles, CA*

MARIELLE GASCON-BARRÉ, PhD, MBA • *Département de pharmacologie, Faculté de médecine, University de Montreal, Centre de recherche, Centre hospitalier de l'Universite de Montreal, Hôpital Saint-Luc, Montréal, Québec, Canada*

IAN J. GRIFFIN, MB, CHB • *Department of Pediatrics, Baylor College of Medicine, USDA/ARS Children's Nutrition Research Center, Texas Children's Hospital, Houston, TX*

EUN-JEONG HA, PhD • *Osteoporosis Prevention and Treatment Center & Bone and Mineral Metabolism Laboratory, Davis Medical Research Center, Ohio State University, Columbus, OH*

SUSAN S. HARRIS, DSC • *Institute for Community Health Studies, New England Research Institutes, Watertown, MA*

ROBERT P. HEANEY, MD • *Osteoporosis Research Center, Creighton University Omaha, NE*

MICHAEL F. HOLICK, PhD, MD • *Vitamin D, Skin, and Bone Research Laboratory, Section of Endocrinology, Diabetes, and Nutrition, Department of Medicine, Dermatology, Physiology, and Biophysics, Boston University School of Medicine, Boston, MA*

SANJEEV KAKAR, MD, MRCS • *Department of Orthopaedic Surgery, Boston University Medical Center, Boston, MA*

DOUGLAS P. KIEL, MD, MPH • *Hebrew Rehabilitation Center for Aged Research and Training Institute, Beth Israel Deaconess Medical Center, Harvard Medical School, Boston, MA*

ANNE KLIBANSKI, MD • *Neuroendocrine Unit, Massachusetts General Hospital and Harvard Medical School, Boston, MA*

ELIZABETH A. KRALL, PhD, MPH • *Department of Health Policy & Health Services Research, Boston University School of Dental Medicine, Boston, MA*

EDITH M. C. LAU, MD, FRCP • *Department of Community and Family Medicine, The Chinese University of Hong Kong, China*

WINNY W. Y. LAU, RD • *Jockey Club Center for Osteoporosis Care Control, The Chinese University of Hong Kong, China*

LEON LENCHIK MD • *Department of Radiology, Wake Forest University School of Medicine, Winston-Salem, NC*

ROBERT LINDSAY, MBChB, PhD, FRCP • *Regional Bone Center, Department of Internal Medicine, Helen Hayes Hospital, West Haverstraw, NY*

BARBARA P. LUKERT, MD, FACP • *Division of Endocrinology, Metabolism, & Genetics, Department of Medicine, University of Kansas School of Medicine, Kansas City, KS*

VELIMIR MATKOVIC, MD, PhD • *Osteoporosis Prevention and Treatment Center & Bone and Mineral Metabolism Laboratory, Davis Medical Research Center, Ohio State University, Columbus, OH*

MADHUSMITA MISRA, MD • *Neuroendocrine and Pediatric Endocrine Units, Massachusetts General Hospital and Harvard Medical School, Boston, MA*

DOROTHY A. NELSON, PhD • *Department of Internal Medicine, Wayne State University, Detroit, MI*

SUSAN A. NEW, BA, MSc, PhD • *Centre for Nutrition & Food Safety, School of Biomedical & Life Sciences, University of Surrey, United Kingdom*

KIMBERLY O. O'BRIEN, PhD • *Bloomberg School of Public Health, Johns Hopkins University, Baltimore, MD*

REGIS J. O'KEEFE, MD, PhD • *Department of Orthopaedics, University of Rochester School of Medicine and Dentistry, Rochester, NY*

J. EDWARD PUZAS, PhD • *Department of Orthopaedics, University of Rochester School of Medicine and Dentistry, Rochester, NY*

LAWRENCE G. RAISZ, MD • *Center for Osteoporosis, Department of Medicine, University of Connecticut Health Center, Farmington, CT*

THOMAS C. REGISTER, PhD • *Department of Pathology, Wake Forest University School of Medicine, Winston-Salem, NC*

SONGYANG REN, MD • *Burns and Allen Research Institute, Division of Endocrinology, Diabetes, and Metabolism, Cedars-Sinai Medical Center, UCLA School of Medicine, Los Angeles, CA*

JOHANN D. RINGE, PhD • *Klinikum Leverkusen, Lehrkrankenhaus der Universität zu Koln, Leverkusen, Germany*

RENÉ RIZZOLI, MD • *Service of Bone Diseases, WHO Collaborating Center for Osteoporosis Prevention, Department of Rehabilitation and Geriatrics, University Hospital, Geneva, Switzerland*

CLIFFORD J. ROSEN, MD • *Maine Center for Osteoporosis Research and Education, St. Joseph Hospital, Bangor, ME*

RANDY N. ROSIER, MD, PhD • *Department of Orthopaedics, University of Rochester School of Medicine and Dentistry, Rochester, NY*

NORMAN J. SAUER, PhD • *Michigan State University, East Lansing, MI*

SUE A. SHAPSES, PhD • *Department of Nutritional Sciences, Rutgers University, New Brunswick, NJ*

BONNY L. SPECKER, PhD • *Ethel Austin Martin Program in Human Nutrition, South Dakota State University, Brookings, SD*

LISA A. SPENCE, PhD, RD • *Food and Nutrition Department, National Dairy Council, Rosemont, IL*

PAUL STEWART, PhD • *Department of Nutrition, Schools of Public Health and Medicine, University of North Carolina, Chapel Hill, NC*

BOYD R. SWITZER, PhD • *Department of Nutrition, Schools of Public Health and Medicine, University of North Carolina, Chapel Hill, NC*

MICHAEL SYMONS, PhD • *Department of Nutrition, Schools of Public Health and Medicine, University of North Carolina, Chapel Hill, NC*

SRIDHAR VATTI, MD • *Department of Radiology, Wake Forest University School of Medicine, Winston-Salem, NC*

MERYL WASTNEY, PhD • *Metabolic Modeling Services, Ltd., Hamilton, New Zealand*

CONNIE M. WEAVER, PhD • *Department of Foods and Nutrition, Purdue University, West Lafayette, IN*

SHAOXING WU, MD • *Burns and Allen Research Institute, Division of Endocrinology, Diabetes, and Metabolism, Cedars-Sinai Medical Center, UCLA School of Medicine, Los Angeles, CA*

I

Basics of Nutrition and Bone Biology

1 Evolutionary Aspects of Bone Health

Development in Early Human Populations

Dorothy A. Nelson, Norman J. Sauer, and Sabrina C. Agarwal

1. INTRODUCTION

The skeleton serves two primary functions: it provides biomechanical support and protection of soft tissue; and it plays a key role in mineral homeostasis. Skeletal health can be affected by a number of factors, including genetics, lifestyle, demographic characteristics, and disease. Skeletal size, strength, and structure can be affected by diet and physical activity, age, body size, ethnicity, and health status. In living persons, most of these factors can be assessed to some extent, and changes can be monitored in individuals over time. Techniques such as bone densitometry, assessment of biochemical markers of bone remodeling, radiography, bone biopsy, and others can be used in the assessment of skeletal status. In contrast, investigations of skeletal health in past populations are limited to various physical characteristics that happen to be preserved at a moment in time for each individual specimen or local population.

For the purposes of examining evolutionary aspects of bone health, it is fortunate that bones (and teeth) are typically preserved in the fossil record. Certain artifacts of culture may also be present in hominid (human) fossil sites, and these provide further information about adaptation. Some of the techniques used for assessing skeletal status in the living, such as bone densitometry and histomorphometry, can also be used in skeletal remains. However, it is impossible to obtain dynamic or longitudinal measurements of physiological processes, or to assess diet and physical activity accurately. Fortunately, anthropological techniques have been developed that allow us to create reasonable models of life and health in past human populations. In this work we will examine aspects of our evolutionary past that may have affected skeletal health in human ancestors and formed the basis for observed skeletal conditions among modern human populations. In order to place bone health in evolutionary perspective, an overview of evolutionary principles and stages will be presented.

From: *Nutrition and Bone Health*
Edited by: M. F. Holick and B. Dawson-Hughes © Humana Press Inc., Totowa, NJ

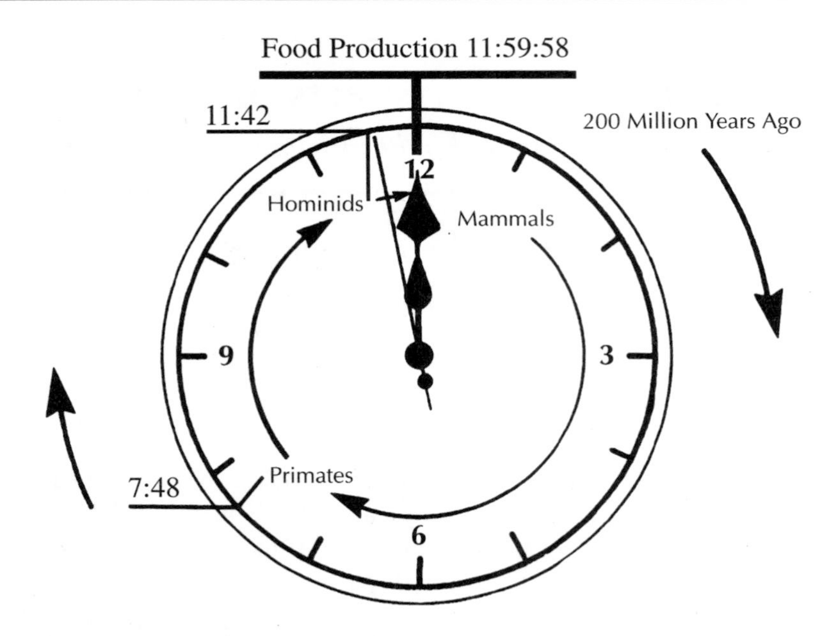

Fig. 1. Analog of 200 million years of evolution depicted on a 12-h clock. (Reprinted with permission from ref. *1,* Fig. 1, p. 326, © Springer-Verlag.)

2. AN OVERVIEW OF EVOLUTION

The period during which our human ancestors evolved is miniscule in relation to the evolutionary record of all living things. The first, simple organism is thought to have appeared around 3.5 billion years ago; evidence of the first vertebrates dates to about 500 million years ago; and, finally, mammals are found in the fossil record some 200 million years ago. The animal order to which we belong, the Primates, appeared approximately 70 million years ago. If we were to fit 200 million years of mammalian evolution into a 12-h clock (Fig. 1), the earliest members of the Primates appear at about 7:48 *(1)*. The first hominids, or members of the human family, appear at 11:42; and food production develops in the final 2 or 3 s. Thus, people are relative newcomers on the earth when compared to other organisms, but the speed with which the species changed is unique among animals.

2.1. Evolutionary Mechanisms

It is important to point out that evolution is a process without direction, and does not imply progress. Thus, no organism is "more evolved" than any other organism. Evolution occurs though genetic change in response to environmental pressures, as well as through relatively random processes. Evolution can be defined as a change in the frequencies of genes in a population over time. It occurs through natural selection, gene flow, genetic drift, and mutation.

Mutations are constant and random in species, providing the genetic raw material through which evolution works. Natural selection favors organisms, in any given environment, that have certain genetic characteristics that allow them to adapt well enough to reproduce successfully. Those individuals or populations that leave more offspring than others will leave their genetic conformation more frequently as well. With each generation, if the environment changes, gene frequencies will change depending on the nature of selective pressures. Microevolution is the change in gene frequencies from one generation to the next. Macroevolution refers to more notable changes over a longer period of time, such as the appearance of new species.

The final two evolutionary forces, gene flow and genetic drift, redistribute existing genes independent of selection and the environment. Gene flow acts through the exchange of genes between populations, affecting the distribution of genes by equalizing frequencies among groups. The higher the rate of gene flow between two populations, the more alike they are. It is gene flow over large spans of time that prevents a species from diversifying and splitting into two. Conversely, the absence of gene flow, isolation, is a critical feature in the formation of new species.

Genetic drift is due to chance. Especially when populations are small, sampling error accounts for random fluctuations in gene frequencies through time. An important example of genetic drift is the founder effect. When a small population splits from a larger one and founds a new group, sampling error dictates that the new group will carry gene frequencies that differ randomly from (and are not representative of) the parent population. Thus, the newly formed population will have a unique set of gene frequencies due entirely to chance factors. Genetic drift is the main reason why certain genetic diseases are unexpectedly high among some populations, such as the Amish in the United States.

2.2. Culture and Adaptation

Our species, *Homo sapiens,* evolved through a combination of these mechanisms just as other species have done. One factor that sets us apart is our reliance on culture, or the set of beliefs, habits, and material goods through which we also adapt to our environment. To some extent, culture interferes with natural selection, but it could also be viewed as one of the environmental forces to which people must adapt. The distinctive feature of human adaptation is the combination of biology and culture that characterizes human evolutionary change.

2.3. Bone Health and Adaptation

The subject of bone health in this chapter will be discussed in the context of adaptation to the environments in which people evolved. The need for specific skeletal characteristics must have changed often as early hominids experienced major shifts in adaptation. Such characteristics include the size, shape, and density

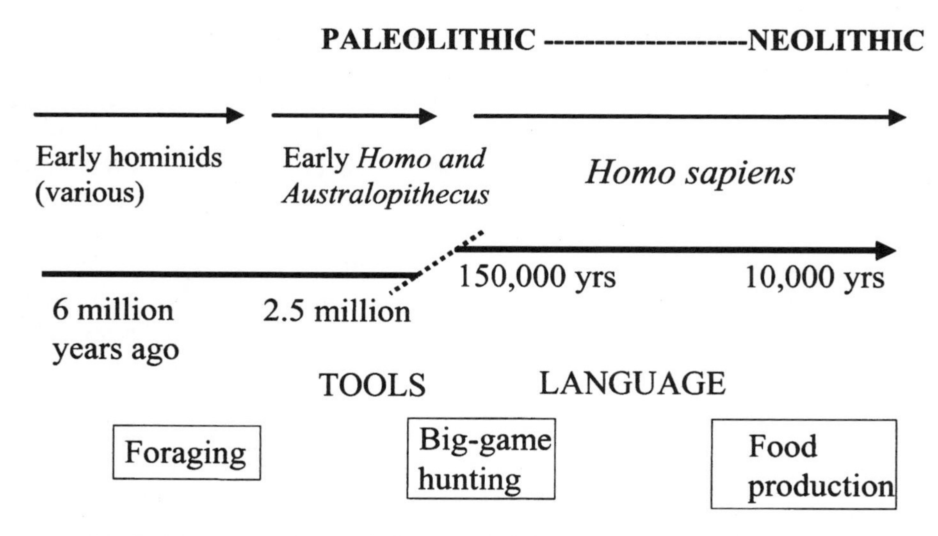

Fig. 2. Time line of events in human evolution over the past 6 million years.

of the skeleton throughout life. Major shifts in adaptation during human evolution include the following:

1. Expansion from the tropics to a wide range of environments
2. Transition from hunting and gathering to food production
3. Change from physically active lifestyles to relative sedentism
4. Increase in life expectancy

These four areas will form the focal points for discussion of bone health over the course of human evolution.

3. THE COURSE OF HUMAN EVOLUTION

It will be helpful to outline the major events in human evolution for the nonanthropologist reader (Fig. 2). Sometime around 6 or 7 million years ago, the hominid clade (the evolutionary line leading to humans) arose when a group of chimpanzee-like creatures began adapting to a terrestrial bipedal niche. Rapidly accumulating fossil evidence for the earliest hominid seems to push this event farther back in time. At the time of the writing of this draft, two candidates, *Orrorin tugenensis* from Kenya and *Ardipithecus ramidus* from Ethiopia, date to about 6 and 4.4 million years, respectively. This is very close to the time of hominid origins suggested by genetic evidence, and there is debate whether *Orrorin* might be a common ancestor to apes as well as to humans *(2,3)*.

The period from the origin of the hominid line to the first evidence for the genus *Homo* was characterized by enormous diversity. Still confined to Africa, hominids were separated into perhaps 4 genera and as many as 10 species, each characterized by unique skeletal and dental features. Probably all primarily vegetarians, their brains were about one-third the size of ours; their molar teeth had about four times the surface area; and forward-jutting faces protruded from great brow ridges. These early members of our evolutionary line probably looked more like upright chimpanzees than modern-day humans, and almost certainly had a diet rich in a variety of plant foods, such as leaves, fruits, seeds, and tubers. Studies of chimpanzees in the wild demonstrate that these primate relatives also occasionally hunt small or young animals as well.

Then, about 2.5 million years ago, a new hominid arose with noticeably smaller teeth, brains about two-thirds the size of ours, and greater stature. Still divided into several species and still confined to Africa, the genus *Homo* is associated with the earliest stone tools. The appearance of stone tools in the archeological record marks the beginning of the "Paleolithic" (Old Stone Age) period (Fig. 2). It is likely that the early *Homo* groups regularly hunted for, or at least scavenged, meat. One species of an earlier genus, *Australopithecus,* lived alongside the genus *Homo,* but by about 1.5 million years ago it died out. Eventually, the early humans spread throughout the continent of Africa, and by about 1.7 million years ago *Homo erectus* migrated north to Eastern Europe and Asia. Evidence for this spread includes not only hominid fossils but also evidence of stone tools and living sites. *Homo erectus* also used fire, apparently for cooking or to keep warm. Scientists debate whether *Homo erectus* or some related species made its way into Western Europe and gave rise to the Neanderthals *(4).*

For many decades, the origin of our species, *Homo sapiens,* has been one of the most hotly debated issues in hominid paleontology. At center stage are the Neanderthals of Western Europe. Having been characterized as brutish, thick-boned, and primitive, some authorities *(5)* assign them to their own species, *Homo neanderthalensis,* and relegate them to an evolutionary dead end. According to the "out of Africa" model, anatomically modern *Homo sapiens* arose in Africa sometime between 150,000 and 200,000 years ago and quickly (in evolutionary terms) spread throughout the Old World, replacing the Neanderthals and any other hominid species they came in contact with. In Asia and Africa, for example, *Homo sapiens* would replace the species *Homo erectus.* The new, less robust species, with smaller teeth and greater stature, soon developed more advanced tools, cave art, and presumably language.

Not all scientists agree on a draconian fate for the Neanderthals. An alternative view, the "multiregional" model, places the Neanderthals and their contemporaries in Asia and Africa, directly in the human evolutionary line *(3).* In fact, proponents of the multiregional model include the Neanderthals in the human species and consider them direct ancestors of modern Europeans. Citing evidence for regional anatomical continuity, these scientists argue that modern *Homo sapiens* evolved from earlier forms in Eastern Europe, Asia, and Africa. Critical to this approach is the understanding that gene flow was sufficient throughout the species range to prevent the diversification necessary for separate species to arise. Whichever hypothesis one supports, virtually everyone agrees that by about 30,000 years ago, a single hominid species, *Homo sapiens,* looking much like we do today, ranged throughout most of the Old World.

This evolutionary period also provides evidence of big-game hunting: the tools that could be used for killing game and preparing carcasses for use; cave art depicting large game animals, indicating their importance to the early cultures; and language, which would have been helpful if not critical in developing cooperative approaches to big-game hunting. The addition of meat to the diet on a regular basis, and the addition of fire to the food-processing regimen, must have dramatically altered the human diet. Dentition steadily reduced in size over evolutionary time as stone tools replaced teeth as tools, and as the texture and type of foods required less vigorous mastication. It can be inferred that bone and mineral metabolism would also have changed in response to these changes in diet and activity.

4. EXPANSION FROM THE TROPICS TO A WIDE RANGE OF ENVIRONMENTS

As discussed above, fossil evidence indicates that our earliest hominid ancestors originated in Africa some 6–7 million years ago. Reconstructions of the physical environment indicate that they lived in subtropical climates, probably in a mixture of grassland/savanna and woodlands. At first, they may have scavenged sources of meat protein or procured sick or young animals to supplement their diet of roots, tubers, seeds, fruits, and other wild plants. There are nonhuman primate examples of this adaptation among baboons and chimpanzees, both of which have been observed to hunt opportunistically but otherwise to subsist mainly on a wide variety of vegetarian food. It can be inferred from this reconstruction of diet that the earliest hominids consumed more calcium than modern humans, given the almost 10-fold higher calcium content in a given unit of wild plant compared with wild game *(6).* The dietary calcium intake levels presumably dropped as *Homo* developed hunting tools and skills and incorporated more meat in their diets.

This decrease in calcium content of the diet most likely accelerated as our ancestors migrated into the northern climates of the Pleistocene, about 1.7 million years ago, and relied even more heavily on hunting as a means of subsistence *(4).* If studies of more recent Arctic populations (Inuit) are relevant to these Ice Age hunters, there is some evidence that a large meat component in the diet may contribute to bone loss *(7).* Skeletal calcium may be resorbed to buffer the effect of the acid load contained in animal proteins *(8).* Additionally, calcium may be bound in the kidney by sulfates and phosphates produced by protein metabolism *(9).* While high-protein diets have been suggested to reduce calcium availability, the influence of high protein intake on bone mineral and bone metabolism is controversial *(10).* For example, some studies show no differences in bone mass with high-protein diets *(11,12).* However, in modern populations, a cross-cultural association has been reported between higher intakes of dietary animal protein and higher hip fractures in an analysis of data from 16 countries *(13).* This offers support for the speculation that our prehistoric ancestors' skeletons may have been adversely affected by an increased reliance on game animals and the associated increase in dietary protein.

The transition to subarctic life and reliance on big-game hunting may have been accompanied by another factor affecting calcium metabolism—decreasing exposure to ultraviolet radiation in northern latitudes, where solar radiation is weaker. Exposure to sunlight may have been reduced even further by the need to wear clothes to keep warm. It is generally assumed that light-colored skin evolved to optimize the synthesis of vitamin D *(14)*. Presumably our earliest ancestors, exposed to high levels of ultraviolet radiation, had dark skin to protect them from the adverse effects of too much sun exposure. However, dark skin would have been maladaptive in northern latitudes, where ultraviolet radiation was weaker, and the colder climate required some type of clothing for warmth (further limiting sun exposure). Under these conditions, it may have been impossible to make enough vitamin D in the skin to allow optimal calcium absorption from the gut—especially when dietary calcium intake may have decreased. Thus, it is assumed that members of *Homo* who lived in subarctic regions must have evolved, through natural selection, fairer skin to allow adequate vitamin D production. It is possible that this adaptation was sufficient in populations that rarely lived past middle age, but not adequate for individuals who lived long enough to experience the well-known degenerative effects of aging on the gut and on nutrient absorption in particular. In modern populations, a high prevalence of vitamin D deficiency has been recognized in elderly populations *(15)*, and this may contribute to the risk of osteoporosis.

5. TRANSITION FROM HUNTING AND GATHERING TO FOOD PRODUCTION

Several studies of past populations suggest that low bone mass was not a problem in human populations until the transition from hunting–gathering to food production (coinciding with the "Neolithic" cultural period) some 10,000–12,000 years ago *(16)*. Factors such as a high infant and childhood mortality rate and a high incidence of injury deaths contributed to the lower life expectancy among prehistoric, technologically simple societies relying on gathering and hunting wild foods. In contrast, in early agricultural societies, infectious disease became a significant factor in limiting life expectancy. Such conditions existed partly because of larger, more sedentary populations, increased interpersonal contact, the accumulation of garbage and contaminants, and the domestication of animals. These problems were often associated with poor nutrition and periods of low calorie intake or starvation. Not surprisingly, some studies of prehistoric populations have found a lower bone mass among transitional agriculturalists compared with hunter–gatherers *(16,17)*.

Various indicators of bone quantity and mass have been measured in skeletal remains of past populations, including cortical thickness, cortical area, bone mineral content, and histomorphometry. Studies of archeological populations are limited by the relative imprecision with which age, sex, and other relevant characteristics can be ascribed to individual skeletons. Reconstructions of past lifeways, including dietary adaptations and physical activity levels, are based on assumptions, fragmentary data, and inherent methodological errors. Bone can also be modified by its burial environment, and such biological and chemical diagenetic changes can affect the reliability of analyses. This is of particular concern with studies that rely on the use of noninvasive methods such as absorptiometry to assess bone mass *(18)*. Furthermore, age- and sex-related changes in bone quality and its role in bone fragility in the past have not been widely considered in archeological populations *(18)*. However, with these caveats in mind, it is still possible to summarize some of the current knowledge gained from studies of bone maintenance in skeletal collections.

Some prehistoric groups appear to have low bone mass in comparison to other groups (either prehistoric or modern) with better overall health indicators. Studies that found a relatively low bone mass in past populations implicate such factors as chronic malnutrition associated with early agricultural adaptations, such as in Nubia (approx 350 BC to 1450 AD) *(19–21)*, and in eastern and southwestern North America (from 2000 BC to the contact period) *(16)*. For example, Nelson *(17)* reported that hunter–gatherers from 6000 years ago in the American midwest had thicker cortices, higher bone mass (measured by single-photon absorptiometry), and better maintenance of bone in late adulthood compared with maize agriculturalists from the same region several millennia later. Ericksen *(22)* also suggested that nutrition was an important determinant of bone loss in her comparative analysis of age-related changes in Eskimo, Pueblo, and Arikara archeological populations. The author found radiographically measured medial-lateral cortical thinning of the humerus and femur to be most pronounced in the Pueblo sample, which relied primarily on a cereal-based diet *(22)*. Ericksen *(23)* also found differences in bone remodeling (based on density of osteons per unit area) between the groups that she suggests reflect dietary differences, as well as differences in physical activity, between the groups. She specifically implicates the high-protein diet of the Eskimo, and the low-protein diet of the sedentary Pueblo, in her explanation of the differences in their remodeling parameters, and a subsequent study of intracortical remodeling by Richman et al. *(24)* of the same skeletal material supports these findings.

Low bone mass has also been reported for some Arctic groups with an unusually heavy intake of animal protein *(7)*. For example, an early comparison of long bone density in U.S. blacks, U.S. whites, and Sadlermiut Inuit (AD 1500–1900), found older Sadlermiut adults to have the earliest and highest loss of bone *(25,26)*. Bone core studies of various archeological Inuit skeletons, when compared to U.S. whites, also show thinner cortices, lower bone mineral content, and increased secondary osteonal remodeling suggestive of an increase in intracortical porosity and subsequent bone loss *(27–29)*.

In summary, low bone mass has been found in some past populations from a variety of geographic regions, representing either early agriculturalists or Arctic hunters. Clearly, these are not just the ancestors of groups currently considered to have the highest risk of osteoporosis *(16),* suggesting a significant contribution from environmental and/or cultural factors. Nutritional models are most commonly used to explain low bone mass in past populations, with or without a physical activity (i.e., biomechanical) component. For example, Ruff et al. have suggested that past populations of agriculturalists were less physically active than hunter–gatherers *(30).* However, the types and intensity of physical activity were almost certainly an important factor as well *(31).*

Despite the findings of low bone mass in some past populations, there is little evidence of osteoporotic fractures in these groups *(18).* The low prevalence of fragility fracture in archeological samples may in part be explained as the result of mortality bias and inaccurate age-at-death estimates. For example, while the low prevalence of age-related or fragility fracture in some past populations may mean that fracture was rare in the past as compared to modern populations, it could also reflect heterogeneity in the oldest age groups, whereby the oldest individuals in skeletal samples may not be developing fragility fractures because they represent an overall "healthier stock" that managed to survive into old age. This is particularly important to consider when comparing old-age individuals in the past and the present, as present-day elderly individuals have benefited from modern medicine and may not be comparable to their historical counterparts.

Perhaps a more concerning problem with using archeological skeletal samples is age-at-death estimations *(32–34).* It is increasingly evident that while some humans in the historic past did likely manage to live into old age, we cannot accurately ascribe age to skeletons older than around 50 years of age. The conservative approach in osteological studies has been to assign only broad age groups with a final open-end age group of, for example, 45 or 50+, to skeletons. However, it has been suggested that if we cannot break down our age estimation after 50 into finer groups, we may not be able to adequately study the rates of degenerative or age-related conditions *(32).* While this may hold true when looking exclusively at age-related bone loss and osteoporosis, certainly the use of broad age categories is still likely adequate to discern broad changes and patterns of bone maintenance in females that are related to menopause. It has also been suggested that bone quality may have been protected, perhaps through the effect of physical activity on bone architecture, thereby reducing the likelihood of fractures due to bone fragility *(18).* Physical activity levels in past populations were almost certainly higher than those of modern populations, were probably high in both sexes, and were probably maintained at a relatively high level throughout the life span.

5.1. Dietary Calcium Intake in Evolutionary Perspective

It follows from the above discussion that the sources and amounts of dietary calcium (and other relevant nutrients) changed over the time period during which our human ancestors evolved. It has been estimated that the dietary intake of Paleolithic populations was at least 1500 mg/d *(6),* which is two or three times

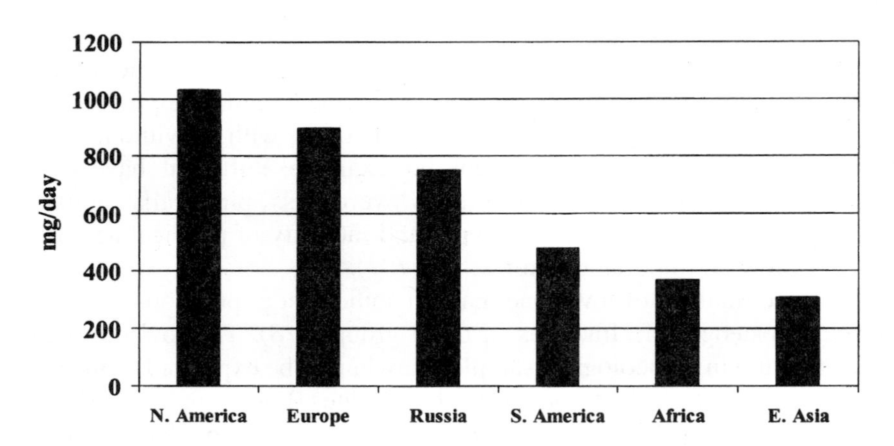

Fig. 3. Estimated average calcium intake (mg/d) in selected geographic regions in 1990 *(35)*.

more than the typical U.S. diet affords. However, calcium intake was only one factor affecting skeletal health over the course of human evolution. The interaction of this nutrient with other dietary components, physical activity levels, exposure to solar radiation, longevity, and general health must be considered in the context of the various biocultural environments in which people lived.

It is clear that a dramatic decline in dietary calcium occurred in our recent evolutionary past with the advent of agriculture *(1)*. Cultivated foods (grains) have a much lower calcium content than uncultivated plant foods *(6)*. Eaton and Nelson *(6)* reported that, on average, cereal grains contain 29 mg of calcium per 100 g of grain, compared with nearly 133 mg/100 g in uncultivated plant sources. Furthermore, grains generally have an undesirable calcium/phosphorus ratio, and may contain phytate (which binds to calcium and reduces its availability). In the modern world, there is a wider variety of foods available, and the dietary intake of nutrients varies widely. Data from the FAO Yearbook, 1990 *(35)*, indicate that dietary calcium intakes in the late 1980s ranged from 300 to 500 mg/d in Asia, Africa, and Latin America to 900–1000 mg/d in some North American and European populations (Fig. 3). This continuum does not necessarily correspond with the prevalence of osteoporosis around the world. Cooper et al. *(36)* estimated that in 1990, half of all hip fractures worldwide occurred in North America and Europe, although this is expected to change as life expectancy increases in the developing countries. Explanations for the apparent paradox that higher dietary calcium intake is associated with more hip fractures include higher protein intakes, and poorer vitamin D status, in Western countries *(8)*. Thus, it is clear that calcium intake must be considered within the context of other factors. Even within a population, subgroups may have differing dietary profiles. For example, nutrient patterns by tertiles of calcium intake were studied in a group of 957 men and women ages 50–79 residing in a community in southern California *(37)*. In both men and women, intakes of protein, vitamin D, magnesium, and phosphorus were significantly higher in the high-calcium tertile *(37)*, providing a complex of nutrients that might affect the skeleton differently from the other two groups. Other lifestyle factors such as physical activity would interact with dietary habits in their effect on the skeleton. Clearly, human behavioral and dietary plasticity have allowed our species to flourish in a wide range of environments, over a wide range of calcium intakes.

6. CHANGE FROM PHYSICALLY ACTIVE LIFESTYLES TO RELATIVE SEDENTISM

6.1. Physical Activity in Prehistoric Times

Perhaps the most dramatic transition in the prehistory of the genus *Homo* was the shift from a hunting and gathering economy to one based primarily on plant domestication. Known as the Neolithic transition in the Old World, about 10,000–12,000 years ago in places as widespread as the Middle East, the Indian subcontintent, and China, human populations gave up their dependence on game and collected wild plant food and adopted agriculture. Soon after, evidence of settled village life appears in the archeological record. Similar events occurred more recently in North and South America.

Intuitively, one might assume that a shift away from life as a nomadic hunter, constantly in search of game for sustenance, to a more predictable dependence on domesticated plants and animals would be accompanied by improved overall health, reduced physical stress, and perhaps better bone quality. Evidence from past skeletal populations, however, reveals a very different scenario. A number of studies confirm that, compared to hunter–gatherers, early agricultural groups suffered increased levels of biological stress, poorer nutrition, and (presumably due to crowded conditions, exposure to contaminants, and association with domesticated animals), elevated levels of infectious disease *(38)*.

Research into the effects of a dependence on plant domestication on human bone strength and activity patterns is inconsistent. From computed tomography (CT) scan cross-sectional analyses of femora and humeri in samples from the Georgia coast, Larsen reports that male and female bone strength was greater and life more mechanically demanding in hunter–gatherers than it was in an agricultural sample *(38)*. In her study of skeletal populations from northwest Alabama, however, Bridges *(39)* reports greater femur strength in male and greater femur and humerus strength in female agriculturalists compared to hunter–gatherers, concluding that "agriculture was more physically demanding than hunting-and-gathering in that region, and that, . . . females took over the majority of the chores" (*40,* p. 116). According to Bridges *(40)*, similar disparities are found in Old World populations as well. Cross-sectional CT scan analysis suggests that bone strength is related to multiple factors and that regional differences, such as terrain, may play a role.

6.2. Physical Activity in Historic Times

Physical activity can play an important role in the severity of osteoporosis, affecting both the achievement of peak bone mass and the subsequent rate of bone loss and deterioration of bone quality. The reduction in habitual physical activity in modern Western populations has been suggested as a primary explanation for the increasing incidence in osteoporotic fracture *(41,42)*.

Previous studies of bone mineral density in archeological skeletons, particularly from historical times, have suggested that physically active lifestyles may have reduced bone loss in the past. For example, a study by Lees et al. *(43)* of femoral bone density in female archeological remains from Spitalfields, England, dated between 1729 and 1852, found no evidence of premenopausal bone loss and less severe postmenopausal loss compared to modern females, which they suggest to be the result of physical activity and possibly unidentified environmental factors. Another study of bone mineral density by Ekenman et al. *(44)* of medieval skeletons from Stockholm, dated between 1300 and 1530 AD, found an absence of low bone density in older age groups, and a higher diaphyseal bone density in the lower extremities as compared to modern reference values, which they also suggest could be the result of environmental factors and physical demands, such as walking and standing.

7. INCREASE IN LIFE EXPECTANCY

The expansion of the lifespan past reproductive age is a relatively recent phenomenon among humans, and is uncommon among free-living nonhuman primates *(45)*. Data from living hunting–gathering groups studied in the past century indicate that life expectancy at birth in these groups was, on average, roughly 20–40 years—much shorter than among people living in technologically advanced modern cultures *(46)*. However, there is some evidence that early agricultural populations had a lower mean age at death than hunter–gatherers, although this may be related to higher birth rates and not higher mortality *(47)*. Some estimates suggest that the average lifespan has tripled since prehistoric times *(48)*. Rapid increases in life expectancy at birth that began in the early 20th century were due largely to drops in mortality among infants and children *(48)*. In the case of females, life expectancy is related not only to infant mortality, but also to risks associated with childbirth. Furthermore, there is no reason to believe that human *longevity* has changed over time, and there is evidence that people did indeed live into old age, at least in historical periods *(32)*. Jackes *(32)* suggests that estimates of a 10% survival beyond age 60 would actually be conservative, highlighting the demographic data of Russell *(49)*, which notes that a number of individuals were expected to live beyond 60 across Europe and North Africa in the first 1500 years AD, and the work of Sjovold *(50)*, who notes a significant number of deaths between the ages of 70 and 80 in an Austrian village in the 250 years prior to 1852. In the past few decades, however, increases in life expectancy have been due largely to reduced death rates among the elderly *(48)*.

Despite the fact that life expectancy was shorter in prehistoric populations, and in hunter–gatherers (until very recently), age-related bone loss has been documented in some prehistoric populations *(16,18)*. However, it is rarely associated with osteoporotic fractures, in part because not many individuals lived long enough to be affected. However, it should also be noted that even in modern times, fracture risk is not tied exclusively to life expectancy. Today, there is a secular trend whereby the increment in the population over the age of 80 has and will continue to rise exponentially as compared to the overall population growth *(41)*. However, the change in demographics does not account entirely for the present increased incidence of several types of fragility fracture. For example, Kanis *(41)* notes that hip fracture incidence in Oxford, England, doubled in the 27 years since the 1950s, and similar increases have been documented in other parts of the world. Clearly, life expectancy is not the only factor involved in the increasing incidence of osteoporosis. There is some evidence to suggest that bone quality may have been maintained better in past populations than in modern populations experiencing low bone mass after mid-adulthood *(18)*. It is conceivable that sustained physical activity levels throughout life may have contributed to the maintenance of bone quality.

Another factor that may have contributed to better bone maintenance in past populations relates to the socioeconomic status of elders in egalitarian (hunter–gatherer) societies. Older individuals held a respected position within their group, and were usually supported by their families except in extreme conditions of food scarcity *(51)*. Thus, they could expect to enjoy relatively good general health and nutrition status into older age. This is not always the case with the elderly in modern Western society, where longevity coupled with worsening health and nutrition contribute to osteoporosis.

One paradox of longevity that is, in large part, the result of modern medical care is that it allows chronic disease to develop and be maintained in modern societies. When captive nonhuman primates live to old age, unlike their wild relatives, they also develop diseases of aging such as degenerative joint disease, atherosclerosis, and osteoporosis *(52)*. Thus, one of the disadvantages of an increased lifespan in human society is that it must bear the socioeconomic cost of diseases associated with aging. These diseases become more prevalent as populations age, as with osteoporosis, which is recognized as a public health burden that could soon reach critical proportions.

From an evolutionary perspective, the increasing prevalence of osteoporosis suggests that this condition either does not impact reproductive success (i.e., it is not subject to natural selection), or that the importance of some other, related characteristic was greater than the "cost" of osteoporosis. Longevity itself may contribute to the reproductive success of individuals or populations, perhaps through the contribution of elders in a society. For example, Hawkes et al. *(53)* propose that older members of a population, and grandmothers in particular, make important contributions to the survival and reproductive success of their lineal descendants.

Although osteoporosis is a debilitating condition that could reduce some individuals' ability to help younger generations, it could be offset by the contributions of the unaffected individuals.

In summary, an increased life expectancy in the modern human species is associated with an increase in osteoporosis. According to Stanley Garn's classic study *(54),* bone loss after middle age is a universal phenomenon in the human species, and this has been corroborated by numerous studies since then. Because bone loss has also been observed in captive primates with a long lifespan, it seems likely that human ancestors might have faced the problem of osteoporosis if they had lived longer. Poor nutrition and decreased physical activity among the modern elderly population almost certainly contribute to the prevalence of osteoporosis, in contrast with the apparent continuation of skeletal health in past populations whose few elderly continued to enjoy good nutrition and adequate activity. Despite the well-known burdens of osteoporosis, individual and societal, longevity itself may be more beneficial to society overall by allowing elders to contribute to the success of the next generations.

8. CONCLUSION

In our relatively short existence on Earth, our species has undergone dramatic changes in adaptation. These include worldwide expansion into diverse environments, the development of food production, changes in physical activity types and levels, and an increasing life expectancy. All of these changes can be related in some way(s) to the modern problem of osteoporosis. In evolutionary perspective, the advantages of these changes for our species must have outweighed the potential, as well as the actual, disadvantages. However, skeletal health in modern populations appears to be at increasingly greater risk from modern lifestyles and environments. An understanding of our evolutionary past can hold some important lessons and provide insight into safeguarding this aspect of health as we move into the new millennium.

REFERENCES

1. Nelson DA. An anthropological perspective on optimizing calcium consumption for the prevention of osteoporosis. Osteopor Int 1996; 6:325–328.
2. Relethford JH. The Human Species, 5th ed. McGraw-Hill, New York, 2003.
3. Wolpoff MH. Paleoanthropology, 2nd ed. McGraw-Hill, New York, 1999.
4. Jurmain R, Kilgore L, Trevathan W, Nelson H. Introduction to Physical Anthropology, 9th ed. Wadsworth, Belmont, CA, 2003.
5. Stringer CB. The emergence of modern humans. Sci Am 1990; 12:98–104.
6. Eaton SB, Nelson DA. Calcium in evolutionary perspective. Am J Clin Nutr 1991; 54:281S–287S.
7. Mazess RB, Mather W. Bone mineral content of North Alaskan Eskimos. Am J Clin Nutr 1974; 27:916–925.
8. FAO/WHO Expert Group: Joint FAO/WHO expert consultation on human vitamin and mineral requirements, Chapter 11, Calcium, 2002. www.fao.org/docrep/004/y2809e/y2809e0h.htm.

9. Schuette SA, Hegsted M, Zemel MB, Linkswiler HM. Renal acid, urinary cyclic AMP, and hydroxyproline excretion as affected by level of protein, sulfur amino acid, and phosphorus intake. J Nutr 1981; 111:2106–2116.
10. Orwoll ES. The effects of dietary protein insufficiency and excess on skeletal health. Bone 1992; 13:343–350.
11. Bell NH, Shary J, Stevens J, Garza M, Gordon L, Edwards J. Demonstration that bone mass is greater in black than in white children. J Bone Miner Res 1991; 6:719–723.
12. Grynpas M. Age and disease-related changes in the mineral of bone. Calcif Tissue Int 1993; 53(suppl 1):S57–S64.
13. Abelow BJ, Holford TR, Insogna KL. Cross-cultural association between dietary animal protein and hip fracture: a hypothesis. Calcif Tissue Int 1992; 50:14–18.
14. Jablonski NG, Chaplin G. The evolution of human skin coloration. J Hum Evol 2000; 39:57–106.
15. Eriksen EF, Glerup H. Vitamin D deficiency and aging: implications for general health and osteoporosis. Biogerontology 2002; 3:73–77.
16. Pfeiffer SK, Lazenby RA. Low bone mass in past and present aboriginal populations. In: Draper HH, ed. Advances in Nutritional Research, Vol. 9. Plenum, New York, 1994, pp. 35–51.
17. Nelson DA. Bone density in three archaeological populations. Am J Phys Anthropol 1984; 63:198.
18. Agarwal SC, Grynpas MD. Bone quantity and quality in past populations. Anat Rec 1996; 246:423–432.
19. Martin DL, Armelagos GJ. Morphometrics of compact bone: an example from Sudanese Nubia. Am J Phys Anthropol 1979; 51:571–578.
20. Martin DL, Armelagos GJ, Goodman AH, Van Gerven DP. The effects of socioeconomic change in prehistoric Africa: Sudanese Nubia as a case study. In: Cohen MN, Armelagos GJ, eds. Paleopathology at the Origins of Agriculture. Academic, New York, 1984, pp. 193–214.
21. Martin DL, Armelagos GJ. Skeletal remodelling and mineralization as indicators of health: an example from prehistoric Sudanese Nubia. J Hum Evol 1985; 14:527–537.
22. Ericksen MF. Cortical bone loss with age in three native American populations. Am J Phys Anthropol 1976; 45:443–452.
23. Ericksen MF. Patterns of microscopic bone remodelling in three aboriginal American populations. In: Brownman DL, ed. Early Native Americans: Prehistoric Demography, Economy, and Technology. Houton, The Hague, 1980, pp. 239–270.
24. Richman EA, Ortner DJ, Schulter-Ellis FP. Differences in intracortical bone remodeling in three aboriginal American populations: possible dietary factors. Calcif Tissue Int 1979; 28:209–214.
25. Mazess RB. Bone density in Sadlermiut Eskimo. Hum Biol 1966; 38:42–48.
26. Mazess RB, Jones R. Weight and density of Sadlermiut Eskimo long bones. Hum Biol 1972; 44:537–548.
27. Thompson DD, Guinness-Hey M. Bone mineral-osteon analysis of Yupik-Inupiaq skeletons. Am J Phys Anthropol 1981; 55:1–7.
28. Thompson DD, Posner AS, Laughlin WS, Blumenthal NC. Comparison of bone apatite in osteoporotic and normal Eskimos. Calcif Tissue Int 1983; 35:392–393.
29. Thompson DD, Salter EM, Laughlin WS. Bone core analysis of Baffin Island skeletons. Arctic Anthropol 1981; 18:87–96.
30. Ruff CB, Larsen CS, Hayes WC. Structural changes in the femur with the transition to agriculture on the Georgia coast. Am J Phys Anthropol 1984; 64:125–136.
31. Bridges PS. Bone cortical area in the evaluation of nutrition and activity levels. Am J Hum Biol 1989; 1:785–792.
32. Jackes M. Building the bases for paleodemographic analyses: adult age determination. In: Katzenberg MA, Saunders SR, eds. Biological Anthropology of the Human Skeleton. Wiley Liss, New York, 2000, pp. 417–466.
33. Milner GR, Wood JW, Boldsen JL. Paleodemography. In: Katzenberg MA, Saunders SR, eds. Biological Anthropology of the Human Skeleton. Wiley Liss, New York, 2000, pp. 467–497.

34. Saunders SR, Hoppa RD. Growth deficit in survivors and non-survivors: biological mortality bias in subadult skeletal samples. Yrbk Phys Anthropol 1993; 36:127–151.
35. FAO of the United Nations. Production Yearbook Vol. 44. FAO, Rome, 1991.
36. Cooper C, Campion G, Melton LJ III. Hip fractures in the elderly: a worldwide projection. Osteoporos Int 1992; 2:285–289.
37. Holbrook TL, Barrett-Connor E. Calcium intake: covariates and confounders. Am J Clin Nutr 1991; 53:741–744.
38. Larsen CS. Bioarchaeology: Interpreting Behavior from the Human Skeleton. Cambridge University Press, Cambridge, UK, 1977.
39. Bridges PS. Skeletal evidence of changes in subsistence activities between the Archaic and Mississippian time periods in northwestern Alabama. In: Powell ML, Bridges PS, Mires AMW, eds. What Mean These Bones: Studies in Southeastern Bioarchaeology. University of Alabama Press, Tuscaloosa, AL, 1991, pp. 89–101.
40. Bridges PS. Skeletal biology and behavior in ancient humans. Evol Biol 1995; 4:112–120.
41. Kanis JA. Osteoporosis. Blackwell Science, Oxford, UK, 1994.
42. Mosekilde L. Osteoporosis and exercise. Bone 1995; 17:193–195.
43. Lees B, Molleson T, Arnett TR, Stevenson JC. Differences in proximal femur bone density over two centuries. Lancet 1993; 341:673–675.
44. Ekenman I, Eriksson SA, Lindgren JU. Bone density in medieval skeletons. Calcif Tissue Int 1995; 56:355–358.
45. Katz SH, Armstrong DF. Cousin marriage and the X-chromosome: evolution of longevity and language. In: Crews DE, Garruto RM, eds. Biological Anthropology and Aging. Oxford University Press, New York, 1994, pp. 101–123.
46. Cohen MN. Health and the Rise of Civilization. Yale University Press, New Haven, CT, 1989.
47. Larsen CS. Biological changes in human populations with agriculture. Ann Rev Anthropol 1995; 24:185–213.
48. Wilmoth JR. Demography of longevity: past, present, and future trends. Exp Gerontol 2000; 35:1111–1129.
49. Russell JC. The Control of Late Ancient and Medieval Populations. American Philosophical Society, Philadelphia, 1985.
50. Sjovold T. Inference concerning the age distribution of skeletal populations and some consequences for paleodemography. Anthrop Kozl 1978; 22:99–114.
51. Klinghardt G. Hunter-gatherers in southern Africa. Iziko Museums of Capetown. 2001. www.museums.org.za/sam/resource/arch/hunters.htm.
52. Sumner DR, Morbeck ME, Lobick JJ. Apparent age-related bone loss among adult female Gombe chimpanzees. Am J Phys Anthropol 1989; 79:25–234.
53. Hawkes K, O'Connell JF, Blurton-Jones NG. Hadza women's time allocation, offspring provisioning, and the evolution of long postmenopausal life spans. Curr Anthropol 1997; 38:551–577.
54. Garn SM. The Earlier Gain and the Later Loss of Cortical Bone in Nutritional Perspective. Charles C Thomas, Springfield, IL, 1970.

2 Genetics, Nutrition, and Bone Health

Serge Ferrari

Positive health requires a knowledge of man's primary constitution and of the powers of various foods, both those natural to them and those resulting from human skill. But eating alone is not enough for health. There must also be exercise, of which the effects must likewise be known.

— *Hippocrates, 480* BC

1. INTRODUCTION

Nutritional, lifestyle, and genetic factors all influence bone mass development during growth *(1)*. Heritability explains up to 80% of the population variance for peak bone mass, and the influence of these genetic factors is expressed well before puberty *(2)*. Genome-wide linkage studies in humans and mice have started to reveal the multitude of quantitative trait loci (QTLs) potentially contributing to bone mass, although most of the specific gene variants that influence discrete traits for bone strength, such as bone size, cortical thickness, and trabecular architecture, remain to be identified *(3)*. Despite this strong genetic determination, genotypes associated with bone mineral density (BMD) in a given cohort have not necessarily been found to be associated with BMD in other cohorts with a similar genetic background but with different nutrient intakes and/or lifestyle factors. Moreover, numerous intervention trials using calcium supplements and/or dairy food products have proven beneficial to improve bone mass gain in children, particularly in those with a spontaneously inadequate calcium intake *(4)*. Thus, the major genetic determination of peak bone mass does not preclude nutrients to modify the "tracking" of bone mass during growth. In fact, there are clear suggestions that nutritional and genetic factors may interact to influence bone modeling, that is, changes in BMD, bone size, and architecture, and mineral homeostasis during the years of peak bone mass acquisition *(5)*. Likewise, gene–environment interactions have been found to influence bone remodeling and the maintenance of bone mass in postmenopausal women *(6)*. These observations open the way for a novel approach in osteoporosis prevention, based on

From: *Nutrition and Bone Health*
Edited by: M. F. Holick and B. Dawson-Hughes

genetic profiling associated with the response to nutritional factors and drugs (i.e., nutrigenetics and pharmacogenetics).

Over the past few years, hundreds of studies have been published in the field of osteoporosis genetics. Many of these studies have led to inconsistent results, in part because areal bone mineral density (aBMD), the most commonly used trait in both linkage and association studies, and to an even greater extent fracture, are complex phenotypes. Fracture in particular is a rare and stochastic event that depends on one side on bone strength *(7)*, and on another side on a number of extrinsic factors related to failure load, such as the propensity to falls, protecting responses, soft tissue padding, etc. *(8)*. The latter factors may have their own heritable and nonheritable components *(9)*. Not surprisingly, then, the apparent contribution of genetic factors to the liability for fracture is less than 30%, as compared to 60–80% for aBMD *(3)*. In order to understand the genetic basis of bone health, and ultimately of osteoporotic fractures, a more precise definition and evaluation of the discrete phenotypes contributing to bone strength is required. Since osteoporosis is a disorder characterized by low bone mass and microarchitectural deterioration of bone tissue, one eventually needs to identify the specific genes associated with both quantitative and "qualitative" bone traits, including volumetric bone density, bone size, geometry, microarchitecture (such as trabecular connectivity), bone turnover, and material properties (including microdamage and collagen cross-linking) *(10)*. In addition, genetic influences on endocrine and paracrine pathways for calcium and phosphate homeostasis, bone formation (osteoblastic function) and resorption (osteoclastic function), as well as on indirect determinants of bone strength, such as body weight, height, lean mass, and muscle strength *(11)*, should all be explored. By using such "proximal phenotypes," that is, by dissecting the multiple causes of a complex disease such as osteoporosis into observably distinct traits that can be mapped individually, stronger signals will be obtained in future genetic studies *(12)*.

2. INHERITANCE AND HERITABILITY OF BONE MASS AND STRENGTH

It has been shown that daughters of osteoporotic women have low BMD *(13)* and that both women and men with a family history of osteoporosis have significantly decreased BMD compared to subjects without such history *(14,15)*. BMD was also found to be decreased among relatives of middle-aged men with severe idiopathic osteoporosis *(16)*. These studies clearly suggest that the risk of osteoporosis is at least partly inherited in both genders. However, daughters of women with vertebral fractures have a BMD deficit at the spine that is already half the deficit of their mothers, whereas daughters of women with hip fractures have only a small BMD deficit at the femur neck *(17)*. These observations suggest that the influence of genetic factors on peak bone mass might affect the risk of osteoporotic fractures differentially at spine and hip, i.e., would be more pronounced for vertebral fractures, whereas hip fracture risk would be more affected by age-related changes in BMD *(17)*.

Heritability (h^2, %) is defined as the proportion of the total variance for a trait across the population that is attributable to the average effects of genes *(18)*. By comparing within-pairs correlations for BMD between monozygotic (MZ) twins, who by essence share 100% of their genes, and dizygotic (DZ) twins, who have 50% of their genes in common, genetic factors have been estimated to account for as much as 80% of the population variance for lumbar spine and proximal femur BMD *(19)*. However, a bias in the estimate of heritability using twins as well as other familial models can be introduced by underestimating environmental sources of covariance *(20,21)*, which may then inflate the apparent additive genetic effects on bone mass. Some studies also suggest that additive genetic covariance exists between bone mass and lean body mass, i.e., that bone and muscle share some common (genetic) determinants *(10)*. This finding is consistent with the influence of mechanical loading on bone structure, particularly cortical thickness, and also with the fact that some gene products may be implicated in the regulation of both bone and muscle metabolism, such as the vitamin D receptor (VDR) (see below). In contrast, the Sydney Twin Study of Osteoporosis has found that the 80% and 65% of variance for lean mass and fat mass, respectively, that was attributable to genetic factors had only little influence on BMD at the lumbar spine or femoral neck *(22)*. Hence, it remains presently unclear whether the relation between lean mass and bone mass is most significantly determined by common environmental or genetic influences. Nevertheless, these observations suggest that the influence of some genes on bone mass might further depend on lifestyle factors, particularly the level of physical exercise, which will at first allow particular gene variants to expressed their differences at the muscle level.

In addition to BMD, twins studies have also reported heritability estimates ranging from 55% to 82% using quantitative ultrasound to evaluate bone properties in the phalanges and/or calcaneum, which were similar to or just slightly lower than BMD values estimated by dual-energy X-ray absorptiometry (DXA) at the lumbar spine and femoral neck *(23,24)*. Moreover, cross-trait correlations suggest that specific genes unrelated to BMD explain at least half of the heritability of the skeletal phenotype(s) measured by ultrasound *(24)*. The major difficulty in interpreting these results comes from our poor current understanding of the actual bone properties measured by ultrasounds, besides BMD. Several parameters of bone geometry, such as femoral cross-sectional area, femoral axis length, and the height and width of vertebral bodies, have also been shown to have a major genetic determination using twins and/or sib-pairs analysis *(3)*.

Parents–offspring correlations for BMD have also been significant, albeit heritability estimates have been somewhat lower in this case (in the range of 50–60%) compared to the twins model *(25,26)*. The heritability of aBMD, bone mineral content (BMC), volumetric bone mineral density (vBMD), and bone area in the lumbar spine and femur (neck, trochanter, and diaphysis) has also been evaluated in 8-yr-old prepubertal girls and their premenopausal mothers *(27)*. In this study, regressions were adjusted for height, weight, and calcium intake in order to minimize the contribution of genetic covariance and shared nutritional factors. Despite great disparities in the maturity of the various bone traits before puberty, ranging from only 30% for BMC to nearly 100% for vBMD in children compared to their mothers' (peak) values, heredity by maternal descent was detectable at all skeletal sites and affected all traits, including bone size and vBMD (h^2 range: 52–76%). Moreover, all bone parameters were reevaluated 2 yr later in the female children, showing high correlation with prepubertal values (all $r > 0.80$) and similar heritability estimates in mothers and daughters as evaluated earlier, despite considerable increase in bone mass in children during this period. These results indicate that a major proportion of the population variance for peak bone mass is explained by genetic influences that are already expressed before puberty, and suggest "tracking" of the genotypic values (i.e., the mean phenotype observed among individuals with a given genotype) during growth. Concerning vBMD in particular, taking together the evidence of a significant heritability in prepubertal children and the constancy of vBMD measurements during growth, it appears that genetic determination might be already established before birth *(17)*. The prepubertal and early pubertal expression of genes accounting for a vast proportion of peak bone mass variance has been independently confirmed using both parents–offspring *(28)* and the twins model *(21)*. The latter study as well as another analysis of variance components for BMD in nuclear families *(29)*, however, suggested that heritability estimates could further increase (up to 84%) after peak bone mass is achieved. This might be explained by a few specific genes being expressed after puberty, perhaps through some interactions with nutritional/lifestyle factors that are specific to young adults.

Information concerning bone mass heritability by paternal descent is scarce. Overall, father–son correlations for BMD at various skeletal sites appear to be slightly lower than mother–daughter correlations *(26,30)*. In addition, mother–son correlations for bone density are similar or lower compared to mother–daughter estimates, whereas father–daughter correlations appear to be the lowest. Altogether, these data suggest a predominant effect on bone mass of genes inherited by maternal descent, which might at least partly be explained by gene imprinting effects and/or by interactions with maternal environmental factors *in utero*.

In contrast to the clear heritability of peak bone mass, the contribution of genetic factors to the population variance for bone turnover and age-related bone loss remains unclear. A small twins study including both females and males suggested better correlations for bone loss among MZ compared to DZ twins *(18)*, whereas another small study in male twins did not find a significant heritability for bone loss *(31)*. The heritability of bone turnover markers seems to be lower in postmenopausal compared to younger women, i.e., in the range of 30% *(32)*. Moreover, BMD heritability estimates between mothers and daughters are lower in post- compared to premenopausal daughters *(33)*. On another side, the age at which cessation of the ovarian function occurs, obviously a major determinant of osteoporosis risk in later years, seems to be genetically determined (h^2, 63%) *(34)*. Even if the overall genetic determination of bone remodeling in later years is less prominent than it is on bone modeling and peak bone mass acquisition, it does not preclude some specific gene variants to play an important role in modulating bone turnover and bone loss, particularly as a result of interactions with hormonal, nutritional, and other lifestyle factors (see below). These gene variants may either be the same that are associated with peak BMD or different genes, mostly associated with bone mass and strength in aging subjects.

3. QUANTITATIVE TRAIT LOCI FOR BONE MASS

The actual number of genes contributing to bone health and, conversely, osteoporosis risk, is currently unknown. It has been hypothesized that the determination of bone mass involves dozens of genes with relatively small additive effects, so-called modulator genes, and a few genes with rather large effects *(35,36)*. Recent segregation studies in populations defined by a very homogeneous genetic background and environment indeed suggest that analytical models accounting for a major gene effect could be the most appropriate to describe BMD heritability *(37–39)*. Genome-wide screening approaches search for loci flanked by DNA microsatellite markers that co-segregate with the phenotype of interest in a population of related individuals. These pedigrees can be constituted by large kindreds and sibships with extreme phenotypes (such as high or low bone mass) or from the population at large *(36.)* Although linkage studies for bone mass and other determinants of bone strength in both humans and mice have identified a large number of QTLs linked to these traits *(3)*, located on virtually every chromosome, some of these QTLs have shown better consistency across various populations, suggesting that gene(s) in those loci could be major contributors to the population variance for bone mineral density (Table 1). In addition, many other QTLs for spine and femur bone geometry have been identified *(3,40)*, as well as for bone ultrasound properties *(41)*. A major advantage of genome-wide scanning is that it makes no assumptions about the genes potentially governing the trait, which can potentially lead to the identification of novel, previously unsuspected, genes contributing to bone mass and strength. Mapping novel gene(s) in the QTLs thus identified may in turn have a major impact on our understanding of the pathophysiology of osteoporosis and other skeletal disorders, such as bone dysplasias. A recent example is the linkage of three Mendelian skeletal disorders, i.e., the autosomal recessive syndrome of juvenile-onset osteoporosis-pseudoglioma (OPPG), familial high bone mass (HBM), and autosomal dominant osteopetrosis Type I (ADOI) to chromosome 11q12-13 *(42–44)*. Later on, genes mapping in this region and coding for low density lipoprotein-receptor related protein 5 *(LRP5)* and the osteoclast-specific subunit of the vacuolar proton pump, *ATP6i* (*TCIRG1* gene), respectively, have been identified to be responsible for the OPPG and HBM syndromes *(LRP5)*

Table 1
Quantitative Trait Loci (QTLs) for Bone Mineral Density (BMD) in Humans

QTL	*Koller (115)*	*Devoto (116)*	*Karasik (117)*	*Deng (118)*	*Wilson (119)*	*Candidate Genes*
1p36		FN			WB	*MTHFR, TNFRSF1B**
1q21–23	LS					*IL6R, BGLAP*
2p23–24		LS				
3p22–14				DR	LS	*PTHR1*
4q32		FN		LS, DR		*PDGF(C), NPY1R, IL-15*
5q33–35	FN					*IL-4, GR*
6p11–12	LS					
6p21.2			FN, LS			*HLA DRB1, BMP6, TNFA*
7p22				LS		*IL6, TWIST*
8q24			Ward			*TNFRSF11b*[§]
9p24				DR		
10q26				FN		*FGFR2*
11q12–13	FN, LS					*LRP5, TCIRG*[¶]
12q24			LS	LS		*IGF-1*
13q33–34				(LS)		*COL4A1, A2*
14q31–34	FT		LS			*TSHR, TRAF3*[#]
16p12–q23			(Ward)		LS	*IL4R, MMP2, CDH11*[†]
17p13				DR (FN)		
21q22.2-qter			FT			*COL6A1*
22q12–13	LS					

QTLs with LOD scores for linkage >1.8 from 5 independent genome-wide screening studies in Caucasians are summarized. In bold, QTLs that have been identified in more than one study. In parentheses, LODs < 1.75. Candidate osteoporosis genes mapped near the identified QTLs are also mentioned. LS, lumbar spine; FN, femoral neck; FT, trochanter; WB, whole body; DR, distal radius; Ward, Ward's area.

* TNF receptor superfamily/1β (TNF receptor 2); [§]osteoprotegerin (OPG); [¶]osteoclast-specific subunit of the vacuolar proton pump, ATP6i; [#]TNF receptor-associated factor 3; [†]osteoblast-cadherin (cadherin-11).

(45–47) and for malignant autosomal recessive osteopetrosis *(TCIRG1) (48)*. Since the 11q12-13 locus has also been linked to femur and spine BMD in pairs of healthy Caucasian-American sisters *(49)* (Table 1) as well as with hand BMD in Russians *(50)*, the *LRP5* and *TCIRG1* genes have become obvious candidates for bone mass determination in the population. A first study suggests that genetic variation in *TCIRG1* may not be associated with peak bone mass in healthy premenopausal women *(51)*. In contrast, preliminary studies suggest that *LRP5* polymorphisms might actually be associated with lumbar spine BMD, size, and the risk of osteoporosis in men.

The discrepancy of QTLs mapping across several studies (Table 1) illustrates some of the limitations of genome scanning for bone mass in humans. On one side, there is the limited power of this analysis, which may require thousands of individuals to achieve the required statistical power, unless sib pairs extremely discordant for the trait or large pedigrees are gathered *(36)*. On another side, it is limited by the density of microsatellites markers used, typically distant 10 or more centi-Morgans (cM) (≥10 mio base pairs). The QTLs thus identified may contain hundreds of genes, whereas chromosomal regions that have not been formally identified by linkage (logorithm of the odds [LODs] below 1.8) may still harbor clear candidate genes for osteoporosis (Table 2). One study in 115 probands with osteoporosis and 499 of their relatives somewhat circumvented the problem of leaving obvious candidate genes unidentified by using a limited number of microsatellites in the vicinity of specific genes implicated in the control of BMD and/or bone metabolism *(52)*. The candidate genes studied coded for structural components, such as type I collagen A1 and A2, type II collagen A1, fibrillin type 1, and osteopontin; for growth factors and cytokines, such as colony-stimulating factor 1, epidermal growth factor, interleukin (IL)-1α, IL-4, IL-6, IL-11, transforming growth factor-β_1, tumor necrosis factors-α and -β; and for components of endocrine systems, such as androgen receptor, VDR, calcium-sensing receptor, estrogen receptor-α (ESR1), insulin-like growth factor (IGF)-1, parathyroid hormone (PTH), PTH-related protein, and PTH receptor type 1. The strongest linkage with BMD was detected with the PTH receptor type 1 gene, whereas ESR1 gene and the IL-6 gene were among the few other loci to be significantly associated with BMD, although with lower LOD scores.

4. POPULATION-BASED ASSOCIATION STUDIES

Population-based association studies have mostly tested the relationship between BMD and/or bone turnover markers in unrelated individuals and polymorphic candidate genes coding for bone structural molecules, hormones, and/or their receptors implicated in calcium/phosphate and bone metabolism, cytokines involved in bone remodeling and, more recently, transcription factors (Table 2) *(53)*. Association studies based on one or a few single nucleotide polymorphisms (SNPs) have their limitations, because they are by definition unable to identify new susceptibility genes for osteoporosis; true associations may be missed because of

Table 2
Candidate Genes in Osteoporosis Association Studies

Protein	*Candidate Gene*	*Chromosome*
Receptors		
Vitamin D	*VDR*	12q13
Estrogen	*ESR1* (α)	6q25.1
	ESR2 (β)	14q23
Parathyroid hormone	*PTH1R*	3p22-21.1
Calcitonin	*CALCR*	7q21.3
Calcium-sensing	*CASR*	3q21-24
Androgen	*AR*	Xq11.2-q12
Glucocorticoid	*GR*	5q31
Tumor necrosis factor (receptor 2)	*TNFRSF 1b*	1p36
Osteoprotegerin	*TNFRSF 11b*	8q24
Growth factors and cytokines		
Transforming growth factor-β	*TGFB1*	19q13.2
Interleukin-6	*IL6*	7p21
Insulin-like growth factor 1	*IGF1*	12q22-23
Interleukin-1 receptor antagonist	*IL1RN*	2q14.2
Tumor necrosis factor-α	*TNFA*	6p21
Enzymes		
Aromatase	*CYP19*	15q21.1
Methylenetetrahydrofolate reductase	*MTHFR*	1p36
Bone-associated proteins		
Collagen type I, α1	*COL1A1*	17q21-22
Collagen type I, α2	*COL1A2*	7q22
Osteocalcin	*BGLAP*	1q25-31
Miscellaneous		
Apolipoprotein E	*APOE*	19q13
α2-HS-glycoprotein	*AHSG*	3q27

the incomplete information provided by individual SNPs; negative results do not rule out association involving nearby SNPs; and positive results may not indicate the discovery of the causal SNP but simply a marker in *linkage disequilibrium* with a true causal SNP located some distance (perhaps several genes) away. Moreover, many association studies with BMD were poorly designed in terms of power to detect true differences between genotypes, and in terms of cohort homogeneity for age, gender, and genetic background *(54)*. By investigating functional gene variants located in gene regulatory (promoter) and coding regions (exons), rather than synonymous or intronic SNPs, and by taking into account potential interactions with other genes and environmental factors, the consistency of population-based association studies can, however, be markedly improved. The following sections discuss genetic variation in three genes, namely, VDR, estrogen receptor-α (ESR1), and IL-6 genes, in order to illustrate the importance of interactions between genetic and nutritional/lifestyle factors on both peak bone mass acquisition and maintenance.

4.1. Interaction of Calcium Intake With VDR Gene Polymorphisms

The VDR mediates the effects of calcitriol *(55)* on the intestinal absorption of calcium and phosphate and on bone mineralization. The VDR gene, whose nine exons and multiple promoters expand over more than 80 kb, is highly polymorphic *(56)*. The best known and actually first variants ever described in association with bone mass are the VDR 3′-UTR alleles (intron 8/exon9 *Bsm 1, Taq,* and *Apa 1,* the first two being in nearly complete *linkage disequilibrium*) *(57,58)*. The original study found that *Bsm 1* polymorphisms (B-allele frequency equal to 0.4 and 0.1 among Caucasians and Asians, respectively) were associated with BMD and postmenopausal bone loss. However, it later became clear that this study had methodological problems, as a number of investigators failed to confirm differences in BMD and markers of bone turnover between *Bsm1* genotypes *(59–61)*. Nevertheless, a large-scale study in 55+-yr-old men and women using VDR haplotypes *(62)* and a meta-analysis combining 16 separate studies *(63)* provided some support for modest differences (2–3%) in BMD between VDR-3′ alleles. A common variant in the VDR first start codon (ATG, *Fok 1*) has also been reported (*f*-allele frequency close to 0.4 among both Caucasians and Asians), wherein the *ff* genotype was associated with a moderately lower BMD compared to *FF* in postmenopausal women *(64,65)*. However, the original cohort in which this association was described was small and non-white, and these results were not confirmed in Caucasians *(66–68)*, an unexpected result considering the fact that *Fok1* variants are predicted to code for VDR molecules differing by three amino acids in length. Nevertheless, it was later shown that VDR-3′ and -5′ alleles might interact on their association with BMD *(67)*. Other studies have shown that significant BMD differences between VDR-3′ *Bsm*I genotypes could be detected in children *(69,70)*, but not in premenopausal women from the same genetic background. These observations led to suggest that age-related factors, such as levels of gonadal steroids and nutritional/lifestyle variables, may have a profound influence on the association of VDR alleles with bone mass.

Consistent with its prominent effects on bone mass growth during childhood and the maintenance of bone mass in aging women, dietary calcium intake seems to play an important role in modulating the association of VDR polymorphisms with BMD in both age groups. Conversely, genetic variation at the VDR may contribute to the highly variable skeletal response to calcium supplementation. In a cohort of 144 prepubertal girls, 1-yr BMD gain at the distal radius and proximal femur was 50–70% higher in those receiving calcium supplements (850 mg/d) provided as milk extracts (containing phosphorus) compared to placebo, whereas lumbar spine BMD was barely affected *(4)*. Not surprisingly, these effects were most prominent in children with inadequate dietary calcium intake for age (below 800 mg/d). In this case, calcium supplements likely influenced bone modeling, as demonstrated by an increase in both vertebral height and stature, and by the maintenance of positive calcium supplements effects on bone mass years after the end of the intervention *(71)*. In this cohort, baseline BMD at lumbar spine and femur neck was significantly lower in subjects with VDR *BsmI* BB genotype compared to heterozygotes and *bb*. BMD gain in response to calcium supplements was increased at several skeletal sites among *BB* and *Bb*, whereas it remained apparently unaffected in *bb* girls, who had a trend for spontaneously higher BMD accumulation on their usual calcium diet *(69)*. Another calcium-intervention trial with a similar design was recently carried out in 235 prepubertal boys with an inadequate calcium intake. In contrast to the above results, calcium supplementation did not have prominent effects on BMD gain, nor was a significant association or interaction with VDR genotypes observed. On one side, these findings may reveal gender-related differences in the influence of VDR polymorphisms on bone mass gain and its response to calcium. More likely, they may indicate further levels of interaction between nutrients themselves, such as proteins and calcium *(72)*, and between these nutrients and other genetic factors. Indeed, BMD in these boys was better correlated with protein than with calcium intake, the former being on average quite high in this cohort (mean, 1.7 g/kg body weight) *(73)*. By its proper effects on IGF-1 expression and bone formation *(74)*, this high protein intake may have offset part of the deficit in spontaneous calcium intake, thereby preventing calcium supplements from being efficient. Accordingly, calcium supplements significantly increased BMD gain in boys with the lowest protein intake. From a genetic standpoint, the high protein intake may have allowed these boys to achieve a BMD gain close to their best genetic potential, thereby obscuring differences among VDR genotypes. In this particular case, it would be interesting to test whether genetic variation in the protein metabolic pathway, such as IGF-1 polymorphisms *(75)*, might be associated with BMD gain during growth.

Another interesting study comparing the distribution of VDR genotypes in 105 rachitic children from Nigeria (calcium intake, 200 mg/d) and 94 healthy controls found that VDR-3′ genotypes were similarly distributed among cases and controls, whereas the *FOKI ff* genotype was significantly underepresented among cases *(76)*, suggesting that the latter genotype might be protective against osteomalacia induced by dietary calcium deficiency. In contrast, in 72 Caucasian, African-American, and Mexican children with adequate calcium intake (above 1000 mg/d), Ames et al. found that carriers of the VDR genotype *ff* had significantly *decreased* intestinal calcium absorption *(77)*. How the VDR genotype *BB* could be associated with both lower BMD and increased BMD gain after calcium supplementation [in prepubertal girls, *(69)*] and the *ff* genotype be associated with protection against rachitism while being associated with decreased intestinal calcium absorption in healthy kids will be explained at the end of this section.

A significant interaction between VDR-3′ genotypes and calcium intake on BMD and BMD changes has also been shown in postmenopausal women by a number of investigators *(78–81)*. In a pioneering study including elderly subjects (90% women, mean age 73 yr) with a high prevalence of osteoporosis and a low calcium intake (590 mg/d), *bb* subjects apparently did not lose bone at the lumbar spine over 18 mo, whereas spine BMD decreased 2% in heterozygotes and *BB* during this time *(78)*. Calcium supplements (800 mg/d) reversed bone loss in *Bb* subjects after 18 mo, but did not significantly alter BMD changes among the other genotypes. Another prospective study in younger postmenopausal women (mean age 59 yr) whose mean calcium intake was very low (400 mg/d) also found that lumbar spine and hip bone loss was significantly higher in *BB* subjects. In subjects receiving calcium supplements (500 mg/d), however, bone mass changes were similar in all genotypic groups, indicating that the response to calcium was actually greater among *BB (79)*. Similar to the previous two studies, a long-term follow-up (6.3 yr) study in postmenopausal women (mean age 69 yr) reported that among individuals with low calcium intake (below 456 mg/d), *TT* homozygotes for VDR *Taq 1* polymorphism (same as *bb*) had a significantly lower rate of bone loss at both the femoral neck and lumbar spine compared to *tt* (same as *BB*). In contrast, among those with a higher dietary calcium intake (above 705 mg/d), there were no more significant differences in BMD changes between genotypes *(82)*. Cross-sectional observations in the Framingham Osteoporosis cohort also reported a significant interaction between VDR *Bsm 1* alleles and calcium intake on bone mass in the elderly. Thus, in men and women aged 69–90 yr, BMD at the femur trochanter and ultra-distal radius (two regions rich in cancellous bone) was significantly higher in *bb* compared to *BB* in subjects whose calcium intake was greater than 800 mg/d *(81)*. In addition, a case-control study reported that the relative risk of hip and wrist fractures among participants to the Nurses Health Study (mean age 60 yr) was significantly higher in *BB* compared to *bb* in a subgroup with calcium intake below 1078 mg/d (odds ratio, 4.3), but not in the subgroup with higher calcium intake (odds = 1) *(83)*. In summary, the VDR allele *B* seems to be associated with increased bone loss after the menopause but also with a better response to calcium supplementation.

Evidence for functional differences among VDR genotypes comes from a few studies on fractional calcium absorption and on calcium/phosphate homeostasis in response to dietary changes, as well as on patients with primary hyperparathyroidism *(55)*. Thus, in postmenopausal women following a calcium-restriction period, fractional absorption of calcium increased significantly less in *BB* compared to *bb* women, despite a trend for higher calcitriol levels among the former *(84)*. Similarly, a short-term (2-wk) dietary modification trial in young healthy males found that *BB* had a subtle resistance to calcitriol while on a low dietary calcium and phosphate diet for several days, leading to significantly higher levels of circulating PTH, decreased tubular reabsorption of phosphate and lower serum phosphate levels *(85)*. Although these two studies are internally consistent and may explain why many investigators found lower BMD in *BB* subjects on low-calcium diets, they remain to be reconciled with the many observations of a better skeletal response to calcium supplementation in carriers of the *B* allele (see above).

To answer this question, we propose a model in which the VDR *BB* (and *FF*) genotype is characterized by *low efficiency–high capacity,* whereas the VDR *bb* (and *ff*) ff genotype is characterized by *high efficiency–poor capacity,* in promoting calcium absorption and/or bone mineralization. This means that carriers of the VDR genotype *ff* may be capable to better extract calcium from a calcium-poor environment (as suggested by the rachitic kids from Nigeria), whereas carriers of the genotype *BB* would be at a disadvantage in similar conditions (the metabolic studies above); on the opposite, in a calcium-rich environment, the capacity of *ff* carriers to utilize large amounts of calcium would be limited (the calcium absorption study in healthy kids), whereas *BB* could do so (based on their better response in calcium supplementation trials). This might also explain why the *BB* genotype is virtually absent among Asians, whose diet is traditionally poor in dairy products. Accordingly, the relationship between BMD and dietary calcium might be better described by sigmoidal curves for which the point of inflection occurs at different calcium intake thresholds depending on VDR genotypes *(86).* In keeping with our interpretation (above), further developments of this model have proposed that the calcium intake–BMD curves might not run parallel to each other but actually cross over *(2).* In this case, the VDR genotypes associated with decreased bone mass at low calcium intake *(BB, FF)* might actually be the ones associated with increased bone mass at higher calcium intakes (hence with a better response to calcium supplements), whereas the opposite would be true for carriers of the alternate homozygous genotypes *bb* and *ff* (Fig. 1).

4.2. Estrogen Receptor Gene Polymorphisms

Female sex hormones appear to be mandatory not only for the acquisition of peak bone mass in both females and males *(87,88),* but also for the maintenance of bone mass in both genders *(89).* They control bone remodeling during reproductive life in females and later on in aging men *(90).* Genotypes identified by *Pvu*II and *Xba*I restriction fragment length polymorphisms in the first intron of the estrogen receptor α (ERα) gene (*ESR1*) (Table 1) were originally found to be significantly associated with BMD in postmenopausal Japanese women, but not with markers of bone turnover *(91).* In contrast, a similar study from Korea reported no significant BMD differences among ERα genotypes in postmenopausal women receiving hormone replacement therapy (HRT) *(92).* Another study from Japan including 173 premenopausal to late postmenopausal women indicated a predominant association between ERα genotypes and adult bone mass, which disappeared with advancing age *(93).* Several investigators have examined ERα gene polymorphisms and bone mass in Caucasian populations as well. In one study, a significant association was found between either the *Pvu*II or the *Xba*I genotypes and lumbar spine BMD in 253

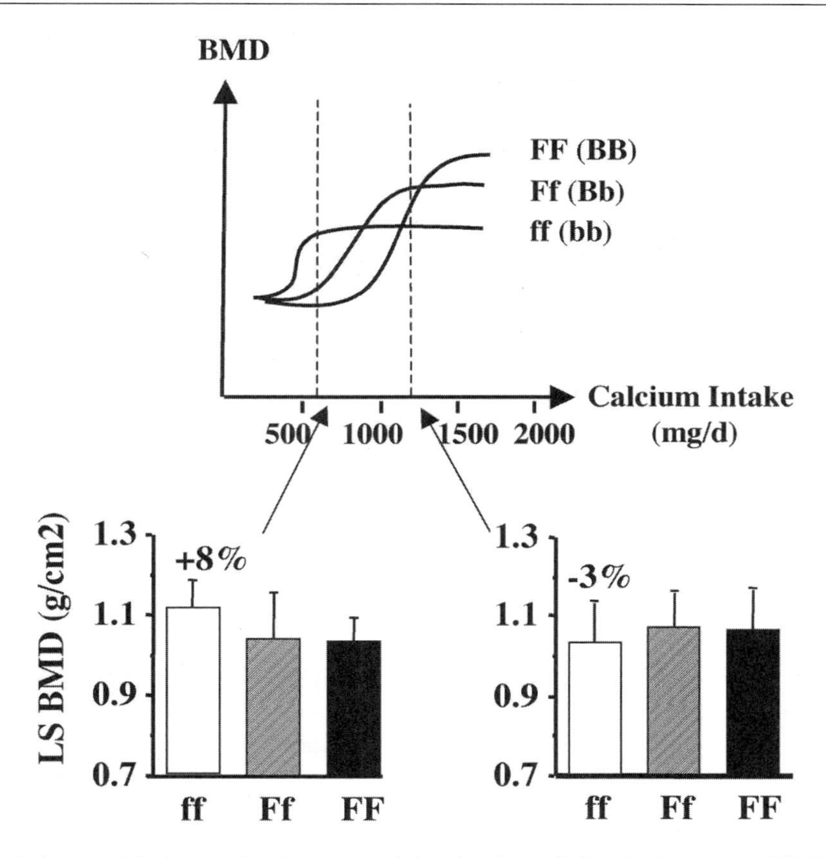

Fig. 1. Scheme of the interaction between calcium intake and vitamin D receptor (VDR) polymorphisms on bone mineral density (BMD). The scheme on top illustrates the relationship between BMD and dietary calcium intake according to VDR polymorphisms (adapted from ref. *2*), with *ff (bb)* being the "high efficiency–low capacity" genotype and *FF (BB)* the "low efficiency–high capacity" genotype. The lumbar spine BMD data shown in the two diagrams below were obtained in 177 healthy premenopausal women genotyped for VDR *Fok 1* polymorphisms (*FF, Ff,* and *ff*), who were further subdivided into two groups for dietary calcium intake (respectively below and above the median for the whole cohort) (adapted from ref. *67*). The arrows and dotted lines indicate the expected BMD differences among VDR genotypes in these two groups of women (mean calcium intake, 550 and 1200 mg/d, respectively).

pre- and perimenopausal women, those with the *Pvu*II *pp* genotype having a 6.4% lower BMD at this site compared to *PP (94)*. However, there were no differences in BMD changes, nor in several biochemical markers of calcium and bone metabolism, including PTH and osteocalcin, over a 3-yr period in this cohort. One limitation of this study was a low rate of bone loss in this cohort over 3 yr (≤1%). In contrast, a very recent study that prospectively investigated the 5-yr bone loss in early postmenopausal women receiving either HRT or placebo in addition to calcium and vitamin D found no significant differences in BMD among ERα polymorphisms at baseline, but significant differences in lumbar spine BMD changes between genotypes *PP* (–6.4%) and *pp* (–2.9%) in the absence of HRT *(95)*. In women receiving HRT, these differences were no more apparent.

A significant gene-by-gene interaction between VDR and ER gene polymorphisms has been suggested by several authors. In the study by Willing et al. *(94)*, BMD at all skeletal sites was lower in subjects with the VDR *Bsm*I genotype *BB*, as compared to *Bb* and *bb*, in the subgroup of women carrying the ERα *Pvu*II genotype *PP*. Of note, however, there were only five *BB/PP* subjects in this cohort. An interaction between VDR-3′ and ERα polymorphisms has also been found in relation to BMD in a cohort of 426 normal and osteoporotic women *(96)*. Subjects carrying the *BB/PP* genotypes had a significantly lower BMD at the lumbar spine compared to alternate homozygotes *bb/pp*. VDR/ ERα polymorphisms have also been related to the rate of postmenopausal bone loss in a small cohort of women (n = 108) with or without HRT *(97)*. These results, however, remain controversial, as a recent study in 313 late postmenopausal women with a low average calcium intake (approx 600 mg/day), including 142 women with a history of osteoporotic fractures, found no significant association between ERα polymorphisms alone or in combination with VDR polymorphisms on BMD, nor on biochemical markers of bone and mineral metabolism *(98)*.

A meta-analysis on the association of ERα genetic variation with BMD and fracture risk in more than 5000 women from 30 studies concludes to the absence of significant differences in lumbar spine or hip BMD between *Pvu*II alleles, whereas homozygotes for the *Xba*I *XX* genotype have a significantly higher BMD at these two sites *(99)*. Since the *Xba*I and *Pvu*II sites are very nearly located and in strong *linkage disequilibrium*, this may explain why some authors found an association with the latter. Moreover, the meta-analysis found a trend for more prominent differences in pre- compared to postmenopausal women, suggesting that ERα genotypes might exert their influence prominently on peak bone mass acquisition. Despite standardized BMD differences of only approximately Z = 0.1 between *XX* and *xx* genotypes, differences in the risk of fracture were disproportionately high (odds ratio, 0.66 in *XX* vs *xx (99)*. The latter observation suggests that ERα genotypes might be implicated in the determination of bone microarchitecture, which is poorly accounted for by DXA measurements.

The molecular mechanisms by which ERα polymorphisms may modulate the actions of estrogens on bone modeling and remodeling remain to be elucidated. A consistent finding, however, is that HRT appears to alleviate BMD differences among these genotypes (see above), indicating that carriers of the x allele might particularly benefit from HRT after the menopause. HRT has also been found to alter the association with BMD of other genes belonging to the biological pathway by which estrogens exert their activity on the skeleton, such as IL-6 (see below).

Thus, future studies should consider the possible interaction of ERα allelic variants not only with estrogens, but also with IL-6 and VDR alleles.

4.3. IL-6 Gene Promoter Polymorphisms

IL-6 is a pleiotropic cytokine playing a central role in the activation of osteoclasts (the bone-resorbing cells) and bone turnover *(100)*. Since IL-6 gene expression is normally repressed by endogenous estrogens, increased IL-6 production is an important factor for postmenopausal bone loss *(101)*. Moreover, IL-6 expression in bone is triggered by parathyroid hormone and it is therefore implicated in the age-related bone loss associated with poor calcium and vitamin D intake *(101,102)*. Several studies have identified the IL-6 gene locus to be linked to BMD in postmenopausal women *(103,104)* and in families of osteoporotic probands *(52,105)*, whereas no linkage between the IL-6 gene locus and bone mass was found in young sib pairs *(106)*. These observations therefore suggested that IL-6 genetic variation might contribute to the population variance in bone loss rather than peak bone mass. More recently, several functional allelic variants have been identified in the IL-6 gene promoter region *(107,108)*. Among them, a common –174G>C polymorphism (frequency of the C allele, 0.4 among Caucasians) is located close to a binding site for a transcription factor (NF-IL-6) that is under the dependency of estrogens. There is some evidence that this variant may produce a functional mutation in that the C allele is associated with lower IL-6 gene transcriptional activity in vitro *(107)*. In addition, a rare G>C allelic polymorphism at position –573 (frequency of the C allele, 0.06 among Caucasians), which is closely related to two glucocorticoid responsive elements, has very recently been identified in this region. The –573 alleles also seem to influence the level of IL-6 transcriptional activity in vitro *(109)*. Further proof of functionality of IL-6 genetic variation comes from the fact that –174CC homozygotes have circulating IL-6 concentrations approximately half those of –174GG homozygotes *(107)*, whereas both the –174C and –573G alleles are associated with significantly lower levels of the IL-6-dependent C-reactive protein (CRP) in serum *(109)*.

In 434 healthy, community-dwelling, white U.S. postmenopausal women (mean age ± SD, 71.7 ± 5.7 yr), C-terminal crosslinks of Type 1 collagen (CTx), a marker of bone resorption, were significantly lower among –174CC and –573GG compared to the other genotypes *(110)*. Postmenopausal women with genotype –174CC (15% of the population) had levels of bone resorption similar to those of premenopausal women. Interestingly, the Il-6 –573G>C and –174 G>C allelic variants appear to cooperate in the regulation of IL-6 gene transcriptional activity *(108)*. Accordingly, homozygous carriers of the two IL-6 variants associated with low IL-6 activity (namely, –174C and –573G) had CTx levels that were 30% lower compared to those not carrying this allelic combination *(109)*. Among these women, women with the –174CC genotype had BMD at the hip and forearm (distal radius) that was 1.5% to 4.7% (nonsignificantly) higher as compared to GG homozygotes *(110)*. However, differences were larger at the trochanter and ultradistal radius, consistent with the predominant effects of estrogen deficiency (and increased IL-6 activity) on cancellous bone. When the cohort was further divided in two groups of early and late postmenopausal women, BMD was found to be significantly lower in the older compared to younger women, but more prominently so in subjects carrying the IL-6 –174 GG and GC genotypes (–9% to –10% at the various skeletal sites) compared to CC women (–5% to 6.1%). Taken together with the lower level of bone resorption associated with the –174C allele, these observations indicate that postmenopausal women with the IL-6 –174CC genotype may be "slow bone losers." Two studies also reported an association of IL-6 polymorphisms with peak bone mass, one in young males *(111)* and the other in premenopausal females *(112)*. However, the latter study failed to detect a lower rate of bone resorption or bone loss associated with the –174CC genotype in 234 postmenopausal women (mean age 64 yr). These data suggest that IL-6 alleles may contribute to peak bone mass, but this association may be blunted in early postmenopausal women by the dramatic hormonal changes occurring at the menopause. Eventually, IL-6 genetic variation might again exert detectable effects on bone turnover and bone mass 20 yr after the menopause, i.e., at a time when nutritional and lifestyle factors, such as poor calcium and vitamin D intake, further modulate IL-6 gene expression.

To test this hypothesis, the interaction between IL-6 promoter polymorphisms and factors known to affect bone turnover, namely, years since menopause, estrogen status, physical activity, smoking, dietary calcium, vitamin D, and alcohol intake, was examined in the Offspring Cohort of the Framingham Heart Study *(113)*. This cohort comprises 1574 unrelated men and women (mean age 60 yr) with bone mineral density measurements at the hip. Consistent with the study of Garnero et al. *(112)*, in models that considered only the main effects of IL-6 polymorphisms, no significant association with bone mineral density was observed in either women or men. In contrast, interactions were found between IL-6 –174 genotypes and years since menopause, estrogen status, dietary calcium, and vitamin D intake in women. Thus, bone mineral density was significantly lower with genotype –174 GG compared to CC, and intermediate with GC, in women above 15 yr past menopause, in those without estrogens or with calcium intake below 940 mg/d. In estrogen-deficient women with poor calcium intake, hip bone mineral density differences between IL-6 –174 genotypes CC and GG were as high as 16%. In contrast, no such interactions were observed in men. These data therefore suggest that age, HRT, and dietary calcium all influence the association between IL-6 alleles and bone mass. Accordingly, HRT and adequate calcium intake could be better targeted to population subgroups genetically identified to be at otherwise increased risk of accelerated bone resorption and low bone mass with aging, such as IL-6 –174GG homozygotes.

5. CONCLUSION

Genetically speaking, humans today live in a nutritional environment that differs from that for which our genetic constitution was selected *(114)*. Thus, gene polymorphisms that appeared a few ten thousand years ago in the human genome as an adaptation to a changing environment (hunter-gatherer, then agricultural), but have become *mal*-adapted to our current nutritional and lifestyle habits, may in turn contribute highly to a number of common disorders including osteoporosis, but also diabetes, hypertension, etc. The identification of such genetic variations and interactions through various approaches combining association studies with candidate genes involved in bone metabolic pathways and genome-wide mapping of QTLs linked to discrete traits for bone strength is ongoing. In the future, advances in the osteoporosis genetics field may allow for individualized Recommended Dietary Allowances for various nutrients, primarily calcium, and for targeted interventions aiming at improving lifestyle factors for better bone health.

ACKNOWLEDGMENTS

I thank Dr. Jean-Philippe Bonjour, Dr. René Rizzoli, and Dr. Thierry Chevalley (Division of Bone Diseases, Departments of Geriatrics and Internal Medicine, Geneva University Hospital, Switzerland) for their contribution to this work.

REFERENCES

1. Ferrari S, Rizzoli R, Bonjour JP. Heritable and nutritional influences on bone mineral mass. Aging (Milano) 1998; 10:205–213.
2. Eisman JA. Genetics of osteoporosis. Endocr Rev 1999; 20:788–804.
3. Peacock M, Turner CH, Econs MJ, Foroud T. Genetics of osteoporosis. Endocr Rev 2002; 23:303–326.
4. Bonjour JP, Carrie AL, Ferrari S, et al. Calcium-enriched foods and bone mass growth in prepubertal girls: a randomized, double-blind, placebo-controlled trial. J Clin Invest 1997; 99:1287–1294.
5. Ferrari S, Rizzoli R, Bonjour J. Vitamin D receptor gene polymorphisms and bone mineral homeostasis. In: Econs MJ, ed. The Genetics of Osteoporosis and Metabolic Bone Disease. Humana, Totowa, NJ, 2000, pp. 45–60.
6. Eisman JA. Vitamin D polymorphisms and calcium homeostasis: a new concept of normal gene variants and physiologic variation. Nutr Rev 1998; 56:s22–s29; discussion s54–s75.
7. Bouxsein MLB. Biomechanics of age-related fractures. In: Marcus RFD, Kelsey J, eds. Osteoporosis. Vol. 1. Academic, San Diego, CA, 2001, pp. 509–526.
8. Pinilla TP, Boardman KC, Bouxsein ML, Myers ER, Hayes WC. Impact direction from a fall influences the failure load of the proximal femur as much as age-related bone loss. Calcif Tissue Int 1996; 58:231–235.
9. Nguyen TV, Eisman JA. Genetics of fracture: challenges and opportunities [editorial] [In Process Citation]. J Bone Miner Res 2000; 15:1253–1256.
10. Chesnut CH 3rd, Rose CJ. Reconsidering the effects of antiresorptive therapies in reducing osteoporotic fracture. J Bone Miner Res 2001; 16:2163–2172.
11. Seeman E, Hopper JL, Young NR, Formica C, Goss P, Tsalamandris C. Do genetic factors explain associations between muscle strength, lean mass, and bone density? A twin study. Am J Physiol 1996; 270:E320–E327.
12. Lee C. Irresistible force meets immovable object: SNP mapping of complex diseases. Trends Genet 2002; 18:67–69.
13. Seeman E, Hopper JL, Bach LA, et al. Reduced bone mass in daughters of women with osteoporosis. N Engl J Med 1989; 320:554–558.

14. Soroko SB, Barrett-Connor E, Edelstein SL, Kritz-Silverstein D. Family history of osteoporosis and bone mineral density at the axial skeleton: the Rancho Bernardo Study. J Bone Miner Res 1994; 9:761–769.
15. Barthe N, Basse-Cathalinat B, Meunier PJ, et al. Measurement of bone mineral density in mother-daughter pairs for evaluating the family influence on bone mass acquisition: a GRIO survey. Osteopor Int 1998; 8:379–384.
16. Cohen-Solal ME, Baudoin C, Omouri M, Kuntz D, De Vernejoul MC. Bone mass in middle-aged osteoporotic men and their relatives: familial effect. J Bone Miner Res 1998; 13:1909–1914.
17. Seeman E. Pathogenesis of bone fragility in women and men. Lancet 2002; 359:1841–1850.
18. Kelly PJ, Morrison NA, Sambrook PN, Nguyen TV, Eisman JA. Genetic influences on bone turnover, bone density and fracture. Eur J Endocrinol 1995; 133:265–271.
19. Pocock NA, Eisman JA, Hopper JL, Yeates MG, Sambrook PN, Eberl S. Genetic determinants of bone mass in adults. A twin study. J Clin Invest 1987; 80:706–710.
20. Slemenda CW, Christian JC, Williams CJ, Norton JA, Johnston CC Jr. Genetic determinants of bone mass in adult women: a reevaluation of the twin model and the potential importance of gene interaction on heritability estimates. J Bone Miner Res 1991; 6:561–567.
21. Hopper JL, Green RM, Nowson CA, et al. Genetic, common environment, and individual specific components of variance for bone mineral density in 10- to 26-year-old females: a twin study [see comments]. Am J Epidemiol 1998; 147:17–29.
22. Nguyen TV, Howard GM, Kelly PJ, Eisman JA. Bone mass, lean mass, and fat mass: same genes or same environments? [see comments]. Am J Epidemiol 1998; 147:3–16.
23. Arden NK, Baker J, Hogg C, Baan K, Spector TD. The heritability of bone mineral density, ultrasound of the calcaneus and hip axis length: a study of postmenopausal twins. J Bone Miner Res 1996; 11:530–534.
24. Howard GM, Nguyen TV, Harris M, Kelly PJ, Eisman JA. Genetic and environmental contributions to the association between quantitative ultrasound and bone mineral density measurements: a twin study. J Bone Miner Res 1998; 13:1318–1327.
25. Tylavsky FA, Bortz AD, Hancock RL, Anderson JJ. Familial resemblance of radial bone mass between premenopausal mothers and their college-age daughters. Calcif Tissue Int 1989; 45:265–272.
26. Krall EA, Dawson-Hughes B. Heritable and life-style determinants of bone mineral density. J Bone Miner Res 1993; 8:1–9.
27. Ferrari S, Rizzoli R, Slosman D, Bonjour JP. Familial resemblance for bone mineral mass is expressed before puberty. J Clin Endocrinol Metab 1998; 83:358–361.
28. Jones G, Nguyen TV. Associations between maternal peak bone mass and bone mass in prepubertal male and female children. J Bone Miner Res 2000; 15:1998–2004.
29. Gueguen R, Jouanny P, Guillemin F, Kuntz C, Pourel J, Siest G. Segregation analysis and variance components analysis of bone mineral density in healthy families. J Bone Miner Res 1995; 10:2017–2022.
30. Jouanny P, Guillemin F, Kuntz C, Jeandel C, Pourel J. Environmental and genetic factors affecting bone mass. Similarity of bone density among members of healthy families. Arthritis Rheum 1995; 38:61–67.
31. Christian JC, Yu PL, Slemenda CW, Johnston CC Jr. Heritability of bone mass: a longitudinal study in aging male twins. Am J Hum Genet 1989; 44:429–433.
32. Garnero P, Arden NK, Griffiths G, Delmas PD, Spector TD. Genetic influence on bone turnover in postmenopausal twins. J Clin Endocrinol Metab 1996; 81:140–146.
33. Danielson ME, Cauley JA, Baker CE, et al. Familial resemblance of bone mineral density (BMD) and calcaneal ultrasound attenuation: the BMD in mothers and daughters study. J Bone Miner Res 1999; 14:102–110.

34. Snieder H, MacGregor AJ, Spector TD. Genes control the cessation of a woman's reproductive life: a twin study of hysterectomy and age at menopause. J Clin Endocrinol Metab 1998; 83:1875–1880.
35. Rogers J, Mahaney MC, Beamer WG, Donahue LR, Rosen CJ. Beyond one gene-one disease: alternative strategies for deciphering genetic determinants of osteoporosis [editorial]. Calcif Tissue Int 1997; 60:225–228.
36. Nguyen TV, Blangero J, Eisman JA. Genetic epidemiological approaches to the search for osteoporosis genes. J Bone Miner Res 2000; 15:392–401.
37. Cardon LR, Garner C, Bennett ST, et al. Evidence for a major gene for bone mineral density in idiopathic osteoporotic families. J Bone Miner Res 2000; 15:1132–1137.
38. Deng HW, Livshits G, Yakovenko K, et al. Evidence for a major gene for bone mineral density/content in human pedigrees identified via probands with extreme bone mineral density. Ann Hum Genet 2002; 66:61–74.
39. Livshits G, Karasik D, Pavlovsky O, Kobyliansky E. Segregation analysis reveals a major gene effect in compact and cancellous bone mineral density in 2 populations. Hum Biol 1999; 71:155–172.
40. Koller DL, Liu G, Econs MJ, et al. Genome screen for quantitative trait loci underlying normal variation in femoral structure. J Bone Miner Res 2001; 16:985–991.
41. Karasik D, Myers RH, Hannan MT, et al. Mapping of quantitative ultrasound of the calcaneus bone to chromosome 1 by genome-wide linkage analysis. Osteopor Int 2002; 13:796–802.
42. Johnson ML, Gong G, Kimberling W, Recker SM, Kimmel DB, Recker RB. Linkage of a gene causing high bone mass to human chromosome 11 (11q12–13) [see comments]. Am J Hum Genet 1997; 60:1326–1332.
43. Gong Y, Vikkula M, Boon L, et al. Osteoporosis-pseudoglioma syndrome, a disorder affecting skeletal strength and vision, is assigned to chromosome region 11q12–13. Am J Hum Genet 1996; 59:146–151.
44. Van Hul E, Gram J, Bollerslev J, et al. Localization of the gene causing autosomal dominant osteopetrosis type I to chromosome 11q12–13. J Bone Miner Res 2002; 17:1111–1117.
45. Gong Y, Slee RB, Fukai N, et al. LDL receptor-related protein 5 (LRP5) affects bone accrual and eye development. Cell 2001; 107:513–523.
46. Little RD, Carulli JP, Del Mastro RG, et al. A mutation in the LDL receptor-related protein 5 gene results in the autosomal dominant high-bone-mass trait. Am J Hum Genet 2002; 70:11–19.
47. Boyden LM, Mao J, Belsky J, et al. High bone density due to a mutation in LDL-receptor-related protein 5. N Engl J Med 2002; 346:1513–1521.
48. Sobacchi C, Frattini A, Orchard P, et al. The mutational spectrum of human malignant autosomal recessive osteopetrosis. Hum Mol Genet 2001; 10:1767–1773.
49. Koller DL, Rodriguez LA, Christian JC, et al. Linkage of a QTL contributing to normal variation in bone mineral density to chromosome 11q12–13. J Bone Miner Res 1998; 13:1903–1908.
50. Livshits G, Trofimov S, Malkin I, Kobyliansky E. Transmission disequilibrium test for hand bone mineral density and 11q12–13 chromosomal segment. Osteopor Int 2002; 13:461–467.
51. Carn G, Koller DL, Peacock M, et al. Sibling pair linkage and association studies between peak bone mineral density and the gene locus for the osteoclast-specific subunit (OC116) of the vacuolar proton pump on chromosome 11p12–13. J Clin Endocrinol Metab 2002; 87:3819–3824.
52. Duncan EL, Brown MA, Sinsheimer J, et al. Suggestive linkage of the parathyroid receptor type 1 to osteoporosis [see comments]. J Bone Miner Res 1999; 14:1993–1999.
53. Rizzoli R, Bonjour JP, Ferrari SL. Osteoporosis, genetics and hormones. J Mol Endocrinol 2001; 26:79–94.
54. Econs MJ, Speer MC. Genetic studies of complex diseases: let the reader beware. J Bone Miner Res 1996; 11:1835–1840.

55. Carling T, Kindmark A, Hellman P, et al. Vitamin D receptor genotypes in primary hyperparathyroidism. Nat Med 1995; 1:1309–1311.
56. Uitterlinden AG, Van Leuuwen J, Pols HA. Genetics and genomics of osteoporosis. In: Marcus RFD, Kelsey J, ed. Osteoporosis. Vol. 1. Academic, San Diego, CA, 2001, pp. 639–668.
57. Morrison NA, Yeoman R, Kelly PJ, Eisman JA. Contribution of trans-acting factor alleles to normal physiological variability: vitamin D receptor gene polymorphism and circulating osteocalcin. Proc Natl Acad Sci USA 1992; 89:6665–6669.
58. Morrison NA, Qi JC, Tokita A, et al. Prediction of bone density from vitamin D receptor alleles [see comments] [published erratum appears in Nature 1997 May 1; 387(6628):106]. Nature 1994; 367:284–287.
59. Garnero P, Borel O, Sornay-Rendu E, Delmas PD. Vitamin D receptor gene polymorphisms do not predict bone turnover and bone mass in healthy premenopausal women. J Bone Miner Res 1995; 10:1283–1288.
60. Garnero P, Borel O, Sornay-Rendu E, Arlot ME, Delmas PD. Vitamin D receptor gene polymorphisms are not related to bone turnover, rate of bone loss, and bone mass in postmenopausal women: the OFELY Study. J Bone Miner Res 1996; 11:827–834.
61. Hustmyer FG, Peacock M, Hui S, Johnston CC, Christian J. Bone mineral density in relation to polymorphism at the vitamin D receptor gene locus. J Clin Invest 1994; 94:2130–2134.
62. Uitterlinden AG, Pols HA, Burger H, et al. A large-scale population-based study of the association of vitamin D receptor gene polymorphisms with bone mineral density. J Bone Miner Res 1996; 11:1241–1248.
63. Cooper GS, Umbach DM. Are vitamin D receptor polymorphisms associated with bone mineral density? A meta-analysis [see comments]. J Bone Miner Res 1996; 11:1841–1849.
64. Gross C, Eccleshall TR, Malloy PJ, Villa ML, Marcus R, Feldman D. The presence of a polymorphism at the translation initiation site of the vitamin D receptor gene is associated with low bone mineral density in postmenopausal Mexican-American women [see comments]. J Bone Miner Res 1996; 11:1850–1855.
65. Harris SS, Eccleshall TR, Gross C, Dawson-Hughes B, Feldman D. The vitamin D receptor start codon polymorphism (FokI) and bone mineral density in premenopausal American black and white women. J Bone Miner Res 1997; 12:1043–1048.
66. Eccleshall TR, Garnero P, Gross C, Delmas PD, Feldman D. Lack of correlation between start codon polymorphism of the vitamin D receptor gene and bone mineral density in premenopausal French women: the OFELY study. J Bone Miner Res 1998; 13:31–35.
67. Ferrari S, Rizzoli R, Manen D, Slosman D, Bonjour JP. Vitamin D receptor gene start codon polymorphisms (FokI) and bone mineral density: interaction with age, dietary calcium, and 3′-end region polymorphisms. J Bone Miner Res 1998; 13:925–930.
68. Langdahl BL, Gravholt CH, Brixen K, Eriksen EF. Polymorphisms in the vitamin D receptor gene and bone mass, bone turnover and osteoporotic fractures [see comments]. Eur J Clin Invest 2000; 30:608–617.
69. Ferrari SL, Rizzoli R, Slosman DO, Bonjour JP. Do dietary calcium and age explain the controversy surrounding the relationship between bone mineral density and vitamin D receptor gene polymorphisms? J Bone Miner Res 1998; 13:363–370.
70. Sainz J, Van Tornout JM, Loro ML, Sayre J, Roe TF, Gilsanz V. Vitamin D-receptor gene polymorphisms and bone density in prepubertal American girls of Mexican descent [see comments]. N Engl J Med 1997; 337:77–82.
71. Bonjour JP, Chevalley T, Ammann P, Slosman D, Rizzoli R. Gain in bone mineral mass in prepubertal girls 3.5 years after discontinuation of calcium supplementation: a follow-up study. Lancet 2001; 358:1208–1212.
72. Dawson-Hughes B. Interaction of dietary calcium and protein in bone health in humans. J Nutr 2003; 133:852S–854S.

73. Chevalley T, Ferrari S, Hans D, et al. Protein intake modulates the effect of calcium supplementation on bone mass gain in prepubertal boys. J Bone Miner Res 2002; 17(suppl 1):S172.
74. Bonjour JP, Ammann P, Chevalley T, Rizzoli R. Protein intake and bone growth. Can J Appl Physiol 2001; 26(suppl):S153–S166.
75. Rosen CJ, Donahue LR. Insulin-like growth factors and bone: the osteoporosis connection revisited. Proc Soc Exp Biol Med 1998; 219:1–7.
76. Fischer PR, Thacher TD, Pettifor JM, Jorde LB, Eccleshall TR, Feldman D. Vitamin D receptor polymorphisms and nutritional rickets in Nigerian children. J Bone Miner Res 2000; 15:2206–2210.
77. Ames SK, Ellis KJ, Gunn SK, Copeland KC, Abrams SA. Vitamin D receptor gene Fok1 polymorphism predicts calcium absorption and bone mineral density in children. J Bone Miner Res 1999; 14:740–746.
78. Ferrari S, Rizzoli R, Chevalley T, Slosman D, Eisman JA, Bonjour JP. Vitamin-D-receptor-gene polymorphisms and change in lumbar-spine bone mineral density [see comments]. Lancet 1995; 345:423–424.
79. Krall EA, Parry P, Lichter JB, Dawson-Hughes B. Vitamin D receptor alleles and rates of bone loss: influences of years since menopause and calcium intake. J Bone Miner Res 1995; 10:978–984.
80. Salamone LM, Ferrell R, Black DM, et al. The association between vitamin D receptor gene polymorphisms and bone mineral density at the spine, hip and whole-body in premenopausal women [published erratum appears in Osteopor Int 1996; 6(3):187–188]. Osteopor Int 1996; 6:63–68.
81. Kiel DP, Myers RH, Cupples LA, et al. The BsmI vitamin D receptor restriction fragment length polymorphism (bb) influences the effect of calcium intake on bone mineral density. J Bone Miner Res 1997; 12:1049–1057.
82. Brown MA, Haughton MA, Grant SF, Gunnell AS, Henderson NK, Eisman JA. Genetic control of bone density and turnover: role of the collagen 1alpha1, estrogen receptor, and vitamin D receptor genes. J Bone Miner Res 2001; 16:758–764.
83. Feskanich D, Hunter DJ, Willett WC, et al. Vitamin D receptor genotype and the risk of bone fractures in women. Epidemiology 1998; 9:535–539.
84. Dawson-Hughes B, Harris SS, Finneran S. Calcium absorption on high and low calcium intakes in relation to vitamin D receptor genotype. J Clin Endocrinol Metab 1995; 80:3657–3661.
85. Ferrari S, Manen D, Bonjour JP, Slosman D, Rizzoli R. Bone mineral mass and calcium and phosphate metabolism in young men: relationships with vitamin D receptor allelic polymorphisms. J Clin Endocrinol Metab 1999; 84:2043–2048.
86. Ferrari S, Bonjour J, Rizzoli R. The vitamin D receptor gene and calcium metabolism. Trends Endocrinol Metab (TEM) 1998; 9:259–264.
87. Carani C, Qin K, Simoni M, et al. Effect of testosterone and estradiol in a man with aromatase deficiency. N Engl J Med 1997; 337:91–95.
88. Rizzoli R, Bonjour JP. Hormones and bones. Lancet 1997; 349(suppl 1):s120–s123.
89. Amin S, Zhang Y, Sawin CT, et al. Association of hypogonadism and estradiol levels with bone mineral density in elderly men from the Framingham study. Ann Intern Med 2000; 133:951–963.
90. Riggs BL, Khosla S, Melton LJ 3rd. A unitary model for involutional osteoporosis: estrogen deficiency causes both type I and type II osteoporosis in postmenopausal women and contributes to bone loss in aging men. J Bone Miner Res 1998; 13:763–773.
91. Kobayashi S, Inoue S, Hosoi T, Ouchi Y, Shiraki M, Orimo H. Association of bone mineral density with polymorphism of the estrogen receptor gene. J Bone Miner Res 1996; 11:306–311.
92. Han KO, Moon IG, Kang YS, Chung HY, Min HK, Han IK. Nonassociation of estrogen receptor genotypes with bone mineral density and estrogen responsiveness to hormone replacement therapy in Korean postmenopausal women. J Clin Endocrinol Metab 1997; 82:991–995.

93. Mizunuma H, Hosoi T, Okano H, et al. Estrogen receptor gene polymorphism and bone mineral density at the lumbar spine of pre- and postmenopausal women. Bone 1997; 21:379–383.
94. Willing M, Sowers M, Aron D, et al. Bone mineral density and its change in white women: estrogen and vitamin D receptor genotypes and their interaction. J Bone Miner Res 1998; 13:695–705.
95. Salmen T, Heikkinen AM, Mahonen A, et al. Early postmenopausal bone loss is associated with PvuII estrogen receptor gene polymorphism in Finnish women: effect of hormone replacement therapy. J Bone Miner Res 2000; 15:315–321.
96. Gennari L, Becherini L, Masi L, et al. Vitamin D and estrogen receptor allelic variants in Italian postmenopausal women: evidence of multiple gene contribution to bone mineral density. J Clin Endocrinol Metab 1998; 83:939–944.
97. Deng HW, Li J, Li JL, et al. Change of bone mass in postmenopausal Caucasian women with and without hormone replacement therapy is associated with vitamin D receptor and estrogen receptor genotypes. Hum Genet 1998; 103:576–585.
98. Vandevyver C, Vanhoof J, Declerck K, et al. Lack of association between estrogen receptor genotypes and bone mineral density, fracture history, or muscle strength in elderly women. J Bone Miner Res 1999; 14:1576–1582.
99. Ioannidis JP, Stavrou I, Trikalinos TA, et al. Association of polymorphisms of the estrogen receptor alpha gene with bone mineral density and fracture risk in women: a meta-analysis. J Bone Miner Res 2002; 17:2048–2060.
100. Manolagas SC, Jilka RL. Bone marrow, cytokines, and bone remodeling. Emerging insights into the pathophysiology of osteoporosis. N Engl J Med 1995; 332:305–311.
101. 1Manolagas SC. The role of IL-6 type cytokines and their receptors in bone. Ann N Y Acad Sci 1998; 840:194–204.
102. Dawson-Hughes B, Harris SS, Krall EA, Dallal GE. Effect of calcium and vitamin D supplementation on bone density in men and women 65 years of age or older [see comments]. N Engl J Med 1997; 337:670–676.
103. Murray RE, McGuigan F, Grant SF, Reid DM, Ralston SH. Polymorphisms of the interleukin-6 gene are associated with bone mineral density. Bone 1997; 21:89–92.
104. Tsukamoto K, Yoshida H, Watanabe S, et al. Association of radial bone mineral density with CA repeat polymorphism at the interleukin 6 locus in postmenopausal Japanese women. J Hum Genet 1999; 44:148–151.
105. Ota N, Hunt SC, Nakajima T, et al. Linkage of interleukin 6 locus to human osteopenia by sibling pair analysis. Hum Genet 1999; 105:253–257.
106. Takacs I, Koller DL, Peacock M, et al. Sib pair linkage and association studies between bone mineral density and the interleukin-6 gene locus. Bone 2000; 27:169–173.
107. Fishman D, Faulds G, Jeffery R, et al. The effect of novel polymorphisms in the interleukin-6 (IL-6) gene on IL-6 transcription and plasma IL-6 levels, and an association with systemic-onset juvenile chronic arthritis. J Clin Invest 1998; 102:1369–1376.
108. Terry CF, Loukaci V, Green FR. Cooperative influence of genetic polymorphisms on interleukin 6 transcriptional regulation. J Biol Chem 2000; 275:18138–18144.
109. Ferrari SL, Ahn-Luong L, Garnero P, Humphries SE, Greenspan SL. Two promoter polymorphisms regulating interleukin-6 gene expression are associated with circulating levels of C-reactive protein and markers of bone resorption in postmenopausal women. J Clin Endocrinol Metab 2003; 88:255–259.
110. Ferrari SL, Garnero P, Emond S, Montgomery H, Humphries SE, Greenspan SL. A functional polymorphic variant in the interleukin-6 gene promoter associated with low bone resorption in postmenopausal women. Arthritis Rheum 2001; 44:196–201.
111. Lorentzon M, Lorentzon R, Nordstrom P. Interleukin-6 gene polymorphism is related to bone mineral density during and after puberty in healthy white males: a cross-sectional and longitudinal study. J Bone Miner Res 2000; 15:1944–1949.

112. Garnero P, Borel O, Sornay-Rendu E, et al. Association between a functional interleukin-6 gene polymorphism and peak bone mineral density and postmenopausal bone loss in women: the OFELY study. Bone 2002; 31:43–50.
113. Ferrari S, Karasik D, Liu J, et al. Interleukin-6 genetic variation and the effects of estogens and dietary calcium on bone mass: The Framingham Osteoporosis Study. J Bone Miner Res 2002; 17(suppl 1):S140.
114. Neel JV. When some fine old genes meet a 'new' environment. World Rev Nutr Diet 1999; 84:1–18.
115. Koller DL, Econs MJ, Morin PA, et al. Genome screen for QTLs contributing to normal variation in bone mineral density and osteoporosis. J Clin Endocrinol Metab 2000; 85:3116–3120.
116. Devoto M, Shimoya K, Caminis J, et al. First-stage autosomal genome screen in extended pedigrees suggests genes predisposing to low bone mineral density on chromosomes 1p, 2p and 4q. Eur J Hum Genet 1998; 6:151–157.
117. Karasik D, Myers RH, Cupples LA, et al. Genome screen for quantitative trait loci contributing to normal variation in bone mineral density: the Framingham Study. J Bone Miner Res 2002; 17:1718–1727.
118. Deng HW, Xu FH, Huang QY, et al. A whole-genome linkage scan suggests several genomic regions potentially containing quantitative trait Loci for osteoporosis. J Clin Endocrinol Metab 2002; 87:5151–5159.
119. Wilson SG, Reed PW, Bansal A, et al. Comparison of genome screens for two independent cohorts provides replication of suggestive linkage of bone mineral density to 3p21 and 1p36. Am J Hum Genet 2003; 72:144–155.

3 Bone Physiology

Bone Cells, Modeling, and Remodeling

Lawrence G. Raisz

1. INTRODUCTION

There has been a remarkable expansion of the understanding of growth and remodeling of the skeleton based on advances in cell and molecular biology and genetics, as well as identification of the many local and systemic factors that regulate bone cell function. Despite this, there remain many unanswered questions concerning the regulation of these processes. This chapter focuses on the interplay of local and systemic factors in bone remodeling, particularly those that are most likely to be affected by changes in nutritional state. For an extensive review on this topic, see ref. *1.*

2. SKELETAL DEVELOPMENT

The skeleton is initially formed by a series of programmed mesenchymal condensations followed by formation of a cartilaginous template for all of the major skeletal structures. The conversion of the cartilaginous template to bone involves either endochondral bone formation in which the cartilage is initially calcified or membranous bone formation in which the bone cells arise adjacent to, but separate from, the cartilage anlage, which does not calcify but simply is resorbed.

Many genes have been identified that act early in embryonic development to determine the specific form of skeletal structures. Remarkably, there is a single' transcriptioned pathway that is critical for the conversion of the cartilaginous to a bony skeleton. The transcription factors RUNX-2/Cbfa-1 and a downstream factor, Osterix, are apparently necessary and sufficient for mesenchymal precursors to differentiate into osteoblastic cells *(1,2)*. The knockout of either of these factors results in a failure to form a bony skeleton, whereas in humans heterozygous for RUNX-2 there is a specific pattern of deformities termed cleido-cranial dysplasia *(3)*. The shape of a skeleton is determined genetically, but the size is determined by local growth factors, particularly insulin-like growth factors 1 and 2 (IGF-1 and IGF-2) and their receptors and binding proteins. Thus, knockouts of IGF-1 or

From: *Nutrition and Bone Health*
Edited by: M. F. Holick and B. Dawson-Hughes © Humana Press Inc., Totowa, NJ

IGF-2 result in decreased skeletal size with an extreme decrease occurring when both genes are deleted *(4,5)*. Growth hormone (GH)-dependent IGF secretion is important in early postnatal growth, but appears to be less important in prenatal growth when local IGF production, independent of GH, predominates. Maternal imprinting of the IGF-2 scavenger receptor limits skeletal growth, while paternal imprinting of IGF-2 enhances growth *(6,7)*.

Other factors may have both stimulatory and inhibitory effect on skeletal size. For example, fibroblast growth factor (FGF) can activate a specific receptor, FGF-3, to produce cessation of cartilage growth *(8)*. Activating mutations of this receptor result in dwarfism. Cartilage growth and differentiation is under complex local regulation by parathyroid hormone-related peptide (PTHRP) and genes involved in local signaling such as Indian hedgehog (Ihh) *(9)*. Insulin also acts on the IGF receptors and increased glucose-stimulated insulin production in fetuses of mothers with poorly controlled diabetes may be responsible for their increased birth weight *(10)*.

3. MODELING AND REMODELING

The skeleton undergoes constant change through the processes of modeling and remodeling. Modeling is defined as change in the size or shape of skeletal elements that occurs when formation and resorption are independent—for example, the enlargement of long bones by periosteal apposition and endosteal resorption. Remodeling is defined as resorption and formation that occur essentially at the same site. Remodeling may either be in balance, that is, the amount of new bone formed at the site is equal to the amount previously resorbed, or there may be remodeling imbalance with either gain or loss of bone because the amount of new formed is greater or less than the amount resorbed. Both modeling and remodeling are subject to systemic and local regulation. Trabecular bone remodeling is probably the predominate mechanism for fulfilling the homeostatic role of the calcium-regulating hormones, whereas cortical bone remodeling is probably of particular importance in the response to loading, the repair of microdamage, and maintenance of cell viability. Although modeling is greatest during skeletal growth, it continues throughout life, largely in response to mechanical forces. For example, increased strain due to weakening of the skeleton by endosteal resorption can produce new periosteal apposition even late in life.

4. THE BONE REMODELING CYCLE

The remodeling cycle involves a series of linked cellular events that has been turned the bone multicellular unit (BMU). Bone remodeling occurs on trabecular surfaces in the form of shallow irregular Howship's lacunae, whereas in cortical bone it occurs as a cylinder to form the Haversian canal, which is then replaced to form an osteon. There are four phases: activation, resorption, reversal, and formation (Fig. 1). Recent histological studies suggest that the BMU is compartmentalized, that is, there are lining cells that separate the BMU from the rest of

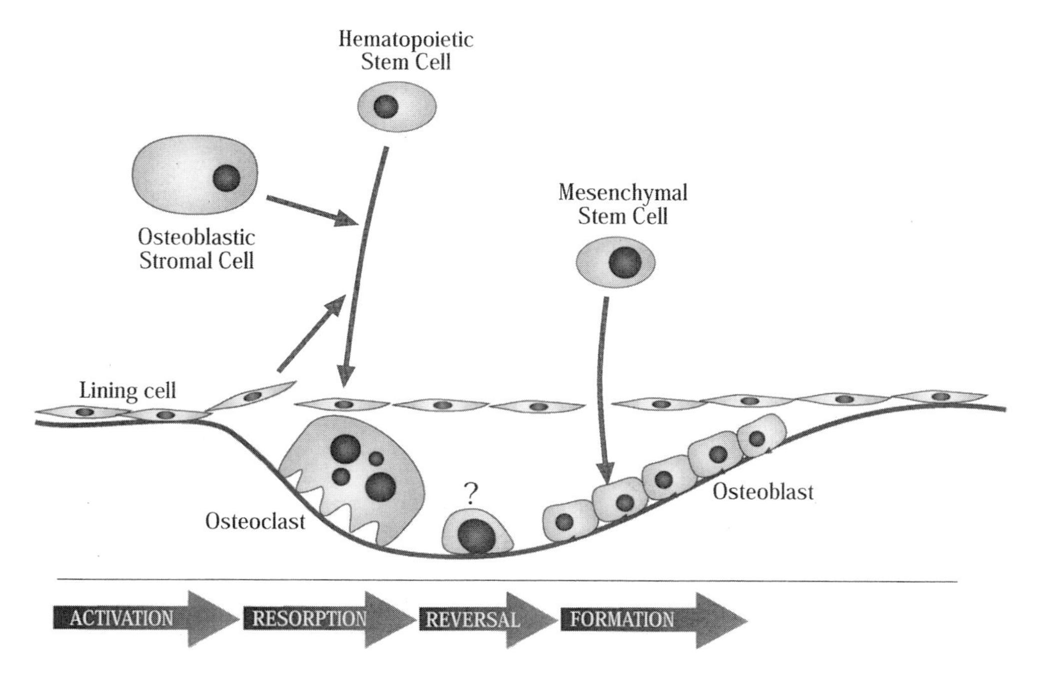

Fig. 1. Bone remodeling. The current concept of trabecular bone remodeling is illustrated here. Either osteoblastic stromal cells or lining cells may activate the resorption process. The bone remodeling units may be compartmentalized. The mechanism of entry of osteoclasts is not clear; in Haversian remodeling, the osteoclasts are presumable carried in by vessels. Although a single layer of osteoblasts is illustrated, multiple layers are required to refill the remodeling space.

the marrow space *(11,12)*. If correct, this could produce a substantially different local milieu for BMU, with levels of local factors and ions that differ from those in the general extracellular fluid.

4.1. Activation

Our understanding of the activation of bone resorption has advanced greatly in recent years with the discovery of a ligand-receptor system that explains the old observation that stimulators of bone resorption act largely on cells of the osteoblastic lineage and that this indirectly results in activation of osteoclasts *(13)* (Fig. 2). Stimulation of resorption results in the production of receptor activator of NFκB ligand (RANKL) on stromal cells and osteoblastic cells. RANKL then interacts with a receptor, RANK, on precursor cells of the hematopoietic lineage to initiate differentiation of these cells to form multinucleated osteoclasts and maintain their resorbing activity. This system is held in balance by the production of a decoy receptor, osteoprotegerin (OPG), which is also produced by osteoblasts and which may be downregulated by stimulators of resorption. While this system

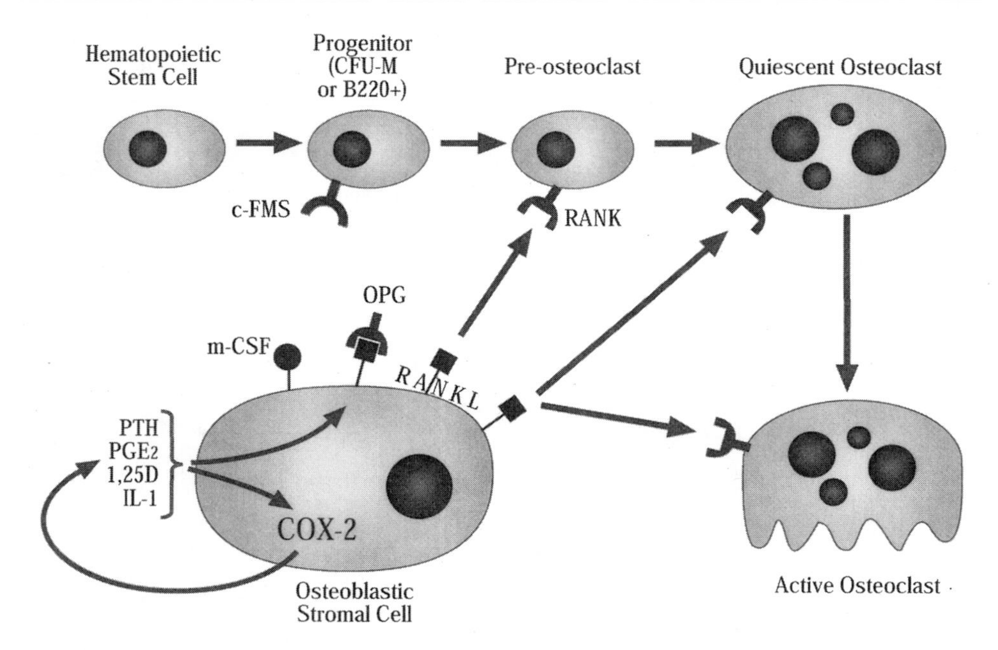

Fig. 2. Osteoblast–osteoclast interaction. The interaction of osteoblasts and osteoclasts involves macrophage colony-stimulating factor (MCSF) and its receptor c-fms, receptor activator of nuclear κB ligand (RANKL) and its receptor RANK. Activation of osteoblasts may be amplified by a number of factors. One mechanism illustrated here is induction of COX-2, but induction of other cytokines such as interleukin (IL)-6 could also be involved. Prostaglandins and cytokines may also act directly on the hematopoietic lineage to enhance osteoclastogenesis.

is clearly important in initiating and regulating osteoclast activity, additional steps may be necessary that are not as well understood. In addition to delineating the BMU compartment, the lining cells may be involved in removal of a protein layer covering bone that is necessary for osteoclast access *(14)*. If the lining cells do form a compartment, then the osteoclast precursors probably need to traverse the lining cell membrane, presumably through penetrating vessels, and this may also be a site for regulation.

4.2. Resorption

The size, duration, and depth of the resorption phase are presumably controlled at least in part by RANKL, which maintains osteoclast viability. However, control of the movement of osteoclasts along the bone surface or in Haversian canals cannels is less well understood. Moreover, we do not know the precise mechanism by which osteoclast resorption is stopped. High concentrations of calcium in the ruffled border area may result in separation and inactivation of osteoclasts *(15)*.

Transforming growth factor-β (TGFβ) released from the bone matrix may also cause osteoclast apoptosis, but TGFβ may also increase osteoclastic activity *(16,17)*. Calcitonin rapidly stops osteoclastic activity partly by blocking cell attachment, but this is probably an emergency brake on osteoclasts to protect against hypercalcemia rather than a general mechanism for stopping osteoclast activity *(18,19)*.

4.3. Reversal

The process of reversal is the least well understood phase of the remodeling cycle. Mononuclear cells are present on the bone surface that may complete the removal of matrix, deposit the mucopolysaccharide-rich material, the so-called cement line, between old and new bone, and possibly produce factors that attract and activate osteoblast precursors. It is not clear whether these mononuclear cells are derived from the mesenchymal or hematopoietic lineage cells or how they are regulated. A recent study of the reversal phase *(14)* suggests that bone lining cells rather than cells of the monocyte–macrophage lineage are present and that they digest protruding collagen fibrils and then deposit a thin osteopontin-rich cement line that is necessary before osteoblastic bone formation can begin.

4.4. Formation

The formation phase involves the replication, migration, and differentiation of mesenchymal cells and may also involve activation of previously dormant lining cells. The formation phase lasts for several months, compared to the few weeks of the resorption and reversal phases. Many factors are known to affect the magnitude of bone formation as discussed below, but the specific signals that lead to initiation and cessation of formation and that result in positive, equal, or negative balance between resorption and formation under different circumstances are still not well understood.

5. THE OSTEOCLAST LINEAGE

The concept that osteoclasts are derived from an alternative differentiation pathway for cells of the monocyte–macrophage lineage may require some revision *(20–22)*. Recent studies have demonstrated that pre-B-cells (B220-positive) cells can differentiate into osteoclasts if their normal pathway into mature B-cells is blocked. B220-positive plus cells may also differentiate to macrophages. Presumably all of these cells are derived initially from multipotent hematopoietic stem cells. Osteoclasts differentiate initially as mononuclear cells that express tartrate resistant acid phosphatase (TRAP) and may even express calcitonin receptors even before the critical process of fusion occurs. The large multinucleated osteoclast is a highly differentiated cell that is able to resorb bone because it adheres to the mineralized surface through a sealing zone and develops within that zone a ruffled bordered area in which the secretion of hydrogen ions and proteolytic enzymes can remove mineral and degrade matrix. The highly convoluted cell surface can form vacuoles, which contain the breakdown products that are transported thought the cell and release these products to the extracellular fluid. The osteoclast has a number of specialized features that permit it to function, including integrins for binding to bone, carbonic anhydrase and proton pumps to facilitate hydrogen ion secretion, lysosomal enzymes, particularly cathepsin K, to break down matrix, and ion-transport systems to permit the transport of calcium, phosphate, and excess bicarbonate to the extracellular fluid. Osteoclasts are terminally differentiated, relatively short-lived cells in which the nuclei undergo apoptosis. It is not clear whether all of the nuclei undergo apoptosis simultaneously or whether there are nuclei with different lifespans within a single multinucleated cell.

6. OSTEOBLASTS AND OSTEOCYTES

The differentiation of mesenchymal cells to fully functional osteoblasts that can deposit bone matrix and promote mineralization is a complex, multistep process. Freidenstein *(23)* coined the terms inducible and determined osteoprogenitor cells (IOPCs and DOPCs) to indicate that there were stromal that were pluripotential and could be induced to form osteoblasts, while other cells further along the differentiation pathway were irreversibly differentiated. This concept has undergone some revision recently because cells that have many of the characteristics of DOPCs can be diverted in culture to other phenotypes such as adipocytes or cartilage cells. Thus it seems more likely that the process of differentiation is a continuum with many different stages. Moreover, the differentiated osteoblasts probably does not represent a single phenotype. There appear to be variations in the expression of genes such as osteocalcin among osteoblasts. Moreover, the morphology and rate of collagen synthesis of osteoblasts may vary from very high in the largest columnar cells to low in more flattened osteoblasts. Osteoblasts not only produce a collagenous matrix, but regulate its posttranslational modification and mineralization through the production of noncollagen proteins. Once an osteoblast has completed its task of matrix formation, it can undergo several fates. Some osteoblasts become buried as osteocytes, while others form lining cells. Probably the majority of osteoblasts undergo apoptosis. The regulation of these alternative pathways undoubtedly has considerable impact on the amount of newly formed bone and its structure.

The osteoblasts, lining cells, and osteocytes in each BMU form a syncytium in which each cell is connected to other by multiple cytoplasmic processes. This syncytium probably fulfills many functions in bone. One clearly important function is to detect and transmit the skeletal response to mechanical forces. This impact loading of the skeletal is amplified by fluid shear stress around the canalicular processes, and the osteocyte cell bodies produces signals involving change in ion flux, nitric oxide production, and prostaglandin production, which can initiate modeling and remodeling responses *(24,25)*. Microdamage as a consequence of repetitive loading may also provide a signal for these responses.

7. OTHER CELL TYPES

Angiogenesis plays a critical role in bone modeling and remodeling *(26)*. Resting and proliferative cartilage are normally resistant to vascular invasion, but hypertrophic cartilage, particularly after calcification of the matrix, appears to induce vascular invasion. This brings chondroclasts that resorb the calcified cartilage and stromal cells that initiate the bone formation in the primary spongiosa. Haversian remodeling also depends on the entry of new blood vessels into bone. Epidermal growth factor (EGF) produced either by endothelial cells or by cells of the osteoblastic lineage plays a critical role in this process. Endothelial cells may also produce cytokines and other regulatory factors such as nitric oxide and prostaglandin, which influence bone cell function.

Interactions between marrow and bone are critical in bone modeling and remodeling. Not only is the marrow the source of stromal cells and osteoclast precursors, other marrow elements may play an important regulatory role. For example, T-cells are an important source of both membrane-associated and soluble RANKL *(26–28)*. This is likely to be of particular importance in inflammatory bone loss and malignancy, but may also be important in physiological remodeling. Increased hematopoiesis is associated with expansion of the marrow cavity at the expense of bone in disorders such as thalassemia *(29)*. There is an ill-defined association between mast cells and bone loss. Patients with mastocytosis develop osteoporosis associated with increased interleukin (IL)-6 production *(30)*, and increased mast cell numbers have been described in idiopathic osteoporosis as well. Finally, there appears to be a reciprocal relationship between adipocytes and osteoblasts in bone. There is an increase in adipocyte numbers in the marrow with aging, which may be exaggerated in osteoporotic patients.

8. SYSTEMIC HORMONES REGULATING BONE REMODELING

Metabolic functions of the skeleton are largely controlled by two calcium-regulating hormones, PTH and 1,25-$(OH)_2D$. Calcitonin appears to play a lesser role—indeed, calcium homeostasis is not impaired in patients with either calcitonin excess such as those with medullary carcinoma of the thyroid or with calcitonin deficiency after total thyroidectomy. Whereas the primary function of PTH is to maintain serum ionized calcium concentration, the level of PTH also establishes the overall rate of bone turnover. Thus, increased PTH levels in hyperparathyroidism result in increased bone remodeling. After an initial period of bone loss, this process often appears to be in balance *(31)*.

Recent studies suggest that both PTH and PTHRP are critical for early skeletal development *(32)*. Stimulation of bone formation by PTH and also by PTHRP is probably critical in establishing optimal skeletal mass and structure. Once the skeleton is fully formed, PTH deficiency, that is, hypoparathyroidism, is associated with low bone turnover and high bone mass.

The relative dominance of the catabolic or anabolic effects of PTH appears to depend on both the level and duration of increased hormone levels. In rats that are given the same dose of PTH intermittently or continuously, the former administration results in increased bone formation with no increase in bone resorption, while the latter results in increased bone resorption with no increase in bone formation *(33)*. A plausible explanation is that the ability of PTH to increase RANKL and decrease OPG levels depends on more prolonged stimulation of osteoblasts, while only a transient stimulation is sufficient to induce growth responses, probably related to increased production of local growth factors including IGF and possibly TGFβ, IL-6 fibroblast growth factor (FGF), and prostaglandin E_2 (PGE_2). PTH effects also differ for different types of bone. In hyperparathyroidism, loss of cortical bone usually predominates while trabecular bone is preserved, although there may be a subset, particularly of postmenopausal women, in whom PTH excess increases spinal trabecular bone loss.

PTH excess may play a role in the pathogenesis of osteoporosis. Patients with calcium and vitamin D deficiency will show an increase in PTH levels, and this will increase bone resorption. Interestingly, the age-related increase in PTH levels may be related not only to calcium and vitamin D deficiency but also to estrogen deficiency, since women given estrogen replacement do not show this increase.

The most important role of the vitamin D hormone system is to maintain the intestinal absorption of calcium and phosphate and supply these ions for bone mineralization. Indeed, skeletal growth and development can be relatively normal in the absence of 1,25-$(OH)_2D$ or its receptor if calcium and phosphate supplies are adequate. In humans and experimental animals lacking the vitamin D receptor, but given sufficient amounts of calcium and phosphate, the major phenotypic abnormality is alopecia *(34)*. 1,25-$(OH)_2D$ can also directly stimulate bone resorption at high concentration. This effect may be particularly important in conditions of marked calcium and phosphate deficiency when 1,25-$(OH)_2D$ synthesis is maximal. Resorption takes these essential ions from the skeleton to provide adequate supplies for soft tissues.

There is considerable debate about the role of 1,25-$(OH)_2D$ in osteoblast function. An anabolic effect of this hormone and particularly of certain analogs has been described, but nothing comparable to effect of PTH has been demonstrated. Vitamin D receptors have been found in many other cells types, particularly hematopoietic cells. Abnormalities of vitamin D metabolism may also affect immune responses *(35)*. Vitamin D may play a role in neuromuscular function, and impaired balance and muscular strength are associated with vitamin D depletion. This may be in part related to changes in calcium and phosphate availability to nerve and muscle.

9. GROWTH-REGULATING HORMONES

The major hormones that regulate tissue growth and metabolism all have a major influence on skeletal growth and remodeling, including the growth hormone–insulin-like growth factor (GH–IGF) system, thyroid hormones, and adrenal glucocorticoids. Recently discovered nutritional regulators, leptin and ghrelin, may also affect the skeleton, although current data are limited and conflicting.

The GH–IGF system determines body size, including the size of the skeleton before epiphyseal closure, and regulates the distribution of body fat, lean body mass, and bone modeling and remodeling after epiphyseal closure. This system can be considered as both a systemic and local regulator of bone metabolism. GH can stimulate IGF production not only in the liver but also in other target organs, including bone. Thus, skeletal growth is not markedly impaired by targeted deletion of IGF-1 production in the liver *(36)*. However, deleting liver IGF-1 and its binding protein does impair growth *(37)*. The GH–IGF system stimulates both resorption and formation. In human studies, GH treatment has resulted in transient decreases in bone mass due to a more rapid increase in resorption with a subsequent gradual and progressive increase in bone formation resulting in an overall increase in bone mass after 2–3 yr of treatment *(38)*. The function of this system is markedly influenced by a number of binding proteins *(39)*. Among these, IGFBP2-4 all appear to have inhibitory effects on IGF action, whereas IGFBP5 appears to be stimulatory and may even have direct anabolic effects. The main circulating binding protein, IGFBP3, can inhibit the effects of IGF-1 in vitro, but prolongs its half-life in the circulation and enhances the in vivo anabolic response to IGF-1.

IGF-2 plays a critical role in skeletal development, but its role in the adult skeleton is less clear. Both IGF-1 and IGF-2 are subject to nutritional regulation, and a decrease in these factors associated with protein-calorie malnutrition probably mediates the associated impairment of growth and loss of bone mass.

Thyroid hormone has an effect similar to PTH on bone. In hyperthyroidism, bone resorption and formation are both increased while PTH levels are decreased *(40)*. However, thyroid hormone does not stimulate vitamin D activation, and calcium absorption may be decreased in hyperthyroid patients. Conversely, in hypothyroid patients, bone turnover is decreased and PTH levels may be increased. Thyroid hormone excess, either as primary hyperthyroidism or due to excessive thyroid hormone therapy, can contribute to bone loss. Both a history of hyperthyroidism and hypothyroidism *(41)* are associated with an increased risk for fractures. Hyperthyroidism may accelerate growth in children, while hypothyroidism clearly diminishes growth. The latter may be the result of an impairment of the GH–IGF axis.

Adrenal glucocorticoids have complex effects on bone cell function *(42,43)*. Glucocorticoids enhance bone cell differentiation. Thus, decreased glucocorticoid production or increased inactivation of glucocorticoids may impair fetal skeletal growth and development. In children, glucocorticoid excess can markedly impair skeletal growth. This may not require high concentrations of glucocorticoid, but depends more on the fact that diurnal rhythm of glucocorticoid secretion is lost so that there is no period in the afternoon and night of extremely low secretion. This diurnal rhythm is essential for skeletal growth. During the period of low glucocorticoid secretion, the GH–IGF-1 axis is activated and there appears to be an increase in bone formation as indicated by a nocturnal rise in the bone formation marker, osteocalcin. Osteocalcin is particularly sensitive to glucocorticoids. There is a glucocorticoid response element (GRE) in the osteocalcin promoter whereby glucocorticoid receptor binding can inhibit transcription *(44)*. Glucocorticoid responses can be controlled not only by the circulating levels, but also by the level of glucocorticoid receptors in the cell and by the rate at which glucocorticoids are inactivated by the enzyme 11β dehydrogenase *(45)*.

The biphasic effects of glucocorticoids can be demonstrated in organ culture. There is an initial increase in collagen synthesis, probably related to an increase in the cell responsiveness to IGF-1, followed by a prolonged decrease associated with decreased IGF-1 production. Glucocorticoid-induced osteoporosis is the most common form of secondary osteoporosis and results largely from the direct inhibition of bone formation by glucocorticoids. However, there may also be an increase in bone resorption, probably as an indirect result of decreased sex hormone production and decreased calcium absorption and secondary hyperparathyroidism. PTH levels are not consistently increased in patients with glucocorticoid excess *(46)*.

Two recently identified hormones that are critical nutritional regulators may also play a role in bone. Leptin is a product of adipose tissues that decreases appetite. There are animal models of obesity with either leptin resistance or leptin deficiency. In these models hypogonadism may occur, yet despite this bone mass is increased. This led to the hypothesis that leptin might inhibit bone formation. Animal studies suggested that this was a indirect central effect and that there was a central nervous system hormone, possibly a β-adrenergic agonist that stimulates bone formation, that is downregulated by leptin *(47)*. However, studies of the peripheral effect of leptin suggest that it is a direct stimulator of bone formation an inhibitor of bone resorption *(48,49)*. These paradoxical central and peripheral pathways have not yet been fully resolved.

Ghrelin was recently identified as a hormone produced by the stomach that stimulates appetite *(50)*. However, the initial results that led to its discovery came from studies of an exogenous growth hormone-releasing peptide (GHRP). Ghrelin was found to be the natural ligand for the GHRP receptor. At this point there is little information on any direct effects of ghrelin on bone. There is likely to be an indirect effect through stimulation of the GH–IGF axis. Another osteotropic hormone from the gastrointestinal tract has been established.

10. SEX HORMONES

Sex hormones play a critical role in the regulation of skeletal development and bone remodeling *(50)*. Estrogen deficiency is probably the single most important factor in the pathogenesis of osteoporosis in both men and women. Estrogen acts on the skeleton not only through its direct effects on cartilage and bone cells, but also through its effects on other hematopoietic cell lineages and possibly also on endothelial cells in bone *(51–55)*. A large number of pathways by which estrogen might act have been identified, but their relative importance in skeletal physiology is still not established. Analysis of estrogen action is complicated by the fact that there are two estrogen receptors ERα and ERβ, and that there may also be estrogen effects on the cell membrane that are independent of either receptor *(56)*. Studies of men with estrogen deficiency, due either to loss of ERα or to a defect in aromatase the enzyme that converts androgen to estrogen, show a similar phenotype. They have failure of epiphyseal closure, increased bone turnover, and low BMD. Current studies also suggest that activation of ERα is the most important pathway for skeletal affects *(57,58)*. Estrogen may inhibit resorption by altering the production of cytokines in marrow cells, increasing OPG production or inducing osteoclast apoptosis, possibly by increasing TGFβ *(16,59–61)*. High concentrations of estrogen can clearly stimulate bone formation in animal models and may enhance the bone formation response to mechanical forces *(58,62)*. In estrogen deficiency, bone formation is increased, but the increase is clearly inadequate to compensate for the increase in bone resorption. Hence there is a relative defect in bone formation. The orphan receptor, estrogen receptor-related receptor α, has recently been identified, which may modulate estrogen responses *(63)*.

The effects of estrogen on bone may involve different pathways and a different dose–response relation from its effects on classic target organs such as the uterus and breast. Selective estrogen receptor modulators (SERMs), which do not stimulate the breast or uterus, can still inhibit bone resorption and prevent bone loss after ovariectomy *(64)*. Bone turnover can be decreased by low doses of estrogen, and fracture risk is increased in women with extremely low serum estradiol concentrations *(65,66)*. Indeed, the estrogens produced by aromatase in fat tissue may be sufficient to protect the skeleton in postmenopausal women, hence the association of low body weight, low bone mass, and increased fracture risk in this population. Estrogens have multiple effects on hematopoietic cells. In animal models, estrogen deficiency is associated with an increased number of B-cells in the marrow, and B-cell precursors may be a source of osteoclasts *(22)*. Both estradiol and raloxifene can affect β-lymphopoiesis *(67)*. Estrogen can modulate activation and cytokine production of T- and B-cells as well as cells of the monocyte–macrophage lineage.

Our understanding of the role of testosterone in bone metabolism is complicated by the fact that testosterone can be converted to estrogen by many tissues, probably including bone *(68)*. Studies using dihydrotestosterone, which cannot be aromatized, suggests that androgens can both stimulate bone formation and inhibit bone resorption *(69,70)*. Whether these effects are mediated only through the androgen receptor or by binding to the estrogen receptor as well is not established. Androgens are probably responsible for the stimulation of bone growth that occurs at puberty and can probably stimulate bone formation in adults as well. However, studies in humans suggest that estrogen is far more important than androgen in inhibiting bone resorption *(71)*. Androgen effects on muscle may indirectly produce skeletal responses by altering mechanical forces exerted on bone *(72)*. The role of progesterone in bone is controversial. A number of authors have suggested that progesterone may stimulate bone formation. However, markers of bone formation were not increased by administration of progestins to postmenopausal women, either untreated or on estrogen therapy *(73–75)*. Other hormones of the pituitary-reproductive system have also been implicated in bone remodeling. In particular, inhibin may suppress bone remodeling *(76)*. Inhibin deficiency could play a role in the increased bone remodeling and bone loss that occurs in the perimenopause, at a time when estrogen levels are not generally decreased but inhibin production declines as indicated by a rising follicle-stimulating hormone level.

11. LOCAL REGULATORY FACTORS

The recognition that there are large number of local factors that regulate bone remodeling has been one of the major advances in bone biology in recent decades *(77,78)*. An adequate review of this topic is beyond the scope of this chapter, not only because of the number of factors, but because of the complexity of their actions and interactions. Cytokines, such as IL-1, and tumor necrosis factor-α (TNFα), growth factors, such IGF-1 and the TGFβ and the related bone morphogenetic protein (BMP) family, small molecules such as prostaglandins, leukotrienes, and nitric oxide, and neuropeptides have all been found to affect bone formation and resorption and have been implicated in meta-physiological and pathological skeletal responses.

The concept that cytokines play a role in bone began with an observation made 30 yr ago that supernatants of human mononuclear cells cultured with a activator, such as an agglutinin or an antigen to which the patient had cell-mediated immunity, could cause bone resorption. This was initially called osteoclast activating factor. Subsequent studies indicated that a major component was IL-1 but TNFα and other cytokines were shown to have quite similar activities. These cytokines can also affect bone formation, although both inhibitory and stimulatory responses have been reported. Other cytokines, such as IL-6 and IL-11, can stimulate both bone resorption and formation. A number of cytokines can inhibit bone resorption, including IL-4, IL-13, IL-18, and interferon (IFN)-γ or IFN-β. Recently it was found that IFN-β expression was increased in cells of the osteoclastic lineage treated with RANKL *(79)*. This appears to be a second balancing system for preventing excessive osteoclastogenesis, analogous to the RANKL-OPG system. Studies with cytokine inhibitors and knockouts of the cytokines or their receptors have confirmed these effects in murine models *(80)*. The relative importance of these cytokines in human bone remodeling and particularly in the pathogenesis of osteoporosis has not been established. It is possible that a deficiency of inhibitory cytokines as well as an excess of those that stimulate resorption could play a pathogenetic role in bone loss.

Prostaglandins, particularly prostaglandin E_2 (PGE_2), are potent multifunctional regulators of bone metabolism *(78)*. The predominant response to PGE_2 is stimulation of bone resorption and formation, although there are also inhibitory responses under certain conditions. Stimulation of bone turnover by PGE_2 is likely to be involved in inflammatory bone loss and appears to play an important role in the response to mechanical forces. The systemic and local factors that stimulate bone resorption increase the production of PGE_2 via the inducible enzyme cyclooxygenase-2 (COX-2). As a result, bone resorptive responses both in vivo and in vitro may be blunted in animals who cannot form prostaglandins because COX-2 is inhibited or deleted or because specific receptors for PGE_2, particularly the EP2 and EP4 receptors, are inactivated *(81–84)*. PTH, 1,25-(OH2)D, thyroid hormone, IL-1 and TNFα, BMP-2 and TGFβ *(85,86)* can all induce COX-2. Osteoclastogenesis in response to these factors is diminished in bone marrow cultures from knockout mice. However, these factors also have the capacity to stimulate bone resorption by prostaglandin-independent pathways. PGE_2 can itself stimulate COX-2 *(87)*. This auto-amplification pathway is probably important in the skeletal response to mechanical forces.

A number of other small molecules have been implicated in bone remodeling. Nitric oxide can act as both an inhibitor of bone resorption and a stimulator of bone formation of PGE_2 production *(88)*. Leukotrienes, neuropeptides, and nucleotides have also been shown to act on bone *(89–92)*.

Bone cells and produce and/or respond to many different growth factors. As discussed above, IGF-1 is a potent stimulator of bone formation, which also increases bone resorption. IGF-1 is not only produced by bone cells but is probably stored in bone matrix together with binding proteins and may be released when these binding proteins are degraded *(93)*. The large family of TGFβ-related proteins including several of the BMPs may also be stored in the matrix. There is a specific release pattern for TGFβ involving release from a binding protein and activation of the precursor molecule *(94)*. This may occur during bone remodeling and has been implicated in the inhibition of resorption and the initiation of formation during the reversal phase of the bone remodeling cycle. Other growth factors, such as vascular endothelial growth factor (VEGF), EGF platelet-derived growth factor (PDGF), and FGF, are also likely to be local regulators of bone formation and possibly also bone resorption *(95–98)*. There may be an amplification system for these factors involving PGE_2. VEGF can stimulate prostaglandin production in endothelial cells, while EGF, PDGF, and FGF can stimulate prostaglandin production in bone. PGE_2 has been shown to stimulate the production of the VEGF and FGF in bone cells *(99)*.

Recently another potential local regulatory pathway has been discovered through the genetic analysis of unusual families that have high bone density and of a rare genetic form of bone loss, the osteoporosis-pseudoglioma syndrome *(100–102)*. High bone density is associated with an activating mutation of low-density lipoprotein receptor related-5 (LRP-5), while deletion of this receptor results in the osteoporosis-pseudoglioma syndrome. This receptor is associated with the Wnt signaling pathway, which is critical for craniofacial development.

Neither the specific ligand that activates the LRP-5 receptor nor the mechanisms by which it affects bone mass has yet been identified.

12. REGULATION BY CALCIUM, PHOSPHORUS, AND OTHER IONS

Calcium is not only essential for regulation of neuromuscular activity and cellular function throughout the body, it also may play a specific role as a local regulator in bone. High concentrations of calcium are likely to develop in the ruffled border area, where the hydrogen ion concentration is high and mineral is being dissolved. These high concentrations of calcium have been shown to inhibit osteoclast function, probably largely by causing a loss of cell adhesion. This effect may also require a increase in intracellular calcium ion concentration. There is controversy concerning the mechanism by which osteoclasts sense calcium. There may be a specific calcium receptor, but this appears to differ from the calcium-sensing receptor of the parathyroid glands. High local concentrations of calcium may also affect osteoblast function *(103)*.

Although extracellular ionized calcium concentration is tightly regulated, the extracellular concentration of phosphate shows wide variation, not only during different stages of skeletal growth and maturation in humans, but also among different mammalian species. In general, higher phosphate concentrations are associated with more rapid skeletal growth and mineralization; for example, serum phosphate concentrations are elevated during the rapid growth phases of early infancy and puberty. In organ culture, increasing phosphate concentration has been shown to enhance both matrix formation and mineralization *(104)*. There is also a reciprocal relation between calcium and phosphate concentrations, particularly because of the effect of parathyroid hormone to decrease renal tubular absorption of phosphate and lower serum levels while increasing serum calcium concentration. Phosphate has complex effects on osteoclastic bone resorption. Phosphate depletion is associated with increased serum calcium and increased resorption rates in organ culture, which may be in part due to a change in the physical chemical gradent for removal of mineral *(105)*. Defects in osteoclast phosphate transport, on the other hand, are associated with impaired osteoclastic bone resorption *(106)*.

Magnesium can affect bone remodeling both directly and indirectly. In severe magnesium deficiency, parathyroid hormone secretion in impaired, bone resorption is decreased, and hypocalcemia can develop *(107)*. The low magnesium levels may also directly affect both osteoblast and osteoclast function *(108)*. The skeleton is a reservoir not only for calcium and phosphate, but also for sodium and for hydroxyl ions. The skeleton forms a second line defense after the kidney in acid–base balance, providing both phosphate and hydroxyl ions to buffer hydrogen ion excess. Severe chronic acidosis can cause bone loss. It has been postulated that the typical Western high-protein diet represents a acid load that may cause mineral loss *(109)*.

Fluoride is a potent stimulator of osteoblasts and can cause increased bone formation *(110,111)*. Unlike the anabolic effect of PTH, the response to fluoride can produce bone that is structurally abnormal, resulting in increased fragility rather than increased strength. Fluoride also is incorporated into the hydroxyapatite mineral, producing an alteration in crystal structure. In humans, low doses of fluoride may increase bone strength and decrease fracture risk, but the use of fluoride for the treatment of osteoporosis has been largely discontinued because of the relatively narrow range between therapeutic and adverse effects.

Strontium is incorporated into the bone mineral and may have an anabolic effect on osteoblasts. Recent studies using a novel strontium salt, strontium enanthate, show not only an increase in bone density but a decrease in fractures in osteoporotic patients *(112)*.

13. CONCLUSION

The major goal of this chapter has been to provide the reader with help in asking appropriate questions concerning possible pathways that might influence bone resorption and formation and that might in turn be influenced by nutrition. While this field is evolving extremely rapidly and many new regulatory factors and new interactions are likely to be discovered, the regulatory mechanisms outlined here are important to understand not only because they are affected by nutrition but also because they may be the key to understanding interactions between nutrition and genetics.

REFERENCES

1. Komori T. Runx2, a multifunctional transcription factor in skeletal development. J Cell Biochem 2002; 87:1–8.
2. Nakashima K, Zhou X, Kunkel G, et al. The novel zinc finger-containing transcription factor osterix is required for osteoblast differentiation and bone formation. Cell 2002; 108:17–29.
3. Yoshida T KH, Osato M, Yanagida M, Miyawaki T, Ito Y, Shigesada K. Functional analysis of RUNX2 mutations in Japanese patients with cleidocranial dysplasia demonstrates novel genotype-phenotype correlations. Am J Hum Genet 2002; 71:724–738.
4. Baker J, Liu JP, Robertson EJ, Efstratiadis A. Role of insulin-like growth factors in embryonic and postnatal growth. Cell 1993; 75:73–82.
5. Liu JP, Baker J, Perkins AS, Robertson EJ, Efstratiadis A. Mice carrying null mutations of the genes encoding insulin-like growth factor I (Igf-1) and type 1 IGF receptor (Igf1r). Cell 1993; 75:59–72.
6. DeChiara TM, Robertson EJ, Efstratiadis A. Parental imprinting of the mouse insulin-like growth factor II gene. Cell 1991; 64:849–859.
7. Reik W, Davies K, Dean W, Kelsey G, Constancia M. Imprinted genes and the coordination of fetal and postnatal growth in mammals. Novartis Found Symp 2001; 237:19–31; discussion 31–42.
8. Vajo Z, Francomano CA, Wilkin DJ. The molecular and genetic basis of fibroblast growth factor receptor 3 disorders: the achondroplasia family of skeletal dysplasias, Muenke craniosynostosis, and Crouzon syndrome with acanthosis nigricans. Endocr Rev 2000; 21:23–39.
9. Kobayashi T, Chung UI, Schipani E, et al. PTHrP and Indian hedgehog control differentiation of growth plate chondrocytes at multiple steps. Development 2002; 129:2977–2986.

10. Lepercq J, Taupin P, Dubois-Laforgue D, et al. Heterogeneity of fetal growth in type 1 diabetic pregnancy. Diabetes Metab 2001; 27:339–344.
11. Hauge EM, Qvesel D, Eriksen EF, Mosekilde L, Melsen F. Cancellous bone remodeling occurs in specialized compartments lined by cells expressing osteoblastic markers. J Bone Miner Res 2001; 16:1575–1582.
12. Parfitt AM. The bone remodeling compartment: a circulatory function for bone lining cells. J Bone Miner Res 2001; 16:1583–1585.
13. Khosla S. Minireview: the OPG/RANKL/RANK system. Endocrinology 2001; 142:5050–5055.
14. Everts V, Delaisse JM, Korper W, et al. The bone lining cell: its role in cleaning Howship's lacunae and initiating bone formation. J Bone Miner Res 2002; 17:77–90.
15. Lorget F, Kamel S, Mentaverri R, et al. High extracellular calcium concentrations directly stimulate osteoclast apoptosis. Biochem Biophys Res Commun 2000; 268:899–903.
16. Hughes DE, Dai A, Tiffee JC, Li HH, Mundy GR, Boyce BF. Estrogen promotes apoptosis of murine osteoclasts mediated by TGF-beta. Nat Med 1996; 2:1132–1136.
17. Fuller K, Lean JM, Bayley KE, Wani MR, Chambers TJ. A role for TGFbeta(1) in osteoclast differentiation and survival. J Cell Sci 2000; 113(Pt 13):2445–2453.
18. Zikan V, Stepan JJ. Plasma type 1 collagen cross-linked C-telopeptide: a sensitive marker of acute effects of salmon calcitonin on bone resorption. Clin Chim Acta 2002; 316:63–69.
19. Sexton PM, Findlay DM, Martin TJ. Calcitonin. Curr Med Chem 1999; 6:1067–1093.
20. Rolink AG, Melchers F. Precursor B cells from Pax-5-deficient mice—stem cells for macrophages, granulocytes, osteoclasts, dendritic cells, natural killer cells, thymocytes and T cells. Curr Top Microbiol Immunol 2000; 251:21–26.
21. Grcevic D, Katavic V, Lukic IK, Kovacic N, Lorenzo JA, Marusic A. Cellular and molecular interactions between immune system and bone. Croat Med J 2001; 42:384–392.
22. Sato T, Shibata T, Ikeda K, Watanabe K. Generation of bone-resorbing osteoclasts from B220+ cells: its role in accelerated osteoclastogenesis due to estrogen deficiency. J Bone Miner Res 2001; 16:2215–2221.
23. Friedenstein AJ, Piatetzky S, II, Petrakova KV. Osteogenesis in transplants of bone marrow cells. J Embryol Exp Morphol 1966; 16:381–390.
24. Klein-Nulend J, Burger EH, Semeins CM, Raisz LG, Pilbeam CC. Pulsating fluid flow stimulates prostaglandin release and inducible prostaglandin G/H synthase mRNA expression in primary mouse bone cells. J Bone Miner Res 1997; 12:45–51.
25. Wadhwa S, Godwin SL, Peterson DR, Epstein MA, Raisz LG, Pilbeam CC. Fluid flow induction of cyclo-oxygenase 2 gene expression in osteoblasts is dependent on an extracellular signal-regulated kinase signaling pathway. J Bone Miner Res 2002; 17:266–274.
26. Collin-Osdoby P. Role of vascular endothelial cells in bone biology. J Cell Biochem 1994; 55:304–309.
27. Kotake S, Udagawa N, Hakoda M, et al. Activated human T cells directly induce osteoclastogenesis from human monocytes: possible role of T cells in bone destruction in rheumatoid arthritis patients. Arthritis Rheum 2001; 44:1003–1012.
28. Nosaka K, Miyamoto T, Sakai T, Mitsuya H, Suda T, Matsuoka M. Mechanism of hypercalcemia in adult T-cell leukemia: overexpression of receptor activator of nuclear factor kappaB ligand on adult T-cell leukemia cells. Blood 2002; 99:634–640.
29. Dresner Pollack R, Rachmilewitz E, Blumenfeld A, Idelson M, Goldfarb AW. Bone mineral metabolism in adults with beta-thalassaemia major and intermedia. Br J Haematol 2000; 111:902–907.
30. Theoharides TC, Boucher W, Spear K. Serum interleukin-6 reflects disease severity and osteoporosis in mastocytosis patients. Int Arch Allergy Immunol 2002; 128:344–350.
31. Silverberg SJ. Natural history of primary hyperparathyroidism. Endocrinol Metab Clin N Am 2000; 29:451–464.
32. Miao D, He B, Karaplis AC, Goltzman D. Parathyroid hormone is essential for normal fetal bone formation. J Clin Invest 2002; 109:1173–1182.

33. Dobnig H, Turner RT. The effects of programmed administration of human parathyroid hormone fragment (1-34) on bone histomorphometry and serum chemistry in rats. Endocrinology 1997; 138:4607–4612.
34. Li YC, Amling M, Pirro AE, et al. Normalization of mineral ion homeostasis by dietary means prevents hyperparathyroidism, rickets, and osteomalacia, but not alopecia in vitamin D receptor-ablated mice. Endocrinology 1998; 139:4391–4396.
35. Panda DK, Miao D, Tremblay ML, et al. Targeted ablation of the 25-hydroxyvitamin D 1alpha-hydroxylase enzyme: evidence for skeletal, reproductive, and immune dysfunction. Proc Natl Acad Sci USA 2001; 98:7498–7503.
36. Sjogren K, Liu JL, Blad K, et al. Liver-derived insulin-like growth factor I (IGF-I) is the principal source of IGF-I in blood but is not required for postnatal body growth in mice. Proc Natl Acad Sci USA 1999; 96:7088–7092.
37. Yakar S, Rosen CJ, Beamer WG, et al. Circulating levels of IGF-1 directly regulate bone growth and density. J Clin Invest 2002; 110:771–781.
38. Blackman MR, Sorkin JD, Munzer T, et al. Growth hormone and sex steroid administration in healthy aged women and men: a randomized controlled trial. JAMA 2002; 288:2282–2292.
39. Mohan S. IGF-binding proteins are multifunctional and act via IGF-dependent and independent mechanisms. J Endocrinol 2002; 175:19–31.
40. Akalin A, Colak O, Alatas O, Efe B. Bone remodelling markers and serum cytokines in patients with hyperthyroidism. Clin Endocrinol (Oxf) 2002; 57:125–129.
41. Vestergaard P, Mosekilde L. Fractures in patients with hyperthyroidism and hypothyroidism: a nationwide follow-up study in 16,249 patients. Thyroid 2002; 12:411–419.
42. Canalis E, Delany AM. Mechanisms of glucocorticoid action in bone. Ann N Y Acad Sci 2002; 966:73–81.
43. Harrison JR, Woitge HW, Kream BE. Genetic approaches to determine the role of glucocorticoid signaling in osteoblasts. Endocrine 2002; 17:37–42.
44. Meyer T, Gustafsson JA, Carlstedt-Duke J. Glucocorticoid-dependent transcriptional repression of the osteocalcin gene by competitive binding at the TATA box. DNA Cell Biol 1997; 16:919–927.
45. Woitge H, Harrison J, Ivkosic A, Krozowski Z, Kream B. Cloning and in vitro characterization of alpha 1(I)-collagen 11 beta-hydroxysteroid dehydrogenase type 2 transgenes as models for osteoblast-selective inactivation of natural glucocorticoids. Endocrinology 2001; 142:1341–1348.
46. Rubin MR, Bilezikian JP. Clinical review 151: the role of parathyroid hormone in the pathogenesis of glucocorticoid-induced osteoporosis: a re-examination of the evidence. J Clin Endocrinol Metab 2002; 87:4033–4041.
47. Takeda S, Elefteriou F, Levasseur R, et al. Leptin regulates bone formation via the sympathetic nervous system. Cell 2002; 111:305–317.
48. Cornish J, Callon KE, Bava U, et al. Leptin directly regulates bone cell function in vitro and reduces bone fragility in vivo. J Endocrinol 2002; 175:405–415.
49. Holloway WR, Collier FM, Aitken CJ, et al. Leptin inhibits osteoclast generation. J Bone Miner Res 2002; 17:200–209.
50. Wang G, Lee HM, Englander E, Greeley GH Jr. Ghrelin—not just another stomach hormone. Regul Pept 2002; 105:75–81.
51. Shevde NK, Bendixen AC, Dienger KM, Pike JW. Estrogens suppress RANK ligand-induced osteoclast differentiation via a stromal cell independent mechanism involving c-Jun repression. Proc Natl Acad Sci USA 2000; 97:7829–7834.
52. Roggia C, Gao Y, Cenci S, et al. Up-regulation of TNF-producing T cells in the bone marrow: a key mechanism by which estrogen deficiency induces bone loss in vivo. Proc Natl Acad Sci USA 2001; 98:13960–13965.

53. Perry MJ, Samuels A, Bird D, Tobias JH. Effects of high-dose estrogen on murine hematopoietic bone marrow precede those on osteogenesis. Am J Physiol Endocrinol Metab 2000; 279:E1159–E1165.
54. Evans MJ, Harris HA, Miller CP, Karathanasis SK, Adelman SJ. Estrogen receptors alpha and beta have similar activities in multiple endothelial cell pathways. Endocrinology 2002; 143:3785–3795.
55. Chambliss KL, Shaul PW. Estrogen modulation of endothelial nitric oxide synthase. Endocr Rev 2002; 23:665–686.
56. Kousteni S, Chen JR, Bellido T, et al. Reversal of bone loss in mice by nongenotropic signaling of sex steroids. Science 2002; 298:843–846.
57. Windahl SH, Andersson G, Gustafsson JA. Elucidation of estrogen receptor function in bone with the use of mouse models. Trends Endocrinol Metab 2002; 13:195–200.
58. McDougall KE, Perry MJ, Gibson RL, et al. Estrogen-induced osteogenesis in intact female mice lacking ERbeta. Am J Physiol Endocrinol Metab 2002; 283:E817—E823.
59. Horowitz MC. Cytokines and estrogen in bone: anti-osteoporotic effects. Science 1993; 260:626–627.
60. Rogers A, Saleh G, Hannon RA, Greenfield D, Eastell R. Circulating estradiol and osteoprotegerin as determinants of bone turnover and bone density in postmenopausal women. J Clin Endocrinol Metab 2002; 87:4470–4475.
61. Khosla S, Atkinson EJ, Dunstan CR, O'Fallon WM. Effect of estrogen versus testosterone on circulating osteoprotegerin and other cytokine levels in normal elderly men. J Clin Endocrinol Metab 2002; 87:1550–1554.
62. Lanyon L, Skerry T. Postmenopausal osteoporosis as a failure of bone's adaptation to functional loading: a hypothesis. J Bone Miner Res 2001; 16:1937–1947.
63. Bonnelye E, Kung V, Laplace C, Galson DL, Aubin JE. Estrogen receptor-related receptor alpha impinges on the estrogen axis in bone: potential function in osteoporosis. Endocrinology 2002; 143:3658–3670.
64. Dunn BK, Anthony M, Sherman S, Costantino JP. Conclusions: considerations regarding SERMs. Ann N Y Acad Sci 2001; 949:352–365.
65. Cummings SR, Browner WS, Bauer D, et al. Endogenous hormones and the risk of hip and vertebral fractures among older women. Study of Osteoporotic Fractures Research Group. N Engl J Med 1998; 339:733–738.
66. Prestwood KM, Kenny AM, Unson C, Kulldorff M. The effect of low dose micronized 17ss-estradiol on bone turnover, sex hormone levels, and side effects in older women: a randomized, double blind, placebo-controlled study. J Clin Endocrinol Metab 2000; 85:4462–4469.
67. Erlandsson MC, Jonsson CA, Lindberg MK, Ohlsson C, Carlsten H. Raloxifene- and estradiol-mediated effects on uterus, bone and B lymphocytes in mice. J Endocrinol 2002; 175:319–327.
68. Takayanagi R, Goto K, Suzuki S, Tanaka S, Shimoda S, Nawata H. Dehydroepiandrosterone (DHEA) as a possible source for estrogen formation in bone cells: correlation between bone mineral density and serum DHEA-sulfate concentration in postmenopausal women, and the presence of aromatase to be enhanced by 1,25-dihydroxyvitamin D3 in human osteoblasts. Mech Ageing Dev 2002; 123:1107–1114.
69. Vandenput L, Ederveen AG, Erben RG, et al. Testosterone prevents orchidectomy-induced bone loss in estrogen receptor-alpha knockout mice. Biochem Biophys Res Commun 2001; 285:70–76.
70. Chen Q, Kaji H, Sugimoto T, Chihara K. Testosterone inhibits osteoclast formation stimulated by parathyroid hormone through androgen receptor. FEBS Lett 2001; 491:91–93.
71. Falahati-Nini A, Riggs BL, Atkinson EJ, O'Fallon WM, Eastell R, Khosla S. Relative contributions of testosterone and estrogen in regulating bone resorption and formation in normal elderly men. J Clin Invest 2000; 106:1553–1560.
72. Liegibel UM, Sommer U, Tomakidi P, et al. Concerted action of androgens and mechanical strain shifts bone metabolism from high turnover into an osteoanabolic mode. J Exp Med 2002; 196:1387–1392.

73. Schmidt IU, Wakley GK, Turner RT. Effects of estrogen and progesterone on tibia histomorphometry in growing rats. Calcif Tissue Int 2000; 67:47–52.
74. Ikram Z, Dulipsingh L, Prestwood KM. Lack of effect of short-term micronized progesterone on bone turnover in postmenopausal women. J Womens Health Gend Based Med 1999; 8:973–978.
75. Onobrakpeya OA, Fall PM, Willard A, Chakravarthi P, Hansen A, Raisz LG. Effect of norethindrone acetate on hormone levels and markers of bone turnover in estrogen-treated postmenopausal women. Endocr Res 2001; 27:473–480.
76. Gaddy-Kurten D, Coker JK, Abe E, Jilka RL, Manolagas SC. Inhibin suppresses and activin stimulates osteoblastogenesis and osteoclastogenesis in murine bone marrow cultures. Endocrinology 2002; 143:74–83.
77. Rodan GA RL, Bilezikian JP. Pathophysiology of osteoporosis. In: Bilezikian JP, Raisz LG, Rodan GA, eds. Principles of Bone Biology. Academic, San Diego, CA, 2001.
78. Pilbeam C, Harrison JR, Raisz LG. Prostaglandins and bone metabolism. In: Bilezikian JP, Raisz LG, Rodan GA, eds. Principles of Bone Biology. Academic, San Diego, CA, 2001, pp. 979–994.
79. Takayanagi H, Kim S, Matsuo K, et al. RANKL maintains bone homeostasis through c-Fos-dependent induction of interferon-beta. Nature 2002; 416:744–749.
80. Lorenzo JA, Naprta A, Rao Y, et al. Mice lacking the type I interleukin-1 receptor do not lose bone mass after ovariectomy. Endocrinology 1998; 139:3022–3025.
81. Li X, Okada Y, Pilbeam CC, et al. Knockout of the murine prostaglandin EP2 receptor impairs osteoclastogenesis in vitro. Endocrinology 2000; 141:2054–2061.
82. Okada Y, Lorenzo JA, Freeman AM, et al. Prostaglandin G/H synthase-2 is required for maximal formation of osteoclast-like cells in culture. J Clin Invest 2000; 105:823–832.
83. Tomita M, Li X, Okada Y, et al. Effects of selective prostaglandin EP4 receptor antagonist on osteoclast formation and bone resorption in vitro. Bone 2002; 30:159–163.
84. Sakuma Y, Tanaka K, Suda M, et al. Crucial involvement of the EP4 subtype of prostaglandin E receptor in osteoclast formation by proinflammatory cytokines and lipopolysaccharide. J Bone Miner Res 2000; 15:218–227.
85. Pilbeam C, Rao Y, Voznesensky O, et al. Transforming growth factor-beta1 regulation of prostaglandin G/H synthase-2 expression in osteoblastic MC3T3-E1 cells. Endocrinology 1997; 138:4672–4682.
86. Chikazu D, Li X, Kawaguchi H, et al. Bone morphogenetic protein 2 induces cyclo-oxygenase 2 in osteoblasts via a Cbfal binding site: role in effects of bone morphogenetic protein 2 in vitro and in vivo. J Bone Miner Res 2002; 17:1430–1440.
87. Pilbeam CC, Raisz LG, Voznesensky O, Alander CB, Delman BN, Kawaguchi H. Autoregulation of inducible prostaglandin G/H synthase in osteoblastic cells by prostaglandins. J Bone Miner Res 1995; 10:406–414.
88. Buttery L, Mancini L, Moradi-Bidhendi N, O'Shaughnessy MC, Polak JM, MacIntyre I. Nitric oxide and other vasoactive agents. In: Bilezikian JP, Raisz LG, Rodan GA, eds. Principles of Bone Biology. Academic, San Diego, CA, 2001.
89. Flynn MA, Qiao M, Garcia C, Dallas M, Bonewald LF. Avian osteoclast cells are stimulated to resorb calcified matrices by and possess receptors for leukotriene B4. Calcif Tissue Int 1999; 64:154–159.
90. Lundberg P, Lerner UH. Expression and regulatory role of receptors for vasoactive intestinal peptide in bone cells. Microsc Res Tech 2002; 58:98–103.
91. Imai S, Matsusue Y. Neuronal regulation of bone metabolism and anabolism: calcitonin gene-related peptide-, substance P-, and tyrosine hydroxylase-containing nerves and the bone. Microsc Res Tech 2002; 58:61–69.
92. Bowler WB, Buckley KA, Gartland A, Hipskind RA, Bilbe G, Gallagher JA. Extracellular nucleotide signaling: a mechanism for integrating local and systemic responses in the activation of bone remodeling. Bone 2001; 28:507–512.

93. Hakeda Y, Kawaguchi H, Hurley M, et al. Intact insulin-like growth factor binding protein-5 (IGFBP-5) associates with bone matrix and the soluble fragments of IGFBP-5 accumulated in culture medium of neonatal mouse calvariae by parathyroid hormone and prostaglandin E2-treatment. J Cell Physiol 1996; 166:370–379.
94. Dallas SL, Rosser JL, Mundy GR, Bonewald LF. Proteolysis of latent transforming growth factor-beta (TGF-beta)-binding protein-1 by osteoclasts. A cellular mechanism for release of TGF-beta from bone matrix. J Biol Chem 2002; 277:21352–21360.
95. Lorenzo JA, Quinton J, Sousa S, Raisz LG. Effects of DNA and prostaglandin synthesis inhibitors on the stimulation of bone resorption by epidermal growth factor in fetal rat long-bone cultures. J Clin Invest 1986; 77:1897–1902.
96. Harada S, Thomas, KA. Vascular Enothelial Growth Factors. In: Bilezikian JP, Raisz LG, Rodan GA, eds. Principles of Bone Biology. Academic, San Deigo, CA, 2001.
97. Montero A, Okada Y, Tomita M, et al. Disruption of the fibroblast growth factor-2 gene results in decreased bone mass and bone formation. J Clin Invest 2000; 105:1085–1093.
98. Franchimont N, Durant D, Rydziel S, Canalis E. Platelet-derived growth factor induces interleukin-6 transcription in osteoblasts through the activator protein-1 complex and activating transcription factor-2. J Biol Chem 1999; 274:6783–6789.
99. Sabbieti MG, Marchetti L, Abreu C, et al. Prostaglandins regulate the expression of fibroblast growth factor-2 in bone. Endocrinology 1999; 140:434–444.
100. Gong Y, Slee RB, Fukai N, et al. LDL receptor-related protein 5 (LRP5) affects bone accrual and eye development. Cell 2001; 107:513–523.
101. Little RD, Carulli JP, Del Mastro RG, et al. A mutation in the LDL receptor-related protein 5 gene results in the autosomal dominant high-bone-mass trait. Am J Hum Genet 2002; 70:11–19.
102. Boyden LM, Mao J, Belsky J, et al. High bone density due to a mutation in LDL-receptor-related protein 5. N Engl J Med 2002; 346:1513–1521.
103. Yamaguchi T, Chattopadhyay N, Kifor O, Sanders JL, Brown EM. Activation of p42/44 and p38 mitogen-activated protein kinases by extracellular calcium-sensing receptor agonists induces mitogenic responses in the mouse osteoblastic MC3T3-E1 cell line. Biochem Biophys Res Commun 2000; 279:363–368.
104. Bingham PJ, Raisz LG. Bone growth in organ culture: effects of phosphate and other nutrients on bone and cartilage. Calcif Tissue Res 1974; 14:31–48.
105. Ivey JL, Morey ER, Baylink DJ. The effects of phosphate depletion on bone. Adv Exp Med Biol 1978; 103:373–380.
106. Gupta A, Tenenhouse HS, Hoag HM, et al. Identification of the type II Na(+)-Pi cotransporter (Npt2) in the osteoclast and the skeletal phenotype of Npt2–/– mice. Bone 2001; 29:467–476.
107. Leicht E, Biro G. Mechanisms of hypocalcaemia in the clinical form of severe magnesium deficit in the human. Magnes Res 1992; 5:37–44.
108. Rude RK, Gruber HE, Wei LY, Frausto A, Mills BG. Magnesium deficiency: effect on bone and mineral metabolism in the mouse. Calcif Tissue Int 2002.
109. Frassetto L, Morris RC Jr, Sellmeyer DE, Todd K, Sebastian A. Diet, evolution and aging—the pathophysiologic effects of the post-agricultural inversion of the potassium-to-sodium and base-to-chloride ratios in the human diet. Eur J Nutr 2001; 40:200–213.
110. Balena R, Kleerekoper M, Foldes JA, et al. Effects of different regimens of sodium fluoride treatment for osteoporosis on the structure, remodeling and mineralization of bone. Osteopor Int 1998; 8:428–435.
111. Chae HJ, Chae SW, Kang JS, Kim DE, Kim HR. Mechanism of mitogenic effect of fluoride on fetal rat osteoblastic cells: evidence for Shc, Grb2 and P-CREB-dependent pathways. Res Commun Mol Pathol Pharmacol 1999; 105:185–199.
112. Meunier PJ, Slosman DO, Delmas PD, et al. Strontium ranelate: dose-dependent effects in established postmenopausal vertebral osteoporosis—a 2-year randomized placebo controlled trial. J Clin Endocrinol Metab 2002; 87:2060–2066.

4 Interpretation of Bone Mineral Density As It Relates to Bone Health and Fracture Risk

Leon Lenchik, Sridhar Vatti,
and Thomas C. Register

1. INTRODUCTION

Examinations measuring bone mineral density (BMD) provide essential information about bone health and fracture risk and have made a significant impact on osteoporosis research as well as on patient management. Yet care must be exercised when interpreting the results of these examinations, as pitfalls are common and often overlooked.

2. TECHNIQUES FOR MEASURING BMD

Standard X-rays provide subjective assessment of bone mineralization characterized by poor inter- and intraobserver reproducibility as well as low correlation with quantitative measurement of BMD *(1–4)*. In the absence of fracture, the ability of standard X-rays to assess bone health and fracture risk is limited; thus, the use of quantitative techniques for measuring BMD has become widespread.

Techniques for measuring BMD are commonly divided into central and peripheral. Central methods allow measurement of the spine and proximal femur and include dual X-ray absorptiometry (DXA) and quantitative computed tomography (QCT). Peripheral methods allow measurement of the phalanges, forearm, or calcaneus and include peripheral dual X-ray absorptiometry (pDXA) and peripheral quantitative computed tomography (pQCT). Although it does not measure BMD, quantitative ultrasound (QUS) is often included with peripheral methods.

The technical characteristics of densitometric devices are typically summarized in terms of precision (i.e., reproducibility based on multiple measurements) and accuracy (i.e., comparison of measured value with "true" mineral content). When used properly, bone densitometry devices have very low precision error *(5–8)*. In

From: *Nutrition and Bone Health*
Edited by: M. F. Holick and B. Dawson-Hughes © Humana Press Inc., Totowa, NJ

particular, the coefficient of variation of posterioanterior (PA) spine measurement with DXA approaches 0.5% *(5,6)*.

2.1. DXA

DXA allows measurement of bone mineral content (BMC) and BMD of virtually any bone in the skeleton. DXA scanners consist of an X-ray source, X-ray collimators, and X-ray detectors. On central DXA devices, the X-ray source is usually located below the scanner table and is coupled via a C-arm to X-ray detectors located above the table. During scan acquisition, the scanner C-arm moves but does not contact the patient.

The physical basis for DXA measurements may be summarized as follows: X-ray photons are differentially attenuated by the patient based in part on their energy and on the density of the tissue through which they pass *(9)*. Dual-energy X-rays are required to determine how much of the attenuation of X-ray photons is attributable to bone rather than soft tissue. Manufacturers of DXA equipment use different approaches for producing and detecting dual-energy X-rays. One approach uses K-edge filtering to split the polyenergetic X-ray beam into high- and low-energy components (used by manufacturers GE-Lunar and Norland). These systems require the use of energy-discriminating detectors and an external calibration phantom. The other approach uses voltage switching between high and low kVp during alternate half-cycles of the main power supply (used by manufacturer Hologic). These systems require current-integrating detectors and an internal calibration drum (or wheel).

DXA scanners also differ according to the size and the orientation of the X-ray beam. Pencil beam scanners have a collimated X-ray beam and a single detector that move in tandem. Fan beam scanners have an array of X-rays and detectors. There are two types of fan beam, wide-angle and narrow-angle. Wide-angle fan beam is oriented transverse to the long axis of the body (used by Hologic), whereas narrow-angle is parallel to it (used by GE-Lunar). In general, fan beam scanners have shorter scan acquisition times and higher image resolution.

There are several limitations of DXA scanning. Perhaps the most important is the fact that DXA provides an *areal* measurement of BMD (in g/cm^2) rather than the *volumetric* measurement (in mg/cm^3) provided by QCT. Areal measurements do not take into account bone thickness and are influenced by body size and bone size. Young men typically have higher areal BMD (aBMD) than young women (who have smaller skeletons). Various investigators *(10–16)* have tried to address this issue by adjusting areal BMD for bone size either by (1) dividing aBMD by height (or height squared, or square root of height), (2) estimating vertebral volume from PA and lateral spine scans, or (3) calculating the volumetric bone mineral apparent density (BMAD). BMAD is calculated as follows: BMC/A$^{3/2}$ for the spine and as BMC/A^2 for the femoral neck, where A is the projected area. Unfortunately, in clinical practice, there is no simple way to account for this limitation (i.e., the DXA software does not perform a volumetric adjustment). Determination of BMD in individual bone compartments is also limited by the two-dimensional nature of DXA, which integrates cortical and trabecular BMD in the path of the X-ray beam *(5,9)*. In contrast, QCT allows differentiation of trabecular and cortical compartments of bone *(5)*.

Despite these limitations, central DXA is generally considered the "gold standard" for clinical measurement of BMD. This is justified by the fact that this technique has been the most widely studied. In particular, because DXA has been used in most epidemiological studies, it is well known how aBMD relates to fracture risk *(17–37)*. DXA has also been used in most pharmaceutical trials for selection and monitoring of subject populations for therapy *(38–50)*. Perhaps the main reason why DXA is considered the clinical gold standard relates to the emerging consensus on its use for the diagnosis of osteoporosis and for monitoring of therapy *(51,52)*. For these reasons as well as the low cost and broad availability of DXA, it is likely that central DXA will continue to be the most widely used approach to clinical BMD measurement. With respect to research applications of BMD measurement, QCT and peripheral devices offer some advantages over central DXA.

2.2. *QCT*

QCT allows measurement of BMD in the spine and proximal femur. Measurements are obtained with a standard imaging CT scanner but using a dedicated software package and calibration phantom. The phantom is scanned either separately or simultaneously with the patient to obtain a linear regression correction curve from which BMD values are derived *(5)*.

QCT has the unique advantage of selectively measuring trabecular BMD and providing a volumetric BMD *(5)*. Another advantage of QCT over DXA is that measurements of BMD with QCT are less influenced by the patient's height and weight or by the presence of degenerative disease *(5)*.

Early studies of QCT were plagued by insufficient precision and high radiation dose, both of which resulted in low acceptance among clinicians *(53,54)*. However, recent improvements in QCT technology, including more accurate phantoms, faster scanners, and more convenient software, have improved precision and lowered radiation doses. The use of spiral CT, in particular, allows faster data acquisition and improves precision by decreasing repositioning errors *(7,8)*.

Clinical utility of QCT is limited by the fact that no agreement exists regarding appropriate thresholds to use for diagnosis of osteoporosis (however, most experts agree that the World Health Organization diagnostic criteria are not appropriate for QCT). The fact that prospective data on the ability of QCT to predict fracture risk are limited further complicates the establishment of an alternative diagnostic threshold. Although QCT is valuable for monitoring changes in BMD, because it measures the more metabolically active trabecular bone, it is rarely used for that purpose in clinical practice, in which most patients' initial scan is a DXA.

Despite the limitations of QCT for clinical use, its use in research has been growing and should be encouraged for a number of reasons *(55–59)*. QCT studies will increase our understanding of how volumetric rather than aBMD relates to fracture risk, drug efficacy, and a variety of other outcome variables relevant to the clinical and research bone communities. Wider acceptance of CT scanning for research in other fields (e.g., obesity and cardiovascular disease) will facilitate broader use in the bone field.

2.3. Peripheral Devices

Peripheral devices include peripheral DXA, peripheral QCT, and QUS and provide greater portability, ease of use, and lower radiation than central devices. The main differences among peripheral DXA devices are the number and the type of skeletal sites that are measured (i.e., phalanges, forearm, or calcaneus) as well as the X-ray beam geometry that is used (i.e., pencil beam or cone beam). Peripheral QCT devices typically measure the forearm or the tibia. Like QCT of the spine, pQCT is able to differentiate trabecular from cortical bone.

Although QUS devices do not measure BMD, they are usually included in discussion of peripheral methods. QUS devices measure the speed of sound (SOS) and/or the broadband ultrasound attenuation (BUA). SOS is expressed in meters per sec (m/s). BUA is expressed in decibels per megahertz (dB/MHz). SOS and BUA values are lower in patients with osteoporosis when compared to controls. From SOS and BUA, additional parameters such as "stiffness" or "quantitative ultrasound index" may be calculated. Differences among QUS devices are related to measurement site (i.e., phalanges, calcaneus, or multiple sites), sound transmission (i.e., transverse or axial), transducer coupling (i.e., water- or gel-based), data acquisition (i.e., fixed single-point or variable imaging), definitions of velocity and attenuation, and calibration methods *(60)*. In general, correlation of calcaneal QUS values with calcaneal BMD (by pDXA) is lower than the correlation with spine or hip BMD (by DXA) *(60)*.

As is the case with QCT, the clinical utility of peripheral devices is limited by the fact that no agreement exists on appropriate diagnostic criteria *(61)*. However, the peripheral devices provide a cheaper alternative to mass screening of the population, and development of these diagnostic criteria through more research should be encouraged. The use of QUS in the pediatric population is especially promising because it provides an opportunity to detect skeletal disease prior to attainment of peak BMD and without exposure to ionizing radiation *(62–64)*.

3. RATIONALE FOR BONE DENSITOMETRY

3.1. Fracture Trials

Bone densitometry is a powerful clinical tool due to its ability to predict the risk of fracture. Since the critical measures of the impact of osteoporosis are fractures and their outcomes, the ability to identify individuals at high risk for fractures is paramount. Many cross-sectional and longitudinal studies have shown that BMD measured at various skeletal sites is highly associated with osteoporotic fractures

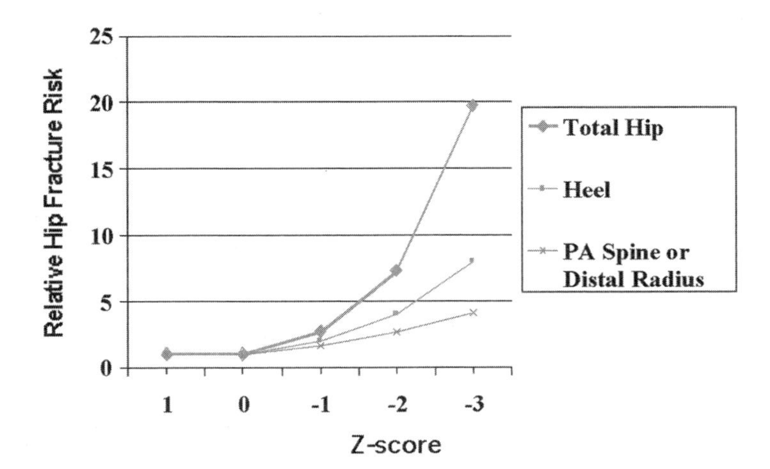

Fig. 1. Relative risk of hip fracture compared to individuals of the same age and with an average BMD obtained from measurements of BMD at various skeletal sites using DXA. In the Study of Osteoporotic Fractures the relative risk of hip fracture per 1 SD decrease in BMD was 2.7 (CI 2.0–3.6) when BMD was measured at the proximal femur, 2.0 (CI 1.5–2.7) when BMD was measured at the heel, and 1.6 (CI 1.2–2.2) when BMD was measured at the PA spine or distal radius *(17)*. Unlike the case of predicting hip fracture risk, in which hip BMD measurement appears superior to other methods, in predicting any osteoporotic fracture, all BMD measurements (i.e., central and peripheral) had comparable relative risk. PA, posterior–anterior.

(17–37). In general, for each standard deviation decrease in BMD, the risk of osteoporotic fracture doubles *(65)* (Fig. 1). DXA measurements at the hip predict hip fracture better than measurements at other skeletal sites (relative risk [RR] range 1.9–3.8) *(26,65)*. However, there is growing evidence that calcaneal and forearm measurement with DXA and calcaneal measurement with QUS are predictive of hip fractures (RR range 1.4–2.7) *(26,30,65)*.

The majority of evidence concerning the ability of BMD measurements to predict fracture risk has been obtained in postmenopausal Caucasian women. As such, the ability of BMD to predict fracture risk in men, non-Caucasian individuals, and younger patients requires further study.

3.2. Biomechanical Studies

It is generally accepted that osteoporotic fractures result from diminished bone strength, which is in turn determined by both BMD and "bone quality." Biomechanical studies have shown that BMD is a powerful predictor of bone strength. Material properties of trabecular bone specimens vary according to BMD and anatomic site. It has been estimated that 60–90% variability in elastic modulus is explained by apparent density *(66–70)*. There is high inverse association between femoral BMD and failure load. Correlations between vertebral BMD and failure load are higher for DXA ($r = 0.80–0.94$) than QCT ($r = 0.30–0.66$), reflecting the contributions of both the size of the bone (which influences areal BMD) as well as the cortical component of bone to biomechanical strength *(71–73)*.

4. CLINICAL USE OF BMD MEASUREMENTS

In clinical practice, bone densitometry is typically used to diagnose osteoporosis, assess risk of fracture, help select patients for pharmacological therapy, and monitor that therapy *(74–76)*.

4.1. Diagnosis of Osteoporosis

Measurement of BMD with bone densitometry has become an essential part of clinical practice, ideally providing the diagnosis of osteoporosis prior to the occurrence of fracture, which is analogous to diagnosis of hypertension prior to the occurrence of stroke.

The diagnosis of osteoporosis is made according to the T-score, a standardized score that is unique to bone densitometry. T-scores are calculated by subtracting mean BMD of a young-normal reference population from the subject's measured BMD and dividing by the standard deviation of a young-normal reference population. In postmenopausal Caucasian women, the World Health Organization (WHO) criteria of osteoporosis (T-score less than or equal to –2.5), osteopenia (T-score between –1.0 and –2.5), and normal (T-scores equal to or above –1.0) are widely used *(51,77)*.

The reason that T-scores rather than absolute BMD values are used to make the diagnose osteoporosis is explained in part by the fact that manufacturers of DXA scanners use different approaches to BMD measurement. Each manufacturer uses a different approach for producing dual-energy X-rays and detecting them, calibrating the scanners, and in some cases defining the regions of interest where the BMD is measured. Thus, the same absolute BMD value cannot be used for diagnosis of osteoporosis using different devices. In contrast, the use of a diagnostic threshold based on a T-score enables the same diagnostic criteria to be used regardless of the central DXA manufacturer.

On most densitometric devices, BMD measurements are also expressed in terms of a Z-score. The Z-score is calculated similarly to the T-score except that an age-matched reference population is used. With the exception of their use in children, the Z-scores are not used for diagnosis. This is appropriate because using Z-score in adults would result in prevalence of osteoporosis remaining constant with age despite an overall loss of bone in the population. Because fracture incidence increases dramatically with age, such an approach would result in many osteopenic and osteoporotic patients being classified as normal. For this reason, diagnosis using a T-score approach has become standard.

It is widely accepted that the WHO diagnostic thresholds are appropriate only for DXA measurement at particular skeletal sites (e.g., PA spine, hip, and forearm) *(51)*. The diagnostic criteria for other densitometric methods (e.g., QCT, QUS) and skeletal sites (e.g., lateral spine, heel, phalanges) are controversial. Furthermore, although the WHO criteria are commonly applied to premenopausal women, men, and non-Caucasian individuals, such an approach is not universally accepted *(78)*. Certainly, the peak bone mass as well as the change in BMD with aging and disease vary between the sexes and among ethnic groups *(58,79)*. Change in BMD with age also depends on gender, BMD measurement site (e.g., hip vs spine), and measurement technique (e.g., DXA vs QCT) *(58,79)*. By definition, the T-score is dependent on the peak bone mass of young normal adults.

4.1.1. Diagnosis in Men

DXA measured BMD is approx 10% higher in young men than in young women *(80)*. This is primarily related to the fact that men have larger bones than women, and aBMD measurement by DXA is influenced by bone size *(80)*. All DXA manufacturers calculate T- and Z-scores in men based on the male reference data, hence at a given BMD value the T-scores in men and women are different. However, an alternative approach of using female reference data for the diagnosis of osteoporosis in men may be worthwhile, because several large studies have shown that men fracture at the same BMD as women. *(81,82)*. Other studies showed that men fracture at higher BMD values than women, arguing for retaining the current approach to T-score calculation in men *(83–86)*. In addition to the controversy about the appropriate reference population to use in men, there is also debate about the appropriate diagnostic threshold to use in men. Currently, most clinicians apply the same WHO diagnostic criteria (T-score ≤–2.5) to men as they do to women *(78)*.

4.1.2. Diagnosis in Non-Caucasians

As with young men, aBMD in young normal African-American women is higher than in young Caucasian women. However, the approach to diagnosis of osteoporosis in non-Caucasians is complicated by the fact that some DXA manufacturers adjust their T-scores for race, whereas others do not (i.e., Hologic adjusts in women and men, Norland adjusts in women but not men, GE-Lunar does not adjust) *(78)*. Because there are many races and ethnicities (and because admixing of races is common), prospective studies to determine the relationship between BMD and fracture risk in non-Caucasians have not been performed and the best approach to diagnosis of osteoporosis in these individuals is unknown. If studies show that fractures occur at the same aBMD in Caucasians as non-Caucasians, then a common database for T-score calculation would be appropriate. Conversely, if studies show different levels of risk at the same BMD level, a race-specific database would be more appropriate. Based on their 2001 position development conference, the International Society for Clinical Densitometry recommends using the Caucasian reference data for non-Caucasians *(78)*.

4.1.3. Diagnosis With Peripheral Devices

T-scores obtained on peripheral devices are not comparable to those obtained with central DXA because these techniques use different physical principles, measure different skeletal sites with different rates and patterns of bone loss, and use different normative databases *(87)*. At the present time, there is no consensus regarding appropriate diagnostic criteria for osteoporosis using peripheral devices *(87)*. Most experts consider the WHO criteria not appropriate for peripheral devices *(87)*. It would be ideal to have unique diagnostic thresholds for each technique based on similar risk profiles for future osteoporotic fractures. Although several such approaches have been proposed, a consensus has not been reached regarding which is best *(87)*.

4.1.4. Other Limitations of the WHO Criteria

The WHO diagnostic criteria have several other important limitations. The use of any threshold for diagnosing osteoporosis may be misleading. The fact that any diagnostic threshold may be confused with a "fracture threshold" is problematic, since the relationship between decreasing BMD and increasing risk of fracture is continuous rather than threshold based *(65)*. Because of substantial overlap in BMD among fracture and nonfracture patients, it is impossible to define a threshold BMD value for a population below which everyone will fracture or above which no one will fracture. It is more appropriate to consider osteoporosis as a continuum of BMD, with the patients with the lowest BMD values having the greatest risk of fracture. Despite that limitation in the use of a diagnostic threshold, such an approach is essential for clinical practice.In the current health care environment (at least in the United States), disease *risk* is generally is not interchangeable with the *diagnosis* of a disease. Instead, a defined level of disease risk is linked with a particular diagnosis. For example, various threshold-based diagnostic approaches are used for hypertension, hypercholesterolemia, and type 2 diabetes, where similar continuous relationships between measured and outcome variables exist.

Another reason why the T-score approach to diagnosis is problematic relates to the fact that T-scores are dependent on an "appropriate" reference database (i.e., young adult reference means and standard deviations). However, there is poor agreement among reference databases of different manufacturers as well as between reference data from various study populations *(87–91)*. Different manufacturers may use different inclusion and exclusion criteria when gathering normative data. Furthermore, the same manufacturer may use different reference populations at different skeletal sites and regions of interest. If the standard deviations are different, the resultant T-scores are different, even when the mean BMD values for two normative populations are the same *(90)*. For these reasons, the same patient measured on different devices is likely to have different T-scores.

Finally, the T-score approach to diagnosis does not adequately account for the discordance among skeletal sites and regions of interest. In fact, patterns of bone loss vary according to skeletal site and region. Using the same T-score cutoff for the same skeletal site, e.g. total hip, femoral neck, trochanter, identifies different populations of patients with different risk of fracture *(91)*. It may be ideal to have unique diagnostic thresholds for each skeletal site based on similar risk profiles for future osteoporotic fractures *(89)*.

For these reasons, the use of T-scores for diagnosis of osteoporosis has been widely debated. In the future, T-scores may be abandoned in favor of diagnostic thresholds based on measured BMD and associated fracture risk.

4.2. Prognosis

Prognosis in the context of osteoporosis typically relates to assessment of fracture risk. Many epidemiological trials have shown that for each standard deviation decrease in BMD there is a 1.5–3-fold increase in risk of fracture *(65)*. Although QUS does not measure BMD, several prospective trials indicate that QUS parameters are predictive of hip fracture rates *(26,30)*. However, in clinical practice, it is difficult to assign numerical fracture risk to an individual patient. For many types of patients (i.e., younger women, non-Caucasians, patients with secondary osteoporosis, and patients on therapy), the relationship between BMD and fracture risk is not known. In addition, non-BMD risk factors contribute substantially to overall fracture risk. Age, in particular, is a powerful predictor of fracture risk (e.g., an 80-yr-old with a T-score of –3 has a greater current risk for fracture than a 55-yr-old with a T-score of –3). Similarly, history of previous fracture further increases fracture risk, regardless of BMD level. In determining an individual patient's prognosis, clinicians must be able to assess these and other non-BMD risk factors (e.g., beginning menopause, discontinuing estrogen at any age, beginning steroids, immobility). In clinical practice, careful history should provide essential information that can be used in combination with BMD measurement to provide a more complete assessment of fracture risk.

4.3. Selection of Patients for Therapy

Using BMD thresholds to help select patients for therapy is recommended by several professional organizations including the National Osteoporosis Foundation (NOF) and American Association of Clinical Endocrinologists (AACE) *(92,93)*. Such an approach is appropriate because evidence for fracture reduction exists mainly in subjects enrolled in pharmaceutical trials based on the presence of low BMD and/or the presence of a vertebral fracture.

4.4. Monitoring of Therapy

Monitoring of therapy using BMD measurements is possible as long as the devices used have low precision errors and measure skeletal sites that respond well to therapy *(52,94,95)*. Because of excellent precision and greatest responses to therapy, measurement of the spine BMD with DXA or QCT is preferable to other densitometric methods and measurement sites. To determine if interval change is statistically significant, the precision of the device must be known and the study must be technically valid. Least significant change is calculated from the precision error and used when monitoring patients *(52)*.

When monitoring patients the following considerations for site selection apply: PA spine is preferred because it has the lowest precision error and because it is most responsive to therapy *(94)*. If PA spine cannot be used, total hip, femoral neck, trochanter regions may provide a viable alternative. In general, BMD measurement with peripheral devices shows smaller levels of response to therapy as central sites. For this reason, monitoring with peripheral devices is not recommended *(52)*.

Monitoring time interval depends on the precision error and the expected change in BMD with therapy *(52)*. Some therapies are associated with large increases in BMD, while others may have modest or no significant BMD changes *(38–50)*. Patient factors (i.e., glucocorticoid therapy, hyperparathyroidism, etc.) also influence expected changes in BMD. However, in most cases, clinicians monitor therapy not to identify patients who gain BMD, but to identify those who lose BMD on therapy *(52)*. These patients may require workup for secondary causes of osteoporosis. For that purpose, a 1–2-yr follow-up interval is generally appropriate. In contrast, patients receiving glucocorticoid treatment may be followed as early as 6–12 mo *(52)*.

5. APPROACH TO CLINICAL DXA INTERPRETATION

5.1. Site Selection

Typically, two skeletal sites are measured with DXA, the PA lumbar spine and proximal femur *(51,96)*. The rationale for measuring both spine and hip is as follows: approx 20–30% of patients have significant spine–hip discordance, in which T-scores at one site are of a different diagnostic category than the other site *(97,98)*. It is desirable to find the site with the lower BMD. Another reason for measuring both relates to fracture prediction, spine BMD is a better determinant predictor of spine fractures, whereas hip BMD is a better determinant predictor of hip fractures. Finally, in patients with severe degenerative disease of the spine, hip BMD offers more accurate monitoring. Various causes of spine–hip discordance have been described, including differences in achievement of peak bone mass and rate of bone loss at different skeletal sites. For example, in perimenopausal women, rate of bone loss is greater in cancellous bone (e.g., spine) than cortical bone (e.g., hip). Body weight, obesity, exercise, and other parameters may also differentially influence BMD in these sites.

In patients in whom the spine and/or the proximal femur scans are invalid, a forearm scan may be obtained. For example, patients with spine instrumentation, fractures, severe degenerative disease, or scoliosis may benefit from a forearm scan *(51)*. Patients with bilateral hip replacements (or other instrumentation), severe hip arthritis, or patients who exceed the weight limit of the table are candidates. Finally, forearm scan is useful in patients with hyperparathyroidism, since the mid-forearm measures primarily cortical bone *(51)*. Cortical bone loss exceeds trabecular bone loss in patients with hyperparathyroidism.

Lateral spine scanning is less influenced by degenerative changes than PA spine scanning. All DXA scanners that are capable of scanning the spine in the PA direction can also scan from the lateral direction. Shortcomings of this method are the frequent overlap of L2 and L4 by ribs and pelvis, respectively. The use of lateral spine scans for diagnosis of osteoporosis is not recommended. However, they may be of value for monitoring therapy.

Recently, lateral scans have been modified for lateral vertebral assessment (LVA™) or instantaneous vertebral assessment (IVA™) and are used not for BMD measurement but for detection of vertebral fracture. Given the importance of vertebral fractures, the use of IVA/LVA should be encouraged.

5.2. Scan Acquisition and Analysis

It is important to weigh the patient and measure his or her height, because these parameters may influence the selection of appropriate scan mode and in some cases influence the Z-score. Changes in weight and height loss should be recorded, as they may influence DXA results.

DXA examinations must be obtained using the manufacturer's recommendations for patient positioning, scan protocols, and scan analysis. The patient should change into a hospital gown (or equivalent) to remove any potential artifacts related to street clothing.

The lumbar spine scans are acquired with the subject's body aligned with the scanner table and legs elevated using the standard positioning block. The scan window extends from the mid portion of L5 to the mid portion of T12 vertebra. BMD and BMC of lumbar vertebrae 1–4 are measured. The proximal femur scans are acquired with the subject's left leg internally rotated using the standard positioning device. BMD and BMC are measured at the femoral neck, total hip, and trochanteric region. The forearm scans are acquired with the subject sitting in a chair adjacent to the scanner table and his or her nondominant forearm placed in the standard positioning device. BMD and BMC are measured at the mid ($^1/_3$ or 33%) radius region, ultra distal (UD) radius region, and total forearm region. Total body scans are acquired with the subject supine and aligned with the scanner table.

5.3. Understanding DXA Printouts

DXA printouts vary according to the manufacturer and software version; common features include a summary of patient demographics, an image of the skeletal site scanned, a plot of patient age vs BMD, and numerical results. The presentation of numerical results is usually configurable by the DXA operator. Typically, the BMD values (in g/cm^2), T-scores, Z-scores, and other data (e.g., BMC, area, %BMD, vertebral height, etc.) for various regions of interest are presented.

When interpreting DXA scans it is important to check whether correct patient demographics were entered into the DXA computer. Incorrect age, gender, and race may influence T-scores and/or Z-scores.

The next step is to evaluate the DXA image for proper patient positioning, scan analysis, and artifacts. On properly positioned PA spine scans, the spine appears straight (aligned with the long axis of scan table), both iliac crests are visible, and the scan starts in the middle of L5 and ends in the middle of T12 (Fig. 2). On

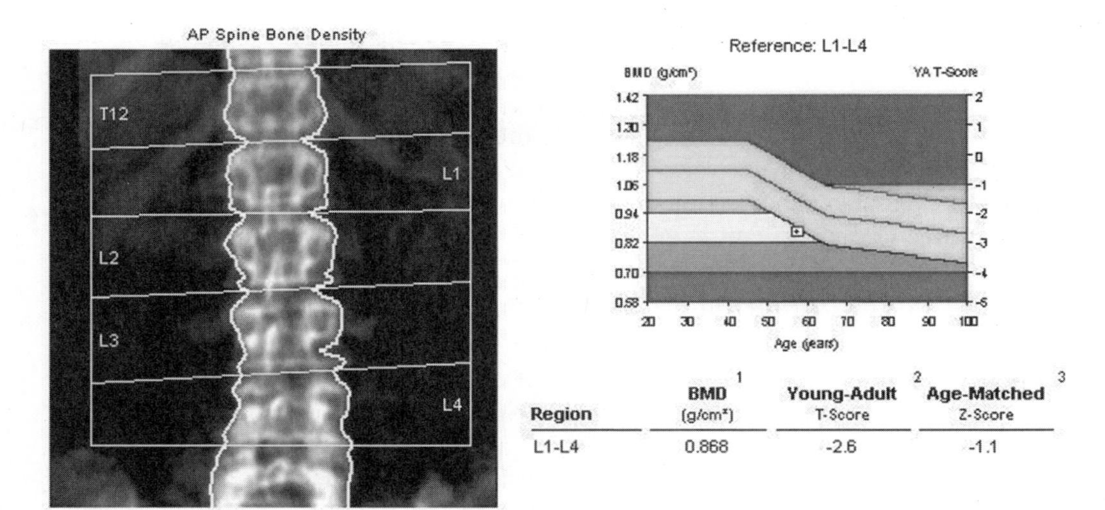

Fig. 2. DXA scan of the PA spine in a 56-yr-old Caucasian woman not receiving any pharmacological therapy. Note proper patient positioning and scan analysis. The T-score is within the WHO range for osteoporosis.

properly positioned hip scans, the femoral shaft is straight (aligned with the long axis of scan table), the hip is internally rotated, and the scan includes the ischium and the greater trochanter (Fig. 3). The lesser trochanter is a posterior structure, and its size is the best indication of the degree of rotation of the proximal femur during a DXA study. BMD values are affected by the degree of rotation of the proximal femur and the position of the femoral neck (region of interest [ROI]), and internal or external rotation will cause an increase in the measured BMD. On properly positioned forearm scans, the forearm is straight and the distal ends of the radius and ulna are visible. On properly positioned total body scans, the patient is centered and the entire body is within scan limits.

It is important to rescan the patient when positioning mistakes are made. For example, if on images of PA spine scans the spine is tilted or not centered, one or both iliac crests are not visible, or the image does not include T12 or L5, the patient should be rescanned. On the image of the hip scan, if the femoral shaft is angled (adducted or abducted) or the leg is not properly rotated (too much lesser trochanter is visible), the patient should be rescanned. On images of forearm scans, if the forearm is not centered, radius and ulna are angled, or the distal cortex of radius and ulna are not visible, the patient should be rescanned.

The images should also show appropriate scan analysis (e.g., region of interest size and location). In the PA spine scan, the vertebral bodies should be numbered correctly (Fig. 1). This is especially true in patients with four or six lumbar vertebrae, in whom numbering should begin at the level of the iliac crest—typically,

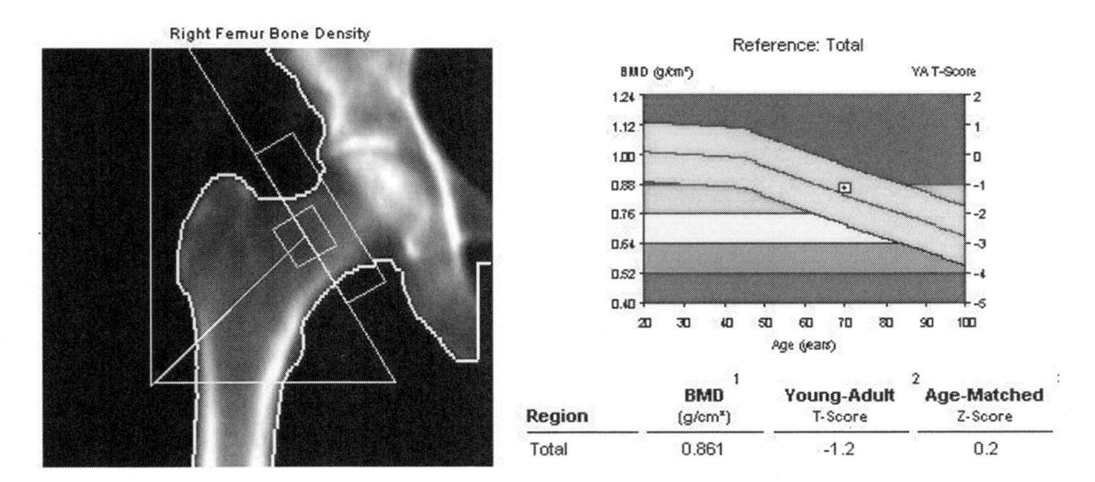

Region	BMD[1] (g/cm²)	Young-Adult[2] T-Score	Age-Matched Z-Score
Total	0.861	-1.2	0.2

Fig. 3. DXA scan of the right proximal femur in a 69-yr-old Caucasian woman on estrogen replacement therapy. Note proper patient positioning and scan analysis. The T-score is within the WHO range for osteopenia.

corresponding to the L4–5 disk space. On lateral spine scans, overlying ribs and pelvis should be excluded, and the regions of interest should be within the individual vertebral bodies. On hip scans, the femoral neck region must not include the greater trochanter or the ischium (Fig. 3). Femoral neck placement is highly manufacturer-specific (GE-Lunar measures the midportion of the femoral neck, Hologic measures the base of the femoral neck). For this reason, it is important to follow the manufacturer's recommendations. On forearm scans, the distal radius region should not include the articular surface of the distal radius.

It is important to be able to recognize artifacts on DXA images. Common artifacts include degenerative disease and fractures (Fig. 4). Degenerative disease typically manifests as disk space narrowing, subchondral sclerosis, osteophytes, and facet osteoarthritis. Spinal degenerative disease often increases BMD *(99,100)*. Severe osteoarthritis of the hip may increase BMD in the femoral neck or total hip (because of buttressing of medial femoral neck), whereas BMD in the trochanteric region is relatively unaffected *(101)*. When degenerative disease of the spine is limited to several vertebra, it should be excluded from the region of interest used for diagnosis or for monitoring. When disease is diffusely distributed, a forearm scan should be added. In general, vertebral fractures increase BMD. Vertebral compression fractures often appear as having decreased height compared to adjacent vertebrae. In some cases, the PA spine DXA image will not show a known fracture. In such cases, discrepant BMD values for individual vertebrae may indicate the presence of a compression fracture. Comparison to a lateral DXA image or lumbar spine X-ray is necessary to identify such fractures definitively. Fractures

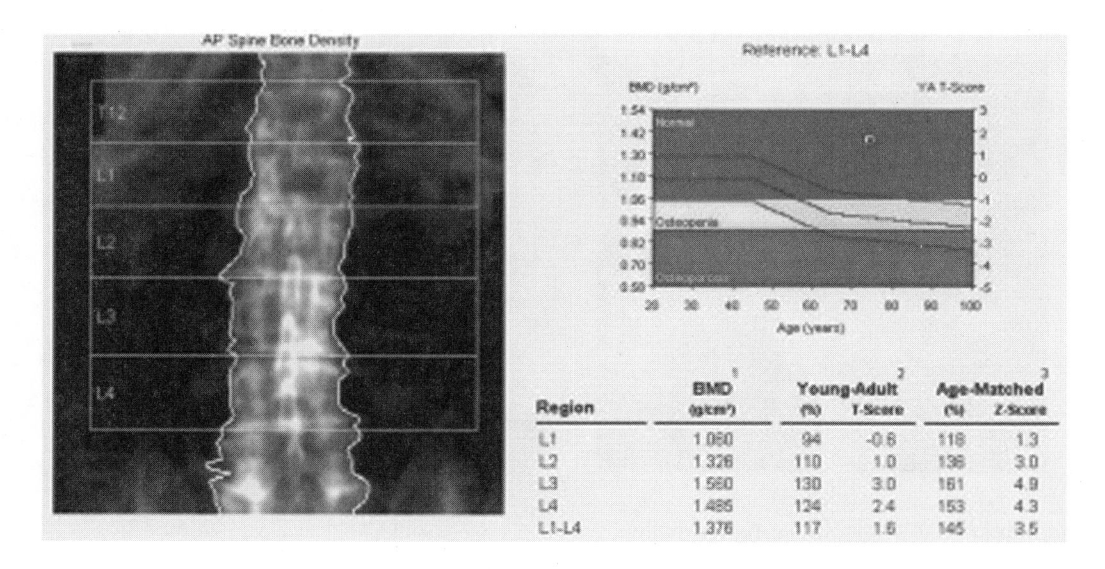

Region	BMD[1] (g/cm²)	Young-Adult[2] (%)	Young-Adult T-Score	Age-Matched[3] (%)	Age-Matched Z-Score
L1	1.060	94	-0.6	118	1.3
L2	1.326	110	1.0	136	3.0
L3	1.560	130	3.0	161	4.9
L4	1.486	124	2.4	153	4.3
L1-L4	1.376	117	1.6	145	3.5

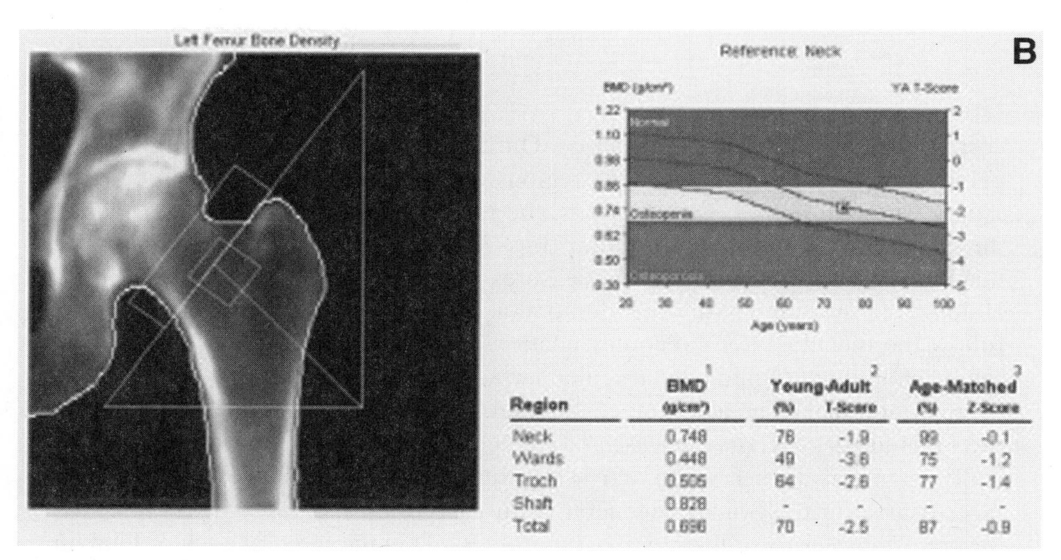

Region	BMD[1] (g/cm²)	Young-Adult[2] (%)	Young-Adult T-Score	Age-Matched[3] (%)	Age-Matched Z-Score
Neck	0.748	76	-1.9	99	-0.1
Wards	0.448	49	-3.6	75	-1.2
Troch	0.506	64	-2.6	77	-1.4
Shaft	0.828	-	-	-	-
Total	0.696	70	-2.5	87	-0.9

Fig. 4. **(A)** DXA scan of the PA spine in a 74-yr-old Caucasian woman not receiving therapy. Note asymmetric sclerosis and osteophytes involving multiple vertebra indicating degenerative disease. Corresponding X-ray showed a fracture of L1 vertebral body. Although the T-score is within the WHO range for normal, the BMD is falsely elevated by degenerative disease and fracture. **(B)** DXA scan of the left proximal femur in the same woman shows the T-score in the total hip and trochanter within the WHO range for osteoporosis. This case illustrates a common occurrence in elderly patients, in which the spine is normal due to the presence of artifacts and yet the hip is osteoporotic.

must be excluded from the region of interest used for scan analysis to prevent overestimation of BMD. Nonvertebral fractures may also increase BMD *(102)*. For this reason, avoid scanning the hip or forearm with an old fracture.

Other artifacts include postoperative changes (i.e., laminectomy defects, spinal instrumentation and/or fusion), vascular calcifications, gastrointestinal contrast, calcium tablets, gallstones, renal stones, pancreatic calcifications, and various metallic devices *(103)*. External artifacts, including buttons, zippers, bra clips, wallets, and jewelry, should be removed prior to scanning.

Patients should not be scanned if there has been recent gastrointestinal contrast, because contrast in overlying tissues invalidates BMD result; the patient is uncooperative and cannot remain still throughout the examination (motion artifacts invalidate BMD result); the patient is obese (DXA scanner tables have a weight limit of 250–450 lb).

The final step is to evaluate the numerical results. DXA results are expressed in absolute BMD units (g/cm^2), T-scores, and Z-scores. The absolute BMD is used to monitor a patient's response to therapy. T-score is used to make a diagnosis of osteoporosis. Z-score is used to guide laboratory workup to exclude secondary causes of osteoporosis; the Z-score is not used for diagnosis (except in children). On PA spine scans there should be an incremental increase in BMD from L1 to L4 *(104)*. Individual vertebral T-scores should be within 1 SD. When these two conditions are not met, the DXA image should be scrutinized to detect an artifact that should be excluded from analysis. Occasionally, a radiograph is needed for clarification.

5.4. Diagnosis

In making the diagnosis of osteoporosis the following approach is used *(51)*: Upon ensuring proper positioning and analysis, and if necessary, exclusion of artifacts, the lower of the T-scores of the PA spine and hip is used. In the spine, using the weighted average of L1-L4 is preferred (Fig. 2). Only those vertebra affected by focal structural abnormalities (i.e., fracture, focal degenerative disease, or surgery) should be excluded from analysis. In the hip, using the lowest T-score of the total hip (total femur), femoral neck, and trochanter is recommended (Fig. 3). Ward's region should not be used to diagnose osteoporosis. Ward's region is a triangular region in the femoral neck that is created by the intersection tensile, compressive, and intertrochanteric trabecular bundles. On DXA scans Ward's region is a square that is variably placed (depending on the manufacturer), and because of its small area has high precision error. It is important to note that discrepancies between sites may indicate a true difference in BMD, but they may also be falsely elevated due to degenerative joint disease or fractures, or falsely decreased due to osteolytic bone tumors or postoperative changes

In making the diagnosis of osteoporosis using DXA, it is essential to recognize that low BMD does not explain its etiology. In particular, low BMD in postmenopausal women does not always imply postmenopausal osteoporosis due to estrogen deficiency but can also be due to secondary causes of osteoporosis.

Furthermore, a single low BMD result may evolve in different ways, resulting, for example, from a low peak BMD followed by normal rate of loss or a normal peak BMD with accelerated rate of loss.

5.5. Monitoring

When monitoring patients with DXA it is important that follow-up scans demonstrate consistent patient positioning and scan analysis *(52)*. Typically, the scan area (in cm^2) should be within 2%. The most important principle of monitoring using DXA is that the same bone (or ROI) must be measured and analyzed the same way. DXA images on the two comparison studies should be inspected to make sure that the ROI is the same size and position. If the measured area differs by more than 2%, the ROI should be reexamined for improper positioning, incorrect scan analysis, and/or artifacts (fractures, degenerative changes, etc.) that may explain the discrepancy. When monitoring patients, BMD values rather than T-scores should be compared, since the T-scores depend on a normative database that may change with software upgrades.

5.6. Clinical Reporting

Reports should include a statement on the technical validity of the scan (note any artifacts such as fracture or degenerative disease that may invalidate numerical results) *(96)*. When appropriate, the diagnosis of osteoporosis (T-score less than or equal to –2.5), osteopenia (T-score between –1.0 and –2.5), or normal (T-scores equal to or above –1.0) should be assigned based on WHO criteria. Statements regarding further monitoring (generally 1–2 yr) if pharmacological therapy is being initiated, laboratory tests to exclude secondary causes for osteoporosis, X-rays to confirm the presence of fracture or to evaluate an artifact seen on the DXA image, should be included when appropriate. According to the NOF guidelines, initiate therapy in women with T-scores less than –2.0 in the absence of other risk factors, or in women with T-scores below –1.5 when risk factors are present. Note that the decision to initiate pharmacological therapy is influenced by BMD as well as other factors.

6. BMD MEASUREMENT IN RESEARCH

Traditional and emerging methods for measurement of bone mass and quality provide important new information in many areas of importance to human health. In clinical trials and cross-sectional studies in people, these studies facilitate a greater understanding of relationships between these skeletal parameters and clinically important diseases and conditions of the skeleton (e.g., osteoporosis, osteoarthritis, and osteopetrosis). Likewise, these methods are useful in the study of the effects on the skeleton of other diseases and conditions, such as disorders of glucose metabolism (i.e., diabetes, insulin resistance, and metabolic syndrome), obesity and weight loss, sarcopenia, aging, amenorrhea, menopause, andropause, and other endocrine disorders. Investigations into the efficacy and/or adverse effects of novel drugs and dietary regimens, endogenous and exogenous hormone therapies, lifestyle parameters such as exercise, dietary composition, behavior and stress, weight loss and gain, occupations, etc., are also being investigated. In the postgenomic era, these measurements will be useful in identifying novel genes that are involved in characteristics of the skeleton. In some cases these studies have already led to the identification of genes that are associated specifically with BMD, one example being the *LRP5* gene *(105,106)*. Exploitation of these methodologies in animal models will allow the further exploration of the skeletal effects (and relevant underlying mechanisms) of emerging investigational drugs, treatments, and dietary regimens that can be tightly controlled in animal models, as well as determining the influence of individual genes on the skeleton in genetically modified animals.

7. CONCLUSION

Despite growing interest in experimental techniques, such as quantitative magnetic resonance imaging, examinations measuring bone mineral density are vital to osteoporosis research as well as to patient management. When interpreted properly, the results of these examinations provide essential information about bone health and fracture risk.

REFERENCES

1. Jergas M, Uffmann M, Escher H, et al. Interobserver variation in the detection of osteopenia by radiography and comparison with dual X-ray absorptiometry of the lumbar spine. Skeletal Radiol 1994; 23(3):195–199.
2. Finsen V, Anda S. Accuracy of visually estimated bone mineralization in routine radiographs of the lower extremity. Skeletal Radiol 1988; 17:270.
3. Haller J, Andre MP, Resnick D, et al. Detection of thoracolumbar vertebral body destruction with lateral spine radiography. Part II. Clinical investigation with computed tomography. Invest Radiol 1990; 25:523.
4. Haller J, Andre MP, Resnick D, et al. Detection of thoracolumbar vertebral body destruction with lateral spine radiography. Part I: Investigation in cadavers. Invest Radiol 1990; 25:517.
5. Genant HK, Engelke K, Fuerst T, et al. Noninvasive assessment of bone mineral and structure: state of the art. J Bone Miner Res 1996; 11(6):707–730.
6. Lilley J, Walters BG, Heath DA, Drolc Z. In vivo and in vitro precision for bone density measured by dual-energy X-ray absorption. Osteopor Int 1991; 1(3):141–146.
7. Lang TF, Li J, Harris ST, Genant HK. Assessment of vertebral bone mineral density using volumetric quantitative CT. J Comput Assist Tomogr 1999; 23(1):130–137.
8. Braillon PM. Quantitative computed tomography precision and accuracy for long-term follow-up of bone mineral density measurements: a five year in vitro assessment. J Clin Densitom 2002; 5(3):259–266.
9. Blake GM, Fogelman I. Technical principles of dual energy x-ray absorptiometry. Semin Nuclear Med 1997; 27(3):210–228.
10. Nevill AM, Holder RL, Maffulli N, et al. Adjusting bone mass for differences in projected bone area and other confounding variables: an allometric perspective. J Bone Miner Res 2002; 17(4):703–708.

11. Melton LJ 3rd, Khosla S, Achenbach SJ, O'Connor MK, O'Fallon WM, Riggs BL. Effects of body size and skeletal site on the estimated prevalence of osteoporosis in women and men. Osteopor Int 2000; 11(11):977–983.
12. Taaffe DR, Cauley JA, Danielson M, et al. Race and sex effects on the association between muscle strength, soft tissue, and bone mineral density in healthy elders: the Health, Aging, and Body Composition Study. J Bone Miner Res 2001; 16:1343–1352.
13. Fieldings KT, Backrach LK, Hudes ML, Crawford PB, Wang MC. Ethnic differences in bone mass of young women vary with method of assessment. J Clin Densitom 2002; 5(3):229–238.
14. Reid IR, Evans MC, Ames RW. Volumetric bone density of the lumbar spine is related to fat mass but not lean mass in normal postmenopausal women. Osteopor Int 1994; 4:362–367.
15. Martini G, Valenti R, Giovani S, Nuti R. Age-related changes in body composition of healthy and osteoporotic women. Maturitas 1997; 27:25–33.
16. Nguyen TV, Howard GM, Kelly PJ, Eisman JA. Bone mass, lean mass, and fat mass: same genes or same environments? Am J Epidemiol 1998; 147:3–16.
17. Cummings SR, Black DM, Nevitt MC, et al. Bone density at various sites for prediction of hip fractures. Lancet 1993; 341:72–75.
18. Huang C, Ross PD, Yates AJ, et al. Prediction of fracture risk by radiographic absorptiometry and quantitative ultrasound: a prospective study. Calcif Tissue Int 1998; 63(5):380–384.
19. Porter RW, Miller CG, Grainger D, et al. Prediction of hip fracture in elderly women: a prospective study. Br Med J 1990; 301(6753):638–641.
20. Melton LJ 3rd, Atkinson EJ, O'Fallon WM, et al. Long-term fracture prediction by bone mineral assessed at different skeletal sites. J Bone Miner Res 1993; 8(10):1227–1233.
21. Black DM, Cummings SR, Genant HK, et al. Axial and appendicular bone density predict fractures in older women. J Bone Miner Res 1992; 7(6):633–638.
22. Cummings SR, Black DM, Nevitt MC, et al. Appendicular bone density and age predict hip fracture in women. The Study of Osteoporotic Fractures Research Group. JAMA 1990; 263(5):665–668.
23. Hui SL, Slemenda CW, Johnston CC Jr. Baseline measurement of bone mass predicts fracture in white women. Ann Intern Med 1989; 111(5):355–361.
24. Hui SL, Slemenda CW, Johnston CC Jr. Age and bone mass as predictors of fracture in a prospective study. J Clin Invest 1988; 81(6):1804–1809.
25. Cleghorn D, Polley K, Bellon M, et al. Fracture rates as a function of forearm mineral density in normal postmenopausal women: retrospective and prospective data. Calcif Tissue Int 1991; 49:161–163.
26. Bauer DC, Gluer CC, Cauley JA, et al. Broadband ultrasound attenuation predicts fractures strongly and independently of densitometry in older women. A prospective study. Study of Osteoporotic Fractures Research Group. Arch Intern Med 1997; 157(6):629–634.
27. de Laet CE, Van Hout BA, Burger H, et al. Hip fracture prediction in elderly men and women: validation in the Rotterdam study. J Bone Miner Res 1998; 13(10):1587–1593.
28. Duboeuf F, Hans D, Schott AM, et al. Different morphometric and densitometric parameters predict cervical and trochanteric hip fracture: the EPIDOS Study. J Bone Miner Res 1997; 12(11):1895–1902.
29. Garnero P, Dargent-Molina P, Hans D, et al. Do markers of bone resorption add to bone mineral density and ultrasonographic heel measurement for the prediction of hip fracture in elderly women? The EPIDOS Prospective Study. Osteopor Int 1998; 8(6):563–569.
30. Hans D, Dargent-Molina P, Schott AM, et al. Ultrasonographic heel measurements to predict hip fracture in elderly women: the EPIDOS Prospective Study. Lancet 1996; 348(9026):511–514.
31. Kroger H, Huopio J, Honkanen R, et al. Prediction of fracture risk using axial bone mineral density in a perimenopausal population: a prospective study. J Bone Miner Res 1995; 10(2):302–306.

32. Mele R, Masci G, Ventura V, et al. Three-year longitudinal study with quantitative ultrasound at the hand phalanx in a female population. Osteopor Int 1997; 7(6):550–557.
33. Nevitt MC, Johnell O, Black DM, et al. Bone mineral density predicts non-spine fractures in very elderly women. Study of Osteoporotic Fractures Research Group. Osteopor Int 1994; 4(6):325–331.
34. Schott AM, Cormier C, Hans D, et al. How hip and whole-body bone mineral density predict hip fracture in elderly women: the EPIDOS Prospective Study. Osteopor Int 1998; 8(3):247–254.
35. Stewart A, Torgerson DJ, Reid DM. Prediction of fractures in perimenopausal women: a comparison of dual energy x-ray absorptiometry and broadband ultrasound attenuation. Ann Rheum Dis 1996; 55:140–142.
36. Vecht-Hart CM, Zwamborn AW, Peeters PH, et al. Prediction of peripheral fracture risk by quantitative microdensitometry. Prevent Med 1997; 26(1):86–91.
37. de Laet CE, van Hout BA, Burger H, et al. Bone density and risk of hip fracture in men and women: cross sectional analysis. Br Med J 1997; 315(7102):221–225.
38. Black DM, Cummings SR, Karpf DB, et al. Randomised trial of effect of alendronate on risk of fracture in women with existing vertebral fractures. Lancet 1996; 348:1535–1541.
39. Liberman UA, Weiss SR, Broll J, et al. Effect of oral alendronate on bone mineral density and the incidence of fractures in postmenopausal osteoporosis. N Engl J Med 1995; 333(22):1437–1443.
40. Cummings SR, Black DM, Thompson DE. Effect of alendronate reduces on risk of fracture in women with low bone density but without vertebral fractures: results from the Fracture Intervention Trial. JAMA 1998; 280(24):2077–2082.
41. Orwoll ES, Oviatt SK, McClung MR, Deftos LJ, Sexton G. The rate of bone mineral loss in normal men and the effects of calcium and cholecalciferol supplementation. Ann Intern Med 1990; 112:29–34.
42. Saag KG, Emkey R, Schnitzer TJ, et al. Alendronate for the prevention and treatment of glucocorticoid-induced osteoporosis. N Engl J Med 1998; 339:292–299.
43. Chesnut CH 3rd, Silverman S, Andriano K, et al. A randomized trial of nasal spray salmon calcitonin in post-menopausal women with established osteoporosis: the prevent recurrence of osteoporotic fractures study. PROOF Study Group. Am J Med 2000; 109(4):267–276.
44. Dawson-Hughes B, Harris SS, Krall EA, Dallal GE. Effect of calcium and vitamin D supplementation on bone density in men and women 65 years of age or older. N Engl J Med 1997; 337:670–676.
45. Delmas PD, Bjarnason NH, Mitlak BH, et al. Effects of raloxifene on bone mineral density, serum cholesterol concentrations, and uterine endometrium in postmenopausal women. N Engl J Med 1997; 337:1641–1647.
46. Felson DT, Zhang Y, Hannan MT, Kiel DP, Wilson PW, Anderson JJ. The effect of postmenopausal estrogen therapy on bone density in elderly women. N Engl J Med 1993; 329:1141–1146.
47. Francis RM. The effects of testosterone on osteoporosis in men. Clin Endocrinol 1999; 50:411–414.
48. Lufkin EG, Wahner HW, O'Fallon WM, et al. Treatment of postmenopausal osteoporosis with transdermal estrogen. Ann Intern Med 1992; 117:1–9.
49. McClung MR, Geusens P, Miller PD, et al. Effect of risedronate on the risk of hip fracture in elderly women. Hip Intervention Program Study Group. N Engl J Med 2001; 344:333–340.
50. Neer RM, Arnaud CD, Zanchetta JR, et al. Effect of parathyroid hormone (1–34) on fractures and bone mineral density in postmenopausal women with osteoporosis. N Engl J Med 2001; 344(19):1434–1441.
51. Hamdy RC, Petak SM, Lenchik L. Which central dual X-ray absorptiometry skeletal sites and regions of interest should be used to determine the diagnosis of osteoporosis? J Clin Densitom 2002; 5(suppl):S11–S18.

52. Lenchik L, Kiebzak GM, Blunt BA. What is the role of serial bone mineral density measurements in patient management? J Clin Densitom 2002; 5(suppl):S29–S38.
53. Ott SM, Kilcoyne RF, Chesnut CH 3rd. Longitudinal changes in bone mass after one year as measured by different techniques in patients with osteoporosis. Calcif Tissue Int 1986; 39(3):133–138.
54. Rosenthal DI, Ganott MA, Wyshak G, Slovik DM, Doppelt SH, Neer RM. Quantitative computed tomography for spinal density measurement. Factors affecting precision. Invest Radiol 1985; 20(3):306–310.
55. Lang T, Augat P, Majumdar S, Ouyang X, Genant HK. Noninvasive assessment of bone density and structure using computed tomography and magnetic resonance. Bone 1998; 22(5 suppl):149S–153S.
56. Rauch F, Schoenau E. Changes in bone density during childhood and adolescence: an approach based on bone's biological organization. J Bone Miner Res 2001; 16(4):597–604.
57. Baroncelli GI, Saggese G. Critical ages and stages of puberty in the accumulation of spinal and femoral bone mass: the validity of bone mass measurements. Horm Res 2000; 54(suppl)1:2–8.
58. Yu W, Qin M, Xu L, et al. Normal changes in spinal bone mineral density in a Chinese population: assessment by quantitative computed tomography and dual-energy X-ray absorptiometry. Osteopor Int 1999; 9(2):179–187.
59. Carr JJ, Shi R, Lenchik L, Langefeld C, Lange L, Bowden DW. Validation of quantitative computed tomography for measurement of bone mineral density in the thoracic spine during cardiac gated protocol for coronary vascular calcium. Radiology 2001; 221:380.
60. Gluer CC. Quantitative ultrasound techniques for the assessment of osteoporosis: expert agreement on current status. The International Quantitative Ultrasound Consensus Group. J Bone Miner Res 1997; 12(8):1280–1288.
61. Miller PD, Siris ES, Barrett-Connor E, et al. Prediction of fracture risk in postmenopausal white women with peripheral bone densitometry: evidence from the National Osteoporosis Risk Assessment. J Bone Miner Res 2002; 17(12):2222–2230.
62. Pluskiewicz W, Halaba Z. First prospective report with the use of quantitative ultrasound (QUS) in children and adolescents. J Clin Densitom 2001; 4(2):173.
63. van den Bergh JP, Noordam C, Ozyilmaz A, Hermus AR, Smals AG, Otten BJ. Calcaneal ultrasound imaging in healthy children and adolescents: relation of the ultrasound parameters BUA and SOS to age, body weight, height, foot dimensions and pubertal stage. Osteopor Int 2000; 11(11):967–976.
64. Falk B, Sadres E, Constantini N, Eliakim A, Zigel L, Foldes AJ. Quantitative ultrasound (QUS) of the tibia: a sensitive tool for the detection of bone changes in growing boys. J Pediatr Endocrinol Metab 2000; 13(8):1129–1135.
65. Marshall D, Johnell O, Wedel H. Meta-analysis of how well measures of bone mineral density predict occurrence of osteoporotic fractures. Br Med J 1996; 312:1254–1259.
66. Carter DR, Hayes WC. The compressive behavior of bone as a two-phase porous structure. J Bone Joint Surg Am 1977; 59(7):954–962.
67. Gibson LJ. The mechanical behaviour of cancellous bone. J Biomech 1985; 18(5):317–328.
68. Hvid I, Jensen NC, Bunger C, Solund K, Djurhuus JC. Bone mineral assay: its relation to the mechanical strength of cancellous bone. Eng Med 1985; 14(2):79–83.
69. Hvid I, Hansen SL. Trabecular bone strength patterns at the proximal tibial epiphysis. J Orthoped Res 1985; 3(4):464–472.
70. Linde F, Hvid I, Pongsoipetch B. Energy absorptive properties of human trabecular bone specimens during axial compression. J Orthoped Res 1989; 7(3):432–439.
71. Moro M, Hecker AT, Bouxsein ML, Myers ER. Failure load of thoracic vertebrae correlates with lumbar bone mineral density measured by DXA. Calcif Tissue Int 1995; 56(3):206–209.

72. Cheng XG, Nicholson PH, Boonen S, et al. Prediction of vertebral strength in vitro by spinal bone densitometry and calcaneal ultrasound. J Bone Miner Res 1997; 12(10):1721–1728.
73. Eriksson SA, Isberg BO, Lindgren JU. Prediction of vertebral strength by dual photon absorptiometry and quantitative computed tomography. Calcif Tissue Int 1989; 44(4):243–250.
74. Bates DW, Black DM, Cummings SR. Clinical use of bone densitometry: clinical applications. JAMA 2002; 288(15):1898–1900.
75. Cummings SR, Bates D, Black DM. Clinical use of bone densitometry: scientific review. JAMA 2002; 288(15):1889–1897.
76. Lenchik L, Sartoris DJ. Current concepts in osteoporosis. AJR Am J Roentgenol 1997; 168(4):905–911.
77. Kanis JA. Assessment of fracture risk and its application to screening for postmenopausal osteoporosis: synopsis of a WHO report. WHO Study Group. Osteopor Int 1994; 4(6):368–381.
78. Binkley NC, Schmeer P, Wasnich RD, Lenchik L. What are the criteria by which a densitometric diagnosis of osteoporosis can be made in males and non-Caucasians? J Clin Densitom 2002; 5(suppl):S19–S27.
79. Hui SL, Zhou L, Evans R, et al. Rates of growth and loss of bone mineral in the spine and femoral neck in white females. Osteopor Int 1999; 9(3):200–205.
80. Melton LJI, Khosla S, Achenbach SJ, et al. Effects of body size and skeletal site on the estimated prevalence of osteoporosis in women and men. Osteopor Int 2000; 11:977–983.
81. De Laet DH, Van Hout BA, Burger H, Hofman A, Pols HAP. Bone density and risk of hip fracture in men and women: cross sectional analysis. Br Med J 1997; 221–225.
82. Kanis JA, Johnell O, Oden A, De Laet C, Mellstrom D. Diagnosis of osteoporosis and fracture threshold in men. Calcif Tissue Int 2001; 69:218–221.
83. Orwoll E. Perspective: assessing bone density in men. J Bone Miner Res 2000; 15:1867–1870.
84. Selby PL, Davies M, Adams JE. Do men and women fracture bones at similar bone densities? Osteopor Int 2000; 11:153–157.
85. Kudlacek S, Schneider B, Resch H, Freudenthaler O, Willvonseder R. Gender differences in fracture risk and bone mineral density. Maturitas 2000; 36:173–180.
86. Cauley JA, Zmuda JM, Palmero L, Stone KL, Black DM, Nevitt MC. Do men and women fracture at the same BMD level. J Bone Miner Res 2000; 15:S144.
87. Miller PD, Njeh CF, Jankowski LG, Lenchik L. What are the standards by which bone mass measurement at peripheral skeletal sites should be used in the diagnosis of osteoporosis? J Clin Densitom 2002; 5(suppl):S39–S45.
88. Faulkner KG, Roberts LA, McClung MR. Discrepancies in normative data between Lunar and Hologic DXA systems. Osteopor Int 1996; 6:432–436.
89. Black D. A proposal to establish comparable diagnostic categories for bone densitometry based on hip fracture risk among Caucasian women over age 65 years. J Bone Miner Res 2001; 16:S342.
90. Faulkner KG, Von Stetten E, Miller PD. Discordance in patient classification using T scores. J Clin Densitom 1999; 2:343–350.
91. Grampp S, Genant HK, Mathur A, et al. Comparisons of noninvasive bone mineral measurements in assessing age-related loss, fracture discrimination, and diagnostic classification. J Bone Miner Res 1997; 12(5):697–711.
92. Hodgson SF, Watts NB, Bilezikian JP, et al. American Association of Clinical Endocrinologists 2001 medical guidelines for clinical practice for the prevention and management of postmenopausal osteoporosis. Endocr Pract 2001 July; 7(4):293–312.
93. Kanis JA, Torgerson D, Cooper C. Comparison of the European and USA practice guidelines for osteoporosis. Trends Endocrinol Metab 2000; 11(1):28–32.
94. Bonnick SL, Johnston CC Jr, Kleerekoper M, et al. Importance of precision in bone density measurements. J Clin Densitom 2001; 4:105–110.

95. Lenchik L, Watts NB. Regression to the mean: what does it mean? Using bone density results to monitor treatment of osteoporosis. J Clin Densitom 2001; 4(1):1–4.
96. Lenchik L, Rochmis P, Sartoris DJ. Optimized interpretation and reporting of dual X-ray absorptiometry (DXA) scans. Am J Roentgenol 1998; 171(6):1509–1520.
97. Varney LF, Parker RA, Vincelette A, Greenspan SL. Classification of osteoporosis and osteopenia in postmenopausal women is dependent on site-specific analysis. J Clin Densitom 1999; 3:275–283.
98. Woodson G. Dual X-ray absorptiometry T score concordance and discordance between the hip and spine measurement sites. J Clin Densitom 2000; 3:319–324.
99. Yu W, Gluer CC, Fuerst T, et al. Influence of degenerative joint disease on spinal bone mineral measurements in postmenopausal women. Calcif Tissue Int 1995; 57:169–174.
100. Drinka PJ, DeSmet AA, Bauwens SF, Rogot A. The effect of overlying calcification on lumbar bone densitometry. Calcif Tissue Int 1992; 50(6):507–510.
101. Preidler KW, White LS, Tashkin J, et al. Dual-energy X-ray absorptiometric densitometry in osteoarthritis of the hip. Influence of secondary bone remodeling of the femoral neck. Acta Radiol 1997; 38:539–542.
102. Akesson K, Gardsell P, Sernbo I, Johnell O, Obrant KJ. Earlier wrist fracture: a confounding factor in distal forearm bone screening. Osteopor Int 1992; 2(4):201–204.
103. Smith JA, Vento JA, Spencer RP, Tendler BE. Aortic calcification contributing to bone densitometry measurement. J Clin Densitom 1999; 2:181–183.
104. Peel NF, Johnson A, Barrington NA, Smith TW, Eastell R. Impact of anomalous vertebral segmentation on measurements of bone mineral density. J Bone Miner Res 1993; 8(6):719–723.
105. Johnson ML, Gong G, Kimberling W, Recker SM, Kimmel DB, Recker RB. Linkage of a gene causing high bone mass to human chromosome 11 (11q12–13) Am J Hum Genet 1997; 60:1326–1332.
106. Boyden LM, Mao J, et al. High bone density due to a mutation in LDL-receptor-related protein 5. N Engl J Med 2002; 346(20):1513–1521.

5 Importance of Nutrition in Fracture Healing

Sanjeev Kakar and Thomas A. Einhorn

1. INTRODUCTION

Injuries to the musculoskeletal system are common and place a significant burden on modern-day health care systems. Within the United States alone, there are more than 6 million fractures a year *(1)*, with hip fractures making up a significant proportion *(2)*. These tend to be concentrated among the elderly osteoporotic population and are a major cause of morbidity and mortality *(3)*.

Fracture healing is a specialized reparative process of bony regeneration. It is highly orchestrated with the restoration of skeletal integrity and functional recovery. There are however, 5–10% of injuries that fail to heal normally and result in delayed or nonunion *(1)*. This may be in part related to the numerous local and systemic factors that affect the repair process *(4)*. Local causes include degrees of periosteal stripping and the presence of infection at the fracture site. In analyzing the various systemic factors that influence fracture healing, an individual's nutritional status has been shown to have an important effect.

In trauma patients with long bone injuries, a hypermetabolic response occurs. The greater the severity of the injury, the larger the catabolism, with a marked loss in body nitrogen, sulfur, and phosphorous on the first day after injury *(5)*. The resultant increase in calorific intake required to keep up with the metabolic demands can exceed 6000 cal/d, nearly three times that of a normal active adult *(6)*. If this nutritional demand is not met, patient recovery is adversely affected.

Bastow et al. *(7)* conducted a study investigating the effects of supplemental tube feeding on the postoperative recovery of patients with fractured femurs. Patients who were deemed to be malnourished upon their anthropometric measurements were entered postoperatively into a randomized, controlled trial of overnight supplemental feeding in addition to their normal diet. Clinical outcome improved in the supplemented group in terms of rehabilitation times and hospital stays.

Nutritional status has been shown to affect fracture repair. Lindholm *(8,9)* demonstrates that calluses in rats fed a calcium- and vitamin D-deficient diet lacked

From: *Nutrition and Bone Health*
Edited by: M. F. Holick and B. Dawson-Hughes © Humana Press Inc., Totowa, NJ

consolidation and calcification and instead were enriched in fibrous and adipose tissues. A further study by Pollak et al. *(10)* investigated the effects of different animal feeds on the biomechanical properties of fracture calluses in rats. By altering the protein and calorific content of the feeds, reductions in tensile strength and stiffness were seen in the fracture calluses of animals fed on low-protein diets. These results have real clinical significance in the face of a high prevalence of malnutrition seen within injured orthopedic surgery patients.

In this chapter, we review the events of fracture healing and the complex interplay of nutritional elements that are known to affect this process.

2. FRACTURE HEALING REPAIR PROCESSES

Fracture healing is a highly orchestrated process comprising a series of biological repair stages intimately linked with each other. The result is near-complete biochemical and biomechanical restoration of the original bony substance. The repair process involves both *primary* and *secondary* healing *(11)*.

Primary fracture healing is the process by which the cortex directly attempts to reestablish its own continuity with the aid of "cutting cones." These are remodeling units of osteoclasts that resorb cortical bone, thereby permitting angiogenesis and stem cell invasion of the fracture site. These cells differentiate into osteoblasts, which leads to the formation of bone, bridging the fracture gap *(12)*.

The response of the periosteum and neighboring soft tissues to bony injury forms the basis of secondary fracture healing through which the majority of fractures heal. It involves both intramembranous and endochondral ossification that proceed concurrently. Intramembranous bone formation occurs at the proximal and distal ends of the disrupted periosteal tissue and is mediated by committed osteoprogenitor cells. The endochondral response is largely dependent on undifferentiated cells in the periosteum *(13)* and neighboring soft tissues *(14,15)* that are induced to differentiate in response to injury.

Bony injury triggers an inflammatory cascade, critical to regulating the repair process. Hematoma formation, necrosis of fracture segments, and local blood vessel disruption occur. Inflammatory mediators such as interleukin (IL)-1 and IL-6 are involved in the chemotaxis of scavenger cells including neutrophils and macrophages, which remove necrotic tissue associated with the injury. Cytokines such as tumour necrosis factor-α (TNF-α) facilitate the repair process by acting on mesenchymal cell recruitment and differentiation *(16)*. Angiogenesis is driven by soft tissue/periosteal responses to low pO_2 that exist within the injury site as a result of high cellular metabolism. It peaks at about 2 wk after injury, before normalizing at 14 wk *(17,18)*.

Once the inflammatory process subsides, callus bridges the fracture gap. This has been described as comprising of a soft (cartilage) and hard (bone) callus. Rather than being discrete entities, a degree of overlap exists between the two. Initially, a rapid mitotic response of the chondrocytes occurs 7–10 d post-injury with type II collagen fiber formation. These fibrils are stabilized by type IX collagen. After 2 wk, the chondrocytes hypertrophy and matrix vesicles are seen budding off them. These contain the necessary enzymes to permit mineralization of soft callus into hard callus enriched with types I and II collagen. Chondroclasts and osteoblasts then convert the callus into woven bone. This results in an increased strength and stiffness of the repair tissue.

The remodeling process marks the last stage of fracture healing, with woven bone slowly transformed into mechanically robust lamellar bone through the coordinated activities of osteoblasts and osteoclasts.

3. NUTRITION IN ORTHOPEDIC PATIENTS

Several studies have shown a high prevalence of malnutrition among surgical patients. In one series of 129 patients undergoing orthopedic procedures, the average prevalence of clinical and subclinical malnutrition was 42.4% *(19)*. Preexisting undernutrition is often not recognized by health care providers and is further exacerbated during hospitilization *(20–24)*.

Malnutrition is a particular problem within the elderly, with up to 40% of patients significantly underweight *(25)*. Several factors account for this, such as physical and mental diseases, medications with appetite-suppressing side effects, and inadequate access to food due to poverty or disabilities *(25)*. Surgical insult further compounds this problem by inducing a catabolic response. Patients' energy requirements increase from 2500 cal of a normal active adult to more than 6000 cal per day *(6)*. Fuel reserves are mobilized within the body in times of metabolic stress. Glycogen stores are exhausted in less than 12 h. Protein breakdown occurs at the expense of vital organs and muscle mass. Its depletion has been linked to impaired wound healing, progressive weakness, and delayed physical rehabilitation. Host resistance is further compromized, with impaired humoral and cell-mediated immunity *(26)*. Patterson et al. *(3)* conducted a prospective study examining the effects of protein depletion in 63 elderly patients admitted with fractured hips. Fifty-nine percent of patients were found to be protein-deficient during their hospital stay. They were found to have a higher rate of complications, prolonged rehabilitation times, and a lower probability of 1-yr survival after their hip fracture. Bonjour et al. *(27)* conducted a prospective randomized, controlled study that investigated the effects of nutritional supplementation on functional recovery. Patients with hip fractures were subdivided into groups that were either given dietary supplementation in terms of increased intakes of energy, protein, and calcium, or fed a standard diet. Patient morbidity and mortality were improved in the supplemented group, with patients experiencing a fourfold reduction in complications vs the control patients. From this it can be concluded that appropriate nutritional replenishment can reverse these trends and improve patient prognosis *(28)*.

Bone mineral density, a determinant that provides 70% of bone strength, is adversely affected by poor nutrition *(29,30)*. This has been shown in several studies that have investigated rates of hip fracture in malnourished patients. Matkovic et al. *(31)* examined the rates of hip fractures in two communities who differed in their calcium intake. Having adjusted for confounding factors, they found a 50% reduction in hip fracture rates among those who were on high-dairy diets as opposed to those on reduced intakes. This was related to the former having a higher bone mineral density (BMD). Similarly, a case-control study by Ray et al. *(32)* investigated the influence of thiazide and other antihypertensive medication on hip fracture rates. Risk of fracture decreased significantly with increasing duration of thiazide use. In contrast, no such trend was seen in patients on other medications. The reduction in age-dependent bone loss and hip fracture rates may be related to the hypocalciuric effect of thiazide medication *(32)*.

Malnutrition can lead to an increase in musculoskeletal injuries by impairing patient coordination *(33)*. In addition, Pruzansky et al. *(34)* showed that underweight women had a significantly higher risk of hip fracture than their normal- and overweight counterparts. The mechanism by which this occurs is unknown but may be related to the effect of Wolff's law *(35)* on bone metabolism. The thinner the patient, the less force is transmitted through the proximal femur, resulting in a reduction in bone growth in that area.

Effective monitoring of the nutritional health of patients is clearly important. This may take the form of a multidisciplinary approach involving surgeons, dietitians, visiting nurses, and home physical therapists. The elements of a nutritional support program should include assessment of the patient's dietary status preoperatively, calculation of caloric requirements, and clinical/biochemical monitoring to ensure that adequate nutritional support is met *(36)*.

4. NUTRITONAL FACTORS AND FRACTURE HEALING

4.1. Calcium

The mineral content of bone accounts for approx 70% of its weight. It exists mainly as calcium hydroxyapatite, which plays an important role in regulating bone's elastic stiffness properties. This has been demonstrated by work of Burstein et al. *(37)*, in which the contribution of collagen and mineral were correlated to the elastic–plastic properties of bone. By mechanically testing bone tissue under tensile loading after it had undergone progressive decalcification, they showed that the mineral phase of bone mainly contributes to its tensile strength.

Calcium is found in a number of foodstuffs including dairy products such as milk, green vegetables, and tofu *(38)*. Its bioavailability is affected by various dietary factors. Oxalates, which are present in large quantities in spinach, zinc, and vitamin A have been shown to lower serum calcium concentrations *(38)*. Calcium absorption rates are also affected by age, with less efficient absorption seen in the elderly *(39,40)*. This may be related to either a deficiency of vitamin D or impairment of the synthesis of its most active metabolite, 1,25-dihydroxyvitamin D [$1,25(OH)_2D_3$] *(41)*.

Calcium requirements are higher than other nutrients because of less efficient intestinal absorption. The recommended dietary Adequate Intake (AI) is set at 1000 and 1200 mg/d for adults 18–50 yr and 51+ yr, respectively *(42)*. Men meet their calcium requirements more readily than women, who ingest less than the AI *(43,44)*.

Calcium supplementation directly affects peak bone mass by reducing cortical bone loss. This is most evident in the elderly population, with the risk of vertebral fractures reduced by as much as 50% and hip fractures by 20% *(45)*. Several mechanisms are thought to account for this. Heaney et al. *(46)* demonstrated suppressed parathyroid hormone (PTH) levels in response to increase dietary calcium intake. This in turn reduced bone resorption. Calcium supplementation also affects bone's remodeling. Orwoll et al. *(47)* investigated the effects of calcium carbonate supplementation, either alone or with 25-hydroxyvitamin D [25(OH)D] on osteoporotic women who had severe trabecular osteopenia and prolonged remodeling phases. After 2 yr of treatment, they found an increase in the rate of matrix mineralization and a reduction in the duration of bone remodeling by over 50%.

The role of dietary calcium in fracture healing has been extensively studied. In 1955, Key and Odell *(48)* reported that fractures healed normally regardless of additional dietary calcium. It appeared that a mechanism independent of dietary calcium existed that provided minerals for fracture healing. Subsequent work by Singh et al. *(49)* investigated the role of the diet and skeleton as a potential source of calcium in fracture repair. Using ^{35}Sr as a tracer, their findings suggested that the skeleton acted as a calcium source in the first 2 wk after fracture. Thereafter, the diet provides the necessary calcium needed for fracture repair.

The biomechanical properties of fracture callus is affected by its calcium concentration. Einhorn et al. investigated the contributions of dietary protein and mineral to the mechanical properties of fracture callus in rats *(50)*. Three animal groups were studied—a control group, a group in which the right femurs were treated with an intramedullary pin, and a group in which the pinned femurs were fractured. Animals in each group were further subdivided in terms of their dietary regimens, which varied in their mineral/protein contents. Results showed that excess protein and/or mineral intake had no adverse effect on fracture healing. The stiffness of the fracture callus was independent of dietary protein and mineral contents. The strength, however, was found to be related to its mineral composition.

4.2. Vitamin D

Vitamin D is an essential micronutrient required for efficient calcium absorption and mineralization of bone. Its levels decrease with age, resulting in impaired calcium absorption due to reductions in 1α-hydroxylation of $25(OH)D_3$ and lowered responsiveness of the intestinal mucosa to $1,25(OH)_2D_3$ *(51)*.

Vitamin D-enriched food products include dairy products, fish, mushrooms, and eggs *(52,53)*. AI is 10 μg (400 IU)/d for those 51–70 yr of age and 15 μg (600 IU)/d for those over 70 yr of age.

Vitamin D malnourishment is very common among institutionalized and hip fracture elderly patients *(54)*. It is estimated that a quarter of the elderly aged up to 79 yr and up to three-quarters of people aged 80 yr or over are vitamin D-deficient *(55)*.

Hypovitaminosis has been linked to an increase in fracture rates, especially in postmenopausal women and the elderly *(56,57)*. These patients have been shown to have elevated PTH levels in response to vitamin D deficiency *(58,59)*. The secondary hyperparathyroidism leads to an increase in skeletal bone turnover with increased remodeling space, increased bone loss, and cortical thinning *(60)*. These changes result in lower BMD and an increase in fracture risk *(54,56)*. Supplementation with calcium and vitamin D can reduce this fracture risk *(56,61)*. In an effort to elucidate the mechanism by which this occurs, Lips et al. *(55)* conducted a trial investigating the effects of vitamin D on serum PTH concentrations. One hundred and forty-two patients living in elderly or nursing homes were randomized to receive vitamin D supplements or placebo. Results showed that supplementation led to a small but significant increase in plasma 25(OH)D with a reduction in PTH levels *(55)*. Further work by the same group investigated the effect of vitamin D supplementation on the incidence of hip fractures. More than 2500 patients, aged 70 yr or over, participated in this double-blind prospective study. They received either vitamin D or placebo. After 1 yr, rises in plasma 25(OH)D levels with reductions in serum PTH were seen in the experimental group. This resulted in a significant increase in bone mineral density at the femoral neck in these patients. This was still present after 2 yr, corresponding to a 15% reduction in hip fracture rates *(61)*. This positive effect in reducing fracture rates is not confined to the hip. Chapuy et al. *(56)* investigated the impact of vitamin D and calcium supplementation on nonvertebral fractures. They found a 32% reduction in these fractures in the experimental group compared to those on placebo.

Vitamin D affects the biology of fracture healing with deficiencies delaying callus formation for up to 9 wk after injury *(62)*. Many investigators have examined the mechanism by which this occurs. Lindolm *(63)* evaluated the properties of fracture calluses in calcium- and vitamin D-deficient rats. He found that the consolidation of fractures were slower in these animals, with a reduction in new bone formation and an increased loss of trabecular bone. Histologically, the callus was qualitatively and quantitatively inferior to that seen in controls. It was primarily composed of fibrous and adipose elements with minimal calcified tissue.

Further research has shown that individual vitamin D metabolites affect the fracture healing process in different ways. Brumbaugh et al. *(64)* investigated the effects of $1,25(OH)_2D_3$ treatment in chicks raised on vitamin D-deficient diets. Following humeral fracture, animals devoid of supplementation exhibited prolonged fracture healing, abnormal endochondral bone formation, and delays in woven bone remodeling. In trying to elucidate the mechanism by which this occurs, Jingushi et al. *(65)* injected radiolabeled $1,25(OH)_2D_3$ into rats and subjected them to femur fractures.

Plasma concentrations rapidly decreased on d 3 after fracture and continued to decline an additional 7 d. Radioactivity in the fractured femurs was higher than in the controls and was concentrated within the callus. 1,25$(OH)_2D_3$ receptor gene expression was also detected in the callus just after fracture. The investigators suggested that serum 1,25$(OH)_2D_3$ was diverted to the callus to regulate the repair process as 1,25$(OH)_2D_3$ stimulates osteoblast and chondrocyte proliferation via the 1,25$(OH)_2D_3$ receptor *(66–68)*. It may be hypothesized that 1,25$(OH)_2D_3$ regulates both cartilage and bone formation during fracture repair.

Much is known about the role of 1,25$(OH)_2D_3$ in fracture repair. Dekel et al. *(69)* examined whether other vitamin D metabolites are involved in this process. Tibial fractures were performed in chicks on a vitamin D-deficient diet. In tandem, they were dosed with radioactively labeled cholecalciferol. Analysis of the callus 9 d after injury revealed a preferential accumulation of 24,25$(OH)_2D_3$ and 1,25$(OH)_2D_3$. Based on their data they concluded that 24,25$(OH)_2D_3$ may have an important effect on fracture healing. In trying to determine its role in bone repair, Seo et al. *(70)* tested natural 24,25$(OH)_2D_3$ and its synthetic epimer alone or in combination with 1,25$(OH)_2D_3$ on the mechanical properties of healed tibiae after injury. Results showed that natural 24,25$(OH)_2D_3$ is an essential vitamin D metabolite for normal bone integrity and healing of fractures in chicks.

Fracture callus is able to discriminate between natural and synthetic versions of 24,25$(OH)_2D_3$. This led many to try and characterize the nature of 24,25$(OH)_2D_3$ receptor. Kato et al. *(71)* found no evidence of a classic nuclear/cytosol receptor for 24,25$(OH)_2D_3$ in fracture callus. Instead, a specific receptor/binding protein for this metabolite was found in the callus membrane fraction, which exhibited selective ligand-binding properties.

Vitamin D metabolites have contrasting effects on the mechanical properties of fracture callus *(72)*. Bone wax impregnated with either 1,25$(OH)_2D_3$ or 24,25$(OH)_2D_3$ was implanted into chick tibial fractures of vitamin D-fed chicks. The animals were harvested at a series of time points and the mechanical and biochemical properties of the tibia were evaluated. 24,25$(OH)_2D_3$ strengthened the callus and increased alkaline phosphatase levels in the first week postfracture. In contrast, 1,25$(OH)_2D_3$ weakened the fracture callus, with a reduction of calcium incorporation. However, when administered together, 24,25$(OH)_2D_3$ was shown to have a positive action on 1,25$(OH)_2D_3$ in terms of enhancing the mechanical strength of callus and the mineral content of fractured bones *(71,73)*.

When do each of the metabolites exert their effects during fracture healing? 24,25$(OH)_2D_3$ is seen to accumulate to a greater extent than 1,25$(OH)_2D_3$ in the early stages of callus formation *(69,74)*. The peak concentration of 24,25$(OH)_2D_3$ is on d 3 post-injury, during which rapid proliferation of mesenchymal cells and their differentiation into chondrocytes occurs. This represents the chondrogenic phase of callus formation. As cartilage is converted into bone, levels of 24,25$(OH)_2D_3$ decline whereas concentrations of 1,25$(OH)_2D_3$ rise, indicating its major role in the remodeling phase of fracture healing *(74)*.

4.3. Phosphorous

Food sources rich in phosphorous include meat, poultry, fish, and grain products. AI equates to 100 mg/d for adults *(75),* with intakes varying with the population group. It has been shown that up to 25% of elderly living in nursing homes have insufficient dietary consumption *(76).*

Phosphorous consumption has been shown to have an effect on bone mineralization. A low intake impairs osteoblast function, whereas too high a consumption reduces serum calcium levels. This results in secondary hyperparathyroidism with cortical bone loss *(51,77–80).* This latter effect is an acute response with no long-term adverse effects on calcium balance *(43).*

Phosphate malnutrition has a negative effect on fracture healing. Urist and McLean *(81)* showed that, in rats with low-phosphate rickets, calcification of callus was impaired. Invasion of the fibrocartilaginous callus by new intramembranous osseous tissue was significantly delayed, sometimes up to 12 d after injury.

4.4. Vitamin K

Vitamin K-rich foods include dark green vegetables such as broccoli and spinach *(82).* The current RDA for vitamin K is 1 μg/kg *(83).* Recent studies, however, suggest that the average U.S. diet for younger adults does not meet this current RDA *(83).*

The clinical observation of bone defects and nasal dysplasia in the offspring of women taking warfarin, a vitamin K antagonist, during their pregnancy led many to investigate the role of vitamin K on bone metabolism. An in vitro study investigating the effects of vitamin K on osteoblasts was the first report demonstrating that it could modulate the proliferative state of these bone cells *(84).* Vitamin K increased alkaline phosphatase activity, a marker of synthetic function within the osteoblasts. It also stimulated collagen synthesis and mineralization. In contrast, warfarin had opposing effects, indicating a vitamin K-dependent mechanism existed that modulated bone forming cells' activation. The identification of γ-carboxylated bone proteins, in particular osteocalcin (OC), has shed light on the role of vitamin K in bone metabolism *(85,86).* Osteocalcin is the most abundant noncollagenous protein in bone matrix. It contains three Gla residues in positions 17, 21, and 24, which enable the protein to bind to calcium ions and hydroxyapatite. Synthesized exclusively by osteoblasts, OC is considered to be a specific marker of bone turnover *(87).* Its exact function, however, is not fully known. It is believed that it may modulate bone remodeling, as studies have shown that OC acts as a chemotactic agent (mediated by its C-terminal fragment) for osteoclast precursors and monocytes, and enhances their attachment to bone *(88,89).*

Decarboxylated OC is a good marker of the body's vitamin K nutritional status *(27,90,91).* With age, its levels are seen to increase, indicating that an age-dependent impairment of γ-carboxylation of OC occurs *(92).* Elevated decarboxylated OC concentrations have been linked to low BMD *(91,93)* and an increase in hip fractures *(94).* This was demonstrated by Szulc et al. *(90),* who examined serum decarboxylated OC levels in elderly women and related the values to the risk of hip fracture. In 195 subjects enrolled in the study, just under 25% had serum decarboxylated OC levels above the upper limit of the normal range for young women. During an 18-mo follow-up, women who sustained a hip fracture had a baseline undercarboxylated (uc) OC higher than in the nonfracture group. In fact, the relative risk of hip fracture was six times higher in women with high decarboxylated OC levels. This trend can be reversed by vitamin K supplementation, which promotes OC carboxylation *(95–98)*. This modifies the protein's three-dimensional structure, enabling hydroxyapatite to bind during bone mineralization *(99,100)*.

Vitamin K has a direct effect on fracture healing. In a study by Dodds et al. *(101)*, the effects of the anticoagulant dicumarol were investigated on fracture repair. Closed metatarsal fractures were produced in the experimental animal group taking this drug. At 12 d after injury, results showed that treatment with dicumarol resulted in a significant decrease in the amount of bone produced. In addition, the callus showed large areas of cartilage and fibrous tissue compared to that seen within the control population. This effect has been similarly reported by others *(102)*.

Investigating the effects of dietary vitamin K, Einhorn et al. *(103)* tested the hypothesis that alterations in OC concentrations have an effect on fracture healing. Thirty rats were divided into a control group, which received vitamin K supplementation, and an experimental group that was fed a vitamin K-deficient diet. Closed femoral fractures were produced in the animals. The bones were harvested at 6 wk and subjected to mechanical testing before the OC content of the calluses was determined. Results showed no difference in the mechanical properties of calluses between the control and experimental groups. In addition, the amount of OC and its degree of carboxylation in the callus's were the same. This suggested that in vitamin K-deficient states, fracture callus is able to maintain its normal biomechanical properties and that its OC concentrations are protected despite its depletion in the rest of the body. These findings raise the intriguing possibility of osteoblast cells possessing a mechanism for γ-carboxylation of glutamic acids that functions at low levels of vitamin K. In trying to explain this, observations made by Szulc et al. *(90)* highlighted the role of vitamin D as being able to influence serum decarboxylated OC concentrations. They noticed that additional vitamin D and calcium significantly decreased serum decarboxylated OC levels. This lead the authors to comment that vitamin D may have a direct effect on the γ-carboxylation of OC, possibly by influencing microsomal γ-carboxylase activity. Lian and co-workers *(104)* investigated whether this effect of vitamin D may be related to the structure of the osteocalcin gene. They found that vitamin D-deficient rats had reduced osteocalcin mRNA levels as well as bone and serum OC concentrations. By injecting the rats with $1,25(OH)_2D_3$, OC mRNA levels increased, indicating a physiological response of the OC gene to vitamin D. Analyzing the genetic structure, they found that this could be accounted for by the presence of a nucleotide sequence, sensitive to $1,25(OH)_2D_3$, located upstream from the transcription site that stimulated transcription of the OC gene.

4.5. Vitamin A

Vitamin A-rich foodstuffs include liver, fish oil, eggs, dairy products, and fortified foods such as cereals *(105)*. The Institute of Medicine recommends 800 μg/d for men and 700 μg/d for women *(106)*. With higher consumption of fortified food products and supplements, 5–10% of the elderly population in the United States ingest excessive amounts in relation to their recommended daily intake *(107,108)*.

Hypervitaminosis A has been linked to an increased risk of hip fracture. This is prevalent in Scandanavian countries, where vitamin A intake has been six times higher compared to southern Europe *(109)*. Melhus and coworkers *(110)* investigated this relationship by conducting a cross-sectional and nested case-control study in Sweden. They demonstrated that increased vitamin A consumption had a negative effect on the bone mineral density of the lumbar spine and femoral neck. For every 1-mg increase in daily retinol intake, a biologically active vitamin A metabolite, the risk of hip fracture increased by 68%. Michaelsson et al. *(111)* showed that serum retinol levels can be used to evaluate the risk of fracture. In a long-term prospective study of 2322 men, serum levels higher than 86 μg/dL (normal range: 20.1–80.2 μg/dL) were related to an increased rate of hip fractures.

In trying to elucidate the mechanism by which this may occur, many have examined the effects of vitamin A on bone biology. Johansson et al. *(112)* studied the effects of subclinical hypervitaminosis A in mature rats. After 12 wk, they noticed that rats in the experimental group exhibited cortical thinning and reduction in the diameter of the long bones. Biomechanical properties were adversely affected with a reduction in three-point bending breaking forces of the femora by more than 10%.

Retinoic acid receptors have been isolated on both osteoblasts and osteoclasts *(113)*. Their activation on osteoblasts results in the secretion of cytokines that stimulate osteoclasts as well as the recruitment of bone marrow precursor cells *(105)*. This leads to an imbalance between bone forming and resorbing cells, with resultant bone tissue loss. Saneshige et al. *(114)* demonstrated that retinoic acid regulates the expression of cathepsin K/OC 2, a dominant cysteine protease, at the transcriptional level in mature osteoclasts isolated from rabbits. Furthermore, elevated levels of retinoic acid receptor mRNA were detected in the osteoclasts. These findings demonstrate that retinoic acid has a negative influence on bone metabolism via its interactions with these bone resorbing cells.

4.6. Protein

Protein malnutrition is commonly seen in surgical patients. More than 25% of elderly patients admitted for orthopedic care have been found to be protein-deficient *(19)*. This degree of malnourishment has significant adverse effects on patient prognosis. Patterson and coworkers *(3)* prospectively evaluated 63 hip fracture patients to determine the effect of protein depletion and postoperative nutritional status on their outcome. Parameters used to assess the degree of protein depletion included levels of prealbumin, albumin, and transferrin, total lymphocyte count, and nitrogen balance. Fifty-eight percent were found to be protein-malnourished. Nitrogen balance studies showed that they needed at least 8 d before returning to an anabolic state. If calorie deficiencies were not replaced, a patient's catabolism was prolonged into the postoperative recovery period, resulting in a higher rate of complications like prolonged hospitalization and inferior survival analysis compared to the non-protein-malnourished group 1 yr after their fracture.

Protein has many dietary sources, including meat, nuts, and soya products *(115)*. RDA in adults is 0.75 g/kg/d, with the average American consuming more than 100 g of protein per day *(116,117)*.

Protein supplementation has a beneficial clinical effect by readdressing negative nitrogen balance that occurs with inadequate intake *(5)*. Bonjour et al. *(27)* demonstrated this relationship by comparing the clinical outcome of 62 elderly patients with hip fractures receiving two different diets that differed in their protein content. One group received an oral supplement of proteins, mineral salts, and vitamin A. The controls received the same diet but without additional protein. Clinical outcome was much better in the patients receiving protein, in terms of reduced complication and mortality rates. Similar positive effects of protein supplementation have been reported by others *(3,19,31,34)*.

Proteins play an important role in the process of fracture healing. This was initially shown by the experimental work of Rhoads and Kasinskas *(118)*. Using a low-protein diet and repeated plasmapharesis, the authors produced a state of hypoproteinemia in dogs. The ulna was then fractured and the repair was followed radiographically. The results showed that up to 74 d post-injury, callus formation was retarded in the experimental group. Further work by Pollak et al. *(10)* revealed that in animals on a low-protein diet, the biomechanical properties of the calluses were significantly lower compared to animals on high-protein diets. Einhorn et al. *(50)* characterized which biomechanical properties of fracture callus were dependent on dietary protein. The animal model comprised a control group, a group in which the rats underwent intramedullary pin insertion of the right femur, and an experimental group in which closed femoral fractures were produced in the pinned femurs. The animals received a regular diet, protein- and/or mineral-free feeds. Results showed that the stiffness of the fracture callus, which is acquired early in the course of normal fracture healing, developed independently of alterations in dietary protein or mineral. The strength of the fracture callus, which is acquired later in the repair process, showed a much greater dependence on protein as compared to minerals, as evidenced by reductions in both torsional strength and absorption of energy in rats that were fed a diet low in protein.

If protein malnutrition is recognized at the time of fracture, Day and Deheer *(119)* showed that dietary intervention in the immediate post-injury period can reverse the detrimental effects of protein deprivation on fracture healing. Rats were maintained on a diet containing either a normal or reduced protein concentration. After sustaining a closed femoral fracture, the animals were divided into a control group that was maintained on a 20% protein diet, a malnourished group that was maintained on a protein-depleted diet, and a third renourished group that received additional protein. When compared to well-nourished rats, the callus (periosteal and external) from the renourished animals resembled that seen in the control groups in terms of histological appearance. Mechanical testing revealed that the strength and stiffness of the fracture callus in the renourished group was greater than that seen in the malnourished and well-fed animals. This finding of replenishing protein at the time of fracture has important clinical relevance in the management of musculoskeletal injuries in protein-depleted patients.

In patients with protein malnutrition, there is a reduction in insulin-like growth factor (IGF) levels *(120)*. This may be due to an increase in IGF plasma clearance *(121)*, impairment of growth hormone (GH) action, and reduced IGF transcription/translation at the hepatic level *(122–124)*. Studies have shown IGF levels can be used as an index of nutritional status. Current available methods such as anthropometric indices, albumin, and transferrin levels lack sensitivity and specificity *(125,126)*. As an alternative, IGF values have been shown to be more informative in determining nutritional states in a group of hospitalized patients *(127)*.

The effects of protein on fracture healing can be mediated by the actions of IGF. Studies have shown that among hip fracture patients, protein supplementation has a beneficial clinical outcome by increasing IGF levels *(128)*. IGFs are anabolic polypeptides that mediate the growth-promoting effects of GH. GH is released by the anterior pituitary in response to hypothalamic gonadotrophin-releasing hormone. Once released into the systemic circulation, GH binds to specific cell surface receptors on target tissues such as the liver. It then stimulates the production of IGF, which in turn is released into the circulation to act on its target cells such as chondrocytes and osteoblasts. In this way, IGF mediates the growth-promoting effects of GH by promoting cell proliferation and matrix synthesis of chondrocytes and osteoblasts *(129–131)*. In contrast to its stimulatory effect on bone-forming cells, IGF negatively affects osteoclasts. By continuously infusing IGF into the arterial supply of the hind limb of ambulatory rats, Spencer et al. *(131)* showed that both cortical and trabecular bone formation were significantly increased. These findings were related to stimulation of osteoblasts and inhibition of osteoclasts.

The quantity of callus formed during fracture repair is influenced by IGF. Schmidmaier and colleagues *(132)* used intramedullary wires coated with IGF to study its local effects on fracture healing. They showed that IGF had a positive effect on the repair process, with greater biomechanical stability and enhanced callus consolidation seen in the experimental group compared to controls.

4.7. Alcohol

Alcohol has a negative effect on several aspects of bone metabolism and repair. It accounts for a reduction in BMD *(133)*, bone mass, and increases risk of fractures *(134)*. Brown et al. investigated the effects of chronic ethanol exposure in a rat osteotomy model. They found that the tibiae in rats exposed to alcohol had a significantly lower load to failure in three-point bending tests compared to control animals. In addition, the experimental animals that underwent distraction osteogenesis or fracture repair had reductions in bone formation compared to controls. These findings indicate that ethanol has an adverse effect on bone repair by inhibiting intramembranous bone formation *(135)*. This may be related to the toxic effects of alcohol on osteoblast function *(136,137)*.

5. CONCLUSION

Nutritional status can have a major impact on a patient's ability to withstand a physiological insult. In trauma patients who are malnourished before, during, or after injury, clinical outcome is adversely affected by malnutrition. Appreciation of this by health care providers is important to ensure that appropriate attention be paid, and that supplementation be provided to optimize nutritional state.

Fracture healing involves the complex interplay of various nutritional elements, thereby necessitating the need for balanced nutrition, as deficiencies can result in impairments of this repair process.

REFERENCES

1. Praemer A, Furner S, Price OP. Musculoskeletal conditions in the United States. Am Acad Orthoped Surg 1992; 85–91.
2. U.S. Congress, Office of Technology Assessment. Hip fracture outcomes in people aged 50 and over—background paper. OTA-BP-H-120. U.S. Government Printing Office, Washington, DC, July 1994.
3. Patterson BM, Cornell CN, Carboni B, Levine B, Chapman D. Protein depletion and metabolic stress in elderly patients who have a fracture of the hip. J Bone Joint Surg 1992; 74A(2):251–260.
4. Einhorn TA. Enhancement of fracture healing. Current concept review. J Bone Joint Surg 1995; 77A(6):940–956.
5. Cuthbertson DP. Further observations on the disturbance of metabolism caused by injury, with particular reference to the dietary requirements of fracture cases. Br J Surg 1936; 23:505–520.
6. Smith TK. Prevention of complications in orthopaedic surgery secondary to nutritional depletion. Cin Orth Rel Res 1987; 222:91–97.
7. Bastow MD, Rawlings J, Allison SP. Benefits of supplementary tube feeding after fractured neck of femur: a randomized controlled trial. Br Med J 1983; 287:1589–1592.
8. Lindholm TS. Histological components of tibial fracture callus in growing osteopenic rats. Acta Chir Scand Suppl 1974; 449:7–18.
9. Lindhom TS. Studies of the skeletal mass and bone formation of tibial fracture callus and femoral bone in growing osteopaenic rats. Acta Chir Scand Suppl 1974; 449:37–42.
10. Pollak D, Floman Y, Simkin A, Avinezer A, Freund HR. The effect of protein malnutrition and nutritional support on the mechanical properties of fracture healing in the injured rat. J Parenter Enteral Nutr 1986; 10(6):564–567.
11. Einhorn TA. The cell and molecular biology of fracture healing. Clin Orthoped Res 1998; 355S:S7–S21.
12. McKibbin B. The biology of fracture healing in long bones. J Bone Joint Surg 1978; 60B:150–162.

13. Nakahara H, Bruder SP, Haynesworth SE, et al. Bone and cartilage formation in diffusion chambers by subcultured cells derived from the periosteum. Bone 1990; 11:181–188.
14. Iwata H, Sakano S, Itoh T, Bauer TW. Demineralized bone matrix and native bone morphogenetic protein in orthopaedic surgery. Clin Orthoped 2002; 395:99–109.
15. Jingushi S, Urabe K, Okazaki K, et al. Intramuscular bone induction by human recombinant bone morphogenetic protein 2 with beta tricalcium phosphate as a carrier: in vivo bone banking for muscle pedicle autograft. J Orthoped Sci 2002; 7:490–494.
16. Gerstenfeld LC, Cho TJ, Kon T, et al. Impaired intramembranous bone formation during bone repair in the absence of tumor necrosis factor alpha signaling. Cells Tissue Organs 2001; 169:285–294.
17. Bolander ME. Regulation of fracture repair by growth factors. Proc Soc Exp Biol Med 1992; 200:165–170.
18. Ostrum RF, Chao EYS, Bassett CAL, et al. Bone injury, regeneration and repair. In: Simon SR, ed. Orthopaedic Basic Science. American Academy of Orthopaedic Surgeons, Rosemont, IL, 1994, pp. 277–323.
19. Jensen JE, Jensen TG, Smith TK, Johnston DA, Dudrick SJ. Nutrition in orthopaedic surgery. J Bone Joint Surg 1982; 64A:1263–1272.
20. Bollet AJ, Owens S. Evaluation of nutritional status of selected hospitalized patients. Am J Clin Nutr 1973; 26:931–938.
21. Bistrian BR, Balckburn GL, Halowell E, Heddle R. Protein status of general surgical patients. JAMA 1974; 230:858.
22. Hill GL, Blackett RL, Pickford I, et al. CJ, Morgan DB. Malnutrition in surgical patients: an unrecognized problem. Lancet 1977; 1:689.
23. Reinhardt GF, Myscofski JW, Wilkens DB, Dobrin PD, Mangan JE Jr, Stannard RT. Incidence and mortality of hypoalbuminaemic patients in hospitalized veterans. J Parent Ent Nutr 1980; 4:357–359.
24. Willard GD, Gilsdorf RB, Price TA. Protein calorie malnutrition in a community hospital. JAMA 1980; 243:1720–1722.
25. Lipschitz DA. Nutritional assessment and interventions in the elderly. In: Buckhardt P, Heaney RP, eds. Nutritional Aspects of Osteoporosis 1994. Challenges of Modern Medicine. Vol 7. Ares-Seromo, Rome, 1995, pp. 177–191.
26. Law DK, Dudrick SJ, Abdou NI. The effects of protein calorie malnutrition on immune competence of the surgical patient. Surg Gynecol Obstet 1968; 139:459–474.
27. Bonjour JP, Schurch MA, Rizzoli R. Nutritional aspects of hip fractures. Bone 1996; 18:3:I39S–I44S.
28. Mullen JL, Gertner MH, Buzby GP, Goodhart GL, Rosato EF. Implications of malnutrition in the surgical patient. Arch Surg 1979; 114:121–125.
29. Dalen N, Hellstrom LG, Jacobson B. Bone mineral content and mechanical strength of the femoral neck. Acta Orthoped Scand 1976; 47:503–508.
30. Delmi M, Rapin CH, Bengoa JM, Delmas PD, Vasey H, Bonjour JP. Dietary supplementation in elderly patients with fracture neck of femur. Lancet 1990; i:1013–1016.
31. Matkovic V. Kostial K, Simonovic I, Buzina R, Brodarec A, Nordin BEC. Bone status and fracture rates in two regions of Yuogoslavia. Am J Clin Nutr 1979; 32:540–549.
32. Ray WA, Griffin MR, Downey W, Melton LJ III. Long-term use of thiazide diuretics and risk of hip fracture. Lancet 1989; i:687–690.
33. Bastow MD, Rawlings J, Allison SP. Undernutrition, hypothermia and injury in elderly women with fractured femur; an injury response to altered metabolism? Lancet 1983; i:143–146.
34. Pruzansky ME, Turano M, Luckey M, Senie R. Low body weight as a risk factor for hip fracture in both black and white women. J Orthoped Res 1989; 7:192–197.
35. Wolff J. In: Hirchwald A, ed. Das Gaesetz der Transformation der Knochen. Berlin, 1892.

36. Braun RM, Schorr R. Surgical nutrition in the patient with multiple injuries. Report of a case. J Bone Joint Surg 1983; 65A:1.
37. Burstein AH, Zika JM, Heiple KG, Klein L. Contribution of collagen and mineral to the elastic properties of Bone. J Bone Joint Surg 1975; 57A:956–961.
38. Einhorn TA, Levine B, Michel P. Nutrition and bone. Orthoped Clin N Am 1990; 21:1:43–50.
39. Nordin BEC, Wilkinson R, Marshall DH, Gallagher JC, Williams A, Peacock M. Calcium absorption in the elderly. Calcif Tissue Res 1976; 21:442–451.
40. Peacock M, McClintock R, Schaefer C, Johnston CC, Hu M. Age related changes in calcium regulating hormones, bone turnover and calcium absorption in men and women over the age of 60. J Bone Miner Res 1993; 8(suppl):891 (abstr).
41. Nordin BEC, Horsman A, Marshall DH, Simpson M, Waterhouse GM. Calcium requirement and calcium therapy. Clin Orthoped 1979; 140:216–239.
42. National Research Council Subcommittee on the Tenth Edition of the RDAs. Recommended Dietary Allowances: Subcommittee on the Tenth Edition of the RDAs, Food and Nutrition Board, Commision of Life Sciences, National Research Council, 10th rev ed. National Academy Press, Washington, DC, 1989.
43. Heaney RP, Gallagher JC, Johnston CC, Neer R, Parfitt AM. Calcium nutrition and bone health in the elderly. Am J Clin Nutr 1982; 36:986–1013.
44. Heaney RP. The role of nutrition in prevention of and management of osteoporosis. Clin Obstet Gynecol 1987; 50:833–846.
45. Lindsay R, Cosman F. Prevention of osteoporosis. In: Favus MJ, ed. Primer on the Metabolic Bone Diseases and Disorders of Mineral Metabolism. 4th ed. Lippincott, Williams and Wilkins, Philadelphia, Baltimore, New York, London, Buenos Aires, Hong Kong, Sydney, Tokyo, 1994, pp. 264–270.
46. Heaney RP, Saville PD, Recker RR. Calcium absorption as a function of calcium intake. J Lab Clin Med 1975; 85:881–890.
47. Orwoll ES, McClung MR, Oviatt SK, Recker RR, Weigel RM. Histomorphometric effects of calcium or calcium plus 25 hydroxyvitamin D3 therapy in senile osteoporosis. J Bone Miner Res 1989; 4:81–88.
48. Key JA, Odell RT. Failure of excess minerals in diet to accelerate the healing of experimental fractures. J Bone Joint Surg 1955; 37A:37.
49. Singh LM, Della Rosa RJ, Dunphy JE. Mobilization of calcium in fractured bones in rats. Surg Gynecol Obstet 1968; 126:2:243–248.
50. Einhorn TA, Bonnarens F, Burstein AH. The contributions of dietary protein and mineral to the healing of experimental fractures. A biomechanical study. J Bone Joint Surg 1986; 68A:9:1389–1395.
51. Heaney RP. Nutrition and osteoporosis. In: Favus MJ, ed. Primer on the Metabolic Bone Diseases and Disorders of Mineral Metabolism. 4th ed. 1994, pp. 270–273.
52. McKenna MJ. Differences in vitamin D status between countries in young adults and the elderly. Am J Med 1992; 93:69–77.
53. Nakamura K, Nashimoto M, Okuda Y, Ota T, Yamamoto M. Fish as a major source of vitamin D in the Japanese diet. Nutrition 2002; 18(5):415–416.
54. Parfitt AM, Gallagher JC, Heaney RP, Johnston CC, Neer R, Whedon GD. Vitamin D and bone health in the elderly. Am J Clin Nutr 1982; 36:1014–1031.
55. Lips P, Wiersinga A, van Ginkel FC, et al. The effect of vitamin D supplementation on vitamin D status and parathyroid function in elderly patients. J Clin Endocrinol Metab 1988; 67:644–650.
56. Chapuy MC, Arlot ME, Duboeuf F, et al. Vitamin D3 and calcium to prevent hip fractures in elderly women. N Engl J Med 1992; 327:1637–1642.
57. LeBoff MS, Kohlmeier L, Hurwitz S, Franklin J, Wright J, Glowacki J. Occult vitamin D deficiency in postmenopausal US women with acute hip fracture. JAMA 1999; 16 (281):1505–1511.

58. Baker MR, McDonald H, Peacock M, Nordin BEC. Plasma 25 hydroxyvitamin D concentrations in patients with fractures of the femoral neck. Br Med J 1979; 1:598.
59. Boonen S, Aerssens J, Dequeker J. Age related endocrine deficiencies and fractures of the proximal femur, II: implications of vitamin D deficiency in the elderly. J Endocrinol 1996; 149:13–17.
60. Lips P, Netelenbos JC, Jongen MJM, et al. Histomorphometric profile and vitamin D status in patients with femoral neck fracture. Metab Bone Dis Rel Res 1982; 4:85–93.
61. Lips P, Ooms ME. The effect of vitamin D supplementation in the elderly. Chall Mod Med 1995; 7:311–315.
62. Hey H, Lund B, Sorensen OH, Lund B. Delayed fracture healing following jejunoileal bypass surgery for obesity. Calcif Tissue Int 1982; 34:13–15.
63. Lindholm TS. Histological components of the femoral bone in growing osteopenic rats. Acta Chir Scand Suppl 1974; 449:19–26.
64. Brumbaugh PF, Speer DP, Pitt MJ. 1α,25 Dihydroxyvitamin D3. A metabolite of vitamin D that promotes bone repair Am J Pathol 1982; 106:171–179.
65. Jingushi S, Iwaki A, Higuchi O, et al. Serum 1α,25-Dihydroxyvitamin D3 accumulates into the fracture callus during rat femoral fracture healing. Endocrinology 1998; 139(4):1467–1473.
66. Beresford JN, Gallagher JA, Russell RGG. 1,25 Dihydroxyvitamin D3 and human bone derived cells in vitro: effects on alkaline phosphatase, type I collagen and proliferation. Endocrinology 1986; 119:4:1776–1785.
67. Schwartz Z, Schlader DL, Ramirez V, Kennedy MB, Boyan BD. Effects of vitamin D metabolites on collagen production and cell proliferation of growth zone and resting zone cartilage cells in vitro. J Bone Miner Res 1989; 4:2:199–207.
68. Klaus G, Eing MH, Hugel U, et al. $1,25(OH)_2D_3$ receptor regulation and $1,25(OH)_2D_3$ effects in primary cultures of growth cartilage cells of the rat. Calcif Tissue Int 1991; 49:340–348.
69. Dekel S, Ornoy A, Sekeles E, Noff D, Edelstein S. Contrasting effects on bone formation and on fracture healing of cholecalciferol and of 1α hydroxycholecalciferol. Calcif Tissue Int 1979; 28:245–251.
70. Seo EG, Einhorn TA, Norman AW. 24R,25 dihydroxyvitamin D3: and essential vitamin D3 metabolite for both normal bone integrity and healing of tibial fractures in chicks. Endocrinology 1997; 138:9:3864–3872.
71. Kato A, Seo EG, Einhorn TA, Bishop JE, Norman AW. Studies on 24R,25 dihydroxyvitamin D3: evidence for a nonnuclear membrane receptor in the chick tibial fracture healing callus. Bone 1998; 23(2):141–146.
72. Lidor C, Dekel S, Meyer MS, Blaugrund E, Hallel T, Edelstein S. Biochemical and biomechanical properties of avian callus after local administration of dihydroxylated vitamin D metabolites. J Bone Joint Surg 1990; 72B:137–140.
73. Dekel S, Salama R, Edelstein S. The effect of vitamin D and its metabolites on fracture repair in chicks. Clin Sci 1983; 65:429–436.
74. Lidor C, Dekel S, Edelstein S. The metabolism of vitamin D3 during fracture healing in chicks. Endocrinology 1987; 120(1):389–393.
75. http://www.hoptechno.com/book29o.htm.
76. Faisant C, Lauque S, Stebenet M, Rouillon M, Vellas B, Albarede JL. Calcium and phosphorous intake in different groups of elderly persons with good or poor nutritional status. Chall Mod Med 1995; 7:209–212.
77. Laflamme GH, Jowsey J. Bone and soft tissue changes with oral phosphate supplements. J Clin Invest 1972; 51:2834–2840.
78. Sie TL, Draper HH, Bell RR. Hypocalcaemia, hyperparathyroidism and bone resorption induced by dietary phosphate. J Nutr 1974; 104:1195–1201.
79. Bell RR, Draper HH, Tzeng DYM, Shin HK, Schmidt GR. Physiological responses of human adults to foods containing phosphate additives. J Nutr 1977; 107:42–50.

80. Draper HH, Bell RR. Nutrition and osteoporosis. Adv Nutr Res 1979; 2:79–106.
81. Urist MR, McLean FC. Calcification and ossification; control of calcification in fracture callus or rachitic rats. J Bone Joint Surg 1941; 23A:283.
82. Booth SL, O'Brien-Morse ME, Dallal GE, Davidson KW, Gundberg CM. Response of vitamin K status to different intakes and sources of phylloquinone-rich foods: comparison of younger and older adults. Am J Clin Nutr 1999; 70(3):368–377.
83. Booth SL, Suttie JW. Dietary intake and adequacy of vitamin K. J Nutr 1998; 128:(5):785–788.
84. Akedo Y, Hosoi T, Inoue S, et al. Vitamin K2 modulates proliferation and function of osteoblastic cells in vitro. Biochem Biophys Res Commun 1992; 187:814–820.
85. Binkley NC, Suttie JW. Vitamin K nutrition and osteoporosis. J Nutr 1995; 125:1812–1821.
86. Vermeer C, Jie KSG, Knapen MHJ. Role of vitamin K in bone metabolism. Annu Rev Nutr 1995; 15:1–22.
87. Lian JB, Grundberg CM. Osteocalcin: biochemical considerations and clinical applications. Clin Orthoped 1988; 226:267.
88. Mundy GR, Poser JW. Chemotactic activity of the γ carboxyglutamic acid containing protein in bone. Calcif Tissue Int 1983; 35:164–168.
89. Lian JB, DunnK, Key LL Jr. In vitro degradation of bone particles by human monocytes is decreased with the depletion of the vitamin K dependent bone protein from the matrix. Endocrinology 1986; 118(4):1636–1642.
90. Szulc P, Chapuy MC, Meunier PJ, Delmas PD. Serum undercarboxylated osteocalcin is a marker of the risk factor of hip fracture in elderly women. J Clin Invest 1993; 91:1769–1774.
91. Szulc P, Arlot M, Chapuy MC, Duboeuf F, Meunier PJ, Delmas PD. Serum undercarboxylated osteocalcin correlates with hip bone mineral density in elderly women. J Bone Miner Res 1994; 9:1591–1595.
92. Plantalech L, Guillaumont M, Leclercq M, Delmas PD. Impaired carboxylation of serum osteocalcin in elderly women. J Bone Miner Res 1991; 6:1211–1216.
93. Jie KSG, Bots ML, Vermeer C, Witteman JCM, Grobbee DR. Vitamin K status and bone mass in women with and without aortic atherosclerosis: a population based study. Calcif Tissue Int 1996; 59:352–356.
94. Vergnaud P, Garnero P, Meunier PJ, Breart G, Kamihagi K, Delmas PD. Undercarboxylated osteocalcin measured with a specific immunoassay predicts hip fracture in elderly women: the EPIDOS study. J Clin Endocrinol Metab 1997; 82:719–724.
95. Knapen MHJ, Hamulyak K, Vermeer C. The effect of vitamin K supplementation on circulating osteocalcin and urinary calcium excretion. Ann Intern Med 1989; 111:1001–1005.
96. Akjba T, Kurihara S, Tachibana K. Vitamin K increased bone mass in hemodialysis patients with low turnover bone disease. J Am Soc Nephrol 1991; 608:42P (abstr).
97. Douglas AS, Robins SP, Hutchison JD, Porter RW, Stewart A, Reid DM. Carboxylation of osteocalcin in post menopausal osteoporotic women following vitamin K and D supplementation. Bone 1995; 17:15–20.
98. Feskanich D, Weber P, Willett WC, Rockett H, Booth SL, Colditz GA. Vitamin K intake and hip fractures in women: a prospective study. Am J Clin Nutr 1999; 69:74–79.
99. Hauschka PV, Lian JB, Cole DEC, Gundberg CM. Osteocalcin and matrix gla protein: vitamin K dependent proteins in bone. Physiol Rev 1989; 69:990–1047.
100. Ducy P, Desbois C, Boyce B, et al. Increased bone formation in osteocalcin deficient mice. Nature 1996; 383:448–452.
101. Dodds PA, Caterall A, Bitensky L, Chayen J. Effects on fracture healing of an antagonist of the vitamin K cycle. Calcif Tissue Int 1984; 36:233–238.
102. Stinchfield FE, Sankaran B, Samilson R. The effect of anticoagulant therapy on bone repair. J Bone Joint Surg 1956; 38A:270–282.

103. Einhorn TA, Gundberg CM, Devlin VJ, Warman J. Fracture healing and osteocalcin metabolism in vitamin K deficiency. Clin Orthoped Rel Res 1987; 237:219–225.
104. Lian J, Stewart C, Puchacz E, et al. Structure of the rat osteocalcin gene and regulation of vitamin D dependent expression. Proc Natl Acad Sci USA 1989; 86:1143–1147.
105. Anderson JJB. Oversupplementation of vitamin A and osteoporotic fractures in the elderly: to supplement or not to supplement with vitamin A. J Bone Miner Res 2002; 17:8:1359–1362.
106. Food and Nutrition Board, Institute of Medicine. Dietary Reference Intakes of Vitamin A, Vitamin K, Arsenic, Boron, Chromium, Copper, Iodine, Iron, Manganese, Molybdenum, Nickel, Silicon, Vanadium, and Zinc. National Academy Press, Washington, DC, 2001, pp. 65–126.
107. Berner LA, Clydesdale FM, Douglass JS. Fortification contributed greatly to vitamin and mineral intakes in the United States, 1989–1991. J Nutr 2001; 131:2177–2183.
108. Norris J, Harnack L, Carmichael S, Pouane T, Wakimoto P, Block G. US trends in nutrient intake: the 1987 and 1992 national health interview surveys. Am J Publ Health 1997; 87:740–746.
109. Cruz JA, Moreiras-Varela O, van Staveren WA, Trichopoulou A, Roszkowski W. Intake of vitamins and minerals. Euronut SENECA investigators. Eur J Clin Nutr 1991; 45(suppl 3):121–138.
110. Melhus H, Michaelsson K, Kindmark A, et al. Excessive dietary intake of vitamin A is associated with reduced bone mineral density and increased risk for hip fracture. Ann Intern Med 1998; 129:770–778.
111. Michaelsson K, Lithell H, Vessby B, Melhus H. Serum retinol levels and the risk of hip fracture. N Engl J Med 2003; 348:4:287–294.
112. Johansson S, Lind PM, Hakansson H, Oxlund H, Orberg J, Melhus H. Subclinical hypervitaminosis A causes fragile bones in rats. Bone 2002; 31(6):686–689.
113. Kindmark A, Torma H, Johansson A, Ljunghall S, Melhus H. Reverse transcription polymerase chain reaction assay demonstrates that the 9 cis retinoic acid receptor alpha is expressed in human osteoblasts. Biochem Biophys Res Commun 1993; 192:1367–1372.
114. Saneshige S, Mano H, Tezuka K, et al. Retinoic acid directly stimulates osteoclastic bone resorption and gene expression of cathepsin K/OC-2. Biochem J 1995; 1;309(Pt3):721–724.
115. http://www.asante.org/StandardPage.asp?MenuID=469.
116. Pellet PL. Protein requirements in humans. Am J Clin Nutr 1990; 51:723–737.
117. http://www.calstatela.edu/faculty/lcalder/ucsb/tsld018.htm.
118. Rhoads JE, Kasinskas BA. The influence of hypoproteinaemia on the formation of callus in experimental fracture. Surgery 1942; 11:38–44.
119. Day SM, DeHeer DH. Reversal of the detrimental effects of chronic protein malnutrition on long bone fracture healing. J Orthoped Trauma 2001; 15(1):47–53.
120. Golden NH, Kreitzer P, Jacobson MS, et al. Disturbances in growth hormone secretion and action in adolescents with anorexia nervosa. J Pediatr 1994; 125:655–660.
121. Thissen JP, Davenport ML, Pucilowska J, Miles MV, Underwood LE. Increased serum clearance and degradation of 125 labeled IGF1 in protein restricted rats. Am J Physiol 1992; 262:E406–E411.
122. Thissen JP, Triest S, Maes M, Underwood LE, Ketelslegers JM. The decreased plasma concentrations in insulin like growth factor I in protein restricted rats is not due to decreased number of growth hormone receptors on isolated hepatocytes. J Endocrinol 1990; 124:159–165.
123. Thissen JP, Triest S, Moats-Statts BM, et al. Evidence that pretranslational and translational defects decrease serum IGF1 concentrations during dietary protein restriction. Endocrinology 1991; 129:429–435.
124. Thissen JP, Underwood LE, Maiter D, Maes M, Clemmons DR, Ketelslegers JM. Failure of IGF1 infusion to promote growth in protein restricted rats despite normalization of serum IGF1 concentrations. Endocrinology 1991; 128:885–890.
125. JeeJeebhoy KN, Baker JP, Wolman SL, et al. Critical evaluation of the role of clinical assessment and body composition studies in patients with malnutrition and after total parenteral nutrition. Am J Clin Nutr 1982; 35:1117–1127.

126. Klein S. The myth of serum albumin as a measure of nutritional status. Gastroenterology 1990; 99:1845–1851.
127. Unterman TG, Vazquez RM, Sla AJ, Martyn PA, Phillips LS. Nutrition and somatomedin. XII. Usefulness of somatomedin C in nutritional assessment. Am J Med 1985; 78:228–234.
128. Schurch MA, Rizzoli R, Slosman D, Vadas L, Vergnaud P, Bonjour JP. Protein supplements increase serum insulin like growth factor 1 levels and attenuate proximal femur loss in patients with recent hip fracture. A randomized, double blind, placebo controlled trial. Ann Intern Med 1998; 128:801–809.
129. Goldring MB, Goldring SR. Skeletal response to cytokines. Clin Orthoped 1990; 258:245–278.
130. Skottner A, Arrhenius NV, Kanje M, Fryklund L. Anabolic and tissue repair functions of recombinant insulin like growth factor I. Acta Pediatr Acad 1990; suppl 367:63–66.
131. Spencer EM, Liu CC, Si EC, Howard GA. In vivo actions of insulin like growth factor I on bone formation and resorption in rats. Bone 1991; 12:21–26.
132. Schmidmaier G, Wildeman B, Heeger J, et al. Improvement of fracture healing by systemic administration of growth hormone and local application of insulin like growth factor 1 and transforming growth factor β1. Bone 2002; 31:1:165–172.
133. Turner RT. Skeletal response to alcohol. Alcohol Clin Exp Res 2000; 24:1693–1701.
134. Purohit V. Introduction to the alcohol and osteoporosis symposium. Alcohol Clin Exp Res 1997; 21:383–384.
135. Brown EC, Perrien DS, Fletcher TW, et al. Skeletal toxicity associated with chronic ethanol exposure in a rat model using total enteral nutrition. Pharmacol Exp Ther 2002; 301(3):1132–1138.
136. Wezeman FH, Emanuele MA, Emanuele NV, et al. Chronic alcohol consumption during male rat adolescence impairs skeletal development through effects on osteoblast gene expression, bone mineral density and bone strength. Alcohol Clin Exp Res 1999; 23:1534–1542.
137. Maran A, Zhang M, Spelsberg TC, Turner RT. The dose response effects of ethanol on the human fetal osteoblastic cell line. J Bone Miner Res 2001; 16:270–276.

6 Nutritional Assessment of Nutrients for Bone Health

Edith M. C. Lau and Winny W. Y. Lau

1. OBJECTIVES

An ideal nutrition intake is a very important determinant of bone health for all ages. The objectives of this chapter are to describe the various methods available for nutrition assessment for bone health, and to discuss the applicability of such methods in the clinical and research settings.

2. ASSESSMENT OF OVERALL NUTRITIONAL STATUS

The objective of an overall nutritional assessment in relation to health is identifying subjects who are malnourished, thus predisposing them to osteoporosis. Practical strategies using readily available markers for malnourishment pertaining to energy, protein, calcium, and vitamin D could be applied for this purpose.

Assessment of overall nutritional status should begin with a comprehensive medical history inquiring about altered oral intake, lack of specific food groups, use of supplements, ability to chew and swallow, vomiting, bowel habits, and symptoms related to vitamin and trace-element deficiency.

Systematic methods for quantifying diet history are also feasible, using such instruments as the Nutrition Screening Initiative Checklist for Nutrition Risk *(1)*. This checklist is known to predict nutrition-related complications in both retrospective and prospective studies *(2,3)*. The simplest assessment of nutritional status is body weight, especially as compared to previous data. Such measurements could be compared to ideal body weight charts to ascertain normality. Alternatively, the body mass index (BMI) could be used. The BMI has the advantage of wide applicability across ethnic groups. For instance, a BMI of less than 18.5 kg/m^2 could be considered as underweight, hence predisposing to osteoporosis and its related fractures.

Other measurements of general nutrition status include mid-arm muscle circumference, as an indicator of muscle mass; and triceps and subscapular skinfold thickness, which are good indices for body fat.

From: *Nutrition and Bone Health*
Edited by: M. F. Holick and B. Dawson-Hughes © Humana Press Inc., Totowa, NJ

Biochemical indices for nutritional assessment could include measurements of albumin, transferrin, insulin-like growth factor (IGF), and urinary urea nitrogen.

3. FOOD-FREQUENCY METHODS

The food-frequency questionnaire is a suitable method for assessing nutrition with respect to bone health. The underlying principle of the food-frequency approach is based on the belief that intake over weeks, months, or years is important, particularly for chronic and degenerative diseases. This concept is certainly applicable to osteoporosis. For instance, long-term calcium intake, rather than recent intake, is associated with bone health. In studies on osteoporosis, it is hence much more appropriate to use the food-frequency method to quantify long-term calcium intake, rather than to use the food diary method to quantify immediate intake.

The basic food-frequency questionnaire consists of two components, a food list and a frequency response section to assess how often each food was eaten.

There are several approaches to compiling the food list. The most basic approach is to examine published food composition tables and identify the foods that contain substantial amounts of the nutrients of interest. Alternatively, one can start with a long list of foods that are potentially important nutrient sources and systemically reduce this list. For a food item to be included, it should be used quite frequently, it must have a substantial content of the nutrient, and intake must not be homogenous across a population. It is particularly important for the food list to be culturally specific. For instance, green vegetables are a major source of calcium in the Chinese diet, and use of an unadapted food list from the United States to assess calcium intake in Chinese will be of a low validity.

The frequency of intake could be categorized as follows *(4):*

Never
Once a month or less
Two to three times per month
Once per week
Two to four times per week
Five to seven times per week
More than one, less than two times a day
Two to three times per day
Four to six times per day
More than six times per day

Figure 1 *(5),* an excerpt from the Nurses' Health Study Dietary Questionnaire, further illustrates the details for the food-frequency method. Using this approach, food-frequency questionnaires could be adapted for different cultures by inclusion of essential source of nutrients.

The major advantages of the food-frequency method are low cost and applicability to large-scale studies. However, there are several disadvantages to the

For each food listed, fill in the circle indicating how often <u>on average</u> you have used the amount specified <u>during the past year.</u>	AVERAGE USE LAST YEAR								
DIARY FOODS	Never, or less than once per month	1-3 per mo.	1 per week	2-4 per week	5-6 per week	1 per day	2-3 per day	4-5 per day	6+ per day
Skim or low fat milk (8 oz. Glass)	○	○	Ⓦ	○	○	Ⓓ	○	○	○
Whole milk (1 oz. Glass)	○	○	Ⓦ	○	○	Ⓓ	○	○	○
Cream, e.g. coffee, whipped (Tbs.)	○	○	Ⓦ	○	○	Ⓓ	○	○	○
Sour cream (Tbs)	○	○	Ⓦ	○	○	Ⓓ	○	○	○
Non-dairy coffee whitener (tsp.)	○	○	Ⓦ	○	○	Ⓓ	○	○	○
Sherbet or ice milk ($^1/_2$ cup)	○	○	Ⓦ	○	○	Ⓓ	○	○	○
Ice-cream ($^1/_2$ cup)	○	○	Ⓦ	○	○	Ⓓ	○	○	○
Yogurt (1 cup)	○	○	Ⓦ	○	○	Ⓓ	○	○	○
Cottage or ricotta cheese ($^1/_2$ cup)	○	○	Ⓦ	○	○	Ⓓ	○	○	○
Cream cheese (1 oz.)	○	○	Ⓦ	○	○	Ⓓ	○	○	○
Other cheese, e.g. American, cheddar, etc., plain or as part of a dish (1 slice or 1 oz. Serving)	○	○	Ⓦ	○	○	Ⓓ	○	○	○
Margarine (pat), added to food or bread; exclude use in cooking	○	○	Ⓦ	○	○	Ⓓ	○	○	○
Butter (pat), added to food or bread; exclude use on cooking	○	○	Ⓦ	○	○	Ⓓ	○	○	○

Fig. 1. Nurses' Health Study Dietary Questionnaire *(5)*.

food-frequency method: it is only semiquantitative, and calorie intake cannot be obtained directly. This is not inconsequential.

Table 1 *(6)* is based on comparisons of average intake of selected specific foods reported on a compressed 61-item questionnaire, with intake measured by 28 d of dietary record among 173 US women. It can be seen that the correlation is fairly good.

4. APPLICATION OF THE FOOD-FREQUENCY METHOD TO EPIDEMIOLOGICAL STUDIES IN OSTEOPOROSIS RESEARCH

The food-frequency method is the method used for assessing dietary intake in most of the cohort studies in the field of osteoporosis. In the Study of Osteoprotic Fractures, a validated food-frequency questionnaire developed from the Second National Health and Nutrition Examination Survey (NHANES-II) was used. Subjects were asked how often in the last year they usually ate each of 20 foods. Food models were used to assist subjects to rate their usual portion size per each food as small, medium, or large. Inclusion of foods on the questionnaire was based on their contribution to total calcium and protein intake in US adults. The

Table 1
Comparison of Average Intake of Selected Specific Foods Reported on a Compressed 61-Item Questionnaire With Intake Measured by 28 d of Dietary Record Among 173 US Women

	Diet Record	Questionnaire	*Record vs Questionnaire (Pearson Correlation)*	
	Mean	*Mean*	*Crude*	*Adjusted*
Low-fat milk (cups)	0.28	0.53	0.79	0.81
Whole milk (cups)	0.27	0.22	0.62	0.62
Margarine (pats)	1.24	1.50	0.71	0.76
Butter (pats)	0.97	0.64	0.79	0.85
Spinach, other greens ($^1/_2$ cup)	0.06	0.28	0.08	0.17
Broccoli ($^1/_2$ cup)	0.07	0.17	0.49	0.69
Apples (1 fruit)	0.20	0.33	0.66	0.80

Source: Ref. *6.*

foods accounted for about 80% of calcium intake, and the correlation between calcium intake in elderly women assessed with the questionnaire and 7-d food records was 0.76. *(7).*

The Rancho Bernardo Study *(8)* is a population-based cohort study on elderly, upper middle-class Caucasian residents of Rancho Bernardo, a southern California community. Dietary information was obtained using the Harvard-Willett diet assessment questionnaires *(9).* Application of such information to the food-frequency method has enabled the investigators to quantify the dietary calcium and protein intake of all participants, so that the relationship between dietary protein intake and bone mineral density (BMD) could be described *(10).*

The NHANES-II survey is one of the most comprehensive nutritional studies conducted in US Caucasians to date. Dietary data from 11,658 adult respondents were used to provide quantitative information regarding the contribution of specific food to the total population intake of key nutrients. Data on nutrients pertaining to bone health are available and include protein intake, calorie intake, calcium intake, and alcohol consumption. Using these data, the food-frequency method was successfully applied to study the role of diet in hip fracture *(11).*

The Dubbo Osteoporosis Study was a large-scale cohort study to document risk factors for osteoporotic fractures in Australian Caucasians. The food-frequency method was used to measure dietary calcium intake, and this was found to be applicable and valid *(12,13).*

In the European Vertebral Osteoporosis Study (EVOS) *(14),* dietary calcium intake was assessed by the number of days per week each subject ate hard cheese, soft cheese, yogurt, milk and "other diary produce." Historical calcium intake was based solely on the amount of milk drunk in three time periods in life. This sim-

Table 2
Comparison of Percentage Differences Between Burke History and 24-h Recall and Between 24-h Recall and 7-d Record

	Difference		
Nutrient	*Massachusetts (n = 28)*	*New York (n = 51)*	*Rhode Island (n = 87)*
Burke history and 24-h recall (%)			
Energy	+21.1	+23.3	+9.7
Protein	+23.8	+20.1	–7.2
Calcium	+20.6	+21.5	+0.1
Phosphorus	+23.8	+20.9	–3.7
Iron	+32.2	+17.3	–10.9
7-d record and 24-h recall (%)			
Energy	+2.4	+6.5	+0.9
Protein	+1.1	+1.9	+4.3
Calcium	+13.1	+2.5	+11.9
Phosphorus	+3.1	+1.3	+8.2
Iron	+7.7	+1.7	+3.4

Source: Ref. *17.*

plified food-frequency approach was found to be applicable in assessing dietary calcium intake across European countries, to allow the relationship between diet, lifestyle, and the prevalence of vertebral deformity to be studied.

Similar methods have been applied, primarily in the Mediterranean Osteoporosis Study, to quantify calcium intake in several European countries, so that risk factors for hip fracture in European women could be determined *(15).*

5. TWENTY-FOUR-HOUR DIET RECALL

The 24-h recall was designed to assess recent nutrient intake quantitatively *(16–19).* Intuitively, 24-h recall should be more accurate than recall of the remote diet. However, long-term intake has more implications on bone health, and such may not be represented in the 24-h recall. For instance, 24-h recall may not be useful in studies on the relationship between long-term calcium, sodium, vitamin K intake, and BMD in the elderly. To the contrary, 24-h recall may be applicable to studies on how recent dietary intake may affect the progress of clinical outcomes.

As discussed by Barrett-Connor *(20),* the degree of reproducibility or representativeness of the 24-h recall method depends on the study populations' monotony of diet, and with the method used for validation. Barrett-Connor *(20)* quoted data from Young *(17)* to show the difference between 24-h recall and 7-d record (*see* Table 2.)

Table 3
Average Daily Intake (J) of 400 Women Based on Diet-Assessment Method

Method	*Energy (KJ/d)*
24-h recall	6760
Current diet history	9084
Past diet history	9561
4-d record	7451

Source: Ref. *16.*

6. FOOD RECORD OR DIARY

In the food record or diary method, subjects are advised to record their exact intake for 3–7 d. The disadvantages of this method include low compliance and possible change in eating pattern during the record period. As shown by Morgan *(16),* intake from a 7-d food record does not always parallel the intake based on a food history.

7. FOOD HISTORY

The food history method includes a 24-h diet recall, a history of usual foods, and data on food preparation. It requires a 1–2-h interview by a specially trained nutritionist. As illustrated in Table 3 *(16),* the results of diet history do not always concur with those from other methods. It is nevertheless used as the "gold standard" for assessing the validity of other methods.

8. CONCLUSION AND FUTURE PROSPECTS

Of all the methods just described, the food-frequency method has been most frequently applied in epidemiological studies on nutrition and bone health. This method also has the greatest potential in investigating new hypotheses between diet and bone. For instance, recently the modified Block food-frequency method *(21)* has been successfully applied to study the relationship between phytoestrogen intake and BMD in various ethnic groups across the United States *(22).* This entailed the establishment of a culturally sensitive database for the source of phytoestrogen, which was then used in the food-frequency method. In the future, similar approaches could be applied to address new research questions on nutritional and bone health.

REFERENCES

1. The Nutrition Screening Initiative. Nutrition Interventions Manual for Professionals Caring for Older Americans. The Nutrition Screening Initiative, Washington, DC, 1992.
2. Kant AK, Schatzkin A, Harris TB, et al. Dietary diversity and subsequent mortality in the First National Health and Nutrition Examination Survey Epidemiologic Follow-up Study. Am J Clin Nutr 1993; 54:434–440.

3. Posner BM, Jette AM, Smith KW, et al. Nutrition and health risks in the elderly: the nutrition screening initiative. Am J Publ Health 1993; 83:972–978.
4. Stefanik PA, Trulson MF. Determining the frequency of foods in large group studies. Am J Clin Nutr 1962; 11:335–343.
5. Willett W. Food-frequency methods. In: Nutritional Epidemiology. Oxford University Press, New York, 1998, pp. 74–100.
6. Salvini S, Hunter DJ, Sampson L, et al. Food-based validation of a dietary questionnaire: the effects of week-to-week variation in food consumption. Int J Epidemiol 1989; 18:858–867.
7. Cumming RG, Cummings SR, Nevitt MC, et al. Calcium intake and fracture risk: results from the study of osteoportoic fractures. Am J Epidemiol 1997; 145:926–927.
8. Criqui MH, Barrett-Connor E, Austin M. Differences between respondents and non-respondents in a population-based cardiovascular disease study. Am J Epidemiol 1978; 108:367–372.
9. Willett WC, Sampson L, Stampfer MJ, et al. Reproducibility and validity of a semiquantitative food frequency questionnaire. Am J Epidemiol 1985; 122:51–65.
10. Promislow J, Goodman-Gruen D, Slymen D, Barrett-Connor E. Protein consumption and bone mineral density in the elderly. Am J Epidemiol 2002; 155:636–644.
11. Mussolino ME, Looker AC, Madans JH, Langlois JA, Orwoll ES. Risk factors for hip fracture in white men: the NHANES I Epidemiologic Follow-up Study. J Bone Miner Res 1998; 13:918–924.
12. Nguyen T, Sambrook P, Kelly P, et al. Prediction of osteoporotic fractures by postural instability and bone density. Br Med J 1993; 307:1111–1115.
13. Nguyen TV, Sambrook PN, Eisman JA. Bone loss, physical activity, and weight change in elderly women: the Dubbo Osteoporosis Epidemiology Study. J Bone Miner Res 1998; 13:1458–1467.
14. Lunt M, Masaryk P, Scheidt-Nave C, et al. The effects of lifestyle, dietary dairy intake and diabetes on bone density and vertebral deformity prevalence: the EVOS Study. Osteopor Int 2001; 12:688–698.
15. Johnell O, Gullberg BO, Kanis JA, et al. Risk factors for hip fracture in European women: the MEDOS Study. J Bone Miner Res 1995; 10:1802–1815.
16. Morgan RW, Jain M, Miller AB, et al. A comparison of dietary methods in epidemiologic studies. Am J Epdemiol 1978; 107:488–498.
17. Young CM, Hagan GC, Tucker RE, Foster WD. A comparison of dietary study methods. II. Diet history vs seven-day record vs 24-hour record. Am J Dietet Assoc 1952; 28:218–221.
18. Balogh M, Kahn H, Medalie JH. Random repeat 24-hour dietary recalls. Am J Clin Nutr 1971; 24:304–310.
19. Beaton GH, Milner J, Corey P, et al. Sources of variance in 24-h dietary recall data: implications for nutrition study design and interpretation. Am J Clin Nutr 1979; 32:2456–2559.
20. Barret-Connor E. Nutrition epidemiology: how do we know what they ate? Am J Clin Nutr 1991; 54:182S–187S.
21. Block G, Hartman AM, Dresser CM, et al. A data-based approach to diet questionnaire design and testing. Am J Epidemiol 1986; 124:453–469.
22. Greendale GA, FitzGerald G, Huang MH, et al. Dietary soy isoflavones and bone mineral density: results from the study of women's health across the nation. Am J Epidemiol 2002; 155:746–754.

7 Nutritional Assessment

Analysis of Relations Between Nutrient Factors and Bone Health

John J. B. Anderson, Boyd R. Switzer, Paul Stewart, and Michael Symons

1. INTRODUCTION

The application of nutritional assessment tools for the study of diet–bone relationships has been greatly advanced by the computerized handling of large datasets. The multifactorial nature of osteoporosis makes it especially difficult to identify major determinants of the diet–bone linkage. However, recent studies using computer-based methods have received widespread attention because of the statistical strategies that permit improved estimation of the strength of relationships between various nutritional factors and bone health.

These nutritional factors are discussed in this chapter as part of a broader review of nutritional assessment methods as they apply to bone health. A few representative published reports have been selected to illustrate different approaches to the assessment of calcium and vitamin D intakes used in association with bone measurements. Epidemiological and biostatistical methods provide the general basis for the analysis of nutrient–bone associations.

2. NUTRITIONAL ASSESSMENT: VARIABLES OF INTEREST

The main nutritional methods of assessment are specified in this review, but related aspects deserve brief mention. For example, nutritional assessment typically includes more than simply obtaining nutrient intakes on a study group or population; other information on lifestyle and other characteristics of a population help in the analyses of diet–bone associations. A few of these other variables, such as age, reproductive status (females), physical activity, lean body mass, and alcohol consumption, are noted.

From: *Nutrition and Bone Health*
Edited by: M. F. Holick and B. Dawson-Hughes © Humana Press Inc., Totowa, NJ

2.1. ABCDs of Assessment

The major domains of measurement include anthropometric (A), biochemical (B), clinical (C), and dietary (D). Information from each domain is used to model diet–bone relationships statistically. This point is illustrated by examples in a later section.

2.2. Nutrients of Interest to Bone Health

Several commonly consumed nutrients generate interest for nutrient–bone relationships, but the focus of this review is on calcium and vitamin D. A few other nutrient variables, such as vitamins A and K, plus energy (calories), are mentioned because of their known roles in bone formation or metabolism, especially as reported from studies of animal models. The emphasis in this review is on food sources of nutrients, usually from dietary intakes. Intakes of nutrient supplements, such as calcium and vitamin D, that may be given in large quantities, are given less coverage. (*See* Chapter 8.)

2.3. Food Composition Tables and Databases

The foods composition databases have been generated by different agencies: the US Department of Agriculture (USDA), the National Heart, Lung, and Blood Institute, and others. These bases provide reasonably accurate estimates of macronutrient and certain micronutrient content of foods, both raw and processed, consumed in the United States. Industries (i.e., specific companies) analyze their food products and usually make their data available to compilers of the large U.S. datasets, such as the Computerized National Data Set (http://www.ars.usda.gov/fnic/foodcomp). The selection of the computerized database depends on several factors, especially the overall accuracy of determination of macronutrients and other nutrient variables of interest.

3. STATISTICAL CONSIDERATIONS

The critical questions in studies of nutrient–bone analyses relate to the relative contribution of one nutrient, compared with others, to bone density (or other bone measurements) in multiple-regression models *(1)*. For example, how important is dietary calcium to total body bone measurements or to specific bone regions of interest, such as bone mineral density (BMD) of the hip in postmenopausal females? Adjustments for energy or nutrients, as well as of nonnutrient variables, in the models may be appropriate. See the text by Willett for additional insights *(2)*.

3.1. Study Designs

The study design selected determines to a large extent the comparisons possible and the variables to be assessed, including the nutrient variables. Often the ideal study would be a prospective (longitudinal) randomized controlled trial (RCT) with double blinding and a placebo control group, but such investigations generally require large numbers of subjects, making them expensive. Such prospective trials are preferred over cross-sectional studies or retrospective case-control approaches, mainly because of the higher quality of information obtained. In the 1980s and early 1990s, many of the diet–bone studies were cross-sectional or retrospective in nature. Since the early 1990s, most diet–bone investigations have been prospective.

3.2. An Example of a Study Protocol

An approach used for a prospective investigation of diet–bone linkages is given in this hypothetical study assessing usual intakes of calcium, vitamin D, and other nutrient variables.

3.2.1. Overview

Consider a hypothetical cohort in which longitudinal nutritional intakes from 200 premenarcheal girls 7 to 12 yr of age will be assessed. The two cohorts of this observational study are designated A (those consuming recommended healthy diet) and B (those established to be on a less than optimal diet, typical "diet"). The primary objective is to compare the two treatments in terms of gains as indicated by measurements of bone mineral content (BMC) (also known as mass) and BMD. Measurements are typically obtained at baseline and after 1 yr, and subsequently at 2- and 3-yr anniversaries (Table 1). A secondary objective is to characterize, by the cross-sectional baseline data, the relationships between BMD and characteristics, such as serum 25-hydroxyvitamin D (25[OH]D), serum parathyroid hormone (PTH), body mass index (BMI), and self-reported calcium intake. Typical outcome variables, predictor variables, identifiers, and covariates of interest are listed in Table 1.

3.2.2. Plans for Statistical Analysis

The plans should address the general and specific aims of the research, and provide control of the error rates associated with testing of hypotheses. The plans include formal inferences about key *a priori* hypotheses along with informal exploratory and supportive analyses. This approach takes full advantage of the fact that, when performed correctly, tests of *a priori* hypotheses are more powerful than tests of hypotheses that are suggested by patterns observed in the data collected *(3)*. Supportive analyses will include simple descriptive tables and graphs for each of the cohorts for each of the efficacy variables listed in Table 1.

3.2.3. Primary Analysis of Efficacy

The primary analysis will focus on mean change from baseline (Δ) in eight primary response variables (Y, 8×1): mass and density in hip, forearm, spine, and total skeleton. Comparison of the two cohorts (A vs B) will rely on a multivariate linear model (MANOVA) for $Y = [Y_1, \ldots, Y_8]$. Graphical and tabular methods will be used to describe the results. Model-based estimates of Δ_A and Δ_B will be tabulated

Table 1
Key Measurements for Hypothetical Study Design

Domain	*Name of Variable*	*Scale*[a]	*Description of the Variable*	*When Recorded*[b]
Identifiers	patient_ID	L	unique patient ID for this study	0,1
	date	I	date of each occasion	0,1
	treatment arm	L	treatment regimen assigned (A, B)	0,1
Safety	AEs	N	list adverse events (e.g., fractures)	0,1
Diet	calcium	X	calcium intake via dietary self-reports	0,1
	vitamin_D	X	vitamin D intake via dietary self-reports	0,1
	diet profile	*	calculated nutrients from self-reports	0,1
Serum profile	25(OH)D	X	25-hydroxyvitamin D	0,1
	PTH	X	parathyroid hormone	0,1
	bone marker	B	abnormal calc/phos/ALP (yes,no)	0,1
Bone response	bone mass total	X	total bone mass (g) via DXA scans	0,1
	bone mass spine	X	spine bone mass (g) ″	0,1
	bone mass arm	X	arm bone mass (g) ″	0,1
	bone mass hip	X	hip bone mass (g) ″	0,1
	BMD total	X	total bone mineral density (g/cm_3) ″	0,1
	BMD spine	X	spine bone mineral density (g/cm_3) ″	0,1
	BMD arm	X	arm bone mineral density (g/cm_3) ″	0,1
	BMD hip	X	hip bone mineral density (g/cm_3) ″	0,1
Development	Tanner stage	C	Tanner stage (0,1,2,3,4)	0,1
	menarche	B	occurrence of menarche during study	0,1
	age	X	age (years)	0,1
	height	X	height (cm)	0,1
	weight	X	weight (kg)	0,1
	BMI	X	body mass index (kg/cm_2)	0,1
	lean percent	X	lean mass component of DXA scan (%)	0,1
	fat percent	X	fat mass component of DXA scan (%)	0,1
Questionnaires	physical activity	*	self-reported levels of physical activity	0,1
	usual health	*	self-reported recent health history	0,1
Compliance	questionnaire	C	Adherence to protocol?	0,1
	pill count	N	Number of pills taken	0,1

[a] Scales: B, binary; C, ordinal categories; I, integer codes; L, nominal labels; N, count; X, continuous; *, self-reported data.

[b] Occasions: 0 = baseline, 1 = after 1 yr, and any years thereafter. ID, identification; AE, adverse event; PTH, parathyroid hormone; ALP, alkaline phosphatase; DXA, dual energy X-ray absorptiometry; BMD, bone mineral density; BMI, body mass index.

with 95% confidence regions. The primary null hypothesis that will be tested is "$\Delta_A = \Delta_B$." A conclusion that the two cohorts differ will be made only if the corresponding appropriate F-test is statistically significant at the 5% level of significance ($\alpha \leq 0.05$). The MANOVA results will be reported as confirmatory results and will be accompanied by supporting comments about auxiliary, diagnostic, and exploratory analyses.

3.2.4. Power Computations

Based on our conjectures about Δ_A and Δ_B, the proposed study should have at least a 90% probability of rejecting a false null hypothesis; i.e., a 10% chance exists that the test will be inconclusive. Sample power computations are shown in Table 2.

3.2.5. Choosing a Sample Size

In selecting a sample size of $N = 100$ per cohort, for example, primary considerations are (1) the precision of the statistical estimators of interest (e.g., the standard errors of means), (2) statistical power for the primary hypothesis tests of interest, (3) costs, and (4) feasibility. Power *and* precision increase with sample size, while costs increase and feasibility decreases with sample size.

Statistical power is the probability that a false null hypothesis test will yield a p-value smaller than the level of significance ($\alpha = 0.05$). It is an inherent characteristic of the procedure used (e.g., t-test procedure) to test a particular hypothesis of interest. Statistical power is influenced by the study design, the sample size, and can be greater than 0.05 only for a false null hypothesis. Each hypothesis test has its own level of power.

Even if the sample size is large enough to provide at least 90% power levels for the tests, and even if our research hypotheses are correct, the proposed study may or may not support a correct research theory. The results will be influenced by many sources of variation, such as which subjects happen to be recruited, temporal variation within each subject, measurement error variation, laboratory/assay measurement error, and other possibilities. Hypothetically, if the study were repeated many times, some replications of the study would yield results that support our research theories and some replications would not. Conducting the study is in this way analogous to a roll of the dice. One purpose of power analysis and sample size computations is to reduce the investigators' research risks to an acceptable level or at least to achieve some awareness of the levels of risk involved in the proposed undertaking.

The nature of statistical hypothesis testing is such that if the null hypothesis is not rejected, i.e., the p-value turns out to be larger than the level of significance ($\alpha = 0.05$), then that test does not support the alternative hypothesis. In other words, a large p-value supports, but does not establish, that the null hypothesis is true.

Some choices of study design, sample size, and statistical procedures are inherently more efficient and accurate than other choices. The investigator's research risks can thus be reduced by careful comparison of these choices. Toward achieving acceptable levels of precision and power, information available from previous

Table 2
Example of Power Computations for Proposed Tests of Efficacy[a]
(for all measurements $\alpha^f = 0.05/8$ by the Bonferroni correction)

Variable	*Y[b]*	*SD of Y[c]*	*Difference[d]* $\Delta_A - \Delta_B$	*Power (%)[e]*	*N / group[g]*
Bone mass (g)	total	—	—	—	—
	spine	0.36	1.950	90	4
	arm	—	—	—	—
	hip	0.36	0.400	90	29
Bone mineral density (g/cm^2)	total	0.06	0.015	90	518
	spine	0.07	0.035	90	131
	arm	0.06	0.035	90	97
	hip	0.07	0.020	90	397

[a] Bone measurements are the outcome (response) variables; the rest are predictor variables in table.
[b] Name of the change from baseline variable *(Y)*.
[c] Anticipated magnitude of the standard deviation (SD) of change *(Y)* from baseline.
[d] Anticipated difference between regimens.
[e] The probability of rejecting a false null hypothesis.
[f] Significance level of the test, with Bonferroni correction.
[g] Sample size per group.

studies about the standard deviations and correlations should be considered in the proposed study. The analysis of the power levels that are characteristic of a particular study design and analysis plan is always a conjectural exercise. By making conjectures about the underlying distributions of key variables, as well as about enrollment and retention, useful sample-size computations may be made. Accuracy of the computations is more sensitive to the factual accuracy of the conjecture than to the computational formulas used. Power calculations should be made early in the design phase.

Apart from hypothesis testing, the study will also yield informative statistical estimates of means, differences, and measures of association. Descriptions of change are accompanied by measures of precision in the form of standard errors and confidence intervals. The level of precision provided by the proposed sample size should also be a consideration.

After taking into account costs and the need for this study and, potentially, other published studies in this area of investigation, such considerations may or may not suggest that a particular sample size and study design are appropriate.

3.2.6. Randomization Issues

Randomization issues, though not relevant to the proposed cohort investigation, are critical for many nutritional studies. Eligibility for inclusion in the study should be *verified* by the study coordinator(s) *prior to* enrollment into the study and *prior to* randomization to a treatment group. Randomization can be stratified with respect to one or more selected factors in order to ensure comparability of the treatment groups with respect to this/these factor(s). If required, stratified randomization involves separate applications of a selected method of randomization within each stratum *(4)*.

The randomization is typically implemented as follows. A computer program is written to generate a sequence of ID numbers and treatment assignments, keeping the number of subjects in each treatment nearly equal. Usually the algorithm will rely on a permuted-blocks method: the sequence of treatment assignments is segmented into blocks of assignments. The order of assignments within a block is randomly permuted. The sizes of the blocks can be randomly assigned. For a two-treatment study (A vs B), blocks will typically be of size 2, 4, or 6 *(4)*.

Under ideal circumstances, a trusted agent (e.g., a coordinating center or some other research database management unit) will securely store the assignment key in a blinded manner, but with provision for ready access in case of medical emergency. This step needs to be completed before enrollment begins. A study coordinator (who is blind to the measurements on the variables in Table 1 for each patient) assigns each new eligible subject to the next randomization ID number.

4. METHODS OF DIETARY ASSESSMENT

The selection of the method used to estimate nutrient intakes depends, in part, on personal philosophy and comfort regarding the specific tool that has typically been employed in previous studies by the same investigative team. For example, the qualitative food frequency instrument that has been validated in different ways has been used in many studies that enroll large numbers of subjects *(5)*. Other investigators have preferred to use the quantitative food frequency tool originally developed by Block and co-workers *(6)* and later modified by Subar and colleagues *(7)*. The National Health and Nutrition Examination Survey (NHANES) used only the 24-h recall at one time, now it is planning to use a combination of methods. Still other researchers use combinations of 24-h recall and 2 d of dietary records (USDA Household Food Consumption Survey). The basic aspects of each method are described next.

4.1. 24-H Recall

This method obtains information on all foods and beverages consumed during the previous day (24 h from midnight to midnight). Although only a 1-d slice in time is captured, good approximations of population means can be determined. Typically, standard deviations are quite broad for nutrient variables (energy, nutrients, fiber, and other components of the diets). If multiple 1-d recalls are made, variability tends to decrease for most nutrients. The number of 1-d recalls necessary to estimate well the usual intake mean varies from 3 to more than 100 d for the various nutrients *(8)*.

The 24-h dietary recall method is most commonly used in dietary assessments, because of its ease of administration. For an individual, much variation exists in intake from one day to the next, especially on weekends and holidays. This intraindividual variation has been shown to be greater than the variation in intake between individuals on the same day. Thus, 24-h recalls are more appropriate for reference about the population average than individual intakes.

4.2. Dietary Record or Diary

Records (diaries) of dietary intakes can be made for as little as 1 d or long as 10 d, either consecutively or spaced over time. The relative validity of the number of days used has been analyzed for many nutrients in an attempt to get at the "true" population's mean intake of a given nutrient *(8)*. An example of three consecutive days of records at an interval of 1 yr is given next.

Software for the analysis of 3-d food records, kept at two periods (baseline and 12 mo) during the study, will incorporate the USDA database *(13)* (http://www.ars.usda.gov/fnic/foodcomp). The analytic software for nutrient intakes, such as the ESHA Food Processor (ESHA Research, Salem OR) software or similar computer software, is used to obtain reasonable estimates of intakes. Research nutritionists conduct an interview with each subject at baseline and upon receipt of each food diary to obtain a diet history regarding usual intake and the use of nutrient and/or herbal supplements. Any allergies or diseases that adversely affect nutritional status will be recorded. Any part of the diet history that may impact on skeletal growth will be reported to the investigators and the data safety monitoring board for evaluation as to whether the subject should continue in the study.

4.3. Qualitative Food Frequency

This form of assessment has been widely applied by Willett and collaborators in several studies involving thousands of participants *(2,5)*. Such large population studies need shortened tools that are consistently applied over time for valid comparative analyses of intakes. Although this instrument may underestimate overall energy intake for one day or a collection of days, it usually places the relative intakes of specific nutrients of individuals in their right order, or rank, for comparisons between or among groups. It also permits relatively easy comparisons of the same individuals over several years of investigation.

4.4. Quantitative Food Frequency

The Block quantitative food frequency questionnaire *(6)* and the adapted version of Subar and colleagues *(7)* have been commonly used by nutritionists because each of these quantitative instruments captures intakes within the context of the past 12 mo. The downsides of this tool are the extensive length of time needed to collect an individual's intake in the context of 12 mo, the inaccurate estimations of portion sizes and daily intakes of nutrients. In addition, this method overestimates total caloric intake per day compared to "true" population's intake as estimated by using doubly labeled water method *(9)*, even though the rank order of nutrient intakes by individuals is probably correct. (Although this assumption seems reasonable, it has not been tested for these questionnaires, as it has for the Willett questionnaire *[5]*.)

In recent years, validation studies of quantitative dietary instruments have used selected biomarkers to assess the accuracy of intakes *(10)*. The validation of calcium or vitamin D intakes with biomarkers has been limited, but this approach is worthy of pursuit.

4.5. Duplicate Analyzed Meal Samples

This method is probably the most accurate and reliable of the methods, but it can only be used for relatively small numbers of study subjects. A duplicate meal (weighed portions) is homogenized, aliquots are stored in plastic containers, and frozen for later analyses. Aliquots of meals are chemically analyzed for the nutrients of interest, but the chemical analysts must also consider the food matrices, chemical diluents, and analytic instruments, as well as other aspects of the analytes. This method has been used in the Dalby, Sweden, investigations of an elderly population *(11)*, but not in many population studies because of the cost.

4.6. Summary

The choice of the specific assessment instrument for a study depends on the nature of the investigation, the design, and the cost. The frequency instrument of Willett has become by far the most widely used tool, but each method has limitations that need to be considered during the planning stages of a study. New investigators, especially those planning a large population-based project, are advised to consider having a collaborator with experience in the use of assessment instruments and nutrient data handling. Keep in mind that many studies in the past have employed retrospective assessments of previous nutrient intake patterns rather than prospective changes in dietary intakes.

5. EXAMPLES OF BONE STUDIES EMPLOYING DIETARY ASSESSMENT METHODS

A few examples of different types of bone studies that utilized dietary assessment methods illustrate the detailed approaches to obtain useful data in identifying nutrient risk factors for bone health. These epidemiological studies primarily focused on older women because of the greater risk of osteoporotic fractures in this gender. The choice of these studies is not so inclusive, but the cited examples show how these nutritional methods—largely retrospective—can be applied to many investigations searching for relationships between specific nutrients and bone, especially with BMD as the response (dependent) variable in regression models. Only a few of the studies cited here were prospective in design.

5.1. Retrospective Studies

5.1.1. J. M. Lacey et al. *(12)*

Significant dietary correlates of cortical bone mass among premenopausal and postmenopausal Japanese women were identified using a quantitative food frequency tool developed for Japanese foods *(12)*. Protein was positively associated with BMC in both pre- and postmenopausal women. The retrospective cross-sectional study design of two age groups of women found several nutrient variables that were correlated to cortical bone variables, but the significant variables differed quite markedly in the two groups. For example, mid-radial (cortical) BMC was highly correlated with vegetable and milk consumption in the postmenopausal women, but not in the premenopausal women. Protein intake, however, was positively correlated with mid-radial BMC in both groups of women.

5.1.2. J. A. Metz et al. *(13)*

Calcium intake and physical activity level were positively related to radial bone mass in young adult women (20s), but intakes of phosphorus and protein were negatively associated in this cross-sectional study. Dietary intakes were determined using a quantitative food frequency instrument *(14)*. Care was needed in the regression analysis of each of these nutrients because of collinearity among the variables. The main finding in these healthy young women, who were quite physically active, was the negative protein association. (Several reports reviewed later in this section also assessed the protein–bone linkage, but different conclusions among the reports suggest that protein effects may be age-specific—more negative at younger ages and more positive at older ages—at least in females.)

5.1.3. F. A. Tylavsky et al. *(15)*

Adequate calcium intakes and moderate or high physical activity patterns during adolescence were found to be associated with greater measurements of radial bone mass than low intakes and less physical activity in a group of more than 800 white college-age females (18–22 yr old) using a cross-sectional design. Retrospective and current dietary assessments were made using a quantitative food frequency instrument *(14)*.

5.1.4. J. W. Nieves et al. *(16)*

Teenage and current calcium intakes were each related to bone mineral density of the hip and forearm in women aged 30–39. The major outcome of this retrospective analysis was that high calcium consumption during the teenage period results in greater BMD of the women in the fourth decade of life. The dietary assessment instrument permitted reasonable statistically significant calcium–bone associations for both the teenage years and the current period *(6)*.

5.1.4. S. A. New et al. *(17)*

In this cross-sectional investigation of premenopausal women (45–49 yr), significant associations with BMD were found for several nutrients with the use of a food frequency questionnaire *(18)*. Specifically, higher intakes of zinc, magnesium, potassium, and fiber were found to contribute to significantly greater mean measurements of lumbar BMD. Higher vitamin C intake was less robust in its relationship to lumbar BMD. (A significant calcium-bone relationship was not found.) The authors suggested that high intakes of nutrients in milk and fruit beginning early in adult life may benefit bone mass and density because of the greater acid-buffering capacity of these nutrients.

5.1.5. T. A. Outila et al. *(19)*

Vitamin D status was speculated to influence parathyroid hormone concentrations during winter in female adolescents, using a cross-sectional design. Dietary intakes of calcium and vitamin D were assessed by a food frequency questionnaire coupled with photos of portion sizes *(20)*. Mean intake of calcium was greater than 1200 mg/d, but vitamin D intake was very low. Milk is not fortified with vitamin D in Finland, so little vitamin D is consumed in foods except for fish. About 13% of the adolescent girls (14–16 yr) had low vitamin D status and elevated parathyroid hormone concentrations. When 25(OH)D was lower than 25 mmol/L during the winter months, the girls also demonstrated suboptimal forearm bone measurements. Adequate vitamin D intakes during winter months at northernmost latitudes appear to be critical for skeletal development in growing adolescent females.

5.2. Prospective Studies

5.2.1. J. A. Reed et al. *(21)*

Changes in radial BMD of elderly white female lacto-ovo-vegetarians (LOVs) and omnivores over a 5-yr period were compared. Because changes in BMD of omnivorous and LOV women were similar, despite higher mean calcium in the omnivores, calcium may not be the variable that influenced the loss of BMD. Calcium intakes changed little in subgroups of omnivores and LOVs, either up or down, over this period, but calcium intake apparently had little effect on slowing the bone loss, even in high-calcium consumers. All subjects lost bone, and those with the greater loss of BMD over 5 yr also had the greater loss of lean body mass. Calcium intake in this study was assessed using a quantitative food frequency instrument *(14)*. The findings of this prospective cohort study suggest that calcium intake by consumers of the different types of diets over the 5-yr period was less critical for maintaining BMD than one other variable, namely, lean body mass.

5.2.2. D. Feskanich et al. *(22)*

The results of this 12-yr prospective study of pre- and early postmenopausal women, aged 34–59 yr, counter the current thinking that higher intakes of milk and other dairy products rich in calcium protect against fractures, especially fractures of the hip. A short food frequency questionnaire (61 food items) was used to estimate calcium intakes *(23)* of women in the Nurses Health Study who had hip (proximal femur) or distal radial fractures. Higher intakes of milk or calcium were not associated with reductions of fractures at either site 12 yr after baseline intake measurements.

5.2.3. A. C. Looker et al. *(24)*

The effect of dietary calcium on the risk of hip fracture in men and women was examined in a prospective study, the NHANES I Epidemiologic Follow-Up Study. For the postmenopausal women who had hip fractures ($n = 122$), hip fracture rates were practically the same in all calcium quartiles (high to low calcium intake) who were not taking estrogen replacement therapy. The authors suggested that a high calcium intake may lower the risk of hip fracture only in late menopausal women, consistent with an earlier study by Dawson-Hughes and colleagues *(25)*. Men were not affected by calcium intake at all in this small cohort. A 24-h recall method was used for assessing calcium intake.

5.2.4. J. Promislow et al. *(26)*

This report assessed nutrient intakes in relation to BMD in elderly women and men living in Rancho Bernardo, California. A prospective design permitted assessment of BMD changes and fractures over a period of 12 yr. The Willett self-administered dietary questionnaire, which consists of 128 common food items, was used *(5)*. Other information on nutrient supplements and alcohol consumption was also obtained. High vitamin A levels, provided mainly by supplements, had a negative effect on BMD, and vitamin A alone was considered responsible for the higher hip fracture rate in the supplement users. The increase in hip fractures among elderly subjects who took vitamin A supplements was a surprising result, especially considering that the total intake of vitamin A was not much greater than the currently recommended intake.

5.2.5. M. T. Hannan et al. *(27)*

A 4-yr prospective study of elderly men and women in the Framingham Osteoporosis Study found that protein intake was positively associated with BMD. Lower protein intakes correlated with increased bone loss of the hip and lumbar spine, suggesting that reasonable amounts, e.g., approx 70 g/d, were necessary for maintaining bone mass and density. Protein source, either animal or plant, made no difference in the elderly subjects of this study. A semiquantitative food frequency questionnaire was used for the Framingham subjects.

5.3. Prospective vs Retrospective

5.3.1. K. Michaelsson et al. *(28)*

This prospective investigation (control study nested within a cohort) actually compared intakes using both retrospective and prospective designs in 123 women before and after a hip fracture. The investigators demonstrated that misclassification commonly occurs in retrospective analyses but not in prospective assessments of dietary intakes. Dietary calcium intakes from dairy foods was almost halved in cases after their fractures using a current assessment tool, but with the use of a retrospective questionnaire no change in calcium intake was reported for the same women! The message is that assessment of retrospective dietary intakes appears to be too inaccurate to rely on it for correctly classifying intakes earlier in adult or adolescent life.

5.4. Questionnaire Development

5.4.1. J. A. Pasco et al. *(29)*

A cross-sectional design based on current eating patterns in Geelong, Australia, revealed that most (76%) of the 1045 women consumed less calcium than recommended, and that young women (20–54 yr) had the lowest mean calcium intakes of all adult subjects. Calcium supplement use was also found in approx 10% of the women. A calcium-specific food frequency questionnaire for Australian women, developed by testing population-based estimates of calcium intakes from common foods (standard servings) against estimated calcium intakes from 4 d of weighed records, was used *(29)*.

5.5. Meta-Analysis

5.5.1. D. C. Welten et al. *(30)*

This meta-analysis of several published reports assessed calcium intakes in relation to bone measurements. This report exemplifies the increased credibility of analysis based on similar findings of several studies, effectively increasing the numbers of study subjects. This approach, while using a common yardstick for the outcome variable, permits the making of stronger conclusions when the combined number of subjects is large, especially when a variety of experimental settings (similar and different) are included. The authors found that calcium intakes by adolescents and young adults of approx 1000 mg/d were necessary to result in optimal bone mass *(30)*.

6. DISCUSSION

The specific instrument selected for nutritional intake estimations yields information on usual dietary intakes, but the data may be limited by the instrument, e.g., the 128 food choices in the Willett questionnaire may not adequately capture the intakes of some of the micronutrients such as vitamin D or zinc *(5)*. The quantitative food frequency questionnaires *(6,7)* overestimate intakes of many nutrients. No perfect assessment method exists, so that investigators need to be aware of the limitations of any method of nutritional assessment selected. In fact, Bingham has reported the use of several dietary methods in the same study subjects to the weighted record method, and her group reported that the food frequency questionnaires are no better than the 24-h recall because of inaccuracies of estimating frequency of food intake *(31)*. However, biases in intakes typically cancel when differences between groups are found.

This review highlights selected published studies that have used a variety of dietary research methods to tease out significant nutrient–bone linkages. These reports have focused more on calcium and vitamin D than on other nutrients, such as protein and vitamin A, which may also have important roles in bone tissue at one or more stages of the life cycle. The assessment of usual calcium intakes has proved to be not only difficult because of wide variability in daily food consumption, but also much less accurate and precise than desirable. In studies in which supplements are used, some of this variability can be overcome by knowing the amount administered (e.g., of calcium), hence the total calcium consumed each day.

Factors other than dietary variables may be confounding many of the prospective studies. Two such factors are lean body mass and early vs late menopausal status. The loss of lean body mass has been positively correlated in several investigations with the loss of bone mass and density *(21)*. Also women who are within 5 yr of the menopause respond poorly to calcium when given as a supplement, but later postmenopausal women respond quite favorably to calcium supplements *(25)*. Another factor that is poorly understood is how older women have been able to adapt—at least reasonably well—to very low habitual calcium intakes for decades without experiencing a hip fracture. These and other factors confound our understandings of the associations between dietary variables and bone health.

7. CONCLUSIONS

The use of computer-assisted methods to obtain intake data of calcium, vitamin D, and other nutritional variables has enabled bone researchers to uncover unsuspected nutrients as important positive or negative risk factors for bone health. An example of such a nutrient not previously suspected of having adverse effects on bone fractures among the elderly is vitamin A (i.e., high retinol intakes), which is now implicated in three prospective population studies. These nutritional assessment methods, whether quantitative or qualitative, provide less precision or accuracy in estimating mean intakes of specific nutrients than investigators would desire. Part of the reason for the wide variabilities of mean nutrient intakes resides in physiological differences. Methodological issues that reduce precision and accuracy include correlated variables, variability in eating patterns from day to day, and incomplete information on nutrient content of foods, especially processed foods. The manuscripts cited in this review represent a variety of approaches to assessing nutrient–bone associations, and they are grouped according to the type of study.

ACKNOWLEDGMENTS

The authors wish to acknowledge the assistance of Agna Boass, PhD, in the preparation of this manuscript.

REFERENCES

1. Freidman L, Furberg C, DeMets D. Fundamentals of Clinical Trials, 2nd ed. PSG, Littleton, MA, 1985.
2. Willett W, ed. Nutritional Epidemiology, 2nd ed. Oxford University Press, New York, 1998.

3. Cox DR. Planning of Experiments. Wiley, New York, 1958.
4. Lachin J, Matts J, Wei L. Randomization in clinical trials: conclusions and recommendations. Controlled Clin Trials 1988; 9:365–374.
5. Willett WC, Sampson L, Browne ML, et al. The use of a self-administered questionnaire to assess diet four years in the past. Am J Epidemiol 1988; 127:188–199.
6. Block G, Hartman AM, Dresser CM, Carroll MD, Gannon J, Gardner L. A data-based approach to diet questionnaire design and testing. Am J Epidemiol 1986; 124:453–469.
7. Subar A, Thompson FE, Kipnis V, et al. Comparative validation of the Block, Willett, and National Cancer Institute Food Frequency Questionnaires. Am J Epidemiol 2001; 154:1089–1099.
8. Beaton GH, Milner J, Corey P, et al. Sources of variance in 24-hour dietary recall data: implications for nutrition study design and interpretation. Am J Clin Nutr 1979; 32:2546–2549.
9. Schoeller DA, Bandini LG, Dietz WH. Inaccuracies in self-reported intake identified by comparison with the doubly labelled water method. Can J Physiol Pharmacol 1990; 68:941–949.
10. Day NE, McKeown N, Wong MY, Welch A, Bingham S. Epidemiological assessment of diet: a comparison of a 7-day diary with a food frequency questionnaire using urinary markers of nitrogen, potassium, and sodium. Int J Epidemiol 2001; 30:309–317.
11. Borgstrom B, Norden A, Akesson B, Jagerstad M. A study of the food consumption by the duplicate portion technique in a sample of the Dalby population. Scand J Social Med 1975; Suppl 10:1–98.
12. Lacey JM, Anderson JJB, Fujita T, et al. Correlates of cortical bone mass among premenopausal and postmenopausal Japanese women. J Bone Miner Res 1991; 7:651–659.
13. Metz JA, Anderson JJB, Gallagher PN Jr. Intakes of calcium, phosphorus, and protein, and physical activity level are related to radial bone mass in young adult women. Am J Clin Nutr 1993; 58:537–542.
14. Tylavsky FA, Anderson JJB. Dietary factors in bone health of elderly lactoovovegetarian and omnivorous women. Am J Clin Nutr 1988; 48:842–850.
15. Tylavsky FA, Anderson JJB, Talmage RV, Taft TN. Are calcium intakes and physical activity during adolesence related to radial bone mass of white college-age females? Osteopor Int 1992; 2:232–240.
16. Nieves JW, Grisso JA, Kelsey JL. A case-control study of hip fracture: evaluation of selected dietary variables and teen-age physical activity. Osteopor Int 1992; 2:122–127.
17. New SA, Bolton-Smith C, Grubb DA, Reid DM. Nutritional influences on bone mineral density: a cross-sectional study in premenopausal women. Am J Clin Nutr 1997; 65:1831–1839.
18. Lanham SA, Bolton-Smith C. Development of a food frequency questionnaire abstr. Proc Nutr Soc 1993; 52:330A.
19. Outila TA. Vitamin D status affects serum parathyroid hormone concentrations during winter in female adolescents: associations with forearm bone mineral density. Am J Clin Nutr 2001; 74:206–210.
20. Outila TA, Karkkainen MU, Seppanen RH, Lamberg-Allardt CJ. Dietary intake in premenopausal women, healthy vegans was insufficient to maintaine concentrations of serum 25-hydroxyvitamin D and intact parathyroid hormone within normal ranges during winter in Finland. J Am Dietet Assoc 2000; 100:434–441.
21. Reed JA, Anderson JJB, Tylavsky FA, Gallagher PN. Comparative changes in radial-bone density of elderly female lacto-ovo-vegetarians and omnivores. Am J Clin Nutr 1994; 59(suppl):1197S–1202S.
22. Feskanich D, Willett W, Stampfer MJ, Colditz GA. Milk dietary calcium, and bone fractures: a 12-year prospective study. Am J Publ Health 1997; 87:992–997.
23. Salvini S, Hunter DJ, Sampson L, et al. Food-based validation of a dietary questionnaire: the effects of week-to-week variation in food consumption. Int J Epidemiol 1989; 18:858–867.
24. Looker AC, Harris TB, Madans JH, Sempos CT. Dietary calcium and hip fracture risk: the NHANES I Epidemiologic Follow-Up Study. Osteopor Int 1993; 3:177–184.

25. Dawson-Hughes B, Dallal GE, Krall EA, Sadowski L, Sahyoun N, Tannenbaum S. A controlled trial of the effect of calcium supplementation on bone density in postmenopausal women. N Engl J Med 1990; 323:878–884.
26. Promislow JHE, Goodman-Gruen D, Slymen DJ, Barrett-Connor E. Retinol intake and bone mineral density in the elderly: the Rancho Bernardo Study. J Bone Miner Res 2002; 17:1349–1358.
27. Hannan MT, Tucker KL, Dawson-Hughes B, Cupples LA, Felson DT, Kiel DP. Effect of dietary protein on bone loss in elderly men and women: the Framingham Osteoporosis Study. J Bone Miner Res 2000; 15:2504–2512.
28. Michaelsson K, Holmberg L, Ljunghall S, Mallmin H, Persson P-G, Wolk A. and The Study Group of the Multiple Risk Survey on Swedish Women for Eating Assessment. Effect of prefracture versus postfracture dietary assessment on hip fracture risk assessments. Int J Epidemiol 1996; 25:403–410.
29. Pasco JA, Nicholson GC, Sanders KM, Seeman E, Henry MJ, Kotowicz MA. Calcium intakes among Australian women: Geelong Osteoporosis Study. Austral NZ J Med 2000; 30:21–27.
30. Welten DC, Kemper HCG, Post GB, van Staveren WA. A meta-analysis of the effect of calcium intake on bone-mass in young and middle aged females and males. J Nutr 1995; 125:2802–2813.
31. Bingham SA, Gill C, Welch A, et al. Comparison of dietary assessment methods in nutritional epidemiology: Weighed records v. 24 h recalls, food-frequency questionnaires and estimated-diet records. Br J Nutr 1994; 72:619–643.

8 Nutrition and Oral Bone Status

Elizabeth A. Krall

1. INTRODUCTION

Tooth loss and periodontal diseases are two clinically important outcomes of oral bone loss. They result from the interaction of multiple factors that range from dental care access to genetic factors. However, it is the loss of the tooth-supporting bone that is a key diagnostic feature of periodontal disease and a risk factor for tooth loss. Nutritional factors related to bone turnover elsewhere in the body, particularly calcium and vitamin D, have roles in maintaining oral bone, but their importance relative to the inflammatory process that causes periodontal disease is uncertain.

2. PERIODONTAL DISEASES AND TOOTH LOSS

The teeth are supported in sockets within the maxilla and mandible by periodontal ligament, alveolar bone, and cementum. Alveolar bone is a plate of compact bone that lines the tooth sockets. Periodontal ligament consists of bundles of collagenous fibers that attach and anchor the tooth root to alveolar bone. Cementum is a calcified tissue that lines the roots of the teeth, but does not undergo remodeling as bone tissue does. In contrast, alveolar bone and the cancellous bone of the jaws are identical to bone tissue in other parts of the body and continuously undergo cycles of resorption and formation. The gingiva is soft periodontal tissue that covers the bone and roots of the teeth and reduces contact between bone and the contents of the mouth.

The most common forms of periodontal diseases, gingivitis and adult periodontitis, are inflammatory diseases initiated by bacteria found in the plaque that coats the teeth. In gingivitis, only the soft periodontal tissue is affected. In periodontitis, there is also loss of periodontal ligament and alveolar bone. The bone loss is primarily vertical; i.e., the height of the bony tooth socket declines as the cortical alveolar bone is eroded (Fig. 1). Along with the loss of bone height, the gingiva recede and pockets develop between the teeth and gums in which bacteria can accumulate and further aggravate the destruction of periodontal tissues. Loss

From: *Nutrition and Bone Health*
Edited by: M. F. Holick and B. Dawson-Hughes © Humana Press Inc., Totowa, NJ

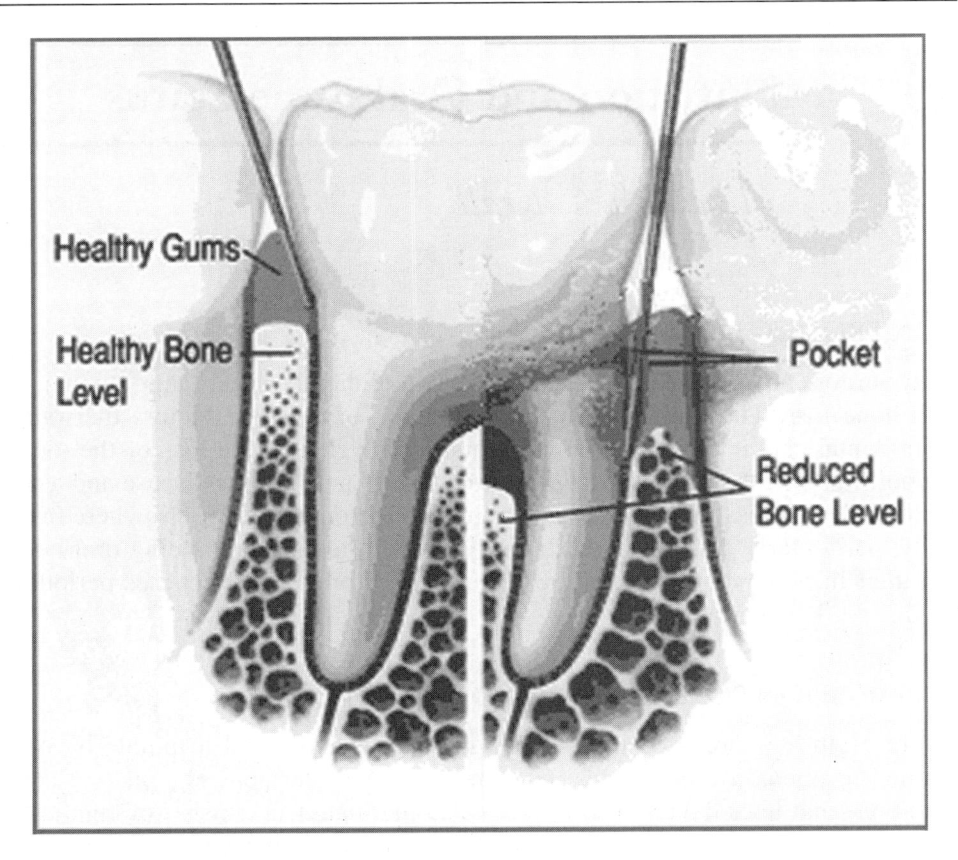

Fig. 1. Illustration showing alveolar bone and gingival tissues in a healthy individual (left side) and in periodontal disease (right side).

of alveolar bone is generally related to localized factors that increase bone resorption or decrease formation. These factors include lipopolysaccharide and cytokines such as interleukin (IL)-1 and tumor necrosis factor (TNF), and are produced by bacteria themselves or by the host immune system in response to periodontal inflammation *(1)*.

As bone support declines, the teeth become loose, mobile, and uncomfortable, and may have to be extracted. Gingival recession also increases the risk of root caries as the roots of the teeth are exposed to acids produced by bacteria. In the general population, caries likely is the major contributor to tooth loss. However, in the elderly more than one-fourth of missing teeth were extracted because of periodontal disease *(2)*.

In the US population aged 30–90 yr, 35% of dentate adults have some form of periodontitis, defined as at least one tooth with a probing pocket depth greater than or equal to 3 mm *(3)*. Most of the cases are considered mild (22%), with a few affected teeth. Only 13% of adults have more teeth involved and/or deeper pockets and are classified as moderate or severe. The prevalence of periodontitis increases sharply with age. Among those aged 65 and older, the prevalence of mild periodontal disease is almost double that of adults aged 30–34, whereas moderate and severe periodontitis are more than four times more common in the elderly than in young adults.

The prevalence of tooth loss also increases dramatically with age. In the third National Health and Nutrition Examination Survey (NHANES III) survey in 1988–1991, nearly 70% of US adults had lost one or more teeth *(4)*. The average number of teeth declined from 27 at ages 18–24, to 15 at ages 60–64, and to 9 at ages 75 and older. The prevalence of edentulism (total tooth loss) increased from 0 to 44% over these same age intervals. Retention of teeth is expected to improve in future cohorts of the elderly. In addition to caries and periodontal disease, other factors that contribute to periodontal disease and tooth loss include poor dental hygiene, lack of access to dental care, genetics, and systemic diseases such as diabetes. Osteoporosis has also been associated with increased periodontal disease and tooth loss *(5)*.

3. RELATIONSHIP BETWEEN SYSTEMIC BONE AND ALVEOLAR BONE LOSS OR TOOTH LOSS

In a 2-yr prospective study, Payne et al. *(6)* compared changes in alveolar bone height between a group of postmenopausal women with spinal osteoporosis and a group of women with normal bone mineral density (BMD). They found that osteoporotic women exhibited significantly more sites with alveolar bone loss than the normal-BMD group. Other studies of the relationship have been cross-sectional in design and included both pre- and postmenopausal women. The correlations between alveolar bone height and BMD of the hip, spine, and forearm were modest, in the range of $r = 0.2–0.5$ *(5,7–10)*. In contrast, one of the larger cross-sectional studies was conducted by Elders et al. *(11)*, who reported no association between spine BMD and loss of alveolar bone height in healthy, early postmenopausal women between the ages of 46 and 55.

Erosion of the alveolar bone and jaw is frequently a consequence of tooth extraction. Findings from several *(10,12–14)* studies of individuals who lost teeth also suggest a correlation between remaining bone height of the jaw and BMD at systemic sites, but not all studies agree *(15)*.

Studies in which tooth loss is the dependent variable tend to be more supportive of a positive association between systemic bone status and oral status. In a prospective, 7-yr observational study of postmenopausal women, rates of bone loss at the hip, spine, and total body over a 7-yr period were three times faster than in women who lost no teeth *(16)*. The risk of tooth loss increased by at least 50% for each 1% per year increment in the rate of systemic bone loss. However, there was no information on other oral factors such as caries burden that cause tooth loss. Several cross-sectional studies of elderly subjects and postmenopausal women report moderate correlations between BMD and the number of remaining teeth *(17–19)*. However, these findings are not in complete agreement either. Two studies of early postmenopausal women and one of elderly women *(20–22)* reported no associations of dentition status with BMD.

4. NUTRITION AND ORAL BONE STATUS

Although it is not certain if the link between systemic bone status and poor oral health is causal, enough studies have shown significant correlations to raise the question of whether nutritional status affects periodontal bone loss and tooth loss. In further support of this possibility is the fact that alveolar bone is metabolically active and responds to the same resorption and formation stimuli as elsewhere in the skeleton. However, the consensus has been that alveolar process is relatively independent of the rest of the skeleton and does not respond quickly to nutritional imbalances. Therefore, nutritional therapy is thought to have little value in managing periodontal diseases *(1)*.

Prior to the 1980s, animal experiments from one group suggested that a low-calcium diet led to alveolar bone resorption, tooth mobility, and tooth loss *(23)*, which was reversed by supplying adequate amounts *(24)*. The authors proposed that calcium deficiency was the initial insult to periodontal bone tissue and local inflammation had a secondary role. This view has been discounted, as later studies failed to replicate the findings *(25–27)* and the role of inflammatory mediators in alveolar bone loss was clarified. Diets with a low calcium content or high P:Ca ratio do produce osteoporotic changes in alveolar bone and the jawbone—porosity and thinning of the cortex—but not the loss of vertical height seen in periodontal disease.

Earlier studies of calcium supplementation in patients with periodontal disease also suggested that increasing calcium intake could prevent periodontal bone loss. A 6-mo regimen of 1000 mg of calcium per day was reported to reduce tooth mobility and pocket depth, and improve the status of alveolar bone in 10 periodontitis patients *(28)*. A 1-yr study of calcium and vitamin D supplements (750 mg calcium and 375 IU vitamin D per day) in 33 patients concluded that supplements were associated with improved periodontal disease *(29)*. However, these studies had serious limitations, such as lack of a control group *(28)*, small sample sizes, and subjective measures of improvement in periodontal disease signs and symptoms. Uhrbom and Jacobson *(30)* conducted a study of 66 patients who took 1000 mg/d of calcium for 6 mo. Using established methods for measuring gingival inflammation, pocket depth, and tooth mobility, they found no differences in periodontal disease status between the supplemented and placebo groups at the end of the study period.

Interest in the role of nutrition in periodontal disease has reemerged with the recent publication of several studies showing a relationship between calcium intake and periodontal disease and tooth loss *(31–33)*. These latter studies have the advantages of large sample size and long-term follow-up *(32,33)* that provide greater statistical power to detect relationships. Nishida et al. *(31)* published the results of an analysis of dietary intake surveys and periodontal examinations from more than 12,000 adults from NHANES III. Periodontal disease, defined as average attachment loss of 1.5 mm or more, was present in about 25% of the population. An inverse association was found between dietary calcium intake level and periodontal disease, controlling for smoking and age. The odds of periodontal disease were 30 and 60% higher in men and women, respectively, with calcium intake below 500 mg compared to those with calcium intake above 800 mg. In a prospective study of older men, those with a calcium intake below 1000 mg/d had more teeth exhibiting rapid progression of alveolar bone loss compared to men with calcium intakes above this level *(32)*.

Calcium supplementation and calcium intake were associated with a reduction in the risk of tooth loss in elderly men and women in a series of two studies that spanned 5 yr *(33)*. During the initial 3 yr, subjects were randomly assigned to take either placebos or supplements containing 500 mg of calcium and 700 IU of vitamin D per day. The odds of losing any teeth were reduced by more than half in the supplemented group at the end of 3 yr. In a 2-yr follow-up of the same individuals, the odds of tooth loss were again about 50% lower in subjects whose total calcium intake was at least 1000 mg/d compared to those who consumed less than 1000 mg. Oral examinations performed at the end of the study suggested the periodontal and caries status of the two groups were similar and the major difference was level of calcium intake.

Vitamin D has a role in maintaining calcium balance. Recent studies have identified another function that has implications for periodontal health. In animal models, vitamin D has been shown to act as an immunosuppressant through regulation of T-cell-mediated immune reactions *(34)*. Vitamin D selectively stimulates transforming growth factor (TGFβ-1) and IL-4, but suppresses gene expression of interferon-γ and TNFα, which is destructive to both bony and connective periodontal tissues.

Vitamin C is necessary for the hydroxylation of amino acids that are used in the synthesis of collagen. Collagen in bone tissue provides the matrix onto which minerals are deposited, and comprises one of the major connective tissue fibers of the gingiva. In the NHANES III survey of adults, low vitamin C intake levels were associated with an increased risk of periodontal disease *(35)*. The association observed in the general population was similar to that of low calcium. Individuals consuming less than 30 mg/d had a 30% increased risk of periodontal disease than those consuming 180 mg or more.

5. CONCLUSION

Periodontal disease and tooth loss have multiple causes. Although the presence of bacteria is necessary to initiate an inflammatory response, it is not known how much of an influence nutritional status may have on the susceptibility to and progression of periodontal disease. The nutrient that has been studied most is calcium because of its obvious importance to maintaining healthy bone tissue. The existing studies generally support the hypothesis that osteoporosis and osteopenia are associated with oral health, particularly tooth loss in the elderly, but it is not clear whether nutritional therapies such as calcium supplements have benefits for oral bone. More prospective studies and clinical trials are needed.

REFERENCES

1. Genco RJ. Pathogenesis and host responses in periodontal disease. In Genco RJ, Goldman HM, Cohen DW, eds. Contemporary Periodontics. Mosby, St. Louis, MO, 1990, pp. 184–193.
2. Brown LJ, Brunelle JA, Kingman A. Periodontal status in the United States, 1988–1991: prevalence, extent, and demographic variation. J Dent Res 1996; 75(spec no):672–683.
3. Albandar JM, Brunelle JA, Kingman A. Destructive periodontal disease in adults 30 years of age and older in the United States, 1988–1994. J Periodontol 1999; 70:13–29.
4. Marcus SE, Drury TF, Brown LJ, Zion GR. Tooth retention and tooth loss in the permanent dentition of adults: United States, 1988–1991. J Dent Res 1996; 75(spec issue):684–695.
5. Wactawski-Wende J, Grossi SG, Trevisan M, et al. The role of osteopenia in oral bone loss and periodontal disease. J Periodontol 1996; 67(suppl):1076–1084.
6. Payne JB, Reinhardt RA, Nummikoski PV, Patil KD. Longitudinal alveolar bone loss in postmenopausal osteoporotic/osteopenic women. Osteopor Int 1999; 10:34–40.
7. Streckfus CF, Johnson RB, Nick T, Tsao A, Tucci M. Comparison of alveolar bone loss, alveolar bone density and second metacarpal bone density, salivary and gingival crevicular fluid interleukin-6 concentrations in healthy premenopausal and postmenopausal women on estrogen therapy. J Gerontol A: Biol Sci Med Sci 1997; 52:M343–M351.
8. Engel MB, Rosenberg HM, Jordan SL, Holm K. Radiological evaluation of bone status in the jaw and in the vertebral column in a group of women. Gerodontology 1994; 11:86–92.
9. Southard KA, Southard TE, Schlechte JA, Meis PA. The relationship between the density of the alveolar processes and that of post-cranial bone. J Dent Res 2000; 79:964–969.
10. Kribbs PJ, Smith DE, Chesnut CH III. Oral findings in osteoporosis. Part II: Relationship between residual ridge and alveolar bone resorption and generalized skeletal osteopenia. J Prosthet Dent 1983; 50:719–724.
11. Elders PJ, Habets LL, Netelenbos JC, van der Linden LW, van der Stelt PF. The relation between periodontitis and systemic bone mass in women between 46 and 55 years of age. J Clin Periodontol 1992; 19:492–496.
12. Kribbs PJ, Chesnut CH III, Ott SM, Kilcoyne RF. Relationships between mandibular and skeletal bone in an osteoporotic population. J Prosthet Dent 1989; 62:703–707.
13. Klemetti E, Vainio P. Effect of bone mineral density in skeleton and mandible on extraction of teeth and clinical alveolar height. J Prosthet Dent 1993; 70:21–25.
14. Klemetti E, Vainio P, Lassila V, Alhava E. Cortical bone mineral density in the mandible and osteoporosis status in postmenopausal women. Scand J Dent Res 1993; 101:219–223.
15. Mercier P, Inoue S. Bone density and serum minerals in cases of residual alveolar ridge atrophy. J Prosthet Dent 1981; 46:250–255.
16. Krall EA, Garcia RI, Dawson-Hughes B. Increased risk of tooth loss is related to bone loss at the whole body, hip, and spine. Calcif Tissue Int 1996; 9:433–437.
17. May H, Reader R, Murphy S, Khaw KT. Self-reported tooth loss and bone mineral density in older men and women. Age Ageing 1995; 24:217–221.
18. Krall EA, Dawson-Hughes B, Papas A, Garcia RI. Tooth loss and skeletal bone density in healthy postmenopausal women. Osteopor Int 1994; 4:104–109.

19. Daniell HW. Postmenopausal tooth loss. Contributions to edentulism by osteoporosis and cigarette smoking. Arch Intern Med 1983; 143:1678–1682.
20. Weyant RJ, Pearlstein ME, Churak AP, Forrest K, Famili P, Cauley JA. The association between osteopenia and periodontal attachment loss in older women. J Periodontol 1999; 70:982–991.
21. Klemetti E, Collin HL, Forss H, Kkanen H, Lassila V. Mineral status of skeleton and advanced periodontal disease. J Clin Periodontol 1994; 21:184–188.
22. Earnshaw SA, Keating N, Hosking DJ, et al. Tooth counts do not predict bone mineral density in early postmenopausal Caucasian women. EPIC study group. Int J Epidemiol 1998; 27:479–483.
23. Henrikson PA. Periodontal disease and calcium deficiency. An experimental study in the dog. Acta Odontol Scand 1968; 26(suppl 50):1–132.
24. Krook L, Lutwak L, Henrikson PA, et al. Reversibility of nutritional osteoporosis; physicochemical data on bones from and experimental study in dogs. J Nutr 1971; 101:233–246.
25. Bissada NF, DeMarco TJ. The effect of a hypocalcemic diet on the periodontal structures of the adult rat. J Periodontol 1974; 45:739–745.
26. Oliver WM. The effect of deficiencies of calcium, vitamin D or calcium and vitamin D and of variations in the source of dietary protein on the supporting tissues of the rat molar. J Periodont Res 1969; 4:56–69.
27. Ferguson HW, Hartles RL. The effects of diets deficient in calcium or phosphorus in the presence and absence of supplements of vitamin D on the secondary cementum and alveolar bone of young rats. Arch Oral Biol 1964; 9:647–658.
28. Krook L, Lutwak L, Whalen JP, Henrikson PA, Lesser GV, Uris R. Human periodontal disease. Morphology and response to calcium therapy. Cornell Vet 1972; 62:32–53.
29. Spiller WF. A clinical evaluation of calcium therapy for periodontal disease. Dental Digest 1971; 77:522–526.
30. Uhrbom E, Jacobson L. Calcium and periodontitis: clinical effect of calcium medication. J Clin Periodontol 1984; 11:230–241.
31. Nishida M, Grossi SG, Dunford RG, Ho AW, Trevisan M, Genco RJ. Calcium and the risk for periodontal disease. J Periodontol 2000; 71:1057–1066.
32. Krall EA. The periodontal-systemic connection: implications for treatment of patients with osteoporosis and periodontal disease. Ann Periodontol 2001; 6:209–213.
33. Krall EA, Wehler C, Harris SS, Garcia RI, Dawson-Hughes B. Calcium and vitamin D supplements reduce tooth loss in the elderly. Am J Med 2001; 111:452–456.
34. DeLuca HF, Cantorna MT. Vitamin D: its role and uses in immunology. FASEB J 2001; 15:2579–2585.
35. Nishida M, Grossi SG, Dunford RG, Ho AW, Trevisan M, Genco RJ. Dietary vitamin C and the risk for periodontal disease. J Periodontol 2000; 71:1215–1223.

II

Nutrition and Bone: Effects of Life Stages and Race

9 Nutrition in Pregnancy and Lactation

Bonny L. Specker

1. INTRODUCTION

Significant changes in maternal calcium metabolism occur during pregnancy and lactation. These changes provide a sufficient calcium supply to the fetus for skeletal growth and to the newborn infant in the form of adequate maternal milk production. There is increasing evidence that calcium is mobilized from the maternal skeleton for milk production and is replaced upon resumption of menses, and that dietary calcium does not prevent the bone loss that occurs during lactation. This chapter summarizes the literature on the effects of human pregnancy, lactation, and weaning on calcium metabolism and bone health, the role of dietary calcium on these changes, the epidemiological evidence relating parity and lactation to osteoporosis and fracture risk, and the role of maternal calcium and vitamin D intake on neonatal vitamin D and bone development.

2. CALCIUM AND BONE METABOLISM DURING PREGNANCY, LACTATION, AND WEANING

Significant calcium demands are placed on the mother during pregnancy and lactation. A term infant contains approx 25–30 g of calcium, the majority of which is accumulated during the last trimester *(1)*. It is estimated that fetal calcium accretion increases from about 50 mg/d at 20 wk of gestation to a maximum of approx 330 mg/d at 35 wk gestation *(2)*. Calcium demands on the mother continue to be high during lactation, when approx 250 mg/d of calcium are lost to breast milk *(3)*. Important physiological adaptations occur during pregnancy and lactation to meet these increased demands. Figures 1–3 summarize the physiological adaptation that occurs during pregnancy, lactation, and weaning in well-nourished women.

2.1. Pregnancy

2.1.1. Calcium and Bone Metabolism

There are three possible calcium sources that may supply the mother with the necessary calcium to support fetal growth: increased intestinal calcium absorption

From: *Nutrition and Bone Health*
Edited by: M. F. Holick and B. Dawson-Hughes © Humana Press Inc., Totowa, NJ

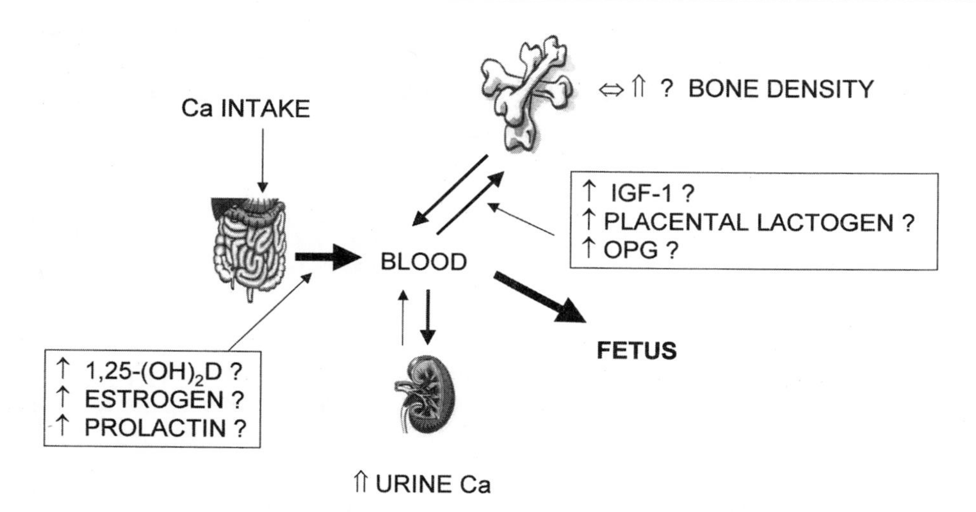

Fig. 1. Schematic illustration of the role of intestinal calcium absorption, renal calcium excretion, and bone turnover in providing calcium to the fetus. The size of the arrows indicates the relative magnitude of the flux of calcium.

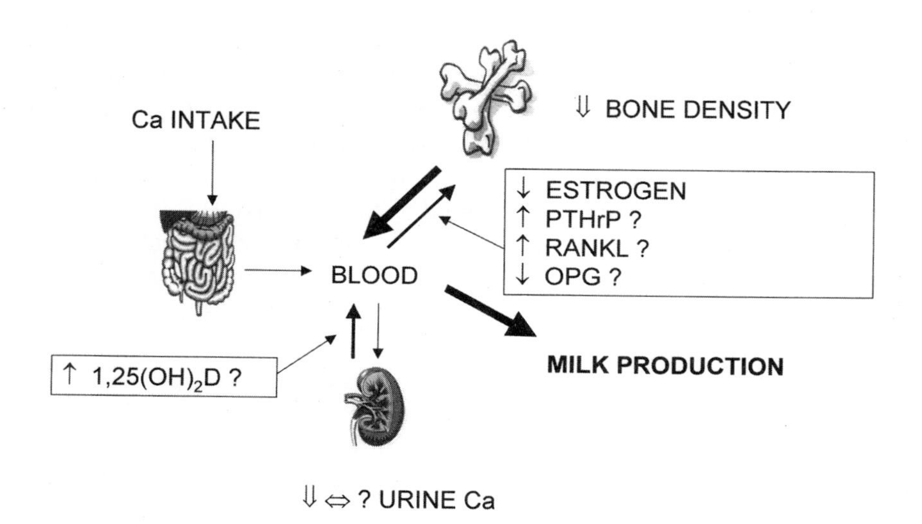

Fig. 2. Schematic illustration of the role of intestinal calcium absorption, renal calcium excretion, and bone turnover in providing calcium for milk production. The size of the arrows indicates the relative magnitude of the flux of calcium.

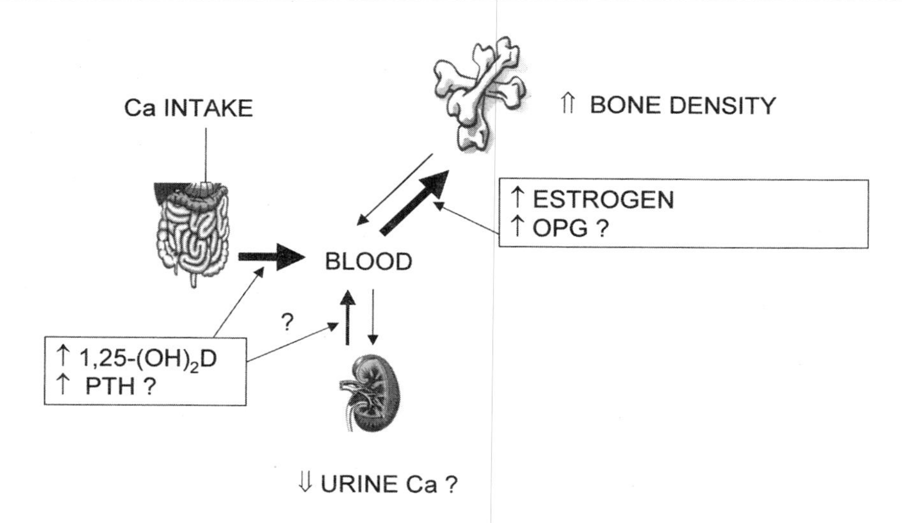

Fig. 3. Schematic illustration of the role of intestinal calcium absorption, renal calcium excretion, and bone turnover in providing the mother with sufficient calcium during weaning. The size of the arrows indicates the relative magnitude of the flux of calcium.

from the diet, increased renal calcium conservation, and increased bone calcium mobilization (Fig. 1). Under normal circumstances, the active metabolite of vitamin D, 1,25-dihydroxyvitamin D [1,25$(OH)_2$D], increases the efficiency of intestinal calcium absorption, decreases renal calcium excretion, and, in conjunction with parathyroid hormone (PTH), mobilizes calcium from bone. Increased fractional calcium absorption appears to be an important mechanism for obtaining extra calcium during pregnancy *(4)*. Fractional calcium absorption, which is typically about 35% in the nonpregnant state, increases to approx 60% during the third trimester *(5,6)*. This represents an additional 300 mg/d of calcium that is obtained solely from increased intestinal absorption among women consuming 1200 mg calcium/d. Calcium absorption is positively associated with serum 1,25$(OH)_2$D concentrations in late human gestation *(5,6)*. Although low calcium intake may result in increased serum concentrations of 1,25$(OH)_2$D during lactation *(7)*, it is not known whether this adaptation occurs during pregnancy. The events that lead to increased serum concentrations of 1,25$(OH)_2$D during pregnancy are not clear. PTH, which is usually considered the stimulus for the increased renal hydroxylation of 25-hydroxyvitamin D [25(OH)D] to 1,25$(OH)_2$D, has not been shown to be high during pregnancy in women consuming adequate amounts of calcium (~1000 mg/d) *(5,6)* or low amounts of calcium (~500 mg/d) *(8)*. It has been speculated that the 1,25$(OH)_2$D present in the maternal circulation may be of placental origin *(9)*. Other suggested mechanisms for the increase in calcium absorption involve estrogen, placental lactogen, or prolactin concentrations as controlling factors, all of which are known to increase during pregnancy *(5,10)*. Renal calcium conservation does not occur during pregnancy and renal calcium excretion increases even when calcium intakes remain relatively constant *(5,6)*, and is likely a result of both increased absorptive load of calcium along with an increase in glomerular filtration rate that occurs during pregnancy *(5,6)*.

Bone is continuously broken down and formed. The sequence of events involved in bone turnover includes activation of osteoclast precursors, followed by osteoclastic bone resorption and then osteoblastic bone formation. Markers of bone turnover are increased in late gestation *(5,6,8)*. Both markers of bone formation (bone-specific alkaline phosphatase and procollagen 1 carboxypeptides) and bone resorption (tartrate-resistant acid phosphatase, deoxypyridinoline) are elevated above nonpregnant concentrations. Although studies consistently find an increase in serum concentrations of biological markers of bone turnover, few have reported changes in bone density (see below). Increased concentrations of insulin-like growth factor (IGF-1) and placental lactogen have been suggested as the possible mechanism behind increased bone turnover during pregnancy *(11)*. It also has been suggested that perhaps some of the increase in bone turnover markers is due to increased turnover of soft tissue collagen of the uterus and skin or possibly may be of placental or fetal origin *(12)*.

Receptor activator of nuclear factor-κ B ligand (RANKL) is important in osteoclast differentiation and also in development of alveolar structures of the mammary gland during pregnancy *(13)*. The cytokine osteoprotegerin (OPG) acts as a decoy receptor for RANKL and prevents its function, thereby decreasing bone resorption. A recent study looked at the changes in calcium-regulating hormones and OPG during pregnancy and found that maternal serum OPG concentrations steadily increased with gestational age *(14)*. The authors speculated that higher OPG concentrations, possibly of placental origin, may play a role in the control of bone metabolism during pregnancy.

2.1.2. Changes in Bone Mineral

Studies of bone changes during pregnancy are difficult to conduct due to the radiation exposure. One study measured bone density of the distal radius *during* pregnancy using single-photon absorptiometry (SPA) *(6)*, whereas other longitudinal studies have measured bone density before and after pregnancy using methods that are more sensitive for detecting small bone changes (dual-energy X-ray absorptiometry [DXA] and quantitative computed tomography [QCT]). It is important to note that the timing of the final postpartum measurement is critical due to rapid bone changes that occur following delivery in lactating women. Ultrasound measurements throughout pregnancy also have been used to investigate bone changes. The effect of pregnancy on BMD changes was recently reviewed *(15)*. Unfortunately, many of the studies measured bone during the 6 wk following delivery, a time when lactation-induced bone loss may have already occurred.

Studies by Cross et al. (5) and Ritchie et al. (6) did not find significant bone changes during pregnancy. However, Naylor and co-workers found a significant decrease in spine bone mineral density (BMD) *(11)*. These contradictory findings may be partly explained by differences in sample sizes and the type of bone measured. All three studies had small sample sizes. Cross and co-workers measured only the ultradistal (predominantly trabecular bone) and one-third distal radius (predominantly cortical bone) in nine women using SPA and found no change *during* pregnancy. Ritchie and co-workers obtained QCT and DXA measurements before and after pregnancy (1–2 wk postpartum) and found no significant changes in spine (predominantly trabecular) or total body BMD among 14 women. However, Naylor and co-workers found a significant decrease in spine BMD measurements obtained from a total-body DXA scan in a study of 16 women who were measured prior to pregnancy and within 2 wk of delivery *(11)*. If a change in BMD were to occur, then likely it would be at a trabecular bone site, such as the spine because trabecular bone is more sensitive to hormonal changes. Spine BMD measurements from a total-body scan have greater variation than QCT or spine regional DXA scans (coefficient of variation 3.6% for spine BMD from a total body scan vs <1% for spine BMD from a regional scan or QCT). Changes in fat distribution before and after pregnancy theoretically may affect BMD measurements of the spine, although this has been reported to be more of a problem with lateral measurements than anterior–posterior spine BMD measurements *(16)*. Whether changes in fat distribution affect spine measurements from a whole-body scan is not known.

Sowers and co-workers used bone ultrasound measurements to determine whether there were differences between adolescents and adults in bone changes during pregnancy *(17)*. Adolescent mothers had greater decreases in speed of sound (speed of signal [SOS] transmission through the heel), broadband ultrasound attenuation (BUA, degree of attenuation of the high-frequency sound waves), and quantitative ultrasound index (combination of SOS and BUA into a single measure) than adult mothers. However, baseline measurements were made at approx 16 wk gestation and at 6 wk postpartum. Significant bone changes could have occurred between birth and 6 wk postpartum if the mothers were lactating. No information was provided on whether any of these women were breast-feeding their infant and whether breast-feeding rates differed between the two groups (adolescent vs adult). In addition, no dietary information was provided and it is not clear whether this may have affected the results. Bezarra and co-workers found that pregnant adolescents consuming a low-calcium diet had greater serum PTH concentrations and lower urinary calcium excretion than pregnant adults consuming a similarly low-calcium intake *(8)*. However, serum deoxypyridinoline and bone-specific alkaline phosphatase concentrations were no different between the adolescent and adult mothers and the authors suggested that pregnant adolescents protect their bones during pregnancy, possibly by decreasing urinary calcium excretion. Whether bone changes during pregnancy differ between adolescent and adult mothers is not known.

2.2. Lactation

2.2.1. Calcium and Bone Metabolism

Similar adaptation in calcium metabolism (increased intestinal calcium absorption from the diet, increased renal calcium conservation, increased bone mobilization) that may occur during pregnancy also may allow the mother to adapt to the calcium needs of milk production (Fig. 2). Numerous studies have shown that intestinal calcium absorption is unchanged during lactation, is higher in the postpartum period in both lactating and nonlactating women, and is not influenced by the maternal calcium intake *(6,18–20)*. However, findings on renal calcium excretion are less consistent. Some studies report a decrease in calcium excretion compared to either prepregnancy concentrations or to a nonlactating control group *(6,18,21)*, whereas others report no effect *(5,22)*. Because urinary calcium excretion is influenced by dietary calcium intake, it is difficult to determine whether differences in excretion are due to differences in calcium intake or renal calcium handling. However, a study of calcium kinetics in seven women, measured both while lactating and not lactating, found lower rates of urinary calcium excretion during lactation that was independent of calcium intake *(18)*. Although some studies have found serum $1,25(OH)_2D$ to be higher during lactation *(7,23,24)*, other studies find no association *(5)*. If $1,25(OH)_2D$ is elevated during lactation the mechanism is not clear because most studies find that serum intact PTH concentrations are low *(25,26)* or do not differ with lactation *(5,21)*.

Biochemical markers of bone resorption and formation are elevated during lactation *(21,25,27,28)*, and a study of calcium kinetics found that bone resorption exceeds bone formation during early lactation *(18)*. The increases in serum concentrations of bone turnover markers are not influenced by either dietary calcium or physical activity *(28)*, and many individuals have suggested that the estrogen deficiency that occurs with lactation may be responsible. Other factors that are associated with estrogen also may be important. PTH-related peptide (PTHrP), which has similar biological actions as PTH, is synthesized in mammary tissue and secreted into milk. Although its role in lactation is not clear, PTHrP concentrations have been reported to be inversely associated with estradiol concentrations and positively associated with the degree of bone loss during lactation *(29)*. However, not all studies have observed a relationship between lactation-induced bone loss and PTHrP *(30)*. PTHrP and prolactin also enhance production of mRNA for RANKL *(13)*, and serum OPG concentrations are positively associated with serum estradiol concentrations *(31)*. High PTHrP and low estrogen levels during lactation theoretically would lead to high RANKL and low OPG, resulting in increased bone resorption. The recent study by Uemura and co-workers reported a significant drop in serum OPG concentrations immediately following delivery to 1 mo postpartum *(14)*, a time when bone turnover increases significantly.

2.2.2. Changes in Bone Mineral

Significant bone loss during lactation, especially at axial bone sites, has been reported in numerous prospective studies *(3,21,25,32–36)*. Bone loss occurs early in lactation and begins to recover once menses resumes *(35,37–39)*. The amount of bone lose that occurs appears to be related to the length of amenorrhea, but on average ranges from 3 to 5%. This amount of loss is considerable given that women typically lose 1–3% per year during the postmenopausal period.

Several trials of calcium supplementation during lactation have found that the lactation-induced bone loss, breast milk calcium concentrations, and intestinal calcium absorption are not modified by the mother's calcium intake *(27,39–42)*. However, one study in adolescent mothers, using older SPA methodology, found that increased calcium intake prevented lactation-induced bone loss *(43)*. In this study, one group of adolescent mothers received routine dietary counseling (control group), whereas the other group received counseling to increase calcium intake to 1600 mg/d. However, the study was not randomized and the group with the higher calcium intake also had significantly higher intakes of calories, protein, vitamin D, and phosphorus. The control group had a significant decrease in bone mineral content (BMC; 10% loss), whereas the experimental group did not (3% loss in BMC). It is not stated whether the change in BMC differed between the two groups. Several concerns have been raised concerning the results of this study, including a higher rate of bone loss than seen in other studies and that 2 of the women had bone measurements greater than 4 SD above normal at baseline *(44,45)*. It is possible, however, that adolescent mothers may respond to lactation differently than adult mothers while consuming a low-calcium diet. Bezerra and co-workers found that among women with low calcium intakes (<500 mg/d), urinary deoxypyridinoline concentrations were higher among nonlactating, nonpregnant (NLNP) adolescents compared to NLNP adult women. Urinary deoxypyridinoline concentrations were higher in both adolescent and adult lactating mothers compared to NLNP, but there was no difference between the adolescent and adult mothers. Serum PTH and bone-specific alkaline phosphatase concentrations were higher, and urinary calcium excretion lower, in lactating adolescent vs adult mothers, findings similar to what was observed in NLNP.

A large randomized, double-blind study of calcium supplementation among women with habitually low calcium intake (<800 mg/d) found that calcium intake did not modify the bone changes observed during lactation *(46)*. Although calcium supplementation (1 g/d) augmented the bone loss among the lactating women, the effect was similar among both lactating and nonlactating women (Fig. 4). There was no difference in breast-milk calcium concentrations between those mothers receiving calcium supplements and those receiving placebo. Prentice and co-workers also conducted a randomized calcium supplementation trial among Gambian women and found no effect of calcium intake on breast-milk calcium concentrations *(40)*. These results indicate that calcium supplementation in women with habitually low calcium intake does not prevent the

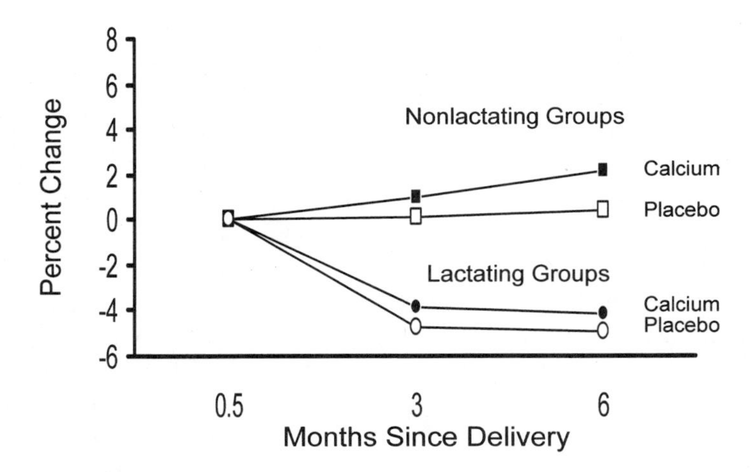

Fig. 4. Effects of calcium supplementation and lactation on the mean (+ SEM) percent change in the bone mineral density of the lumbar spine during the first 6 mo postpartum. Values are adjusted for baseline bone mineral density, height, weight, change in weight, dietary intake of calcium, and level of physical activity. $p = 0.01$ for the effect of calcium; $p < 0.001$ for the effect of lactation; and $p = 0.23$ for the interaction between calcium supplementation and lactation. (From ref. *41*. Copyright © 1997 Massachusetts Medical Society. All rights reserved.)

bone loss that is observed during lactation and does not influence breast-milk calcium concentrations.

2.3. Weaning

2.3.1. Calcium and Bone Metabolism

During weaning the mother recovers much, if not all, of the bone calcium she lost during lactation by increasing intestinal calcium absorption and decreasing renal calcium excretion (Fig. 3). Intestinal calcium absorption is higher in women who are weaning vs postpartum women who either did not breast-feed or were breast-feeding *(19)*. The greater fractional calcium absorption is associated with higher serum $1,25(OH)_2D$ concentrations that also are observed at this time *(19,47)*. Serum PTH concentrations also have been reported to be higher in weaning vs nonweaning women and is probably the stimulus for the higher $1,25(OH)_2D$ serum concentrations *(47)*. The higher PTH concentrations also may be responsible for the decrease in renal calcium excretion *(21)* that also is observed during weaning.

2.3.2. Changes in Bone Mineral

The bone loss that occurs early in lactation begins to recover once menses resumes *(35–38)*. The earlier the resumption of menses, the less bone loss that occurs during lactation and the greater the bone gain during weaning *(35)*. The length of time to complete recovery is not known, but is probably greater than

6 mo postweaning *(25,46)*. Factors that may be associated with the bone gain during weaning include closely spaced pregnancies, maternal age, and possibly calcium intake.

An early study in the 1970s found that women who had small families (0–2 children) had similar bone density as women who had larger families (7 or more children), supporting the hypothesis that closely spaced pregnancies are not detrimental to bone *(48)*. Sowers and co-workers found that women who breast-fed for at least 6 mo and became pregnant within 18 mo of initiating breast feeding had BMD recovery similar to women who breast-fed for a similar amount of time but did not become pregnant *(33)*. A more recent study using DXA found that 30 multiparous women (>5 children) who lactated for at least 6 mo per child had similar BMD as 6 nulliparous women *(49)*. Therefore, it does not appear that closely spaced pregnancies, which may interfere with bone recovery during weaning, are detrimental to long-term bone health.

Maternal age at the time of lactation also may be important. As described previously, adolescent mothers may adapt to the calcium needs differently than older mothers, especially if their calcium intake is low. Whether there are adaptation differences in adolescent vs adult mothers during weaning is not known. Hopkinson and co-workers reported that the net changes in bone mass following lactation were negatively associated with age: older mothers had less of a recovery during weaning than younger mothers *(38)*. No other studies have reported differences in bone recovery between younger vs older adult women.

A randomized study of calcium supplementation among women with habitually low calcium intake (<800 mg/d) studied the effect of calcium supplementation on bone changes during weaning *(35)*. A group of 95 lactating women and 92 non-lactating women were enrolled at approx 6 mo postpartum and were followed for 6 mo. The lactating women were breasting feeding on average 5.5 times per day when enrolled and weaned their infants during the next 2 mo. Within each group, half of the women were randomized to placebo and half to calcium supplements (1 g/d). The percent increase in spine BMD in women who were weaning and receiving calcium was of similar magnitude as the percent increase in women who were not weaning but receiving calcium (Fig. 5). These results indicate that calcium supplementation in women with habitually low calcium intakes increases BMD to the same extent in women, whether or not they are weaning. Others have recently reported similar findings *(39)*.

3. EPIDEMIOLOGICAL STUDIES OF PARITY AND LACTATION ON OSTEOPOROSIS INCIDENCE AND FRACTURES

Although there are case reports of pregnancy-induced osteoporosis *(50,51)*, this is a pathological condition and does not occur during normal pregnancy. The preceding sections focused on changes in calcium metabolism and bone mineral status that enable the mother to adapt to the calcium needs of the growing fetus and

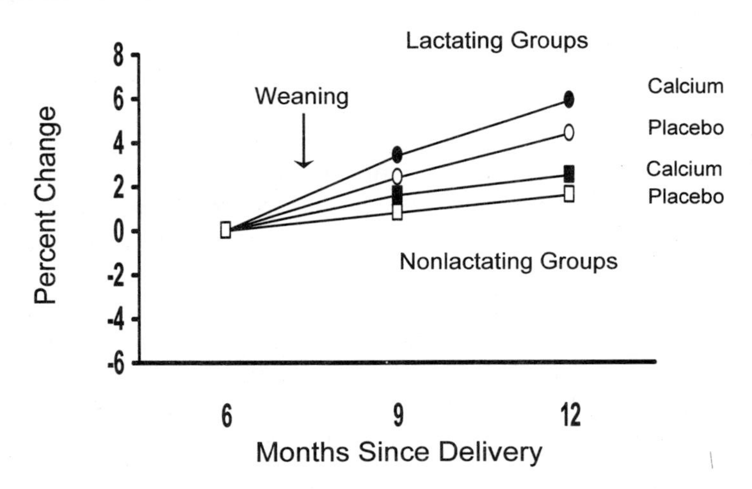

Fig. 5. Effects of calcium supplementation and lactation on the mean (+ SEM) percent change in the bone mineral density of the lumbar spine during the second 6 mo postpartum. Values are adjusted for baseline bone mineral density, height, weight, change in weight, dietary intake of calcium, and level of physical activity. $p < 0.001$ for the effect of calcium; $p < 0.001$ for the effect of weaning; and $p = 0.36$ for the interaction between calcium supplementation and weaning. The lactating women were fully breast feeding at baseline, and the arrow indicates the average time at which breast feeding was completely ended. (From ref. *41*. Copyright © 1997 Massachusetts Medical Society. All rights reserved.)

of milk production for the infant. However, there also may be consequences of reproduction on long-term bone health.

Retrospective studies have found lactation history to be associated with both increased BMD *(52–55)* and decreased BMD *(56,57)* later in life, but there are also studies that report no relationship *(48,58,59)*. Numerous methodologies for assessing bone status have been used in these studies, ranging from radiographs with aluminum wedges for estimation of bone density to DXA measurements in older women. Investigators measure BMD because it is used to determine whether someone is osteoporotic and is also considered a major predictor of future fracture risk *(60)*. Osteoporosis is currently defined by the World Health Organization as "a disease characterized by low bone mass and micro-architectural deterioration of bone tissue leading to enhanced bone fragility and a consequent increase in fracture risk" *(61)*. Osteoporosis is defined clinically by a BMD measurement that is 2.5 SD or more below the gender-specific mean value for peak bone mass. As expected, studies of lactation and fracture risk typically find similar results as those studies that have measured BMD as the outcome.

Studies on the relationship between fracture risk and lactation history have found similar risk of fracture among women who lactated compared to those who did not lactate *(62–65)*. One study found no association between fracture risk and parity

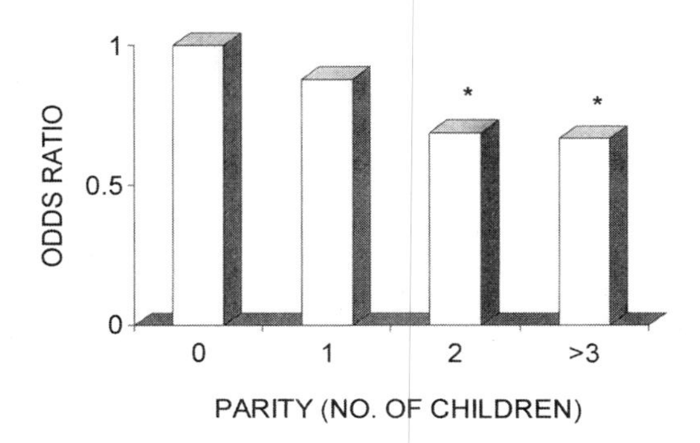

Fig. 6. The odds ratio for hip fracture risk was significantly less than 1 at parities of 2 or greater (*). Odds ratios were calculated using nulliparous women as the reference. (Data from ref. *67.*)

but did find that fracture cases had breast-fed fewer times and for fewer months than controls, even when age and parity were included as covariates in the statistical analysis *(66)*. A recent study of 1328 incident cases with hip fracture and 3312 randomly selected controls found that the hip fracture risk among all women was reduced by 10% per child and the odds ratio was statistically significant at parities of 2 or greater (Fig. 6) *(67)*. These results confirm previous findings. Hoffman and co-workers found that each birth was associated on average with a 9% reduction in the odds of hip fracture *(64)*. In neither study was the fracture risk associated with duration of lactation once parity was taken into account.

4. EFFECTS OF MATERNAL DIET ON THE NEONATE

Maternal nutrition during pregnancy or lactation may have effects on other outcomes other than maternal calcium metabolism and bone mineral status. As described above, dietary calcium intake during lactation does not influence breast milk calcium concentrations. However, there are some reports that maternal calcium intake during pregnancy may influence neonatal bone. Maternal vitamin D intake during pregnancy and lactation also may influence both neonatal vitamin D and calcium metabolism and bone mineral status.

4.1. Maternal Calcium Intake and Neonatal Bone

Maternal calcium intake during pregnancy does not influence the maternal bone changes. However, low maternal calcium intake during pregnancy may influence neonatal bone density. In one study in undernourished pregnant mothers, supplementation with 300 or 600 mg calcium/d did not increase maternal metacarpal bone density compared to unsupplemented mothers, but maternal calcium supplementation did increase neonatal bone density *(68)*. Similar results were recently reported in a large randomized trial of maternal

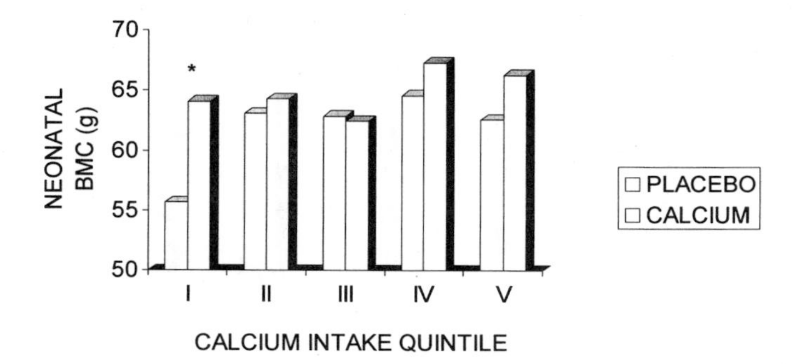

Fig. 7. Neonatal bone mineral content (BMC) in infants whose mothers received placebo (open bars) and those who received supplemental calcium (2 g/d) (dark bars) by quintile of maternal calcium intake. *Neonates of mothers in the lowest quintile of calcium intake (<600 mg/d) who were randomized to placebo had lower BMC compared to infants in the lowest quintile whose mothers were randomized to calcium supplement. (Data from ref. *69.*)

calcium supplementation for the prevention of preeclampsia *(69)*. A total of 256 women were enrolled in the randomized, double-blind, placebo-controlled trial. Newborn infants of mothers in the lowest quintile of calcium intake (<600 mg/d) who were randomized to placebo had lower BMC compared to newborns in the lowest quintile whose mothers were randomized to calcium supplement (Fig. 7). There was no difference in neonatal BMC between placebo and supplemented maternal groups in the upper quintiles of calcium intake.

4.2. Maternal Vitamin D Intake and Neonatal Vitamin D and Calcium Status and Bone Mineral

Vitamin D is generally considered a prohormone. It may be derived endogenously from the skin or supplied from exogenous sources. Endogenously, provitamin D_3 is converted to vitamin D_3 by ultraviolet radiation and subsequent thermal isomerization. Vitamin D_3 is then transformed in the liver to 25(OH)D and further hydroxylated in the kidney to 1,25$(OH)_2$D. A drop in serum concentrations of calcium or phosphorus, or increased serum PTH concentrations, facilitate this transformation to 1,25$(OH)_2$D. Serum 25(OH)D concentrations serve as the most reliable indicator of vitamin D status in the body, whereas 1,25$(OH)_2$D is the hormonally active metabolite, and its concentrations reflect the body's need for calcium and phosphorus.

Maternal vitamin D deficiency during pregnancy can have significant effects on neonatal calcium metabolism and possibly bone development. Placental transfer of 25(OH)D occurs and serum 25(OH)D concentrations of newborn infants are approx 50% of the mother's concentrations *(70,71)*. Maternal vitamin D deficiency is associated with secondary hyperparathyroidism, and vitamin D deficiency and hyperparathyroidism during pregnancy may lead to neonatal hypocalcemia or tetany *(72,73)*. In the early 1970s, Purvis and co-workers noted that the occurrence of neonatal tetany among 112 infants was inversely related to the amount of sunlight exposure the mothers had during the last trimester of pregnancy *(74)*. The authors speculated that the mothers developed hyperparathyroidism secondary to vitamin D deficiency that resulted in a transitory hypoparathyroidism in the newborn infants. This neonatal hypoparathyroidism may subsequently lead to hypocalcemia. However, Delvin and co-workers conducted a randomized trial of vitamin D supplementation during pregnancy and found that cord serum PTH concentrations were similar in infants of mothers supplemented with 1000 IU vitamin D/d during the last trimester vs those mothers not supplemented *(75)*. However, neonates of mothers who were supplemented with vitamin D had less of an increase in serum PTH and less of a decrease in serum calcium concentrations between birth (cord blood) and 4 d of age than infants of mothers not receiving vitamin D. These results indicate that adequate maternal vitamin D status during pregnancy ensures appropriate neonatal handling of calcium.

Maternal vitamin D deficiency during pregnancy also may lead to impaired fetal growth and bone development. Occurrence of vitamin D deficiency is high among Asians from the Indian subcontinent living in Britain *(73)*. A trial of vitamin D supplementation (1000 IU/d) among pregnant Asian women found that a higher percent of the infants in the placebo group were small-for-gestational-age compared to infants in the supplemented group *(76)*. In addition, infants of mothers who were not supplemented had larger fontanelles than infants of mothers who were given vitamin D, which is consistent with impaired ossification of the skull *(76)*. Low cord 25(OH)D concentrations also are observed in infants with congenital craniotabes *(77)*, and a more recent study conducted in China found that the presence of wrist ossification centers is related to cord serum 25(OH)D concentrations *(78)*. These findings are suggestive of a role of maternal vitamin D in fetal bone development.

Vitamin D deficiency leads to rickets in children. Infant formula is routinely fortified with vitamin D, but very low vitamin D concentrations have been found in human milk *(79,80)*. Most of the reported cases of rickets have been of black infants, supporting the contention that persons with dark skin have difficulty synthesizing adequate concentrations of vitamin D due to the relative inability of sunlight to penetrate darker-pigmented skin. In addition, the diet of mothers of rachitic infants appears to be low in vitamin D, and the mothers may be vitamin D deficient themselves. Specker and co-workers found that, based on conservative estimates, exclusively breast-fed infants residing in Cincinnati could maintain serum 25(OH)D concentrations above the lower limit of normal (11 ng/mL) with 2 h of sunshine exposure per week if fully clothed except for the face *(81)*. The cutoff for defining low 25(OH)D (11 ng/mL) is based on the concentration that nutritional rickets have been observed, although this may be variable. Other factors such as latitude, season, weather conditions, and use of sunscreens may affect vitamin D status. Large seasonal differences in sunlight exposure and serum 25(OH)D concentrations over the first year of life have been observed in infants followed longitudinally *(82)*. These findings indicate that infants' sunlight exposure plays a more dominant role in determining their vitamin D status than the mothers' vitamin D status or milk vitamin D.

Although maternal intake of vitamin D is correlated with breast milk vitamin D concentrations, mothers who consume 600–700 IU vitamin D/d have total vitamin D concentrations ranging from only 5 to 136 IU/L *(80)*. Supplementing lactating mothers with 1000 IU vitamin D/d during winter months did not result in increased serum 25(OH)D concentrations among the infants *(83)*. Infant serum 25(OH)D is correlated with maternal vitamin D status early in the neonatal period and is probably a result of placental vitamin D transfer and fetal stores. Beyond the neonatal period, the breast-fed infant's serum 25(OH)D concentrations are correlated neither with breast milk vitamin D concentrations nor maternal serum 25(OH)D concentrations *(81)*, and the infant is dependent on endogenous synthesis or other dietary sources for vitamin D.

5. CONCLUSION

In summary, adaptive changes occur during pregnancy, lactation, and weaning to ensure an adequate calcium supply for fetal growth, milk production, and maternal bone recovery. During pregnancy, serum $1,25(OH)_2D$ concentrations increase, resulting in increased intestinal calcium absorption. Urinary calcium excretion also increases and is probably owing to increased glomerular filtration and increased absorptive calcium load. Although bone turnover markers are elevated, it is not apparent whether there are significant changes in bone mass during pregnancy. During lactation, changes in calcium homeostasis are independent of the mother's calcium intake and are more dependent on return of ovarian function. As ovarian function returns, serum $1,25(OH)_2D$ concentrations increase, intestinal calcium absorption increases, a higher renal calcium retention persists, and biochemical markers of bone turnover return to normal concentrations as bone is regained.

Although maternal vitamin D intake during lactation is correlated with breast-milk vitamin D concentrations, milk concentrations are too low to provide the infant with sufficient vitamin D. During pregnancy, low maternal calcium intake is associated with low neonatal BMC and maternal vitamin D deficiency influences fetal bone development and neonatal calcium homeostasis. However, these effects are seen at maternal dietary intake levels well below the current recommended amounts.

REFERENCES

1. Widdowson EM. Changes in body composition during growth. In: Davis JA, Dobbings J, eds. Scientific Foundations of Pediatrics. Heinemann, London, 1981, pp. 330–342.
2. Forbes GB. Letter: Calcium accumulation by the human fetus. Pediatrics 1976; 57:976–977.
3. Laskey MA, Prentice A, Hanratty LA, et al. Bone changes after 3 mo of lactation: influence of calcium intake, breast-milk output, and vitamin D-receptor genotype. Am J Clin Nutr 1998; 67:685–692.

4. Kent GN, Price RI, Gutteridge DH, et al. The efficiency of intestinal calcium absorption is increased in late pregnancy but not in established lactation. Calcif Tissue Int 1991; 48:293–295.
5. Cross NA, Hillman LS, Allen SH, Krause GF, Vieira NE. Calcium homeostasis and bone metabolism during pregnancy, lactation, and postweaning: a longitudinal study. Am J Clin Nutr 1995; 61:514–523.
6. Ritchie LD, Fung EB, Halloran BP, et al. A longitudinal study of calcium homeostasis during human pregnancy and lactation and after resumption of menses. Am J Clin Nutr 1998; 67:693–701.
7. Specker BL, Tsang RC, Ho ML, Miller D. Effect of vegetarian diet on serum 1,25-dihydroxyvitamin D concentrations during lactation. Obstet Gynecol 1987; 70:870–874.
8. Bezerra FF, Laboissiere FP, King JC, Donangelo CM. Pregnancy and lactation affect markers of calcium and bone metabolism differently in adolescent and adult women with low calcium intakes. J Nutr 2002; 132:2183–2187.
9. Seki K, Makimura N, Mitsui C, Hirata J, Nagata I. Calcium-regulating hormones and osteocalcin levels during pregnancy: A longitudinal study. Am J Obstet Gynecol 1991; 164:1248–1252.
10. Heaney RP, Skillman TG. Calcium metabolism in normal pregnancy. J Clin Endocrinol 1971; 33:661–670.
11. Naylor KE, Iqbal P, Fledelius C, Fraser RB, Eastell R. The effect of pregnancy on bone density and bone turnover. J Bone Miner Res 2000; 15:129–137.
12. Kent GN, Price RI, Gutteridge DH, et al. Effect of pregnancy and lactation on maternal bone mass and calcium metabolism. Osteopor Int 1993; 1:S44–S47.
13. Martin TJ, Gillespie MT. Receptor activator of nuclear factor kappa B ligand (RANKL): another link between breast and bone. Trends Endocrinol Metabol 2001; 12:2–4.
14. Uemura H, Yasui T, Kiyokawa M, et al. Serum osteoprotegerin/osteoclastogenesis-inhibitory factor during pregnancy and lactation and the relationship with calcium-regulating hormones and bone turnover markers. J Endocrinol 2002; 174:353–359.
15. Ensom MHH, Liu PY, Stephenson MD. Effect of pregnancy on bone mineral density in healthy women. Obstetr Gynecol 2002; 57:99–111.
16. Tothill P, Avenell A. Errors in dual-energy X-ray absorptiometry of the lumbar spine owing to fat distribution and soft tissue thickness during weight change. Br J Radiol 1994; 67:71–75.
17. Sowers MF, Scholl T, Harris L, Jannausch M. Bone loss in adolescent and adult pregnant women. Obstet Gynecol 2000; 96:189–193.
18. Specker BL, Vieira NE, O'Brien KO, et al. Calcium kinetics in lactating women with low and high calcium intakes. Am J Clin Nutr 1994; 59:593–599.
19. Kalkwarf HJ, Specker BL, Heubi JE, Vieira NE, Yergey AL. Intestinal calcium absorption of women during lactation and after weaning. Am J Clin Nutr 1996; 63:526–531.
20. Moser-Veillon PG, Mangels AR, Vieira NE, et al. Calcium fractional absorption and metabolism assessed using stable isotopes differ between postpartum and never pregnant women. J Nutr 2001; 131:2295–2299.
21. Kent GN, Price RI, Gutteridge D, et al. Human lactation: forearm trabecular bone loss, increased bone turnover, and renal conservation of calcium and inorganic phosphate with recovery of bone mass following weaning. J Bone Miner Res 1990; 5:361–369.
22. Kalkwarf HJ, Specker BL, Ho M. Effects of calcium supplementation on calcium homeostasis and bone turnover in lactating women. J Clin Endocrinol Metab 1999; 84:464–470.
23. Kumar R, Cohen WR, Silva P, Epstein FH. Elevated 1,25-dihydroxyvitamin D plasma levels in normal human pregnancy and lactation. J Clin Invest 1979; 63:342–344.
24. Hillman L, Sateesha S, Haussler M, Wiest W, Slatopolsky E, Haddad J. Control of mineral homeostasis durig lactation: interrelationships of 25-hydroxyvitamin D, 24,25-dihydroxyvitamin D, 1,25-dihydroxyvitamin D, parathyroid hormone, calcitonin, prolactin, and estradiol. Am J Obstet Gynecol 1981; 139:471–476.

25. Affinito P, Tommaselli GA, DiCarlo C, Guida F, Nappi C. Changes in bone mineral density and calcium metabolism in breast-feeding women: a one year follow-up study. J Clin Endocrinol Metabol 1996; 81:2314–2318.
26. Krebs NF, Reidinger CJ, Robertson AD, Brenner M. Bone mineral density changes during lactation: maternal, dietary, and biochemical correlates. Am J Clin Nutr 1997; 65:1738–1746.
27. Prentice A, Jarjou LMA, Stirling DM, Buffenstein R, Fairweather-Tait S. Biochemical markers of calcium and bone metabolism during 18 months of lactation in Gambian women accustomed to a low calcium intake and in those consuming a calcium supplement. J Clin Endocrinol Metab 1998; 83:1059–1066.
28. Sowers MF, Eyre D, Hollis BW, et al. Biochemical markers of bone turnover in lactating and nonlactating postpartum women. J Clin Endocrinol Metab 1995; 80:2210–2216.
29. Sowers MF, Hollis BW, Shapiro B, et al. Elevated parathyroid hormone-related peptide associated with lactation and bone density loss. JAMA 1996; 276:549–554.
30. Dobnig H, Kainer F, Stepan V, et al. Elevated parathyroid hormone-related peptide levels after human gestation: relationship to changes in bone and mineral metabolism. J Clin Endocrinol Metab 1995; 80:3699–3707.
31. Szulc P, Hofbauer LC, Heufelder AE, Roth S, Delmas PD. Osteoprotegerin serum levels in men: correlation with age, estrogen, and testosterone status. J Clin Endocrinol Metab 2001; 86:3162–3165.
32. Hayslip CC, Dlein TA, Wray L, Duncan WE. The effects of lactation on bone mineral content in healthy postpartum women. Obstet Gynecol 1989; 73:588–592.
33. Sowers MF, Randolph J, Shapiro B, Jannausch M. A prospective study of bone density and pregnancy after an extended period of lactation with bone loss. Obstet Gynecol 1995; 85:285–289.
34. Sowers MF, Corton G, Shapiro B, et al. Changes in bone density with lactation. JAMA 1993; 269:3130–3135.
35. Kalkwarf HJ, Specker BL. Bone mineral loss during lactation and recovery during weaning. Obstet Gynecol 1995; 86:26–32.
36. Lopez JM, Gonzalez G, Reyes V, Campino C, Diaz S. Bone turnover and density in healthy women during breastfeeding and after weaning. Osteopor Int 1996; 6:153–159.
37. Kolthoff N, Eiken P, Kristensen B, Nielsen SP. Bone mineral changes during pregnancy and lactation: a longitudinal cohort study. Clin Sci 1998; 94:405–412.
38. Hopkinson JM, Butte NF, Ellis K, Smith EO. Lactation delays postpartum bone mineral accretion and temporarily alters its regional distribution in women. J Nutr 2000; 130:777–783.
39. Polatti F, Capuzzo E, Viazzo F, Collconi R, Klersy C. Bone mineral changes during and after lactation. Obstet Gynecol 1999; 94:52–56.
40. Prentice A, Jarjou LM, Cole TJ, Stirling DM, Dibba B, Fairweather-Tait S. Calcium requirements of lactating Gambian mothers: effects of a calcium supplement on breast-milk calcium concentration, maternal bone mineral content, and urinary calcium excretion. Am J Clin Nutr 1995; 62:58–67.
41. Kalkwarf HJ, Specker BL, Bianchi D, Ranz J. Randomized trial of calcium supplementation on bone changes during lactation and after weaning. N Engl J Med 1997; 337:523–528.
42. Cross NA, Hillman LS, Allen SH, Krasue GF. Changes in BMD and markers of bone remodeling during lactation and postweaning in women consuming high amounts of calcium. J Bone Mineral Res 1995; 10:1312–1320.
43. Chan GM, McMurry M, Westover K, Englebert-Fenton K, Thoman R. Effects of increased dietary calcium intake upon the calcium and bone mineral status of lactating adolescent and adult women. Am J Clin Nutr 1987; 46:319–323.
44. Greer FR, Garn SM. Loss of bone mineral content in lactating adolescents. J Pediatr 1982; 101:718–719.
45. Cunningham AS, Mazess RB. Bone mineral loss in lactating adolescents. J Pediatr 1983; 101:338–339.

46. Kalkwarf HJ, Specker BL, Bianchi DC, Ranz J, Ho M. The effect of calcium supplementation on bone density during lactation and after weaning. N Engl J Med 1997; 337:523–528.
47. Specker BL, Tsang RC, Ho ML. Changes in calcium homeostasis over the first year postpartum: effect of lactation and weaning. Obstetr Gynecol 1991; 78:56–62.
48. Walker ARP, Richardson B, Walker F. The influence of numerous pregnancies and lactations on bone dimensions in South African Bantu and Caucasian mothers. Clin Sci 1972; 42:189–196.
49. Henderson PH, Sower M, Kutzko KE, Jannausch ML. Bone mineral density in grand multiparous women with extended lactation. Am J Obstet Gynecol 2000; 182:1371–1377.
50. Khastgir G, Studd JWW, King H, et al. Changes in bone density and biochemical markers of bone turnover in pregnancy-associated osteoporosis. Br J Obstetr Gynaecol 1996; 103:716–718.
51. Gruber HE, Gutteridge DH, Baylink DJ. Osteoporosis associated with pregnancy and lactation: bone biopsy and skeletal features in three patients. Metab Bone Dis & Rel Res 1984; 5:159–165.
52. Aloia JF, Vaswani AN, Yeh JK, Ross P, Ellis K, Cohn SH. Determinants of bone mass in postmeopausal women. Arch Intern Med 1983; 143:1700–1704.
53. Feldblum PJ, Zhang J, Rich LE, Forney JA, Talmage RV. Lactation history and bone mineral density among perimenopausal women. Epidemiology 1992; 3:527–531.
54. Hreshchyshyn MM, Hopkins A, Zylstra S, Anbar M. Associations of parity, breast-feeding, and birth control pills with lumbar spine and femoral neck bone densities. Am J Obstet Gynecol 1988; 159:318–322.
55. Melton III LJ, Bryant SC, Wahner HW, et al. Influence of breastfeeding and other reproductive factors on bone mass later in life. Osteopor Int 1993; 3:76–83.
56. Lissner L, Bengtsson C, Hansson T. Bone mineral content in relation to lactation history in pre- and postmeopausal women. Calcif Tissue Int 1991; 48:319–325.
57. Wardlaw GM, Pke AM. The effect of lactation on peak adult shaft and ultra-distal forearm bone mass in women. Am J Clin Nutr 1986; 44:283–286.
58. Koetting CA, Wardlaw GM. Wrist, spine, and hip bone density in women with variable histories of lactation. Am J Clin Nutr 1988; 48:1479–1481.
59. Wasnich R, Yano K, Vogel J. Postmenopausal bone loss at multiple skeletal sites: relationship to estrogen use. J Chron Dis 1983; 36:781–790.
60. Cummings SR, Black DM, Nevitt MC, et al. The Study of Osteoporotic Fractures Research Group: bone density at various sites for prediction of hip fractures. Lancet 1993; 341:72–75.
61. Assessment of Fracture Risk and Its Application to Screening for Postmenopausal Osteoporosis. World Health Organization Technical Report Series 843. World Health Organization, Geneva, 1994.
62. Alderman BW, Weiss NS, Daling JR, Ure CL, Ballard JH. Reproductive history and postmenopausal risk of hip and forearm fracture. Am J Epidemiol 1986; 124:262–267.
63. Cummings SR, Nevitt MC, Browner WS, et al. Risk factors for hip fracture in white women. N Engl J Med 1995; 332:767–773.
64. Hoffman S, Grisso JA, Kelsey JL, Gammon DM, O'Brien LA. Parity, lactation and hip fracture. Osteopor Int 1993; 3:171–176.
65. Cumming RG, Klineberg RJ. Breastfeeding and other reproductive factors and the risk of hip fractures in elderly women. Int J Epidemiol 1993; 22:684–691.
66. Kreiger N, Kelsey JL, Holford TR, O'Connor T. An epidemiologic study of hip fracture in postmenopausal women. Am J Epidemiol 1982; 116:141–148.
67. Michaelsson K, Baron JA, Farahmand BY, Ljunghall S. Influence of parity and lactation on hip fracture risk. Am J Epidemiol 2001; 153:1166–1172.
68. Raman L, Rajalakshmi K, Krishnamachari KAVR, Sastry JG. Effect of calcium supplementation to undernourished mothers during pregnancy on the bone density of the neonates. Am J Clin Nutr 1978; 31:466–469.
69. Koo W, Walters J, Esterlitz J, Levine R, Bush A, Sibai B. Maternal calicum supplementation and fetal bone mineralization. Obstet Gynecol 1999; 94:577–582.

70. Markestad T, Aksnes L, Ulstein M, Aarskog D. 25-Hydroxyvitamin D and 1,25-dihydroxyvitamin D of D2 and D3 origin in maternal and umbilical cord serum after vitamin D2 supplementation in human pregnancy. Am J Clin Nutr 1984; 40:1057–1063.
71. Gertner JH, Glassman MC, Coustan DR, Goodman DB. Fetomaternal vitamin D relationships at term. J Pediatr 1980; 97:637–640.
72. Daaboul J, Sanderson S, Kristensen K, Kitson H. Vitamin D deficiency in pregnant and breast-feeding women and their infants. J Perinatol 1997; 17:10–14.
73. Okonofua F, Menon RK, Houlder S, et al. Parathyroid hormone and neonatal calcium homeostasis: evidence for secondary hyperparathyroidism in the Asian neonate. Metabolism 1986; 35:803–806.
74. Purvis RJ, MacKay GS, Cockburn F, et al. Enamel hypoplasia of the teeth associated with neonatal tetany: a manifestation of maternal vitamin D deficiency. Lancet 1973; ii:811–814.
75. Delvin EE, Salle BL, Glorieux FH, Adeleine P, David LS. Vitamin D supplementation during pregnancy: effect on neonatal calcium homeostasis. J Pediatr 1986; 109:328–334.
76. Brooke DG, Brown IRF, Bone CDM, et al. Vitamin D supplements in pegnant Asian women: effects on calcium status and fetal growth. Br Med J 1980; 280:751–754.
77. Reif S, Katzir Y, Eisenberg Z, Weisman Y. Serum 25-hydroxyvitamin D levels in congenital craniotabes. Acta Paediatr Scand 1988; 77:167–168.
78. Specker B, Ho M, Oestreich A, et al. Prospective study of vitamin D supplementation and rickets in China. J Pediatr 1992; 120:733–739.
79. Hollis BW, Roos BA, Draper HH, Lambert PW. Vitamin D and its metabolites in human and bovine milk. J Nutr 1981; 111:1240–1248.
80. Specker BL, Tsang RC, Hollis BW. Effect of race and diet on human milk vitamin D and 25-hydroxyvitamin D. Am J Dis Child 1985; 139:1134–1137.
81. Specker B, Valanis B, Hertzberg V, Edwards N, Tsang R. Sunshine exposure and serum 25-hydroxyvitamin D concentrations in exclusively breast-fed infants. J Pediatr 1985; 107:372.
82. Specker B, Tsang R. Cyclical serum 25-hydroxyvitamin D concentrations paralleling sunshine exposure in exclusively breast-fed infants. J Pediatr 1987; 110:744–747.
83. Ala-Houhala M. 25-Hydroxyvitamin D levels during breast-feeding with or without maternal or infantile supplementation of vitamin D. J Pediatr Gastro Nutr 1985; 4:220–226.

10 Nutritional Requirements for Fetal and Neonatal Bone Health and Development

Stephanie A. Atkinson

1. INTRODUCTION

The trajectory for bone mineral deposition through the periods of fetal and infant development is becoming better defined as more data emerge on whole-body and regional analysis of bone mineral content (BMC) using dual-energy X-ray absorptiometry (DXA). Previously, knowledge of fetal accretion of nutrients was based on analysis of the chemical composition of fetal and neonatal bodies *(1),* and more recently for minerals by neutron activation analysis *(2).* Accretion of nutrients after birth was derived primarily from metabolic balance studies that yielded retention of mineral. Beginning in the early 1980s, BMC of the radius or humerus using single-photon absorptiometry (SPA) was possible to quantitate in small premature infants, but it was not until the early 1990s that the measurement of whole-body BMC by DXA was sufficiently sophisticated to be validated as a tool *(3,4)* for estimating bone, lean and fat mass in small infants. All three sources of information—fetal tissue analysis, metabolic balance studies, and estimation of BMC by DXA—have contributed to the determination of recommended intakes of minerals and vitamin D to support healthy bone growth. In addition to nutrients, early skeletal development is dependent on genetics, an appropriate endocrine environment including parathyroid hormone (PTH), 1,25-dihydroxycholecalciferol, growth hormone (GH), insulin and insulin-like growth factors (IGFs), as well as physical activity. The integration of these factors is important to skeletal development *per se* and also the response of such factors to influences in the environment of the growing fetus and child.

2. NUTRIENT NEEDS FOR FETAL SKELETAL DEVELOPMENT

Deposition of the key essential minerals for fetal bone development—calcium (Ca), phosphorus (P), and magnesium (Mg)—occurs predominantly in the third

From: *Nutrition and Bone Health*
Edited by: M. F. Holick and B. Dawson-Hughes © Humana Press Inc., Totowa, NJ

trimester of pregnancy, during which approx 80% of the mineral of the infant born at term is accrued in the skeleton *(5)*. A maternal source of vitamin D appears to be important for the transplacental transfer of Ca (and presumably P) to the fetus, and the synthesis of the active metabolite (1,25-dihydroxyvitamin D) that functions in this transport role appears to occur not only in maternal tissues but also in placental tissue and the fetal kidney (reviewed in ref. *6*). After birth, infants are dependent on a normal circulating concentration of 25-hydroxyvitamin D to produce the active metabolite.

2.1. Estimates of Intrauterine Accretion of Bone Minerals

Biochemical analysis of body composition of aborted fetuses or infants dying shortly after birth have formed the basis of estimations of accretion of nutrients in the third trimester of pregnancy *(1,7)*. From the compositional data, and using weight growth curves of more recent studies, mineral accretion over the period of 24–36 wk of gestation was estimated to be 90–120 mg/kg fetal body weight/d for Ca, and possibly higher from 36 to 38 wk of gestation, which is the time of peak fetal accretion of bone mineral *(7)*. During this intrauterine growth period, P is deposited in amounts of about 60–75 mg/kg/d and Mg accretion is deposited in amounts of about 2.5–3.4 mg/kg/d *(7)*. The exact amount of maternal vitamin D intake required to optimize fetal bone accretion is unknown. The 25-hydroxyvitamin D accumulates in the fetus and serum concentrations at term birth will reflect maternal vitamin D status, race (African-American being lower than Caucasian), and season of the year *(8)*. In preterm infants, cord blood concentration of this vitamin D metabolite is variable, with some lower and some higher than the normal reference range (50–215 pmol/L) *(9)*.

2.2. Accretion of Bone As a Measure of Calcium Needs During Fetal Life

In vivo intrauterine estimates of fetal bone mineral accretion have not been measured even with DXA technology, due to the small exposure to radiation (of the order of 0.3 mrem per scan). Instead estimates of BMC at sequential gestational ages have been inferred from cross-sectional measures in preterm infants taken shortly after birth using DXA. For example, in a cross-sectional study of German infants born at a gestational age of 25–42 wk, lumbar spine BMC increased 5.5-fold and mid-humerus BMC increased 2.4-fold *(10)*. In a similar cross-sectional study of U.S. infants *(11)*, accretion of Ca *in utero* in the third trimester of pregnancy was estimated to be 23.2 ± 35 g (mean ± SD), assuming that bone mineral contains 32.2% Ca. Reference intrauterine values for BMC in Belgian infants were converted to Ca content of bone and found to compare favorably with the Ca content measured by chemical analysis or with neutron activation analysis *(12)*.

2.3. Factors Influencing Fetal Bone Development

Bone mineral accretion during fetal life is influenced by both genetics (measured as parental size at birth) and environmental factors such as the diet and lifestyle habits of the mother. The genetic determinants of BMC in premature infants were recently explored in Finnish infants *(13)*. Associations were observed between specific vitamin D receptor genotypes and BMC at single sites at 3 mo of age and at 9–11 yr. Because of the small number of infants studied (n = 37), no definitive conclusions with respect to genotype and BMC could be drawn.

Maternal dietary intake of Ca during pregnancy was demonstrated to influence fetal bone mineral accretion. For women with low dietary Ca intakes (<600 mg/d), a supplement of 2 g of Ca from before 22 wk of gestation resulted in higher BMC of the total body in infants born at term *(14)*.

Other maternal lifestyle factors such as smoking and physical activity were identified as determinants of fetal accretion of bone in a population-based cohort study. Infants born at term whose mothers smoked during pregnancy had significantly lower (by 11%) whole-body BMC than infants of nonsmokers *(15)*. Maternal thinness as reflected in low triceps skin fold thickness and more frequent and vigorous activity in late pregnancy were also associated with a lower BMC in the infants *(15)*. These maternal predictors of newborn bone mass were independent of placental weight and thus not likely a result of reduced placental delivery of nutrients. However, intrauterine growth restriction leading to small infant size at birth may also be an important marker of ultimate skeletal development, since birth weight is a recognized predictor of bone mass in later life *(16,17)*.

It has been hypothesized that variations in fetal skeletal development related to environmental influences via the mother occur as a result of perturbations in a maternal/fetal hormone regulatory pathway such as the GH–IGF axis *(18)*. In infants of diabetic mothers, whole-body BMC was greater at birth than normal weight infants with an average z-score of 1.3 ± 0.9 *(19)*. The authors proposed that since insulin stimulates production of IGF-1 in the liver, which in turn promotes bone collagen and matrix synthesis, the hyperinsulinemic state in infants of diabetic mothers contributed to an enhanced bone mineralization *in utero (19)*. Such observations emphasize the relevance of the interrelationships between insulin, IGF-1, and bone formation.

3. NUTRIENT NEEDS FOR SKELETAL DEVELOPMENT IN TERM-BORN INFANTS

3.1. Dietary Reference Intake (DRI) for Calcium, Phosphorus, and Vitamin D in Infants

The most recent nutrient-based recommendations for mineral and vitamin D intakes are those of the Food and Nutrition Board, Institute of Medicine (IOM) *(20)*, which are intended for use by Americans and Canadians. In the 1997 report on calcium and other bone nutrients, the recommended intakes or Adequate Intakes (AI) for minerals were based on observed intakes of breast-fed infants. The value for an AI for infants from birth to 12 mo was derived

Table 1
Comparison of Recommendations by Various Countries for Mineral and Vitamin D Intakes of Term and Premature Infants

	Term Infants 0–6/7–12 mo (mg/d) (20)	*Preterm Infants—US (mg/kg.d) (31)*	*Preterm Infants—Europe (mg/kg.d) (32)*	*Preterm Infants—Canada (mg/kg.d) (33)*
Calcium	210/250	200–230	84–168	160–220
Phosphorus	100/255	110–123	60–104	78–118
Magnesium	30/75	–	7.2–14.4	4.8–9.6
Vitamin D (IU/d)	200	400	800–1600	400

Note. See text for details.

from the intake of Ca, P, and Mg from observed volumes of breast milk intake (for 0–6 mo) or breast milk in addition to published values for intake of solid foods (for 7–12 mo) *(20)*. The recommended adequate intakes for term infants are summarized in Table 1 (in the left-hand column for 0–6 and 7–12 mo of age). For vitamin D, it was determined that a dietary (or supplement) intake of 100 IU would likely prevent rickets but not maintain normal circulating concentrations of 25-hydroxyvitamin D. Thus, assuming that most infants obtain minimal or no vitamin D via exposure to sunlight, an AI of 200 IU (5 μg) was established. This amount of vitamin D was also recommended for infants of 7–12 mo assuming that most infants can maintain normal vitamin D status with this intake. Recently, the AI for vitamin D set by the IOM was adopted by the American Academy of Pediatrics (AAP) *(21)*. The AAP recommended a minimum intake of 200 IU vitamin D per day for all infants beginning during the first 2 mo of life *(21)*, in recognition of the risk in vitamin D deficiency rickets in the United States, especially among infants who are breast-fed for a number of months without vitamin D supplementation.

3.2. Bone Mineral Content in Breast- and Formula-Fed Infants

In term infants, BMC of the distal radius as measured by SPA doubles over the first year of life and is similar between infants fed human milk, cow milk or soy-based formula *(22)*. The latter observation was not consistent across studies, with some evidence for lower radial BMC in infants fed human milk compared to infant formulas (Fig. 1a). In extremely low-birth-weight (ELBW) prematurely born infants, BMC of the distal radius increased threefold or more from about 35 wk gestational age to 1 yr corrected age (Fig. 1b). Compared to reported radial BMC values in term infants who were formula-fed *(23)*, the ELBW infants demonstrated catch-up in BMC in those who had received enriched formula to 3 mo *(24)*, reaching the mean value for term-born infants by 6 mo (Fig. 1b).

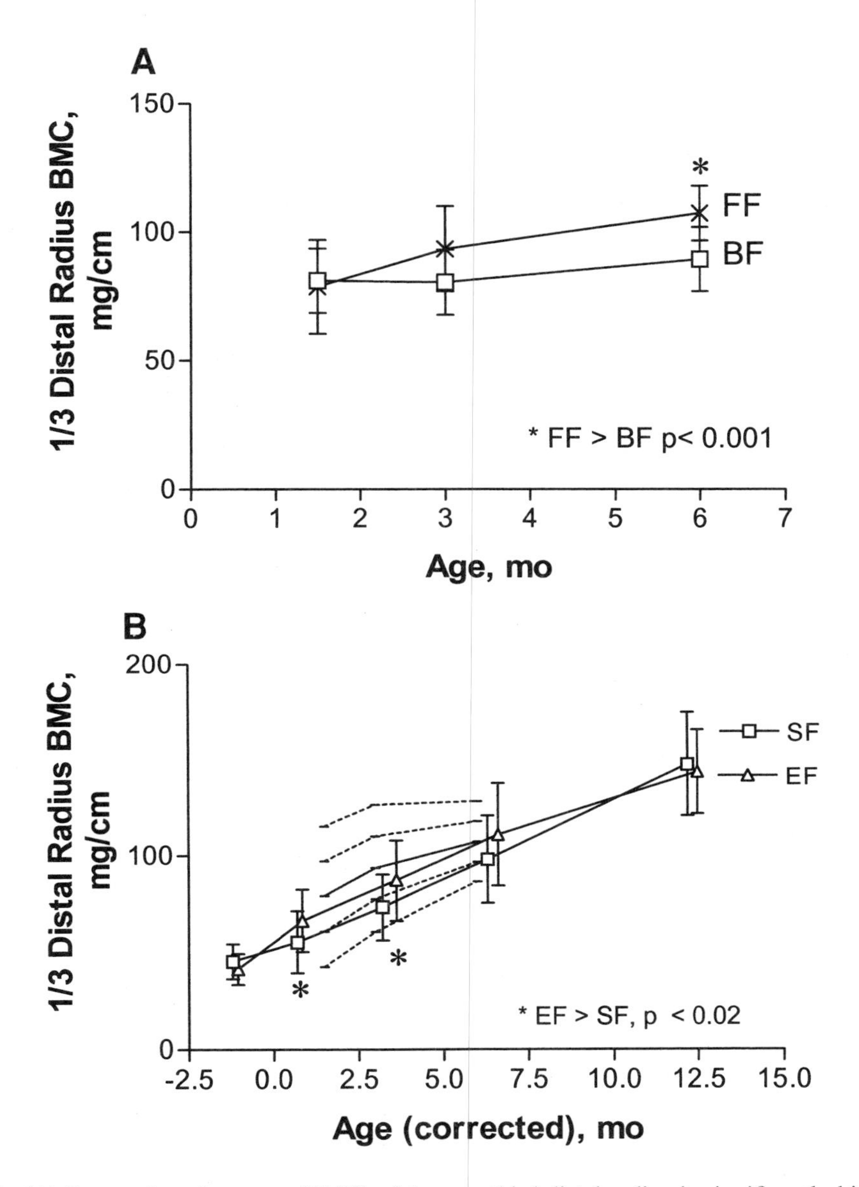

Fig. 1. **(A)** Bone mineral content (BMC) of the one-third distal radius is significantly higher ($p < 0.001$) in term-born infants at 6 mo of age if they were formula-fed (FF) than if breast-fed (BF) with supplemental vitamin D (400 IU/d) *(22)*. The gain in BMC over the study period of 4.5 mo was 36% for FF and 17% for BF. **(B)** BMC of the one-third distal radius in prematurely born infants of extremely low birth weight (n = 56, birth weight = 866 ± 16 g, gestational age = 24 ± 1.5 wk) fed formula with standard protein and mineral content as in term infant formula (SF) or nutrient-enriched formula (EF) to 3 mo corrected age *(24)* and then followed to 12 mo corrected age while receiving standard infant formula and foods as selected by parents (Brunton, Saigal, and Atkinson, unpublished). The infants fed EF had significantly higher BMC at 3 mo (p = 0.003) and 6 mo (p = 0.02), but by 12 mo corrected age the BMC was similar in the two diet groups. The data from term infants fed standard formula *(22)* are shown as mean (in solid line) ± 2 SD (in hatched lines) as a reference population. Radial BMC in the EF preterm infants reached the mean value for term infants by 6 mo corrected age.

While measure of BMC by SPA reflects cortical bone, DXA technology provides measures of whole-body and regional analysis of BMC, thus allowing for assessment of the more metabolically active trabecular bone such as in the lumbar spine, or in combination with cortical bone in the whole body. Since whole-body BMC does not correlate with BMC by SPA *(25)*, it is not surprising that observed responses in infants in BMC measures using DXA have diverged from previous observations of BMC measured by SPA. For instance, longitudinal measures of whole-body BMC in term-born infants demonstrated that BMC increases by 2.5- to 3.6-fold over the first year of life *(26,27)*, a greater velocity than observed for radial BMC *(23)*. Based on reports of DXA measures of body composition in term infants, Koo *(28)* estimated that BMC increased by 400% during infancy while body weight increased by 330%. However, body weight appears to be the strongest determinant of BMC in growing healthy infants *(11,12,29)*. The independent influence of diet—such as vitamin D or mineral intake or dietary practice of breast feeding compared to formula feeding—on bone mineral accretion in infancy has been addressed in only a few studies.

Whole-body BMC in breast-fed term infants has consistently been observed to be lower than in formula-fed infants *(27,30)*. In one study *(27)*, a significantly lower BMC was observed at 3 and 6 mo of age in breast-fed infants (average milk Ca and P of 300 and 150 mg/L) compared to those fed formula containing moderate amounts of Ca and P (510 and 390 mg/L) but not low Ca and P (430 and 220 mg/L). However, from 6 to 12 mo of age, the previously breast-fed infants were fed formula with moderate or high (1350 mg Ca and 900 mg P/L) mineral content and demonstrated a greater rate of accretion of BMC (81 ± 161 mg/6 mo) than the infants fed formula in the first 6 mo (73 ± 15 and 71 ± 15 mg/6 mo). As a result, by 12 mo of age there were no differences in whole-body BMC between feeding groups. If infants were breast-fed beyond 6 mo, a lower whole-body BMC than formula-fed infants was maintained to 12 mo of age, but such differences were not apparent by 2 yr of age *(30)*. A lower intake of protein and macrominerals from exclusive feeding with breast milk compared to standard infant formulas is the likely explanation for observed variations in growth patterns in early life *(30)*. Taken together, the available studies indicate that slower accretion of bone mass in early life may represent the biological norm and is not predictive of lower ultimate bone mass in early childhood. Whether differences in patterns of skeletal accretion of bone mineral influence metabolic programming of growth in later life and final adult bone mass must await future research.

4. NUTRIENT NEEDS FOR SKELETAL DEVELOPMENT IN PREMATURELY BORN INFANTS

4.1. Approaches to Setting Nutrient Requirements for Bone Growth

As noted under Subheading 3, estimates of nutrient intake of breast-fed infants represent the "gold standard" for setting recommended intakes for term infants.

This paradigm is not applicable to preterm infants, due to the greater nutrient intakes required to support the more rapid velocity of growth as patterned after intrauterine deposition of nutrients. Recommendations for intakes of Ca, P, Mg, and vitamin D in premature infants are greater than for term infants (on a body-weight basis) and vary considerably among international sources (Table 1). These recommendations represent the goals for the stable growing period when a reasonable target for oral intake of Ca, P, and Mg is to achieve retention of mineral that is similar to intrauterine accretion.

The recommendations for nutrient intakes of preterm infants were initially based on intrauterine accretion of nutrients as derived from body composition studies of aborted fetuses or infants dying soon after birth *(1,5)*. Using such data, Ziegler et al. (1981) *(34)* were the first to develop a factorial approach to setting recommended intakes by summing the accretion of nutrient, loss of nutrient via the feces and skin, and correcting this derived value for efficiency of absorption. Over the past two decades the mineral needs of preterm infants have been estimated from metabolic balance studies to determine accretion of nutrients, specific biochemical indices of endocrine or bone mineral status, and the measurement of bone mineral content as a functional outcome.

4.1.1. Calcium and Phosphorus

Unfortified mother's milk provides only about 6.4 mmol (236 mg) Ca/L and 4.5 mmol (140 mg) P/L and thus intakes of mineral could not exceed 1 mmol Ca and 0.77 mmol P/kg/d. Such amounts of Ca and P are considered inadequate, since preterm infants fed unfortified mother's milk in early life develop hypophosphatemia, with or without hypercalcemia *(35)* and hypercalciuria *(36)*. While normal serum and urinary biochemistry can be attained on Ca intakes as low as 2.5 mmol (107 mg)/kg/d and as high as 5.5 mmol (220 mg) /kg/d *(37)*, intrauterine accretion of Ca will only be achieved at the higher intakes. Consequently, for preterm infants, fortification of mother's milk with Ca and P is a widely accepted practice, although without consensus as to the amount or form of mineral salt that is optimal. Bovine-based human milk fortifiers or single supplements of Ca and P are both used and provide up to 3 mmol (120 mg) Ca and 2.4 mmol (75 mg) P/kg body weight/day.

Measurement of nutrient balance in infants offers an estimate of the amount of nutrient being deposited in the body as well as the efficiency or bioavailability of varying amounts and sources of mineral salts that are added as fortifiers to human milk or to preterm formulas. From such metabolic studies it is known that absorption of Ca from human milk (when expressed as percent of intake absorbed), with Ca and P salts added (73–79%), is usually somewhat greater than from human milk fortified with a multinutrient supplement (50–82%), or from preterm formula (40–72%) *(38)*. Absorption of Ca varies with the type of mineral salt, and the amount and quality of fat, protein, and phosphorus in the diet *(39,40)*. Phosphorus is absorbed more efficiently than Ca, usually at least 90% of intake,

Table 2
Recommended Reasonable Range of Intakes of Calcium (Ca), Phosphorus (P), Magnesium (Mg) and Vitamin D in Preterm Infants *(38)*

	Ca	*P*	*Mg*	*Vitamin D*[a]
Enteral:				
mmol/kg/d	3.0–5.0[b]	2.0–3.5	0.3–0.4	μg/d 5–23
mg/kg/d	120–200[b]	70–120	7.2–9.6	IU/d 200–1000

[a] The recommended amounts of vitamin D represent the total daily intake. On a body weight basis the minimum recommended oral intake is 2.23 μg (90 IU)/kg/d.

[b] For enteral nutrition the ratio Ca:P should be maintained between 1.4 and 1.6 M (1.7–2.0 mg).

and is not influenced greatly by composition of the feeding. The most recent estimates for recommended intakes of Ca and P in the neonatal period *(38)* are summarized in Table 2.

4.1.2. Vitamin D

An exogenous source of vitamin D may be required in the growing preterm infant to support optimal accretion of bone mineral, since birth at an early gestation would preclude sufficient transplacental transfer of 25-hydroxyvitamin D [25(OH)D] from the mother. This concept has been challenged based on evidence that the absorption and retention of Ca and P are independent of the availability of the active hormone of vitamin D in early neonatal life *(41)*. If provided with supplemental vitamin D, preterm neonates appear to have the metabolic maturity to absorb and hydroxylate vitamin D to produce 25(OH)D and the active metabolite 1,25-dihydroxyvitamin D [1,25$(OH)_2$D] from the first day of life *(6)*. Thus, the postnatal age at which a dependency on vitamin D-regulated transport of Ca and P begins is uncertain.

In North America and Australia, supplementation with a single daily dose of vitamin D of 400 IU/d is standard practice. This recommendation is supported by the findings from a clinical trial in preterm infants (birth weight = 1200 g) randomized to daily vitamin D intakes of 200 IU (90 IU/kg), 400 (180 IU/kg), or 800 IU (360 IU/kg) from 16 d of age for 1 mo *(42)*. All groups maintained normal vitamin D status and no radiological differences were observed between groups. In a similar study in Finnish preterm infants (n = 39, birth weight = 735–2230 g) *(9)*, plasma 25(OH)D was maintained from early neonatal life to 3 mo with administration of supplemental vitamin D in a dose of 200 IU/kg/d to a total of 400 IU/d. No benefit to vitamin D status or forearm BMC was observed at a higher dose of 960 IU/d *(9)*. In Europe, the Nutrition Committee for the European Society of Pediatric Gastroenterology, Hepatology and Nutrition *(32)* recommends a vitamin D intake up to 800–1600 IU/d (Table 1), although this recommendation is thought by some *(9)* to be too high, especially when extra minerals are provided. Based on the noted studies, the minimum recommended oral intake of vitamin D is 90 IU/kg/d and the maximum amount as 1000 IU/day (Table 2).

4.1.3. Magnesium

Of total body Mg, 50–60% is found in bone or other cartilaginous tissues. Magnesium functions as a cofactor in many biological processes, including the release of PTH, thus making it part of the calciotropic hormone axis. Although Mg absorption in preterm infants is generally greater from human milk than bovine-based formulas *(35,43)*, the net amount absorbed is marginally adequate to meet intrauterine accretion *(36,39,44)*. Accordingly, most human milk fortifiers contain Mg. The range of recommended intake is set at about 7–10 mg/kg/d (Table 2).

4.2. Nutrition and Bone Mineralization in the Neonatal Period

In ELBW premature infants, BMC of the distal radius increased threefold or more (Fig. 1b) from term to 1 yr corrected age, although it did not reach the reported mean value for term-born infants by 12 mo of age in our study (Fig. 1b) or that of Greer and McCormick *(45)*. Whether the observations in a single bone site representing primarily cortical bone reflected bone mineral status in the whole body in preterm infants was unknown until DXA technology became available.

The impact of mineral fortification of mother's milk or specialized formulas for preterm infants on accretion of bone mineral measured by bone densitometry has been inconsistent. In a systematic review *(46)*, the effectiveness of multicomponent fortification of human milk on the promotion of growth and bone mineralization in preterm infants was evaluated. Ten trials (*N* = 596 infants) were included in the analysis, which represented random or quasi-random allocation to supplementation of human milk with multiple nutrients or no supplementation within a nursery setting. Unfortunately, only half of the reported trials measured BMC, and there was inconsistency in the use of radial or whole-body measures of bone. The main results of the review showed that BMC was increased by nutrient fortification of formula. However, in four of five studies in which BMC was measured, there were no statistical differences between control and treatment groups. A meta-analysis of BMC measures from the 5 studies (in which one study contributed 59 of 79 infants) showed a positive effect on BMC of fortification of mother's milk with a human milk fortifier (weighted mean difference [WMD] 8.3 mg/cm, 95% confidence interval (CI) 3.8–12.8 mg/cm). Plasma alkaline phosphatase, a marker of bone turnover, was not different between treatment groups.

Whether rapid accretion of bone in early life in preterm infants is of positive benefit to ultimate skeletal BMC has been questioned. One proposition is that lower intakes of Ca and P in early life may be important in the programming of subsequent regulation of bone mineralization *(47)*. This is based on observations that the feeding of unfortified mother's milk to low birth weight (LBW) infants in early life was associated with a greater bone mineral content at 5 yr of age compared to infants who had been fed formula with higher mineral content *(47)*.

The short-term impact of early nutrition (while in hospital) on whole-body measures of BMC at term-adjusted age have been evaluated in relation to feeding of fortified mother's milk compared to preterm formulas. In some studies in infants less than 1500 g birth weight and gestational age less than 32 wk, no differences in whole-body BMC were observed at or near term age between those fed fortified mother's milk or preterm formula in hospital *(48–50)*. However, the mean BMC of the preterm infants was lower than –1 SD from the mean value for the term infants. In preterm infants of larger birth weight (<1750 g), feeding of preterm formula compared to fortified human milk for 3 wk resulted in higher whole-body BMC at 37 wk gestational age *(51)*. However, BMC related to bone area or to body weight were similar between the feeding groups and represented about a –2 *z*-score compared to term-born reference infants *(51)*. Taken together, the reported studies provide evidence that bone mineral deposition equivalent to that which occurs *in utero* in the third trimester of pregnancy is not achieved by term age in infants born prematurely, despite many advances in the delivery of nutrients in early life in this population.

Some of the observed differences in BMC at term age between reported studies may relate to differences in birth size, particularly if the population studied included infants of extremely LBW or small for gestational age. As published previously *(52)*, mean values for whole-body BMC at term-corrected age for preterm infants of varying birth weight and in-hospital feeding regimens fall below (< –1.5 SD) (see values at 0 [term] age in Fig. 2) that of term reference infants. Mean absolute BMC for preterm infants who were appropriate for gestational age (AGA) at term-corrected age was 16–30% lower, and for those who were small for gestational age (SGA) was 36% lower than BMC in term-born infants *(52)*. Since most preterm infants are smaller in both weight and length at term-corrected age than term-born infants, BMC values were expressed as a function of weight or length. Compared to a whole-body BMC for body weight of 20 ± 2 g/kg for term infants, SGA infants (17 ± 3 g/kg) and AGA infants (17 ± 2 g/kg) had lower bone mass at term-corrected age but represented only 15% of BMC for weight of term infants. Expressed as a function of length, BMC g/cm was 1.5 ± 0.2, 1.0 ± 0.3, and 1.1 ± 0.2 for the term, preterm SGA, and preterm AGA infants, respectively. Thus, even with correction for body size, the premature infants did not attain a bone mass comparable to term infants at birth.

Physical activity, in particular weight-bearing exercise, positively influences bone mass accretion even in preterm infants in early neonatal life as measured using SPA *(53)* or DXA *(54)*. Using quantitative ultrasound of the tibia early (about 2 wk postnatal), intervention with brief daily passive range of motion exercise reduced the usually observed postnatal decline in tibial speed of sound (SOS) measures *(55)*. The interactive effects of diet and physical activity on accretion of bone mass is just beginning to be addressed in term infants *(56)*, but in preterm infants they remain to be elucidated.

4.3. Nutrition and Bone Growth From Term-Corrected Age Through Adolescence

The expected timing for "catch-up" growth and bone accretion and nutrient needs to support skeletal development in preterm infants after discharge home has

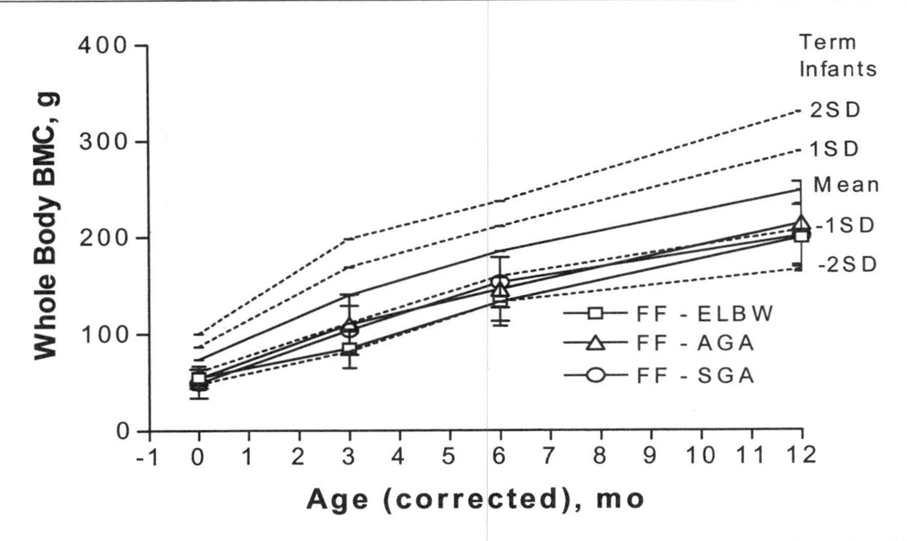

Fig. 2. Whole-body bone mineral content (BMC) in prematurely born infants of varying birth weight groups plotted against reference values for term-born infants fed standard infant formula over the first year of life. The premature infants were all fed standard term infant formula (FF) from hospital discharge to 1 yr corrected age. The preterm infant groups are ELBW, extremely low birth weight (n = 18, birth weight = 845 ± 189 g, gestational age = 24 ± 2 wk); AGA, appropriate for gestational age (n = 34, birth weight = 1322 ± 313 g, gestational age = 30 ± 2 wk); and SGA, small for gestational age (n = 26, birth weight = 1411 ± 461 g, gestational age = 33.4 ± 3.2 wk). By 12 mo corrected age mean BMC in all preterm groups remained at or below –1 SD of the mean for term-born infants (Atkinson SA, McMaster University).

not been delineated. Radial bone mass using SPA was observed to be lower in preterm infants fed either human milk or a standard term formula compared to infants fed a fortified formula up to 1 yr of age *(47,57,58)*. However, such differences in radial BMC were no longer evident by 2 yr of age *(59)*.

Longitudinal measures of whole-body BMC from term date to 1 yr of age demonstrated that both term-born *(26)* and prematurely born infants *(60)* experience an increase in BMC of 3.6- to 4-fold over the 12-mo period. However, the velocity of accretion of BMC and capacity for catch-up bone mineral accretion (that which crosses centile or standard deviation lines) in the whole body of preterm infants appears to vary with birth size and nutritional management over the first year of life (Fig. 2). By 12 mo, "catch-up" in BMC to above –1 SD was achieved only in infants of mean birth weight over 1000 g who were AGA (Fig. 2). Those infants who were SGA or of ELBW had whole-body BMC at 12 mo corrected age between –2 and –1 SD compared to term-born reference infants (Fig. 2). Breast-fed premature infants who received more than 80% of milk intake as breast milk, also did not catch up in bone mineral content when compared to published values for breast-fed term-born infants *(26)*. At 6 mo their BMC was close to –2 SD compared to the reference term values for BMC (Fig. 3).

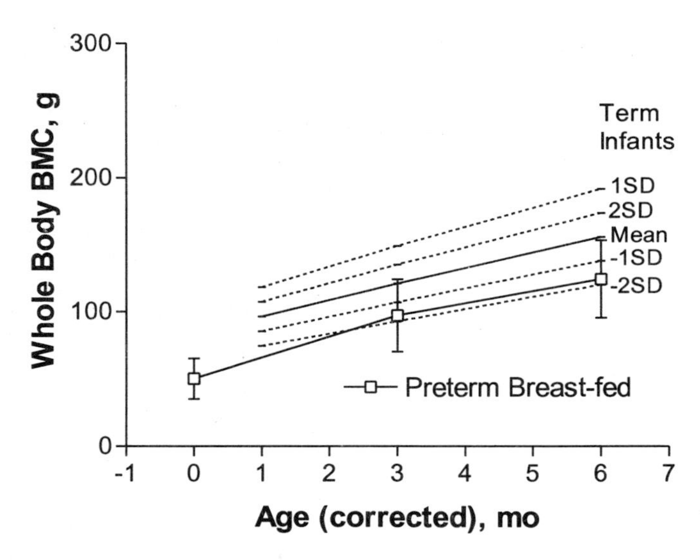

Fig. 3. Whole-body bone mineral content (BMC) of preterm breast-fed infants (n = 25, birth weight = 1187 ± 232 g, gestational age = 29 ± 3 wk) at term (noted as 0 on X axis), 3, and 6 mo corrected age plotted against published values for the mean ± 2 SD for term-born breast-fed infants *(26)*. Mean BMC for preterm infants remained close to –2 SD of the mean for term infants at 6 mo of age.

Nutrient-fortified formulas have been proposed as a means to promote catch-up growth in preterm infants after discharge from hospital. In two randomized trials of a dietary intervention to 3 mo corrected age, nutrient-enriched formula that included supplemental dietary protein, Ca, and P after hospital discharge had a positive immediate benefit to BMC when the intervention was continued to 3 mo *(23)* or 9 *(61)* mo corrected age. However, such a positive effect of early nutrition on BMC is not always sustained *(62,63)*. In addition, the whole-body BMC attained in the preterm infants does not catch up to that of term infants at 1 yr of age, regardless of receiving protein and mineral intakes as currently recommended *(63)*.

The goal for growth as set out in the 4th edition of the *Pediatric Handbook* by the AAP (1998) *(64)* is that premature infants should attain growth and body composition similar to that of term-born infants. Based on the data reviewed above, this goal is not attainable with the nutritional regimens adopted for preterm infants over the past decade. Intrauterine accretion of bone minerals is not achievable by term-corrected age, and "catch-up" does not occur for many infants even by 1 yr of age despite intervention with nutrient-enriched formulas.

Does early nutrition of preterm infants influence bone health in childhood and adolescence? The long-term outcome of bone growth and attainment of peak bone mass in preterm infants has only recently been addressed through observational studies of former preterm infants at various stages of childhood or adolescence. At 8–12 yr of age, whole-body BMC in prematurely born infants was significantly lower than for children of similar age born at term *(65)*. However, the preterm infants were also shorter and lighter, so that their BMC was appropriate for body mass when compared with children born at term. Neither diet in early life (breast milk compared to formula) nor current calcium intake or weight-bearing physical activity was a significant determinant of BMC at this peripubertal age *(65)*. Measures of bone mass of the radius and lumbar spine in 70 preterm infants followed at 9–11 yr of age were also lower and height shorter than in term infants; and neonatal diet interventions of Ca, P, or vitamin D were not associated with BMC in childhood *(66)*. Preliminary evidence suggests that catch-up does not occur during the pubertal growth spurt. Former preterm infants ($n = 26$) studied at 16–19 yr were also found to be shorter and with lower BMC in the whole body, hip, and lumbar spine, although the BMC was appropriate for body size compared to 23 adolescents who were born at term *(67)*. In the latter study, early life variables such as the amount of human milk fed and duration of time until regain of birth weight was achieved were inversely associated with BMC at adolescence *(67)*.

5. CONCLUSION

In summary, the trajectory of bone mass accretion in preterm infants during infancy through to adolescence, and the determinants of the peak bone mass achieved by the end of the second decade in this special infant population, are just beginning to be appreciated. While greater amounts of minerals and protein provided in human milk fortifiers or formulas may benefit BMC when fed in early life, this immediate benefit of enhanced nutrition to BMC is not always sustained after the dietary intervention period. Long-term follow-up of preterm infants on varying dietary intakes will be required in order to determine whether a specific nutritional prescription, as yet not defined, is required to support skeletal mineralization and achievement of peak bone mass to parallel that of infants born at term out to the age of late adolescence when the latter occurs.

ACKNOWLEDGMENTS

The contributions of collaborators, research assistants, and graduate students to the research cited in this chapter from the author's laboratory are gratefully acknowledged.

REFERENCES

1. Widdowson EM, Dickerson JWT. Chemical composition of the body as a whole. In: Mineral Metabolism: An Advanced Treastise. Vol 2. Part A. (Comar CL & Bronne F, eds). Academic, New York, NY, 1964, pp. 1–247.
2. Ellis KJ, Shypailo RH, Schanler RJ, et al. Body composition analysis of the neonate: new reference data. Am J Hum Biol 1993; 3:323–330.
3. Brunton JA, Weiler HA, Atkinson SA. Improvement in the accuracy of dual energy X-ray absorptiometry for whole body and regional analysis of body composition. Validation using piglets and methodological considerations in infants. Pediatr Res 1997; 41:1–7.

4. Koo WWK, Walters J, Bush AJ. Technical considerations of dual-energy X-ray absorptiometery-based bone mineral measurements for pediatric studies. J Bone Miner Res 1995; 10:1998–2004.
5. Widdowson EM, Southgate DA, Hey E. Fetal growth and body composition. In: Linblad BS ed. Perinatal Nutrition. Academic, New York, 1988, pp. 3–14.
6. Salle BL, Delvin EE, Lapillone A, Bishop NJ, Glorieux FH. Perinatal metabolism of vitamin D. Am J Clin Nutr 2000; 71:1317S–1322S.
7. Ziegler EE, O'Donnell Am, Nelson SE, Fomon SJ. Body composition of the reference fetus. Growth 1976; 40:329–341.
8. Hollis BW, Pittard WB. Evaluation of the total fetomaternal vitamin D relationships at term: evidence for racial differences. J Clin Endocrinol Metab 1984; 59:652–657.
9. Backstrom MC, Maki R, Kuusela AL, et al. Randomised controlled trial of vitamin D supplementation on bone density and biochemical indices in preterm infants. Arch Dis Child Fetal Neonatal Ed 1999; 80:F161–F166.
10. Pohlandt F, Mathers N. Bone mineral content of appropriate and light for gestational age preterm and term newborn infants. Acta Paediatr Scand 1989; 78:835–839.
11. Koo WWK, Walters J, Bush AJ, Chesney RW, Carlson SE. Dual energy X-ray absorptiometry studies of bone mineral status in newborn infants. J Bone Miner Res 1996; 11:997–1002.
12. Rigo J, Nyamugabo K, Picaud JC, et al. Reference values of body composition obtained by dual energy X-ray absorptiometery in preterm and term neonates. J Pediatr Gastroenterol Nutr 1998; 25:184–190.
13. Backstrom MC, Mahonen A, Ala-Houhala M, et al. Genetic determinants of bone mineral content in premature infants. Arch Dis Child Fetal Neonatal Ed 2001; 85:F214–F216.
14. Koo WW, Walters JC Esterlitz J, Levine RJ, Bush AJ, Sibai B. Maternal calcium supplementation and fetal bone mineralization. Obstet Gynecol 1999; 94(4):577–582.
15. Godfrey K, Walker-Bone K, Robinson S, et al. Neonatal bone mass: influence of parental birth weight, maternal smoking, body composition, and activity during pregnancy. J Bone Miner Res 2001; 16(9):1694–1703.
16. Jones G, Dwyer T. Birth weight, birth length, and bone density in prepubertal children: evidence for an association that may be mediated by genetic factors. Calcif Tissue Int 2000; 67:304–308.
17. Gale CR, Martyn CN, Kellingray S, Eastell R, Cooper C. Intrauterine programming of adult body composition. J Clin Endocrinol Metab 2001; 86:247–252.
18. Fall C, Hindmarsh P, Dennison E, Kellingray S, Barker D, Cooper C. Programming of growth hormone secretion and bone mineral density in elderly men: a hypothesis. J Clin Endocrinol Metab 1998; 83:135–139.
19. Lapillonne A, Guerin S, Braillon P, Claris O, Delmas PD, Salle BL. Diabetes during pregnancy does not alter whole body bone mineral content in infants. J Clin Endocrinol Metab 1997; 82(12):3993–3997.
20. Institute of Medicine, Food and Nutrition Board. Dietary Reference Intakes for Calcium, Phosphorus, Magnesium, Vitamin D, and Fluoride. National Academy Press, Washington, DC, 1997.
21. Lawrence M, Gartner MD, Frank R, Greer MD, Section on Breastfeeding and Committee on Nutrition. Prevention of rickets and vitamin D deficiency: new guidelines for vitamin D intake. Pediatrics 2003; 111(4):908–910.
22. Mimouni F, Campaigne B, Neylan M, Tsang RC. Bone mineralization in the first year of life in infants fed human milk, cow-milk formula or soy-based formula. J Pediatr 1993; 122:348–354.
23. Greer FR, Marshall S. Bone mineral content, serum vitamin D metabolite concentrations, and ultraviolet B light exposure in infants fed human milk with and without vitamin D2 supplements. J Pediatr 1989; 112:204–212.
24. Brunton JA, Saigal S, Atkinson SA. Growth and body composition in infants with bronchopulmonary dysplasia up to 3 months corrected age: a randomized trial of a high-energy nutrient-enriched formula fed after hospital discharge. J Pediatr 1998; 133:340–345.

25. Venkataraman PS, Ahluwalia BW. Total bone mineral content and body composition by X-ray densitometry in newborns. Pediatrics 1992; 90:767–770.
26. Koo WWK, Bush AJ, Walters J, Carlson SE. Postnatal development of bone mineral status during infancy. J Am Coll Nutr 1998; 17:65–70.
27. Specker BL, Beck A, Kalkwarf H, Ho M. Randomized trial of varying mineral intake on total body bone mineral accretion during the first year of life. Pediatrics 1999; 99:e12.
28. Koo WWK. Body composition measurements in infants. In: Yasumura S, ed. 5th International Symposium on In Vivo Body Composition Studies, New York. Ann N Y Acad Sci 2000; 904:383–392.
29. Lapillonne A, Braillon P, Claris O, Chatelain PG, Delmas PD, Salle BL. Body composition in appropriate and in small for gestational infants. Acta Pediatr 1997; 86:196–200.
30. Butte NF, Wong WW, Hopkinson JM, Smith EO, Ellis KJ. Infant feeding mode affects early growth and body composition. Pediatrics 2000; 106(6):1355–1366.
31. American Academy of Pediatrics. Committee on Nutrition. Nutritional needs of low birth weight infants. Pediatrics 1985; 75:976–983.
32. European Society for Pediatric Gastroenterology, Committee on Nutrition of the Preterm Infant. Nutrition and feeding of preterm infants. Acta Paediatr Scand Suppl 1987; 111:122–126.
33. Canadian Pediatric Society—Nutrition Committee. Nutrient needs and feeding of premature infants. Can Med Assoc J 1995; 152:1765–1783.
34. Ziegler EE, Biga RL, Fomon SJ. Nutritional requirements of the premature infant. In: Suskind RM, ed. Textbook of Pediatric Nutrition. Raven, New York, 1981, p. 29.
35. Atkinson SA, Radde IC, Anderson GH. Macromineral balances in premature infants fed their own other's milk or formula. J Pediatr 1983; 102:99–106.
36. Atkinson SA, Chappell JA, Clandinin MT. Calcium supplementation of mothers' milk for low birth weight infants: problems related to absorption and excretion. Nutr Res 1987; 7:813–823.
37. Mize CE, Uauy R, Waidelich D, Neylan MJ, Jacobs J. Effect of phosphorus supply on mineral balance at high calcium intakes in very low birth weight infants. Am J Clin Nutr 1995; 62:385–391.
38. Atkinson SA, Tsang RC. Calcium, magnesium, phosphorus and vitamin D. In: Uauy R, Tsang RC, Zlotkin S, eds. Nutrient Needs of Preterm Infants. Digital Publishing, Cincinnati, OH, 2003, in press.
39. Schanler RJ, Abrams SA. Postnatal attainment of intrauterine macromineral accretion rates in low birth weight infants fed fortified human milk. J Pediatr 1995; 124:441–447.
40. Rigo J, De Curtis M, Pieltrain C, Picaud JC, Salle BL, Senterre J. Bone mineral metabolism in the micropremie. Clin Perinatol 2000; 25(1):147–170.
41. Bronner F, Salle BL, Putet G, Rigo J, Senterre J. Net calcium absorption in premature infants: results of 103 metabolic balance studies [published erratum appears in Am J Clin Nutr 1993; 57(3):451]. Am J Clin Nutr 1992; 56:1037–1044.
42. Koo WWK, Krug-Wispe S, Neylan M, Succop P, Oestreich AE, Tsang RC. Effect of three levels of vitamin D intake in preterm infants receiving high mineral-containing milk. J Pediatr Gastroenterol Nutr 1995; 21:182–189.
43. Schanler RJ, Rifka M. Calcium, phosphorus and magnesium needs for the low birth weight infant. Acta Pediatr Suppl 1994; 405:111–116.
44. Giles MM, Laing IA, Elton RA, Robins JB, Sanderson M, Hume R. Magnesium metabolism in preterm infants: effects of calcium, magnesium, and phosphorus, and of postnatal and gestational age. J Pediatr 1990; 117:147–154.
45. Greer FR, McCormick A. Bone growth with low bone mineral content in very low birth weight premature infants. Pediatr Res 1986; 20:923–926.
46. Kuschel CA, Harding JE. Multicomponent fortified human milk to promote growth in preterm infants (Cochrane Review). In: The Cochrane Library, Issue 2, 1999. Update Software, Oxford.
47. Bishop J, Dahlenburg S, Fewtrell M, Morley R, Lucas A. Early diet of preterm infants and bone mineralization at age five years. Acta Paediatr 1996; 85:230–236.

48. Wauben I, Atkinson SA, Grad TL, Shah JK, Paes B. Moderate nutrient supplementation to mother's milk for preterm infants supports adequate bone mass and short-term growth. A randomized controlled trial. Am J Clin Nutr 1998; 67:465–472.
49. Lapillone AA, Glorieux FH, Salle B, et al. Mineral balance and whole body bone mineral content in very-low-birth-weight infants. Acta Pediatr Suppl 1994; 405:117–122.
50. Faerk J, Petersen S, Peitersen B, Fleischer Michaelsen K. Diet and bone mineral content at term in premature infants. Pediatr Res 2000; 47:148–156.
51. Pieltain C, De Curtis M, Gerard P, Rigo J. Weight gain composition in preterm infants with dual energy X-ray absorptiometry. Pediatr Res 2001; 49:120–122.
52. Atkinson SA, Randall-Simpson J. Factors influencing body composition of premature infants at term-adjusted age. In: Yasumura S, ed. 5th International Symposium on In Vivo Body Composition Studies, New York. Ann N Y Acad Sci 2000; 904:393–400.
53. Moyer-Mileur L, Leutkemeler M, Boomer L, Chan GM. Effect of physical activity on bone mineralization in premature infants. J Pediatr 1995; 125:620–623.
54. Moyer-Milleur L, Brunstetter V, McNaught TP, Gill G, Chan GM. Daily physical activity program increases bone mineralization and growth in preterm very low birth weight infants. Pediatrics 2000; 106:1082–1092.
55. Nemet D, Dolfin T, Litmanovitz I, Shainkin-Kestenbaum R, Lis M, Eliakim A. Evidence for exercise-induced bone formation in premature infants. Int J Sports Med 2002; 23:1–4.
56. Specker B, Binkley T. Randomized trial of physical activity and calcium supplementation on bone mineral content in 3- to 5-year-old children. J Bone Miner Res 2003; 18:885–892.
57. Chan GM. Growth and bone mineral status of discharged very low birth weight infants fed different formulas or human milk. J Pediatr 1993; 123:439–443.
58. Abrams SA, Schanler RJ, Garza C. Bone mineralization in former very low birth weight infants fed either human milk or commercial formula J Pediatr 1988; 112:956–960.
59. Schanler RJ. Burns PA, Abrams SA, Garza C. Bone mineralization outcomes in human milk fed preterm infants. Pediatr Res 1992; 31:583–586.
60. Wauben IPM, Atkinson SA, Shah JK, Paes B. Growth and body composition of preterm infants: influence of nutrient fortification of mother's milk in hospital and breast feeding post-hospital discharge. Acta Paediatr 1998; 87:780–785.
61. Bishop NJ, King FJ, Lucas A. Increased bone mineral content of preterm babies fed with a nutrient enriched formula after discharge from hospital. Arch Dis Child 1993; 68:573–578.
62. Rubinacci A, Sirtori P, Moro G, Galli L, Minoli I, Tessari L. Is there an impact of birth weight and early life nutrition on bone mineral content in preterm born infants and children? Acta Pediatr 1993; 82:711–713.
63. Brunton JA, Saigal S Atkinson S. Nutrient intake similar to recommended values does not result in catch-up growth by 12 mo of age in very low birth weight infant (VLBW) with bronchopulmonary dysplasia (BPD). Am J Clin Nutr 1997; 55(i), Abstr. 102.
64. American Academy of Pediatrics. Committee on Nutrition. Nutritional needs of preterm infants. In: Kleinmen RE, ed. Pediatric Handbook, 4th ed. American Academy of Pediatrics, Elk Grove Village, IL, 1998, pp. 55–87.
65. Fewtrell MS, Prentice A, Jones SC, et al. Bone mineralization and turnover in preterm infants at 8–12 years of age: the effect of early diet. J Bone Miner Res 1999; 14:810–820.
66. Backstrom MC, Maki R, Kuusela AL, et al. The long-term effect of early mineral, vitamin D, and breast milk intake on bone mineral status in 9- to 11-year-old children born prematurely. J Pediatr Gastroenterol Nutr 1999; 29:575–582.
67. Weiler HA, Yuen CK, Seshia MM. Growth and bone mineralization of young adults weighing less than 1500 g birth weight. J Early Hum Develop 2002; 67:101–112.

11 Nutrition and Bone Health in Children and Adolescents

Velimir Matkovic, Nancy Badenhop-Stevens, Eun-Jeong Ha, Zeljka Crncevic-Orlic, and Albert Clairmont

1. INTRODUCTION

Adult skeletons evolved from a single cell with a programmed system of constraints on development and mineralization which is under strict genetic control. It has been speculated that genetics contributes about 80% of the variance in bone mass and the remaining 20% is affected by one's environment, although the exact contribution of each major determinant of bone mass is unknown. Research data support the hypothesis that peak bone size, bone mass, and to a lesser extent the distribution of bone tissue within the bone as an organ (volumetric bone density) in young individuals are strongly influenced by genetic information by both parents (Fig. 1) *(1)*. This indirectly suggests that bone candidate genes responsible for bone modeling drifts along the longitudinal and periosteal axes in interaction with nutritional factors and physical exercise have an important impact on skeletal development and peak bone mass.

Skeletal tissue not only provides the best example of general growth progress through maturity but also gives us our best approximation of biological age of humans. The rate of this process and the time required to achieve the mature adult skeleton is variable. Skeletal tissue goes through several developmental stages from fetal life to peak bone mass of young adulthood. Because bone fractures are closely related to diminished peak bone mass, we have to identify all underlying causes responsible for inadequate accumulation of bone tissue during skeletal growth and consolidation *(2,3)*.

The purpose of this chapter is to define some of the most important nutritional determinants of bone mass during childhood and adolescence (Fig. 2) *(4)*, and to discuss the strategy of primary prevention of osteoporosis in the search for better bone health. Over the last few decades, the focus of nutrition research and clinical

From: *Nutrition and Bone Health*
Edited by: M. F. Holick and B. Dawson-Hughes © Humana Press Inc., Totowa, NJ

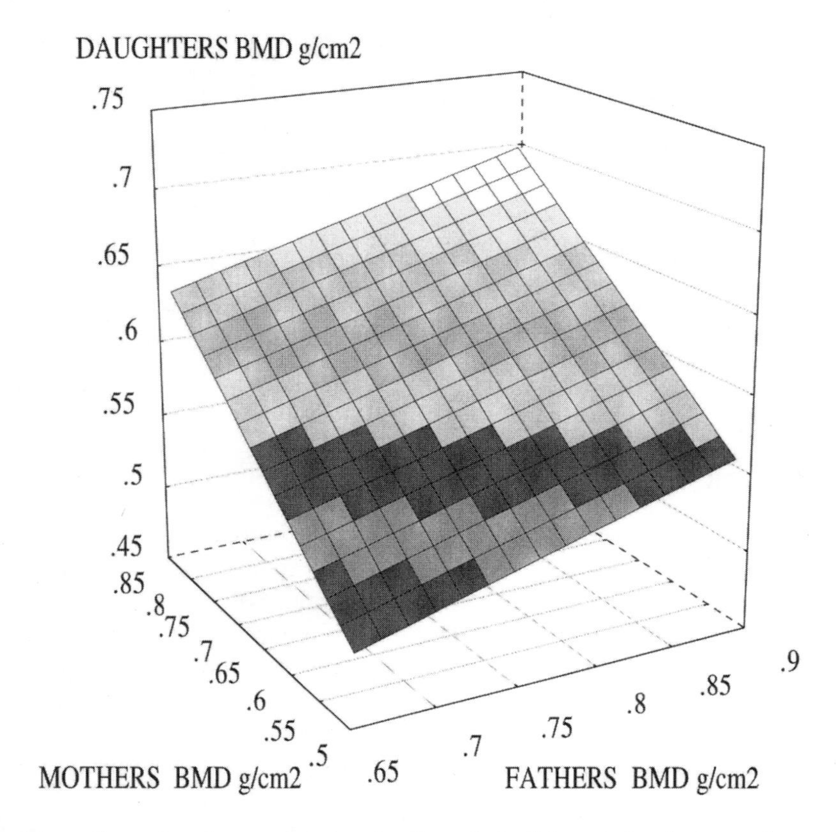

Fig. 1. Three-dimensional representation (surface plot) of the relation between forearm bone mineral areal density (BMD) of daughters *(z)* and their fathers *(y)* and mothers *(x)*. Mothers and fathers with higher BMD have daughters with higher BMD. (Adapted from ref. *1*.)

practices in pediatrics has shifted from the prevention of nutritional deficiencies in young individuals to the establishment of recommended diets to prevent chronic diseases later in life. These priorities may lead eventually to dietary guidelines for the prevention and treatment of osteoporosis by targeting predisposed individuals early in life, as emphasized in the *Healthy People 2010: National Health Promotion and Disease Prevention Objectives (5)*.

2. SKELETAL DEVELOPMENT AND PEAK BONE MASS

Peak bone mass is defined as the highest level of bone mass achieved as a result of normal growth *(6)*. Peak bone mass is important because it determines resistance or susceptibility to fracture *(2)*. This was first shown in the study of bone mass and hip fracture rates in two farming communities with different dietary habits but the same high level of physical activity over a lifetime. It appeared that

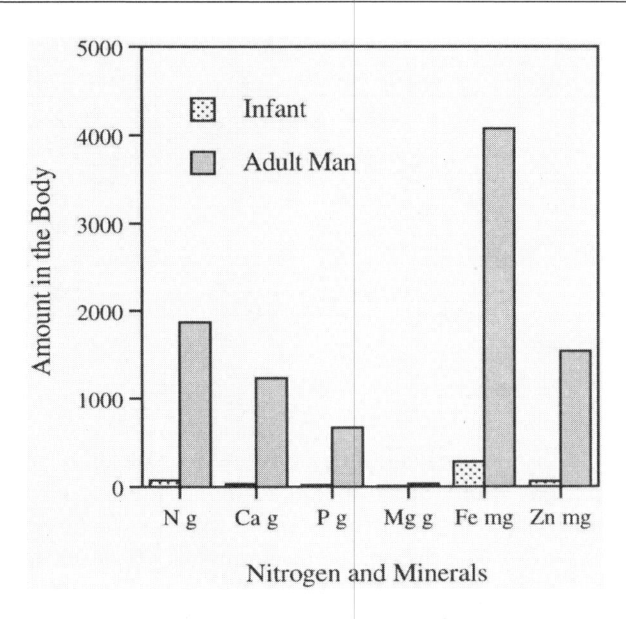

Fig. 2. Total body nitrogen and minerals in infants and young adults. The difference between these two phases of life represent the amount of mineral accumulated during childhood and adolescence. (Adapted from ref. *4*.)

both populations were losing bone with age at about the same rate, but those who started with more bone ended up having higher bone mass and lower incidence of hip fractures. It was concluded that, other things being equal, a high peak bone mass provided a larger reserve later in life. The differences in bone mass and fracture rates were attributed primarily to calcium/protein intake. The differences in bone mass between the communities were established at an early age (30 yr), implying that if calcium intake is important, it may be during skeletal growth that it has its greatest impact. This was the first proposal of the hypothesis that increasing peak bone mass by calcium supplementation during skeletal formation may contribute to osteoporosis prevention *(2)*. Results of a similar ecological study conducted in China on populations accustomed to different calcium intakes over the lifetime *(7)* confirmed the above finding and reiterated the importance of adequate nutrition for skeletal formation and peak bone mass. In a retrospective study, Sandler et al. *(8)* also showed that postmenopausal women who consumed more milk and dairy products during adolescence had higher bone mineral density at the forearm than those who did not. Overall, it is likely that variations in calcium nutrition early in life may account for as much as a 5–10% difference in peak adult bone mass. Such a difference, although small, probably contributes to more than 50% of the difference in the hip fracture rate later in life *(2)*.

Most of the skeletal mass is accumulated by the average age of 18 yr *(9–11)*. Thereafter, there is a minimal change in bone mass and density with age up to the

Table 1
Calcium Intake Thresholds and Balances for Growing Individuals

Age group (yr)	*Threshold intake (mg/d)*	*Threshold balance (mg/d)*
0–1	+503	1090
2–8	+246	1390
9–17	+396	1480
18–30	+114	957

Source: Ref. *14.*

time of menopause. Some skeletal sites continue to lose bone immediately after the age of 18 (proximal femur, and trabecular bone in the vertebrae), and the other sites show continuous apposition of bone up to the time of menopause (forearm, total spine, head) *(11,12)*. The difference in volumetric density (mass/volume ratio in g/cm^3) between childhood and adulthood is minimal, indicating that most of the changes we measure during growth using dual-energy X-ray absorptiometry (DXA, areal density) are due predominantly to the change in bone volume, and to a much lesser extent to increases in true bone mineral density (BMD; g/cm^3) *(11)*.

To achieve maximal peak bone mass, dietary calcium and its absorption need to be at or above the threshold level to satisfy skeletal modeling and consolidation and also obligatory losses in urine, feces, and sweat. Calcium-intake thresholds with corresponding threshold balances for growing individuals were reported (Table 1) *(13,14)*. Growing individuals, contrary to adults, therefore need to be in positive calcium balance to meet the extra needs of skeletal growth and consolidation. The net positive balance of bone tissue contributes to the constant demand for calcium and proteins throughout the developmental process. Except for a few isolated reports that calcium deficiency can cause rickets in children, low calcium intake does not have any deleterious effect on bone health of young individuals *(15,16)*. It has been suggested that calcium deficiency during growth could contribute to bone fragility fractures during puberty *(17,18)*, but definitive data to support this hypothesis are still lacking. Calcium deficiency during skeletal formation may negatively impact peak bone mass, therefore reducing fracture resistance among the elderly *(2)*.

3. CHILDHOOD

Bone accretion during childhood (2–8 yr) is proportional to the rate of growth. During this age interval, height velocity is relatively slow and about 5.5 cm/yr, for both boys and girls. Bone mineral areal density of the whole body increases at a rate of about 1%/yr as compared to adolescence, when bone mineral areal density is increasing at about 4%/yr (Fig. 3). As a direct consequence, the retention of

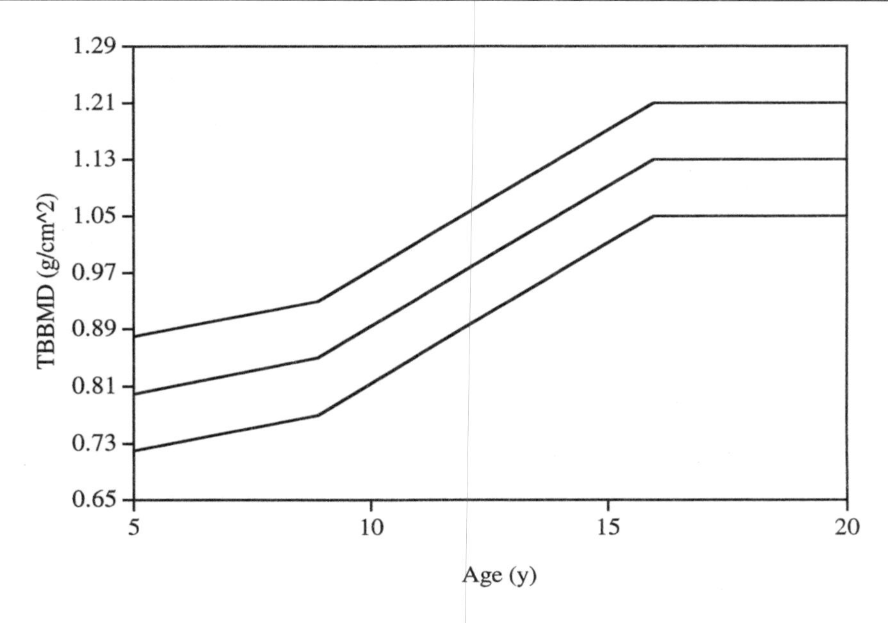

Fig. 3. Bone mineral areal density of the total body during growth. Adapted from GE-Lunar standards for boys (GE-Lunar, Madison, WI). These normative data are based on a cross-sectional study and may not accurately represent the change in total-body bone mineral density (TBBMD) over time, in particular after the inflection point at age 16 yr. Total body bone mass should be steadily increasing during late adolescence and throughout young adulthood but at a lower rate of change.

calcium in the body of an average child is lower than the calcium retention in an adolescent *(13)*. Even so, growing children need two to four times as much calcium per kilogram of body weight compared with adults. Daily required skeletal calcium retention during this period of life has been estimated at about 100 mg/d. On the average calcium intake of about 1100 mg/d, children retain about 200 mg/d of calcium, or 18% of intake. At about 500 mg/d calcium intake, children are still in positive calcium balance for about 60 mg/d *(13,14)*. In one study from India, children were consuming even less calcium (about 300 mg/d) and were still able to maintain a sufficiently positive calcium balance to assure skeletal development *(19)*. The results of this study cannot be applied directly to Caucasian children because Indian people are of different ethnic background, with lower peak bone mass than their Caucasian counterparts. Calcium intake at about 1600 mg/d leads to the positive calcium balance of about 300 mg/d. Dietary surveys in the United States have shown that calcium intake conforms to the current recommended daily intake (RDI) (800 mg/d) for this age group *(20,21)*. With a relatively low urinary calcium excretion, such a calcium intake in this population should provide adequate calcium retention to satisfy the requirements for skeletal growth in children.

It therefore appears likely that most children between infancy and puberty are able to meet the daily calcium requirements necessary for adequate skeletal calcium retention. The fact that children are able to retain more calcium with further increase in calcium intake is very important one, because it could contribute to higher bone mass and density during this age period. Children who were accustomed to a higher level of dairy product consumption during early growth have higher bone mass at the beginning of puberty *(22)*. Assuming that those children will maintain their dietary behavior through adolescence, they can complete their growth phase with a higher peak bone mass, as showed recently in a 7-yr-long observational study *(23)*. In a more recent study conducted in New Zealand among 30 girls and 20 boys ages 3–10 yr, children who avoided drinking milk had low dietary calcium intakes and poor bone health as represented in the bone mineral areal density of the forearm and total body *(24)*. There is evidence that calcium supplements (milk) has improved stature in children. In 1927 a series of tests were carried out in Scotland in which about 1500 children were given additional milk at school for a period of 7 mo. Periodic measurements of the children showed that the rate of growth in those getting the additional milk was faster than in those not getting the supplement. The increased rate of growth was accompanied by a noticeable improvement in health and vigor *(25)*. An assessment of BMD was not available at that time. The great increase in the height of young Japanese adults from 1950 to 1970 coincided with a tripling of the national calcium intake from about the lowest in the world to 600 mg/d *(26)*. In a more recent study, a group of British teenage boys who took calcium supplementation for 12 mo were taller and had a greater bone mass at the end of the intervention than their placebo counterparts *(27)*. However, the same investigators were not able to observe this phenomenon in a cohort of malnourished Gambian children *(28)*, whose bone mineral areal density, but not stature, increased with calcium supplementation. It seems that the overall nutritional status is a determinant of the calcium response with regard to longitudinal bone growth.

4. ADOLESCENCE

Adolescence is characterized by accelerated muscular, skeletal, and sexual development. By the age of 10 the mean height velocity is 6 cm/yr in girls and increases to an average peak of 9 cm/yr by the age of 12. Peak height velocity for boys starts at 12 yr of age (5 cm/yr) and reaches a maximum by age of 14 (10 cm/yr). Mean height velocity will be close to zero by the age of 15 in girls, and by the age of 17 in boys. The average gain in height in girls between the age of 8 and 17 is about 23% of the mean adult value. This gain in height is, however, dramatically outpaced by the gain in total body calcium: 132% vs 19%, respectively (Fig. 4) *(11)*. Bone size, bone mass, and BMD of the regional skeletal sites increase on average by about 4%/yr from childhood (age 8 yr) to late adolescence and young adulthood, when most of the bone mass will be accumulated. This

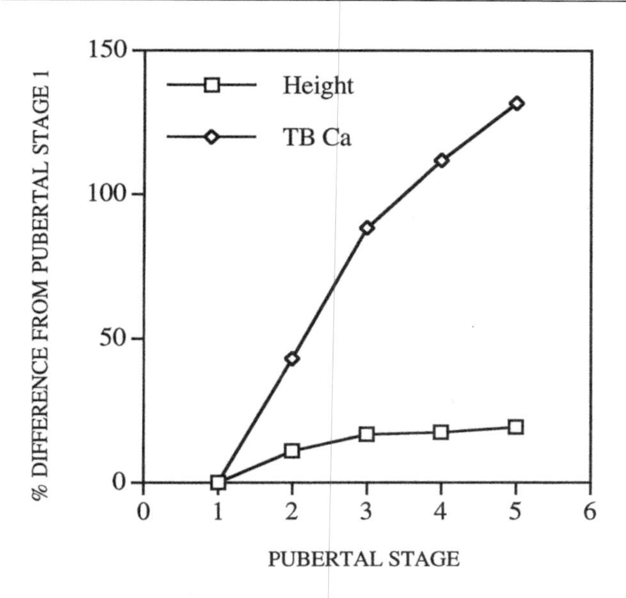

Fig. 4. Relative difference in whole-body calcium and stature between pubertal stage 1 and subsequent stages. Based on a cross-sectional study conducted in 80 healthy adolescent Caucasian females, ages 8–17 yr. (Adapted from ref. *11*.)

ranged from 1.2% for the estimate of true density of the body of L_3 vertebra to 6.6% for the femoral neck *(11)*.

When relating bone mass to pubertal developmental stage, it becomes obvious that most of the bone mass (37%) is being accumulated between pubertal stages 2 and 4 *(11)*. This rapid accumulation of bone mass correlates with the rate of growth and probably also requires the concerted action of growth hormone, insulin-like growth factor-I (IGF-I), and sex steroids and its receptors. The increase in circulating IGF-I at early puberty correlates with sexual development and results from the interaction between sex steroids and growth hormone. Specifically, the surge in sex steroids in turn increases the secretion of growth hormone, which stimulates the production of IGF-I *(29)*. The amount of estrogen required to stimulate longitudinal bone growth is very small. Doses of 100 ng/kg/d produce maximal growth in agonadal individuals. This doses seem to be insufficient to cause either the development of secondary sexual characteristics or an increase in sex hormone-binding globulin. These low-dose effects are consistent with the observation that girls attain peak height velocity early in puberty at serum estradiol levels of less than 30 pg/ML which is one-fifth the mean level found in young adult women *(30)*. During this phase of rapid skeletal modeling (growth), bone mass is not yet consolidated and bone mass per bone volume ratio is relatively low *(31)*. Presumably, this results in the high incidence of the bone

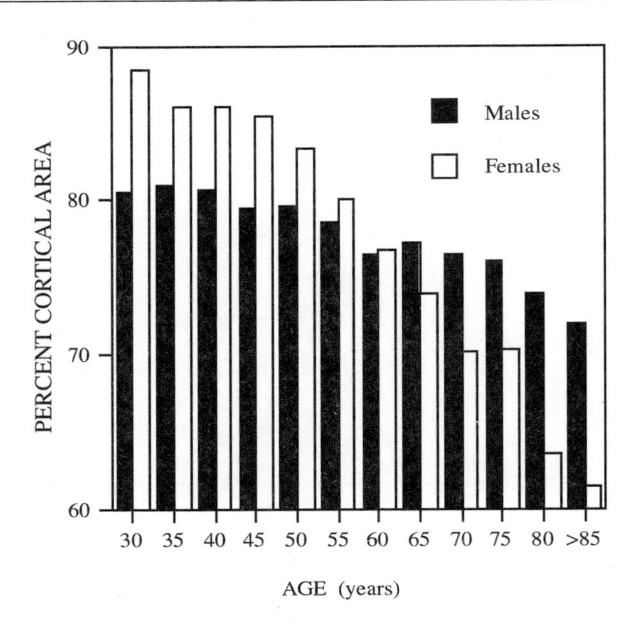

Fig. 5. Bar graph of percent cortical area (%CA) of the second metacarpal bone in males and females according to age. Note higher relative cortical bone mass in premenopausal women comparable to men (age < 50 yr). (Adapted from ref. *33*.)

fragility fractures (distal forearm) in children reaching the levels observed in women after menopause *(32)*.

Bone consolidation proceeds by the cessation of longitudinal bone growth. This coincides with the increase in estradiol secretion by the beginning of menarche. Time since menarche, therefore, is the best predictor of bone events in young females, and comparable to the time since menopause in older women. Estrogen-driven endosteal apposition of bone is responsible for the increase in the relative amount of cortical bone (bone tissue within the bone volume) in premenopausal women as compared to men (higher cortical to total area ratio) (Fig. 5) *(33)*. this endosteal apposition of cortical bone starts at menarche and ends at perimenopause. What has been accumulated during the reproductive phase will be lost after menopause, making elderly women more vulnerable to bone fragility fracture in addition to having smaller bones and lower peak bone mass.

Several studies document lower BMD in adult women with a history of late menarche, as indicative of inadequate sex hormone levels during this critical period of skeletal development and/or a short time interval between menarche and menopause. Both young adult women and adolescents with hypothalamic amenorrhea have reduced bone mass at skeletal sites that should not be losing bone *(34)*. This might lead to the reduction in bone mass at maturity and predispose this population to the increased risk of osteoporosis later in life. The above-mentioned

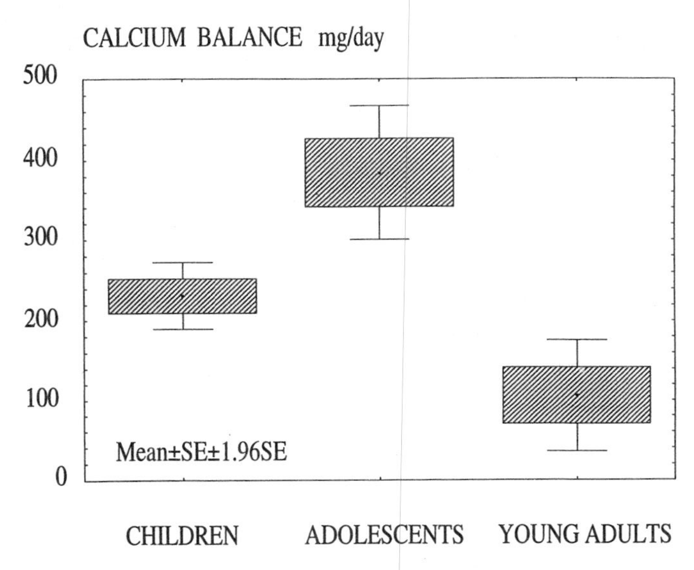

Fig. 6. Box plots of calcium balances in young individuals of different age. Notice high calcium retention during phase of rapid skeletal modeling of adolescence. Calcium balances in all three groups were conducted at intake of 1186 ± 96 mg/d. Subtracting skin losses of about 60 mg/d from an average calcium balance among adolescents of 400 mg/d leaves skeletal retention of about 340 mg/d. This value corresponds to the average daily skeletal accretion of calcium at the peak pubertal growth (time zero since menarche) as assessed by the whole-body bone mineral content measurement by dual-energy X-ray absorptiometry based on a longitudinal study presented in Fig. 7. (Adapted from ref. *13*.)

menstrual disturbances could be in part due to relative or absolute energy deficiency induced by weight loss or inadequate weight gain as seen in protein-calorie malnutrition, young athletic women, anorexics, and other clinical situations. The onset of menarche is related to serum leptin level and body composition *(35)*.

Calcium needs are greater during adolescence (9–17 yr) than in either childhood or adulthood. As a result of the rapid skeletal changes, calcium metabolism in adolescents differs significantly from that in childhood and young adulthood, and it has some similarities with calcium metabolism during infancy. In general, adolescents retain more calcium than either children or young adults (Figs. 6 and 7) *(13)*. According to calcium balance studies, the threshold intake for adolescents is about 1500 mg/d *(14)*. The corresponding average calcium retention that saturates the skeletons of teenagers is about 400 mg/d. Recommendations for calcium nutrition should take into account that the calcium intake threshold is highly variable and depends on body size as well as stage of human development. When metacarpal cortical bone mass was examined separately in a segment of the population from high- and low-calcium districts in Croatia, there was a larger discrepancy in the cortical bone mass per corresponding bone volume for persons of a larger body

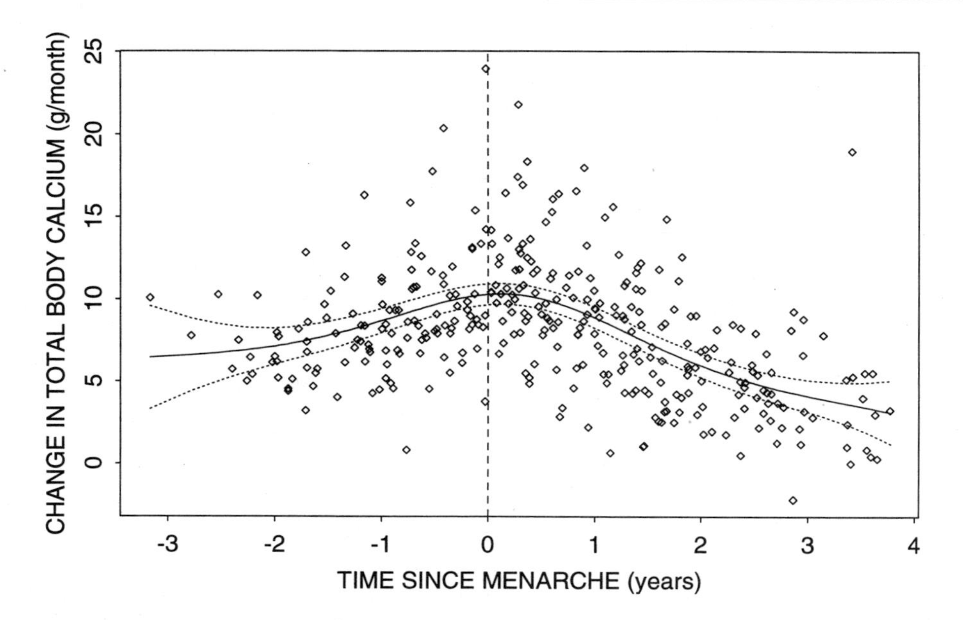

Fig. 7. The rate of change in total body calcium (g/mo, n = 90) with time since menarche (yr) in a cohort of young females followed annually for 4 yr. Total body calcium is considered a fraction (38%) of total body bone mineral content as measured by dual-energy X-ray absorptiometry. Scatterplots shown with cubic splines and 95% confidence intervals. (Adapted from ref. *57*.)

size than for the smaller individuals (Fig. 8) *(36)*. This indicates that calcium deficiency during growth could disproportionally affect individuals who are genetically predestined to reach a higher level in their peak bone mass, as they require more calcium. The importance of a positive calcium balance during adolescent years is further emphasized by the need to meet not only the rapidly expanding skeletal compartment, but also losses of calcium through the skin (not measured in usual balance studies), which may amount to as much as 60 mg/d in adults *(37)*. Young athletes could lose considerable calcium through sweat; this may be up to 60–80 mg/h of intensive training. Low calcium intake may lead to a negative calcium balance and bone loss, as reported for basketball players *(38)*.

Adolescents, in general, absorb more calcium from their diet than either children or young adults (Fig. 9) *(13)*. The concentration of serum calcitriol (1,25-dihydroxy-vitamin D) is highest during peak growth; pubertal stages 2–3 (Fig. 10) *(39)*. Urinary calcium increases during the period of adolescence and reaches its maximum by the age of 15–16 yr, or by the cessation of puberty. The mean urinary calcium output for young boys and girls aged 9–17 yr is about 130 mg/d. Mean urinary calcium excretion at an intake of about 500 mg/d is about 120 mg/d. Further increases in intake up to 1800 mg/d increase urinary calcium excretion

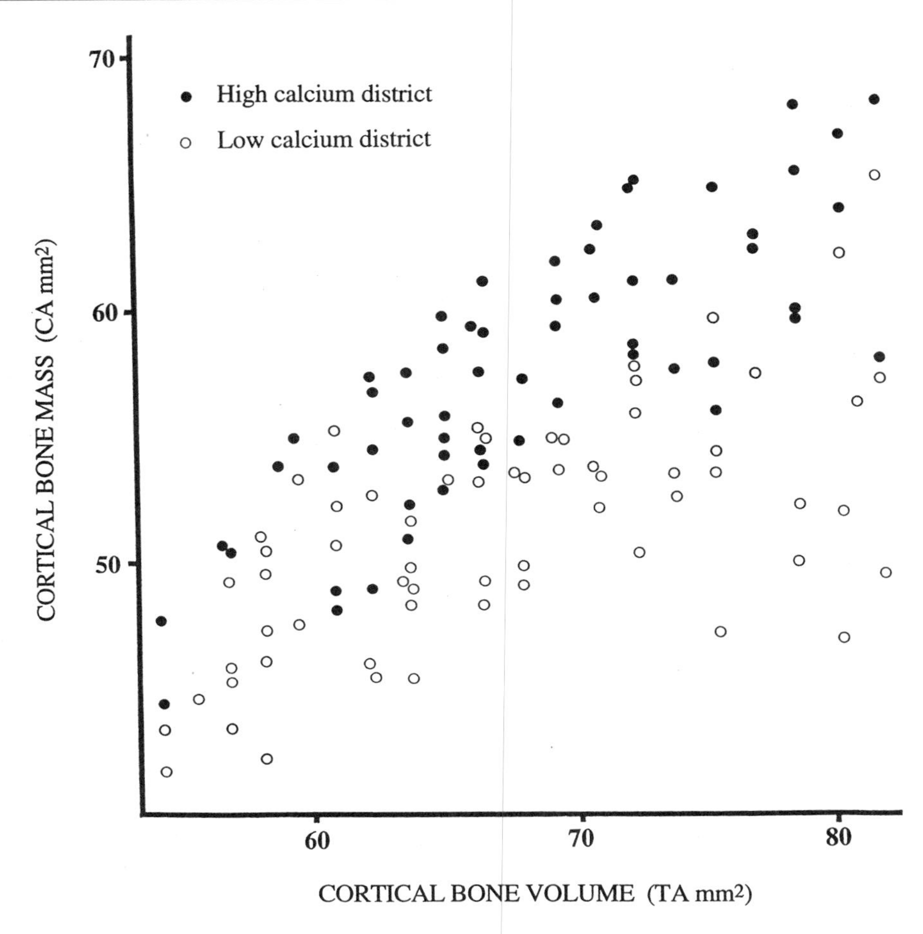

Fig. 8. Relationship between metacarpal cortical bone mass (CA) and bone volume (TA) in 40–44 yr old men (n = 121) accustomed to different calcium intakes over lifetime. Note a greater disparity between bone mass per bone volume in bigger individuals from high- and low-calcium regions. (Adapted from ref. *2*.)

only by 10–20 mg/d. This level of urinary calcium excretion of about 130 mg/d can, therefore, be considered as the mean obligatory urinary calcium loss for the age group 9–17 yr *(13)*. The above indicate that urinary calcium excretion during adolescence is barely related to calcium intake (Fig. 11). Body weight and age seem to be the principal determinants of urinary calcium excretion during this phase of life. More powerful relationships between urinary and dietary calcium definitely exist in adults. The explanation for the above is that adolescents retain the absorbed calcium in the skeleton rather than excreting it in the urine *(13,40)*. Calcium in the urine is expected to rise, as a result of the increase in the filtered load of calcium, only after the skeletal compartment is being saturated with calci-

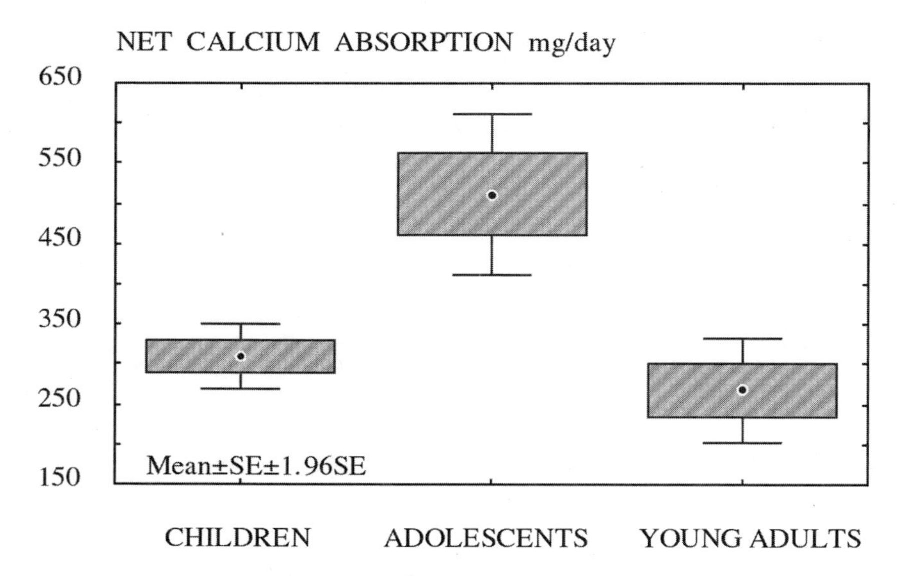

Fig. 9. Net calcium absorption in children, adolescents, and young adults at calcium intake of about 1200 mg/d. (Adapted from ref. *13*.)

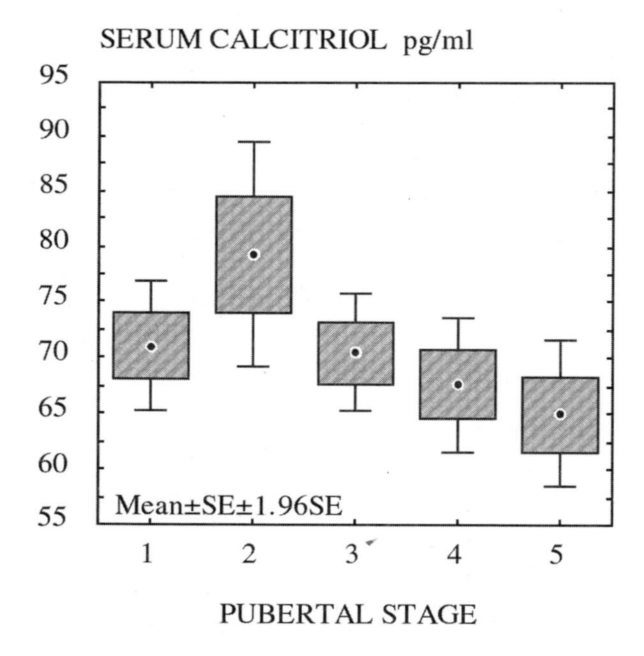

Fig. 10. Box plots of serum calcitriol level in young females according to pubertal developmental stage (breast). Highest concentration found during peak pubertal growth in stage 2. (Adapted from ref. *39*.)

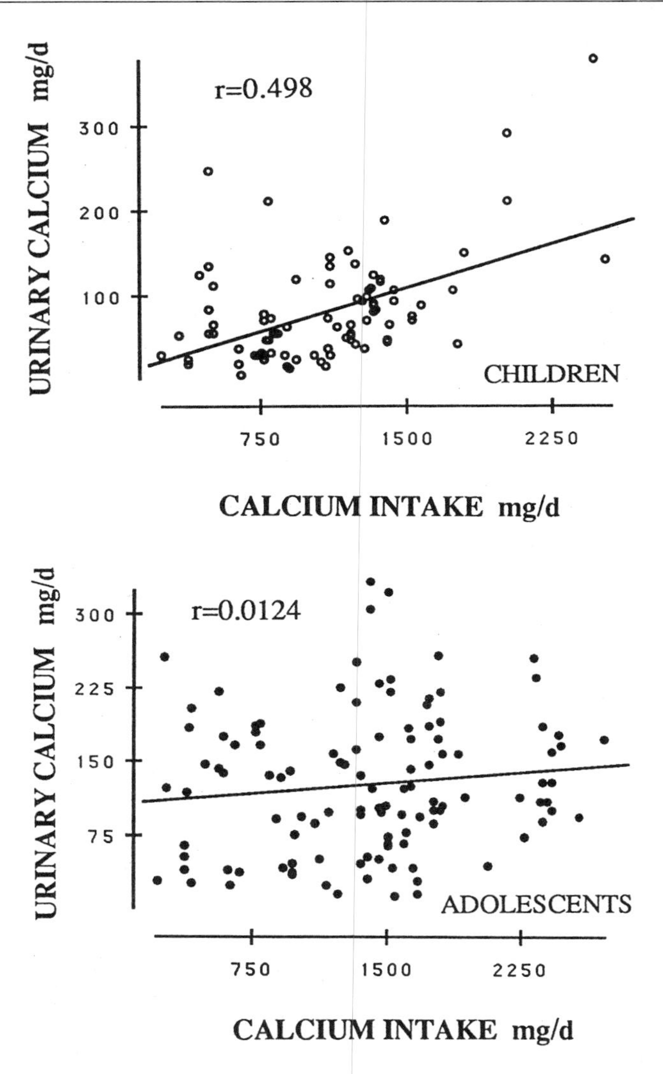

Fig. 11. Relationship between urinary calcium and calcium intake in children (top) and adolescents (bottom). The association is more pronounced during childhood than during adolescence, mimicking a "hungry bone syndrome." (Adapted from ref. *13*.)

um at intakes above the threshold level. Renal excretion of calcium is believed to be regulated by parathyroid hormone and estrogens and is also influenced by sodium intake. High consumption of salt increases the obligatory calcium loss in the urine, which may jeopardize adequate bone accretion during growth (Fig. 12) *(41)*.

Nationwide surveys reveal that adolescents, especially females, consume inadequate amounts of calcium *(20)*. In addition, psychological changes involving the

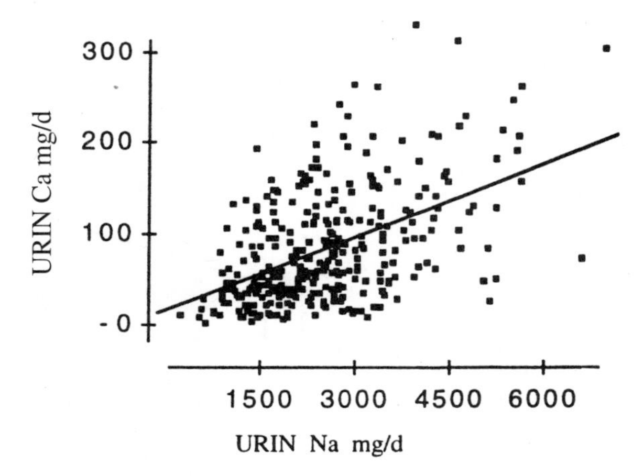

Fig. 12. The relationship between urinary calcium and urinary sodium in a cohort of pubertal females ($N = 325$, $R^2 = 25.0\%$, $p < 0.0001$). (Adapted from ref. *41*.)

adolescent's search for independence and identity, desire for acceptance by peers, and preoccupation with physical appearance may affect eating habits, food choices, nutrient intake, and ultimately nutritional status. As children reach their teen years they drink less milk and their calcium intake declines far below the current standard (1300 mg/d) and even more so in comparison to the calcium intake threshold of 1500 mg/d *(14,21)*. Optimizing the calcium intake of young Americans along with increased energy expenditure (physical activity) is, therefore, of critical importance. Recent improvements in calcium intake have been reported for most age groups, however, young females showed a decrease in calcium intake as compared to a decade earlier. Public health strategies to promote optimal calcium intake should have a broad outreach and should involve educators, health professionals, and the private and public sectors.

The World Health Organization recommendations for dietary calcium intakes for 11- to 15-yr-old adolescents are 600–700 mg/d and for 16- to 19-yr-old adolescents, 500–600 mg/d *(26)*. Although those figures might satisfy some population groups characterized by small body frames, this is not enough for the vast majority of Western populations. It is apparent from these differences in recommended intakes that the amount of dietary calcium needed to sustain growth, as well as to provide maintenance, requires much more study, particularly with regard to different populations. The standards for one ethnic group might not satisfy the requirements of another. Each country should develop its own standards specific for the people living in the region. Factors such as ethnicity, stature and body frame, dietary habits, determinants of calcium economy in the body (sodium intake, sunlight exposure), and activity level all play a role. Ideally, standards should be based on calcium intake thresholds obtained from balance studies and/or

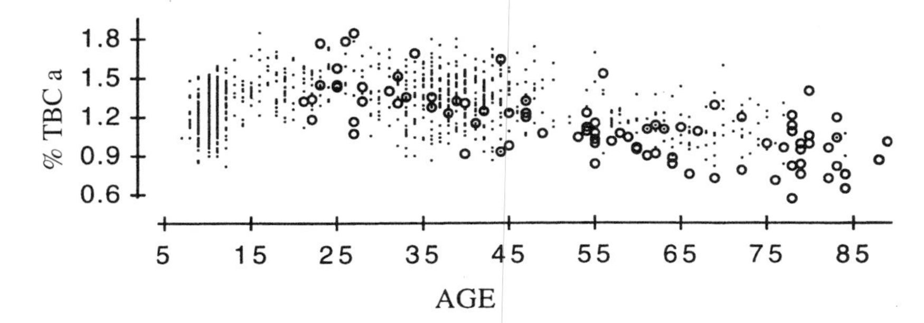

Fig. 13. Relative total body calcium content (% of body weight) in 1218 females (1128 Caucasian females, dots; 90 Japanese females, open circles) ages 7–89 yr. Data based on total body bone mineral content measurements by DXA technology and assuming Ca content is 38% of the BMC. Relative calcium content increases from about 0.9% in childhood to about 1.5% in young adulthood and than declines in elderly women (~1.0%). The relative calcium content per body weight is the same for Japanese and Caucasian women of different ages. The data for Japanese women were kindly supplied by Drs. T. Fujita and A. Tomita.

whole-body bone mass measurements by DXA (38% of bone mineral content is calcium; worldwide accessible technique) at various calcium intakes. In the absence of metabolic wards or densitometry machines, simple but crude estimates could be based on body weight data. The whole body calcium is about 0.8–1.5% of body weight, regardless of ethnicity. Minimal variations are related to age (Fig. 13). The best example is the comparison of the absolute and relative body calcium (derived from DXA) between two ethnic groups known for significant differences in body frames: Caucasians vs Japanese. Japanese have lower amounts of total body calcium but a comparable ratio of calcium to body weight (Fig. 14).

Several studies indicated that children and teenagers, particularly, may benefit from higher calcium intake with further gain in bone mass.This is important not only with regard to peak bone mass acquisition, but also with regard to fracture prevention during growth. The peak incidence of the distal forearm fractures occurs during growth spurt and maximal bone modeling *(42)*. Fractures of the distal end of the forearm in adolescents may satisfy the criteria of bone fragility fracture comparable to the same in adults. One-third of all fractures in children could not be related to a specific activity or environment, but rather to some other contributing factors such as nutrition. Chan et al. (17) were first to suggest that low dietary calcium intake might contribute to bone fragility fractures in children. However, as the results of the study were based on a few cases, this fact remains to be confirmed. In a study in Palma de Mallorca, Spain, a significant difference in the fracture rate was found when cities with a high calcium content in their water (282 mg/L) were compared with those with a lower calcium content (86 mg/L) *(43)*.

All of the clinical trials with calcium supplements in children and adolescents completed to date were relatively short in duration (12 mo to 3 yr) and showed a

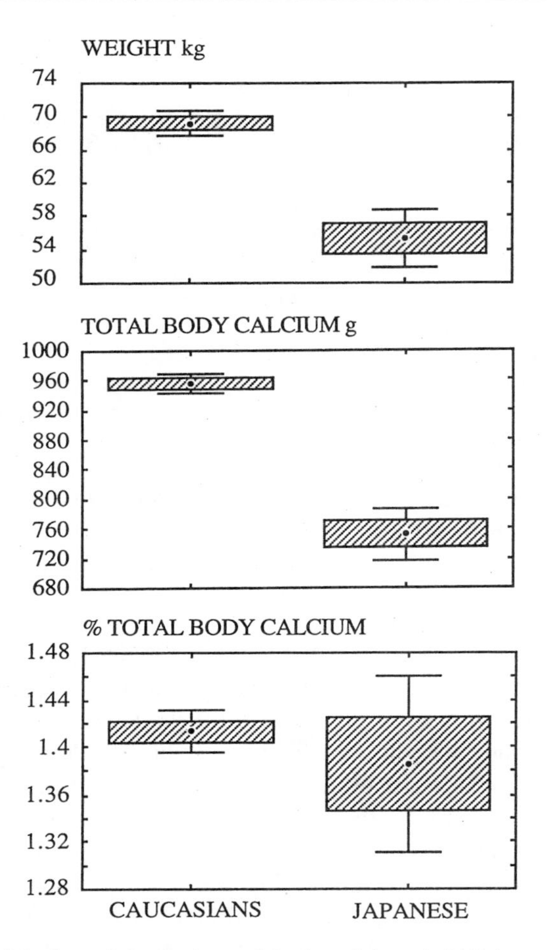

Fig. 14. Box plots of body weight (top), total body calcium (middle), and percent total body calcium (bottom) in the group of young adult (18–50 yr) Caucasian ($n = 437$) and Japanese ($n = 34$) premenopausal women. Results presented as mean – SE – 1.96SE. Significant differences are present for body weight and total body calcium, but not for relative body calcium to weight. Data for Japanese women kindly provided by Drs. T. Fujita and A. Tomita. The average total body calcium in the group of Japanese women is 752 – 107 g and in Caucasians is 956 – 145 g. Subtracting 30 g of skeletal calcium at birth and assuming that most of the bone mass was accumulated by the age 18 yr, the average Japanese and Caucasian women from the group were retaining 109 and 141 mg of Ca/d on average, respectively. (Adapted from ref. *57.*)

positive effect of calcium on bone mass of young individuals *(1,28,44–51)*. The increase in bone mass observed in those studies could be explained to a large extent by the remodeling transient phenomenon emphasized recently by Heaney *(52)*. When calcium intervention was discontinued in some of the studies, the difference in bone mass between the groups diminished *(53,54)*, indicating that some

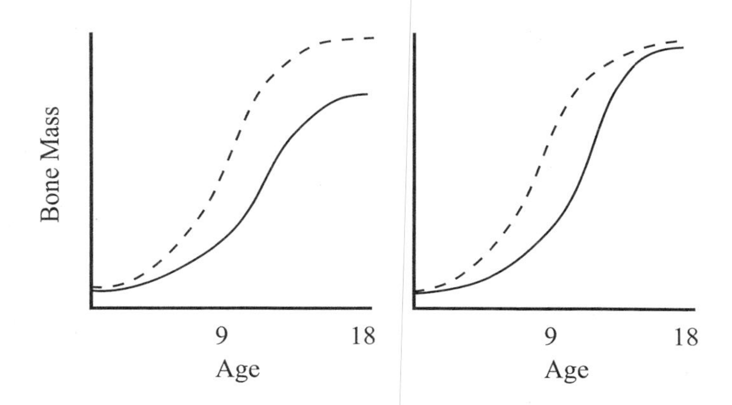

Fig. 15. Hypothetical models of bone mass accumulation with calcium supplementation during growth. A very low dietary calcium intake over time results in a permanent deficit in peak bone mass at the time of skeletal maturity (left). Recovery of the transient deficit in bone mass during growth by the time of skeletal maturity (right). Calcium-supplemented individuals will reach peak bone mass earlier than nonsupplemented subjects. The final level of peak bone mass is the same in the two groups. (Adapted from ref. *56.*)

of the gain was lost as the result of the second transient. In some studies the effect of intervention were still maintained 1–3 yr after discontinuation of dairy products, fortified foods, or calcium supplementation. This may be specific to the dairy products *(47),* the way calcium supplements were given *(49),* and/or the level of calcium intake in the habitual diet *(28,55).*

It is highly unlikely that calcium deficiency alone as found in Western diets could lead to a reduction in bone size. Calcium deficiency during rapid skeletal growth disturbs the balance in the internal bone remodeling, leading to a reduction in bone mass. This defect is transient and can be recovered when bone modeling changes to the bone consolidation phase, with a rapid decline in calcium requirement (the drop in calcium intake threshold from ~1500 mg/d to ~900 mg/d). Permanent reduction in bone mass at skeletal maturity (peak bone mass) is expected in young individuals whose internal bone architecture was not developed appropriately or was sacrificed as a result of a relative calcium deficiency to compensate for longitudinal bone growth and periosteal bone expansion (Fig. 15) *(56).* Preliminary results of a long-term intervention study with calcium supplementation extending from childhood to young adulthood confirmed the above statements, showing that both models are adequate depending on the relative degree of calcium deficiency in the habitual diet *(23).*

In addition to calcium, phosphorus is essential for normal bone and tooth formation and therefore plays a very important role during skeletal development. Of about 700 g of phosphorus contained in the human body, approx 85% is in the bone while the remaining part is in the soft tissues, where it plays an important role in energy storage and release systems. Phosphorus is a ubiquitous element present in almost all the foods we consume. Most of the consumed phosphorus is excreted in the feces and in the urine. Phosphorus balance studies in adults showed that phosphorus output is equal to input at various intake levels from 700 to 1800 mg/d, indicating excellent adaptation. Owing to the lack of balance studies at very low phosphorus intakes, Nordin concluded that it is almost impossible to calculate phosphorus requirements for adults *(26)*. There is presumably an intake below which adult humans go into negative balance; however, this figure is unknown. As phosphorus is an essential component of calcium hydroxyapatite crystal, growing individuals should be in a positive phosphorus balance. As in adults, phosphorus output (urinary and fecal excretion) in adolescents is highly related to phosphorus input (dietary phosphorus). Most young females are in positive phosphorus balance of about 97 ± 17 mg/d, regardless of their phosphorus intake *(57)*. The regression line of phosphorus output on phosphorus intake for the particular intake range (800–2000 mg/d) has a slope of 0.96 and is almost parallel to the line of equality, with an intercept of –58. The difference between the lines is due to phosphorus retention in the body, required primarily for skeletal development. A positive phosphorus balance of about 150–200 mg/d is expected during the pubertal growth spurt, age approx 12. As sodium contributes to calcium excretion in the urine, it can also influence phosphorus excretion and affect phosphorus balance. As milk and dairy products are the main source of calcium in the diet, they are also a good source of phosphorus, with a Calcium/Phosphorus (Ca:P) ratio of 1.3. High calcium intake may negatively affect phosphorus absorption, as documented for adults *(58)*. A similar phenomenon exists during growth, however, the significance of this for young people has not been established (Fig. 16). This may be of importance for adolescents requiring calcium supplements, therefore disturbing Ca:P ratio in the diet with impact on skeletal mineralization due to a relative phosphorus deficiency. The consumption of milk should be encouraged among adolescents because it contains both minerals important for skeletal mineralization in a favorable ratio. The negative Ca:P ratio on the other side has no deleterious effects on bone of young individuals if calcium intakes meet the requirement for growth *(57)*.

Protein calorie malnutrition during childhood can cause growth retardation and decreased formation of cortical bone, and therefore can interfere with peak bone mass acquisition *(59)*. This is probably mediated by IGF-I and leptin through its effect on the reproductive function. Serum IGF-I is considered a biochemical marker of nutritional status, primarily protein intake (Fig. 17). Excessive protein intake and increase in protein consumption above the recommended allowance level can be associated with hypercalciuria; however, this has not been confirmed in children *(41)*. A strong positive nitrogen balance is required for growth, and therefore the amounts of proteins commonly consumed by young Americans have no influence on urinary calcium excretion.

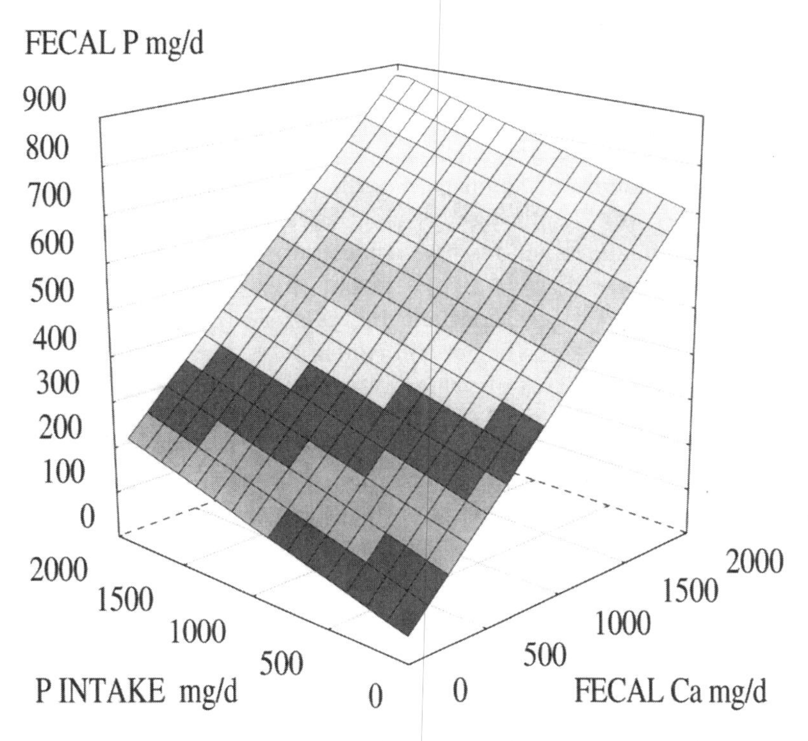

Fig. 16. Three-dimensional representation (surface plot) of the relationship between fecal phosphorus *(z)*, fecal calcium *(y)*, and phosphorus intake *(x)* in a group of adolescent females ($N = 42$). Fecal calcium was the main determinant of fecal phosphorus ($p < 0.0001$), while phosphorus intake had minimal effect ($p < 0.095$). Phosphorus balances were collected from the same sources as calcium balances presented earlier, ref. *13*.

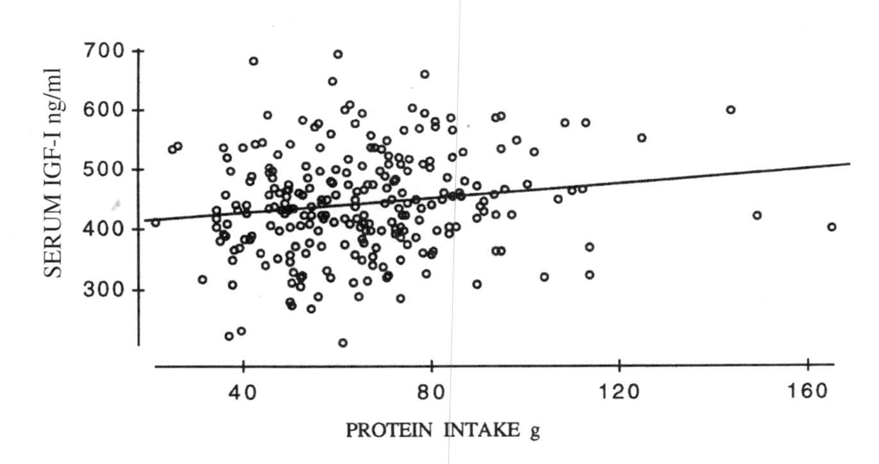

Fig. 17. Scatterplot with the regression line of the relationship between serum insulin-like growth factor (IGF)-I and protein intake in a cohort of healthy adolescent females ($N = 229$). In a stepwise regression analysis along with chronological age, skeletal age, and sexual maturity index, the effect of dietary protein on serum IGF-I was found to be significant at $p < 0.010$, partial-$R = 2.6\%$). IGF-I measured by Nichols-RIA in the laboratory of Dr. C. Rosen, Bangor, ME).

5. NUTRIENT INTERACTIONS DURING CHILDHOOD AND ADOLESCENCE

The National Academy of Sciences, Food and Nutrition Board, as well as the National Institutes of Health Consensus Panel on Optimal Calcium Intake increased calcium intake standards for teenagers to 1300 and 1500 mg/d, respectively *(21,60)*. However, there is a concern that high calcium intake may lead to decreased absorption of other important minerals. However, based on the results of several recent studies, it can be concluded that high dietary calcium intake does not influence nutritional status of some of the major minerals and trace elements (Magnesium, Iron, Zinc, Selenium) in children and adolescents *(61–64)*. Therefore, public health measures to elevate calcium intakes among young Americans to the new standards (up to 1500 mg/d) are safe recommendations and should not trigger a concern for the possible induction of iron deficiency anemia, hypomagnesemia, zinc deficiency, and growth retardation, or predispose young people to cardiomyopathy as the result of selenium deficiency.

6. CONCLUSION

Calcium, phosphorus, and proteins are essential for bone growth and skeletal development. Calcium deficiency has been associated with rickets in children, bone fragility fractures during pubertal growth spurt, and with inadequate peak bone mass formation by young adulthood. Mineral and protein requirements are the highest during pubertal growth spurt to allow for longitudinal bone growth and periosteal bone expansion. Calcium is a threshold nutrient, and dietary intake standards have been established for the various phases of growth and development. Public health measures should assure that all children and teenagers can meet those standards to allow for optimal skeletal health early in life.

ACKNOWLEDGMENTS

This work was supported in part by National Institutes of Health grants RO1 AR40736-01A1 and CRC-NIH M01-RR00034, and NRICGP/U.S. Department of Agriculture grants to The Ohio State University.

REFERENCES

1. Matkovic V, Fontana D, Tominac C, Goel P, Chesnut CH. Factors which influence peak bone mass formation: a study of calcium balance and the inheritance of bone mass in adolescent females. Am J Clin Nutr 1990; 52:878–888.
2. Matkovic V, Kostial K, Simonovic I, Buzina R, Brodarec A, Nordin, BEC. Bone status and fracture rates in two regions of Yugoslavia. Am J Clin Nutr 1979; 32:540–549.
3. Heaney RP, Abrams S, Dawson-Hughes B, et al. Peak Bone Mass. Osteopor Int 2000; 11:985–1009.
4. Widdowson EM. Growth and body composition in childhood. In: Brunser O, Carrazza F, Gracey M, Nichols B, Senterre J, eds. Clinical Nutrition of the Young Child. Raven, New York, 1985, pp. 1–21.

5. U.S. Department of Health and Human Services, Public Health Service. Healthy People 2010. Vol I. understanding and improving health. National Health Promotion and Disease Prevention Objectives. Jones and Bartlett, Boston, 2000.
6. Heaney RP, Matkovic V. Inadequate peak bone mass. In: Riggs BL, Melton LJ, eds. Osteoporosis: Etiology, Diagnosis and Management, 2nd ed. Lippincott-Raven, Philadelphia, 1995, pp. 115–131.
7. Hu JF, Zhao XH, Jia JB, Parpia B, Campbell TC. Dietary calcium and bone density among middle-aged and elderly women in China. Am J Clin Nutr 1993; 58:219–227.
8. Sandler RB, Slemenda C, LaPorte RE, et al. Postmenopausal bone density and milk consumption in childhood and adolescence. Am J Clin Nutr 1985; 42:270–274.
9. Glastre C, Braillon P, David L, Cochat P, Meunier PJ, Delmas PD. Measurement of bone mineral content of the lumbar spine by dual energy X-ray absorptiometry in normal children: correlations with growth parameters. J Clin Endocrinol Metab 1990; 70:1330–1333.
10. Bonjour JP, Theintz G, Buchs B, Slosman D, Rizzoli R. Critical years and stages of puberty for spinal and femoral bone mass accumulation during adolescence. J Clin Endocrinol Metab 1991; 73:555–563.
11. Matkovic V, Jelic T, Wardlaw GM, et al. Timing of peak bone mass in caucasian females and its implication for the prevention of osteoporosis. Inference from a cross-sectional model. J Clin Invest 1994; 93:799–808.
12. Recker RR, Davies KM, Hinders SM, Heaney RP, Stegman MR, Kimmel DB. Bone gain in young adult women. JAMA 1992; 268:2403–2408.
13. Matkovic V. Calcium metabolism and calcium requirements during skeletal modeling and consolidation of bone mass. Am J Clin Nutr 1991; 54:245S–260S.
14. Matkovic V, Heaney RP. Calcium balance during human growth: evidence for threshold behavior. Am J Clin Nutr 1992; 55:992–996.
15. Pettifor JM, Ross FP, Moodley G, DeLuca HF, Travers R, Glorieux FH. Calcium deficiency rickets associated with elevated 1,25-dihydroxyvitamin D concentrations in a rural black population. In: Norman AW, Schaefer K, Herrath DV, et al., eds. Vitamin D, Basic Research and Its Clinical Application. Walter de Gruyter, New York, 1979, pp. 1125–1127.
16. Thacher TD, Fischer PR, Pettifor JM, et al. A comparison of calcium, vitamin D, or both for nutritional rickets in Nigerian children. N Engl J Med 1999; 341:563–568.
17. Chan GM, Hess M, Hollis J, Book LS. Bone mineral status in childhood accident fractures. Am J Dis Child 1984; 139:569–570.
18. Goulding A, Cannan R, Williams SM, Gold EJ, Taylor RW, Lewis-Barned NJ. Bone mineral density in girls with forearm fractures. J Bone Miner Res 1998; 13:143–148.
19. Begum A, Pereira SM. Calcium balance studies on children accustomed to low calcium intakes. Br J Nutr 1969; 23:905–911.
20. Fleming KH, Heimbach JT. Consumption of calcium in the U.S.: food sources and intake levels. J Nutr 1994; 124:1426S–1430S.
21. Dietary Reference Intakes. Food and Nutrition Board, Institute of Medicine, National Academy Press, Washington, DC, 1997.
22. Ilich JZ, Skugor M, Hangartner T, Baoshe A, Matkovic V. Relation of nutrition, body composition, and physical activity to skeletal development: a cross-sectional study in preadolescent females. J Am Coll Nutr 1998; 17:136–147.
23. Matkovic V, Badenhop-Stevens NE, Landoll JD, Goel P, Li B. Long term effect of calcium supplementation and dairy products on bone mass of young females. J Bone Miner Res 2002; 17:S172.
24. Black RE, Williams SM, Jones IE, Goulding A. Children who avoid drinking cow milk have low dietary intakes and poor bone health. Am J Clin Nutr 2002; 76:675–680.
25. Orr JB. Milk consumption and the growth of school children. Lancet 1928; 1:202–203.
26. Nordin BEC. Nutritional consideration. In: Nordin BEC, ed. Calcium, Phosphate and Magnesium Metabolism. Churchill Livingstone, Edinburgh, 1976, pp. 1–35.

27. Prentice A, Stear SJ, Ginty F, Jones SC, Mills L, Cole TJ. Calcium supplementation increases height and bone mass of 16–18 year old boys. J Bone Miner Res 2002; 17:S397.
28. Dibba B, Prentice A, Ceesay M, Stirling DM, Cole TJ, Poskitt EME. Effect of calcium supplementation on bone mineral accretion in Gambian children accustomed to a low-calcium diet. Am J Clin Nutr 2000; 71:544–549.
29. Moll GW, Rosenfield RL, Fang VS. Administration of low dose estrogen rapidly and directly stimulates growth hormone production. Am J Dis Child 1986; 140;124–127.
30. Ross JL, Cassorla FG, Skerda MC, Valk IG, Loriaux L, Culter GB. A preliminary study of the effect of estrogen dose on growth in Turner's syndrome. New Engl J Med 1983; 309:1104.
31. Garn SM. The Earlier Gain and the Later Loss of Cortical Bone. Charles C Thomas, Springfield, IL, 1970.
32. Bailey DA, Wedge JH, McCulloch RG, Martin AD, Bernhardson SC. Eidemiology of fractures of the distal end of the radius in children as ssociated with growth. J Bone Joint Surg 1989; 71-A, 8:1225–1231.
33. Matkovic V, Ciganovic M, Tominac C, Kostial K. Osteoporosis and epidemiology of fractures in Croatia. An international comparison. Henry Ford Hosp Med J 1980; 28:116–126.
34. Rigotti NA, Nussbaum SR, Herzog DB, Neer RM. Osteoporosis in women with anorexia nervosa. N Engl J Med 1984; 311:1601–1606.
35. Matkovic V, Ilich JZ, Skugor M, et al. Leptin is inversely related to age at menarche in human females. J Clin Endocrinol Metab 1997; 82:3239–3245.
36. Matkovic V, Ilich JZ. Calcium requirements during growth. Are the current standards adequate? Nutr Rev 1993; 51:171–180.
37. Charles P, Taagehoj Jensen F, Mosekilde L, Hvid Hansen H. Calcium metabolism evaluated by 47Ca kinetics: estimation of dermal calcium loss. Clin Sci 1983; 65:415–422.
38. Klesges RC, Ward KD, Shelton ML, et al. Changes in bone mineral content in male athletes. Mechanisms of action and intervention effects. JAMA 1996; 276:226–230.
39. Ilich JZ, Badenhop NE, Jelic T, Clairmont AC, Nagode LA, Matkovic V. Calcitriol and bone mass accumulation in females during puberty. Calcif Tissue Int 1997; 61:104–109.
40. Weaver CM, Martin BR, Plawecki KL. Diferences in calcium metaboilism between adolescent and adult females. Am J Clin Nutr 1995; 61:577–581.
41. Matkovic V, Ilich JZ, Andon MB, et al. Urinary calcium, sodium, and bone mass of young females. Am J Clin Nutr 1995; 62:417–425.
42. Alffram PA, Bauer GCH. Epidemiology of fractures of the forearm. J Bone Joint Surg 1962; 44A:105–114.
43. Verd Vellespir S, Dominguez Sanches J, Gonzales Quintial M, et al. Asociacion entre el contenido en calcio de las aguas de consumo y las fracturas en los ninos. An Esp Pediatr 1992; 37:461–465.
44. Johnston CC Jr, Miller JZ, Slemenda CW, et al. Calcium supplementation and increases in bone mineral density in children. N Engl J Med 1992; 327:82–87.
45. Lloyd T, Andon MB, Rollings N, et al. Calcium supplementation and bone mineral density in adolescent girls. JAMA 1993; 270:841–844.
46. Lee WTK, Leung SSF, Wang SF, et al. Double-blind, controlled calcium supplementation and bone mineral accretion in children accustomed to a low-calcium diet. Am J Clin Nutr 1994; 60:744–750.
47. Cadogan J, Eastell R, Jones N, Barker ME. Milk intake and bone mineral acquisition in adolescent girls: randomised, controlled intervention trial. Br Med J 1997; 315:1255–1260.
48. Chan GM, Hoffman K, McMurray M. Effect of dairy products on bone and body composition in pubertal girls. J Ped 1995; 126:551–556.
49. Bonjour JP, Carrie AL, Ferrarri S, et al. Calcium-enriched foods and bone mass growth in prepubertal girls: a randomized, double-blind, placebo-controlled, trial. J Clin Invest 1997; 99:1287–1294.

50. Nowson CA, Green RM, Hopper JL, et al. A co-twin study of the effect of calcium supplementation on bone density during adolescence. Osteopor Int 1997; 7:219–225.
51. Merriles MJ, Smart EJ, Gilchrist NL, et al. Effects of dairy food supplements on bone mineral density in teenage girls. Eur J Nutr 2000; 39:256–262.
52. Heaney RP. Interpreting trials of bone-active agents. Am J Med 1995; 98:329–330.
53. Slemenda C, Reister TK, Peacock M, Johnston CC Jr. Bone growth in children following the cessation of calcium supplementation. J Bone Miner Res 1993; 8:S154.
54. Lee WTK, Leung SSF, Leung DMY, Cheng JCY. A follow-up study on the effects of calcium-supplement withdrawl and puberty on bone acquisition of children. Am J Clin Nutr 1996; 64:71–77.
55. Dibba B, Prentice A, Ceesay M, et al. Bone mineral contents and plasma osteocalcin concentrations of Gambian children 12 and 24 mo after the withdrawl of a calcium supplement. Am J Clin Nutr 2002; 76:681–686.
56. Matkovic V. Can osteoporosis be prevented? Bone mineralization during growth and development. In: Johnston FE, Zemel B, Eveleth PB, eds. Human Growth in Context. Smith-Gordon, London, UK, 1999, pp. 183–193.
57. Matkovic V, Badenhop NE, Ilich JZ. Trace element and mineral nutrition in healthy people: adolescents. In: Bogden JD, Klevay LM, eds. The Clinical Nutrition of the Essential Trace Elements and Minerals—The Guide for Health Professionals. Humana, Totowa, NJ, 2000, pp. 153–182.
58. Heaney RP, Nordin BEC. Calcium effects on phosphorus absorption: implications for the prevention and co-therapy of osteoporosis. J Am Coll Nutr 2002; 21:239–244.
59. Garn SM, Rohmann CG, Behar M, Viteri F, Gozman M. Compact bone deficiency in proteincalorie malnutrition. Science 1964; 145:1444–1445.
60. NIH Consensus Conference: Optimal calcium intake. JAMA 1994; 272:1942–1948.
61. Ilich-Ernst JZ, McKenna AA, Badenhop NE, et al. Iron status, menarche, and calcium supplementation in adolescent girls. Am J Clin Nutr 1998; 68:880–887.
62. Andon MB, Ilich JZ, Tzagournis MA, Matkovic V. Magnesium balance in adolescent females consuming a low or high calcium diet. Am J Clin Nutr 1996; 63:950–953.
63. McKenna AA, Ilich JZ, Andon MB, Wang C, Matkovic V. Zinc balance in adolescent females consuming a low- or high-calcium diet. Am J Clin Nutr 1997; 65:1460–1464.
64. Holben D, Smith AM, Ha EJ, Ilich JZ, Matkovic V. Selenium (Se) absorption, balance, and status in adolescent females throughout puberty. FASEB J 1996; 10:A532.

12 Calcium and Vitamin D for Bone Health in Adults

Bess Dawson-Hughes

Osteoporotic fractures are common and devastating occurrences. Many lifestyle risk factors for osteoporosis have been identified and, generally speaking, their effects are additive. Low calcium and vitamin D intakes are established risk factors for osteoporosis. This chapter reviews the impact of dietary calcium and vitamin D on calcium homeostasis, examines evidence used to define the intake requirements of these nutrients, and considers their impact on fracture rates, their interface with pharmacotherapy for osteoporosis, and current intake recommendations.

1. PHYSIOLOGY

An inadequate intake of calcium and an inadequate level of vitamin D, alone and in combination, influence calcium-regulating hormone levels. Deficiency of either nutrient results in reduced calcium absorption and a lower circulating ionized calcium concentration. The latter stimulates the secretion of parathyroid hormone (PTH), a potent bone-resorbing agent. Over time, a small increase in the circulating level of PTH leads to measurable and significant bone loss and increased risk of fracture (Fig. 1).

1.1. Calcium

Intestinal calcium absorption occurs by active transport and by passive diffusion. In individuals with low to moderate calcium intakes, absorption occurs largely by active transport, a process mediated by 1,25-dihydroxyvitamin D (1,25[OH]$_2$D). As intake increases above approx 500 mg/d, passive diffusion accounts for an increasing proportion of calcium absorbed *(1)*. Estrogen has a direct effect on intestinal responsiveness to 1,25(OH)$_2$D *(2)*. There is an age-related decline in calcium absorption efficiency in men and women *(3)*. This may be related to loss of intestinal vitamin D receptors or resistance of these receptors to the action of 1,25(OH)$_2$D *(4)*. Aging, season, and race also influence calcium absorption efficiency.

From: *Nutrition and Bone Health*
Edited by: M. F. Holick and B. Dawson-Hughes © Humana Press Inc., Totowa, NJ

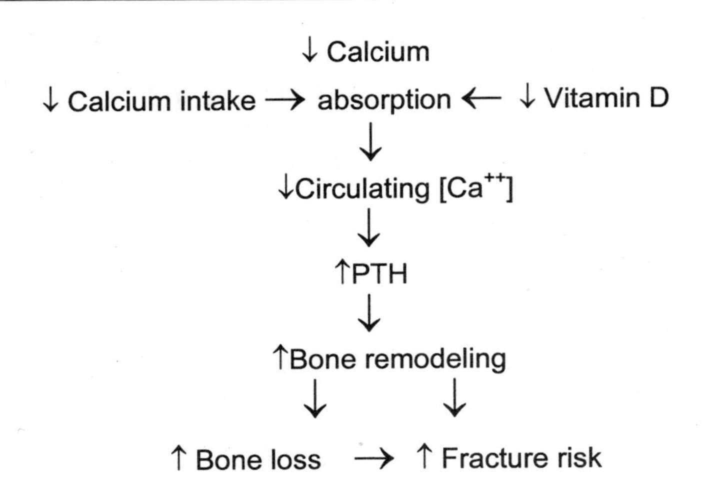

Fig. 1. Consequences of calcium and/or vitamin D insufficiency.

Calcium serves as a substrate to support the bone formation phase of bone remodeling in adults. Typically, about 5 mmol (200 mg) of calcium is removed from the adult skeleton and replaced each day. To supply this, one would need to consume about 600 mg of calcium, because calcium is not very efficiently absorbed. This intake estimate is an approximation, since some of the absorbed calcium would be excreted in sweat, urine, and feces and thus not available for deposition in bone and some of the resorbed calcium would be recycled back into bone at new remodeling sites. The amount of calcium needed to replace daily skeletal resorption losses might be thought of as the subsistence requirement. Several studies have demonstrated a significant impact of increasing calcium intake to the subsistence level in subjects with very low usual calcium diets. For example, supplementation with 500 mg/d of calcium had a greater positive effect on change in bone mineral density (BMD) among postmenopausal women with self-selected calcium intakes under 400 mg/d than among those with intakes in the range 400–650 mg/d *(5)*. In a large case-controlled study of hip fracture risk in women in Europe *(6)*, fracture risk declined as calcium intake rose to an estimated milk score of 4.6 (equivalent to a total calcium intake of 500 mg/d, assuming that milk accounts for about half of all dietary calcium, as has been reported). Notably, only 10% of the women studied had calcium intakes above 500 mg/d *(6)*. Were calcium considered to be solely a substrate for bone formation, the average calcium requirement might be about 600 mg (and the associated recommended dietary intake therefore about 800 mg/d).

The second mechanism by which calcium affects the skeleton is through its impact on the remodeling rate. A high remodeling rate is an independent risk factor for fracture *(7)*. This is perhaps because a high remodeling rate causes more trabecular perforations (or greater architectural deformity). A high remodeling rate may also lead to incomplete mineralization at new remodeling sites. Currently approved treatments for osteoporosis, all antiresorptive agents, may rely more on lowering the remodeling rate than on increasing BMD for their antifracture efficacy. With antiresorptive treatments that lower fracture rates in half, biochemical markers of bone turnover decrease by 40–50%, whereas BMD increases by only 3–5%. Dietary calcium at sufficiently high levels, usually 1000 mg/d or more, lowers the bone remodeling rate by about 10–20% in older men and women *(8–11)*. The degree of suppression appears to be dose related, as illustrated by Elders, who treated postmenopausal Dutch women with either 1000 or 2000 mg of supplemental calcium *(11)*.

1.2. Vitamin D

Vitamin D is acquired from diet and from skin synthesis. Serum 25-hydroxyvitamin D [25(OH)D] levels decline with aging for several reasons. There is a decline in the amount of 7-dehydrocholesterol, the precursor to vitamin D, in the epidermal layer of skin with aging *(12)*. In addition to having a decreased capacity to photosynthesize vitamin D, older people often avoid sun exposure or use sun screens that block access of ultraviolet B rays to the skin. There does not appear to be any impairment in the intestinal absorption of vitamin D with aging *(13)*.

Season is a major determinant of vitamin D levels in people residing in the temperate zone. At 42° North (the latitude of Boston), skin synthesis of vitamin D does not occur between October and March. In healthy postmenopausal women in Boston, mean 25(OH)D levels ranged from a low of 63 nmol/L in March to 95 nmol/L in August *(14)*. In these women, serum PTH levels varied inversely with the serum 25(OH)D levels *(14)*. Higher wintertime levels of PTH raise the possibility that bone loss may be increased during the wintertime. In two prospective studies, bone loss from the spine *(15)* and femoral neck *(16)* was greater in the 6-mo period when PTH levels were highest (winter/spring) than in the 6 mo when PTH levels were lowest *(15,16)*. In both of these studies, supplementation with vitamin D preferentially attenuated wintertime bone loss.

2. DEFINING REQUIREMENTS

2.1. Indicators of Calcium Adequacy

2.1.1. Intake Associated With Maximal Calcium Retention

Historically calcium balance studies have been performed to identify the intake associated with the zero balance, or the intake at which calcium is neither lost nor gained from the body. Balance studies may also be used to identify the intake associated not with zero balance but with the most favorable balance that can be achieved as a result of increasing calcium intake. Increasing intake above the level associated with maximal calcium retention would result in more calcium being absorbed, but that calcium would be excreted rather than retained.

Most balance studies have been conducted in subjects at their self-selected calcium intakes. Because usual intakes of most men and women are low, these studies have not included enough subjects with high enough calcium intakes to be useful for determining the point of maximal retention. However, there is one exception. Spencer *(17)* performed balance studies in 181 men, age 34–71 yr, at six different calcium intake levels ranging from 234 to 2320 mg/d. Diets were supplemented with either calcium gluconate or with milk for periods averaging 20–38 d. Although the zero balance point in this study occurred at an intake of 800 mg/d, calcium retention increased significantly with increasing intake up to a maximum of about 1200 mg/d, in both the calcium gluconate and milk groups.

There are two reports of balance studies with small numbers of women consuming over 1000 mg/d of calcium *(18,19)*. In both, there were positive correlations between intake and balance. These studies support the findings of Spencer *(17)* that increasing intake up to about 1200 mg/d is associated with increasing skeletal calcium retention. They do not, however, exclude the possibility that even higher calcium intakes could result in greater retention in women, as a result of further suppression of the remodeling rate. If short-term balance studies define longer term patterns, then greater retention would signify greater bone mass.

2.1.2. Calcium and Changes in BMD

Many randomized, controlled calcium-intervention trials with change in BMD as the primary endpoint have been reported. In a meta-analysis of 13 trials, calcium induced significant mean gains (or slowed loss) of 0.6% at the distal forearm, 3% at the spine, and 2.6% at the femoral neck *(20)*. Mean differences after the first year were also positive, but smaller. The relatively strong initial response to calcium is attributed to closure of remodeling space. A more recent meta-analysis of 15 trials found that calcium caused positive mean percentage changes from baseline of 1.66% at lumbar spine, 1.64% at the hip, and 1.91% at the distal radius *(21)*.

Time since menopause appears to influence the impact of calcium on changes in BMD. In the first 5 yr after menopause, rapid bone loss occurs as a result of declining estrogen levels. The increased bone resorption that accompanies estrogen deficiency provides calcium to the blood and other extracellular space. As a result, serum PTH levels decline, $1,25(OH)_2D$ levels decline, and thus the signal to absorb calcium declines. Increasing calcium intake does not entirely reverse this sequence and cannot be relied upon to prevent early menopausal bone loss, although it does in some studies attenuate it *(11,22)*. Calcium is generally effective in older postmenopausal women who have completed their skeletal adaptation to the loss of estrogen *(5)*. In large enough doses, calcium can reverse age-related increases in serum PTH and in bone remodeling *(23)*. In one trial, the effects of calcium from food (milk powder) and supplement sources on changes in BMD in older postmenopausal women were compared and found to be similar *(24)*.

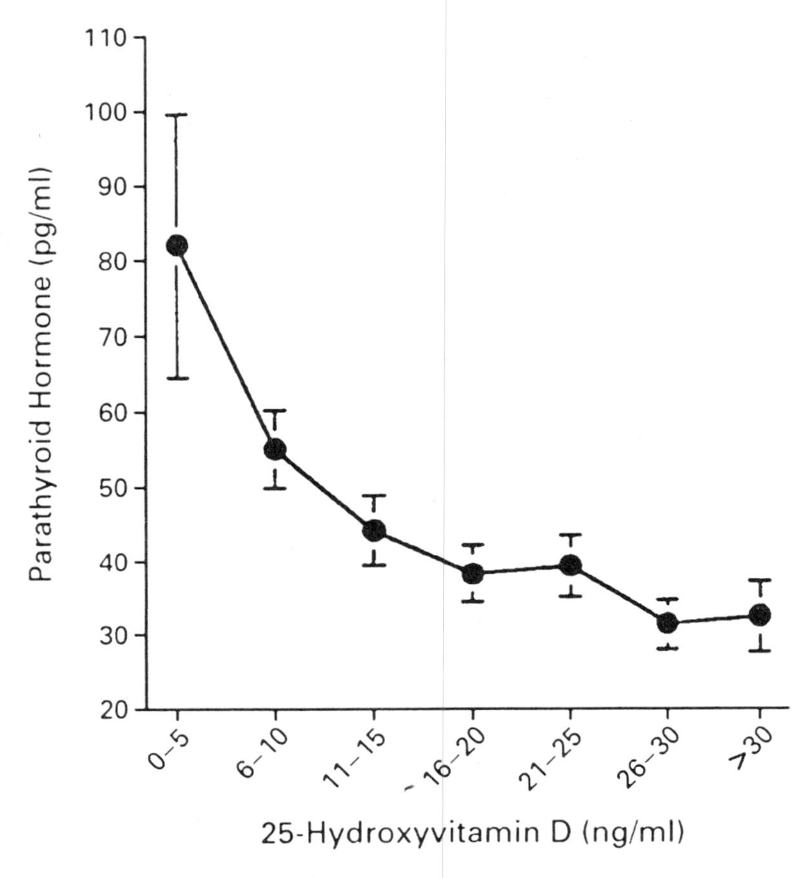

Fig. 2. Relation between serum 25(OH)D and parathyroid hormone (PTH). For PTH, 1 pg/mL = 1 ng/L; for 25(OH)D, 1 ng/mL = 2.5 nmol/L. (From Thomas et al. *[26]*, with permission.)

2.2. Indicators of Vitamin D Adequacy

2.2.1. Maximal Suppression of PTH

Declines in serum 25(OH)D trigger compensatory increases in circulating PTH concentrations. Malabanan et al. *(25)* assessed serum PTH responses to treatment with vitamin D according to the initial 25(OH)D levels of the subjects. In this study, 35 subjects were treated with 50,000 IU/wk of vitamin D for 8 wk. Subjects with 25(OH)D levels below 50 nmol/L (20 ng/mL) had significant declines in serum PTH with treatment, but those with initial 25(OH)D concentrations above this level did not. Vitamin D adequacy in older adults is often defined as the level of 25(OH)D needed to maximally suppress serum PTH levels. Typically, the association is not linear but hyperbolic, as illustrated in Fig. 2 *(26)*. The participants in this study were patients admitted to the medical service at an urban hospital in Boston. Other investigators have documented similar associations between serum 25(OH)D and PTH levels *(14,27–30)*. Analyses were performed in these five cross-sectional data sets to identify the inflection point, or the 25(OH)D level above which serum PTH no longer decreases significantly, have yielded a wide range of estimates. One estimate was 25 nmol/L (10 ng/mL) *(27)* and the other four were within the range 75–110 nmol/L (30 to 44 ng/mL) *(14,28–30)*. Reasons for the rather large discrepancies in threshold 25(OH)D levels are not clear. Some of the factors that may have influenced the estimates include subject age, ambient calcium intake, and 25(OH)D assay differences. The latter, however, does not appear to be fully explanatory *(31)*. Until new evidence emerges indicating a better goal, it is probably wise to maintain 25(OH)D levels at 80 nmol/L (32 ng/mL) or above in the elderly, that is, in the upper half of the normal range. This implies that many subjects in the reference population for the assay had some degree of vitamin D insufficiency, an assumption supported by the evidence in many of the clinical studies discussed above. More recently, a similar inverse association of serum PTH with serum 25(OH)D levels has been documented in adolescent girls *(32)* and in healthy young men and women *(33)*, but inflection points have not been identified in these younger populations.

2.2.2. Maximal BMD

A desirable functional endpoint for assessing the vitamin D requirement would be its effect on the development and preservation of BMD in children and adults and the prevention of fractures in older people. Such evidence is very difficult to obtain from randomized controlled trials because no more than one or two doses can be tested in any given study. A second and less precise option is to examine the association between serum levels of 25(OH)D and BMD in cross-sectional studies. In a Dutch cohort, 25(OH)D and BMD were positively associated up to a 25(OH)D level of 30 nmol/L (12 ng/mL) but not at higher 25(OH)D levels *(34)*. The inflection point of 30 nmol/L is very close to the 25(OH)D level that was associated with maximal suppression of PTH in the same population (25 nmol/L) *(35)*. The association between 25(OH)D and BMD should be examined in other populations. Meanwhile, the 25(OH)D level needed to maximally suppress PTH is being used as the functional end point.

By any of these criteria, vitamin D deficiency is commonplace in older adults. In Boston, for example, 50% of patients admitted to the medical service of a local hospital had 25(OH)D levels below 37 nmol/L (14.8 ng/mL) *(26)*, 50% of patients admitted to another local hospital with acute hip fractures had 25(OH)D levels below 30 nmol/L (12 ng/mL) *(36)*, and 80% of residents of a local nursing home had 25(OH)D levels below 37 nmol/L *(37)*. In contrast, only 14% of patients seen in a local outpatient thyroid clinic had 25(OH)D levels below 40 nmol/L (16 ng/mL) *(38)*.

It is useful to know what vitamin D intake is needed to raise the serum 25(OH)D to the desired level of 80 nmol/L (32 ng/mL) or higher. We examined this question in 376 men and women who participated in a calcium and vitamin D

Table 1
Percentages of Elderly Men and Women (*N* = 376) Whose 25(OH)D Concentrations Reached Selected Levels After 1 yr of Supplementation With 700 IU/d of Vitamin D and 500 mg/d Calcium or Placebo *(39)*

Group	*25(OH)D (nmol/L)*	*Men (%)*	*Women (%)*
Vitamin D + calcium	20	100.0	100.0
	40	100.0	99.0
	60	97.7	96.9
	80	89.8	86.6
	100	75.0	71.1
Placebo	20	100.0	95.1
	40	96.6	84.5
	60	77.3	58.3
	80	56.8	28.2
	100	35.2	15.5

intervention study *(10)*. The men and women in this study consumed an average of 200 IU/d of vitamin D (and 800 mg/d of calcium) in their diets and were supplemented with 700 IU/d of vitamin D (plus 500 mg/d of calcium) or with placebo. Serum 25(OH)D levels were measured throughout the year. After 1 yr of supplementation, 90% of the men and 87% of the women taking 700 IU/d of vitamin D had 25(OH)D levels of 80 nmol/L (32 ng/mL) and above. In contrast, only about half of the men and one-third of the women who were taking placebo had levels of 80 nmol/L and above (Table 1) *(39)*. This indicates that a total intake of 900 IU/d of vitamin D (200 IU from the diet and 700 IU from the supplement) is adequate for most healthy men and women age 65 and older but that an intake greater than 900 IU/d would be required to meet the needs of 98% of the population (i.e., for a Recommended Daily Allowance).

A more refined if impractical approach is to make season-specific vitamin D intake recommendations. In a survey of outpatients conducted in the months of January, February, and March in Boston, when skin synthesis is minimal, 65% of the subjects who took 400 IU/d of supplemental vitamin D had 25(OH)D levels equal to or greater than 80 nmol/L (32 ng/mL), whereas only 30% of the unsupplemented subjects reached this level *(38)*. Clearly, the commonly recommended 400 IU/d is not sufficient for many adults.

3. IMPACT OF CALCIUM AND VITAMIN D ON FRACTURE RATES

3.1. Calcium

Little is known about the impact of calcium on fracture rates *(9,40–42)*. The recent Shea meta-analysis *(21)* found that calcium alone (vs placebo) tended to lower risk of vertebral fractures (RR 0.77 [CI 0.54–1.09]) but not nonvertebral fractures (RR 0.86 [CI 0.43–1.72]). The studies in this analysis range from 18 mo to 4 yr in duration.

3.2. Vitamin D

The effect of vitamin D on fracture incidence has been variable. Heikinheimo et al. *(43)* reported that an annual intramuscular injection of 150,000–300,000 IU of vitamin D reduced fracture rates in older men and women. In contrast, a large perspective controlled trial found that supplementation with 400 IU/d of vitamin D did not alter hip fracture rates in elderly Dutch men and women *(44)*. More recently, supplementation with 400 IU/d of vitamin D as cod liver oil was found to have no effect on hip or other fracture rates in very elderly men and women (mean age 85 yr) in Norway *(45)*. Similarly, supplementation of early post-menopausal Finnish women with 300 IU/d of vitamin D had no impact on non-vertebral fracture rates *(46)*. In the latter study, supplementation increased the mean 25(OH)D level from 47 to 64 nmol/L (18.8 to 25.6 ng/mL), but there was no decline in serum PTH or in serum osteocalcin, a biochemical marker of bone turnover. The mean calcium intake of this study population was only 450 mg/d. It appears that a dose of vitamin D higher than 400 IU/d is needed to lower the PTH stimulus for bone resorption and, thereby, to lower the fracture rate. In addition, the inadequate calcium intake of the Norwegian and Finnish study subjects also may have impaired their skeletal responses to the vitamin D.

3.3. Combination of Calcium and Vitamin D

Combined vitamin D and calcium supplementation has been studied in a cohort of more than 3200 very elderly French nursing home residents who received 1200 mg of calcium as the triphosphate together with 800 IU/d of vitamin D or double placebo *(47)*. The supplements significantly reduced the incidence of hip fractures and other nonvertebral fractures within 18 mo, and the reduction was sustained throughout the 3-yr study. The nursing home residents were at high risk for fracture on the basis of their mean age of 84 yr, their low self-selected calcium intakes (approx 500 mg/d), and their low initial mean 25(OH)D level of 24 nmol/L (9.6 ng/mL) (measured in a subset). These investigators confirmed their findings in another smaller group of 583 women, mean age 85 yr *(48)*. On the same supplement doses (1200 mg/d of calcium and 800 IU/d of vitamin D), the relative risk of hip fracture in the placebo group was 1.7 compared with the supplemented group (95% CI = 0.96, 3.0; trend). A third study of combined calcium and vitamin D supplementation was conducted in a lower-risk group of 389 healthy ambulatory men and women age 65 and older (mean age 71 yr) *(49)*. In this randomized trial, the subjects received 500 mg/d of calcium as the highly absorbable citrate malate plus 700 IU/d of vitamin D or double placebo. Over the 3-yr intervention period, supplementation reduced rates of bone loss from the spine, hip, and total body and significantly lowered the incidence of clinical fractures by 50%. The subjects in this study consumed approx 700 mg/d of calcium and 200 IU/d of vitamin D in their diets at entry and during the study. The positive effects of combined calcium and vitamin D supplementation in these three studies may be due to the fact that higher doses of vitamin D (compared with the vitamin D intervention studies described above) were used and/or to the concurrent use of calcium.

3.4. Discontinuation of Calcium and Vitamin D

One of the cohorts just described *(10)* was followed for an additional 2 yr after they discontinued the study supplements *(49)*. During that interval the subjects were free to take any supplements they chose and were reminded of the recommendations to consume 1200 mg/d of calcium and 600 IU/d of vitamin D. The subjects returned for BMD measurements at annual intervals. Over the 2-yr follow-up period, all gains in BMD at the spine and hip were lost, so that the two treatment arms were indistinguishable in BMD at the end of 5 yr. This occurred despite the fact that half of the subjects reported taking calcium and vitamin D supplements some of the time during the 2-yr follow-up period. The reversal in BMD was accompanied by a reversal in suppression of bone turnover markers, indicating the reopening of remodeling space. In order to sustain the benefits offered by increased calcium and vitamin D intakes, the higher intakes need to be maintained.

4. CALCIUM, VITAMIN D, AND PHARMACOTHERAPY

In recent randomized, controlled trials testing the antifracture efficacy of the antiresorptive therapies alendronate *(50)*, risedronate *(51)*, raloxifene *(52)*, and calcitonin *(53)* and the anabolic drug, PTH 1-34 *(54)*, calcium, and vitamin D have been given to both the control and intervention groups. This allows one to define the impact of the drug in calcium and vitamin D-replete patients and to conclude that any efficacy of the drug is beyond that associated with calcium and vitamin D alone. Based on the evidence that follows, one cannot conclude that these drugs would have the same efficacy in calcium- and vitamin D-deficient patients.

Nieves et al. *(55)* examined BMD responses of early postmenopausal women to hormone replacement therapy (HRT) in relation to calcium intake. Among the 31 HRT intervention trials that provided information about calcium intake, 20 added calcium to both study arms (bringing mean total intake to 1183 mg/d) and 11 did not (mean calcium intake 563 mg/d). The BMD gains at the spine, hip, and forearm were several-fold greater in the women who increased their calcium intakes than in those who took the HRT without added calcium. From this it appears that calcium enables estrogen to be more effective in building BMD. In a large, observational study, the Mediterranean Osteoporosis study (MEDOS) in southern Europe, calcium and bone medication use was estimated by questionnaire and incident fractures were documented *(56)*. Use of the antiresorptive drug nasal calcitonin was associated with a nonsignificant decrease in vertebral fracture risk ($rr = 0.78$ [CI_{95} 0.48, 1.27],

$p = 0.318$), as was use of calcium alone ($rr = 0.82$ [CI_{95} 0.63, 1.07], $p = 0.149$). Use of calcitonin and calcium together, however, was associated with a significant reduction in vertebral fracture risk ($rr = 0.63$ [CI95 0.44, 0.90], $p = 0.012$). This suggests that the effects of calcium and antiresorptive therapy are additive. Little information is available on the interface of other osteoporosis treatments with calcium intake, and evidence is not likely to be forthcoming since it is considered unethical to enroll osteoporotic patients into a trial with a true placebo arm. One could speculate that an adequate calcium intake would be essential for successful treatment with the anabolic drug PTH (1-34). Osteoporotic patients increased their spinal BMD by almost 10% over an 18-mo period with this drug *(54)*. Patients on this anabolic drug would also need an adequate supply of phosphorus to support the extensive bone formation that it induces.

Currently, there is little evidence of an interaction of vitamin D with pharmacotherapy. Komulainen compared the effects of HRT, vitamin D (300 IU/d) and the combination on rates of bone loss in early postmenopausal women *(46)*. The HRT was effective but the vitamin D was not. Furthermore, the combination of HRT with vitamin D was no more effective than HRT alone.

5. INTAKE RECOMMENDATIONS

5.1. Calcium

Calcium intake recommendations vary enormously worldwide. Recommendations by the US National Academy of Sciences are among the highest. The recommended intake of calcium for women and men over age 50 is 1200 mg/d *(57)*. This intake is higher than the Academy's previous recommendation of 800 mg/d. The change was made in 1997 on the basis of evidence that a higher intake is likely to improve calcium retention and bone mass and reduce the risk of fracture. The Safe Upper Limit for calcium is 2500 mg/d *(57)*.

5.2. Vitamin D

The vitamin D intake recommendations are 400 IU/d for men and women age 51–70 yr and 600 IU/d for men and women age 71 and older, and the safe upper limit for vitamin D is 2000 IU/d *(57)*. National survey data do not include information about dietary vitamin D intakes. Based on intake estimates from several convenience samples, however, intakes of vitamin D in the elderly are consistently below the levels now recommended.

6. CONCLUSIONS

In conclusion, calcium and vitamin D are essential for bone health. Current recommendations for calcium are based on a combination of balance data and clinical trial evidence and they appear to be pretty solid. The current recommendations for vitamin D are based on the amount needed to suppress PTH secretion, but variability in this endpoint is very large across study populations. Several large studies have placed the 25(OH)D level needed for maximal PTH suppression in the range 80–110 nmol/L (32–44 ng/mL), whereas another places it as low as 25 nmol/L (10 ng/mL). It is important to have more data on other indicators of vitamin D adequacy such as BMD and change in BMD to corroborate and support the current evidence on PTH suppressibility. The currently recommended vitamin D intake of 600 IU/d for men and women age 71 and older is not adequate to bring a large majority of the elderly population to 25(OH)D levels of 80 nmol/L (32 ng/mL). To meet this target, vitamin D intakes of 1000 IU/d or higher appear to be needed.

ACKNOWLEDGMENTS

This material is based on work supported by a grant (AG10353) from the National Institutes of Health and by the US Department of Agriculture, under agreement No. 58-1950-9001. Any opinions, findings, conclusions, or recommendations expressed in this publication are those of the authors, and do not necessarily reflect the view of the US Department of Agriculture.

REFERENCES

1. Ireland P, Fordtran JS. Effect of dietary calcium and age on jejunal calcium absorption in humans studied by intestinal perfusion. J Clin Invest 1973; 52(11):2672–2681.
2. Gennari C, Agnusdei D, Nardi P, Civitelli R. Estrogen preserves a normal intestinal responsiveness to 1,25-dihydroxyvitamin D3 in oophorectomized women. J Clin Endocrinol Metab 1990; 71(5):1288–1293.
3. Bullamore JR, Wilkinson R, Gallagher JC, Nordin BE, Marshall DH. Effect of age on calcium absorption. Lancet 1970; 2(7672):535–537.
4. Ebeling PR, Sandgren ME, DiMagno EP, Lane AW, DeLuca HF, Riggs BL. Evidence of an age-related decrease in intestinal responsiveness to vitamin D: relationship between serum 1,25-dihydroxyvitamin D3 and intestinal vitamin D receptor concentrations in normal women. J Clin Endocrinol Metab 1992; 75(1):176–182.
5. Dawson-Hughes B, Dallal GE, Krall EA, Sadowski L, Sahyoun N, Tannenbaum S. A controlled trial of the effect of calcium supplementation on bone density in postmenopausal women. N Engl J Med 1990; 323(13):878–883.
6. Johnell O, Gullberg B, Kanis JA, et al. Risk factors for hip fracture in European women: the MEDOS Study. Mediterranean Osteoporosis Study. J Bone Miner Res 1995; 10(11):1802–1815.
7. Garnero P, Hausherr E, Chapuy MC, et al. Markers of bone resorption predict hip fracture in elderly women: the EPIDOS Prospective Study. J Bone Miner Res 1996; 11(10):1531–1538.
8. Riis B, Thomsen K, Christiansen C. Does calcium supplementation prevent postmenopausal bone loss? A double-blind, controlled clinical study. N Engl J Med 1987; 316(4):173–177.
9. Chevalley T, Rizzoli R, Nydegger V, et al. Effects of calcium supplements on femoral bone mineral density and vertebral fracture rate in vitamin-D-replete elderly patients. Osteopor Int 1994; 4(5):245–252.
10. Dawson-Hughes B, Harris SS, Krall EA, Dallal GE. Effect of calcium and vitamin D supplementation on bone density in men and women 65 years of age or older. N Engl J Med 1997; 337(10):670–676.
11. Elders PJ, Netelenbos JC, Lips P, et al. Calcium supplementation reduces vertebral bone loss in perimenopausal women: a controlled trial in 248 women between 46 and 55 years of age. J Clin Endocrinol Metab 1991; 73(3):533–540.

12. MacLaughlin J, Holick MF. Aging decreases the capacity of human skin to produce vitamin D3. J Clin Invest 1985; 76(4):1536–1538.
13. Harris SS, Dawson-Hughes B. Plasma vitamin D and 25(OH)D responses of young and old men to supplementation with vitamin D3. J Am Coll Nutr 2002; 21(4):357–362.
14. Krall EA, Sahyoun N, Tannenbaum S, Dallal GE, Dawson-Hughes B. Effect of vitamin D intake on seasonal variations in parathyroid hormone secretion in postmenopausal women. N Engl J Med 1989; 321(26):1777–1783.
15. Dawson-Hughes B, Dallal GE, Krall EA, Harris S, Sokoll LJ, Falconer G. Effect of vitamin D supplementation on wintertime and overall bone loss in healthy postmenopausal women. Ann Intern Med 1991; 115(7):505–512.
16. Dawson-Hughes B, Harris SS, Krall EA, Dallal GE, Falconer G, Green CL. Rates of bone loss in postmenopausal women randomly assigned to one of two dosages of vitamin D. Am J Clin Nutr 1995; 61(5):1140–1145.
17. Spencer H, Kramer L, Lesniak M, De Bartolo M, Norris C, Osis D. Calcium requirements in humans. Report of original data and a review. Clin Orthoped 1984; (184):270–280.
18. Hasling C, Charles P, Jensen FT, Mosekilde L. Calcium metabolism in postmenopausal osteoporosis: the influence of dietary calcium and net absorbed calcium. J Bone Miner Res 1990; 5(9):939–946.
19. Selby PL. Calcium requirement—a reappraisal of the methods used in its determination and their application to patients with osteoporosis. Am J Clin Nutr 1994; 60(6):944–948.
20. Mackerras D, Lumley T. First- and second-year effects in trials of calcium supplementation on the loss of bone density in postmenopausal women. Bone 1997; 21(6):527–533.
21. Shea B, Wells G, Cranney A, et al. VII. Meta-analysis of calcium supplementation for the prevention of postmenopausal osteoporosis. Endocr Rev 2002; 23(4):552–559.
22. Aloia JF, Vaswani A, Yeh JK, Ross PL, Flaster E, Dilmanian FA. Calcium supplementation with and without hormone replacement therapy to prevent postmenopausal bone loss. Ann Intern Med 1994; 120(2):97–103.
23. McKane WR, Khosla S, Egan KS, Robins SP, Burritt MF, Riggs BL. Role of calcium intake in modulating age-related increases in parathyroid function and bone resorption. J Clin Endocrinol Metab 1996; 81(5):1699–1703.
24. Prince R, Devine A, Dick I, et al. The effects of calcium supplementation (milk powder or tablets) and exercise on bone density in postmenopausal women. J Bone Miner Res 1995; 10(7):1068–1075.
25. Malabanan A, Veronikis IE, Holick MF. Redefining vitamin D insufficiency. Lancet 1998; 351(9105):805–806.
26. Thomas MK, Lloyd-Jones DM, Thadhani RI, et al. Hypovitaminosis D in medical inpatients. N Engl J Med 1998; 338(12):777–783.
27. Lips P, Wiersinga A, van Ginkel FC, et al. The effect of vitamin D supplementation on vitamin D status and parathyroid function in elderly subjects. J Clin Endocrinol Metab 1988; 67(4):644–650.
28. Peacock M. Effects of calcium and vitamin D insufficiency on the skeleton. Osteopor Int 1998; 8(suppl 2):S45–S51.
29. Chapuy MC, Preziosi P, Maamer M, et al. Prevalence of vitamin D insufficiency in an adult normal population. Osteopor Int 1997; 7(5):439–443.
30. Dawson-Hughes B, Harris SS, Dallal GE. Plasma calcidiol, season, and serum parathyroid hormone concentrations in healthy elderly men and women. Am J Clin Nutr 1997; 65(1):67–71.
31. Lips P, Chapuy MC, Dawson-Hughes B, Pols HA, Holick MF. An international comparison of serum 25-hydroxyvitamin D measurements. Osteopor Int 1999; 9(5):394–397.
32. Outila TA, Karkkainen MU, Lamberg-Allardt CJ. Vitamin D status affects serum parathyroid hormone concentrations during winter in female adolescents: associations with forearm bone mineral density. Am J Clin Nutr 2001; 74(2):206–210.

33. Tangpricha V, Pearce EN, Chen TC, Holick MF. Vitamin D insufficiency among free-living healthy young adults. Am J Med 2002; 112(8):659–662.
34. Ooms ME, Lips P, Roos JC, et al. Vitamin D status and sex hormone binding globulin: determinants of bone turnover and bone mineral density in elderly women. J Bone Miner Res 1995; 10(8):1177–1184.
35. Ooms ME. Osteoporosis in elderly women; vitamin D deficiency and other risk factors. PhD thesis, Vrije Universiteit, Amsterdam, 1994.
36. LeBoff MS, Kohlmeier L, Hurwitz S, Franklin J, Wright J, Glowacki J. Occult vitamin D deficiency in postmenopausal US women with acute hip fracture. JAMA 1999; 281(16):1505–1511.
37. Webb AR, Pilbeam C, Hanafin N, Holick MF. An evaluation of the relative contributions of exposure to sunlight and of diet to the circulating concentrations of 25-hydroxyvitamin D in an elderly nursing home population in Boston. Am J Clin Nutr 1990; 51(6):1075–1081.
38. Margiloff L, Harris SS, Lee S, Lechan R, Dawson-Hughes B. Vitamin D insufficiency in both supplemented and unsupplemented ambulatory care patients. Calcif Tissue Int 2001; 69:263–267.
39. Dawson-Hughes B. Impact of vitamin D and calcium on bone and mineral metabolism in older adults. In: Holick MF, ed. Biologic Effects of Light 2001. Kluwer, Boston, 2002, pp. 175–183.
40. Recker RR, Hinders S, Davies KM, et al. Correcting calcium nutritional deficiency prevents spine fractures in elderly women. J Bone Miner Res 1996; 11(12):1961–1966.
41. Reid IR, Ames RW, Evans MC, Gamble GD, Sharpe SJ. Long-term effects of calcium supplementation on bone loss and fractures in postmenopausal women: a randomized controlled trial. Am J Med 1995; 98(4):331–335.
42. Riggs BL, O'Fallon WM, Muhs J, O'Connor MK, Kumar R, Melton LJ III. Long-term effects of calcium supplementation on serum parathyroid hormone level, bone turnover, and bone loss in elderly women. J Bone Miner Res 1998; 13(2):168–174.
43. Heikinheimo RJ, Inkovaara JA, Harju EJ, et al. Annual injection of vitamin D and fractures of aged bones. Calcif Tissue Int 1992; 51(2):105–110.
44. Lips P, Graafmans WC, Ooms ME, Bezemer PD, Bouter LM. Vitamin D supplementation and fracture incidence in elderly persons. A randomized, placebo-controlled clinical trial. Ann Intern Med 1996; 124(4):400–406.
45. Meyer HE, Smedshaug GB, Kvaavik E, Falch JA, Tverdal A, Pedersen JI. Can vitamin D supplementation reduce the risk of fracture in the elderly? A randomized controlled trial. J Bone Miner Res 2002; 17(4):709–715.
46. Komulainen MH, Kroger H, Tuppurainen MT, et al. HRT and Vit D in prevention of non-vertebral fractures in postmenopausal women; a 5 year randomized trial. Maturitas 1998; 31(1):45–54.
47. Chapuy MC, Arlot ME, Duboeuf F, et al. Vitamin D3 and calcium to prevent hip fractures in the elderly women. N Engl J Med 1992; 327(23):1637–1642.
48. Chapuy MC, Pamphile R, Paris E, et al. Combined calcium and vitamin D3 supplementation in elderly women: confirmation of reversal of secondary hyperparathyroidism and hip fracture risk: the Decalyos II study. Osteopor Int 2002; 13(3):257–264.
49. Dawson-Hughes B, Harris SS, Krall EA, Dallal GE. Effect of withdrawal of calcium and vitamin D supplements on bone mass in elderly men and women. Am J Clin Nutr 2000; 72(3):745–750.
50. Black DM, Cummings SR, Karpf DB, et al. Randomised trial of effect of alendronate on risk of fracture in women with existing vertebral fractures. Fracture Intervention Trial Research Group. Lancet 1996; 348(9041):1535–1541.
51. Harris STM, Watts NBM, Genant HKM, et al. Effects of risedronate treatment on vertebral and nonvertebral fractures in women with postmenopausal osteoporosis: a randomized controlled trial. JAMA 1999; 282(14):1344–1352.
52. Ettinger BM, Black DMP, Mitlak BHM, et al. Reduction of vertebral fracture risk in postmenopausal women with osteoporosis treated with raloxifene: results from a 3-year randomized clinical trial. JAMA 1999; 282(7):637–645.

53. Chesnut CHI, Silverman SM, Andriano KP, et al. A randomized trial of nasal spray salmon calcitonin in postmenopausal women with established osteoporosis: the Prevent Recurrence of Osteoporotic Fractures Study. Am J Med 2000; 109(4):267–276.
54. Neer RM, Arnaud CD, Zanchetta JR, et al. Effect of parathyroid hormone (1-34) on fractures and bone mineral density in postmenopausal women with osteoporosis. N Engl J Med 2001; 344(19):1434–1441.
55. Nieves JW, Komar L, Cosman F, Lindsay R. Calcium potentiates the effect of estrogen and calcitonin on bone mass: review and analysis. Am J Clin Nutr 1998; 67(1):18–24.
56. Kanis JA, Johnell O, Gullberg B, et al. Evidence for efficacy of drugs affecting bone metabolism in preventing hip fracture. Br Med J 1992; 305(6862):1124–1128.
57. Standing Committee on the Scientific Evaluation of Dietary Reference Intakes. Dietary Reference Intakes: Calcium, Phosphorus, Magnesium, Vitamin D, and Fluoride. National Academy Press, Washington, DC, 1997, 2001.

13 Nutrition and Bone Health in the Elderly

Clifford J. Rosen

1. INTRODUCTION

Osteoporosis, a common disorder of women and men, is associated with significant morbidity, mortality, and health care costs. Age-related fractures are the most common manifestation of osteoporosis and are usually manifested after menopause in women, and in the seventh or eighth decade in men. Biochemical, biomechanical, and nonskeletal factors contribute to fragility fractures in the elderly. However, nutritional determinants also play a key role in both the pathophysiology and treatment of individuals with osteoporosis. Traditionally, calcium has been considered the key nutrient for optimal skeletal growth and maintenance. However, recent evidence has emerged linking vitamins D, K, and A to the pathogenesis of osteoporosis, especially in elderly individuals. Additionally, adequate protein intake is essential for bone formation, particularly in vulnerable populations residing in nursing homes and assisted-living facilities. Preservation of skeletal architecture is also dependent on maintenance of muscle mass, another soft tissue component that is influenced by nutrient intake. This chapter will focus on the skeletal and nonskeletal pathways that contribute to fractures in the older individual and the role of nutrition in treating and preventing age-related osteoporosis. However, it should be noted that the pathogenesis of skeletal frailty is complex and centers on the entire organism, not just bone. For example, fractures generally result from falls; falling can be related to poor muscle strength, which in turn can be strongly influenced by nutritional intake. In sum, for elderly individuals the confluence of hormonal, environmental, nutritional, and heritable influences all contribute to fracture risk and disease outcome. As such, all these factors work at the tissue level through alterations in bone remodeling. Hence, the first part of this chapter will deal with the bone remodeling cycle, how it is altered in osteoporosis, and how nutritional determinants affect bone turnover.

From: *Nutrition and Bone Health*
Edited by: M. F. Holick and B. Dawson-Hughes © Humana Press Inc., Totowa, NJ

2. BONE HEALTH, BONE REMODELING, AND NUTRITIONAL STATUS

The adult human skeleton is more than just calcium phosphate crystals melded into a protein matrix. It is a metabolically active organ that constantly undergoes remodeling, a deliberate and well-orchestrated process that allows the skeleton to remake itself in a cyclic fashion. Remodeling serves to enhance skeletal integrity without changing mass while maintaining metabolic balance for the rest of the organism, especially in relation to essential ions such as calcium and phosphate. Modeling, on the other hand, is the process of skeletal growth, during which there is both linear expansion from the growth plate by chondrocytes and expansion from the periosteal surface by osteoblasts. Modeling in humans lasts from birth to adolescence, whereas remodeling, which maintains the balance between resorption of old bone and formation of new bone, begins during adolescence and persists across the life of the individual. This chapter will focus predominantly on changes in remodeling that lead to osteoporosis in the adult skeleton.

Every 10 yr the entire skeleton is remodeled, with the greatest turnover noted in the trabecular-rich regions of the thoraco-lumbar spine and several areas of the femur. Because of this huge undertaking, it is not surprising that the mammalian skeleton is highly organized and physiologically active *(1)*. The skeleton serves two purposes: to maintain structure; and to preserve calcium homeostasis for all physiological processes. As such, mammalian bone is uniquely designed for its protective and structural roles. There is an outer surface of cortical bone that surrounds the inner trabecular elements. Marrow bathes the trabecular skeleton, whereas cortical bone is nourished by periosteal vessels and a series of canaliculi connecting osteocytes to lining cells and osteoblasts *(1)*. Gravitary forces help model the cortical skeleton, although periosteal osteoblasts and the underlying growth plate are principally responsible for longitudinal growth. (See Fig. 1.) Both cortical and trabecular bone undergo remodeling, but the frequency of this process is much less in the cortex than in the trabecular components of the spine and femur.

Numerous growth factors and cytokines, each of which contributes to coupling bone dissolution (i.e., resorption) to new bone formation, orchestrate bone remodeling within the skeletal compartments. Preosteoblasts (pre-OBs), derived from mesenchymal stromal cells, and under the influence of a key transcription factor (Cbfa1, i.e., core binding factor I or RUNX2), represent target cells for initiation of the remodeling cycle *(2,3)*. Systemic and local factors enhance pre-OB differentiation, and this, in turn, leads to the synthesis and release of macrophage colony-stimulating factor (M-CSF) and receptor activator of nuclear factor kappa B ligand (RANKL) *(3)*. These two peptides are necessary and sufficient for the recruitment of bone resorbing cells, that is, the osteoclasts. Once bone resorption occurs, calcium, collagen fragments, and growth factors such as the insulin-like growth factors (IGFs), and transforming growth-factors (TGFs), are released from the bony matrix. The latter factors enhance the recruitment of osteoblasts to the bone surface, thereby setting the stage for collagen synthesis and matrix deposition/mineralization *(4)*.

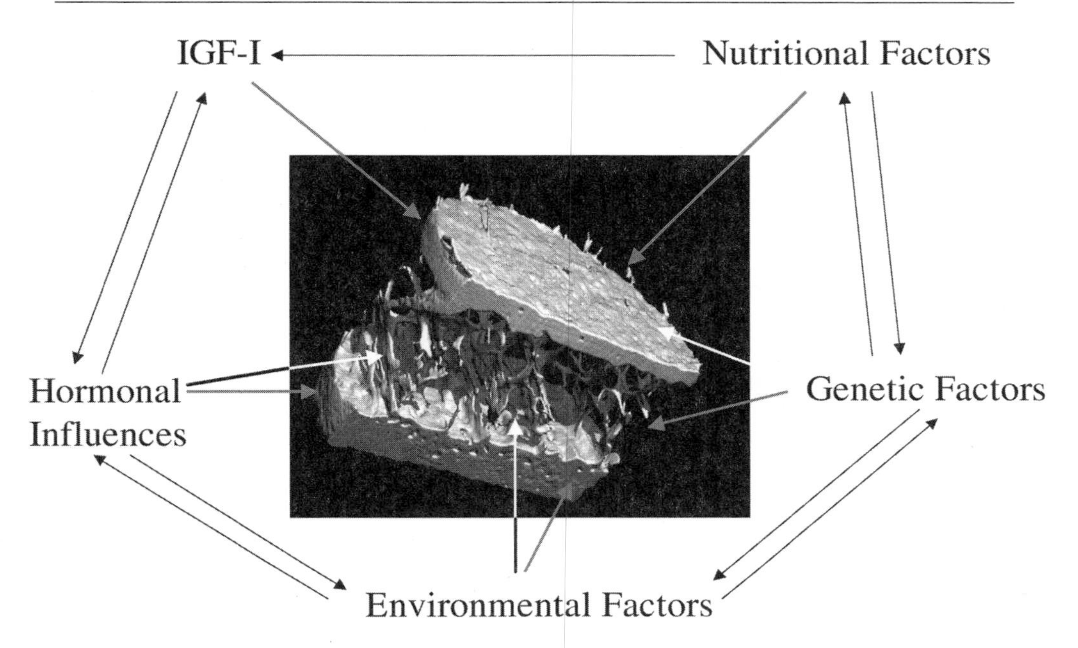

Fig. 1. Interactions among various factors and the adult skeleton. The adult human skeleton is composed of both cortical and trabecular elements. Several factors including nutritional determinants regulate both cortical and trabecular components. With aging, all these factors interact. IGF-I, insulin-like growth factor-I.

The entire remodeling cycle in humans takes approx 90 d, with the majority of time consumed by the elaborate process of bone formation and subsequent mineralization. And, at each step along the way, systemic hormones such as vitamin D, parathyroid hormone (PTH), estrogen, thyroxine, and growth hormone (GH) influence the time and direction of remodeling.

One of the most critical local and systemic growth factors influencing bone remodeling is IGF-I. Not surprisingly, this peptide is directly influenced by nutrient status. (See Fig. 2.) IGF-I and IGF-2 are major components of both the organic skeletal matrix and the circulation. Indeed, the serum of most mammals contains significant concentrations of both IGF-I and IGF-2, bound to high- and low-molecular-weight insulin-like growth factor-binding proteins (IGFBPs) *(5)*. Similarly, the skeletal matrix also is highly enriched with these growth factors and other noncollagenous proteins, including all six IGFBPs and several IGFBP proteases. In addition, the Type I IGF receptor is present on both osteoblasts and osteoclasts. It is now reasonably certain that skeletal IGFs originate from two sources: (1) *de novo* synthesis by bone-forming cells (i.e., pre-OBs, and fully differentiated osteoblasts); and (2) the circulation. In fact, some skeletal IGFs probably make their way into the matrix by way of specialized canaliculi and sinusoids within the bone microcirculation *(5,6)*. IGFs, bound to IGFBPs can also be

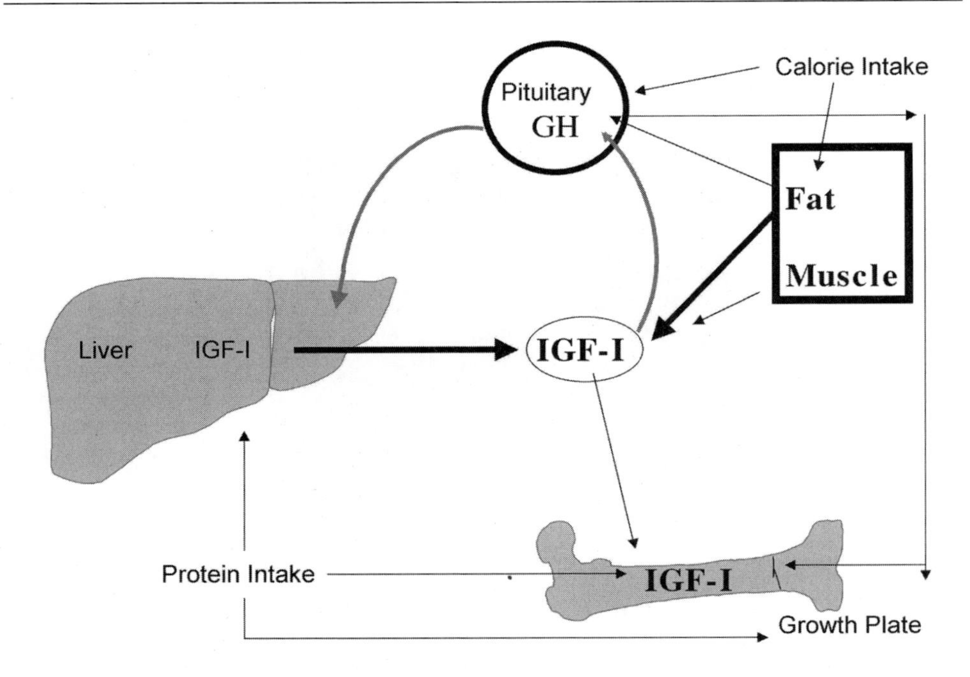

Fig. 2. Nutrient regulation of systemic and skeletal IGF-I. Skeletal insulin-like growth factor (IGF)-I regulates trabecular bone; systemic IGF-I regulates cortical bone. GH, growth hormone.

found within the marrow milieu in close contact with the endosteal surface of bone. However, by most accounts, the vast majority of IGF-I in bone is derived from local osteoblastic synthesis. Yet during active bone resorption, as the matrix is dissolved, significant amounts of IGF-I and IGF-2 are released from storage (i.e., binding to IGFBP-5 and hydroxyapatite). Subsequently, both IGFs recruit precursor osteoblasts, and possibly early osteoclasts to the bone surface, where remodeling is occurring *(5,6)*.

Circulating and skeletal IGF-I are both profoundly influenced by nutritional determinants *(7)*. Growth retardation, a major feature of protein-calorie malnutrition in children, is associated with significant declines in circulating IGF-I despite enhanced GH secretion *(7–9)*. Similarly, elderly individuals with poor protein intake have low serum IGF-I levels *(9,10)*. This, almost certainly, is due to reduced message half-life for IGF-I in the liver, although zinc deficiency has also been associated with reduced serum and hepatic IGF-I synthesis due to an uncharacterized defect in transcriptional regulation *(7–9)*. In a randomized trial of older postmenopausal women, Heaney et al. recently showed that milk supplementation increased serum IGF-I by 10% while at the same time suppressing markers of bone resorption *(11)*. The increase in IGF-I appears to be related to the enhanced protein intake from milk supplementation, further supporting the role of this endogenous growth factor in maintenance of the skeleton. Indeed, new studies suggest that circulating IGF-I may be essential for periosteal growth, development, and maintenance *(5,6,10)*.

Regardless of the mechanism, IGF-I expression is modulated by changes in nutrient intake. And, recent studies of elderly women who sustained a hip fracture (which is the end stage of osteoporosis) support this contention *(12,13)*. Longitudinal studies have demonstrated that, after a hip fracture, there is a profound decline in serum IGF-I that can be partially reversed by oral protein supplementation *(12,13)*. Restoration of circulating IGF-I using recombinant IGF-I associated with IGF-binding protein 3 in elderly patients after hip fractures also resulted in improvement in functional outcomes, similar to what is observed with dietary supplementation immediately after fracture *(14)*.

The remodeling cycle is sensitive to changes in the levels of other nutrients working independent of the IGF regulatory system. High protein intake, for example, leads to an increase in acid production and excessive calcium loss in the urine. Increased hydrogen ions in the milieu of the bone remodeling unit can lead to suppression of bone formation, whereas the loss of calcium in the urine can result in secondary increases in PTH secretion. Excessive sodium intake also increases the risk of hypercalcuria and secondary hyperparathyroidism, which in turn stimulates bone resorption preferentially over formation. Phosphate balance is important for mineralization, and low phosphate levels trigger activation of 1,α-hydroxylase, keying the conversion of 25-hydroxyvitamin D to the active vitamin D compound 1,α 25-dihydroxyvitamin D. Conversely, high levels of phosphorus stimulate PTH secretion, resulting in marked activation of remodeling and enhanced bone resorption. Low calcium intakes also trigger PTH release and almost certainly is a principal factor in the secondary hyperparathyroidism seen in elderly individuals *(15)*.

Vitamin D, which is added to milk, is found in some fish, and is produced endogenously by the skin in response to sunlight, is absolutely essential for the bone remodeling cycle. Besides enhancing calcium intake in the gut, 1,25-dihydroxyvitamin D plays a critical role in osteoclast recruitment and osteoblast differentiation as well as mineralization *(16)*. Vitamin K, which is found only in vegetable sources, is an essential cofactor for γ-carboxylation of osteocalcin, the most common noncollagenous protein in bone. Osteocalcin is produced by bone-forming cells that are highly differentiated, and this protein may also be important for mineralization. Vitamin C (ascorbic acid) is required for collagen modification in bone tissue, whereas vitamin A is essential for the differentiation of both osteoblasts and osteoclasts. Other trace elements such as boron and strontium, which may concentrate in bone, affect bone cell function in vitro, but their direct role in the remodeling cycle remains uncertain. Low magnesium intake has been reported to be associated with reduced bone mineral density (BMD), but the mechanism remains unknown, and the effect of this cation on remodeling is not clear.

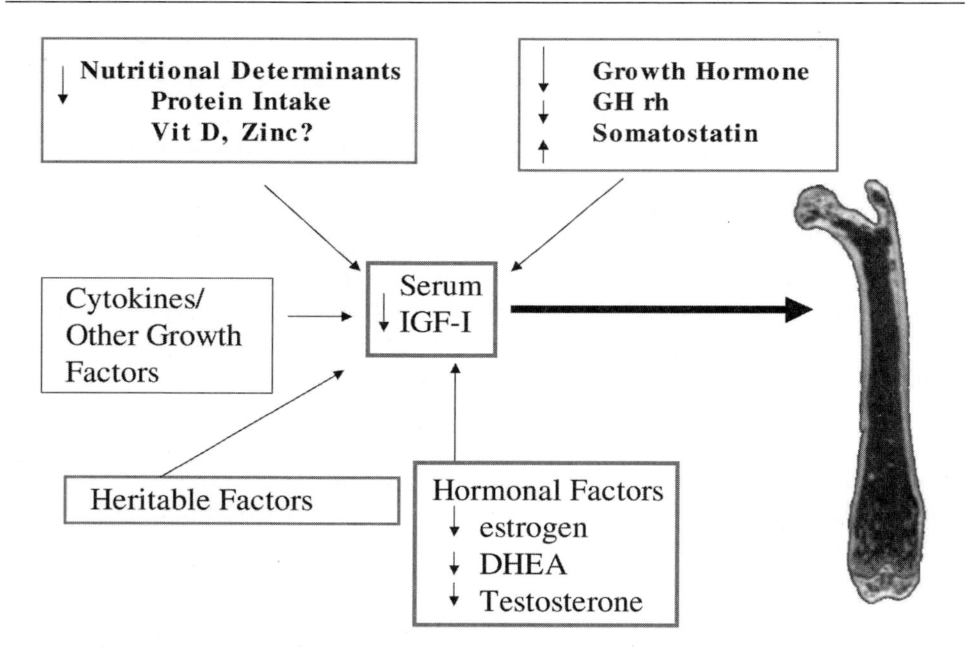

Fig. 3. Aging adversely affects hormonal determinants and nutritional factors. Advanced aging results in poor nutritional status, reduced hormonal function, and an overall reduction in serum insulin-like growth factor (IGF)-I, which in turn can lead to poor bone quality. The eventual outcome is skeletal fragility and fracture. Nutritional factors can also alter endocrine and paracrine growth factor synthesis. GH, growth hormone; DHEA, dehydroepiandrosterone.

3. AGING, ABNORMAL BONE REMODELING, AND NUTRIENT STATUS

Aging is associated with a modest uncoupling in bone resorption compared to bone formation. In general, bone resorption is markedly enhanced by several factors (including, as noted above, calcium/vitamin D-deficient intake), while formation tends to slow with aging. Hence, over a lifespan, women lose approx 42% of their spinal bone mass and 58% of their femoral bone mass *(17)*. Surprisingly, rates of bone loss in the eighth and ninth decades of life may be *comparable to or even exceed* those found in the immediate peri- and postmenopausal period of some women *(18,19)*. This is the result of uncoupling in the bone remodeling cycle of older individuals, resulting in a marked increase in bone resorption but no change or a decrease in bone formation *(20,21)*. However, the mechanisms that lead to an uncoupled bone remodeling unit, especially in the elderly, remain to be elucidated. As noted above, changes in hormonal factors (estrogen, testosterone, and GH), as well as nutrient intake, can clearly affect remodeling cycles. In most cases, however, dramatic changes at the cellular level are the result of multiple determinants. (*See* Fig. 3.)

Recent advances in clinical chemistry have unmasked the remodeling process for investigators by defining changes in bone-specific turnover markers with pathological conditions. Alterations in bone turnover can be detected by several biochemical markers, including bone resorption indices (e.g., urinary and serum *N*-telopeptide, *C*-telopeptide, and urinary-free and total deoxypyridinoline) and bone formation markers (e.g., osteocalcin, procollagen peptide, bone-specific alkaline phosphatase). In general, Bbone turnover markers are significantly higher in older than younger postmenopausal women, and these indices are inversely related to BMD *(224)*. For example, in the Epidemiologie de l'Osteoporose (EPIDOS) trial of elderly European females, the highest levels of osteocalcin, *N*-telopeptide, *C*-telopeptide, and bone-specific alkaline phosphatase were noted for those in the lowest tertile of femoral bone density *(23)*. Also, increased bone resorption indices were associated with a greater fracture risk, independent of BMD *(23)*. For those women in EPIDOS with low BMD and a high bone resorption rate, there was a nearly fivefold greater risk of a hip fracture.

In contrast to a consistent pattern of high bone resorption indices, bone formation markers in the elderly are more variable. Serum osteocalcin levels are high in elderly individuals, but this may be indicative of an increase in bone turnover rather than reflecting a true rise in bone formation *(23)*. On the other hand, bone-specific alkaline phosphatase, and procollagen peptide levels have been reported to be high, normal, or low in elderly men and womens *(24)*. Bone histomorphometric indices in elders are also quite variable. Thus, although there is strong evidence for an age-associated rise in bone resorption, changes in bone formation are inconsistent. Still, there is uncoupling of the remodeling unit that leads to bone loss, altered skeletal architecture, and an increased propensity to fractures.

4. FACTORS THAT CONTRIBUTE TO AGE-RELATED BONE LOSS

4.1. Nutritional Factors

By far the most common cause of increased bone resorption in older individuals is calcium deficiency in a setting of low endogenous estrogen. Very low calcium intakes (i.e., less than 800 mg/d) and vitamin D insufficiency are the rule, rather than the exception in the elderly and are due to a number of factors, including dietary changes, lack of sunlight exposure, malabsorption, and generalized anorexia. The end result of low calcium intake is persistent secondary hyperparathyroidism, which, as noted previously, leads to increased bone resorption, often accentuated by occult vitamin D deficiency, especially in women living in northern latitudes *(25)*. Indeed, most women who present with hip fractures in their latter years are truly vitamin D deficient *(26)*. Furthermore, longitudinal trials of elderly men and women supplemented with calcium and vitamin D have demonstrated preservation of BMD and a reduction in osteoporotic fractures *(27–29)*. These studies led the National Academy of Sciences to recommend an increase in the minimal daily requirement for calcium intake in people over 65 yr of age to 1500 mg/d and 600 IU/d of vitamin D *(30)*.

Besides calcium and vitamin D, other nutritional factors may also play a role in age-related osteoporosis, including the amount of protein and sodium intake and the dietary consumption of vitamins K and A *(31–33)*. Protein-calorie malnutrition stimulates bone resorption and impairs bone formation both directly and through other mechanisms, including a global reduction in serum IGF-I *(13)*. These changes result in low BMD. In addition, low protein intake increases the propensity to fall, due to impaired coordination and muscle weakness (see below). As noted, excessive protein intake can also result in low bone mass, principally through enhanced calcium loss in the urine. Vitamin K deficiency may contribute to an increased risk of osteoporotic fractures, possibly through effects on the carboxylation of bone proteins such as osteocalcin *(34)*. At least one observational study suggested that vitamin A intake might be associated with low BMD *(35)*. Recently a case control study demonstrated a significant relationship between excessive intakes of vitamin A and age-related fractures *(36)*.

4.2. Hormonal Factors

Estrogen deficiency has long been recognized as a major cause of bone loss in the first decade after menopause. More recently, investigators have identified a strong relationship between endogenous estrogen and bone mass in both elderly men and women. In one prospective study, Slemenda et al. noted that both estrogens and androgens were independent predictors of bone loss in older postmenopausal women *(37)*. In the Rancho Bernardo cohort and the Framingham Cohort, estradiol levels were very strongly related to BMD at the spine, hip, and forearm in women *(38,39)*. However, males also suffer from age-related bone loss, and evidence suggests that absolute estrogen levels, rather than testosterone concentrations, are essential for maintenance of BMD. In the Rancho Bernardo cohort, serum estradiol levels in elderly men correlated closely with bone mass at several sites *(38)*. Recently, Falahati et al. demonstrated that small amounts of estradiol were essential for preventing bone resorption in men, in part by upregulating osteoprotogerin (OPG) *(40,41)*. Endogenous testosterone also plays a role in regulating bone turnover, possibly more on the formation side than with respect to resorption. Serum testosterone levels decline with age at a rate of approx 1.2%/yr, whereas sex hormone-binding globulin (SHBG) levels rise. Males treated with androgen antagonists or gonadotropin agonists for prostate cancer metastases rapidly lose bone mass and may be at high risk for subsequent osteoporotic fractures *(42)*. Overall, it seems likely that both androgens and estrogens are important in the elderly male. However, whether changes in male hormone levels are causally related to age-related bone loss in men will have to await large-scale prospective studies.

Changes in the GH–IGF-I axis may also contribute to age-related bone loss. GH secretion declines 14% per decade and, along with reduced nutrient intake, is the principal cause for low serum IGF-I concentrations in both elderly men and women *(43,44)*. Severe protein undernutrition can also impair growth hormone secretion and result in even lower serum IGF-I levels in the elderly. Recent efforts to implement replacement with rh IGF-I or rhGH in elderly individuals has demonstrated only modest increases in BMD *(5,14)*.

4.3. Heritable and Environmental Factors

Age-related bone loss can be dramatic in some individuals, and this decline cannot be attributed solely to hormonal or nutritional factors. Several investigators have hypothesized that there is genetic programming that, when triggered by environmental factors such as low calcium and/or vitamin D intake, may lead to bone loss, especially in the elderly. This postulate has been reinforced by some studies showing that the presence of a particular polymorphism in the vitamin D receptor, when coupled with low calcium intake, may lead to lower bone mass. Some animal models, but not others, have demonstrated a heritable component to age-related bone loss *(45)*. In humans, the multiplicity of environmental factors, including deficiencies in calcium intake, makes the determination of fracture heritability complicated although recent publications suggest a genetic component *(46,47)*. On the other hand, environmental agents such as smoking, alcohol, and medications such as glucocorticoids and anticonvulsants may contribute to an excessive rate of bone loss in some elders.

5. FRACTURES AND FALLS IN THE ELDERLY: THE IMPORTANCE OF ADEQUATE NUTRITION

In elderly individuals decreased bone strength, as reflected by BMD, is only one of many important contributors to overall hip fracture risk *(48,49)*. Other factors include propensity to fall, inability to correct a postural imbalance, characteristics of the faller such as height and muscle activity, the orientation of the fall, adequacy of local tissue shock absorbers, and characteristics of the impact surface. For its part, the resistance of a skeletal structure to failure (i.e., fracture) depends on the geometry of the bone, the material properties of the calcified tissue, and the location and direction of the loads to which the bone is subjected (i.e., during a fall or other activities). Estimations of the forces generated within the bone in response to a given load can be estimated using basic engineering principles. Those forces can then be compared with the strengths of the tissue. The ratio of the impact force expected during a fall to the force required to cause the bone to fail incorporates the two major determinants of fracture risk. When this ratio is close to or more than 1, the structure is at great risk of failure. In the elderly, this ratio is 0.3 at the femoral neck for simple stance and normal ambulation *(48)*.

Falls in older people are rarely due to a single cause and usually occur when a threat to the normal homeostatic mechanisms that maintain postural stability is superimposed on underlying age-related declines in balance, nutrient intake, ambulation, and cardiovascular function. In some cases, this may involve an acute illness such as a fever or infection, or an environmental stress such as a newly initiated drug treatment, unsafe walking surface, or failure to thrive. Nutritional deficiencies contribute to the greater risk of falling by dramatically altering normal homeostasis of both bone and muscle *(50)*. Regardless of the nature of the stress, an elderly person may not be able to compensate because of either age-related declines in function or severe chronic disease. Yet it is unlikely for an extrinsic stress to explain the circumstances of a fall completely. Older persons, by virtue of their age alone, experience declines in physiological function, have greater numbers of chronic diseases, acute illnesses, and hospitalizations, use multiple medications, and have chronic nutrient deficiencies. Superimposed on these age-related characteristics, challenges to postural control may have a greater impact in aged persons according to their risk-taking behavior and opportunity to fall. In sum, multiple factors contribute to falls. Adequate nutritional status helps to maintain muscle function, particularly with respect to vitamin D status. As such, preventive measures to reduce fractures must include attention to nonskeletal factors including nutrient intake, ambulation strength, coordination, and muscle function *(50,51)*.

6. FRACTURE PREVENTION AND TREATMENT IN THE ELDERLY: NUTRITIONAL ASPECTS

Because the elderly have lower BMD to start, are continuing to lose bone, and are in the age group most likely to fracture, interventions would be expected to be most cost-effective when initiated in these individuals. The interventions include dietary supplementation, physical activity to enhance muscle mass, and pharmacological therapy. This chapter will limit the discussion to nutritional interventions that may correct deficiency states or enhance the activity of antiosteoporosis drugs such as the bisphosphonates or estrogens. However, as noted above, improvement in nutrient intake may also affect nonskeletal tissues, thereby reducing falls and improving coordination. As such, these effects should never be overlooked when considering the appropriate therapeutic regimen for an elderly individual.

6.1. Calcium Supplementation As a Therapeutic Intervention

Calcium in various forms has been used for more than half a century, both as a supplement and as a therapeutic agent for postmenopausal osteoporosis. The rationale is relatively straightforward and centers on the thesis that low calcium intake enhances PTH release from the parathyroids, which in turn results in excessive bone turnover and eventual bone loss. Indeed, several studies have demonstrated that calcium supplementation (1000–1500 mg/d) for elderly women is associated with a reduction in bone resorption (as measured by bone turnover indices), a partial suppression of PTH, and maintenance or modest increases in BMD *(52,53)*. In a recent meta-analysis, Shea et al. examined 15 randomized placebo-controlled trials in postmenopausal women that utilized calcium supplementation (minimum of 1000 mg Ca/d) in comparison to regular dietary intake *(54)*. Women who received

Table 1
Summary of an Evidence-Based Meta-Analysis of Calcium and Vitamin D as Therapies for Osteoporosis

Treatment	*No. of subjects*	*RR of Vert Frx*	*RR of non-Vert Frx*	*2-yr Change in BMD—spine*	*2-yr Change in BMD—Hip*
Calcium	576	0.77; (0.54–1.09) $p = 0.14$	0.86; (0.43–1.7) $p = 0.66$	1.66; $p < 0.01$	1.64; $p < 0.01$
Vitamin D	6100	0.63 (0.45–0.88); $p < 0.01$	0.77 (0.57–1.04); $p = 0.09$	S: 0.86 OH: 4.6 P = 0.01	S: 0.98 OH: 2.46 P = 0.03

S; standard vitamin D; OH, hydroxylated vitamin D. Change in BMD represents weighted mean difference between treatment with calcium or vitamin D and placebo. (Adapted from ref. *58.*)

calcium supplementation (calcium citrate or carbonate) had greater increases in bone mass than women receiving placebo after 2 yr or more of treatment. The pooled difference for percentage change in BMD from baseline in supplemented women compared to groups of women with an average dietary calcium intake of approx 700 mg/d was +2.05% for total body BMD, and 1.65% for lumbar spine and hip BMD (*see* Table 1). Thus, by combining randomized trials in a formal meta-analysis, the evidence suggests that calcium supplementation prevents bone loss in susceptible individuals. This is supported by a randomized placebo-controlled trial of elderly Maine women by Storm et al. Those investigators looked at seasonal bone loss in women living in northern latitudes and found that 1000 mg Ca/d for 2 yr prevented significant winter bone loss from the hip *(33)*. However, whether calcium supplementation alone as a therapeutic entity prevents osteoporotic fractures is less certain. In the same meta-analysis by Shea et al., spine and nonspine fracture risk reduction with calcium supplementation was not statistically different from placebo, although the number of subjects and the number of studies used in the meta-analysis were quite small (*see* Table 1) *(54)*. Indeed, for spine fractures, there was a 23% reduction in new fractures in women supplemented with calcium, suggesting the possibility that with more studies, a true effect may become apparent. Similarly, Dawson Hughes et al. showed a 50% reduction in nonspine fractures in elderly New England women supplemented with calcium and vitamin D *(27)*.

Also uncertain is whether calcium supplementation by dietary means affects bone mass. In the Storm et al. trial, milk supplementation (three to four glasses/d) was associated with less bone loss than placebo, but was not nearly as effective as calcium supplements *(53)*. With the paucity of randomized trials, investigators have turned to epidemiological studies, which suggest that very low calcium intake (<250 mg/d) places women at additional risk for osteoporotic fractures *(15)*. On the other hand, in cohort studies, high calcium intake has been shown to reduce fracture risk, but the effect is weak at best *(15)*. A large population-based interventional study in elderly women is currently nearing completion and may provide more evidence about the role of dietary calcium in the prevention of fractures. Notwithstanding that study outcome, it should be noted that calcium and vitamin D are necessary components of any therapeutic regimen for age-related osteoporosis when combined with antiresorptive therapies (calcitonin, bisphosphonates, estrogen, selective estrogen receptor modulators) or anabolics (e.g., PTH).

6.2. Vitamin D As a Therapeutic Supplement

Vitamin D is critical for maintenance of bone mass through its actions on several pathways. First, it promotes calcium absorption in the gut. Second, it facilitates bone mineralization, both at the osteoblast level by enhancing differentiation and through increased absorption of phosphates. Third, the active form of vitamin D (1,25-dihydroxyvitamin D) is important for both osteoblast and osteoclast function *(55)*. As such, it is not surprising that vitamin D deficiency could cause or exacerbate age-related osteoporosis. Moreover, vitamin D deficiency has been associated with muscle weakness, which may result in a greater likelihood of falling. Hence the strong rationale for clinical trials with different forms of vitamin D that have been undertaken for nearly 30 yr. Papadimitropoulos et al. recently performed a meta-analysis of all randomized placebo-controlled trials of vitamin D and its analogs over the last two decades *(56)*. They reported a significant reduction in spine fractures (i.e., 37% risk reduction; $p < 0.01$) in women taking vitamin D (standard or hydroxylated) compared to placebo, and a nearly significant reduction in nonspine fractures (i.e., 23% risk reduction, $p < 0.09$). With respect to bone density, standard vitamin D had virtually no effect on any bone site, whereas hydroxylated vitamin D resulted in a 2–3% increase in BMD at several sites over 3 yr.

Two of the strongest randomized placebo-controlled trials that were part of this meta-analysis are illuminating. Chapuy et al. provided 800 IU cholecalciferol and 1200 mg of calcium to nursing home residents for 18 mo and found a 35% reduction in hip fractures *(28)*. Dawson Hughes et al. provided 700 IU of cholecalciferol plus 500 mg of calcium citrate to elderly men and women and found a nearly 50% reduction in nonvertebral fractures *(27)*. These studies and the meta-analysis provide some insight into potential safe and effective preventive measures to reduce the morbidity and cost of hip fractures. The use of supplemental calcium and vitamin D is safe, noninvasive, inexpensive, and cost-effective in the elderly population. In addition, because it may have some nonskeletal benefits, this combination of nutrients may become standard of care for men and women in assisted-living situations or nursing homes.

6.3. Other Nutritional Interventions

Unfortunately, the level of evidence from randomized placebo-controlled trials with other dietary or nutrient supplements and their role in preventing osteoporotic fractures or preserving bone mass is quite sparse. Some work, principally in Europe, has focused on the role of protein supplementation after hip fracture. For a number of years it has been recognized that adequate protein intake is essential for skeletal growth and maintenance. Recently, more attention has focused on providing adequate protein intake after fractures, to enhance healing and improve the overall metabolic status of the patient. Protein undernutrition can enhance the risk of osteoporotic fractures, especially hip fractures, in part by altering muscle function, and impairing bone formation. Geinoz et al. reported that men and women who sustained hip fractures had low BMD and very low protein intake *(57)*. In a subsequent study, Tkatch et al. used a 20-g protein daily supplement in elderly women with recent hip fractures and demonstrated improvement in their clinical outcomes *(12)*. Schurch et al. performed a double-blind placebo-controlled study of both men and women over the age of 80 who sustained a hip fracture, and compared those given 20 g protein supplement with 200,000 IU of vitamin D vs those just receiving vitamin D *(13)*. Protein repletion after hip fracture was associated with shorter rehabilitation stays, greater biceps and quadriceps strength, and improved functional status. Changes in BMD after the fracture were also less in the group receiving protein supplementation *(13)*. These changes were also associated with a marked increase in serum IGF-I, suggesting that the mechanism of action of protein supplementation after hip fracture could be through the IGF regulatory system.

7. CONCLUSION

Osteoporosis, the most common skeletal disorder of older men and women, is a multifactorial disease that is influenced strongly by nutritional aspects. Protein, calcium, sodium, and phosphorus are all critical macro- and micronutrients that work through the bone remodeling units to sustain skeletal health. Moreover, the fat-soluble vitamins, and in particular, vitamin D, are absolutely essential for the acquisition and maintenance of the skeleton. In fact, many clinicians advocate a minimal requirement of 700 IU of vitamin D/d for all elderly individuals. On the other hand, the role of trace elements in bone remodeling, especially in the elderly, requires further studies. There is no longer any doubt that the successful treatment and prevention of osteoporosis requires, at a minimum, adequate calcium and vitamin D. Even with the advent of newer and more powerful drugs to prevent bone resorption or enhance bone formation, calcium and vitamin D are essential building blocks to restore skeletal health. The challenge in this century for public health experts is to define a greater role for nutritional determinants in preventing age-related bone loss.

REFERENCES

1. Rosen CJ. Restoring aging Bones. Sci Am 2003; 288:70–78.
2. Robey PB, Bianco P. Cellular mechanism of age related bone loss. In: Rosen CJ, Glowacki J, Bilezikian JP, eds. The Aging Skeleton. Academic Press, San Diego, CA, 1999, pp. 145–157.

3. Lacey DL, Timms E, Tan HL, et al. Osteoprotegerin ligand is a cytokine that regulates osteoclast differentiation and activation. Cell 1998; 93(2):165–176.
4. Martin TJ, Ng KW. Mechanisms by which cells of the osteoblast lineage control osteoclast formation and activity. J Cell Biochem 1994; 56:357–366.
5. Donahue LR, Rosen CJ. IGFs and bone: the osteoporosis connection revisited. Proc Soc Exp Biol Med 1998; 219:1–7.
6. Beamer WG, Donahue LR, Rosen CJ. IGF-I and bone: from mouse to man. International Symposium on GH and Growth Factors in Endocrinology and Metabolism; Growth Hormone and IGF Research 2000, 6:B S103–S105.
7. Thissen JP, Ketelslegers JM, Underwood LE. Nutritional regulation of the IGFs. Endocrinol Rev 1994; 15:80–101.
8. Musey VC, Goldstein S, Farmer PK, Moore PB, Phillips LS. Differential regulation of IGF-I and IGFBP-1 by dietary composition in humans. Am J Med Sci 1993; 305:131–138.
9. Ketelslegers JM, Maiter D, Maes M, Underwood LD, Thissen JP. Nutritional regulation of IGF-I. Metabolism 1995; 44:50–57.
10. Rosen CJ, Kessenich C. The role of IGF-I in senesence: implications for therapy. Endocrinologist. 1996; 6:102–106.
11. Heaney RP, McCarron DA, Dawson-Hughes B, et al. Dietary changes favorably affect bone remodeling. J Am Diet Assoc 1999; 99:1228–1233.
12. Tkatch L, Rapin CH, Rizzoli R, et al. Benefits of oral protein supplement in elderly patients with fracture of the proximal femur. J Am Coll Nutr 1992; 11:619–523.
13. Schurch MA, Rizzoli R, Slosman D, Vadas L, Vergnaud P, Bonjour JP. Protein supplements increase serum insulin-like growth factor-I levels and attenuate proximal femur bone loss in patients with recent hip fracture. A randomized, double-blind, placebo-controlled trial. Ann Intern Med 1998; 128(10):801–890.
14. Boonen S, Rosen C, Bouillon R, et al. Musculoskeletal effects of the rhIGF-I/IGFBP-3 complex in osteoporotic patients with proximal femoral fracture. J Clin Endocrinol Metab 2002; 87:1593–1599.
15. Prince RL. The rationale of calcium supplementation in the therapeutics of age related osteoporosis. In: Rosen CJ, Glowacki J, Bilezikian JP, eds. The Aging Skeleton. Academic, San Diego, CA, 1999, pp. 479–494.
16. Gloth FM. Vitamin D. In: Rosen CJ, Glowacki J, Bilezikian JP, eds. The Aging Skeleton. Academic, San Diego, CA, 1999, pp. 185–194.
17. Riggs BL, Wahner W, Seeman E, et al. Changes in bone mineral density of the proximal femur and spine with aging: differences between the postmenopausal and senile osteoporosis syndromes. J Clin Invest 1982; 70:716–723.
18. Ensrud KE, Palermo L, Black DM, et al. Hip and calcaneal bone loss increase with advancing age: longitudinal results from the study of osteoporotic fractures. J Bone Miner Res 1995; 10:1778–1787.
19. Hannan MT, Felson DT, Dawson-Hughes B, et al. Risk factors for longitudinal bone loss in elderly men and women: the Framingham Osteoporosis Study. J Bone Miner Res 2000; 15:710–720.
20. Ensrud KE, Palmero L, Black MD, et al. Hip and calcaneal bone loss increase with advancing age. J Bone Miner Res 1995; 10:1778–1787.
21. Ross PD, Knowlton W. Rapid bone loss is associated with increased levels of biochemical markers. J Bone Miner Res 1998; 13:297–302.
22. Dresner-Pollak R, Parker RA, Poku M, Thompson J, Seibel MJ, Greenspan SL. Biochemical markers of bone turnover reflect femoral bone loss in elderly women. Calcif Tiss Int 1996; 59:328–333.
23. Garnero P, Hausherr E, Chapuy MC, et al. Markers of bone resorption predict hip fracture in elderly women: the EPIDOS prospective study. J Bone Min Res 1996; 11:1531–1538.

24. Bollen AM, Kiyak HA, Eyre DR. Longitudinal evaluation of a bone resorption marker in elderly subjects. Osteopor Int 1997; 7:544–549.
25. Chapuy MC, Schott AM, Garnero P, Hans D, Delmas PD, Meunier PJ. Healthy elderly French women living at home have secondary hyperparathyroidism and high bone turnover in winter. J Clin Endocrinol Metab 1996; 81:1129–1133.
26. LeBoff MS, Kohlmeier L, Hurwitz S, Franklin J, Wright J, Glowacki J. Occult vitamin D deficiency in postmenopausal US women with acute hip fracture. JAMA 1999; 281(16):1505–1511.
27. Dawson-Hughes B, Harris SS, Krall EA, Dallal GE. Effect of calcium and vitamin D on bone density in men and women 65 years of age and older. N Engl J Med 1997; 337:670–676.
28. Chapuy MC, Arlot ME, Duboeuf F, et al. Vitamin D3 and calcium to prevent hip fractures in elderly women. N Engl J Med 1992; 327:1637–1642.
29. Recker RR, Hinders S, Davies M, et al. Correcting calcium deficiency prevents spine fractures in elderly women. J Bone Miner Res 1996; 11:1961–1966.
30. NIH Consensus Conference Optimal Calcium Intake. JAMA 1995; 272:1942–1948.
31. Dawson-Hughes B, Harris SS. Calcium intake influcences the association of protein intake with rates of bone loss in elderly men and women. Am J Clin Nutr 2002; 75:773–779.
32. Hannan MT, Tucker KL, Dawson-Hughes B, et al. Effect of dietary protein on bone loss in elderly men and women: the Framingham Osteoporosis Study. [in process citation] J Bone Miner Res 2000; 15:2504–2512.
33. Tucker KL, Chen H, Hannan MT, et al. Bone mineral density and dietary patterns in older adults: the Framingham Osteoporosis Study. Am J Clin Nutr 2002; 76:245–252.
34. McKeown NM, Jacques PF, Gundberg CM, et al. Dietary and nondietary determinants of vitamin K biochemical measures in men and women. J Nutr 2002; 132:1329–1334.
35. Promislow JH, Goodman-Gruen D., Slymen DJ, Barrett-Connor E. Retinol intake and bone mineral density in the elderly: the Rancho Bernardo Study. J Bone Miner Res 2002; 17(8):1349–1358.
36. Michaelsson K, Lithell H, Vessby B, Melhus H. Serum retinol levels and the risk of fracture. N Engl J Med 2003; 348(4):287–294.
37. Slemenda CW, Longcope C, Zhou L, Hui S, Peacock M, Johnston CC. Sex steroids and bone mass in older men: positive associations with serum estrogens and negative associations with androgens. J Clin Invest 1997; 100:1755–1759.
38. Greendale GA, Edelstein S, Barrett-Connor E. Endogenous sex steroids and bone mineral density in older women and men: the Rancho Bernardo Study. J Bone Miner Res 1997; 12:1833–1843.
39. Amin S, Xhang Y, Sawin CT, et al. Association of hypogonadism and estradiol levels with bone mineral density in elderly men from the Framingham study. Ann Intern Med 2000; 133:951–963.
40. Khosla S, Atkinson EJ, Dunstan CR, O'Fallon WM. Effect of estrogen versus testosterone on circulating osteoprotegerin and other cytokine levels in normal elderly men. J Clin Endocrinol Metab 2002; 87(4):1550–1554.
41. Falahati-Nini A, Riggs BL, Atkinson EJ, O'Fallon WM, Eastell R, Khosla S. Relative contribution of testosterone and estrogen in regulating bone resorption and formation in normal elderly men. J Clin Invest 2000; 106(12):1553–1560.
42. Smith MR, Finkelstein JS, McGovern FJ, et al. Changes in body composition during androgen deprivation therapy for prostate cancer. J Clin Endocrinol Metab 2002; 87(2):599–603.
43. Rosen CJ. Serum IGF-I and IGF binding proteins: clinical implications. Clin Chem 1999; 45:1384–1390.
44. Langlois JA, Rosen CJ, Visser M, et al. The association between IGF-I and bone mineral density in women and men: the Framingham Heart Study. J Clin Endocrinol Metab 1998; 83:4257–4262.
45. Halloran BP, Ferguson VL, Simske SJ, Burghardt A, Venton LL, Majumdar S. Changes in bone structure and mass with advancing age in the male C57BL/6J mouse. J Bone Miner Res 2002; 17(6):1044–1050.

46. Deng HW, Mahaney MC, Williams JT, et al. Relevance of the genes for bone mass variation to susceptibility to osteoporotic fractures and its implications to gene search for complex human disease. Genet Epidemiol 2002; 22(1):12–25.
47. Kiel DP, Myers RH, Cupples LA, et al. The BsmI vitamin D receptor restriction fragment length polymorphism influences the effect of calcium intake on BMD. J Bone Miner Res 1997; 12:1049–1057.
48. Hayes WC. Biomechanics of cortical and trabecular bone: implications for assessment of fracture risk. In: Mow VC, Hayes WC, eds. Basic Orthopaedic Biomechanics. Raven, New York, 1991, pp. 93–142.
49. Cummings SR, Nevitt MC, Browner WS, et al. Risk factors for hip fracture in white women. N Engl J Med 1995; 332:767–773.
50. Leibson CL, Tosteson ANA, Gabriel SE, Ransom JE, Melton LJ. Mortality, disability, and nursing home use for persons with and without hip fracture: a population-based study. J Am Geriatr Soc 2002; 50:1644–1650.
51. Gillespie LD, Gillespie WJ, Robertson MC, Lamb SE, Cumming RG, Rowe BH. Interventions to reduce the incidence of falling in the elderly. The Cochrane Database of Systematic Reviews Issue 4, 2002.
52. Riggs BL, O'Fallon WM, Muhs J, O'Connor MK, Kumar R, Melton LJ. Long term effects of calcium supplementation on serum PTH level, bone turnover and bone loss in elderly women. J Bone Miner Res 1998; 13:168–174.
53. Storm D, Smith-Porter E, Musgrave KO, et al. Calcium supplementaiton prevents seasonal bone loss and changes in biochemical markers of bone turnover in elderly New England women: a randomized placebo-controlled trial. J Clin Endocrinol Metab 1998; 83:3817–3826.
54. Shea B, Wells G, Cranney A, et al. Meta analysis of calcium supplementation for the prevention of postmenopausal osteoporosis. Endocr Rev 2002; 23:552–559.
55. Holick MF. Vitamin D: a millennium perspective. J Cell Biochem 2003; 88:296–307.
56. Papdimitropoulos E, Wells G, Shea B, et al. Meta analysis of the efficacy of vitamin D treatment in preventing osteoporosis in postmenopausal women. Endocr Rev 2002; 23:560–569.
57. Geinoz G, Rapin CH, Tizzoli R, et al. Relationship between BMD and dietary intakes in the elderly. Osteopor Int 1993; 3:242–248.
58. Guyatt GH, Cranney A, Griffith L, et al. Summary of meta-analyses of therapies for postmenopausal osteoporosis and the relationship between bone density and fractures. Endocrinol Metab Clin N Am 2002; 31(3):659–679.

14 Nutrition and Skeletal Health in Blacks

Susan S. Harris

Population groups in Africa are extremely diverse genetically. The focus of this chapter is on African Americans, a group whose African ancestors came predominantly from West Africa and that also reflects a variable admixture of European and other genetic influences. The bone density values and fracture rates of at least some African populations differ from those of US blacks *(1),* and the nutritional and other environmental influences in the United States are quite different from those in many African countries.

It is well established that US blacks have a reduced risk for osteoporosis compared with US whites and others. Medicare data suggest that about 5% of black women and 3% of black men will have a hip fracture by age 90, compared with 16% of white women and 6% of white men *(2).* In addition, the prevalence of all nonvertebral fractures among blacks is about half that of whites *(3).* Although other factors, including bone geometry *(4),* may also play a role, the lower fracture rates of blacks probably result primarily from their higher bone mineral density (BMD) *(5–9),* particularly from achievement of a higher peak density by young adulthood. Racial comparisons of BMD or size-adjusted bone mineral content generally show higher values in prepubertal black compared with white children *(10–13).* In addition, black children appear to have greater increases in BMD than white children during puberty *(14–17).*

BMD of young black adults is reported to be from 4% to 13% higher than that of young white adults at various skeletal sites *(5,18–22).* Early postmenopausal bone loss may be modestly slower in black women than in white women *(9),* but bone loss in later years appears to be about the same for blacks and whites *(9,23).*

This skeletal advantage exists despite nutritional factors that would appear to put blacks at an *increased* risk compared with other groups. The most notable of these differences is a far higher prevalence of vitamin D insufficiency *(6,7,24–31).* For example, data from NHANES-III show that, among residents of the Southern states who were measured in winter, more than 53% of non-Hispanic blacks had

From: *Nutrition and Bone Health*
Edited by: M. F. Holick and B. Dawson-Hughes © Humana Press Inc., Totowa, NJ

25-hydroxyvitamin D [25(OH)D)] levels indicative of vitamin D insufficiency compared with only 8% of non-Hispanic whites *(30)*. This difference may result in part from lower vitamin D intake owing particularly to less use of supplements *(32)* including multivitamins *(33)* and other vitamin D supplements *(33)*, but the principal reason is that skin pigmentation in blacks sharply reduces the amount of vitamin D that is produced during sunlight exposure. Among healthy, free-living young women living in New England, 25(OH)D concentrations in blacks were only half as high as those in whites, and the increases from winter to summer were smaller *(26)*. This difference probably does not result from any absolute limit in the production of vitamin D or 25(OH)D, because, when given adequate exposure to ultraviolet light, black adults can reach mean concentrations of both parent vitamin D *(34)* and 25(OH)D *(35)* that are similar to those of whites. Reduced vitamin D acquisition, from either skin production or diet, is partially compensated for by an increased rate of 25(OH)D production *(28)*, and the extent of this adaptive mechanism appears to be similar in blacks and whites when both have low initial 25(OH)D concentrations *(28)*.

Total calcium intake, although below recommended levels in most adult Americans, is even lower among blacks because of lower dietary calcium intakes (compared with whites and Hispanics) *(36,37)* and also because of less calcium supplement use *(33,36)*. Despite a higher prevalence of lactose intolerance among blacks *(38)*, dairy consumption appears to be only modestly lower in blacks than whites *(33,39)*. Milk is the most important single source of dietary calcium among black adolescents and adults *(40)*, and lactose digestion can apparently be improved with prolonged consumption of dairy products *(41)*.

In the United States, obesity is more prevalent among black women than white women *(42)*, and their reduction in osteoporosis risk despite relatively poor calcium and vitamin D nutrition can be explained in part by differences in body size and body composition. The contributions of these factors to BMD differences is difficult to assess because of the fact that the densitometric methods used to measure bone density are themselves affected by body size variables, including bone size *(43)* and the thickness and composition of soft tissue *(44)*. Nevertheless, it is a fairly consistent finding that *adjusting* for body weight, height, or related measures reduces but does not eliminate the black–white difference in BMD *(8,22)*.

Concentrations of the calcium regulating hormones 1,25-dihydroxyvitamin D [1,25$(OH)_2$D] and parathyroid hormone (PTH) tend to differ by race. Specifically, both young *(25,26,45)* and old *(7,27,29)* blacks have increased PTH concentrations compared with whites. Blacks also tend to have higher concentrations of 1,25$(OH)_2$D, the active form of vitamin D, despite substantially lower concentrations of its precursor, 25(OH)D *(7,45)*. This inverse association is observed in other populations as well *(46)*, and likely results from an increased rate of kidney synthesis of 1,25$(OH)_2$D stimulated by elevated PTH. The higher PTH and higher 1,25$(OH)_2$D of blacks are both consistent with their lower calcium intake and poorer vitamin D status. When young black and white women who had fairly similar calcium intake and 25(OH)D concentrations were observed, no differences in $1,25(OH)_2D$ or PTH were observed *(24)*.

There is extensive evidence in whites and more limited evidence in young *(25,26)* and old *(27,31)* blacks that reduced 25(OH)D is associated with increased PTH. Adequately large studies are not available to determine whether the magnitude of the association of 25(OH)D with PTH is similar in healthy blacks and whites. However, a large study of patients at an osteoporosis clinic demonstrated a greater increase in PTH with vitamin D depletion among blacks than among whites *(31)*, suggesting a potentially *more* harmful effect of vitamin D insufficiency in blacks.

It has been suggested that the higher BMD of blacks despite a higher prevalence of secondary hyperparathyroidism may result from a relative skeletal resistance to the effect of PTH *(47)*. This would theoretically allow them to benefit from the kidney effects of high PTH [increased synthesis of $1,25(OH)_2D$ and increased reabsorption of calcium] without suffering from the increased skeletal calcium release that is also associated with elevated PTH. Adult black women have generally been reported to have reduced rates of bone turnover compared with whites *(7,48,49)*. Consistent with a skeletal resistance to PTH, reduced bone turnover in blacks has been observed despite higher baseline PTH, higher stimulated PTH levels, or similar high PTH concentrations during PTH infusion *(47,50)*. Furthermore, blacks have lower calcium excretion than whites even after multiple factors including calcium intake and PTH are controlled for *(24)*.

Another potential explanation for the higher BMD of blacks is that they may have an intestinal resistance to the actions of $1,25(OH)_2D$. This is supported by a study of calcium retention in black and white women given high- and low-calcium diets *(51)*. The black women had lower 25(OH)D and higher $1,25(OH)_2D$ than the white women throughout the study, and had greater increases in $1,25(OH)_2D$ when calcium intake was decreased. Despite the greater $1,25(OH)_2D$ response of the black women to the diet change, the calcium retention fraction did not differ by race, consistent with an intestinal resistance to the actions of $1,25(OH)_2D$. This hypothesis was further supported by an experiment in which the effect of administered calcitriol on calcium absorption was observed in black and white women. The groups had similar increases in $1,25(OH)_2D$, but the resulting increase in calcium absorption was smaller among the black women *(52)*. Such an adaptation could benefit the skeleton because $1,25(OH)_2D$ may have a positive effect on bone that is independent of its effect on calcium absorption *(53)*, perhaps through stimulation of osteoblastic activity *(54)*. It may be that such a direct effect requires a relatively high $1,25(OH)_2D$ concentration, as may be more readily achieved if intestinal resistance to the hormone protects against hypercalcemia *(55)*.

1. CONCLUSIONS

Little research has been undertaken to determine whether improving nutritional status of blacks would reduce their rates of bone loss and fracture, but some indirect evidence suggests that improved calcium and/or vitamin D status might do so. Vitamin D insufficiency was associated with increased PTH in young black women *(26),* and with higher rates of secondary hyperparathyroidism in elderly blacks *(27).* In the same elderly blacks, secondary hyperparathyroidism was associated with lower calcaneus BMD *(56)* and increased bone turnover *(56).* A small intervention study showed decreases in PTH and bone turnover in older black women supplemented with vitamin D *(57).* Given the high prevalence of vitamin D insufficiency among blacks, a high priority should be given to conducting adequately large and controlled studies to determine the effects of vitamin D supplementation on skeletal and other health outcomes in blacks.

REFERENCES

1. Melton LJ III, Marquez MA, Achenbach SJ, et al. Variations in bone density among persons of African heritage. Osteopor Int 2002; 13(7):551–559.
2. Barrett JA, Baron JA, Karagas MR, Beach ML. Fracture risk in the U.S. medicare population. J Clin Epidemiol 1999; 52(3):243–249.
3. Bohannon AD, Hanlon JT, Landerman R, Gold DT. Association of race and other potential risk factors with nonvertebral fractures in community-dwelling elderly women. Am J Epidemiol 1999; 149(11):1002–1009.
4. Cummings SR, Cauley JA, Palermo L, et al. Racial differences in hip axis lengths might explain racial differences in rates of hip fracture. Study of Osteoporotic Fractures Research Group. Osteopor Int 1994; 4(4):226–229.
5. Looker AC, Wahner HW, Dunn WL, et al. Updated data on proximal femur bone mineral levels of US adults. Osteopor Int 1998; 8(5):468–489.
6. Aloia JF, Vaswani A, Yeh JK, Flaster E. Risk for osteoporosis in black women. Calcif Tissue Int 1996; 59(6):415–423.
7. Kleerekoper M, Nelson DA, Peterson EL, et al. Reference data for bone mass, calciotropic hormones, and biochemical markers of bone remodeling in older (55–75) postmenopausal white and black women. J Bone Miner Res 1994; 9(8):1267–1276.
8. Finkelstein JS, Lee ML, Sowers M, et al. Ethnic variation in bone density in premenopausal and early perimenopausal women: effects of anthropometric and lifestyle factors. J Clin Endocrinol Metab 2002; 87(7):3057–3067.
9. Luckey MM, Wallenstein S, Lapinski R, Meier DE. A prospective study of bone loss in African-American and white women—a clinical research center study. J Clin Endocrinol Metab 1996; 81(8):2948–2956.
10. Nelson DA, Simpson PM, Johnson CC, Barondess DA, Kleerekoper M. The accumulation of whole body skeletal mass in third- and fourth-grade children: effects of age, gender, ethnicity, and body composition. Bone 1997; 20(1):73–78.
11. McCormick DP, Ponder SW, Fawcett HD, Palmer JL. Spinal bone mineral density in 335 normal and obese children and adolescents: evidence for ethnic and sex differences. J Bone Miner Res 1991; 6(5):507–513.
12. Li JY, Specker BL, Ho ML, Tsang RC. Bone mineral content in black and white children 1 to 6 years of age. Early appearance of race and sex differences. Am J Dis Child 1989; 143(11):1346–1349.
13. Bell NH, Shary J, Stevens J, Garza M, Gordon L, Edwards J. Demonstration that bone mass is greater in black than in white children. J Bone Miner Res 1991; 6(7):719–723.
14. Henry YM, Eastell R. Ethnic and gender differences in bone mineral density and bone turnover in young adults: effect of bone size. Osteopor Int 2000; 11(6):512–517.

15. Gilsanz V, Skaggs DL, Kovanlikaya A, et al. Differential effect of race on the axial and appendicular skeletons of children. J Clin Endocrinol Metab 1998; 83(5):1420–1427.
16. Gilsanz V, Roe TF, Mora S, Costin G, Goodman WG. Changes in vertebral bone density in black girls and white girls during childhood and puberty. N Engl J Med 1991; 325(23):1597–1600.
17. Wang MC, Aguirre M, Bhudhikanok GS, et al. Bone mass and hip axis length in healthy Asian, black, Hispanic, and white American youths. J Bone Miner Res 1997; 12(11):1922–1935.
18. Liel Y, Edwards J, Shary J, Spicer KM, Gordon L, Bell NH. The effects of race and body habitus on bone mineral density of the radius, hip, and spine in premenopausal women. J Clin Endocrinol Metab 1988; 66(6):1247–1250.
19. Aloia JF, Vaswani A, Mikhail M, Flaster ER. Body composition by dual-energy X-ray absorptiometry in black compared with white women. Osteopor Int 1999; 10(2):114–119.
20. Meier DE, Luckey MM, Wallenstein S, Lapinski RH, Catherwood B. Racial differences in pre- and postmenopausal bone homeostasis: association with bone density. J Bone Miner Res 1992; 7(10):1181–1189.
21. Nelson DA, Kleerekoper M, Parfitt AM. Bone mass, skin color and body size among black and white women. Bone Miner 1988; 4(3):257–264.
22. Harris SS, Wood MJ, Dawson-Hughes B. Bone mineral density of the total body and forearm in premenopausal black and white women. Bone 1995; 16(4 suppl):311S–315S.
23. Aloia JF, Vaswani A, Feuerman M, Mikhail M, Ma R. Differences in skeletal and muscle mass with aging in black and white women. Am J Physiol Endocrinol Metab 2000; 278(6):E1153–E1157.
24. Meier DE, Luckey MM, Wallenstein S, Clemens TL, Orwoll ES, Waslien CI. Calcium, vitamin D, and parathyroid hormone status in young white and black women: association with racial differences in bone mass. J Clin Endocrinol Metab 1991; 72(3):703–710.
25. Bell NH, Greene A, Epstein S, Oexmann MJ, Shaw S, Shary J. Evidence for alteration of the vitamin D-endocrine system in blacks. J Clin Invest 1985; 76(2):470–473.
26. Harris SS, Dawson-Hughes B. Seasonal changes in plasma 25-hydroxyvitamin D concentrations of young American black and white women. Am J Clin Nutr 1998; 67(6):1232–1236.
27. Harris SS, Soteriades E, Coolidge JA, Mudgal S, Dawson-Hughes B. Vitamin D insufficiency and hyperparathyroidism in a low income, multiracial, elderly population. J Clin Endocrinol Metab 2000; 85(11):4125–4130.
28. Matsuoka LY, Wortsman J, Chen TC, Holick MF. Compensation for the interracial variance in the cutaneous synthesis of vitamin D. J Lab Clin Med 1995; 126(5):452–457.
29. Perry HM 3rd, Horowitz M, Morley JE, et al. Aging and bone metabolism in African American and Caucasian women. J Clin Endocrinol Metab 1996; 81(3):1108–1117.
30. Looker AC, Dawson-Hughes B, Calvo MS, Gunter EW, Sahyoun NR. Serum 25-hydroxyvitamin D status of adolescents and adults in two seasonal subpopulations from NHANES III. Bone 2002; 30(5):771–777.
31. Parikh N ET, Hill J, Phillips E, Rao D. Prevalence of vitamin D depletion among subjects seeking advice on osteoporosis: a five-year cross sectional study with therapeutic and public health implications. J Bone Miner Res 2002; 17(suppl 1):S201.
32. Balluz LS, Kieszak SM, Philen RM, Mulinare J. Vitamin and mineral supplement use in the United States. Results from the Third National Health and Nutrition Examination Survey. Arch Fam Med 2000; 9(3):258–262.
33. Bell RA, Quandt SA, Spangler JG, Case LD. Dietary calcium intake and supplement use among older African American, white, and Native American women in a rural southeastern community. J Am Dietet Assoc 2002; 102(6):844–847.
34. Clemens TL, Adams JS, Henderson SL, Holick MF. Increased skin pigment reduces the capacity of skin to synthesise vitamin D3. Lancet 1982; 1(8263):74–76.
35. Brazerol WF, McPhee AJ, Mimouni F, Specker BL, Tsang RC. Serial ultraviolet B exposure and serum 25 hydroxyvitamin D response in young adult American blacks and whites: no racial differences. J Am Coll Nutr 1988; 7(2):111–1118.

36. Ervin RB, Kennedy-Stephenson J. Mineral intakes of elderly adult supplement and non-supplement users in the third national health and nutrition examination survey. J Nutr 2002; 132(11):3422–3427.
37. Siega-Riz AM, Popkin BM. Dietary trends among low socioeconomic status women of child-bearing age in the United States from 1977 to 1996:a comparison among ethnic groups. J Am Med Womens Assoc 2001; 56(2):44–48, 72.
38. Rao DR, Bello H, Warren AP, Brown GE. Prevalence of lactose maldigestion. Influence and interaction of age, race, and sex. Dig Dis Sci 1994; 39(7):1519–1524.
39. Wiecha JM, Fink AK, Wiecha J, Hebert J. Differences in dietary patterns of Vietnamese, white, African-American, and Hispanic adolescents in Worcester, Mass. J Am Dietet Assoc 2001; 101(2):248–251.
40. Looker AC, Loria CM, Carroll MD, McDowell MA, Johnson CL. Calcium intakes of Mexican Americans, Cubans, Puerto Ricans, non-Hispanic whites, and non-Hispanic blacks in the United States. J Am Dietet Assoc 1993; 93(11):1274–1279.
41. Pribila BA, Hertzler SR, Martin BR, Weaver CM, Savaiano DA. Improved lactose digestion and intolerance among African-American adolescent girls fed a dairy-rich diet. J Am Dietet Assoc 2000; 100(5):524–528; quiz 529–530.
42. Flegal KM, Carroll MD, Ogden CL, Johnson CL. Prevalence and trends in obesity among US adults, 1999–2000. JAMA 2002; 288(14):1723–1727.
43. Looker AC. The skeleton, race, and ethnicity. J Clin Endocrinol Metab 2002; 87(7):3047–3750.
44. Dawson-Hughes B, Deehr MS, Berger PS, Dallal GE, Sadowski LJ. Correction of the effects of source, source strength, and soft-tissue thickness on spine dual-photon absorptiometry measurements. Calcif Tissue Int 1989; 44(4):251–257.
45. Bikle DD, Ettinger B, Sidney S, Tekawa IS, Tolan K. Differences in calcium metabolism between black and white men and women. Miner Electrolyte Metab 1999; 25(3):178–184.
46. Need AG, Horowitz M, Morris HA, Nordin BC. Vitamin D status: effects on parathyroid hormone and 1,25-dihydroxyvitamin D in postmenopausal women. Am J Clin Nutr 2000; 71(6):1577–1581.
47. Fuleihan GE, Gundberg CM, Gleason R, et al. Racial differences in parathyroid hormone dynamics. J Clin Endocrinol Metab 1994; 79(6):1642–1647.
48. Finkelstein JS, Sowers M, Greendale GA, et al. Ethnic variation in bone turnover in pre- and early perimenopausal women: effects of anthropometric and lifestyle factors. J Clin Endocrinol Metab 2002; 87(7):3051–3056.
49. Weinstein RS, Bell NH. Diminished rates of bone formation in normal black adults. N Engl J Med 1988; 319(26):1698–1701.
50. Cosman F, Morgan DC, Nieves JW, et al. Resistance to bone resorbing effects of PTH in black women. J Bone Miner Res 1997; 12(6):958–966.
51. Dawson-Hughes B, Harris S, Kramich C, Dallal G, Rasmussen HM. Calcium retention and hormone levels in black and white women on high- and low-calcium diets. J Bone Miner Res 1993; 8(7):779–787.
52. Dawson-Hughes B, Harris SS, Finneran S, Rasmussen HM. Calcium absorption responses to calcitriol in black and white premenopausal women. J Clin Endocrinol Metab 1995; 80(10):3068–3072.
53. Tilyard MW, Spears GF, Thomson J, Dovey S. Treatment of postmenopausal osteoporosis with calcitriol or calcium. N Engl J Med 1992; 326(6):357–362.
54. Geusens P, Vanderschueren D, Verstraeten A, Dequeker J, Devos P, Bouillon R. Short-term course of 1,25$(OH)_2D_3$ stimulates osteoblasts but not osteoclasts in osteoporosis and osteoarthritis. Calcif Tissue Int 1991; 49(3):168–173.
55. Gallagher JC, Goldgar D. Treatment of postmenopausal osteoporosis with high doses of synthetic calcitriol. A randomized controlled study. Ann Intern Med 1990; 113(9):649–655.
56. Harris SS, Soteriades E, Dawson-Hughes B. Secondary hyperparathyroidism and bone turnover in elderly blacks and whites. J Clin Endocrinol Metab 2001; 86(8):3801–3804.
57. Kyriakidou-Himonas M, Aloia JF, Yeh JK. Vitamin D supplementation in postmenopausal black women. J Clin Endocrinol Metab 1999; 84(11):3988–3990.

III

Effects of Dietary Macronutrients

15 Food Groups and Bone Health

Susan A. New

1. INTRODUCTION

Studies aimed at determining the relationship between key nutrients (especially calcium) and bone health have examined directly the effect of a specific nutrient (or in some cases a variety of nutrients) commonly consumed in the human diet. Consideration of the "foods" we actually consume rather than the nutrients contained within them is an alternative strategy, which has been adopted in other disciplines for examining the relationship between diet and disease. Because site sensory characteristics and the direct hedonic pleasure derived from food are two of the key determinants of food choice, such a "food-orientated" approach is logical way forward *(1)*.

Interestingly, such a procedure has been used to examine associations between dietary factors and a variety of diseases. Examples include breast cancer *(2)*, stomach cancer *(3)*, ischemic heart disease *(4)*, and chronic disease *(5)*.

1.1. The Balance of Good Health Plate/Pyramid

The commonality of "nutrients" that exist within food groups is the critical factor to be considered in determining which "food groups" may be important to the etiology of bone health within populations and groups of individuals. Incredibly, there is a considerable level of agreement within and between countries concerning the proportions with which we should be eating food/food groups. A number of countries use different formats, but essentially the focus of the message is predominantly the same. For example, in the United Kingdom a plate model is used (*see* Fig. 1), and in the United States, a pyramid model is used. The agreed guidelines are shown in Table 1.

1.2. Specified Food Groupings and Bone Health

Five food groups exist, with recommendations to consume in the following proportions in the diet: (1) bread, other cereals, and potatoes (35%, 6–11 servings per day); (2) fruit and vegetables (30%, 5–9 servings per day); (3) meat, fish, and alternatives (15%, 2–3 servings per day); (4) milk and dairy foods (15%, 2–3 servings

From: *Nutrition and Bone Health*
Edited by: M. F. Holick and B. Dawson-Hughes © Humana Press Inc., Totowa, NJ

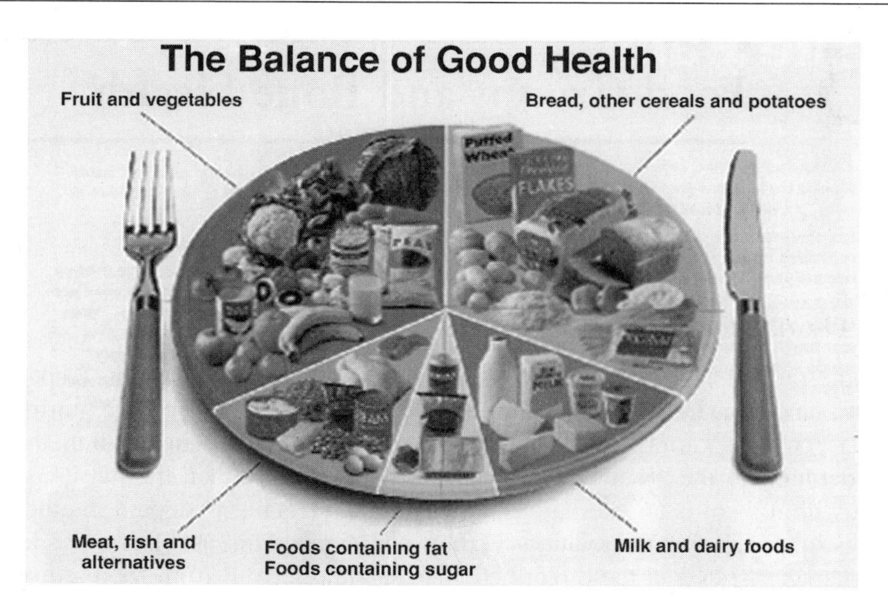

Fig. 1. UK plate model for food group consumption.

Table 1
Specified Guidelines for a Healthy Diet

Eat a variety of different foods.
Eat plenty of foods rich in starch and fiber.
Eat plenty of fruit and vegetables.
Don't eat too many foods that contain a lot of fat.
Don't have sugary foods and drinks too often.

per day); (5) fatty and sugary foods (5%, used sparingly). In reviewing the current evidence for an effect of such food groupings on bone health, emphasis will be given to the following areas: the impact of dietary "quality" in general on bone; the effect of milk and dairy products on the skeleton; the influence of meat/animal products on bone integrity; a fruit and vegetable link to skeletal health; and the impact of other food types on bone. Areas will be reviewed in the context of the available research information, and areas for future research will be discussed.

2. IMPACT OF DIETARY "QUALITY" ON BONE

Only two large population-based studies have examined the specific impact of dietary "quality"/food groups directly on indices of bone health, namely, the

Aberdeen Prospective Osteoporosis Screening Study (APOSS) *(6)* and the Framingham. Offspring Study *(7)*. Cluster analysis on 904 women (mean age 54 yr), pre-, peri-, and postmenopausal, showed that a number of foods groups, including fried foods, cakes, processed meats, and puddings, were associated with worsening hip bone loss (Spearman rank correlation coefficients ranged from –0.07 to –0.08, respectively, $p < 0.05$) *(6)*. In the Framingham study, analyses of food groups were divided into the following categories: (1) fruit, vegetables, milk, and cereal (termed the "healthy group"); (2) soda, pizza, and salty snacks; (3) cheese and other dairy; (4) meat, bread, and potatoes; (5) baked goods and sweets; and (6) alcohol. The hip bone mass in both males ($n = 601$) and females ($n = 905$) was found to be significantly higher in the "healthy group." For a total of three of the four hip sites measured, male subjects in the candy group were found to have significantly lower bone mineral density (BMD) than those in the fruit, vegetable, and cereal group, and women in the candy group were also found to have significantly lower BMD than all but one other group at the radius *(7)*.

These data support the findings of both the original APOSS baseline study *(8,9)* and the older Framingham cohort *(10)* and indicate that a high fruit and vegetable intake is protective to the skeleton, whereas high candy consumption is associated with lower bone mass, regardless of gender. These data also suggest that a high intake of fatty, sugary foods is detrimental to bone health around the time of the menopause *(11)*.

3. MILK AND MILK PRODUCTS AND THE SKELETON

3.1. Milk Consumption: General Comments

Because milk and milk products provide more than 50% of the total calcium in the Western diet as well as a number of other key nutrients including phosphorus, magnesium, and zinc, they have a fundamental role in bone health determination. Milk supplementation has also been shown to improve the nutritional quality of the diet of postmenopausal women to a greater extent than calcium alone *(12)*. As noted by Heaney *(13)*, dairy products are complex, containing many essential nutrients, and thus their effects on bone health are likely "more than can be accounted for by any single constituent and the totality of their effects may be more than the sum of parts" *(13)*.

3.2. Milk Studies and Peak Bone Mass Development

Only a few studies have used food sources of calcium (particularly dairy products) as the supplementary vehicle to investigate the relationship between calcium and peak bone mass attainment. Of those studies that have been published, the results demonstrate a clear positive effect on BMD *(14–18)*. In the prospective, randomized, placebo-controlled, double-blind study published by Bonjour et al. *(16)*, the effects of calcium-fortified foods on bone mass at different skeletal sites were investigated in a cohort of prepubertal girls aged 7–9 yr. The foods were enriched with calcium salts from milk extract (in the form of phosphate salts). Benefits were seen in both bone mass and in height between the supplemented and unsupplemented groups, and the effects were greatest in girls with a spontaneous intake below the median. One year after discontinuation of the intervention, the differences in the gain in BMD and in the size of some bones were still detectable, and after 3.5 yr, this difference was also still there *(19)*. In the milk supplementation trial by Cadogan et al. *(15)*, teenage girls consuming a 300-mL milk supplement every day for 18 mo had significant increases in total body BMD (9.6% vs 8.5%) and total body bone mineral content (BMC) (27% vs 24%) compared with the control group. Raised insulin-like growth factor (IGF)-1 levels were also reported in the children supplemented with milk, which may point to a stimulation of periosteal bone apposition with the resultant effect of a larger skeletal envelope. Follow-up of these subjects 1 yr after discontinuation of the supplement showed a sustained benefit of the milk supplement on bone mass. There is evidence in the literature that milk basic protein (MBP) directly suppresses osteoclastic-mediated bone resorption, resulting in the prevention of bone loss in the animal model *(20)*.

A number of retrospective studies have been reported that show a link between low milk consumption during the childhood and adolescence and lower BMD in a number of age groups. Results are consistent for young women *(21)*, postemenopausal women *(8)*, and older postmenopausal women *(22,23)*. In the most recently published study of Chinese adolescent girls aged 12–14 yr (n = 649), milk intake was found to be positively associated with distal radius and ulna bone mass ($p < 0.05$) *(24)*. Milk intake was found to account for 3.2% of the variation in BMD.

3.3. Milk Studies: Effects on Postmenopausal Bone Loss

Of the milk supplementation studies published, there is evidence of a positive relationship with bone health in pre- and postmenopausal women. In a 24-mo investigation by Prince et al. *(25)*, a daily supplement of fortified skimmed milk powder was found to prevent bone loss in postmenopausal women and was comparable with a calcium supplement. Favorable effects on markers of bone metabolism have been demonstrated in postmenopausal women following supplementation with dairy products *(26)*.

In the most recent supplementation study by Lau et al. *(27)*, the effect of milk supplementation on postmenopausal bone loss in Chinese women accustomed to a low-calcium diet was investigated. A total of 200 Chinese women aged 55–59 yr were randomly assigned to receive 50 g of milk powder per day that contained 800 mg of calcium, or were assigned to a control group. The milk-supplementation group were found to have significantly reduced height loss as well as BMD loss at the three sites measured. Serum parathyroid hormone (PTH) was also found to be lower and serum 25-hydroxyvitamin D [25(OH)D] significantly higher in the milk group compared with the control group at 12 mo.

The mechanisms for reduced bone loss in the postmenopausal stage remain to be fully quantified. A study by Lau and colleagues *(27)* suggests a reduction in

PTH levels, which were found to be significantly lower in the milk-supplemented group, is likely to have resulted in a reduction in bone turnover, although no specific measurements were taken. The findings of this study are particularly relevant given the low-calcium characteristic of the diet consumed by the Asian population and the rising public health problem of osteoporosis in this region.

Other recent cross-sectional studies also suggest an impact of milk intake on bone health in this age group. Lifetime milk consumption was found to be positively associated with bone mass at the lumbar spine, hip, and mid-radius in 581 postmenopausal women *(28)*, and similar positive results for milk and bone mass have been found in 965 Japanese men *(29)*.

3.4. Milk Intake and Risk of Fracture

Only a few studies have been published concerning milk consumption and fracture risk, and the findings are conflicting: Cumming and Klineberg *(30)* found that a higher level of consumption of dairy products at 20 yr of age was associated with an increased risk of hip fracture in both elderly men and women (aged 65 yr and over) *(30)*, and similar results have been found in the USA Nurse's Health Study. However, Johnell et al. *(31)* found that low milk intake was a significant risk factor for fracture in a large study of European women (mean age 78.1 yr) *(31)*, and a low intake of milk and cheese has also been found to be associated with an increased risk of fracture in elderly men *(32)*.

4. IMPACT OF MEAT/ANIMAL PROTEIN ON THE SKELETON

Few data are available on populations consuming a diet highly dependent on animal foods, particularly meat. Mazess and Mather *(33)* examined the BMC of forearm bones in a sample of children, adults, and elderly Eskimo natives of the north coast of Alaska. After the age of 40 yr, the Eskimos of both sexes were found to have a deficit of bone mineral of an order of magnitude between 10% and 15% relative to white standards. In a further study on Canadian Eskimos, ageing bone loss was found to occur at an even greater rate *(34)*.

More recently, the hypothesis that a high dietary ratio of animal to vegetable protein increases bone loss and risk of fracture has been studied in a prospective cohort of 1035 women who participated in the Study of Osteoporotic Fractures (SOF) *(35)*. Community-dwelling white women aged over 65 yr were recruited into the study. Recent dietary history (over the preceding 12 mo) was assessed using a "validated" food-frequency questionnaire. BMD was measured using dual-energy X-ray absorptiometry (DXA) at the total hip and subregions. Two BMD measurements were taken, with an average of 3.6 yr (SD 0.4 yr) between each assessment, and the rate of bone loss was calculated as the percent difference between two BMD measurements in a subset of the participants ($n = 742$). Hip fractures were assessed prospectively for 7 yr (SD 1.5 yr), and fracture data were available for all 1035 women for whom dietary data was collected. Fractures were confirmed with radiographs and a review of radiologists' reports.

Women with a higher ratio of animal to vegetable protein intake had a higher rate of bone loss at the femoral neck than did those with a low ratio, as well as a greater risk of hip fracture (relative risk = 3.7). These findings remained significant after adjustment for important confounding factors including age, weight, estrogen use, tobacco use, physical activity, and total calcium and protein intake. These findings suggest that a reduction in animal protein and an increase in vegetable protein may decrease bone loss and risk of hip fracture, but it remains an area of controversy.

5. FRUIT AND VEGETABLES AND BONE HEALTH

5.1. Early Work Linking Acid–Base Imbalance and the Skeleton

As early as the 1880s, the skeleton was considered a potential a source of buffer, contributing to both the preservation of the body's pH and defense of the system against acid–base disorders *(36)*. Studies more than three decades ago showed the detrimental effects of "acid" from the diet on bone mineral in humans and animals. There is evidence that in natural (e.g., starvation), pathological (e.g., diabetic acidosis), and experimental (e.g., ammonium chloride ingestion) states of acid loading and acidosis, an association exists with both hypercalciuria and negative calcium balance *(37)*. Because the majority of calcium is contained in bone, it is likely that increased urinary calcium excretion is from an osseous source, and it has been demonstrated that acidosis decreases renal calcium reabsorption *(38)*.

The role of bone in acid–base homeostasis is complex. The skeleton has been referred as being "a giant ion-exchange column loaded with an alkali buffer" because 80% of body carbonate, 80% of body citrate, and 35% of body sodium are contained in solution within the hydration shell of bone and released in response to metabolic acid *(39,40)*.

5.1.1. Dietary Link to Osteoporosis: A Hypothesis

In 1968, Wachman and Bernstein put forward a hypothesis linking the daily diet to the development of osteoporosis: "the increased incidence of osteoporosis with age may represent, in part, the results of a life-long utilisation of the buffering capacity of the basic salts of bone for the constant assault against pH homeostasis" *(41)*. On a Western diet, adult humans produce roughly 1 meq of acid per day, and the more acid precursors a diet includes, the greater is the degree of systemic acidity. There is also good evidence to show that with increasing age, overall renal function declines and acidity increases and thus humans become (albeit slightly) more acidic *(42)*.

5.2. Effect of Vegetarianism on Bone Health

Studies of populations following a lacto-ovo-vegetarian diet and their effects on bone mass published prior to 1990 found bone mass higher in the vegetarian group compared with omnivores *(43)*. (*See* Table 2; *44–55*.) Results of several of these studies are likely to have been subject to bias because the data were based on

Table 2
Vegetarianism and Bone Health: Summary of Studies

Author	*Year (Reference)*	*Source*	*Findings*	*Summary*
Ellis et al.	1972 *(44)*	Am J Clin Nutr 25:555–558	BMD ↑ in vegetarian group	✓
Ellis et al.	1974 *(45)*	Am J Clin Nutr 27:769–770	BMD ↓ in vegetarian group	X
Marsh et al.	1980 *(46)*	JAMA 76:148–151	Bone loss ↑ in omnivores	✓
Marsh et al.	1983 *(47)*	Am J Clin Nutr 37:453–456	BMD ↑ in vegetarians	✓
Marsh et al.	1988 *(48)*	Am J Clin Nutr 48:837–841	BMD ↑ in elderly vegetarians	✓
Tylavsky et al.	1988 *(49)*	Am J Clin Nutr 48:842–849	No difference in BMD between groups	–
Hunt et al.	1989 *(50)*	Am J Clin Nutr 50:517–523	No difference in BMD between groups	–
Lloyd et al.	1991 *(51)*	Am J Clin Nutr 54:1005–1010	No difference in BMD between groups	–
Tesar et al.	1992 *(52)*	Am J Clin Nutr 56:699–704	No difference in BMD between groups	–
Reed et al.	1994 *(53)*	Am J Clin Nutr 59:1997–1202	Bone loss rates similar	–
Chui et al.	1997 *(54)*	Calcif Tissue Int 60:245–249	BMD ↓ in vegan group	X
Lau et al.	1998 *(55)*	Eur J Clin Nutr 52:60–64	Hip BMD lower in vegetarian group	X

Source: Adapted from ref. *46,* with permission.

BMD, bone mineral density.

Seventh Day Adventists (SDAs), who had a significantly different lifestyle from that of the omnivorous group (e.g., higher physical activity levels and abstaining from smoking, caffeine, and alcohol). More recently published studies suggest no differences in BMD between vegetarians and omnivores. In the strongest of all the studies published, no differences were seen in bone loss rates between the lacto-ovo-vegetarians and the omnivorous group *(53)* as part of a 5-yr prospective study of changes in radial bone density of elderly white US women (mean age 81 yr) living in residential communities.

5.2.1. Acidity in Foods: Importance of PRAL (Potential Renal Acid Load)

Vegetable-based proteins generate a considerable amount of acid in the urine *(56)*. The potential renal acid load (PRAL) is a useful marker characterizing the acidity of foods, and it has been shown that many grain products and some cheeses have a high PRAL level. The PRAL concept may provide an explanation for the lack of a positive effect on bone health indices in studies comparing vegetarians vs omnivores because these foods are likely to be consumed in large quantities by lacto-ovo-vegetarians (Table 3).

5.3. Fruit and Vegetable Intake and Bone: A Review of Population-Based and Intervention Studies

A variety of population-based studies have recently been published, with remarkable similarities between two of the largest (and most recent) nutrition and bone health surveys *(57)*. A beneficial effect of fruit and vegetable/potassium intake on bone mass has been shown in children; premenopausal, perimenopausal, postmenopausal, and elderly women; on bone loss in men; and on markers of bone metabolism and peripheral skeletal sites in women (Table 4) *(58–64)*. Results of the Dietary Approaches to Stopping Hypertension (DASH) intervention trial lend further support to the view that diets high in fruit and vegetables may be important to bone health. Diets rich in fruit and vegetables were associated with a significant fall in blood pressure compared with baseline measurements. However, of particular interest to the bone fields was the finding that increasing fruit and vegetable intake from 3.6 to 9.5 daily servings decreased urinary calcium excretion from 157 to 110 mg/d *(65)*. More recently, it has been reported that the DASH diet (which emphasizes low-fat dairy products, fruit and vegetables, and a reduced amount of red meat) was found to significantly reduce markers of bone metabolism *(66)*.

6. OTHER FOOD GROUPS: HOT AND COLD BEVERAGES

Other foods remaining on the balance of the good health plate/pyramid have received relatively scant attention in the literature with respect to a direct effect on bone health. Data for alcohol consumption are intriguing: excessive alcohol intake is associated with osteoporosis and osteoporotic fractures *(67)*, and it is

Table 3
Potential Renal Acid Load (PRAL) Values of a Variety of Foods and Food Groups

Food/food group	*PRAL (mEq/100 g edible portion)*	*Food/food group*	*PRAL (mEq/100 g edible portion)*
Fruits and fruit juices		Milk, dairy products and eggs	
Apples	–2.2	Milk (whole, pasteurised)	0.7
Bananas	–5.5	Yoghurt (whole milk, plain)	1.5
Raisins	–21.0	Cheddar cheese (reduced fat)	26.4
Grape juice	–1.0	Cottage cheese	8.7
Lemon juice	–2.5	Eggs (yolk)	23.4
Vegetables		Meat, meat products, and fish	
Spinach	–14.0	Beef (lean only)	7.8
Broccoli	–1.2	Chicken (meat only)	8.7
Carrots	–4.9	Pork (lean only)	7.9
Potatoes	–4.0	Liver sausage	10.6
Grain products		Beverages	
Bread (white wheat)	3.7	Coca Cola	0.4
Oat flakes	10.7	Coffee (infusion)	–1.4
Rice (brown)	12.5	Tea (Indian infusion)	–0.3
Spaghetti (white)	6.5	White wine	–1.2
Cornflakes	6.0	Red wine	–2.4

Source: From ref. *45.*

well known that alcohol is directly toxic to bone-forming cells *(68)* and may disrupt bone metabolism in humans *(69)*. However, moderate alcohol consumption may not be detrimental to bone health. A positive association between moderate alcohol intake and BMD has been reported in premenopausal women *(8)*, postmenopausal women *(70)*, and elderly women *(71)*, as well as men *(70)*. The mechanisms for this positive relationship require further clarification but point to (1) alcohol affects endogenous hormone levels and induces adrenal production of androstenedrone and its adrenal conversion to estrone; and (2) alcohol stimulates the secretion of calcitonin, which is likely to favor an increase in bone mass *(69)*. Varied results have been obtained in studies examining coffee consumption and bone density, but data appear to suggest a negative association only when high caffeine intake is accompanied by very low calcium intake *(72)*. Recently, a positive association has been noted between tea drinking and bone mass in postmenopausal women and may point to the influence of flavonoids contained in tea on bone health, but this is an area requiring further attention *(73)*.

Table 4
Impact of Fruit and Vegetables on Bone: A Review of Population-Based Studies Showing a Positive Link

Author	*Year (Reference)*	*Source*	*Details*	*Findings*
Eaton-Evans et al.	1993 *(58)*	Proc Nutr Soc 52:44A	77 Females, 46–56 yr	✓ Vegetables
Michaelsson et al.	1995 *(59)*	Calcif Tissue Int 57:86–93	175 Females, 28–74 yr	✓ K Intake
New et al.	1997 *(8)*	Am J Clin Nutr 65:1831–1839	994 Females, 45–49 yr	✓ K, Mg, fiber, vitamin C ✓ Past intake: fruit & vegetables
Tucker et al.	1999 *(7)*	Am J Clin Nutr 69:727–736	229 Males, 349 females, 75 yr	✓ K, Mg, fruit & vegetables
New et al.	2000 *(9)*	Am J Clin Nutr 72:142–151	62 Females, 45–54 yr	✓ K, Mg, fiber, vitamin C ✓ Past intake: fruit & vegetables
Jones et al.	2001 *(60)*	Am J Clin Nutr 73:839–844	215 Boys, 115 girls, 8–14 yr	✓ K, urinary K
Chen et al.	2001 *(61)*	J Bone Miner Res 16:S386	668 Females, >48–62 yr	✓ Fruit
Miller et al.	2001 *(62)*	J Bone Miner Res 16:S395	300 Males, 50–91 yr	✓ K, Mg
Stone et al.	2001 *(63)*	J Bone Miner Res 16:S388	1075 Men, 65 yr and over	✓ K, lutein
New et al.	2002 *(64)*	Osteopor Int 13:S77	164 Females, 55–87 yr	✓ K, fruit & vegetables

Source: Adapted from ref. *46,* reproduced with permission.

7. CONCLUSION AND AREAS FOR FURTHER RESEARCH

The approach of using food groups to examine the relationship between diet and disease is an appropriate and logical approach to examining the relationship between diet and osteoporosis. *(74)*. There is somewhat remarkable agreement among countries as to the proportions with which we should be eating food groups. The data suggest that milk and milk products (as providers of more than 50% of total dietary calcium) and fruit and vegetables are beneficial to bone health across the age ranges, although clearly more work on fracture reduction is required. Experimental work is required to determine the exact mechanisms of action of these two specific food groups on the skeleton. Further research is required concerning the effect of other food groups on both indices of bone health and in fracture risk reduction *(75)*.

REFERENCES

1. Kearney J, Gibney M. A pan-European survey of consumer attitudes to food, nutrition and health overview. Food Qual Preference 1998; 9:467–478.
2. Ronco A, De Stefani E, Boffetta P, Denso-Pellegrini H, Mendilaharso M, Leborgne F. Vegetables, fruits and related nutrients and risk of breast cancer: a case-control study in Uruguay. Nutr Cancer 1999; 35:111–119.
3. Risch HA, Jain M, Choi NW. Dietary factors and the incidence of cancer of the stomach. Am J Epidemiol 1985; 122:949–959.
4. Law MR, Morris JK. By how much does fruit and vegetable consumption reduce the risk of ischaemic heart disease? Eur J Clin Nutr 1998; 52:549–556.
5. La Vecchia C, Decarli A, Pagano R. Vegetable consumption and risk of chronic disease. Epidemiology 1998; 9:208–210.
6. Macdonald HM, New SA, Grubb DA, Golden MHN, Reid DM. Impact of food groups on perimenopausal bone loss. In: Burckhardt P, Dawson-Hughes B, Heaney RP, eds. Nutritional Aspects of Osteoporosis 2000 (4th International Symposium on Nutritional Aspects of Osteoporosis, Switzerland, 1997). Challenges of Modern Medicine. Ares-Serono, Academic, New York, 2001, pp. 399–408.
7. Tucker KL, Chen H, Hannan MT, et al. Bone mineral density and dietary patterns in older adults: the Framingham Osteoporosis Study. Am J Clin Nutr 2002; 76:245–252.
8. New SA, Bolton-Smith C, Grubb DA, Reid DM. Nutritional influences on bone mineral density: a cross-sectional study in premenopausal women. Am J Clin Nutr 1997; 65:1831–1839.
9. New SA, Robins SP, Campbell MK, et al. Dietary influences on bone mass and bone metabolism: further evidence of a positive link between fruit and vegetable consumption and bone health? Am J Clin Nutr 2000; 71:142–151.
10. Tucker KL, Hannan MT, Chen H, Cupples A, Wilson PWF, Kiel DP. Potassium and fruit & vegetables are associated with greater bone mineral density in elderly men and women. Am J Clin Nutr 1999; 69:727–736.
19. Bonjour JP, Chevalley T, Ammann P, Slosman D, Rizzoli R. Gain in bone mineral mass in prepubertal girls 3.5 years after discontinuation of calcium supplementation: a follow-up study. Lancet 2001; 358:1208–1212.
20. Toba Y, Takada Y, Yamamura J, et al. Milk basic protein: a novel protective function of milk against osteoporsis. Bone 2000; 27:403–408.
21. Teegarden D, Lyle RM, McCabe R. Dietary calcium, protein and phosphorus are related to bone mineral density and content in young women. Am J Clin Nutr 1998; 68:749–754.

22. Sandler RB, Slemenda CW, LaPorte RE, et al. Postmenopausal bone density and milk consumption in childhood and adolescence. Am J Clin Nutr 1985; 42:270–274.
23. Murphy S, Khaw KT, May H, Compston JE. Milk consumption and bone mineral density in middle-age and elderly women. Br Med J 1994; 308:939–941.
24. Du XQ, Greenfield H, Fraser DR, Ge KY, Liu ZH, He W. Milk consumption and bone mineral content in Chinese adolescent girls. Bone 2002; 30:521–528.
25. Prince R, Devine A, Dick I, et al. The effects of calcium supplementation (milk powder or tablets) and exercise on bone density ini postmenopausal women. J Bone Miner Res 1995; 10:1068–1075.
26. Heaney RP, McCarron DA, Dawson-Hughes B, et al. Dietary changes favourably affect bone remodeling in older adults. J Am Diet Assoc 1999; 99:1228–1233.
27. Lau EMC, Woo J, Lam V, Hong A. Milk supplementation of the diet of postmenopausal Chinese women on a low calcium intake retards bone loss. J Bone Miner Res 2001; 16:1704–1709.
28. Soroko S, et al. Lifetime milk consumption and bone mineral density in older women. Am J Public Health 1994; 84:1319–1322.
29. Sone T, et al. Influence of exercise and degenerative vertebral changes on BMD: a cross-sectional study in Japanese men. Gerontology 1996; 42:57–66.
30. Cumming RG, Klineberg RJ. Case-control study of risk factors for hip fractures in the elderly. Am J Epidemiol 1994; 139:493–503.
31. Johnell O, et al. Risk factors for hip fracture in European women. The MEDOS study. J Bone Miner Res 1995; 10:1802–1815.
32. Kanis J, et al. Risk factors for hip fracture in men from Southern Europe: the MEDOS study. Osteoporosis Int 1999; 9:45–54.
33. Mazess RB, Mather WE. Bone mineral content of North Alaskan Eskimos. Am J Clin Nutr 1974; 27:916–925.
34. Mazess RB, Mather WE. Bone mineral content in Canadian Eskimos. Hum Biol 1975; 47:45.
35. Sellmeyer DE, Stone KL, Sebastian A, Cummings SR, for the Study of Osteoporotic Fractures. A high ratio of dietary animal to vegetable protein increases the rate of bone loss and the risk of fracture in postmenopausal women. Am J Clin Nutr 2001; 73:118–122.
36. Irving L, Chute AL. The participation of the carbonates of bone in the neutralisation of ingested acid. J Cell Comp Physiol 1933; 2:157.
37. Reidenberg MM, Haag BL, Channick BJ, Schuman CR, Wilson TGG. The response of bone to metabolic acidosis in man. Metabolism 1966; 15:236–241.
38. Bushinsky DA. Acid-base imbalance and the skeleton. In: Burckhardt P, Dawson-Hughes B, Heaney RP, eds. Proceedings of the 3rd International Symposium on Nutritional Aspects of Osteoporosis, Switzerland, 1997. Nutritional Aspects of Osteoporosis 1997. Ares-Serono, New York, 1998, pp. 208–217.
39. Green J, Kleeman R. Role of bone in regulation of systematic acid-base balance (editorial review). Kidney Int 1991; 39:9–26.
40. Barzel US. The skeleton as an ion exchange system: implications for the role of acid-base imbalance in the genesis of osteoporosis. J Bone Miner Res 1995; 10:1431–1436.
41. Wachman A, Bernstein DS. Diet and osteoporosis. Lancet I, 958–959.
42. Frassetto LA, Sebastian A. Age and systemic acid-base equilibrium: analysis of published data. J Gerontol 1996; 51A:B91–B99.
43. New SA. Impact of food clusters on bone. In: Dawson-Hughes B, Burckhardt P, Heaney RP, eds. Nutritional Aspects of Osteoporosis 2000. 4th International Symposium on Nutritional Aspects of Osteoporosis, Switzerland, 1997. Challenges of Modern Medicine. Ares-Serono, Academic, New York, 2001, pp. 379–397.
44. Ellis FR, Holesh S, Ellis JW. Incidence of osteoporosis in vegetarians and omnivores. Am J Clin Nutr 1972; 25:555–558.

45. Ellis FR, Holesh S, Sanders TA. Osteoporosis in British vegetarians and omnivores. Am J Clin Nutr 1974; 27:769–770.
46. Marsh AG, Sanchez TV, Micklesen O, Keiser J, Major G. Cortical bone density of adult lactoovovegetarians and omnivorous women. J Am Diet Assoc 1980; 76:148–151.
47. Marsh AG, Sanchez TV, Chaffee FL, Mayor GH, Mickelsen O. Bone mineral mass in adult lactoovovegetarian and omnivorous males. Am J Clin Nutr 1983; 83:155–162.
48. Marsh AG, Sanchez TV, Michelsen O, Chaffee FL, Fagal SM. Vegetarian lifestyle and bone mineral density. Am J Clin Nutr 1988; 48:837–841.
49. Tylavsky F, Anderson JJB. Bone health of elderly lactoovovegetarian and omnivorous women. Am J Clin Nutr 1988; 48:842–849.
50. Hunt IF, Murphy NJ, Henderson C, et al. Bone mineral content in postmenopausal women: comparison of omnivores and vegetarians. Am J Clin Nutr 1989; 50:517–523.
51. Lloyd T, Schaeffer JM, Walker MA, Demers LM. Urinary hormonal concentrations and spinal bone densities of premenopausal vegetarian and nonvegetarian women. Am J Clin Nutr 1991; 54:1005–1010.
52. Tesar R, Notelovitz M, Shim E, Kauwell G, Brown J. Axial and peripheral bone density and nutrient intakes of postmenopausal vegetarian and omnivorous women. Am J Clin Nutr 1992; 56:699–704.
53. Reed JA, Anderson JBB, Tylavsky FA, Gallagher PNJ Jr. Comparative changes in radial bone density of elderly female lactoovovegetarians and omnivores. Am J Clin Nutr 1994; 59:1197S–1202S.
54. Chiu JF, Lan SJ, Yang CY, et al. Long term vegetarian diet and bone mineral density in postmenopausal Taiwanese women. Calcif Tissue Int 1997; 60:245–249.
55. Lau EM, Kwok T, Woo J, Ho SC. Bone mineral density in Chinese elderly female vegetarians, vegans, lactoovovegetarians and omnivores. Eur J Clin Nutr 1998; 52:60–64.
56. Remer T, Manz F. Estimation of the renal net acid excretion by adults consuming diets containing variable amounts of protein. Am J Clin Nutr 1994; 59:1356–1361.
57. New SA. The role of the skeleton in acid-base homeostasis. The 2001 Nutrition Society Medal Lecture. Proc Nutr Soc 2002; 61:151–164.
58. Eaton-Evans J, McIlrath EM, Jackson WE, Bradley P, Strain JJ. Dietary factors and vertebral bone density in perimenopausal women from a general medical practice in Northern Ireland (abstr). Proc Nutr Soc 1993; 52:44A.
59. Michaelsson K, Holmberg L, Maumin H, Wolk A, Bergstrom R, Ljunghall S. Diet, bone mass and osteocalcin; a cross-sectional study. Calcif Tissue Int 1995; 57:86–93.
60. Jones G, Riley MD, Whiting S. Association between urinary potassium, urinary sodium, current diet, and bone density in prepubertal children. Am J Clin Nutr 2001; 73:839–844.
61. Chen Y, Ho SC, Lee R, Lam S, Woo J. Fruit intake is associated with better bone mass among Hong Kong Chinese early postmenopausal women. J Bone Miner Res 2001; 16(S1):S386.
62. Miller DR, Krall EA, Anderson JJ, Rich SE, Rourke A, Chan J. Dietary mineral intake and low bone mass in men: the VALOR Study. J Bone Miner Res 2001; 16(S1):S395.
63. Stone KL, Blackwell T, Orwoll ES, et al. The relationship between diet and bone mineral density in older men. J Bone Miner Res 2001; 16(S1):S388.
64. New SA, Smith R, Brown JC, Reid DM. Positive associations between fruit & vegetable consumption and bone mineral density in late postmenopausal and elderly women. Osteopor Int 2002; 13:S77.
65. Appel LJ, Moore TJ, Obarzanek E, et al. A clinical trial of the effects of dietary patterns on blood pressure. New Engl J Med 1997; 336:1117–1124.
66. Lin P, Ginty F, Appel L, et al. Impact of sodium intake and dietary patterns on biochemical markers of bone and calcium metabolism. J Bone Miner Res 2001; 16(S1):S511.
67. Peris P, Pares A, Guanabens N. Reduced spinal and femoral bone mass and deranged bone mineral metabolism in chronic alcoholics. Alcohol Alcoholism 1992; 27:619–625.

68. Bickle D, Genant H, Cann C, Recker R, Haloran B, Stewler G. Bone disease in alcohol abuse. Ann Intern Med 1985; 103, 42–48.
69. Rico H. Alcohol and bone disease. Alcohol Alcoholism 1990; 25:345–352.
70. Holbrook TL, Barrett-Connor E. A prospective study of alcohol consumption and bone mineral density. Br Med J 1993; 306:1506–1509.
71. Rapuri PB, Gallagher JC, Balhorn KE, Ryschon KL. Alcohol intake and bone metabolism in elderly women. Am J Clin Nutr 2000; 72:1206–1213.
72. Barrett-Connor E, Chun Chang J, Edelstein SL. Coffee-associated osteoporosis offset by daily milk consumption. J Am Med Assoc 1994; 271:280–283.
73. Hegarty VM, May HM, Khaw KT. Tea drinking and bone mineral density in older women. Am J Clin Nutr 2000; 71:1003–1007.
74. Goulding A. Nutritional Strategies for prevention and treatment of osteoporosis in populations and individuals. In: New SA, Bonjour JP, eds. Nutritional Aspects of Bone Health. The Royal Society of Chemistry, Cambridge, UK 2003, pp. 709–732.
75. Iuliano-Burns S., Seeman E. In: New SA, Bonjour JP, eds. Nutritional Aspects of Bone Health. Effect of diet on fracture risk reduction in populations. The Royal Society of Chemistry, Cambridge, UK, 2003, pp. 673–692.

16 Vegetarianism and Bone Health in Women

Susan I. Barr

1. INTRODUCTION

In recent years there has been an increased interest in possible health benefits of plant-based diets. Although vegetarians comprise a small percentage of the population, some data suggest that the prevalence may be increasing, particularly among young women *(1)*. Whether vegetarianism has an impact on bone health has not been studied extensively to date. Accordingly, the purpose of this chapter is to review possible mechanisms whereby dietary and lifestyle factors associated with vegetarianism could influence bone health, and to review the available literature comparing bone health between vegetarians and omnivores. Because very few studies have been conducted with men, this review will be confined primarily to women.

2. VEGETARIAN DIETS

Although vegetarian diets are often described as excluding tissue protein sources (meat, fish, and poultry), there is great variability among the diets followed by individuals who consider themselves to be vegetarian *(1)*. Table 1 shows the characteristics of different types of vegetarian diets. It can be seen that the nutrient composition of the diet could vary considerably depending on the type of vegetarian diet followed. For example, if use of dairy products was increased to replace tissue protein sources, a lacto-ovo-vegetarian (LOV) diet could contain more calcium than a traditional omnivorous (OMNI) diet, whereas a vegan diet would likely contain less calcium. Another source of variability not shown in the table is the extent to which use of plant sources of protein, such as legumes, nuts, and seeds, are incorporated into the diet.

3. POSSIBLE MECHANISMS BY WHICH VEGETARIANISM COULD INFLUENCE BONE MINERAL DENSITY

There are a number of mechanisms whereby components of a vegetarian diet or aspects of a vegetarian lifestyle could influence bone mineral density (BMD) in

From: *Nutrition and Bone Health*
Edited by: M. F. Holick and B. Dawson-Hughes © Humana Press Inc., Totowa, NJ

Table 1
Definitions of Vegetarian Diets

Type of Diet	*Characteristics*
Semivegetarian	No use or infrequent use of red meat
	Some individuals may include chicken (pollo-vegetarians) and/or fish (pesco-vegetarians)
	Dairy products and eggs are included
Lacto-ovo-vegetarian	Do not consume meat, fish, or poultry
	Include dairy products and eggs
	Some individuals may exclude eggs (lacto-vegetarian) or dairy products (ovo-vegetarian)
Vegan	Do not consume meat, fish, or poultry
	Do not consume dairy products or eggs
	Some vegans will consume dairy products and/or eggs as ingredients in other foods (e.g., baked goods), while others also exclude foods containing these ingredients
	Some vegans also exclude foods that are derived from animals (e.g., honey, foods containing gelatin)
Macrobiotic	Diet is based on cereals, pulses, and vegetables
	Small amounts of seaweeds, fermented foods, nuts, seeds, and seasonal fruit are consumed
	Fish may be consumed occasionally, but meat and dairy products are usually avoided

either a positive or negative manner. Most of these possible mechanisms are discussed in depth elsewhere in this volume (e.g., Chaps. 9, 11, 14, 18, 22, 23, 26); accordingly, only brief overviews are presented here.

3.1. Dietary Factors

3.1.1. Calcium

Calcium is a major component of bone mineral, and must be provided in the diet for bone mineralization to occur. It has been suggested that calcium is a threshold nutrient: higher intakes of calcium are related to increased mineralization during growth (or decreased loss during aging) up to the putative threshold, above which higher intakes have no additional effect *(2)*. Thus, diets that differ in calcium content at levels below the threshold could be hypothesized to lead to differences in bone mineralization.

Vegetarian diets have the potential to contain either more, similar amounts, or less calcium than omnivorous diets. Several studies suggest intakes are similar *(3–5)*, although LOVs who increase their use of dairy products to replace tissue proteins may have higher intakes *(6,7)*. Conversely, vegans have calcium intakes that are lower than those of omnivores *(8,9)*.

3.1.2. Vitamin D

The role of vitamin D in prevention of osteoporosis has received increased attention in recent years *(10)*. Naturally occurring dietary sources of vitamin D are limited to only a few foods. The richest sources are fish liver oils; other important sources are fatty fish such as herring and salmon. Small amounts are found in chicken and beef liver, shrimp, egg yolk, and butter. In some countries (e.g., the United States and Canada), fluid milk is fortified with the vitamin, but globally, most people rely on endogenous synthesis to meet their needs. However, sunlight exposure may not be adequate as distance from the equator increases, and sunscreen use blocks endogenous synthesis.

Dietary vitamin D intakes are likely to be very low in North American vegetarians who exclude fluid milk. In countries that do not fortify milk, vegetarians may also have lower intakes because of the exclusion of fish. Low circulating vitamin D levels have been reported in premenopausal vegan women living in Finland *(11)*.

3.1.3. Protein

The role of protein in the maintenance of bone has engendered considerable debate, and evidence exists to suggest adverse effects of both inadequate and excessive protein intakes *(12)*. Clearly, protein is required for synthesis of bone matrix proteins and is also needed to support an anabolic environment. Protein may also act to enhance calcium absorption. On the other hand, excretion of excessive amounts of acidic end products of protein metabolism contributes to increased calciuria. Controversy also exists concerning the roles of animal- vs plant-based proteins *(12)*. Compared to OMNI diets, most vegetarian diets contain lower amounts of protein *(3–5,8,13,14)*.

3.1.4. Acid/Alkali Balance

Bone plays an important role as a metabolic buffer, and responds to an acid load by dissolution of its basic mineral salts *(15)*. Thus, diets that result in a net acid load could lead to loss of bone mineral because of the need to buffer the acid, whereas diets with a low acid load would be less likely to do so. Most high-protein foods lead to acid end products, whereas most fruits and vegetables are high in organic anions and yield net alkali. Thus, although many plant-based foods yield alkali, this is not universally true. Foods such as soybeans and other high-protein legumes, soy-based beverages, most nuts, and many grains are acidogenic.

Further to this hypothesis, diets high in fruits and vegetables have been associated with greater BMD *(16,17)*. To the extent that vegetarian diets contain more fruits and vegetables than do OMNI diets, they have the potential to lead to a lower acid load, and less requirement for use of bone as a buffer.

Table 2
Possible Effect of Dietary Factors on Calcium Balance and/or Bone Mineralization

	May Enhance	*May Enhance or Adversely Affect*	*May Adversely Affect*
Possibly higher in vegetarian diet	Phytosterols Alkali balance		Phytate/oxalate
Possibly lower in vegetarian diet	Calcium Vitamin D	Protein	

3.1.5. Phytates/Oxalates

Calcium may be poorly absorbed from foods rich in oxalic acid or phytic acid, both of which are found in plant foods (oxalates in spinach, sweet potatoes, rhubarb, and beans; phytates in unleavened bread, raw beans, seeds, nuts and grains, and soy isolates). Although specific intake data could not be located, it is likely that vegetarian diets contain higher amounts of phytates and oxalates than OMNI diets.

3.1.6. Phytosterols

Soy products are rich in the isoflavones genistein and daidzein, which act as naturally occurring selective estrogen receptor modulators and thereby have the potential to have effects on bone *(18)*. It is important to note that effects on bone may vary by estrogen status. For example, positive associations between soy isoflavone intake and bone density have been observed in postmenopausal women (e.g., refs. *19,20*), but not premenopausal women *(20)*. A second consideration is that soy product intake and vegetarianism are not synonymous. Although average intakes are likely higher in vegetarians, some omnivorous women may use large amounts of soy products, and not all vegetarians incorporate them in their diets.

3.1.7. Summary of Dietary Factors

Table 2 shows how the above dietary components may differ between vegetarian and omnivorous diets, and the possible impact on bone mineralization. Depending on the type of vegetarian diet followed, either beneficial or adverse effects could occur.

3.2. Lifestyle Factors

A vegetarian lifestyle may be associated with other health-promoting behaviors, such as not smoking and being physically active, that also have the potential to influence bone and could therefore confound comparisons of bone density based on diet. This possibility can be illustrated by comparisons of overall mortality rates between vegetarians and omnivores *(21)*. When comparisons were made to population data, standardized mortality ratios for all causes of death for vegetarians were significantly below the reference value of 100. However, when comparisons were made to controls with similar socioeconomic status and health behaviors, no differences were seen *(21)*. This emphasizes the need to choose appropriate OMNI controls when assessing the possible impact of vegetarianism on bone.

In contrast to the positive lifestyle behaviors referred to above, some weight-conscious young women may adopt a vegetarian diet because of concerns about body weight, in essence using vegetarianism as a socially acceptable way of limiting food intake *(22–24)*. To the extent that dieting or weight concerns can compromise nutrient intake, increase cortisol levels, and/or interfere with menstrual function, the potential for adverse effects on bone exists *(25)*. It should be emphasized that these effects are not specific to vegetarians, and that a vegetarian diet *per se* does not appear to interfere with normal menstrual function in otherwise healthy women *(26)*. Although the association between weight-related concerns and vegetarianism is unlikely to be apparent among those who follow vegetarian dietary patterns for religious reasons (e.g., Seventh-Day Adventists, Buddhists), it may be a factor among some who adopt vegetarian diets for other reasons. Again, attempting to control for this would be important when assessing the effects of vegetarianism on bone.

4. STUDIES ASSESSING BMD IN VEGETARIANS AND OMNIVORES

The previous section indicated that there are several mechanisms whereby vegetarianism could either have a protective or an adverse effect on BMD. In this section the results of available studies comparing bone density between vegetarians and omnivores are presented. Interpretation of these studies is not straightforward: First, many of the vegetarians were members of religious groups that advocate a vegetarian diet. They may differ in several respects, other than diet, from those who adopt vegetarian diets because of concerns about animal rights, for example. Second, particularly for vegetarians whose vegetarianism is not based on membership in a religious group, the characteristics of the vegetarian diet—and even whether a vegetarian diet is followed—may change over time *(27)*. And finally, as indicated earlier, there is great heterogeneity in diets that are considered to be "vegetarian," and in particular, the extent to which intakes of nutrients such as calcium are impacted by the diet.

4.1. Comparison of LOVs to OMNIs

4.1.1. Studies Reporting Higher BMD in LOVs

One of the earliest reports comparing BMD between OMNIs and LOVs is that of Ellis and colleagues *(6)*. Twenty-five British vegetarians (LOVs) aged 53–79, who had followed their diets for 12–77 yr, were compared to height- and weight-matched OMNI controls. Bone density at the center of the third metacarpal medulla and the proximal phalanx, assessed using X-ray densitometry, was significantly higher in vegetarians than omnivores. Unfortunately, dietary intakes were not assessed, although the possibility that calcium intakes may have been higher among the LOVs was mentioned. Differences in dietary acid/alkali balance (also not assessed) were suggested as explaining the difference in bone density.

Supportive results were later obtained by Marsh et al. *(28)*. Women who had followed LOV diets for at least 20 yr were compared with matched OMNI controls. In this cross-sectional study, no differences in mean BMD of the radius were seen for women up to about age 50. However, among women aged 50–87, LOV women had higher BMD than OMNI women. In the initial study, dietary intakes were not reported, but subsequently, 10 LOVs and 10 OMNIs aged 52–87, from the same geographic area as the initial study, kept 7-d weighed diet records to assess whether dietary factors contributed to the observed difference in bone density *(28)*. Mean calcium intakes (LOV = 898 mg/d and OMNI = 712 mg/d) did not differ significantly, and mean phosphorus intakes were almost identical (LOV = 1094 mg/d and OMNI = 1103 mg/d). However, the Ca:P ratio was significantly higher in vegetarians ($p < 0.001$), suggesting that the difference in calcium intakes likely approached significance. The other difference between groups was that the LOV group had a significantly more alkaline diet *(7)*.

4.1.2. Studies Reporting Similar BMD in LOVs and OMNIs

Postmenopausal Women. Several studies have not detected differences in BMD between postmenopausal LOVs and OMNIs *(3–5,29)*. Tylavsky and colleagues reported a comparison of bone density at the mid and distal radial sites between 88 LOV and 287 OMNI postmenopausal women aged 60–98 *(3)*. Vegetarians, who were members of the Seventh-Day Adventist religious group, had adhered to their diets for at least 16 yr. Bone density was assessed using single-photon densitometry, and dietary intakes were assessed using a quantitative food frequency questionnaire. No difference in age-adjusted bone density or bone mineral content (BMC) was observed between groups at either bone site, nor did calcium intakes differ. Five years after the initial bone measurements, follow-up measurements were made on 189 members of this group (49 LOVs and 140 OMNIs) *(29)*. Although all women lost bone, BMD loss rates did not differ between LOVs and OMNIs.

Hunt and colleagues compared BMC and bone width (BW) of the nondominant radius between 146 Methodist OMNI and 144 Seventh-Day Adventist LOV women *(4)*. The two groups of women were similar in age (mean = 66 yr), height, and weight. They also had similar calcium intakes, although protein intakes were higher in OMNIs. BMC/BW did not differ between LOVs and OMNIs, averaging 0.568 ± 0.098 and 0.575 ± 0.092 g/cm^2, respectively.

Both axial and peripheral bone density were compared between 28 matched pairs of postmenopausal LOVs and OMNIs studied by Tesar and colleagues *(5)*. All women were members of religious groups, and vegetarians had followed a vegetarian diet for at least 10 yr (mean = 35 yr). Single-photon absorptiometry was used to measure BMC and BMD at the midshaft and distal radius, and dual-photon absorptiometry was used to measure whole-body, regional, and lumbar spine BMC and BMD. Diet was assessed using a 24-h recall and a 6-d record, and supplement use was recorded. Dietary and supplemental intakes of calcium, phosphorus, and the Ca:P ratio were very similar between groups, although protein intake was higher among omnivores. Twenty comparisons of BMC and BMD were made between groups, and only head BMD differed, being higher among vegetarians ($p < 0.05$). Since p values were not adjusted for multiple comparisons, this difference would be expected based on chance alone.

Premenopausal Women. The vegetarian women in the studies of postmenopausal women just cited were members of religious groups. Fewer studies of BMD in premenopausal vegetarian women have been conducted, and for the most part, sample sizes lacked adequate power to make comparisons. Nevertheless, because the premenopausal vegetarians studied were not members of religious groups, and because the results tend to differ qualitatively from those of postmenopausal women, they are worth considering.

Spinal bone density was compared between 23 vegetarian (all LOV) and 37 nonvegetarian women aged 28–45 yr who reported regular menstrual cycles *(14)*. As assessed using a 3-d diet record, neither calcium nor vitamin D intakes differed between groups, although the vegetarians tended to have higher calcium intakes (mean ± SE = 972.6 ± 86.5 mg/d vs 770.3 ± 52.3 mg/d), and as is usually observed, significantly lower protein intakes (mean ± SE = 62.0 ± 3.6 g/d vs 76.0 ± 3.2 g/d, $p < 0.05$). Bone density, assessed using dual-photon absorptiometry, was nonsignificantly lower among vegetarians (mean ± SE = 1.02 ± 0.02 g/mL vs 1.06 ± 0.02 g/mL).

Nonsignificantly lower spinal BMD was also observed in the premenopausal vegetarian women studied by Barr and colleagues *(30)*. In their study, 22 vegetarians (8 vegans and 15 LOVs) were compared to 23 OMNI women. Women were aged 20–40, weight stable, reported regular menstrual cycles, and had followed their respective diets for at least 2 yr. Spinal BMD was assessed using dual energy X-ray absorptiometry (DXA), and diets were monitored using 3-d records repeated on three occasions. Protein intake was significantly lower among both vegetarian groups, although calcium was significantly lower only in vegans. Spinal BMD in vegetarians was very similar between LOVs and vegans, averaging 1.15 ± 0.11 g/cm^2. This tended to be lower than spinal BMD in OMNIs, which was 1.22 ± 0.13 g/cm^2 ($p = 0.06$, or with body mass index [BMI] as a covariate, $p = 0.14$).

In summary, the majority of studies comparing BMD between postmenopausal LOV and OMNI women have not detected important differences, particularly in studies in which calcium intakes were similar. When higher values for LOV were observed, this occurred in conjunction with diets that were demonstrated or speculated to contain more calcium and a lower acid load. Among premenopausal women whose vegetarianism was not associated with their religion, nonsignificant tendencies to lower BMD have been reported.

4.2. Comparison of Vegans to OMNIs or LOVs

4.2.1. Postmenopausal Women

Two studies, both conducted in postmenopausal Asian Buddhist women, have compared BMD between long-term vegan vegetarians and controls *(13,31)*. Chiu and colleagues *(13)*, who studied 258 healthy postmenopausal Taiwanese women, measured BMD at the lumbar spine and the femoral neck using dual-photon absorptiometry, and assessed diet by 24-h recall. Subjects were defined as "long-term vegan vegetarians" ($n = 71$, those who had followed a strict vegan diet for at least 15 yr) or "others" ($n = 187$, including those who were LOVs, who alternated between vegan and OMNI diets, or who had been vegan for less than 15 yr). A non-vegetarian control group was not included. Compared to "others," long-term vegans were significantly shorter, but had similar BMI. Calcium intake was similarly low in both groups (343 ± 88 mg/d for "others" vs 364 ± 97 mg/d in long-term vegans) and protein intakes were also similar. BMD at the lumbar spine and femoral neck was significantly lower in long-term vegans (0.94 ± 0.19 g/cm^2 vs 0.99 ± 0.17 g/cm^2 at the spine, $p < 0.05$, and 0.69 ± 0.11 g/cm^2 vs 0.75 ± 0.13 g/cm^2 at the femoral neck, $p < 0.05$).

Generally, similar results were obtained by Lau and colleagues *(13)*. They studied 76 Hong Kong Buddhist women aged 70–89 yr who had been vegetarian for more than 30 yr, and compared them to 109 normal volunteers in the same age range. In this study, BMD at the spine and hip was measured by DXA. Diets of the Buddhist women were assessed using a 24-h recall, and were compared to intakes of omnivorous subjects from a previous study. The Buddhist women were characterized as "lacto-vegetarian" or vegan based on whether they consumed milk more than once a month, but both "lacto-vegetarian" and vegans had low calcium intakes (medians were 276 mg/1000 kcal for "lacto-vegetarian" and 282 mg/1000 kcal for vegans). BMD did not differ between these two vegetarian subgroups, so they were combined for comparisons to OMNI. Interestingly, despite the vegetarians' low calcium intakes, their mean intakes were higher than those of the OMNI controls studied previously. Their protein intakes, however, were significantly lower. The vegetarians and OMNIs had similar spinal BMD (0.70 ± 0.13 g/cm^2 vs 0.72 ± 0.14 g/cm^2), but vegetarians had significantly lower BMD at the femoral neck (0.49 ± 0.08 g/cm^2 vs 0.53 ± 0.08 g/cm^2, $p < 0.001$) and Ward's triangle (0.29 ± 0.09 g/cm^2 vs 0.34 ± 0.09 g/cm^2, $p < 0.01$).

Additional data in postmenopausal women are available from the studies of Marsh and colleagues *(7)*. Eleven vegan women were included in approx 300 vegetarians aged 52–90. Calcium intake was not assessed, but milk intake averaged 450 mL/d for lifetime LOVs and 0 mL/d for vegans. Although statistical analysis of the results is not reported, the radial BMD of the vegan women was lower than that of lifetime LOVs (0.465 g/cm^2 vs 0.571 g/cm^2).

4.2.2. Younger Individuals

A small study of premenopausal women provides suggestive evidence that BMD of vegans may be compromised. Outila and colleagues compared BMD at the lumbar spine and femoral neck in 6 vegan, 6 LOV, and 16 OMNI women with a mean age of 34 yr *(11)*. Three BMD measurements were made over the course of a year, and were averaged for group comparisons. Spinal BMD of vegans was significantly lower than that of omnivores (1.03 ± 0.17 g/cm^2 vs 1.18 ± 0.10 g/cm^2, $p = 0.05$) and tended to be lower than that of LOVs (1.03 ± 0.17 g/cm^2 vs 1.14 ± 0.06 g/cm^2, $p = 0.17$). At the femoral neck, BMD of vegans (0.84 ± 0.12 g/cm^2) tended to be lower than that of OMNIs (1.00 ± 0.14 g/cm^2, $p = 0.07$) and of LOVs (0.96 ± 0.06 g/cm^2, $p = 0.15$). Similar results (i.e., lower BMD in premenopausal vegan women compared to LOVs or OMNIs) were reported by Johnston and colleagues (unpublished observations, cited in ref. *32*).

Barr and colleagues measured spinal BMD at baseline and 1 yr in 5 vegan, 6 LOV, and 9 OMNI women with a mean age of 27 yr *(30)*. Over the year, BMD increased in OMNI (from 1.226 ± 0.15 to 1.247 ± 0.15 g/cm^2, $p < 0.05$) but did not change in vegetarian women (initially, 1.170 ± 0.11 g/cm^2 vs 1.176 ± 0.11 g/cm^2 at 1 yr). Vegans and LOVs had generally similar results, but the small sample size precluded detection of differences.

Finally, BMC and bone area for the whole body, lumbar spine, proximal femur, midshaft radius and distal radius were measured in 195 adolescents (103 girls and 92 boys) aged 9–15 *(33)*. Ninety-three children had followed a macrobiotic diet from birth onward for a mean of 6 yr, while the remaining 102 were controls. Previous dietary assessments, conducted while the children were following macrobiotic diets, revealed calcium intakes that were only 30% of the intakes of OMNI controls, very low vitamin D intakes, and lower intakes of energy and protein. Current calcium intake, assessed using a food-frequency questionnaire, remained lower in the children who had previously followed a macrobiotic diet than in controls (for boys, 660 ± 396 mg/d vs 1064 ± 382 mg/d; for girls, 557 ± 324 mg/d vs 1045 ± 367 mg/d). After adjusting for bone area, weight, height, percent body lean, and pubertal status, BMC was significantly lower in both male and female macrobiotic subjects at all bone sites studied except the wrist. Percentage differences from controls ranged from –2.5% to –8.5%.

5. CONCLUSION

The potential exists for characteristics of vegetarian diets to have beneficial or adverse effects on bone density. Following an LOV diet with adequate calcium and vitamin D, and with generous intakes of alkali-producing fruits and vegetables, has potential to favorably affect BMD, although it is possible that an OMNI diet with similar characteristics would be equally favorable. Long-term adherence to a vegan diet, however, appears to be consistently associated with lower bone density. The mechanism underlying this observation is not yet known, but it is reasonable to suggest that suboptimal intakes of nutrients such as calcium, protein, and vitamin D contribute to the lower values. Accordingly, it would be prudent for committed vegans to choose fortified foods and/or supplements to ensure they meet current recommendations for these nutrients.

REFERENCES

1. Sabate J, Ratzin-Turner RA, Brown JE. Vegetarian diets: descriptions and trends. In: Sabate J, ed. Vegetarian Nutrition. CRC Press, Boca Raton, FL, 2001, pp. 3–17.
2. Matkovic V, Heaney RP. Calcium balance during human growth: evidence for threshold behavior. Am J Clin Nutr 1992; 55:992–996.
3. Tylavsky FA, Anderson JJB. Dietary factors in bone health of elderly lactoovovegetarian and omnivorous women. Am J Clin Nutr 1988; 48:842–849.
4. Hunt IF, Murphy NJ, Henderson C, et al. Bone mineral content in postmenopausal women: comparison of omnivores and vegetarians. Am J Clin Nutr 1989; 50:517–523.
5. Tesar R, Notelovitz M, Shim E, Kauwell G, Brown J. Axial and peripheral bone density and nutrient intakes of postmenopausal vegetarian and omnivorous women. Am J Clin Nutr 1992; 56:699–704.
6. Ellis FR, Holesh S, Ellis JW. Incidence of osteoporosis in vegetarians and omnivores. Am J Clin Nutr 1972; 25:555–558.
7. Marsh AG, Sanchez TV, Michelsen O, Chaffee FL, Fagal SM. Vegetarian lifestyle and bone mineral density. Am J Clin Nutr 1988; 48:837–841.
8. Janelle KC, Barr SI. Nutrient intakes and eating behavior scores of vegetarian and nonvegetarian women. J Am Diet Assoc 1995; 95:180–186, 189.
9. Draper A, Lewis J, Malhotra N, Wheeler E. The energy and nutrient intakes of different types of vegetarians: a case for supplements? Br J Nutr 1993; 69:3–19.
10. Gennari C. Calcium and vitamin D nutrition and bone disease of the elderly. Pub Health Nutr 2001; 4:547–559.
11. Outila TA, Karkkainen MUM, Seppanen RH, Lamberg-Allardt CJE. Dietary intake of vitamin D in premenopausal, healthy vegans was insufficient to maintain concentrations of serum 25-hydroxyvitamin D and intact parathyroid hormone within normal ranges during the winter in Finland. J Am Diet Assoc 2000; 100:434–441.
12. Cloutier G, Barr SI. The effect of protein nutriture on bone health: literature review and counselling implications. Can J Diet Pract Res 2003; 64:5–11.
13. Lau EMC, Kwok T, Woo J, Ho SC. Bone mineral density in Chinese elderly female vegetarian vegans, lacto-vegetarians and omnivores. Eur J Clin Nutr 1998; 52:60–64.
14. Lloyd T, Schaeffer JM, Walker MA, Demers LM. Urinary hormonal concentrations and spinal bone densities of premenopausal vegetarian and nonvegetarian women. Am J Clin Nutr 1991; 54:1005–1010.
15. Green J, Kleeman CR. The role of bone in the regulation of systemic acid-base balance. Contrib Nephrol 1991; 81:61–76.
16. Tucker KL, Hannan MT, Chen H, Cupples LA, Wilson PW, Kiel DP. Potassium, magnesium, and fruit and vegetable intakes are associated with greater bone mineral density in elderly men and women. Am J Clin Nutr 1999; 69:727–736.
17. New SA, Robins SP, Campbell MK, et al. Dietary influences on bone mass and bone metabolism: further evidence of a positive link between fruit and vegetable consumption and bone health? Am J Clin Nutr 2000; 71:142–151.
18. Tham DM, Gadner CD, Haskell WL. Potential health benefits of dietary phytoestrogens: a review of the clinical, epidemiological, and mechanistic evidence. J Clin Endocrinol Metab 1998; 83:2223–2235.

19. Greendale GA, FitzGerald G, Huang MH, et al. Dietary soy isoflavones and bone mineral density: results from the study of women's health across the nation. Am J Epidemiol 2002; 155:746–754.
20. Mei J, Yeung SS, Kung AW. High dietary phytoestrogen intake is associated with higher bone mineral density in postmenopausal but not premenopausal women. J Clin Endocrinol Metab 2001; 86:5217–5221.
21. Appleby PN, Key TJ, Thorogood M, Burr ML, Mann J. Mortality in British vegetarians. Pub Health Nutr 2002; 5:29–36.
22. Neumark-Sztainer D, Story M, Resnick MD, Blum RW. Adolescent vegetarians. A behavioral profile of a school-based population in Minnesota. Arch Pediatr Adolesc Med 1997; 151:833–838.
23. Martins Y, Pliner P, O'Connor R. Restrained eating among vegetarians: does a vegetarian eating style mask concerns about body weight? Appetite 1999; 32:145–154.
24. McLean JA, Barr SI. Cognitive dietary restraint is associated with eating behaviors, lifestyle practices, personality characteristics, and menstrual irregularity in college women. Appetite 2003; 40:185–192.
25. McLean JA, Barr SI, Prior JC. Dietary restraint, exercise, and bone density in young women: are they related? Med Sci Sports Exerc 2001; 33:1292–1296.
26. Barr SI. Vegetarianism and menstrual cycle disturbances: is there an association? Am J Clin Nutr 1999; 70:549S–554S.
27. Barr SI. Chapman GE. Perceptions and practices of self-defined current vegetarian, former vegetarian, and nonvegetarian women. J Am Diet Assoc 2002; 102:354–360.
28. Marsh AG, Sanchez TV, Mickelsen O, Keise J, Mayor G. Cortical bone density of adult lacto-ovo-vegetarian and omnivorous women. J Am Diet Assoc 1980; 76:148–151.
29. Reed JA, Anderson JJB, Tylavsky FA, Gallagher PN Jr. Comparative changes in radial-bone density of elderly female lactoovovegetarians and omnivores. Am J Clin Nutr 1994; 59(suppl):1197S–1202S.
30. Barr SI, Prior JC, Janelle KC, Lentle BC. Spinal bone mineral density in premenopausal vegetarian and nonvegetarian women: cross-sectional and prospective comparisons. J Am Diet Assoc 1998; 98:760–765.
31. Chiu J-F, Lan S-J, Yang C-Y, et al. Long-term vegetarian diet and bone mineral density in postmenopausal Taiwanese women. Calcif Tissue Int 1997; 60:245–249.
32. Rajaram S, Wien M. Vegetarian diets in the prevention of osteoporosis, diabetes and neurological disorders. In: Sabate J, ed. Vegetarian Nutrition. CRC Press, Boca Raton, FL, 2001, pp. 109–134.
33. Parsons TJ, Van Dusseldorp M, Van der Vliet M, Van de Werken K, Schaafsma G, Van Staveren WA. Reduced bone mass in Dutch adolescents fed a macrobiotic diet in early life. J Bone Miner Res 1997; 12:1486–1494.

17 Protein Intake and Bone Health

Jean-Philippe Bonjour, Patrick Ammann, Thierry Chevalley, and René Rizzoli

Deficiency in specific nutrients can play a major contributing role in the pathogenesis of osteoporosis and fragility fractures. In addition to calcium and vitamin D, several studies point to the existence of a tight connection between protein intake and bone metabolism. Protein intake below the recommended daily allowance could be particularly detrimental for both the acquisition of bone mass and its conservation throughout adult life. Various studies have found some relationship between the level of protein intake and either calcium–phosphate metabolism, bone mineral mass, or the risk of osteoporotic fracture *(1–3)*. Nevertheless, long-term influence of dietary protein on bone mineral metabolism and skeletal mass has been difficult to document.

1. PROTEIN INTAKE AND BONE GROWTH

Undernutrition during growth, including inadequate supply of energy and protein, can severely impair bone development. Studies in experimental animals indicate that isolated protein deficiency leads to reduced bone mass and strength, i.e., to osteoporosis, without histomorphometric evidence of osteomalacia *(1,2)*. Thus, an inadequate supply of protein appears to play a central role in the pathogenesis of the delayed skeletal growth and reduced bone mass observed in undernourished children *(4)* (Fig. 1). Dietary proteins play an essential role in the hormonal regulation of skeletal growth. This role of dietary proteins should not be considered as merely that of "brick supplier" to the osteogenic cells, thus conferring on them the capacity to lay down the organic bone matrix. The functioning of the endocrine axis connecting growth hormone (GH) to insulin-like growth factor 1 (IGF-1) can be markedly impaired by insufficient intakes of proteins *(3)*. Some amino acids, which remain to be identified, are required for the hepatic production of IGF in response to GH as well as to the action of IGF-1 on bone anabolism. Protein restriction also affects nonhepatic IGF-1 production *(5)*. In vitro, the production of IGF-1 by osteogenic cells can also be stimulated by certain amino acids such as arginine *(6)*. This observation

From: *Nutrition and Bone Health*
Edited by: M. F. Holick and B. Dawson-Hughes © Humana Press Inc., Totowa, NJ

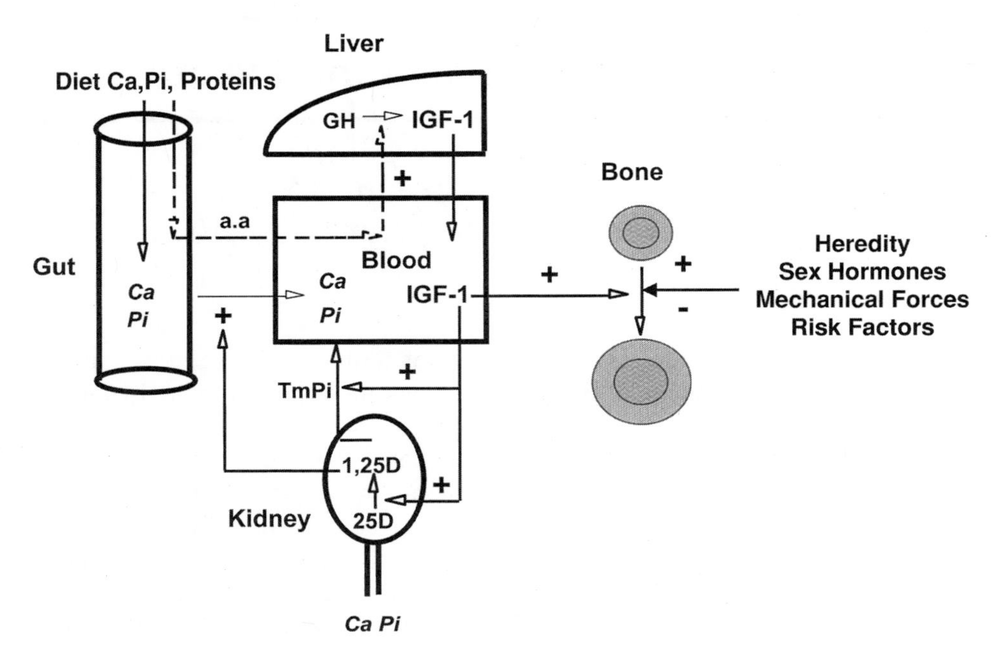

Fig. 1. Role of insulin-like growth factor (IGF)-1 in calcium phosphate metabolism during childhood and pubertal maturation in relation with essential nutrients for bone growth. During the pubertal bone growth spurt there is a rise in circulating IGF-1. The hepatic production of IGF-1 is under the positive influence of growth hormone (GH) and essential amino acids (a.a.). IGF-1 stimulates bone growth. At the kidney level, IGF-1 increases both the 1,25-dihydroxyvitamin D (1,25 D) conversion from 25-hydroxyvitamin D (25D) and the maximal tubular reabsorption of Pi (TmPi). By this dual renal action IGF-1 favors a positive calcium and phosphate balance as required by the increased bone mineral accrual. Heredity, sex hormones, mechanical forces, and risk factors can either positively or negatively influence the bone response to IGF-1. See text for further details.

raises the possibility of a defined modulating role of dietary proteins in the paracrine-autocrine regulation of the proliferation-differentiation of osteogenic cells.

In "well"-nourished children and adolescents, the question arises whether variations in the protein intake within the "normal" range can influence skeletal growth and thereby modulate the influence of genetic factors on peak bone mass attainment *(7)*. In the relationship between bone mass gain at the lumbar and femoral levels and protein intake, it is not surprising to find a positive correlation between these two variables. The association appears to be particularly significant in prepubertal children *(8)*. Indeed, in healthy prepubertal subjects, independent of the intake of calcium, a relatively low-protein diet is associated with a reduced gain in areal bone mineral density (aBMD) or bone mineral content (BMC) at both the femoral and spinal levels (Fig. 2). These results suggest that a relatively high protein intake could favor

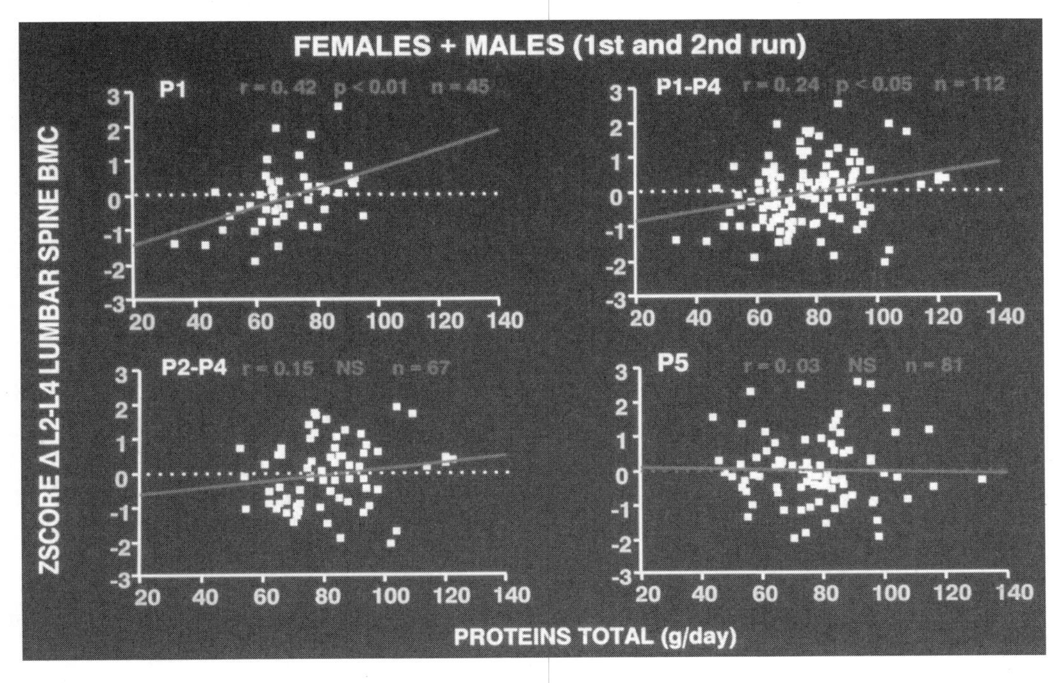

Fig. 2. Relation between protein intake and change in lumbar bone mineral content (BMC) in pre-peri- and postpubertal female and male adolescents. The mean protein intake from dairy, vegetable, and mineral sources was recorded in two 5-d diet diaries at 1-yr intervals. A positive correlation was found in prepupertal (P1), but neither in peripubertal (P2-P4) nor in postpubertal (P5) subjects. Each dot corresponds to the change in BMC adjusted for age and gender (Z-score) in 193 subjects aged from 9 to 19 yr. (Data are from refs *7* and *7a.*)

bone growth accrual during childhood. Nevertheless, these prospective observational results should not be interpreted as evidence for a causal relationship between bone mass gain and protein intake. Indeed, it is quite possible that protein intake, which is related to the overall amount of ingested calories, is to a large extent determined by growth requirements during childhood and adolescence. Only interventional studies testing different levels of protein intake in otherwise isocaloric diets could eventually determine the quantitative relationship between protein intake and bone mass acquisition during childhood and adolescence. Very recent data from our research group suggest that in healthy prepubertal boys the response to calcium supplementation can be influenced by spontaneous protein intake *(9).*

2. PROTEIN INTAKE AND BONE MASS IN YOUNG ADULTS

A positive correlation between protein intake and bone mass has been found in premenopausal women *(10–12).* In women keeping to a low-calorie diet, insufficient protein intake could be particularly deleterious for bone mass integrity.

Anorexia nervosa (AN) is a condition frequently observed in young women. Bone mineral density (BMD) is reduced at several skeletal sites in most women with anorexia nervosa. A recent study suggests that body weight, but not estrogen use, is a significant predictor of BMD in women with AN *(13)*. Another survey also indicates that young women with AN are at increased risk of fracture later in life *(14)*. Abnormally low serum albumin levels ($\leq$ 36 g/L) and a low body weight ($\leq$ 60% of average body weight) at the initial examination were variables best able to predict a lethal course *(15)*. With estrogen and calcium deficiency, low protein intake very likely contributes to the bone loss observed in AN. Experimental evidence obtained in adult female rats indicates that supplying more carbohydrates cannot compensate for the detrimental effect exerted by a low protein intake on bone mass *(16)*.

In athletes or ballet dancers, intensive exercise can lead to hypothalamic dysfunction with delayed menarche and disruption of menstrual cyclicity and bone loss *(17–21)*. Nutritional restriction can play an important role in the disturbance of the female reproductive system resulting from intense physical activity. The propensity to nutritional restriction will be more common when leanness confers an advantage for athletic performance *(19)*. Insufficient energy intake with respect to energy expenditure is supposed to impair the secretion of gonadotropin-releasing hormone and thereby lead to a state of hypoestrogenism. However, the relative contribution of insufficient protein intake frequently associated with low energy intake remains to be assessed. Indeed, experimental studies indicate that bone loss induced by isocaloric protein restriction in adult female rats is mediated by both dependent and independent sex-hormone deficiency mechanisms *(16,22,23)*.

3. PROTEIN INTAKE AND OSTEOPOROSIS IN THE ELDERLY

Among the determinants of osteoporosis in the elderly, nutritional deficiencies certainly play a major contributing role *(24,25)*. Indeed, undernutrition is often observed in the elderly, and it appears to be more severe in patients with hip fracture than in the general aging population *(26–30)*. Low body mass index (BMI), which can be interpreted as a crude "marker" of nutritional state, was found to be a major risk factor in a large prospective study of hip fracture incidence in men aged 50 yr or more living in several countries of southern Europe *(31)*. A variety of evidences also leads to the conclusion that protein intake below RDA could be particularly detrimental for the conservation of bone integrity with aging *(2,3)*. Indeed, undernutrition can accelerate age-dependent bone loss *(24,25,32)*. Protein undernutrition can favor the occurrence of hip fracture by increasing the propensity to fall as a result of muscle weakness and of impairment in movement coordination and by affecting protective mechanisms, such as reaction time, muscle strength, and/or by decreasing bone mass *(33–36)*. Furthermore, a reduction in the protective layer of soft tissue padding decreases the force required to fracture an osteoporotic hip *(28,37–40)*.

4. RELATION BETWEEN PROTEIN INTAKE, BMD, AND OSTEOPOROTIC FRACTURE

Numerous cross-sectional studies have examined the relationship between spontaneous protein intake and bone mass. In hip fracture patients, bone mass was directly proportional to serum albumin, taken as a reflection of nutritional intakes *(41)*. In a survey carried out in hospitalized elderly patients, low protein intake was associated with reduced femoral neck aBMD and poor physical performance *(33)*. The group with high intakes and a greater BMD, particularly at the femoral neck level, also had better improvement of bicepital and quadricepital muscle strength and performance, as indicated by increased capacity to walk and climb stairs, after 4 wk of hospitalization *(33)*. In postmenopausal women there was a positive correlation between protein intake and aBMD *(42–47)* and also in men *(33,46,48)*. In a longitudinal follow-up in the Framingham study, the rate of bone mineral loss was inversely correlated to dietary protein intake in women as well as in men *(49)*. Inconsistent results have been obtained as to the putative detrimental intake of high protein diet. Thus, in a cross-sectional study, a protein intake close to 2 g/kg body weight in young college women was associated with reduced BMD only at one of two forearm sites studied *(50)*. Altogether, these results indicate that a sufficient protein intake is mandatory for bone health *(51)*. Thus, whereas a gradual decline in caloric intakes with age can be considered as an adequate adjustment to the progressive reduction in energy expenditure, the parallel reduction in protein intakes may be detrimental for maintaining the integrity and function of several organs or systems, including skeletal muscles and bone.

The impression that high protein may be harmful for bone stems mainly from studies indicating that urinary calcium was positively correlated with protein intake. This association suggested that high protein intake would induce a negative calcium balance and consequently would favor bone loss *(52)*. However, further studies indicated that a reduction in dietary protein led to a decline in calcium absorption and secondary hyperparathyroidism *(53)*. There is some evidence that the favorable effect of increasing protein intake on bone mineral mass, as repeatedly observed in both genders *(33,42–49)*, is better expressed with an adequate supply of both calcium and vitamin D *(51,54–57)*.

The question of whether the source of proteins, animal vs vegetable, affect calcium metabolism differently has been the object of more emotional belief than serious scientific demonstration *(55)*. First, the belief lies in the hypothesis that animal proteins generate more sulfuric acid from sulfur-containing amino acids than a vegetarian diet. Furthermore, the difference in nutrition-generated acid load would lead in healthy individuals to increased bone dissolution, this by analogy to the classical physicochemical in vitro observation that lowering pH favors the dissolution of calcium phosphate crystals, including those of hydroxyapatite, The theory suggesting that animal protein in contrast to vegetable protein would be more detrimental for bone health is not supported by any robust experimental facts. First, a vegetarian diet with protein derived from grains and legumes would deliver as many millimoles of sulfur per gram of protein as would a purely meat-based diet *(55)*. Second, the net release of proton buffers from bone mineral does not appear to contribute significantly to blood acid–base equilibrium, even in response to a conspicuous proton load, unless renal function is severely impaired *(58)*. Third, several recent human studies do not suggest that the protective effect of protein on either bone loss or osteoporotic fracture is due to vegetarian rather than animal protein (*49,57,59–61; see* Chapters 16 and 18, this volume). In apparently sharp contrast with these very consistent results, an epidemiological study reported that individuals consuming diets with high ratios of animal to vegetal protein lost bone more rapidly than did those with lower ratios, and had a greater risk of hip fracture *(62)*. The physiological meaning, particularly in terms of impact on calcium–phosphate and bone metabolism of animal to vegetal protein ratio remains mechanistically quite obscure. Indeed, variations in this calculated ratio can result from differences in the absolute intake of either animal or vegetable protein. More important, however, in this study *(62)* the statistically negative relationship between the animal-to-vegetable protein ratio and bone loss was obtained only after multiple adjustments, not only for age but also for energy intake, total calcium intake (dietary plus supplements), total protein intake, weight, current estrogen use, physical activity, smoking status, and alcohol intake *(62)*. In sharp contrast, a positive relationship between the animal-to-vegetable protein ratio and baseline BMD was found when the statistical model adjusted only for age *(62)*. This inconsistency according to the way this set of data was analyzed makes difficult the generalization of these findings in terms of nutritional recommendations for bone health and osteoporosis prevention *(55)*.

As already mentioned, various studies have found a relationship between the level of protein intake and calcium–phosphate or bone metabolism *(1,3,6,8,11,50,63)* and came to the conclusion that either a deficient or an excessive protein supply could negatively affect the balance of calcium. An indirect argument in favor of a deleterious effect of high protein intake on bone is that hip fracture appeared to be more frequent in countries with high protein intake of animal origin *(64)*. However, as expected, the countries with the highest hip fracture incidence are those with longest life expectancy, which could increase fracture incidence. In the Nurse Health Study including a large number of subjects followed over 12 yr, a trend toward a hip fracture incidence inversely related to protein intake has been reported *(65)*. In the same study, forearm fracture incidence increased, however, in subjects with high protein intake of animal origin. However, this epidemiological observation was not confirmed in other studies. Thus, in the NHANES I study, hip fracture was higher with low energy intake, low serum albumin levels, and low muscle strength *(66)*. The strongest argument for a favorable influence of protein has been provided by a prospective study carried out on more than 30,000 women in Iowa, in which higher protein intake was associated with a reduced risk of hip fracture (relative risk in the highest protein intake quartile: 0.33, and 0.44 after adjustment for age, body size, parity, smoking, alcohol intake, estrogen use, and physical activity) *(60)*. The association was particularly evident with protein of animal rather than vegetable origin.

Similarly, a reduced relative risk of hip fracture was found with higher intake of milk in both genders *(31,67)*. In another survey, no association between hip fracture and nondairy animal protein intake could be detected *(68,69)*. A low plasma albumin level, which can reflect low nutritional intakes, has been repeatedly found in patients with hip fracture as compared to age-matched healthy subjects or patients with osteoarthritis *(26,29,41,70)*.

5. UNDERNUTRITION, PROTEIN REPLENISHMENT, AND OSTEOPOROSIS

A state of undernutrition on admission, which is consistently documented in elderly patients with hip fracture *(28,30)*, followed by an inadequate food intake during hospital stay, can adversely influence clinical outcome *(28,29,70)*. In hip fracture patients, in whom a lower femoral neck BMD at the level of the proximal femur has been demonstrated *(71)*, a dietary survey based on 50 daily precise measurements of food intake confirmed that nutritional requirements were not met while the patients were in hospital, even though adequate quantities of food were offered *(29)*. Undernutrition can involve all kinds of nutrients, and the specific role of a low protein intake in addition to low calorie consumption can be difficult to appraise in the elderly *(33)*. Intervention studies using supplementary feeding by nasogastric tube or parenteral nutrition *(72)*, or even a simple oral dietary preparation that normalizes protein intake *(29)*, can improve clinical outcome after hip fracture. Oral correction of deficient food intake has obvious practical and psychological advantages over nasogastric tube feeding or parenteral nutrition. It should be emphasized that in the study mentioned above *(29)*, a 20-g protein supplement brought the intake from low to a level still below RDA (0.8 g/kg body weight). Follow-up showed a significant difference in the clinical course in the rehabilitation hospitals, with the supplemented patients doing better. Although the mean duration of dietary supplementation did not exceed 30 d, the significantly lower rate of complication (bedsores, severe anemia, intercurrent lung or renal infections, 44% vs 87%), and deaths was still observed at 6 mo (40% vs 74%) *(29)*. The duration of hospital stay of elderly patients with hip fracture is determined not only by the present medical condition, but also by domestic and social factors *(28,73,74)*. In this study *(29)*, the total length of stay in the orthopedic ward and convalescent hospital was significantly shorter in supplemented patients than in controls (median: 24 vs 40 d). After the favorable outcome shown in hip fracture patients receiving a protein-calorie supplement *(29)*, the question as to whether protein represented the key nutrient responsible for the beneficial effect was addressed by comparing the clinical outcome of elderly patients with hip fracture (mean age 82 yr), receiving two different dietary supplements that differed only by their protein content *(75)*. The clinical course was significantly better in the group receiving protein, with 79% having a favorable course as compared to 36% in the control group during the stay in the recovery hospital, indicating a specific effect of protein supplements on the outcome.

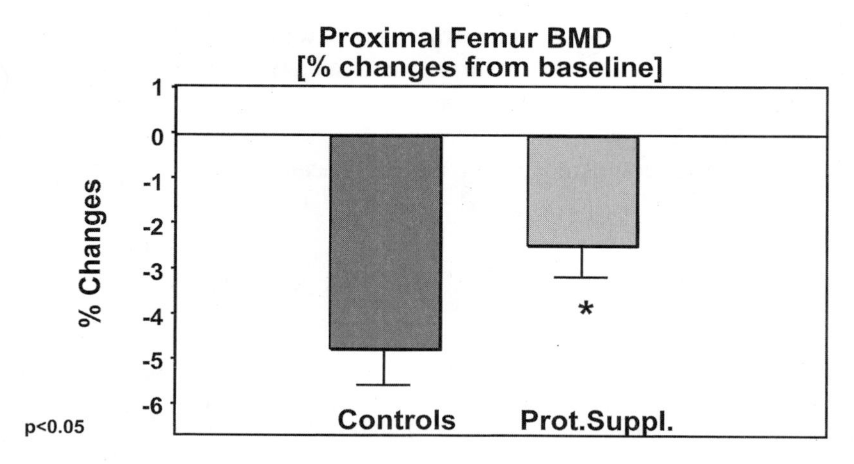

Fig. 3. Protein supplements attenuate proximal femur bone loss in patients with recent hip fracture. Percent change in femoral neck aBMD in patients receiving daily either an oral dietary supplement (Prot.suppl.) containing 20 g of proteins as casein or an isocaloric serving containing no protein (controls) during the 6 mo following hip fracture. Both groups were supplemented in calcium and vitamin D. The study was randomized, double-blind and placebo-controlled. (From ref. *59.*)

In undernourished elderly with a recent hip fracture, an increase in protein intake, from low to normal, can also be beneficial for bone integrity. Indeed, in a double-blind, placebo-controlled study, the effects of protein repletion were investigated in patients with a recent hip fracture *(59)*. Within 1 wk after an osteoporotic hip fracture, 82 patients (80.7 ± 1.2 yr) were randomly allocated to a daily 20-g protein supplement, which nearly corrected protein deficiency, or to an isocaloric placebo, for 6 mo. All were given 200,000 IU vitamin D once at baseline, and 550 mg/d of calcium. As compared with the placebo group, the protein-supplemented patients had significantly greater gains in serum prealbumin, in IGF-1, and in IgM *(59)*. In agreement with previous results *(29,75)*, protein repletion after hip fracture was associated with a more favorable outcome, including shorter rehabilitation hospital stay. In a multiple regression analysis, baseline IGF-1 concentrations and biceps muscle strength, together with the protein supplements, accounted for more than 30% of the variance of the length of stay in rehabilitation hospitals ($r^2 = 0.312$, $p < 0.0005$). These results support the hypothesis that the beneficial effects of protein repletion after hip fracture could be associated with a stimulation of the IGF-1 system in the protein-supplemented patients. Thus, the lower incidence of medical complications observed after such a protein supplement *(29,75)* is also compatible with the hypothesis of IGF-1 improving the immune status, as this growth factor can stimulate the proliferation of immunocompetent cells and modulate immunoglobulin secretion *(76)*. Importantly, the proximal femur BMD decrease observed at 1 yr in the placebo group was attenuated by approx 50% *(59)* (Fig. 3). These results raise the question whether protein repletion of frail elderly could prevent the age-dependent decrease in IGF-1 levels, and help thereby to prevent falls and to increase bone mass. Indeed, a low IGF-1 level has been shown to be a predictor of fracture risk *(77)*. Our intervention study *(59)* stresses the impressive bone loss occurring during the mo following a hip fracture in the patients who did not receive a protein-containing supplement. This may explain why patients with osteoporotic fracture have at least a twofold increased risk of experiencing another fracture. This prompted our group to set up a dedicated clinical pathway for the management of patients with low-trauma fracture *(78)*. This pathway includes a multidisciplinary teaching program on physical therapy, daily living activities, and nutrition, delivered to the patients and their families, as well as nutritional recommendations on the intake required of calcium, vitamin D, and protein, with very practical aspects.

6. RELATION BETWEEN PROTEIN INTAKE AND IGF-1

In association with the progressive age-dependent decrease in both protein intake and bone mass, several reports have documented a decrement in IGF-1 plasma levels *(79–82)*. Experimental and clinical studies suggest that dietary proteins, by influencing both the production and action of growth factors, particularly the GH–IGF system, could control bone anabolism *(83,84)*. The hepatic production and plasma level of IGF-1 is under the influence of dietary proteins *(85,86)*. Protein restriction has been shown to reduce IGF-1 plasma levels by inducing a resistance to the action of GH at the hepatic level *(87,88)*, and by an increase of IGF-1 metabolic clearance rate *(89)*. Decreased levels of IGF-1 have been found in states of undernutrition such as marasmus, AN, celiac disease, or HIV-infected patients *(89,90–92)*. Refeeding these patients led to an increase of IGF-1 *(90,93,94)*. Furthermore, elevated protein intake is able to prevent the decrease in IGF-1 usually observed in hypocaloric states *(85,95)*. In addition, protein restriction could render target organs less sensitive to IGF-1. When IGF-1 was given to growing rats maintained under a low-protein diet at doses normalizing their plasma levels, it failed to restore skeletal longitudinal growth *(96)*.

To evaluate the early cellular responses in both cortical and trabecular bone, we submitted adult female rats to a diet with low protein content but isocaloric to the control diet. Histomorphometric and biochemical analyses were performed after 2 wk of protein deficiency, when plasma IGF-1 was significantly lower in protein-restricted rats *(16,97)*. Thereafter, to investigate the bone cellular response to IGF-1, we administered a pharmacological dose of *rh*IGF-1/IGFBP-3 to 15% and 2.5% casein-fed rats for 10 d and evaluated its effects histologically and biochemically *(97)*. After 14 d of protein restriction, significant drops in periosteal formation and mineral apposition rates were observed, indicating a decreased osteoblast recruitment and activity. In rats fed the 15% casein diet, *rh*IGF/IGFBP-3 increased cancellous and periosteal formation rates, indicating an increased osteoblast recruitment. However, in protein-restricted rats, *rh*IGF/IGFBP-3 failed to increase cancellous or periosteal bone formation. The early response of bone cells activities to isocaloric low protein intake in adult female rats is tissue specific. Indeed, short-term protein restriction impairs bone formation in the cortex at the periosteal level, but not in cancellous bone. In addition, dietary protein restriction induces an osteoblastic resistance to IGF-1 in both envelopes. This may suggest that low plasma IGF-1 and/or osteoblast resistance to IGF-1 in response to low protein intake could play an important role in the impairment of periosteal osteoblasts. Moreover, these results suggest that therapeutic administration of IGF-1 to subjects with a dietary protein deficiency may be ineffective on bone.

We investigated the influence of IGF-1 on BMD in adult rats made osteoporotic by ovariectomy *(98–100)*. BMD was measured by DXA at the levels of lumbar spine, proximal, and total tibia *(101)*. This technique allows a precise and accurate in vivo longitudinal measurement of bone mineral mass, areal bone mineral density, and outer bone dimensions in rats, at various skeletal sites characterized by different proportions of cortical and trabecular bone, and thereby by different rates of bone remodeling and responses to dietary or hormonal manipulations. A 6-wk infusion of IGF-1 induced a dose-dependent increment of BMD at the three scanned skeletal sites *(98)*. The increase in BMD induced by IGF-1 was associated with an increase in the resistance to mechanical strain in relation also with an increase of bone shaft outer dimensions *(93,99)*.

The local production of IGF-1 by osteoblastic cells in relation with amino acid concentrations could also play a role. Indeed, the amino acids arginine or lysine increased IGF-1 production and collagen synthesis in cultured mouse osteoblastic cells, on a time- and concentration-dependent manner *(6)*. This study underlines a possible influence of the local environment in proteins or amino acids on IGF-1 production by bone cells, and suggests a potential role of locally produced IGF-1 under the influence of extracellular amino acid concentration in the regulation of osteoblast function.

To address the issue of a specific influence of protein deficiency in the pathogenesis of osteoporosis, an experimental model in adult female or male rats of selective protein deprivation with isocaloric low-protein diets supplemented by identical amounts of minerals has been developed *(22,23,97,102)*. This model enables the study of bone mineral mass, bone strength, and remodeling. A decrease of BMD was observed at the level of skeletal sites formed by trabecular or cortical bone only in animals fed 2.5% casein. This was associated with a marked and early decrease in plasma IGF-1 by 40%. In this model, sex-hormone deficiency or action was also observed, since estrous cycles disappeared under a long-term low-protein isocaloric diet. The effects of ovariectomy and protein deficiency were additive, suggesting distinct mechanisms of action. Both histomorphometric analysis and biochemical markers of bone remodeling indicate that the low-protein-intake-induced decrease in bone mineral mass and bone strength was related to an uncoupling between bone formation and resorption *(16,22,23)*.

7. OTHER MECHANISMS POTENTIALLY INVOLVED IN PROTEIN-RELATED BONE LOSS

Beside the production and action of the GH–IGF-1 system, protein undernutrition can be associated with alterations of cytokines secretion, such as interferon-γ, tumor necrosis factor-α (TNF-α), or transforming growth factor-β *(102,103)*. TNF-α and interleukin-6 are generally increasing with aging *(104)*. In a situation of cachexia, such as in chronic heart failure, an inverse correlation between bone mineral density and TNF-α levels has been found *(105,106)*, further implicating a possible role of uncontrolled cytokines production in bone loss. Increased TNF-α can be a crucial factor in sex hormone deficiency-induced bone loss *(107)*, but it also plays a role in target organ resistance to insulin, and possibly to IGF-1 *(108)*. Along the same line, certain amino acids given to rats fed a low-protein diet can increase the liver protein synthesis response to TNF-α *(109)*. The amino acid oxidation rate is lower in children with kwashiorkor replete with milk as compared with egg white, and protein breakdown and synthesis correlated inversely with TNF-α levels *(110)*. The modulation by nutritional intakes of cytokines production and action *(111)* and the strong implication of various cytokines in the regulation of bone remodeling *(112)* suggest a possible role of certain cytokines in the nutrition–bone link.

8. CONCLUSION

In the pathogenesis of osteoporosis and fragility fractures it is well established that insufficient supply of specific nutrients can play a major contributing role. In addition to calcium and vitamin D, several studies point to the existence of a tight connection between protein intake and bone metabolism. Protein intake below RDA could be particularly detrimental for both the acquisition of bone mass and its conservation throughout adult life. In healthy prepubertal children, independent of the intake of calcium, a relatively low-protein diet is associated with a reduced gain in aBMD at both femoral and spinal levels. In women keeping to a low-calorie diet, insufficient protein intake could be particularly deleterious for bone mass integrity. With estrogen and calcium deficiency, low protein intake probably contributes to the bone loss observed in anorexia nervosa. In elderly women, reduced protein intake is associated with lower femoral neck aBMD and poor physical performance. Furthermore, protein malnutrition is present in many elderly women with low femoral neck aBMD and hip fracture. More recent North American epidemiological studies corroborate this relationship.

Clinical outcome after hip fracture can be significantly improved by normalizing protein intake. This effect could be, at least in part, mediated by a positive influence on IGF-1, of which the plasma level decreases in both genders with advancing age. As compared with nonfractured controls, patients with hip fracture have a lower IGF-1 plasma level, which is associated with reduced proximal femur aBMD, lower plasma levels of prealbumin, albumin, and lower upper-extremities muscle strength. In undernourished elderly subjects with hip fracture, an increase in the protein intake—in the form of milk proteins—from low to normal, as compared to an isocaloric placebo, induces after 6 mo of intervention a significantly greater gain in plasma prealbumin, IgM, and IGF-1. This intervention also attenuates femoral bone loss and is associated with a shorter stay in rehabilitation hospitals. In order to understand the mechanism whereby protein intake influences bone mineral mass, animal models using either adult female or male rats have been developed. In these models isocaloric protein undernutrition mimics osteoporosis observed in the elderly, in whom both cortical and trabecular skeletal sites are affected with negative uncoupling between bone formation and resorption. Reduced bone formation could be mediated by decreased IGF-1 production and action. Increased bone resorption appears to result from both sex hormone deficiency and increased cytokines such as TNF-α. Thus, dietary protein contributes to maintain bone integrity from early childhood to old age. An adequate intake of proteins should be recommended in the prevention and treatment of postmenopausal and age-dependent osteoporosis.

REFERENCES

1. Orwoll ES. The effects of dietary protein insufficiency and excess on skeletal health. Bone 1992; 13:343–350.
2. Bonjour JP, Schürch MA, Chevalley T, Ammann P, Rizzoli R. Protein intake, IGF-1 and osteoporosis. Osteoporos Int 1997; 7(suppl 3):S36–S42.
3. Bonjour JP, Schürch MA, Rizzoli R. Nutritional aspects of hip fractures. Bone 1996; 18(suppl):S139–S144.
4. Rizzoli R, Ammann P, Chevalley T, Bonjour J-P. Protein intake during childhood and adolescence and attainment of peak bone mass. In: Bonjour J-P, Tsang RC, eds. Nutrition and Bone Development. Lippincott-Raven, Philadelphia, 1999, pp. 231–243.
5. Naranjo WM, Yakar S, Sanchez-Gomez M, Perez AU, Setzer J, LeRoith D. Protein calorie restriction affects non hepatic IGF-1 production and the lymphoid system: studies using the liver-specific IGF-1 gene-deleted mouse model. Endocrinology 2002; 143:2233–2341.
6. Chevalley T, Rizzoli R, Manen D, Caverzasio J, Bonjour JP. Arginine increases insulin-like growth factor-I production and collagen synthesis in osteoblast-like cells. Bone 1998; 23:103–109.
7. Clavien H, Theintz G, Rizzoli R, Bonjour JP. Does puberty alter dietary habits in adolescents living in a Western society? J Adolesc Health 1996; 19:68–75.

7a. Theintz G, Buchs B, Rizzoli R, et al. Longitudinal monitoring of bone mass accumulation in healthy adolescents; evidence for a marked reduction after 16 years of age at the levels of lumbar spine and femoral neck in female subjects. J Clin Endocrinol Metab 1992; 75:1060–1065.

8. Rizzoli R, Bonjour JP. Determinants of peak bone mass and mechanisms of bone loss. Osteoporos Int 1999; 9(suppl 2):S17–S23.
9. Chevalley T, Ferrari S, Hans D, et al. Protein intake modulates the effet of calcium supplementation on bone mass gain in prepubertal boys. J Bone Miner Res 2002; 17(suppl 1):S172.
10. Hirota T, Nara M, Ohguri M, Manago E, Hirota K. Effect of diet and lifestyle on bone mass in Asian young women. Am J Clin Nutr 1992; 55:1168–1173.
11. Cooper C, Atkinson EJ, Hensrud DD, et al. Dietary protein intake and bone mass in women. Calcif Tissue Int 1996; 58:320–325.
12. Teegarden D, Lyle RM, McCabe GP, et al. Dietary calcium, protein, and phosphorus are related to bone mineral density and content in young women. Am J Clin Nutr 1998; 68:749–954.

13. Grinspoon S, Thomas E, Pitts S, et al. Prevalence and predictive factors for regional osteopenia in women with anorexia nervosa. Ann Intern Med 2000; 133:790–794.
14. Lucas AR, Melton LJ 3rd, Crowson CS, O'Fallon WM. Long-term fracture risk among women with anorexia nervosa: a population-based cohort study. Mayo Clin Proc 1999; 74:972–977.
15. Herzog W, Deter HC, Fiehn W, Petzold E. Medical findings and predictors of long-term physical outcome in anorexia nervosa: a prospective, 12-year follow-up study. Psychol Med 1997; 27:269–279.
16. Ammann P, Bourrin S, Bonjour JP, Meyer JM, Rizzoli R. Protein undernutrition-induced bone loss is associated with decreased IGF-1 levels and estrogen deficiency. J Bone Miner Res 2000; 15:683–690.
17. Drinkwater BL, Nilson K, Chesnut CH III, Bremner WJ, Shainholtz S, Southworth MB. Bone mineral content of amenorrheic and eumenorrheic athletes. N Engl J Med 1984; 311:277–281.
18. Marcus R, Cann C, Madvig P, et al. Menstrual function and bone mass in elite women distance runners. Ann Intern Med 1985; 102:158–163.
19. Warren MP, Perlroth NE. The effects of intense exercise on the female reproductive system. J Endocrinol 2001; 17:3–11.
20. Gremion G, Rizzoli R, Slosman D, Theintz G, Bonjour J-P. Oligo-amenorrheic long-distance runners may lose more bone in spine than in femur. Med Sci Sports Exerc 2001; 33:15–21.
21. Beck BR, Shaw J, Snow CM. Physical activity and osteoporosis. In: Marcus R, Feldman D, Kelsey J, eds. Osteoporosis, 2nd ed. Academic, San Diego, CA, 2001, vol. 1, pp. 701–720.
22. Ammann P, Rizzol R, Bonjour JP. Protein malnutrition-induced bone loss is associated with alteration of growth hormone-IGF-1 axis and with estrogen deficiency in adult rats. Osteoporos Int 1998; 8(suppl 3):10.
23. Ammann P, Bourrin S, Bonjour JP, Meyer JM, Rizzoli R. Protein undernutrition-induced bone loss is associated with decreased IGF-1 levels and estrogen deficiency. J Bone Miner Res 1999; 15:683–690.
24. Parfitt AM. Dietary risk factors for age-related bone loss and fractures. Lancet 1983; ii:1181–1184.
25. Schaafsma G, Van Beresteyn ECH, Raymakers JA, Duursma SA. Nutritional aspects of osteoporosis. World Rev Nutr Diet 1987; 49:121–159.
26. Rapin CH, Lagier R, Boivin G, Jung A, MacGee W. Biochemical findings in blood of aged patients with femoral neck fractures: a contribution to the detection of occult osteomalacia. Calcif Tissue Int 1982; 34:465–469.
27. Older MWJ, Edwards D, Dickerson JWT. A nutrient survey in elderly women with femoral neck fractures. Br J Surg 1980; 67:884–886.
28. Bastow MD, Rawlings J, Allison SP. Benefits of supplementary tube feeding after fractured neck of femur: a randomised controlled trial. Br Med J 1983; 287:1589–1592.
29. Delmi M, Rapin CH, Bengoa JM, Delmas PD, Vasey H, Bonjour JP. Dietary supplementation in elderly patients with fractured neck of the femur. Lancet 1990; 335:1013–1016.
30. Jensen JE, Jensen TG, Smith TK, Johnston DA, Dudrick SJ. Nutrition in orthopaedic surgery. J Bone Joint Surg 1982; 64:1263–1272.
31. Kanis J, Johnell O, Gullberg B, et al. Risk factors for hip fracture in men from Southern Europe: The Medos Study. Osteoporos Int 1999; 9:45–54.
32. Garn SM, Guzman MA, Wagner B. Subperiostal gain and endosteal loss in protein-calorie malnutrition. Am J Phys Anthropol 1969; 30:153–155.
33. Geinoz G, Rapin CH, Rizzoli R, et al. Relationship between bone mineral density and dietary intakes in the elderly. Osteoporos Int 1993; 3:242–248.
34. Campbell WW, Barton ML Jr, Cyr-Campbell D, et al. Effects of an omnivorous diet compared with a lactoovovegetarian diet on resistance-training-induced changes in body composition and skeletal muscle in older men. Am J Clin Nutr 1999; 70:1032–1039.

35. Evans WJ. Protein nutrition and resistance exercise. Can J Appl Physiol 2001; 26(suppl):S141–S152.
36. Haub MD, Wells AM, Tarnopolsky MA, Campbell WW. Effect of protein source on resistive-training-induced changes in body composition and muscle size in older men. Am J Clin Nutr 2002; 76:511–517.
37. Grisso JA, Kelsey JL, Strom BL, et al., and the Northeast Hip Fracture Study Group. Risk factors for falls as a cause of hip fracture in women. N Engl J Med 1991; 324:1326–1331.
38. Vellas B, Baumgartner RN, Wayne SJ, et al. Relationship between malnutrition and falls in the elderly. Nutrition 1992; 8:105–108.
39. Vellas BJ, Albarede JL, Garry PJ. Diseases and aging: patterns of morbidity with age: relationship between aging and age-associated diseases. Am J Clin Nutr 1992; 55(suppl 6):1225S–1230S.
40. Schwartz A, Capezuti E, Grisso JA. Falls as risk factors for fractures. In: Marcus R, Feldman D, Kelsey J, eds. Osteoporosis. Academic Press, San Diego, CA, 2001, pp. 795–808.
41. Thiébaud D, Burckhardt P, Costanza M, et al. Importance of albumin, 25(OH)-vitamin D and IGFBP-3 as risk factors in elderly women and men with hip fracture. Osteoporos Int 1997; 7:457–462.
42. Tylavsky FA, Anderson JJ. Dietary factors in bone health of elderly lactoovovegetarian and omnivorous women. Am J Clin Nutr 1988; 48:842–849.
43. Lacey JM, Anderson JJ, Fujita T, et al. Correlates of cortical bone mass among the premenopausal and postmenopausal Japanese women. J Bone Miner Res 1991; 6:651–659.
44. Michaelsson K, Holmberg L, Mallmin H, Wolk A, Bergstrom R, Ljunghall S. Diet, bone mass, and osteocalcin: a cross-sectional study. Calcif Tissue Int 1995; 57:86–93.
45. Chiu JF, Lan SJ, Yang CY, et al. Long-term vegetarian diet and bone mineral density in postmenopausal Taiwanese women. Calcif Tissue Int 1997; 60:245–249.
46. Calvo MS, Barton CN, Park YK. Bone mass and high dietary intake of meat and protein: analyses of data from the Third National Health and Nutrition Examination Survey (NHANES III, 1988–94). Bone 1998; 23(suppl):S290.
47. Lau EM, Kwok T, Woo J, Ho SC. Bone mineral density in Chinese elderly female vegetarians, vegans, lacto-vegetarians and omnivores. Eur J Clin Nutr 1998; 52:60–64.
48. Orwoll ES, Weigel RM, Oviatt SK, Meier DE, McClung MR. Serum protein concentrations and bone mineral content in aging normal men. Am J Clin Nutr 1987; 46:614–621.
49. Hannan MT, Tucker KL, Dawson-Hughes B, Cupples LA, Felson DT, Kiel DP. Effect of dietary protein on bone loss in elderly men and women: the Framingham Osteoporosis Study. J Bone Miner Res 2000; 15:2504–2512.
50. Anderson JJB, Metz JA. Adverse association of high protein intake to bone density. Chall Mod Med 1995; 7:407–412.
51. Bell J, Whiting SJ. Elderly women need dietary protein to maintain bone mass. Nutr Rev 2002; 60:337–341.
52. Heaney RP, Recker RR. Effects of nitrogen, phosphorus, and caffeine on calcium balance in women. J Lab Clin Med 1982; 99:46–55.
53. Kerstetter JE, O'Brien K, Insogna K. Dietary protein and intestinal calcium absorption. Am J Clin Nutr 2002; 73:990–992.
54. Heaney RP. Calcium, dairy product and osteoporosis. J Am Coll Nutr 2000; 19(suppl):83S–99S.
55. Heaney RP. Protein intake and bone health: the influence of belief systems on the conduct of nutritional science. Am J Clin Nutr 2001; 73:5–6.
56. Heaney RP. Protein and calcium: antagonists or synergists? Am J Clin Nutr 2002; 75:609–610.
57. Dawson-Hughes B, Harris SS. Calcium intake influences the association of protein intake with rate of bone loss in elderly men and women. Am J Clin Nutr 2002; 75:773–779.
58. Freudiger H, Bonjour JP. Bisphosphonates and extrarenal acid buffering capacity. Calcif Tissue Int 1989; 44:3–10.

59. Schürch MA, Rizzoli R, Slosman D, Vadas L, Vergnaud P, Bonjour JP. Protein supplements increase serum insulin-like growth factor-I levels and attenuate proximal femur bone loss in patients with recent hip fractureA randomized, double-blind, placebo-controlled trial. Ann Intern Med 1998; 128:801–809.
60. Munger RG, Cerhan JR, Chiu BCH. Prospective study of dietary protein intake and risk of hip fracture in postmenopausal women. Am J Clin Nutr 1999; 69:147–152.
61. Promislow JHE, Goodman-Gruen D, Slymen DJ, Barret-Connor E. Protein consumption and bone mineral density in the elderly. The Rancho Bernardo Study. Am J Epidemiol 2002; 155:636–644.
62. Sellmeyer DE, Stone KL, Sebastian A, Cummings SR, for the Study of Osteoporotic Fractures Research Group. A high ratio of dietary animal to vegetable protein increases the rate of bone loss and the risk of fracture in posmenopausal women. Am J Clin Nutr 2001; 73:118–122.
63. Metz JA, Anderson JJB, Gallagher PN. Intakes of calcium, phosphorus, and protein, and physical-activity level are related to radial bone mass in young adult women. Am J Clin Nutr 1993; 58:537–542.
64. Abelow BJ, Holford TR, Insogna KL. Cross-cultural association between dietary animal protein and hip fracture: a hypothesis. Calcif Tissue Int 1992; 50:14–18.
65. Feskanich D, Willett WC, Stampfer MJ, Colditz GA. Protein consumption and bone fractures in women. Am J Epidemiol 1996; 143:472–479.
66. Huang Z, Himes JH, McGovern PG. Nutrition and subsequent hip fracture risk among a national cohort of white women. Am J Epidemiol 1996; 144:124–134.
67. Johnell O, Gullberg B, Kanis JA, et al. Risk factors for hip fracture in European women: the MEDOS study. J Bone Miner Res 1995; 10:1802–1815.
68. Meyer HE, Pedersen JI, Løken EB, Tverdal A. Dietary factors and the incidence of hip fracture in middle-aged Norwegians. A prospective study. Am J Epidemiol 1997; 145:117–123.
69. Feskanich D, Willett WC, Stampfer MJ, Colditz GA. Milk, dietary calcium, and bone fractures in women: a 12-year prospective study. Am J Public Health 1997; 87:992–997.
70. Patterson BM, Cornell CN, Carbone B, Levine B, Chapman D. Protein depletion and metabolic stress in elderly patients who have a fracture of the hip. J Bone Joint Surg 1992; 74A:251–260.
71. Chevalley T, Rizzoli R, Nydegger V, et al. Preferential low bone mineral density of the femoral neck in patients with a recent fracture of the proximal femur. Osteoporos Int 1991; 1:147–154.
72. Bastow MD, Rawlings J, Allison SP. Undernutrition, hypothermia, and injury in elderly women with fractured femur: an injury response to altered metabolism? Lancet 1983; 1:143–146.
73. Sullivan DH, Patch GA, Walls RC, Lipschitz DA. Impact of nutrition status on morbidity in a select population of geriatric rehabilitation patients. Am J Clin Nutr 1990; 51:749–758.
74. Schürch MA, Rizzoli R, Mermillod B, Vasey H, Michel JP, Bonjour JP. A prospective study on socioeconomic aspects of fracture of the proximal femur. J Bone Miner Res 1996; 11:1935–1942.
75. Tkatch L, Rapin CH, Rizzoli R, et al. Benefits of oral protein supplement in elderly patients with fracture of the proximal femur. J Am Coll Nutr 1992; 11:519–525.
76. Auernhammer CJ, Strasburger CJ. Effects of growth hormone and insulin-like growth factor I on the immune system. Eur J Endocrinol 1995; 133:635–645.
77. Garnero P, Sornay-Rendu E, Delmas PD. Low serum IGF-I and occurrence of osteoporotic fractures in postmenopausal women. Lancet 2000; 355:898–899.
78. Chevalley T, Hoffmeyer P, Bonjour JP, Rizzoli R. A critical pathway for the medical management of osteoporotic fracture: a way to select patients for targeted optimal prevention. Osteoporos Int 2000; 11(suppl 5):S6–S7.
79. Hammerman MR. Insulin-like growth factors and aging. Endocrinol Metab Clin N Am 1987; 16:995–1011.
80. Quesada JM, Coopmans W, Ruiz B, Aljama P, Jans I, Bouillon R. Influence of vitamin D on parathyroid function in the elderly. J Clin Endocrinol Metab 1992; 75:494–501.
81. Goodman-Gruen D, Barrett-Connor E. Epidemiology of insulin-like growth factor-I in elderly men and women. The Rancho Bernardo Study. Am J Epidemiol 1997; 145:970–976.

82. Langlois JA, Rosen CJ, Visser M, et al. Association between insulin-like growth factor I and bone mineral density in older women and men: the Framingham Heart Study. J Clin Endocrinol Metab 1998; 83:4257–4262.
83. Rosen CJ, Donahue LR. Insulin-like growth factors: potential therapeutic options for osteoporosis. Trends Endocrinol Metab 1995; 6:235–241.
84. Canalis E, Agnusdei D. Insulin-like growth factors and their role in osteoporosis. Calcif Tissue Int 1996; 58:133–134.
85. Isley WL, Underwood LE, Clemmons DR. Dietary components that regulate serum somatomedin-C concentrations in humans. J Clin Invest 1983; 71:175–182.
86. Thissen JP, Ketelslegers JM, Underwood LE. Nutritional regulation of the insulin-like growth factors. Endocrine Rev 1994; 15:80–101.
87. Thissen JP, Triest S, Maes M, Underwood LE, Ketelslegers JM. The decreased plasma concentrations of insulin-like growth factor-I in protein-restricted rats is not due to decreased number of growth hormone receptors on isolated hepatocytes. J Endocrinol 1990; 124:159–165.
88. VandeHaar MJ, Moats-Staats BM, Davelport ML, et al. Reduced serum concentrations of insulin-like growth factor-I (IGF-I) in protein-restricted growing rats are accompanied by reduced IGF-I mRNA levels in liver and skeletal muscle. J Endocrinol 1991; 130:305–312.
89. Thissen JP, Davenport ML, Pucilowska J, Miles MV, Underwood LE. Increased serum clearance and degradation of (125I)-labeled IGF-I in protein-restricted rats. Am J Physiol 1992; 262:E406–E411.
90. Pucilowska JB, Davenport ML, Kabir I, et al. The effect of dietary protein supplementation on insulin-like growth factors (IGFs) and IGF-binding proteins in children with shigellosis. J Clin Endocrinol Metab 1993; 77:1516–1521.
91. Hill KK, Hill DB, McClain MP, Humphries LL, McClain CJ. Serum insulin-like growth factor-I concentrations in the recovery of patients with anorexia nervosa. J Am Coll Nutr 1993; 4:475–478.
92. Sullivan DH, Carter WJ. Insulin-like growth factor I as an indicator of protein-energy undernutrition among metabolically stable hospitalized elderly. J Am Coll Nutr 1994; 13:184–191.
93. Clemmons DR, Seek MM, Underwood LE. Supplemental essential amino acids augment the somatomedin-C/insulin-like growth factor-I response to refeeding after fasting. Metabolism 1985; 34:391–395.
94. Clemmons DR, Underwood LE, Dickerson RN, et al. Use of plasma somatomedin-C/insulin-like growth factor-I measurements to monitor the response to nutritional repletion in malnourished patients. Am J Clin Nutr 1985; 41:191–198.
95. Musey VC, Goldstein S, Farmer PK, Moore PB, Phillips LS. Differential regulation of IGF-I and IGF-binding protein-I by dietary composition in humans. Am J Med Sci 1993; 305:131–138.
96. Thissen JP, Triest S, Moats-Staats BM, et al. Evidence that pretranslational and translational defects decrease serum insulin-like growth factor-I concentrations during dietary protein restriction. Endocrinology 1991; 129:429–435.
97. Bourrin S, Ammann P, Bonjour JP, Rizzoli R. Dietary protein restriction lowers plasma insulin-like growth factor I (IGF-1), impairs cortical bone formation, and induces osteoblastic resistance to IGF-1 in adult female rats. Endocrinology 2000; 141:3149–3155.
98. Ammann P, Rizzoli R, Müller K, Slosman D, Bonjour JP. IGF-1 and pamidronate increase bone mineral density in ovariectomized adult rats. Am J Physiol 1993; 265:E770–E776.
99. Ammann P, Rizzoli R, Meyer JM, Bonjour JP. Bone density and shape as determinants of bone strength in IGF-1 and/or pamidronate-treated ovariectomized rats. Osteoporos Int 1996; 6:219–227.
100. Ammann P, Rizzoli R, Caverzasio J, Bonjour JP. Fluoride potentiates the osteogenic effects of IGF-1 in aged ovariectomized rats. Bone 1998; 22:39–43.

101. Ammann P, Rizzoli R, Slosman D, Bonjour JP. Sequential and precise in vivo measurement of bone mineral density in rats using dual-energy X-ray absorptiometry. J Bone Miner Res 1992; 7:311–316.
102. Bourrin S, Toromanoff A, Ammann P, Bonjour JP, Rizzoli R. Dietary protein deficiency induces osteoporosis in aged male rats. J Bone Miner Res 2000; 15:1555–1563.
102. Chan J, Tian Y, Tanaka KE, et al. Effect of protein calorie malnutrition on tuberculosis in mice. Proc Natl Acad Sci USA 1996; 93:14857–14861.
103. Dai G, McMurray DN. Altered cytokine production and impaired antimycobacterial immunity in protein-malnourished guinea pigs. Infect Immun 1998; 66:3562–3568.
104. Spaulding CC, Walford RL, Effros RB. Calorie restriction inhibits the age-related dysregulation of the cytokines TNF-alpha and IL-6 in C3B10RF1 mice. Mech Ageing Dev 1997; 93:87–94.
105. Anker SD, Coats AJ. Cardiac cachexia: a syndrome with impaired survival and immune and neuroendocrine activation. Chest 1999; 115:836–847.
106. Anker SD, Clark AL, Teixeira MM, Hellewell PG, Coast AJ. Loss of bone mineral in patients with cachexia due to chronic heart failure. Am J Cardiol 1999; 83:612–615, A10.
107. Ammann P, Rizzoli R, Bonjour JP, et al. Transgenic mice expressing soluble tumor necrosis factor-receptor are protected against bone loss caused by estrogen deficiency. J Clin Invest 1997; 99:1699–1703.
108. Hotamisligil GS. Mechanisms of TNF-alpha-induced insulin resistance. Exp Clin Endocrinol Diabetes 1999; 107:119–125.
109. Grimble RF, Jackson AA, Persaud C, Wride MJ, Delers F, Engler R. Cysteine and glycine supplementation modulate the metabolic response to tumor necrosis factor alpha in rats fed a low protein diet. J Nutr 1992; 122:2066–2073.
110. Manary MJ, Brewster DR, Broadhead RL, et al. Whole-body protein kinetics in children with kwashiorkor and infection: a comparison of egg white and milk as dietary sources of protein. Am J Clin Nutr 1997; 66:643–648.
111. Grimble RF. Nutritional modulation of cytokine biology. Nutrition 1998; 14:634–640.
112. Jilka RL. Cytokines, bone remodeling, and estrogen deficiency: a 1998 update. Bone 1998; 23:75–81.

18 Acid–Base Balance and Bone Health

David A. Bushinsky

1. DIETARY ACID INTAKE

On a daily basis, humans eat substances that during metabolism generate or consume protons *(1–3)*. Acid may be generated by the following reactions:

$$\text{methionine or cystene} \rightarrow \text{glucose} + \text{urea} + SO_4^{2-} + 2\,H^+$$
$$\text{arginine}^+ \rightarrow \text{glucose} + \text{urea} + H^+$$
$$R\text{-}H_2PO_4 + H_20 \rightarrow ROH + 0.8\,HPO_4^{2-}/0.2\,H_2PO_4 + 1.8\,H^+$$

Acid may be consumed, which is equivalent to base being generated, by the following reactions:

$$\text{glutamate}^- + H^+ \rightarrow \text{glucose} + \text{urea}$$
$$\text{lactate}^- + H^+ \rightarrow \text{glucose} + CO_2$$
$$\text{citrate}^- + 4.5\,O_2 \rightarrow 5CO_2 + 3H_2O + HCO_3^-$$

The net result is that when consuming a typical Western diet adult humans generate approx 1 mEq of acid/kg/d *(4,5)*. It has been demonstrated that the more acid precursors are contained in the diet, the greater is the degree of systemic acidity *(4)*. The maintenance of a stable physiological systemic pH is of critical importance to the survival of mammals *(5–7)*. Although only net loss of hydrogen ions can ultimately correct acidosis *(5)*, bone appears to be instrumental in the maintenance of a stable physiological systemic pH during metabolic acidosis; however, this homeostatic function is often at the expense of its mineral content *(1–3,8–11)*.

2. IN VIVO EFFECTS OF METABOLIC ACIDOSIS ON BONE

During in vivo, acute metabolic acidosis (a primary decrease in bicarbonate concentration [HCO_3]), approx 60% of the administered protons (hydrogen ions) are buffered outside of the extracellular fluid *(12)* by soft tissues *(13–15)* and by bone *(6,7,16–24)*. The in vivo evidence that bone acutely buffers protons, and in the process releases calcium, derives principally from the loss of bone sodium and/or potassium *(25–31)*, carbonate *(19,23,29,32)*, and the increase in serum calcium *(33)*

From: *Nutrition and Bone Health*
Edited by: M. F. Holick and B. Dawson-Hughes © Humana Press Inc., Totowa, NJ

observed during acidosis. Bone sodium (or potassium) loss implies proton for sodium (or potassium) exchange and carbonate loss suggests consumption of this buffer by the administered protons. As at least 98% of body calcium is contained within bone *(34,35)*, the increase in serum calcium is likely to derive from mineral stores.

Chronic metabolic acidosis increases urinary calcium excretion *(36–38)* without an increase in intestinal calcium absorption *(39,40)*, resulting in negative calcium balance *(9,10)*, which appears to reflect proton-mediated dissolution of bone mineral *(5,7,8,16,36,41)*. Indeed, chronic metabolic acidosis appears to decrease mineral content in most in vivo studies *(8–11,42)*.

The independent effect of acidosis to suppress bone formation was elegantly demonstrated in children with renal tubular acidosis *(43–45)*. Children with renal tubular acidosis have stunted linear growth. The provision of base to these children has been demonstrated to increase their growth dramatically. There is ample clinical evidence that acidosis adversely affects bone during renal failure *(46–49)*, which may be corrected by bicarbonate administration *(50–52)*. Bone carbonate is decreased in acidic, uremic patients *(53–55)*. This decrease may represent dissolution of bone carbonate stores or replacement by phosphate resulting in the incorporation of H^+ into the mineral *(19,23,32)*. In a radiographic study, the majority of patients with proximal renal tubular acidosis had rickets or osteopenia *(56)*.

Adults with distal renal tubular acidosis and normal renal function were recently shown to have a lower bone mineral density (BMD) than normal controls *(57)*. Bone histomorphometry demonstrated that these patients with renal tubular acidosis had significantly decreased bone formation rate and an increased osteoid surface and osteoid volume when compared to normal controls. These patients were then treated with potassium citrate, which is metabolized to bicarbonate, to correct the acidosis *(58)*. The treatment resulted in a significant increase in BMD at the trochanter of the femur and in the total femur. There was an increase in bone formation rate. Interestingly, the level of parathyroid hormone (PTH) rose significantly in the treated patients.

Chronic acid ingestion, in the form of the common North American high-protein diet, coupled with the known effects of acid on bone, have led to the suggestion that this acid production may play a role in the etiology of osteoporosis *(8,59–61)*. As we age there is a decrease in overall renal function, including the ability to excrete acid *(62)*. With increasing age, humans become slightly, but significantly, more acidemic leading to bone resorption *(62,63)*. Supporting this hypothesis is the observation that administration of base appears to decrease the negative calcium balance induced by a high-protein diet *(64–66)*.

3. IN VITRO OBSERVATIONS

Although in vivo evidence strongly suggests that bone is involved in the systemic response to acid–base disorders, until fairly recently there was little direct in vitro confirmation *(6)*. Neuman et al. found that a reduction of medium pH produced a marked increase in hydroxyapatite solubility *(67)*. Dominguez and Raisz *(68)* determined that an acid medium induced movement of prelabeled calcium from bone.

We undertook a series of studies to test the hypotheses that cultured bone exposed to a physiologically acidic medium would release calcium into the medium and buffer the increased medium hydrogen ion concentration *(1–3,5–7,19–26,30,31,42,69–91)*. We utilized the model of cultured neonatal mouse calvariae, as the calvariae (frontal and parietal bones of the skull) have functioning osteoclasts and osteoblasts *(92,93)*, respond to hormones, and synthesize DNA and proteins as does bone in vivo *(94)*. Calvariae can be cultured in the physiological carbon dioxide–bicarbonate buffer system *(24)*, in which medium pH can be regulated precisely by independently altering the partial pressure of carbon dioxide or bicarbonate concentration, simulating either "respiratory" or "metabolic" (respectively) acid–base disorders *(1–3,5,24)*.

4. ACUTE ACIDOSIS

4.1. Calcium Release

Calvariae cultured in acidic medium exhibit proton-dependent net calcium efflux during both acute (3 h) and more chronic (>24–99 h) incubations *(1–3,5–7,19–26,30,31,42,69–71)*. During acute incubations there was net calcium efflux from the calvariae when medium pH was decreased to less than the physiological normal of 7.40 by decreasing the [HCO_3], no net flux at a neutral physiological pH, and an influx of calcium into bone when pH was greater than 7.40 *(24)* (Fig. 1).

The hypothesis that the mechanism of proton-mediated net calcium efflux from bone during these acute incubations was direct physicochemical (non-cell-mediated) calcium release was next tested. Calvariae were cultured with agents that would stimulate or suppress bone cell activity but not affect the mineral directly *(69)*. It was found that bone cells contributed a constant, pH-independent net calcium flux from the mineral during these acute (3-h) experiments, thus acute proton-mediated calcium release was the result of physicochemical and not cell-mediated mechanisms *(69)* (Fig. 2). To confirm that acidic medium could alter physicochemical forces and promote dissolution of the bone mineral, synthetic carbonated apatite disks were cultured in physiologically acidic medium *(75)*. The carbonated apatite disks are an accurate, cell-free model of bone mineral *(95–101)*. Net calcium flux was observed from cultured carbonated apatite disks in response to a physiological acidosis, similar to that of cultured calvariae, supporting the hypothesis that acidic medium can induce physicochemical calcium release from bone *(75)*.

The type of bone mineral in equilibrium with the medium, and thus altered by the physicochemical forces, might be carbonate or phosphate in association with calcium. To determine which, calvariae were cultured in medium in which the driving forces for crystallization with respect to the solid phase of the bone mineral were altered by changing medium pH *(23)*. With respect to calcium and

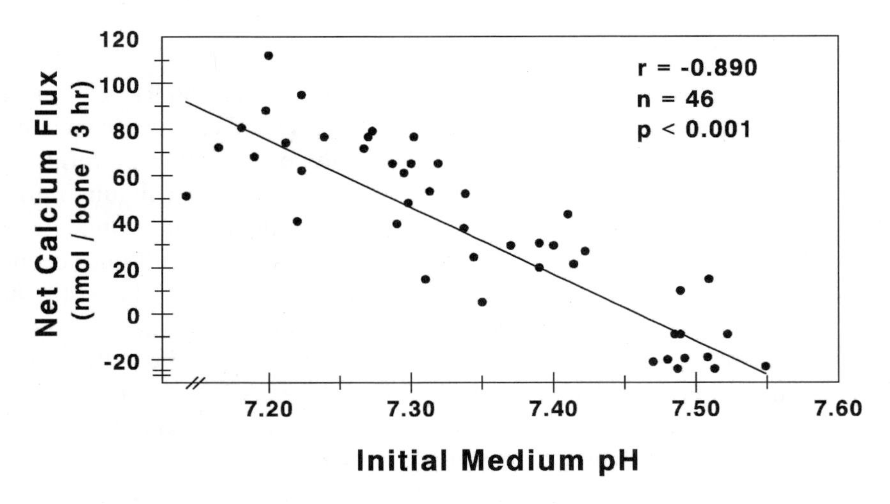

Fig. 1. Effect of initial medium pH on net calcium flux in neonatal mouse calvariae cultured for 3 h. A positive flux indicates net calcium movement from the bone into the medium. Medium pH was adjusted with concentrated HCl or NaOH at a partial pressure of carbon dioxide of 40 mmHg. Calvariae were preincubated in neutral pH medium for 24 h prior to this 3-h incubation. ($r = -0.890$, $n = 46$, $p < 0.001$) (Data from ref. *69*.)

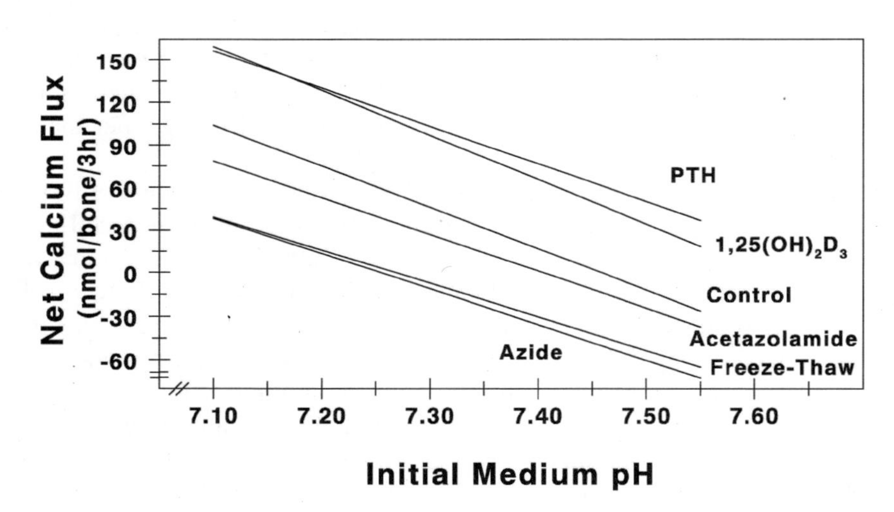

Fig. 2. Comparison of regressions of initial medium pH on net calcium flux for six separate groups of calvariae. Calvariae were incubated for 24 h in similar medium prior to this 3-h incubation. Abbreviations: Control, calvariae incubated in control medium (pH ≈ 7.40); PTH, calvariae incubated in control medium with parathyroid hormone 1×10^{-8} M; $1,25(OH)_2D_3$, calvariae incubated in control medium with 1,25-dihydroxyvitamin D_3 1×10^{-8} M; Acetazolamide, calvariae incubated in control medium with acetazolamide 4×10^{-4} M; Freeze-Thaw, calvariae incubated in control medium after three successive freeze–thaw cycles. Regressions are different due to a difference in intercepts of all groups except PTH and $1,25(OH)_2D_3$, which are similar, and azide and freeze-thaw, which are similar. Slopes are similar in all six groups. (Data from ref. *69*.)

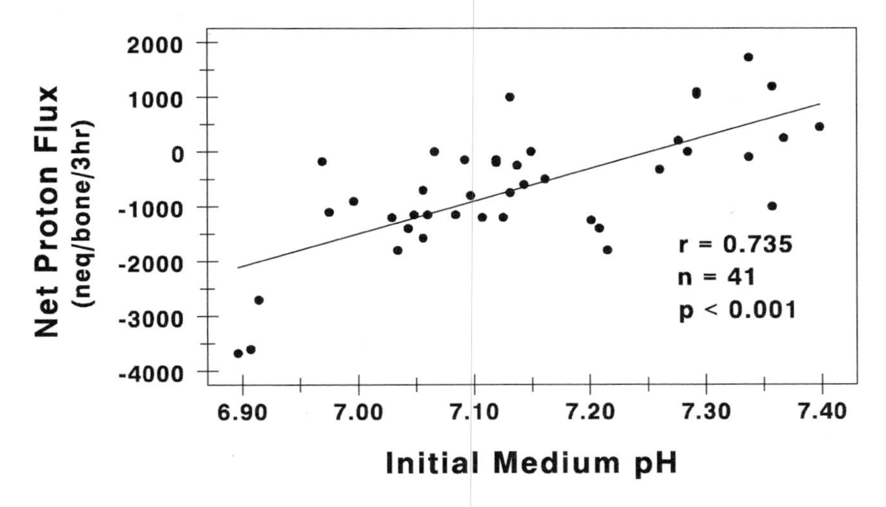

Fig. 3. Effect of initial medium pH on net proton flux in calvariae cultured for 3 h. A positive flux indicates net proton movement from the calvariae into the medium, a negative flux the opposite. pH was adjusted for the 3-h incubations with concentrated HCl or NaOH at a partial pressure of carbon dioxide of 40 mmHg. (Data from ref. *24.*)

carbonate, but not calcium and phosphate, there was bone formation in a supersaturated medium, no change in the bone mineral when cultured in a saturated medium, and bone dissolution into an undersaturated medium. Thus bone carbonate appears to be solubilized during an acute reduction in medium pH leading to a release of calcium. When calvariae are cultured in acidic medium there is a progressive loss of total bone carbonate during a model of metabolic acidosis *(19)*. Further support for the role of carbonate in acid-mediated bone mineral dissolution comes from studies in which it was demonstrated that at a constant pH, whether physiologically neutral or acid, bone net calcium flux is dependent on the medium bicarbonate; the lower the bicarbonate, the greater is the calcium efflux from bone *(74)*. Bone carbonate appears to be in the form of carbonated apatite *(98,102,103)*.

4.2. Hydrogen Ion Buffering

The in vitro evidence for proton buffering by bone is derived from studies of acidosis-induced proton influx into bone *(21–24)* and microprobe evidence for a depletion of bone sodium and potassium during acidosis *(25,26,30,31,71)*. When calvariae are cultured in medium acidified by a decrease in the concentration of bicarbonate (metabolic acidosis), there is a net influx of protons into the bone, decreasing the medium proton concentration and indicating that the additional hydrogen ions are being buffered by bone *(21–24)* (Fig. 3). This leads to an increase in the pH of the culture medium.

4.2.1. Proton for Sodium and/or Potassium Exchange

Bone is a reservoir for sodium and potassium, and its surface has fixed negative sites that normally complex with sodium, potassium, and hydrogen ions; the sodium and potassium appear to exchange freely with the surrounding fluid *(34,35)*. Using a high-resolution scanning-ion microprobe with secondary ion mass spectroscopy, it was found that the surface of the bone is rich in sodium and potassium relative to calcium *(25,26,71,104–106)*. After incubation in acidic medium there is loss of surface sodium and potassium relative to calcium *(25,26,30,31,71)* in conjunction with buffering of the additional protons, suggesting sodium and potassium exchange for hydrogen ions on the bone surface resulting in a decrease in medium acidity *(25,27,28)*. When osteoclastic function is inhibited with calcitonin, microprobe analysis indicates that physicochemical proton buffering by bone causes relatively equal calcium and sodium loss *(26)*. In acidic medium, osteoclastic function is necessary to support the enriched levels of bone potassium *(31)*.

4.2.2. Fall in Bone Carbonate

Bone contains approx 80% of the total carbon dioxide (including CO_3^{2-}, HCO_3^-, and CO_2) in the body *(107)*. Approximately two-thirds of this is in the form of carbonate (CO_3^{2-}) complexed with H^+ (as HCO_3^-), calcium, potassium and sodium and other cations, and is located in the lattice of the bone crystals, where it is relatively inaccessible to the systemic circulation. The other one-third is located in the hydration shell of hydroxyapatite, where it is readily available to the systemic circulation. Acute metabolic acidosis decreases bone total carbon dioxide *(32)*. Acidosis induces the release of calcium and carbonate from bone *(23)*, leading to a progressive loss of bone carbonate during metabolic acidosis *(19)*.

When both the in vitro and in vivo studies are considered together, there is strong evidence that bone is a H^+ buffer capable of maintaining the extracellular fluid pH near the physiological normal. The loss of both bone sodium and carbonate suggests that, in addition to sodium for H^+ exchange, there is a progressive loss of carbonate in response to acidosis.

5. CHRONIC ACIDOSIS

5.1. Calcium Release

Chronic metabolic acidosis induces the release of bone calcium, predominantly by enhanced cell-mediated bone resorption and decreased bone formation *(70,73,76,77,79,80,83,108)*; however, there is a component of direct physicochemical acid-induced dissolution, as in acute metabolic acidosis *(21,23,24,26,69,75)*. In vivo rat studies have shown stimulation of cell-mediated bone calcium resorption during prolonged acidosis *(33,109)*.

There is cell-mediated resorption of bone calcium after 99 h of culture in acidic medium produced by a decrease in medium bicarbonate *(70)* (Fig. 4). Acidosis has

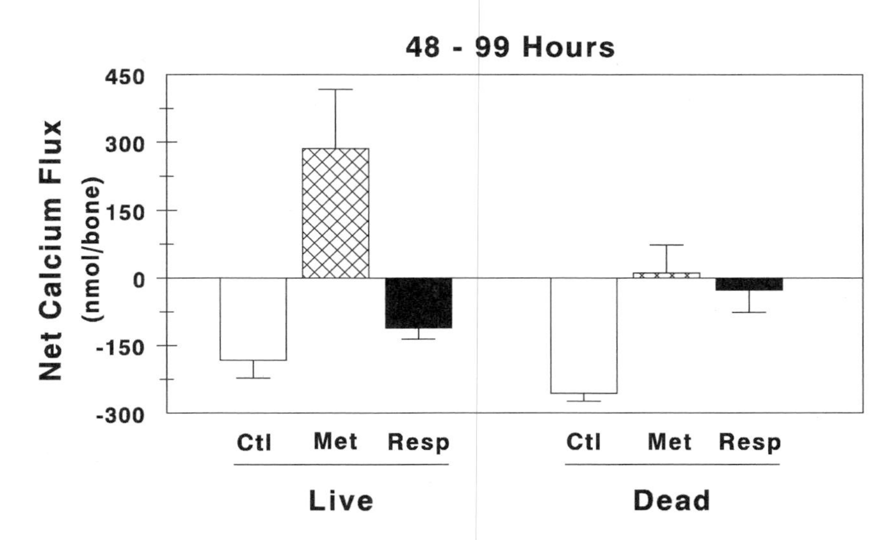

Fig. 4. Net calcium flux during the final 51 h of a 99-h incubation for the six groups of calvariae indicated. Live, calvariae cultured in living state; dead, calvariae subjected to three freeze–thaw cycles before culture; Ctl, calvariae cultured in unaltered medium; Met, medium acidified by lowering the bicarbonate concentration; Resp, medium acidified by increasing the partial pressure of carbon dioxide. Values are mean ± SEM. (Data from ref. *70.*)

also been shown to increase osteoclastic and inhibit osteoblastic activity *(73);* release of the osteoclastic enzyme β-glucuronidase was stimulated (Fig. 5), whereas osteoblastic collagen synthesis (Fig. 6) and alkaline phosphatase were inhibited. Conversely, an increase in [HCO_3], metabolic alkalosis, decreases net calcium efflux from bone through an increase in osteoblastic bone formation and a decrease in osteoclastic bone resorption *(80).* Further evidence that metabolic acidosis inhibits osteoblastic function was obtained utilizing primary osteoblasts in culture. Isolated osteoblasts cultured for 3 wk synthesize collagen and form nodules of apatitic bone *(110–113).* Metabolic acidosis leads not only to fewer nodules, but to decreased calcium influx into the nodules *(76)* (Fig. 7). Thus, it appears that both augmentation of osteoclastic bone resorption and inhibition of osteoblastic bone formation have a prominent role in the hypercalciuria of chronic metabolic acidosis *(73,76).*

During renal failure there is often increased PTH in addition to acidosis *(114,115).* To determine if acidosis and PTH have additive effects on net calcium efflux, mouse calvariae were cultured in acidic medium containing PTH *(79).* Acidosis and PTH were found to independently stimulate net calcium efflux from bone (Fig. 8), inhibit osteoblastic collagen synthesis (Fig. 6), and stimulate osteoclastic β-glucuronidase secretion (Fig. 5), while the combination had a greater effect on each of these parameters than either alone.

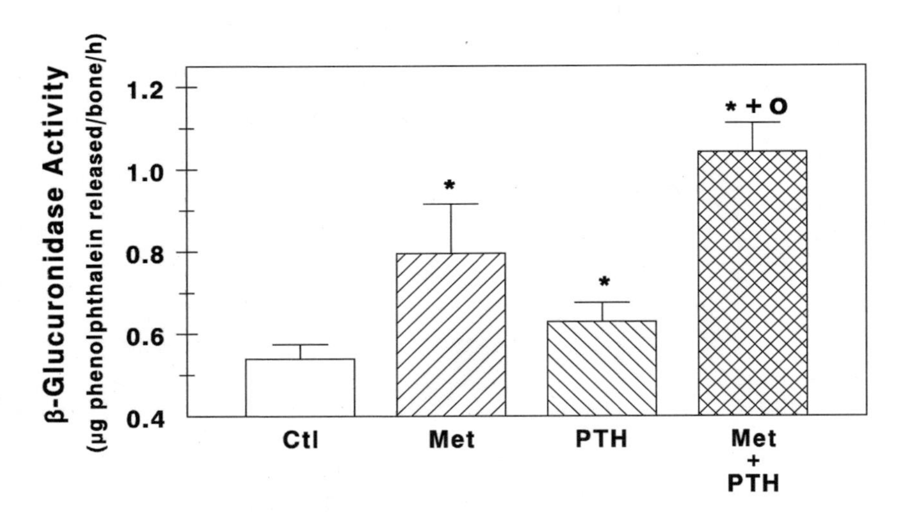

Fig. 5. Effect of acidosis and parathyroid hormone (PTH), alone and in combination, on osteoclastic β-glucuronidase activity. Calvariae were incubated in control medium (Ctl), medium acidified to pH ≈ 7.10 (Met); medium with parathyroid hormone 10^{-10} M final concentration (PTH), or with PTH added to acidic medium (Met+PTH). Calvariae were incubated for 24 h and then transferred to similar fresh medium for an additional 24 h. At the end of the second 24-h incubation, aliquots of medium were removed for assay of β-glucuronidase activity. Values are mean ± SEM. * $p < 0.05$ vs Ctl; +, $p < 0.05$ vs Met; o, $p < 0.05$ vs PTH. (Data from ref. *79.*)

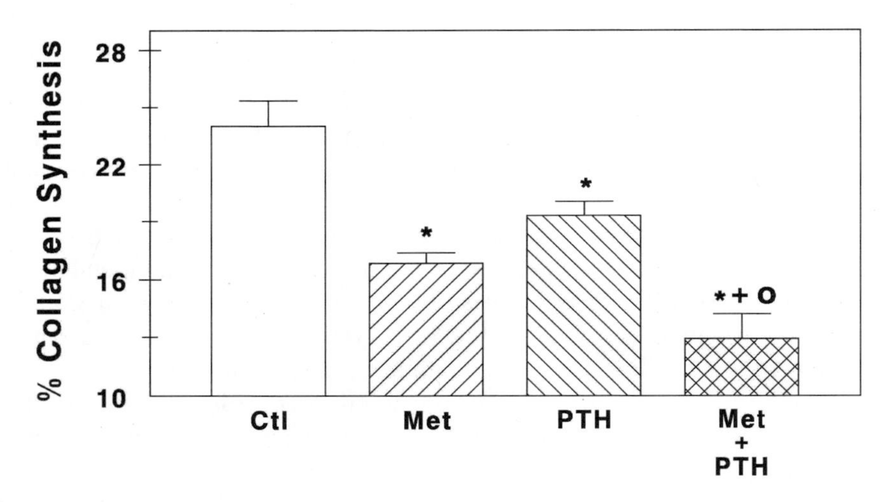

Fig. 6. Effect of acidosis and parathyroid hormone (PTH), alone and in combination, on osteoblastic collagen synthesis. Calvariae were incubated in control medium (Ctl), medium acidified to pH ≈ 7.10 (Met); medium with PTH 10^{-10} M final concentration (PTH), or with PTH added to acidic medium (Met+PTH). Calvariae were incubated for 24 h and then transferred to similar fresh medium for an additional 24 h. Incorporation of [^{3}H]proline into collagenase-digestible protein in calvariae was measured during the final 3 h of the second 24-h incubation. Values are mean ± SEM. *, $p < 0.05$ vs Ctl; +, $p < 0.05$ vs Met; o, $p < 0.05$ vs PTH. (Data from ref. *79.*)

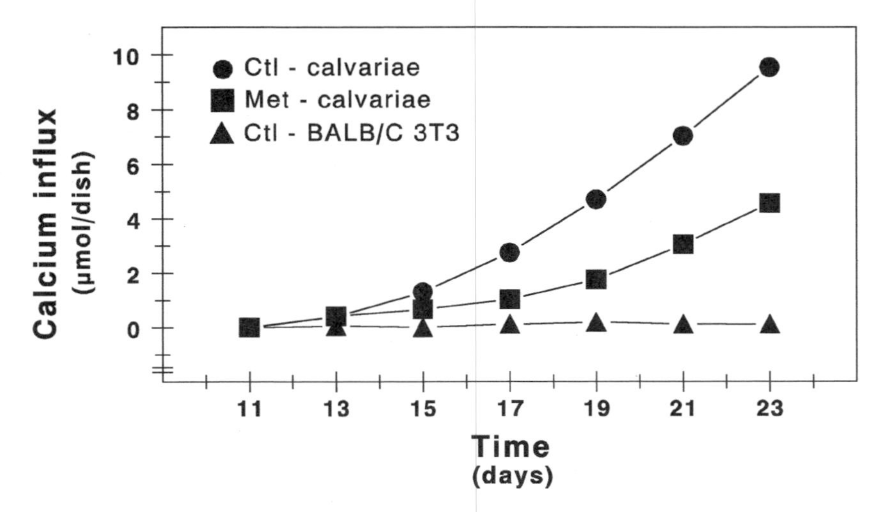

Fig. 7. Cumulative calcium influx as a function of incubation time for cultured neonatal mouse calvarial cells. Cells, which are predominantly osteoblasts, were incubated in control medium until confluent (d 9) and then cultured for an additional 14 d in control medium (Ctl, calvariae), medium acidified by decreasing the medium bicarbonate concentration (Met, calvariae). Balb/C 3T3 mouse fibroblasts were also incubated in control medium (Ctl, BALB/C 3T3). Values are mean ± SE. Changes in medium calcium concentration were calculated by subtracting the final from the initial calcium concentration and correcting for volume. Results are summed over the 14-d incubation period and represent calcium influx into the cultured cells. (Data from ref. *76.*)

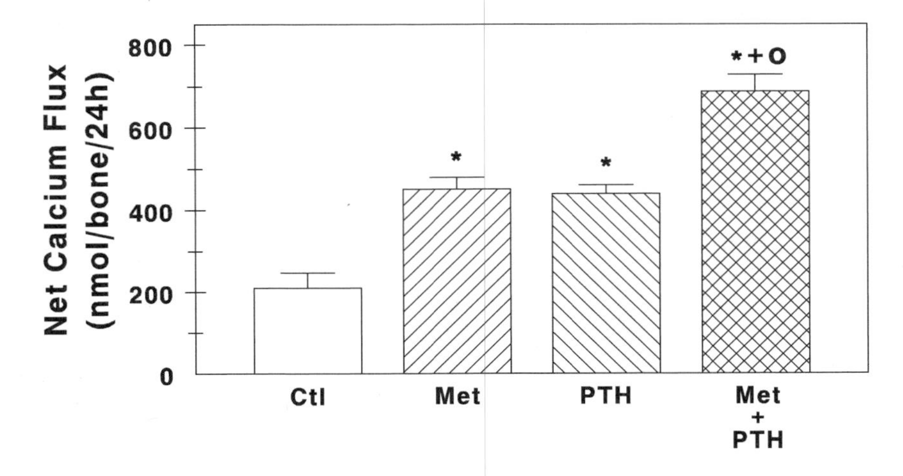

Fig. 8. Effect of acidosis and parathyroid hormone (PTH), alone and in combination, on net calcium efflux from cultured neonatal mouse calvariae. Calvariae were incubated in control medium (Ctl), medium acidified to pH ≈ 7.10 (Met); medium with parathyroid hormone 10^{-10} M final concentration (PTH), or with PTH added to acidic medium (Met+PTH). Calvariae were incubated for 24 h and then transferred to similar fresh medium for an additional 24 h. At the end of the second 24-h incubation, aliquots of medium were removed for assay of net calcium flux. Values are mean ± SEM. *, $p < 0.05$ vs Ctl; +, $p < 0.05$ vs Met; o, $p < 0.05$ vs PTH. (Data from ref. *79.*)

6. ACIDOSIS-INDUCED ALTERATIONS IN GENE ACTIVITY

Based on the H^+-induced increase in osteoclastic bone resorption and decrease in osteoblastic bone formation *(73,77)*, it was hypothesized that acidosis affects the pattern of gene expression in osteoblasts. As a model system, primary neonatal mouse calvarial cells were used, which are principally osteoblasts *(108)*. To assay acute effects of acidosis on gene expression, cells were cultured in physiologically neutral pH medium until confluent and then stimulated with fresh medium at either neutral or acidic pH. RNA was harvested at various times after stimulation. Among a group of immediate early response genes, including *Egr-1, junB, c-jun, junD,* and *c-fos,* only the magnitude of *Egr-1* stimulation was altered by medium pH. A progressive decrease in pH to 6.8 led to a parallel decrease in *Egr-1* stimulation, and an increase in pH to 7.6 led to an increase in *Egr-1* stimulation *(108)*. Osteoblasts express type 1 collagen as the major component of the bone extracellular matrix, which subsequently becomes mineralized. Type I collagen RNA was stimulated approximately three- to fivefold, 40 min after medium change; the stimulation was again decreased by acidosis and increased by alkalosis *(108)*.

Cultured primary mouse calvarial cells differentiate and form sites of mineralization, known as bone nodules *(76,110–113)*. During this process, osteoblasts express a number of matrix proteins distinct to bone, including bone sialoprotein, osteocalcin, osteonectin, osteopontin, and matrix gla protein *(116)*. Metabolic acidosis is known to decrease bone nodule formation and subsequent mineralization *(76)*. It was hypothesized that acidosis would alter the pattern of matrix gene expression in chronic cultures of bone cells, resulting in a matrix that mineralizes less extensively than that from cultures incubated at neutral pH. After 3–4 wk in neutral pH medium there was a dramatic increase in osteopontin RNA (Fig. 9). In contrast, there was no increase in osteopontin RNA in acidic cultures. Osteopontin contains RGD domains and serves as an anchoring protein for macrophages and osteoclasts; osteopontin may also be a chemoattractant for these cell types *(117,118)*. Downregulation of osteopontin expression may serve to limit recruitment of bone-resorbing cells during acidosis, perhaps a cause of low-turnover renal osteodystrophy *(7)*. RNA for matrix Gla protein is also induced by neutral differentiation medium, reaching levels 20- to 30-fold greater than those before differentiation. Again, acidosis almost totally prevents the increase in matrix Gla protein RNA levels (Fig. 10). Although matrix Gla protein expression is not limited to bone, matrix Gla protein comprises about 10% of the carboxyglutamic acid found in bone *(119)*; the Gla residue coordinates with calcium and may serve to direct calcification *(119)*. The levels of RNA for the housekeeping gene GAPDH did not vary with pH, nor did the level of a second bone-associated RNA species, osteonectin, or transforming growth factor (TGF) β_1, indicating that there is not overall cellular toxicity.

To determine if acidosis reversibly impairs cellular production of osteopontin and matrix Gla protein, cultures of primary calvarial bone cells were put in acidic differentiation medium at d 8, then switched to neutral medium at either d 15, 22, or 29 *(83)*. One week of exposure to acidic medium had no lasting effect on osteopontin

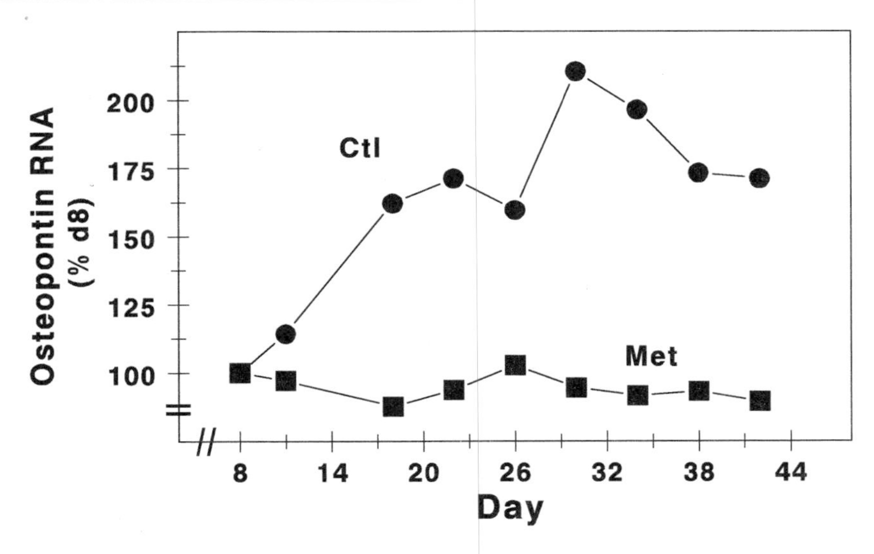

Fig. 9. Time course of osteopontin gene expression. Cells were grown for 8 d in control medium (pH = 7.5), prior to incubation in neutral (pH = 7.5, Ctl) or acidic (pH = 7.1, Met) differentiation medium, and were harvested for RNA at indicated times. After Northern blotting and hybridization with an osteopontin probe, the filter was quantitated using a Molecular Dynamics PhosphorImager. To correct for variations in loading, the filter was stripped and rehybridized with GAPDH. Values expressed on *day 8 (d8)* are the ratio of osteopontin RNA to GAPDH RNA. All values on subsequent days are expressed as the ratio of osteopontin RNA to GAPDH RNA on that day divided by the ratio on *day 8* (predifferentiation). Plot is of a single, representative experiment. (Data from ref. *83.*)

and matrix Gla protein expression, while a 2-wk exposure had a small inhibitory effect. There was partial recovery of RNA for osteopontin and matrix Gla protein after 3 wk of acidosis. In the same samples, osteonectin and GAPDH RNA expression were not affected.

7. EFFECTS OF ACIDOSIS ON PROSTAGLANDIN E_2 AND RANK LIGAND

Prostaglandins, especially prostaglandin E_2 (PGE_2), mediate bone resorption induced by a variety of hormones. To test the hypothesis that acid-induced bone resorption is mediated by prostaglandins, neonatal mouse calvariae were cultured in neutral or physiologically acidic medium with, or without, indomethacin to inhibit prostaglandin synthesis *(87–89)*. Net calcium efflux and medium PGE_2 levels were determined. Compared to neutral pH medium, acid medium led to an increase in net calcium flux and PGE_2 levels after both 48 and 51 h, a time at which acid-induced net calcium flux is predominantly cell-mediated, and indomethacin inhibited the acid-induced increase in both net calcium flux and PGE_2 (Fig. 11). Net calcium flux was correlated directly with medium PGE_2.

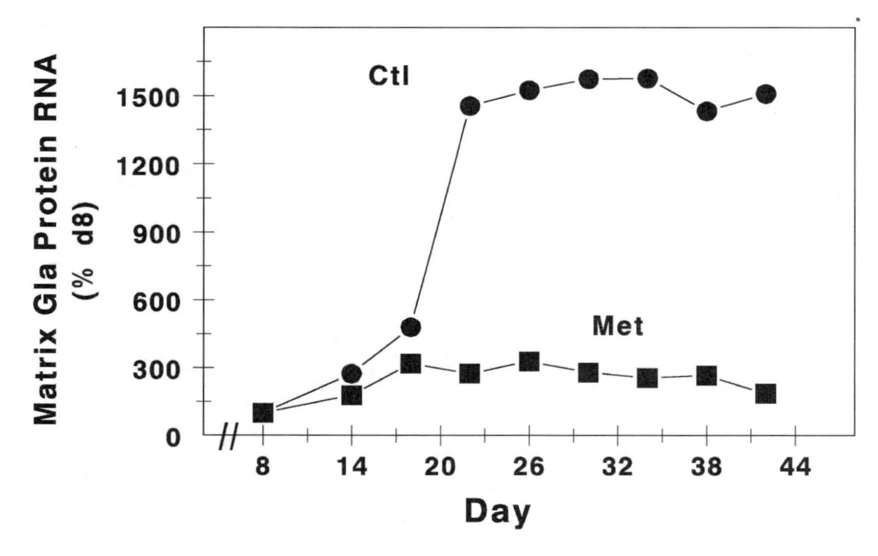

Fig. 10. Time course of Matrix Gla Protein RNA expression. Cells were grown for 8 d in control medium (pH = 7.5), prior to incubation in neutral (pH = 7.5, Ctl) or acidic (pH = 7.1, Met) differentiation medium, and were harvested for RNA at indicated times. After Northern blotting and hybridization with a Matrix Gla Protein probe, the filter was quantitated. To correct for variations in loading, the filter was stripped and rehybridized with GAPDH. Values expressed on *day 8 (d8)* are the ratio of Matrix Gla Protein RNA to GAPDH RNA. All values on subsequent days are expressed as the ratio of Matrix Gla Protein RNA to GAPDH RNA on that day divided by the ratio on *day 8* (predifferentiation). Plot is of a single, representative experiment. (Data from ref. *83*.)

Exogenous PGE_2, at a level similar to that found after acid incubation, induced net calcium efflux in bones cultured in neutral medium. Acid medium also stimulated an increase in PGE_2 levels in isolated bone cells (principally osteoblasts), which was again inhibited by indomethacin. Thus, acid-induced stimulation of cell-mediated bone resorption appears to be mediated by endogenous osteoblastic PGE_2 synthesis.

Growth and maturation of osteoclasts are dependent on the interaction of the osteoclastic cell-surface receptor RANK with a ligand expressed on the surface of osteoblasts, RANKL *(91)*. The RANK/RANKL interaction initiates a differentiation cascade that leads to mature, bone-resorbing osteoclasts and increases their resorptive capacity and survival. To test the hypothesis that metabolic acidosis increases expression of RANKL, neonatal mouse calvariae were cultured in acidic or neutral medium, and the relative expression of RANKL RNA by were determined by reverse transcriptase polymerase chain reaction (RT-PCR) and quantitated by Northern analysis. Metabolic acidosis significantly increased the expression of RANKL RNA at 24 and 48 h compared to respective controls (Fig. 12). Bone calcium efflux was increased into acidic, compared to control,

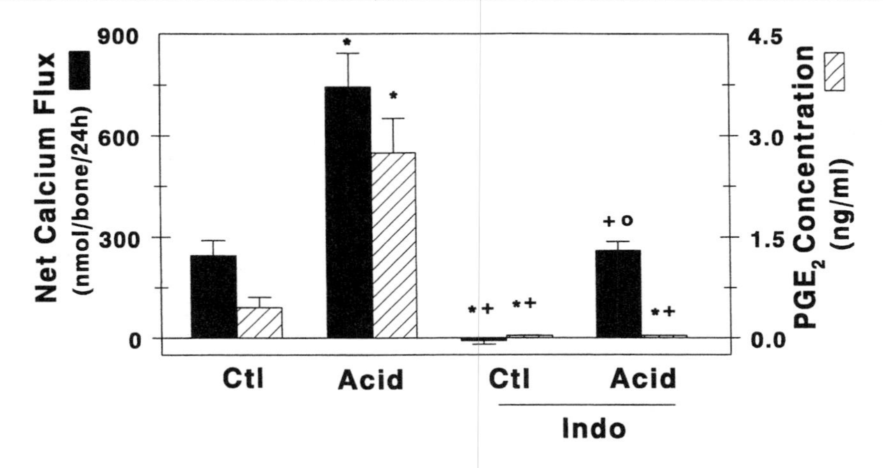

Fig. 11. Acid-induced net calcium efflux and change in medium prostaglandin E_2 levels in mouse calvariae: 24–48 h time period. Calvariae were incubated in neutral (Ctl, pH = 7.5) or acidic (Acid, pH = 7.0) medium for 48 h in the absence or presence of 0.56 μM indomethacin (Indo), with a medium change at 24 h. Net calcium efflux and medium PGE_2 concentration are shown for the 24- to 48-h time period. Results are the mean ± SE for 12 pairs of bones in each group. *, $p < 0.001$ vs Ctl; +, $p < 0.001$ vs acid; o, $p < 0.001$ vs Ctl + Indo.

medium. At 48 h, calcium efflux was correlated directly with RANKL expression. Indomethacin prevented the acid-induced increase in RANKL RNA, indicating a pivotal role for prostaglandins in the induction of RANKL by metabolic acidosis. The acidosis-induced increase in osteoblastic RANKL would augment osteoclastic bone resorption and help explain the acidosis-induced coupling of osteoblastic and osteoclastic activity (Fig. 13) and the increase in bone calcium efflux.

8. ACIDOSIS-INDUCED CHANGES IN BONE ION COMPOSITION

A high-resolution scanning-ion microprobe with secondary ion mass spectroscopy was utilized to determine how $[H^+]$ alters the ion composition of the bone mineral *(25,26,30,31,71,90,104–106,120,121)*. Studies to date have shown that the calvarial surface is rich in sodium and potassium relative to calcium *(25,26,30,31,42,71,90,104–106,120)*. The excess bone potassium is maintained through cell-mediated processes *(105)*. Loss of bone cell function produces an influx of calcium and marked release of bone potassium; there is a fall in the ratio of potassium/calcium, and to a lesser extent sodium/calcium, at the superficial surface of the mineral *(105)*. Metabolic acidosis causes release of mineral calcium and leads to a reduction in the surface ratio of sodium/calcium and potassium/calcium, indicating a greater relative release of mineral sodium and potassium

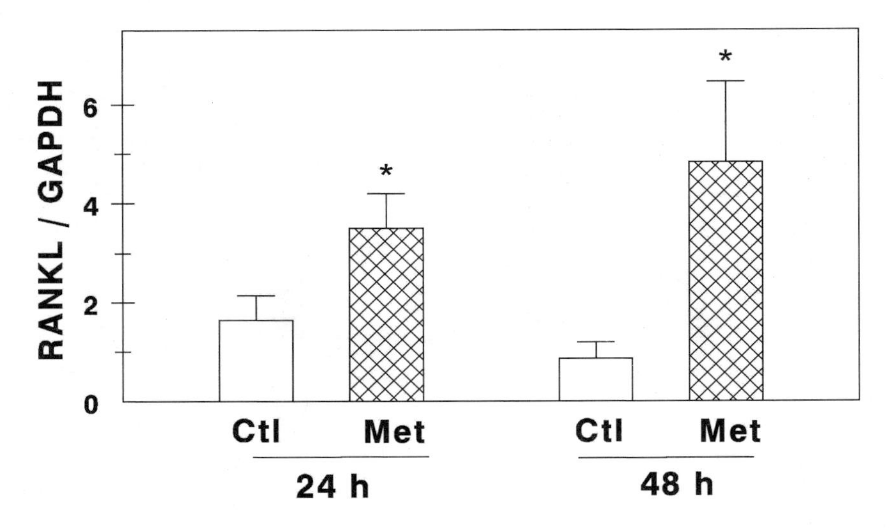

Fig. 12. Effect of metabolic acidosis on RANKL expression. Neonatal mouse calvariae were incubated in neutral medium (Ctl) or in medium acidified by a primary reduction of the bicarbonate concentration, to model metabolic acidosis (Met), either for 24 h or for 48 h. In the 48 h experiments, after the initial 24 h, calvariae were moved to similar fresh preincubated medium for the final 24 h. Calvariae from each culture dish were pooled for isolation of total RNA using a Qiagen RNeasy kit. First-strand cDNA was synthesized from total RNA. Aliquots were amplified using gene-specific primers and then were electrophoresed on agarose. Filters were autoradiographed and quantitated using a densitometer and then were stripped and reprobed for GAPDH. To control for differences in RNA loading, each signal for RANKL was normalized to its respective GAPDH RNA. Data are mean ± SE, $n = 7–11$ pairs of calvariae in each group. Compared to CTL, with MET there was a significant increase in the ratio of RANKL to GAPDH at 24 h and at 48 h. Data are mean ± SE. Abbreviations: RANKL, receptor activator of NFκB ligand; *, different from CTL, $p < 0.05$. (Data from ref. *91.*)

than calcium *(25)*. However, the mineral and medium are in equilibrium *(23)* and there is movement of ions between the two *(67)*, making it difficult to interpret the apparent ion fluxes, especially with respect to potassium and sodium. To help understand better the effects of acidosis on potassium relative to calcium, bone mineral was labeled in vivo with the stable isotope ^{41}K to determine the response of the bone mineral to acidosis. The mineral was found to be rich in potassium relative to calcium; acidosis caused a fall in the ratio of ^{41}K to calcium, indicating loss of this stable isotope from the bone mineral *(31)*.

Because mineral in live bone is rich in potassium relative to calcium, it was unclear if the osteoclasts selectively removed potassium or if they nonselectively removed the surface of the bone mineral. Neonatal mouse bone cells were isolated and cultured on bovine cortical bone slices in the presence of PTH *(30)*. The ion microprobe was then utilized to compare the unresorbed bone to that at the base of the osteoclastic resorption pits. In the presence of parathyroid hor-

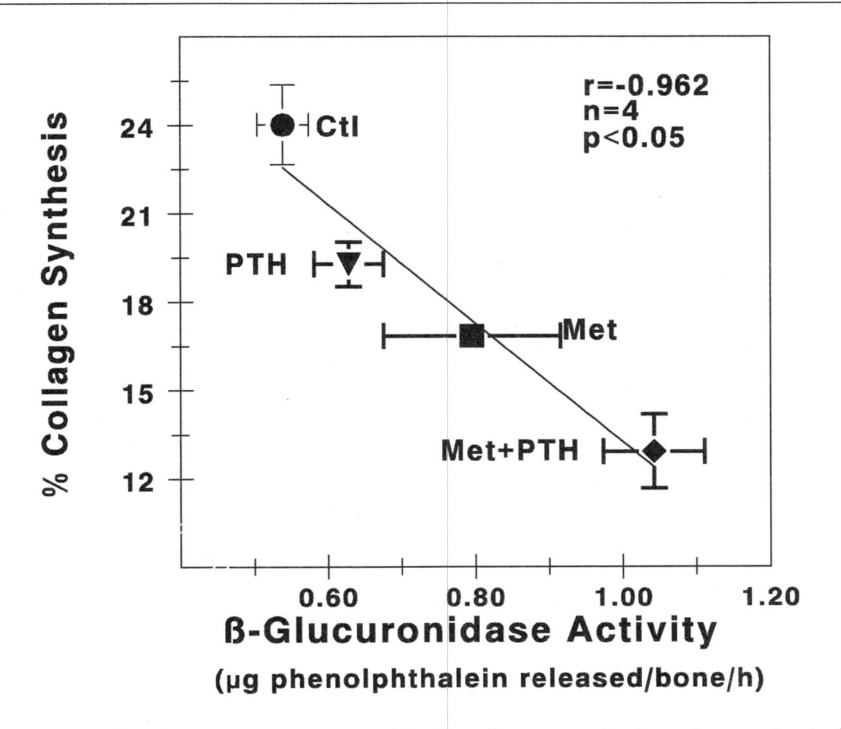

Fig. 13. Correlation between percent osteoblastic collagen synthesis and osteoclastic β-glucuronidase activity. Calvariae were incubated in Ctl, Met, PTH, 10^{-10} M final concentration, or with Met + PTH. Calvariae were incubated for 24 h and then transferred to similar fresh medium for an additional 24-h incubation. Incorporation of [^{3}H] proline into collagenase-digestible protein in calvariae was measured during final 3 h of second 24-h incubation. At end of second 24-h incubation, aliquots of medium were removed for assay of β-glucuronidase activity. (Data from ref. *79*.)

mone the osteoclasts nonselectively removed the potassium-rich surface of the bone mineral *(30)*.

The microprobe was also used to study acute physiochemical bone mineral dissolution caused by acidosis *(26)*. When calvariae were cultured with the osteoclastic inhibitor calcitonin there was a fall in the ratio of sodium/calcium coupled to an influx of calcium into bone, indicating little change in bone sodium. When calvariae were cultured in acidic medium with calcitonin there was calcium release with no change in sodium/calcium, indicating that physicochemical bone mineral dissolution causes relatively equal calcium and sodium release *(26)*.

All of the previous work using the ion microprobe to study the effects of acidosis on bone used bone cultured in vitro *(42)*. To understand better the effects of acid on bone, an in vivo model was established. The microprobe was used to determine the mass spectra of important ion groups from femurs of mice acidified with oral ammonium chloride compared to mice drinking only distilled water. An area in the

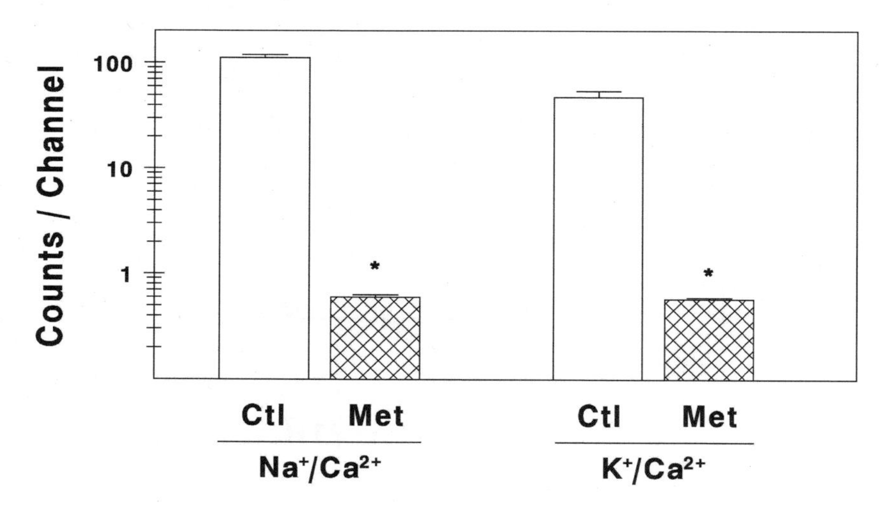

Fig. 14. Ratio of sodium to calcium (Na^+/Ca^{2+}) and potassium to calcium (K^+/Ca^{2+}) in the mid-cortex of neonatal mouse femurs after drinking only distilled water (Ctl) or water with 1.5% NH_4Cl (Met) for 7 d. Values are expressed as mean plus the upper 95% confidence limit. Compared to Ctl, there was a significant fall in the ratios of Na/Ca and K/Ca after acid treatment. *, $p < 0.05$ vs Ctl. (Data from ref. *42.*)

midcortex (midway between the marrow space and the superficial cortex of the longitudinally split femur) midway down the bone shaft was studied. Compared to mice given only oral distilled water, the addition of NH_4Cl to the drinking water led to a marked change in the positive ion spectrum. In control mouse femurs the peak for potassium and sodium is far higher than that for calcium, indicating that there is more potassium and sodium then calcium in the midcortex of the bone (Fig. 14). However, after oral ammonium chloride, there is a fall in the ratios of potassium and sodium relative to calcium. With respect to the negative ions in the midcortex of the control femurs, there was almost as much phosphate as carbon and as much phosphate as the carbon–nitrogen bond. However, oral ammonium chloride led to a fall in the ratios of phosphate to carbon and phosphate to the carbon–nitrogen bond (Fig. 15). Additionally, there was a marked decrease in the ratio of bicarbonate to carbon and bicarbonate to the carbon–nitrogen bond with acidosis (Fig. 16). Thus, it appears that both bicarbonate and phosphate are used as buffers to mitigate the increase in hydrogen ion concentration during in vivo metabolic acidosis.

9. ROLE OF PCO_2 VS $[HCO_3^-]$

Clinically, a decrease in blood pH may be due to either a reduction in bicarbonate concentration ($[HCO_3^-]$, metabolic acidosis) or to an increase in the partial pressure of carbon dioxide (Pco_2, respiratory acidosis). In mammals, metabolic acidosis induces a far greater increase in urine calcium excretion than respiratory

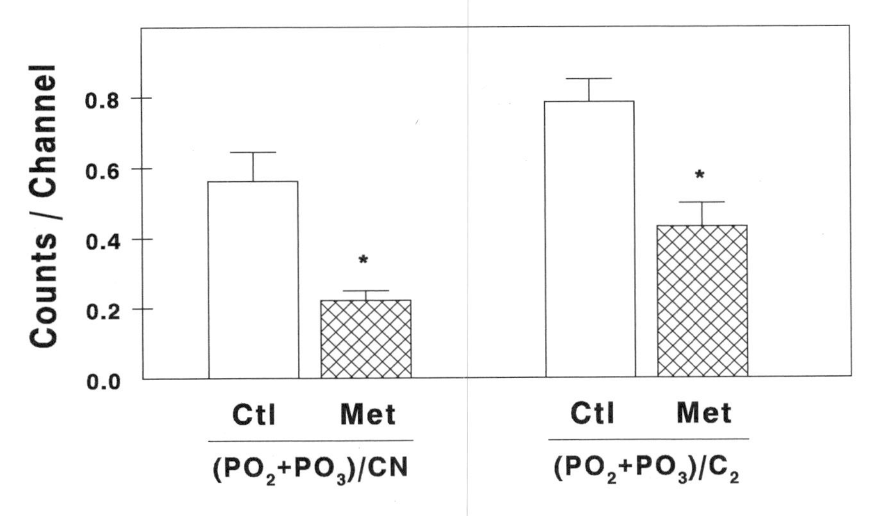

Fig. 15. Ratio of total phosphates (PO_2 + PO_3) to the carbon–nitrogen bond (CN) and total phosphates to carbon (C_2) in the mid-cortex of neonatal mouse femurs after drinking only distilled water (Ctl) or water with 1.5% NH_4Cl (Met) for 7 d. Values are expressed as mean plus the upper 95% confidence limit. Compared to Ctl, there was a significant fall in the ratios of (PO_2 + PO_3)/CN and (PO_2 + PO_3)/C_2 after acid treatment. *, $p < 0.05$ vs Ctl. (Data from ref. *42.*)

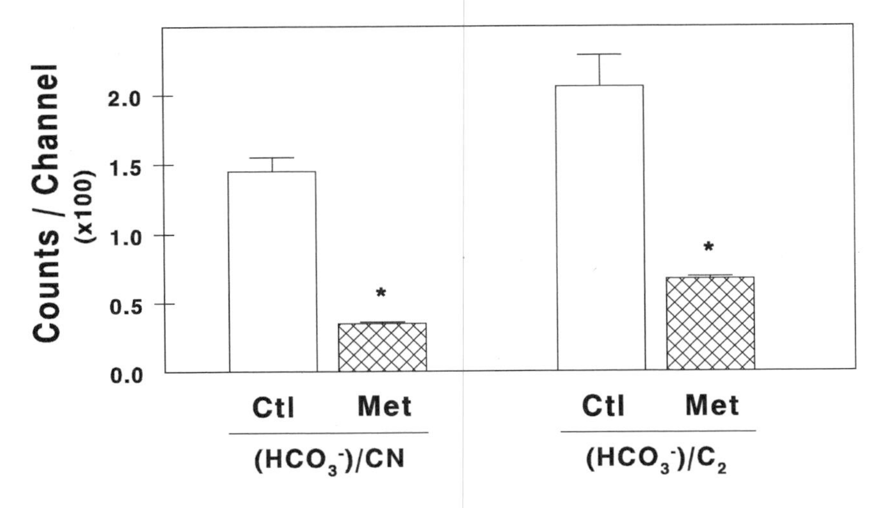

Fig. 16. Ratio of bicarbonate (HCO_3^-) to the carbon–nitrogen bond (CN) and bicarbonate to carbon (C_2) in the mid-cortex of neonatal mouse femurs after drinking only distilled water (Ctl) or water with 1.5% NH_4Cl (Met) for 7 d. Values are expressed as mean plus the upper 95% confidence limit. Compared to Ctl, there was a significant fall in the ratios of HCO_3^-/CN and HCO_3^-/C_2 after acid treatment. *, $p < 0.05$ vs Ctl. (Data from ref. *42.*)

acidosis *(122–125)*, and this increase occurs without an alteration in intestinal calcium absorption, indicating that the additional urinary calcium is derived from the bone mineral *(9–11,38)*.

Most in vivo and in vitro studies have utilized hydrochloric acid or ammonium chloride to decrease bicarbonate as a model of metabolic acidosis. In vitro, the type of acidosis appears to be critical in determining the magnitude of net calcium flux and proton buffering by bone. There are clear distinctions between the effects of metabolic (decreased bicarbonate) and respiratory (increased partial pressure of carbon dioxide) acidosis on cultured bone *(1–3,18,19,21,23,24,70–72,74,76–78)*. In acute studies there was a greater net calcium efflux during culture in decreased-bicarbonate medium than during culture in isohydric acidosis produced by an increase in the partial pressure of carbon dioxide *(21)*. The decreased net calcium efflux during respiratory, compared to metabolic, acidosis is due to decreased unidirectional calcium efflux from the mineral coupled to deposition of medium calcium on the bone surface during hypercapnia *(72)*. There was decreased bone carbonate in response to metabolic, but not respiratory, acidosis *(19)*. These results suggest that over this short time period acidosis affects the physicochemical driving forces for mineral formation and dissolution *(23,26,69,74,75)*. During metabolic acidosis the decreased bicarbonate favors the dissolution, while during respiratory acidosis the increased partial pressure of carbon dioxide and bicarbonate favors the deposition of carbonated apatite. Indeed, there is no net proton influx into bone during respiratory acidosis *(21)*. Extending these studies to compensated metabolic and respiratory acidosis, we found that at a constant pH, whether physiologically neutral or acidic, net calcium efflux from bone is dependent on bicarbonate concentration: the lower the medium bicarbonate, the greater is the calcium efflux from bone *(74)*.

During more chronic incubations there is cell-mediated net calcium efflux from bone during models of metabolic, but not respiratory, acidosis *(70,77)*. A number of studies have shown that metabolic acidosis stimulates osteoclastic resorption *(70,72,73,109,126–128)*. Respiratory acidosis does not alter osteoclastic β-glucuronidase release or osteoblastic collagen synthesis or alkaline phosphatase activity as does metabolic acidosis *(77)*. Medium PGE_2 levels and net calcium efflux from bone were increased with metabolic, but not respiratory, acidosis *(88)*. Respiratory acidosis does not appreciably alter the surface ion composition of bone *(71)*, as does metabolic acidosis *(25,26,30,31,71,90)*.

10. RELATIONSHIP BETWEEN CALCIUM RELEASE AND HYDROGEN ION BUFFERING

During acute metabolic acidosis a reduction in pH causes both bone calcium release and proton buffering by bone. Were all buffering the result of mineral dissolution, there should be a 1:1 ratio of protons buffered to calcium released in the case of calcium carbonate, 5:3 for apatite, and 1:1 for brushite *(129,130)*. However, with cultured calvariae the ratio was found to be 16–21 to 1, indicating that proton

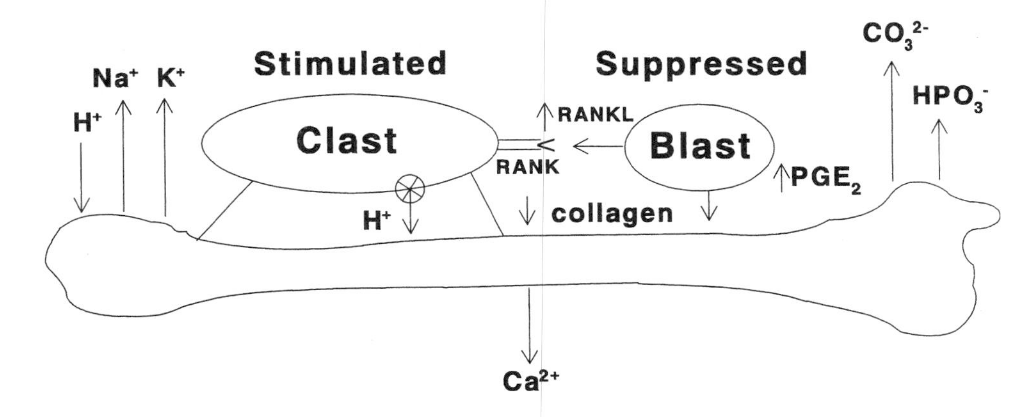

Fig. 17. Schematic of effects of metabolic acidosis on bone. Details as in text. Abbreviations: Clast, osteoclast; Blast, osteoblast. (Modified from refs. *1–3*.)

buffering could not be due simply to mineral dissolution *(24)*. That calcium release is only one component of proton buffering by bone is demonstrated by microprobe studies that show substantial sodium and potassium exchange for protons *(25,26,30,31,71)* and loss of bone phosphate and bicarbonate with acidosis *(90)*.

In vivo studies have shown that metabolic acidosis induces changes in the mid-cortical bone mineral *(42)*, which are consistent with its purported role as a proton buffer. There is a fall in mineral sodium, potassium, carbonate, and phosphate. Each will buffer protons and lead to an increase in systemic pH toward the physiological normal. This apparent protective function of bone will come, in part, at the expense of its mineral stores. Future studies will be necessary to determine if the proton-buffering properties of bone are described by a dynamic equilibrium: protonation of phosphate and carbonate and release of sodium and potassium during acidosis coupled to deprotonation and uptake of sodium and potassium during alkalosis. This attractive hypothetical mechanism has clear survival advantage for mammals *(1–3)*.

11. CONCLUSION

Thus, through a variety of mechanisms, bone appears to decrease the magnitude of fall in serum [HCO_3] and blood pH during metabolic acidosis in conjunction with an increase in net calcium efflux (Fig. 17) *(1–3)*. The acidosis may be mild and secondary to the consumption of food rich in acid precursors. Initially there is physicochemical sodium for hydrogen and potassium for hydrogen exchange on the mineral surface in conjunction with dissolution of carbonate and release of bone calcium. This is followed by stimulation of cell-mediated osteoclastic resorption and inhibition of osteoblastic collagen deposition. The increased resorption releases calcium and the buffers carbonate and phosphate. The fall in bone formation

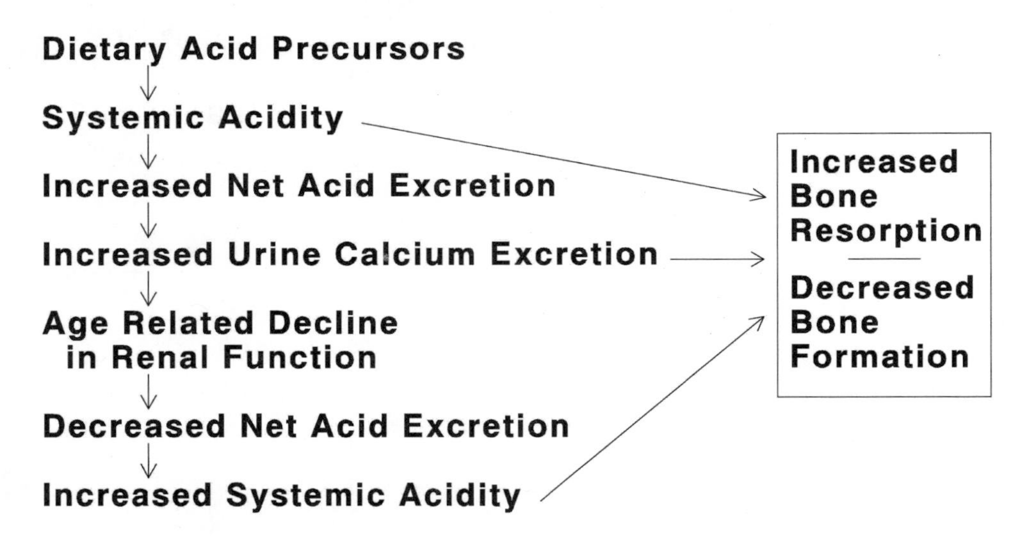

Fig. 18. Hypothesis for the mechanisms leading to dietary acid-induced increased bone resorption and decreased bone formation. Detail as in text. (Modified from ref. *1*.)

prevents calcium uptake and blocks the hydrogen ion release that accompanies bone mineral formation.

With adequate renal function, this mild metabolic acidosis leads to an increase in urine calcium excretion, evidence for bone mineral dissolution and resorption, with buffering of the additional hydrogen ions. However, as we age, renal function slowly deteriorates decreasing the ability of the kidney to excrete the daily acid load (Fig. 18). Exchange of systemic hydrogen ions for bone sodium and potassium and release of carbonate and phosphate all help to normalize the decrease in pH. Osteoclastic bone resorption is further stimulated and osteoblastic bone formation is further suppressed. Bone mineralization continues to decrease, setting the stage for osteoporosis and fracture.

ACKNOWLEDGMENTS

This work was supported in part by grants AR 46289, DK 57716, and DK 56788 from the National Institutes of Health. I thank Nancy S. Krieger, Ph.D., Kevin K. Frick, Ph.D., and Riccardo Levi-Setti, Ph.D., for years of fruitful collaboration.

REFERENCES

1. Bushinsky DA. Acid-base imbalance and the skeleton. In: Burckhardt P, Dawson-Hughes B, Heaney RP, eds. Nutritional Aspects of Osteoporosis. Serono Symposia USA, Norwell, MA, 1998, pp. 208–217.

2. Bushinsky DA. Acid-base imbalance and the skeleton. Eur J Nutrition 2001; 40:238–244.
3. Bushinsky DA, Frick KK. The effects of acid on bone. Curr Opin Nephrol Hypertens 2000; 9:369–379.
4. Kurtz I, Maher T, Hulter HN, Schambelan M, Sebastian A. Effect of diet on plasma acid-base composition in normal humans. Kidney Int 1983; 24:670–680.
5. Bushinsky DA. Metabolic acidosis. In: Jacobson HR, Striker GE, Klahr S, eds. The Principles and Practice of Nephrology. Mosby, St. Louis, MO, 1995, pp. 924–932.
6. Bushinsky DA. Internal exchanges of hydrogen ions: bone. In: Seldin DW, Giebisch G, eds. The Regulation of Acid-Base Balance. Raven, New York, 1989, pp. 69–88.
7. Bushinsky DA. The contribution of acidosis to renal osteodystrophy. Kidney Int 1995; 47:1816–1832.
8. Barzel US. The skeleton as an ion exchange system: implications for the role of acid-base imbalance in the genesis of osteoporosis. J Bone Min Res 1995; 10:1431–1436.
9. Bushinsky DA, Favus MJ, Schneider AB, Sen PK, Sherwood LM, Coe FL. Effects of metabolic acidosis on PTH and $1,25(OH)_2D_3$ response to low calcium diet. Am J Physiol (Renal Fluid Electrolyte Physiol 12) 1982; 243:F570–F575.
10. Lemann J Jr, Litzow JR, Lennon EJ. The effects of chronic acid loads in normal man: further evidence for the participation of bone mineral in the defense against chronic metabolic acidosis. J Clin Invest 1966; 45:1608–1614.
11. Lemann J Jr, Litzow JR, Lennon EJ. Studies of the mechanism by which chronic metabolic acidosis augments urinary calcium excretion in man. J Clin Invest 1967; 46:1318–1328.
12. Swan RC, Pitts RF. Neutralization of infused acid by nephrectomized dogs. J Clin Invest 1955; 34:205–212.
13. Levitt MF, Turner LB, Sweet AY, Pandiri D. The response of bone, connective tissue, and muscle to acute acidosis. J Clin Invest 1956; 35:98–105.
14. Adler S, Roy A, Relman AS. Intracellular acid-base regulation. I. The response of muscle cells to changes in CO_2 tension or extra-cellular bicarbonate concentration. J Clin Invest 1965; 44:8–20.
15. Poole-Wilson PA, Cameron IR. Intracellular pH and K^+ of cardiac and skeletal muscle in acidosis and alkalosis. Am J Physiol 1975; 229:1305–1310.
16. Bushinsky DA, Krieger NS. Regulation of bone formation and dissolution. In: Coe F, Favus M, Pak C, Parks J, Preminger G, eds. Kidney Stones: Medical and Surgical Management. Raven, New York, 1996, pp. 239–258.
17. Bushinsky DA. Acidosis and bone. Miner Electrolyte Metab 1994; 20:40–52.
18. Bushinsky DA, Ori Y. Effects of metabolic and respiratory acidosis on bone. Curr Opin Nephrol Hypertens 1993; 2:588–596.
19. Bushinsky DA, Lam BC, Nespeca R, Sessler NE, Grynpas MD. Decreased bone carbonate content in response to metabolic, but not respiratory, acidosis. Am J Physiol (Renal Fluid Electrolyte Physiol 34) 1993; 265:F530–F536.
20. Bushinsky DA, Krieger NS. Role of the skeleton in calcium homeostasis. In: Seldin DW, Giebisch G, eds. The Kidney: Physiology and Pathophysiology. Raven, New York, 1992, pp. 2395–2430.
21. Bushinsky DA. Net proton influx into bone during metabolic, but not respiratory, acidosis. Am J Physiol (Renal Fluid Electrolyte Physiol 23) 1988; 254:F306–F310.
22. Bushinsky DA. Effects of parathyroid hormone on net proton flux from neonatal mouse calvariae. Am J Physiol (Renal Fluid Electrolyte Physiol 21) 1987; 252:F585–F589.
23. Bushinsky DA, Lechleider RJ. Mechanism of proton-induced bone calcium release: calcium carbonate-dissolution. Am J Physiol (Renal Fluid Electrolyte Physiol 22) 1987; 253:F998–F1005.
24. Bushinsky DA, Krieger NS, Geisser DI, Grossman EB, Coe FL. Effects of pH on bone calcium and proton fluxes in vitro. Am J Physiol (Renal Fluid Electrolyte Physiol 14) 1983; 245:F204–F209.

25. Bushinsky DA, Levi-Setti R, Coe FL. Ion microprobe determination of bone surface elements: effects of reduced medium pH. Am J Physiol (Renal Fluid Electrolyte Physiol 19) 1986; 250:F1090–F1097.
26. Bushinsky DA, Wolbach W, Sessler NE, Mogilevsky R, Levi-Setti R. Physicochemical effects of acidosis on bone calcium flux and surface ion composition. J Bone Miner Res 1993; 8:93–102.
27. Bergstrom WH, Ruva FD. Changes in bone sodium during acute acidosis in the rat. Am J Physiol 1960; 198:1126–1128.
28. Bettice JA, Gamble JL Jr. Skeletal buffering of acute metabolic acidosis. Am J Physiol 1975; 229:1618–1624.
29. Burnell JM. Changes in bone sodium and carbonate in metabolic acidosis and alkalosis in the dog. J Clin Invest 1971; 50:327–331.
30. Bushinsky DA, Gavrilov K, Stathopoulos VM, Krieger NS, Chabala JM, Levi-Setti R. Effects of osteoclastic resorption on bone surface ion composition. Am J Physiol (Cell Physiol 40) 1996; 271:C1025–C1031.
31. Bushinsky DA, Gavrilov K, Chabala JM, Featherstone JDB, Levi-Setti R. Effect of metabolic acidosis on the potassium content of bone. J Bone Miner Res 1997; 12:1664–1671.
32. Bettice JA. Skeletal carbon dioxide stores during metabolic acidosis. Am J Physiol (Renal Fluid Electrolyte Physiol 16) 1984; 247:F326–F330.
33. Kraut JA, Mishler DR, Kurokawa K. Effect of colchicine and calcitonin on calcemic response to metabolic acidosis. Kidney Int 1984; 25:608–612.
34. Widdowson EM, McCance RA, Spray CM. The chemical composition of the human body. Clin Sci 1951; 10:113–125.
35. Widdowson EM, Dickerson JWT. Chemical composition of the body. In: Comar CL, Bronner F, eds. Mineral Metabolism. Academic, New York, 1964, pp. 1–247.
36. Coe FL, Bushinsky DA. Pathophysiology of hypercalciuria. Am J Physiol (Renal Fluid Electrolyte Physiol) 1984; 247:F1–F13.
37. Bushinsky DA, Riera GS, Favus MJ, Coe FL. Response of serum 1,25$(OH)_2D_3$ to variation of ionized calcium during chronic acidosis. Am J Physiol (Renal Fluid Electrolyte Physiol 18) 1985; 249:F361–365.
38. Lemann J Jr, Adams ND, Gray RW. Urinary calcium excretion in human beings. N Engl J Med 1979; 301:535–541.
39. Adams ND, Gray RW, Lemann J Jr. The calciuria of increased fixed acid production in humans: evidence against a role for parathyroid hormone and 1,25(OH)2-vitamin D. Calcif Tissue Int 1979; 28:233–238.
40. Weber HP, Gray RW, Dominguez JH, Lemann J Jr. The lack of effect of chronic metabolic acidosis on 25-OH vitamin D metabolism and serum parathyroid hormone in humans. J Clin Endocrinol Metab 1976; 43:1047–1055.
41. Green J, Kleeman CR. Role of bone in regulation of systemic acid-base balance. Kidney Int 1991; 39:9–26.
42. Bushinsky DA, Chabala JM, Gavrilov KL, Levi-Setti R. Effects of *in vivo* metabolic acidosis on midcortical bone ion composition. Am J Physiol (Renal Physiol 46) 1999; 277:F813–F819.
43. McSherry E, Morris RC. Attainment and maintenance of normal stature with alkali therapy in infants and children with classic renal tubular acidosis. J Clin Invest 1978; 61:509–527.
44. McSherry E. Acidosis and growth in nonuremic renal disease. Kidney Int 1978; 14:349–354.
45. Challa A, Krieg RJ Jr, Thabet MA, Veldhuis JD, Chan JC. Metabolic acidosis inhibits growth hormone secretion in rats: mechanism of growth retardation. Am J Physiol 1993; 265:E547–E553.
46. Goodman AD, Lemann J Jr, Lennon EJ, Relman AS. Production, excretion, and net balance of fixed acid in patients with renal acidosis. J Clin Invest 1965; 44:495–506.
47. Litzow JR, Lemann J Jr, Lennon EJ. The effect of treatment of acidosis on calcium balance in patients with chronic azotemic renal disease. J Clin Invest 1967; 46:280–286.

48. Mora Palma FJ, Ellis HA, Cook DB, et al. Osteomalacia in patients with chronic renal failure before dialysis or transplantation. J Med 1983; S2:332–348.
49. Fletcher RF, Jones JH, Morgan DB. Bone disease in chronic renal failure. Quart J Med 1963; 32:321–339.
50. Bishop MC, Ledingham JG. Alkali treatment of renal osteodystrophy. Br Med J 1972; 4:529.
51. Cochran M, Wilkinson R. Effect of correction of metabolic acidosis on bone mineralization rates in patients with renal osteomalacia. Nephron 1975; 15:98–110.
52. Lefebvre A, de Vernejoul MC, Gueris J, Goldfarb B, Graulet AM, Morieux C. Optimal correction of acidosis changes progression of dialysis osteodystrophy. Kidney Int 1989; 36:1112–1118.
53. Pellegrino ED, Blitz RM, Letteri JM. Inter-relationships of carbonate, phosphate, monohydrogen phosphate, calcium, magnesium, and sodium in uraemic bone: comparison of dialyzed and non-dialyzed patients. Clin Sci Mol Med 1977; 53:307–316.
54. Kaye M, Frueth AJ, Silverman M, Henderson J, Thibault T. A study of vertebral bone powder from patients with chronic renal failure. J Clin Invest 1970; 49:442–453.
55. Pellegrino ED, Biltz RM. The composition of human bone in uremia. Medicine 1965; 44:397–418.
56. Brenner RJ, Spring DB, Sebastian A, et al. Incidence of radiographically evident bone disease, nephrocalcinosis, and nephrolithiasis in various types of renal tubular acidosis. N Engl J Med 1982; 307:217–221.
57. Domrongkitchaiporn S, Pongsakul C, Stitchantrakul W, et al. Bone mineral density and histology in distal renal tubular acidosis. Kidney Int 2001; 59:1086–1093.
58. Domrongkitchaiporn S, Pongskul C, Sirikulchayanonta V, et al. Bone histology and bone mineral density after correction of acidosis in distal renal tubular acidosis. Kidney Int 2002; 62:2160–2166.
59. Lennon EJ, Lemann J Jr, Litzow JR. The effects of diet and stool composition on the net external acid balance of normal subjects. J Clin Invest 1966; 45:1601–1607.
60. Barzel US. The role of bone in acid base metabolism. In: Barzel US, ed. Osteoporosis. Grune & Stratton, New York, 1970, p. 199.
61. Barzel US. Osteoporosis II: an overview. In: Barzel US, ed. Osteoporosis II. Grune & Stratton, New York, 1976, p. 1.
62. Frassetto LA, Morris RC Jr, Sebastian A. Effect of age on blood acid-base composition in adult humans: role of age-related renal functional decline. Am J Physiol (Renal Fluid Electrolyte Physiol 40) 1996; 271:F1114–F1122.
63. Frassetto L, Sebastian A. Age and systemic acid-base equilibrium: analysis of published data. J Gerontol 1996; 51:B91–B99.
64. Sakhaee K, Nicar M, Hill K, Pak CY. Contrasting effects of potassium citrate and sodium citrate therapies on urinary chemistries and crystallization of stone-forming salts. Kidney Int 1983; 24:348–352.
65. Lemann J Jr, Gray RW, Pleuss JA. Potassium bicarbonate, but not sodium bicarbonate, reduces urinary calcium excretion and improves calcium balance in healthy men. Kidney Int 1989; 35:688–695.
66. Sebastian A, Harris ST, Ottaway JH, Todd KM, Morris RC Jr. Improved mineral balance and skeletal metabolism in postmenopausal women treated with potassium bicarbonate. N Engl J Med 1994; 330:1776–1781.
67. Neuman WF, Neuman MW. The Chemical Dynamics of Bone Mineral. University of Chicago Press, Chicago, 1958.
68. Dominguez JH, Raisz LG. Effects of changing hydrogen ion, carbonic acid, and bicarbonate concentrations on bone resorption in vitro. Calcif Tissue Int 1979; 29:7–13.
69. Bushinsky DA, Goldring JM, Coe FL. Cellular contribution to pH-mediated calcium flux in neonatal mouse calvariae. Am J Physiol (Renal Fluid Electrolyte Physiol 17) 1985; 248:F785–F789.

70. Bushinsky DA. Net calcium efflux from live bone during chronic metabolic, but not respiratory, acidosis. Am J Physiol (Renal Fluid Electrolyte Physiol 25) 1989; 256:F836–F842.
71. Chabala JM, Levi-Setti R, Bushinsky DA. Alteration in surface ion composition of cultured bone during metabolic, but not respiratory, acidosis. Am J Physiol (Renal Fluid Electrolyte Physiol 30) 1991; 261:F76–F84.
72. Bushinsky DA, Sessler NE, Krieger NS. Greater unidirectional calcium efflux from bone during metabolic, compared with respiratory, acidosis. Am J Physiol (Renal Fluid Electrolyte Physiol 31) 1992; 262:F425–F431.
73. Krieger NS, Sessler NE, Bushinsky DA. Acidosis inhibits osteoblastic and stimulates osteoclastic activity in vitro. Am J Physiol (Renal Fluid Electrolyte Physiol 31) 1992; 262:F442–F448.
74. Bushinsky DA, Sessler NE. Critical role of bicarbonate in calcium release from bone. Am J Physiol (Renal Fluid Electrolyte Physiol 32) 1992; 263:F510–F515.
75. Bushinsky DA, Sessler NE, Glena RE, Featherstone JDB. Proton-induced physicochemical calcium release from ceramic apatite disks. J Bone Miner Res 1994; 9:213–220.
76. Sprague SM, Krieger NS, Bushinsky DA. Greater inhibition of in vitro bone mineralization with metabolic than respiratory acidosis. Kidney Int 1994; 46:1199–1206.
77. Bushinsky DA. Stimulated osteoclastic and suppressed osteoblastic activity in metabolic but not respiratory acidosis. Am J Physiol (Cell Physiol 37) 1995; 268:C80–C88.
78. Ori Y, Lee SG, Krieger NS, Bushinsky DA. Osteoblastic intracellular pH and calcium in metabolic and respiratory acidosis. Kidney Int 1995; 47:1790–1796.
79. Bushinsky DA, Nilsson EL. Additive effects of acidosis and parathyroid hormone on mouse osteoblastic and osteoclastic function. Am J Physiol (Cell Physiol 38) 1995; 269:C1364–C1370.
80. Bushinsky DA. Metabolic alkalosis decreases bone calcium efflux by suppressing osteoclasts and stimulating osteoblasts. Am J Physiol (Renal Fluid Electrolyte Physiol 40) 1996; 271:F216–F222.
81. Bushinsky DA, Krieger NS. Integration of calcium metabolism in the adult. In: Coe FL, Favus MJ, eds. Disorders of Bone and Mineral Metabolism. Raven, New York, 1992, pp. 417–432.
82. Bushinsky DA, Kittaka MK, Weisinger JR, Langman CB, Favus MJ. Effects of chronic metabolic alkalosis on Ca^{++}, PTH and $1,25(OH)_2D_3$ in the rat. Am J Physiol (Endocrinol Met Physiol 20) 1989; 257:E579–E582.
83. Frick KK, Bushinsky DA. Chronic metabolic acidosis reversibly inhibits extracellular matrix gene expression in mouse osteoblasts. Am J Physiol (Renal Physiol 44) 1998; 275:F840–F847.
84. Frick KK, Bushinsky DA. Effect of simulated matrix on the response of osteoblasts to chronic metabolic acidosis (abstr). J Am Soc Nephrol 1998; 9:546A.
85. Frick KK, Bushinsky DA. In vitro metabolic and respiratory acidosis selectively inhibit osteoblastic matrix gene expression. Am J Physiol (Renal Physiol 46) 1999; 277:F750–F755.
86. Krieger NS, Bushinsky DA. Metabolic acidosis regulates MAP-kinase activity in primary bone cells (abstr). J Am Soc Nephrol 1998; 9:546A.
87. Krieger NS, Parker WR, Alexander KM, Bushinsky DA. Prostaglandins regulate acid-induced cell-mediated bone resorption. Am J Physiol Renal Physiol 2000; 279:F1077–F1082.
88. Bushinsky DA, Parker WR, Alexander KM, Krieger NS. Metabolic, but not respiratory, acidosis increases bone PGE_2 levels and calcium release. Am J Physiol (Renal Fluid Electrolyte Physiol) 2001; 281:F1058–F1066.
89. Krieger NS, Frick KK, Bushinsky DA. Cortisol inhibits acid-induced bone resorption in vitro. J Am Soc Nephrol 2002; 13:2534–2539.
90. Bushinsky DA, Smith SB, Gavrilov KL, Gavrilov LF, Li J, Levi-Setti R. Acute acidosis-induced alteration in bone bicarbonate and phosphate. Am J Physiol Renal Physiol 2002; 283:F1091–F1097.
91. Frick KK, Bushinsky DA. Metabolic acidosis stimulates RANK ligand RNA expression in bone through a cyclogenase dependent mechanism. J Bone Miner Res 2003; 18:1317–1325.

92. Schelling SH, Wolfe HJ, Tashjian AH Jr. Role of the osteoclast in prostaglandin E_2-stimulated bone resorption. A correlative morphometric and biochemical analysis. Lab Invest 1980; 42:290–295.
93. Feldman RS, Krieger NS, Tashjian AH Jr. Effects of parathyroid hormone and calcitonin on osteoclast formation in vitro. Endocrinology 1980; 107:1137–1143.
94. Stern PH, Raisz LG. Organ culture of bone. In: Simmons DJ, Kunin AS, eds. Skeletal Research. Academic, New York, 1979, pp. 21–59.
95. Ellies LG, Nelson DGA, Featherstone JD. Crystallographic structure and surface morphology of sintered carbonated apatites. J Biomed Mater Res 1988; 22:541–553.
96. Ellies LG, Carter JM, Natiella JR, Featherstone JDB, Nelson DGA. Quantitative analysis of early in vivo tissue response to synthetic apatite implants. J Biomed Mater Res 1988; 22:137–148.
97. Nelson DGA, Featherstone JDB, Duncan JF, Cutress TW. Effect of carbonate and fluoride on the dissolution behavior of synthetic apatites. Caries Res 1983; 17:200–211.
98. Nelson DGA, Featherstone JDB. Preparation, analysis and characterization of carbonated apatites. Calcif Tissue Int 1982; 34:S69–S81.
99. Featherstone JDB, Pearson S, LeGeros RZ. An infrared method for quantification of carbonate in carbonated-apatites. Caries Res 1984; 18:63–66.
100. Nelson DGA, Barry JC, Shields CP, Glena R, Featherstone JDB. Crystal morphology, composition, and dissolution behavior of carbonated apatites prepared at controlled pH and temperature. J Colloid Interf Sci 1989; 130:467–479.
101. Featherstone JDB, Nelson DGA. Recent uses of electron microscopy in the study of physicochemical processes affecting the reactivity of synthetic and biological apatites. Scanning Microsc 1989; 3:815–828.
102. Boskey AL, Posner AS. Conversion of amorphous calcium phosphate to microcrystalline hydroxyapatite-pH-dependent, solution mediated, solid-solid conversion. J Phys Chem 1973; 77:2313–2317.
103. Grynpas MD, Bonar LC, Glimcher MJ. Failure to detect an amorphous calcium-phosphate solid phase in bone mineral. A radial distribution function study. Calcif Tissue Int 1984; 36:291–301.
104. Bushinsky DA, Chabala JM, Levi-Setti R. Ion microprobe analysis of bone surface elements: effects of $1,25(OH)_2D_3$. Am J Physiol (Endocrinol Met Physiol 20) 1989; 257:E815–E822.
105. Bushinsky DA, Chabala JM, Levi-Setti R. Ion microprobe analysis of mouse calvariae in vitro: evidence for a "bone membrane." Am J Physiol (Endocrinol Metab 19) 1989; 256:E152–E158.
106. Bushinsky DA, Chabala JM, Levi-Setti R. Comparison of in vitro and in vivo ^{44}Ca labeling of bone by scanning ion microprobe. Am J Physiol (Endocrinol Metab 22) 1990; 259:E586–E592.
107. Pasquale SM, Messier AA, Shea ML, Schaefer KE. Bone CO_2-titration curves in acute hypercapnia obtained with a modified titration technique. J Appl Physiol 1980; 48:197–201.
108. Frick KK, Jiang L, Bushinsky DA. Acute metabolic acidosis inhibits the induction of osteoblastic *egr*-1 and type 1 collagen. Am J Physiol (Cell Physiol 41) 1997; 272:C1450–C1456.
109. Kraut JA, Mishler DR, Singer FR, Goodman WG. The effects of metabolic acidosis on bone formation and bone resorption in the rat. Kidney Int 1986; 30:694–700.
110. Bhargava U, Bar-Lev M, Bellows CG, Aubin JE. Ultrastructural analysis of bone nodules formed in vitro by isolated fetal rat calvaria cells. Bone 1988; 9:155–163.
111. Ecarot-Charrier B, Glorieux FH, van der Rest M, Pereira G. Osteoblasts isolated from mouse calvaria initiate matrix mineralization in culture. J Cell Biol 1983; 96:639–643.
112. Sudo H, Kodama HA, Amagai Y, Yamamoto S, Kasai S. In vitro differentiation and calcification in a new clonal osteogenic cell line derived from newborn mouse calvaria. J Cell Biol 1983; 96:191–198.
113. Sprague SM, Krieger NS, Bushinsky DA. Aluminum inhibits bone nodule formation and calcification in vitro. Am J Physiol (Renal Fluid Electrolyte Physiol 33) 1993; 264:F882–F890.

114. Goodman WG, Coburn JW, Slatopolsky E, Salusky IB. Renal osteodystrophy in adults and children. In: Favus MJ, ed. Primer on the Metabolic Bone Diseases and Disorders of Mineral Metabolism. Lippincott-Raven, Philadelphia, 1999, pp. 347–363.
115. Slatopolsky E, Delmez J. Bone disease in chronic renal failure and after renal transplantation. In: Coe F, Favus M, eds. Disorders of Bone and Mineral Metabolism. Raven, New York, 1992, pp. 905–934.
116. Stein GS, Lian JB, Stein JL, van Wijnen AJ, Montecino M. Transcriptional control of osteoblast growth and differentiation. Physiol Rev 1996; 76:593–629.
117. Davies JE. In vitro modeling of the bone/implant interface. Anat Rec 1996; 245:426–445.
118. Rodan GA. Osteopontin overview. Ann N Y Acad Sci 1995; 760:1–5.
119. Hauschka PV, Lian JB, Cole DE, Gundberg CM. Osteocalcin and matrix gla protein: vitamin K-dependent proteins in bone. Physiol Rev 1989; 69:990–1047.
120. Bushinsky DA, Sprague SM, Hallegot P, Girod C, Chabala JM, Levi-Setti R. Effects of aluminum on bone surface ion composition. J Bone Miner Res 1995; 10:1988–1997.
121. Atkins KB, Simpson RU, Somerman MJ. Stimulation of osteopontin mRNA expression in HL-60 cells is independent of differentiation. Arch Biochem Biophys 1997; 343:157–163.
122. Canzanello VJ, Bodvarsson M, Kraut JA, Johns CA, Slatopolsky E, Madias NE. Effect of chronic respiratory acidosis on urinary calcium excretion in the dog. Kidney Int 1990; 38:409–416.
123. Lau K, Rodriguez Nichols F, Tannen RL. Renal excretion of divalent ions in response to chronic acidosis: evidence that systemic pH is not the controlling variable. J Lab Clin Med 1987; 109:27–33.
124. Schaefer KE, Pasquale S, Messier AA, Shea M. Phasic changes in bone CO_2 fractions, calcium, and phosphorus during chronic hypercapnia. J Appl Physiol 1980; 48:802–811.
125. Schaefer KE, Nichols G Jr, Carey CR. Calcium phosphorus metabolism in man during acclimatization to carbon dioxide. J Appl Physiol 1963; 18:1079–1084.
126. Arnett TR, Dempster DW. A comparative study of disaggregated chick and rat osteoclasts in vitro: effects of calcitonin and prostaglandins. Endocrinology 1987; 120:602–608.
127. Arnett TR, Dempster DW. Effect of pH on bone resorption by rat osteoclasts in vitro. Endocrinology 1986; 119:119–124.
128. Goldhaber P, Rabadjija L. H^+ stimulation of cell-mediated bone resorption in tissue culture. Am J Physiol (Endocrinol Metab 16) 1987; 253:E90–E98.
129. Glimcher MJ. The nature of the mineral component of bone and the mechanism of calcification. Instr Course Lect 1987; 36:49–69.
130. Glimcher MJ. Composition, structure and organization of bone and other mineralized tissues, and the mechanism of calcification. In: Greep RO, Astwood EB, Aurbach GD, eds. Handbook of Physiology, Endocrinology. American Physiology Society, Washington, DC, 1976, pp. 25–116.

IV

Minerals

19 Quantitative Clinical Nutrition Approaches to the Study of Calcium and Bone Metabolism

Connie M. Weaver, Meryl Wastney, and Lisa A. Spence

1. WHY USE QUANTITATIVE CLINICAL NUTRITION APPROACHES?

Understanding the role of nutrients, food components, and diet in bone health is important as a means available to individuals to build and maintain peak bone mass within their genetic potential. There are several approaches an investigator can take to study the relationship between diet and bone, including epidemiology, randomized controlled trials, or metabolic balance and kinetic studies. Epidemiology is conducive to the study of large numbers of subjects, a distinct advantage for power to find an interesting relationship. The outcome measures can be quite precise for studying diet and bone health, primarily bone mineral density and incidence of fracture. However, the assessment of nutrient intake or even dietary patterns is quite poor, owing to an accumulation of errors and intrasubject variation associated with subject recall, inability to estimate portion size accurately, variability of food composition, and availability of this information *(1)*. Moreover, the number of confounders is nearly infinite. The quantitative effect of any one nutrient on some aspect of bone is nearly impossible to determine, as the relationship is dwarfed by other factors such as body weight and age of the subject. Nevertheless, this approach allows hypotheses to be generated. The randomized controlled trial improves the ability to determine the effect of the intervention nutrient or food on the outcome measure of interest, but it is far from quantitative as the background diet is as difficult to assess as for epidemiological studies and compliance with the intervention is unsupervised, and therefore uncertain.

In this chapter, we will review methodologies that allow quantitative nutrition effects on bone to be determined. They are resource intensive, and thus, not feasible to use in large population studies. The necessarily small study group cannot

From: *Nutrition and Bone Health*
Edited by: M. F. Holick and B. Dawson-Hughes © Humana Press Inc., Totowa, NJ

represent the entire population, which is a limitation of this approach. Metabolic balance studies can offer a highly controlled independent variable, that is, diet. When used in a crossover design, the effect of one dietary change on net calcium retention can be quantitated.

Kinetic studies provide additional information over balance studies alone, and analysis of tracers offers greater precision. When balance studies are used in conjunction with isotopic tracers, parameters of calcium metabolism can be studied including absorption, endogenous secretion, excretion, bone formation rates, and bone resorption rates. As 99% of the body's calcium resides in the skeleton, to study calcium metabolism is to study bone metabolism.

2. METABOLIC BALANCE STUDIES

2.1. Application of Balance Studies

Metabolic balance studies have been used at least since the 1920s *(2)*. A thorough accounting of calcium balance methodology was reported in 1945 by a group of authors that included Fuller Albright, for whom the highest award given by the American Society for Bone and Mineral Research was named *(3)*.

Balance studies calculate net retention as intake minus excretion. Balance studies are sufficiently sensitive, when rigorously controlled, to distinguish differences when large effects are expected, as exists when comparing pubertal growth vs adults *(4)*, lactating state vs nonlactating state *(5)*, racial differences *(6)*, the effects of skeletal unloading *(7)*, and some diet effects such as calcium intake *(8)*. The treatment differences in these examples exceed 200 mg calcium retention per day. Power calculations show that sample sizes of five to six subjects per group are sufficient to find significant differences of this magnitude at an α of 0.05 with 80% power even though the variances were large. The ability to determine smaller effects of diet depends on the magnitude of the effect and the specific population. We have been able to show treatment effects on calcium retention of 40 mg/d with 10–15 adolescent subjects in crossover studies. However, for treatment differences in calcium retention of approx 40 mg/d in postmenopausal women using the variance we observed in a recent study, power calculations suggest that 180 subjects would be needed to show significance using a crossover design.

Balance studies can also be used to determine calcium requirements, as the response of calcium retention to calcium intake reaches a plateau when calcium intake is no longer limiting maximal calcium, that is, bone retention. Useful information can be determined about the role of other dietary factors or lifestyle choices in shifting the maximal retention curve, which shifts calcium requirements higher or lower (Fig. 1) *(9)*. This application has the advantage of not putting so much weight on actual values of calcium retention. The maximal retention approach seeks the intake where a plateau occurs rather than an absolute retention. The errors associated with balance are not equally distributed. Errors associated with incomplete consumption of the diet or collection of urine and fecal excretion and failure to measure

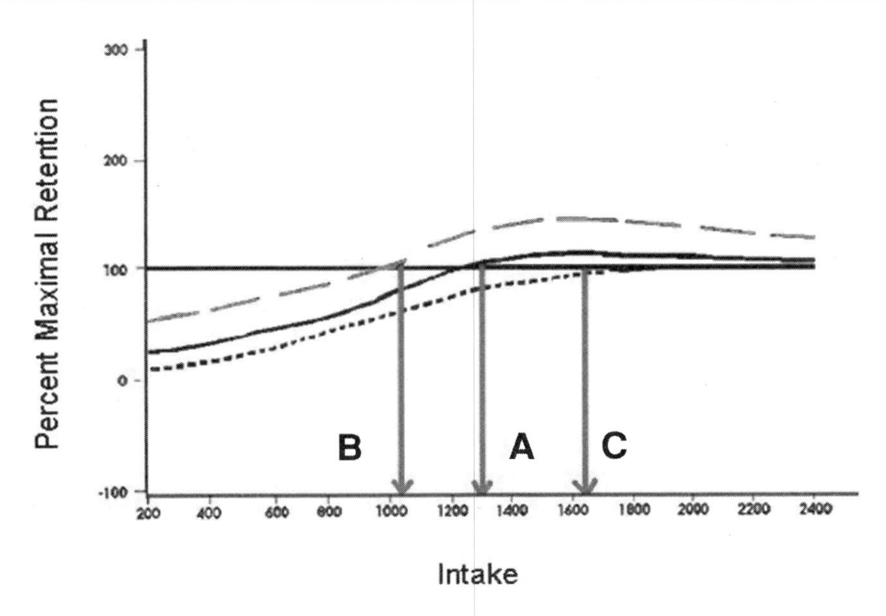

Fig. 1. Maximal calcium retention curve by calcium intake. **A** Reference curve used to determine calcium requirements for adolescents. **(B)** Curve shifted left by a change such as decreasing salt intake. **(C)** Curve shifted to the right by a change such as increasing dietary salt. Curve **A** was developed from data from Jackman et al. *(9)*.

other losses including dermal are often-cited limitations of this method. Also, analytical procedures typically have a coefficient of variation of ≥5%. Some guidance for minimizing these errors is discussed in subsequent sections of this chapter.

2.2. Conducting Balance Studies

Metabolic balance studies involve feeding a controlled diet, collecting excreta, and measuring calcium input and output. Intake cannot be estimated from food composition tables. All foods and beverages containing calcium need to be prepared by weighing ingredients to the nearest 0.1 g. Prepared commercial foods can be used if their composition is homogeneous. Duplicate collections of all of the foods and beverages consumed in a 24-h period are analyzed for calcium and other constituents that influence calcium balance, including protein, phosphorus, fiber, and electrolytes. Diets should be designed to be constant in these constituents throughout the study period. Foods, beverages, and oral health care products that contain calcium, including tap water, inhibitors of calcium absorption such as tea, which contains oxalate, or hypercalciuric ingredients such as salt cannot be allowed *ad libitum.*

If the metabolic study is not conducted in all subjects simultaneously, but rather as a rolling enrollment, it may not be practical to analyze a duplicate sample of each day. In that case, dietary composites representing each cycle day from a dietary intervention should be prepared in intervals throughout a study period to track the variability that occurs over time and the variability between the daily diets. Dietary composites should be measured for calcium and those nutrients that could potentially affect calcium metabolism, that is, protein, sodium, potassium, and phosphorus. Dietary homogenates representing each day of the menu cycle, that is, 7 d for a 7-d menu cycle, should be freeze-dried and aliquoted in triplicate for all nutrient analysis. Variation among these triplicate samples is an indication of homogeneity in the sample and analytical precision. Replicate analysis of dietary composites prepared over the entire study period demonstrates variation due to variability in food items, dietary preparation, and laboratory analysis. Analysis across cycle menus represents daily variability within the diet.

Analysis of dietary composites from a metabolic balance study in our laboratory that used a 7-d menu cycle over a 19-mo period demonstrated 3% variation in calcium from triplicate analysis of dietary samples. The variation in calcium from the replicate analysis of dietary composites, which were collected at quarterly intervals over the 19-mo study period, was 5%. Daily variation in calcium across the 7-d cycle menus was 6%. Protein varied by 8% in both replicate analysis and across cycle menus. Phosphorus, potassium, and sodium varied by 9%, 13%, and 18%, respectively in replicate analysis and by 15%, 16%, and 16%, across cycle menus. The diets were designed with the aid of the computer program, Nutritionist IV. The most careful attention was paid to the calcium and protein content of the diets to test the study hypothesis. Thus, laboratory analysis revealed less variation in the daily analysis of the diets for calcium and protein compared to the daily variation in the other nutrients measured.

Urine and feces are collected in acid-washed containers for later analysis of total calcium by atomic absorption spectrometry or inductively coupled plasma. A 1-d lag is used when calculating intake minus fecal excretion to account for the approx 19-h transit time in the gut. Menstrual losses of calcium can generally be ignored. Dermal losses are often ignored but can be measured by extracting pre-acid-washed clothing worn for 24 h in addition to whole-body wash-down procedures before and after the collection . Using this method we have determined dermal losses of approx 52 mg/d in adolescents. Dermal losses determined by patches over estimated calcium losses by almost eightfold *(10)*. Dermal losses in adults have been estimated to be 60 mg/d by the difference between whole-body retention of ^{47}Ca and excretion in urine and feces *(11)*.

To determine the effect of a variable on calcium retention within a population, the best approach is the use of a crossover design in the same subjects, to minimize such confounding effects that are constant within an individual such as hormonal status, gastrointestinal and kidney function, mucosal mass, transit time, and vitamin D status. Randomized-order assignments of treatment can minimize seasonal effects of vitamin D status. Nevertheless, the presence or absence of an order effect should be tested statistically. Subjects can also be pretreated with vitamin D

Table 1
Fecal Calcium: PEG Ratios (mg/mg) During a 3-wk Balance Period*

Group	*Calcium intake (mg/d)*	*Week 1*	*Week 2*	*Week 3*
Adolescent girls				
	800	0.25 ± 0.13[a]	0.21 ± 0.06[b]	0.19 ± 0.04[b]
	1300	0.82 ± 2.20[a]	0.32 ± 0.06[b]	0.38 ± 0.72[b]
	1800	0.53 ± 0.08[a]	0.48 ± 0.09[b]	0.48 ± 0.07[b]
Adults				
	1300	1.49 ± 5.20[a]	0.36 ± 0.09[b]	0.36 ± 0.08[b]

* Different letter superscripts within rows indicate means are significantly different for each level of calcium intake at $p < 0.05$. Data from ref. *12* and unpublished data.

supplements and continued throughout the study period if they have hypovitaminosis D.

The length of the run-in period needed for a subject to adapt to the study diet and the length of the balance period once steady state is reached must be carefully considered. Misinterpretations have occurred when subjects have been switched from high- to low-calcium intake periods without an appropriate adaptation period, as the higher-calcium intakes spill into the feces during the lower-calcium period for several days. When a nonabsorbable fecal marker such as polyethylene glycol (PEG) is given at every meal, the fecal calcium:PEG ratio can be used to determine when steady state is achieved. We have found that the Ca:PEG ratio becomes constant after 6 d in adolescents and most adults (Table 1, adapted from ref. *12* plus unpublished data). Similarly, when adult black and white women were switched from a diet containing 2000 mg/d for 3 wk to 300 mg/d for 8 wk, whole-body retention of ^{47}Ca varied from wk 1 to wk 2, but not from wk 2 to wk 8 *(13)*. Thus, determining balance during the run-in period can give useful information about when steady state is achieved in calcium balance or another dietary constituent being tested for its effect on calcium retention. Subjects cannot adapt to a low calcium intake to become "in balance," as the homeostatic control mechanisms is inefficient. Malm *(14)* studied prisoners for up to 2 yr and found continued negative balances.

Balance periods should be sufficiently long to evaluate trends over multiple periods. Some investigators make collections in several day pools and monitor multiple periods. We typically make collections in 24-h periods for 2–3 wk after a 1-wk adaptation period. Calculating daily balances for multiple periods allows an error term to be determined so that differences from zero balance can be determined for each individual. In pubertal children, balance periods should not exceed rapid hormonal shifts, which can outweigh the influence of diet on calcium retention *(12)*.

2.3. Monitoring Compliance

Methods to assess compliance of urine and fecal collections and to adjust for discrete 24-h periods are helpful in reducing variation in balance data and in interpreting the quality of data. However, errors can be made in measuring compliance markers, so that corrected data may be less accurate than uncorrected data for any given day. Thus, all components of any calculation should be carefully inspected. Especially troublesome is the apparent overcorrection of fecal calcium using a marker to adjust for low compliance.

Adjustment of urine is usually made with creatinine. Subjects excrete a rather constant level of creatinine proportional to lean body mass. The mean daily creatinine excretion of a subject over the study period can be used to adjust each day to a more precise 24-h period, as it is difficult to completely empty one's bladder at precise regular time periods, especially for children. Twenty-four-hour pools with creatinine values less than 11 mg/kg should be discarded.

Daily fecal calcium output is highly variable despite constant conditions due to variable gut transit times, which does not follow a continuum of discrete periods *(15)*. A number of nonabsorbable fecal markers have been employed to evaluate compliance, transit time, and to convert individual stools collected at irregular intervals to daily fecal calcium output. We use PEG 4000 as a continuously administered, nonabsorbable marker as developed by Wilkinson *(16)*, who demonstrated clearly a reduction in daily variation by correcting stool samples by recovery of this marker. Capsules are prepared containing PEG weighed to the nearest milligram and consumed at each meal throughout the study. The ratio of Ca:PEG in each 24-h pool is multiplied by the amount of PEG consumed during 24-h to determine daily fecal calcium. This marker is superior for water-soluble dietary constituents such as calcium to previously used markers which more closely follows the insoluble pool such as Cr_2O_3 and barium sulfate, although recovery of all three markers was 98–100% *(16)*.

Adjusting fecal calcium as described above supposedly corrects for incomplete stool collections. However, Eastell et al. *(17)* reported a PEG recovery of only 81% of that compared to 95% with ^{51}Cr in the same experiment. Therefore, we suspect that adjusting fecal calcium with PEG may overcorrect, which becomes worse with decreasing compliance. To examine this issue, we used data from one of our studies in which we calculated calcium balance using the PEG adjustment. Each group was studied three times, and we computed residuals for calcium balance by subtracting the group mean from each individual observation. A plot of these residuals vs PEG is given in Fig. 2. The plot includes a centerline at zero (the mean of the residuals) as well as a smooth fit to the data. There appears to be a positive association between the PEG value and the residual. This means that observations with low values of PEG tend to be associated with balance values that are low relative to the group mean and, similarly, high PEG values are associated with balance values that are high relative to the group mean. This association is consistent with a scenario in which the PEG overcorrects the fecal calcium values:

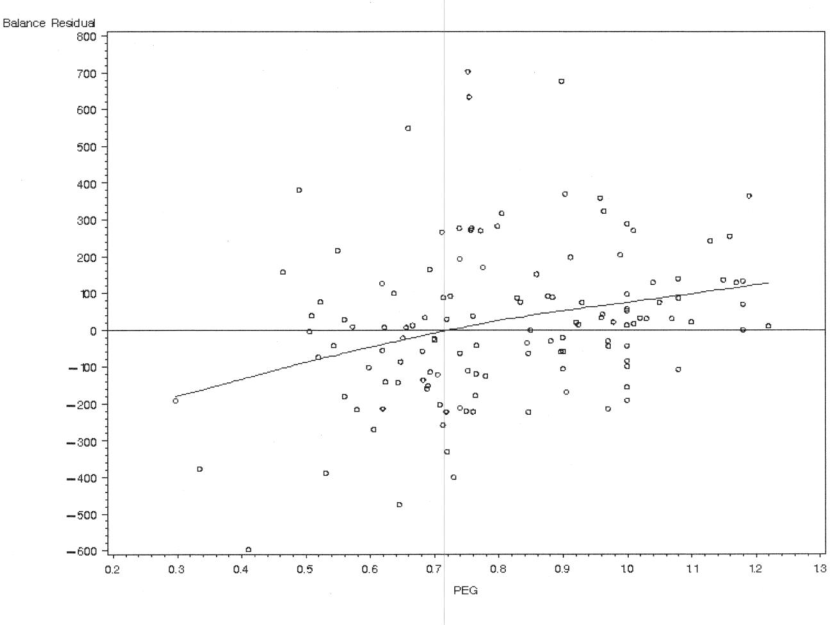

Fig. 2. Calcium balance residuals vs percent PEG recovery in one study of postmenopausal women.

when the PEG is low, the corrected fecal values are too high, and therefore the balance values are too low. More research is needed to understand this issue.

2.4. Feasibility of Free-Living Subjects vs Metabolic Ward

In metabolic studies with free-living subjects or subjects housed in metabolic wards, consumption compliancy and excreta collection compliancy can influence the study results. Classical balance studies have typically been conducted in metabolic wards, where subjects' activities are monitored, particularly, food consumption and excreta collection. Once individuals are allowed to participate in balance studies as free-living individuals, monitoring activities becomes difficult. Food consumption and excreta collection cannot be monitored in free-living individuals as in a metabolic ward. In studies with free-living subjects, all foods and beverages are provided along with instructions to consume each item. When subjects do not consume all food and beverage items due to various reasons, they are instructed to return uneaten portions of the food or beverage. These items are then analyzed in the laboratory for calcium content and calcium balance is corrected based on the calcium content of the uneaten food or beverage.

In a metabolic study with postmenopausal women, our laboratory analysis demonstrated a consistent daily creatinine output and average PEG recovery rate of approx 80%. These indicators of collection compliancy were consistent with results seen in adult subjects participating in in-house metabolic studies *(4,18)*. Collection compliancy remains an obstacle in metabolic studies whether subjects are free-living or maintained in monitored environments. Success depends on committed subjects in either environment. Some populations, such as children, undoubtedly require a supervised environment to be successful. The effect of compliancy on treatment effect can be determined by examining the F statistic when data are evaluated by using various cutoffs for percent PEG recovery as inclusion criteria.

3. TRACER STUDIES

3.1. Application of Tracer Studies

Isotopic tracer data are less variable than balance data. Thus, although fractional absorption determined by tracer studies is similar to net calcium absorption determined by balance studies *(19)*, more subtle differences can be discriminated with isotopic tracer studies. Tracers are required for kinetic studies. Kinetic studies involve the study of movement of calcium using a calcium isotope from one compartment to another, rates of transfer, and body pool sizes. Depending on the route of administration, the number of tracers used, the samples collected, and data analysis, the amount of information gained over balance studies alone can be considerable. Calcium tracer experiments offer insights into the bone microenvironment and transfer at the blood–bone interface. Because most of the calcium in bone is not exchangeable with tracers in short-term experiments, total body pool size cannot be measured, but processes of bone turnover can be measured with tracers. Calcium clearance rates can be used to determine metabolic bone disease. Pitfalls of tracer studies occur if isotopes fail to mix adequately *(20)*. Compartments do not generally equate to anatomically distinct entities and their contents may not be able to be interpreted without further biochemical or physiological investigation.

3.2. Available Calcium Isotopes

A list of isotopic tracers of calcium appears in Table 2. Useful radiotracers of calcium are ^{47}Ca and ^{45}Ca. ^{47}Ca is a γ-emitter, and therefore can be used for whole-body counting in studies of calcium retention in animals or humans in facilities where animal or human γ counters are available. Its short half-life limits the length of the experiment and is the reason for its scarcity and relatively high expense. However, whole-body retention using ^{47}Ca is attractive, as neither compliance nor failure to collect dermal losses are issues as occur with balance studies *(21)*. A limitation of whole-body counting is that mechanisms cannot be investigated because the tissue perturbed, that is, gut, kidney, or bone, cannot be inferred from whole-body retention curves. As a β-emitter, ^{45}Ca is measured in a liquid scintillation counter and is appropriate for biological fluids or samples that can be converted to fluids. Although ^{47}Ca can also be measured in biological fluids, the lower costs and

Table 2
Calcium Isotopic Tracers

Atomic number	Symbol and mass number	Radioisotopes Half-life	Maximum radiation energies β (Mev)	E (Mev)	Stable Isotopes Natural abundance (%)
20	^{41}Ca	10^5 yr	—	—	10^{-15}
	^{42}Ca	—	—	—	0.646
	^{43}Ca	—	—	—	0.135
	^{44}Ca	—	—	—	2.083
	^{45}Ca	164 d	0.255	—	—
	^{46}Ca	—	—	—	0.0033
	^{47}Ca	4.53 d	1.98	1.29	—
	^{48}Ca	—	—	—	0.18

longer half-life typically make ^{45}Ca the preferred radioisotope for tracer studies. Precision of analysis with radioisotopes depend on the counting rate, but samples can be counted to 1–2% precision.

There are many more nonradioactive (stable) isotopes of calcium than radioisotopes. These isotopes of heavier mass than ^{40}Ca, which represents almost 97% of calcium in nature, are measured as isotopic ratios by mass spectroscopy. The methods of choice currently are high-resolution, inductively coupled plasma, mass spectrometry (HR-ICPMS) *(22)* and thermal ionization mass spectrometry (TIMS) *(23)*. The former has the advantage of greater sample throughput and the latter has the advantage of greater precision (1–2% vs <0.1–0.2%). Stable isotopic tracers have the advantage of not exposing subjects to radioactivity and not having to time experiments around a short half-life. They have the disadvantage of being more expensive to purchase and analyze. Use of calcium stable isotopes for clinical studies of calcium metabolism was first proposed in 1983 *(24)*.

The long-lived radioisotope, ^{41}Ca, can be used in such small doses (≤100 nCi) that it can be considered to be radiologically benign. A single dose of this size labels the skeleton for life, which poses a lifetime radiation exposure of less than 2 µrem. The benefits of this tracer are that the tracer can be monitored for long experiments, in contrast to the upper limit of approx 2 wk with other isotopes. Urinary appearance of ^{41}Ca after 100 d from dosing, when the ^{41}Ca can be considered coming from the skeleton, provides a direct, sensitive measure of bone resorption. Changes in bone loss can be accurately measured following an intervention. The disadvantage of this approach is that ^{41}Ca is measured with an accelerator mass spectrometer (AMS), which is not available in most research centers. There are two in the United States, one at Purdue University and one at Lawrence

Livermore National Laboratory. Opportunities with AMS in nutrition have been recently reviewed *(25)*.

3.3. Kinetic Studies

3.3.1. Conducting Kinetic Studies

A comprehensive kinetic study will be described first to present the model and nomenclature. In subsequent sections, individual components of calcium metabolism will be discussed with comments on simplifying experimental designs. The most complete model for calcium metabolism can be developed when subjects participate in a metabolic balance study and have achieved steady state before isotopes are administered. Isotopes are administered orally and intraveneously in doses that should not perturb the normal movement of calcium, that is, less than 10% of the circulating calcium pool. The actual dose administered depends on the precision of the detection method and the length of time the tracers are to be followed, typically limited to 2 wk. If different isotopes are administered orally and intravenously, the isotopes can be administered almost simultaneously. This dual-isotope procedure was described by De Grazia et al. *(26)*. Oral isotopes take longer to enter the plasma pool than intravenous doses, so we give the oral isotope 1 h prior to giving the intravenous isotope. The two isotopes track identically after 20 h *(27)*. Activity of radioisotopes or stable isotope ratio measurements are made on the urine and fecal samples collected for the balance study in addition to total calcium. In addition, plasma or serum samples are collected periodically for isotope measurements. Measurements can also be made on saliva *(28)*. With kinetic studies, the more data collected, the better the model will be. However, there are ethical limits on the volume of blood that can be taken over the study period. More measurements are needed early postadministration, when turnover is more rapid. For subject comfort, frequent collection of blood during the first few hours postadministration is often done through a catheter. Stable isotopes are administered intravenously over several minutes through a catheter and blood collected serially through a separate catheter to avoid contamination.

3.3.2. Mathematical Modeling

Mathematical modeling is the expression of metabolism in terms of equations. Some of the reasons for analyzing data by modeling are to test an hypothesis against data, integrate information from different parts of a system (e.g., absorption and bone deposition), measure attributes of a system that are not measurable directly, calculate a parameter of interest, investigate processes that cannot be studied directly, and to identify changes in metabolism between two treatments.

There are many approaches to modeling. For example, models may be descriptive or mechanistic. Descriptive models tend to be simple and parameter values are arbitrary. Mechanistic models tend to be complex, parameters relate to actual physiological processes of a system, and they can be used for prediction. The approach chosen relates to the questions being addressed by the study *(29)*. A

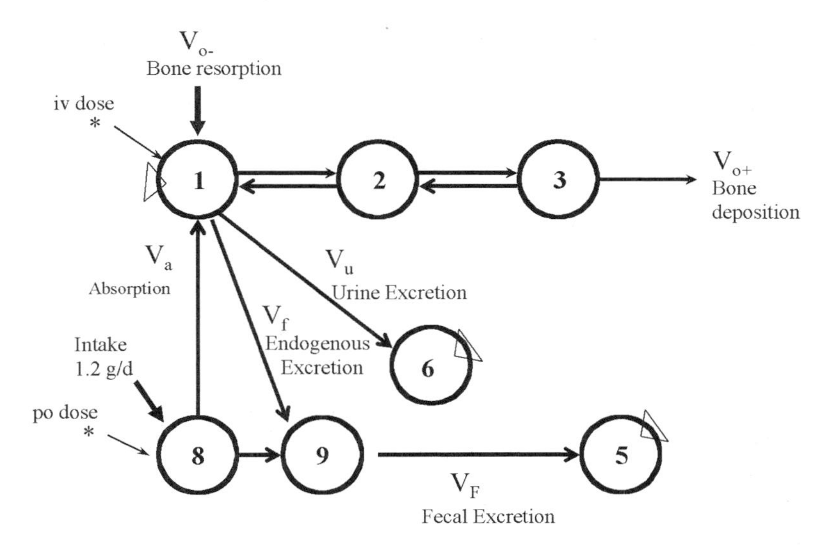

Fig. 3. Model for calcium metabolism. Circles represent compartments, numbers in circles represent compartment number, thin arrows represent movement between compartments, thick arrows represent entry of calcium via the diet or bone resorption (Vo_-). Asterisks indicate entry of tracer and triangles identify sample compartments. Compartment 1 contains blood, compartment 2 soft tissue, and compartment 3 exchangeable calcium on bone. (Adapted from ref. *30.*)

number of modeling packages are available specifically for modeling biological systems, and some of these have been described by Wastney et al. *(29).*

Compartmental models are useful as movement between pools represent known physiological processes, such as absorption, endogenous excretion, and bone deposition. The number of compartments defined from a study can vary based on the sampling frequency and period of the study, but generally three compartments describe the exchange of calcium with serum *(30)* (Fig. 3). The general development of a compartment model has been described *(29),* and some of the challenges in fitting a model to calcium data from human studies have been detailed *(31).* These include the need to simulate multiple tracers per subject, tracer levels in multiple tissues, carryover between studies, and multiple studies, that is, identifying kinetic differences between populations or changes between treatments.

In addition to compartmental approaches, noncompartmental models have been used (e.g., power functions), and results with both approaches have been compared *(20,32).* Weiss et al. *(33)* proposed the use of a non-Markovian model as a new generalized compartment model for calcium kinetics. It differs only in the interpretation of the pathways from other three-compartment models

In fitting any model to data, assumptions made are important for the interpretation of the results. In compartmental modeling these include (1) the system is in steady state (i.e., pool sizes do not change during the study), (2) the tracer does not perturb the system, (3) the sampling period covers both rapid and slow pools, and (4) if the system has been perturbed, a new steady state has been reached.

In addition to the underlying assumptions of a model, an important criterion of modeling is how well the model fits the data. This is determined by lack of consistent deviations between observed and calculated values, low errors associated with fitted parameters, and low correlations between fitted parameters.

4. CALCIUM METABOLIC PARAMETERS

4.1. Absorption

V_a can be determined from balance and kinetic studies as described above (see model Fig. 3). There are many simpler experimental designs that can be used when absorption is the primary metabolism parameter of interest. Eastell et al. *(34)* reported a 1-d method in which oral and intravenous isotopes were administered with all three meals and the ratios determined in the urine. Nevertheless, subjects were adapted to the diets for 7 d prior to administration of the isotope. The method predicted well calcium absorption by balance.

Fractional calcium absorption from a fixed load is useful for determining intrinsic absorptive capacity or for determining bioavailability of calcium sources. There are many study designs that have been used to determine fractional calcium absorption. Most do not adapt subjects to a controlled diet. When absorption is calculated from unabsorbed tracer appearing in the stools, the diet might be controlled long enough to encompass the transit time of the tracer *(35)*. When tracer appearance in blood or urine is used to monitor calcium fractional absorption, often the tracer is given at breakfast following an overnight fast. Typically, the diet is not controlled except for the breakfast when blood is collected or for just 1 d when urine is collected *(36)*. A 24-h urine collection may be sufficient, but when a response delay is expected as occurs in the presence of nondigestible fiber *(37)*, urine might need to be collected for several days. Ideally, oral and intravenous tracers are given and the ratio determined in the blood or urine. This is the most accurate of the simpler methods, aside from whole-body counting, as the oral isotope labels the dietary calcium and its absorption and the intravenous isotope measures the calcium removal from blood. However, a single oral dose may be sufficient. A single 5-h blood draw following an oral dose has been demonstrated to correlate highly with the double-isotope tracer technique *(38,39)*. Good agreement has also been reported between the double-radioisotope method and the fecal recovery method from a single isotope *(26)*. Also, the double stable isotope method and whole-body retention of ^{47}Ca were highly correlated *(40)*.

When determining intrinsic absorption capacity, important considerations are the size of the calcium load and the chemical form of calcium to be administered. As fractional absorption is inversely related to load, all comparisons should be made using the same load. Frequently, loads of between 100 and 300 mg calcium are tested. Some choose the load equivalent to one-third of the daily intake. When fractional absorption is compared across experiments, it is better to include a common source as a reference. Radioisotopes typically are purchased as $CaCl_2$. This soluble isotope can be mixed with milk or juice for consumption or converted to another salt. It is not recommended to give pure $CaCl_2$, as it is a stomach irritant. Alternatively, a capsule of a pre-weighed calcium salt containing the tracer can serve as the oral dose. This is common with stable isotopes of calcium that are purchased as calcium carbonate.

Bioavailability studies are undertaken to determine relative calcium absorption from a calcium-containing food, beverage, or supplement compared to a reference, typically milk or calcium carbonate. Use of a crossover design to compare two or more sources adjusted to the same calcium load eliminates variance associated with factors endogenous to the subject, as described earlier under balance studies. A 5% differences in fractional calcium absorption can usually be detected with 10–15 subjects using a crossover design.

The method chosen to incorporate an isotope into the food being tested for bioavailability deserves thoughtful consideration. Intrinsic labeling techniques, which incorporate isotopes during growth of plants or animals as previously described *(41)* or during the synthesis of a supplement *(42)*, attempts to prepare the label in the same form as endogenous calcium. Extrinsic labeling of calcium sources is simpler and frequently, but not always, allows a good approximation of calcium absorption from intrinsically labeled sources *(43)*. This approach involves premixing a soluble form of the calcium isotope with the food to be tested prior to consumption and assumes the tracer has adequately exchanged with endogenous calcium.

4.2. Urinary Excretion

The major route of obligatory calcium loss is through the urine. Urinary calcium can derive from diet or bone. Thus, it can reflect absorption or bone resorption. Tracers can help distinguish the source of urinary calcium. Typically, urinary calcium is expressed as a 24-h excretion rate.

4.3. Endogenous Excretion

Endogenous fecal excretion is absorbed calcium that has been reexcreted into the gut. Determination of endogenous secretions in 191 perimenopausal women varied inversely with fraction of calcium absorbed and directly with calcium intake *(44)*. The mean was 102 ± 25 mg/d; thus, variance is 25% of the mean. With this amount of variation and dependency on exogenous factors, balances calculated using fecal calcium estimated as unabsorbed calcium from absorption measurements and corrected for estimates of endogenous secretion from the literature without performing stool collections, as proposed by some *(45)*, can give quite different results. These "tracer-assisted" calculated balances reduce variation arising from variation in fecal calcium. However, in a recent metabolic study of postmenopausal women in our laboratory, balances calculated in this way overestimated observed balances by an average of approx 130 mg/d.

Another method employed to determine endogenous excretion is fecal appearance of tracer administered intravenously. However, we observed twice as much intravenous tracer excretion by adult women compared to teen girls, but endogenous excretion rate (calculated by complete kinetic analysis) was shown not to differ between girls and women *(30)*. Fecal intravenous tracer levels were higher in women because serum levels were higher. Serum levels were higher because bone deposition was lower compared to the teens, meaning that tracer remained in serum longer. This shows that tracer approaches that do not utilize data from several tissue sites may lead to erroneous results.

4.4. Bone Turnover

Bone formation rates and bone resorption rates can be determined as Vo_+ and Vo_- using kinetic modeling (Fig. 3). These values are expressed as 24-h rates. To examine the influence of length of study on the value calculated for Vo_+, we fitted various serum profiles of an adolescent subject who participated in a 3-wk balance study reported previously *(30)*. Data were fitted assuming we had 7, 14, and 21 d of blood samples following administration of oral and intravenous stable isotopes (Fig. 4). Model fitting produced different curves if only 7 d of data were available compared to 14 d of data, but the curve did not change appreciably with an additional data point at 21 d. This resulted in considerable differences in calculated Vo_+ and Vo_- if data were available for 2 wk or more compared to only 1 wk (Table 3). The shorter the study, the greater the overestimate of Vo_+ and Vo_-. Assuming 21 d resulted in accurate values for Vo_+ and Vo_-, 7-d data overestimated Vo_+ by 46.6% and Vo_- by 57.1%, whereas 14 d overestimated Vo_+ and Vo_- by only 1.7%. Similarly, the error overestimate of Vo_+ in adults with decreasing length of study compared to 20 d was 3% for 10 d, 16% for 4 d, and 24% for 1 d *(46)*. In contrast, 7 d were sufficient to accurately determine calcium absorption and urinary calcium (Table 3). Calcium retention determined as $Vo_+ - Vo_-$ was greatly underestimated by only 7 d of data compared to 14 d or more. Note that this applies to the use of kinetic studies to calculate balance. Balance estimated by difference between calcium intake and calcium excretion is not time dependent.

A variety of biochemical markers of bone turnover have been used to estimate bone formation and bone resorption rates from serum and urine samples. They are convenient and conducive for monitoring clinical interventions. However, these methods are not specific for either calcium or bone. They are reported in units reflecting enzyme activities of osteoblasts or collagen breakdown products. Attempts have been made to develop regression equations to transpose biochemical marker values into bone formation and bone resorption rates expressed as mass of calcium per day *(47–49)*. This enables quantitative changes in bone turnover to be calculated from qualitative changes. Although biochemical markers and Vo_+ are highly correlated, the variance of biomarkers is greater than for kinetic parameters

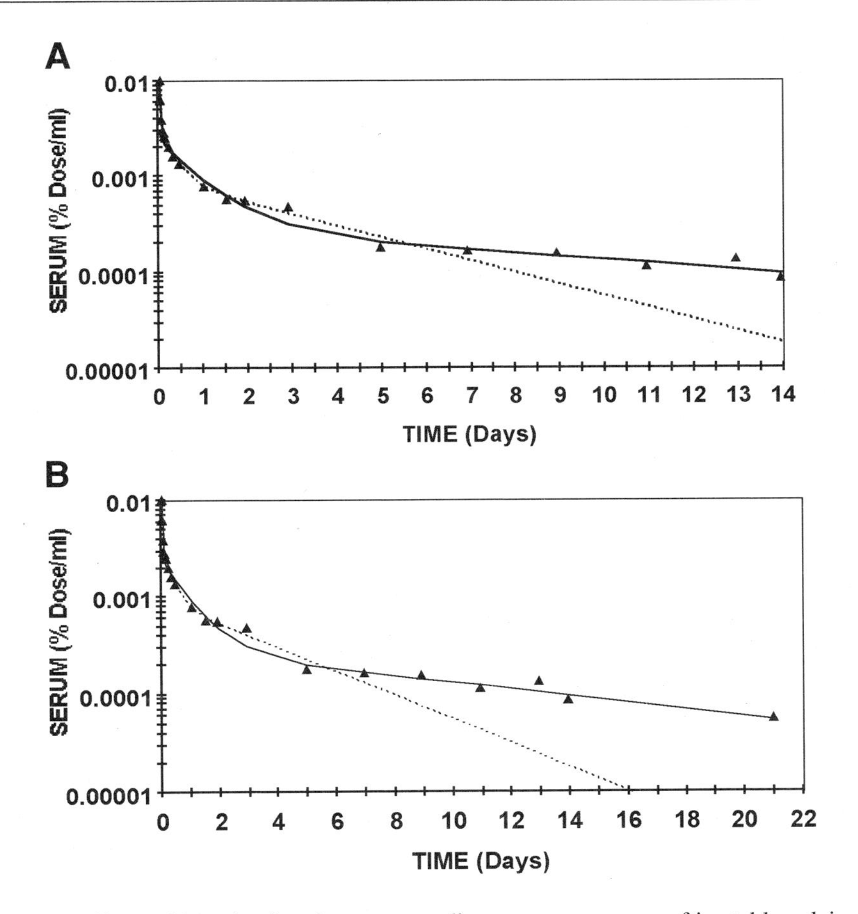

Fig. 4. The effects of length of study on serum disappearance curves of iv stable calcium isotopic tracer in an adolescent girl. Symbols are observed data, lines are values calculated by the model shown in Fig. 2.

Table 3
Results of 7, 14, and 21 d Study in Teen Girl (1300 mg Ca/d intake)

	7 d	*14 d*	*21 d*
L (0,3) fract/d	0.355	0.090	0.085
Absorption (%)	52	49	49
Vo_+ (mg/d)	2282	1583	1557
Vo_- (mg/d)	2273	1472	1447
Balance (mg/d)	8	110	110
V_u (mg/d)	113	113	113

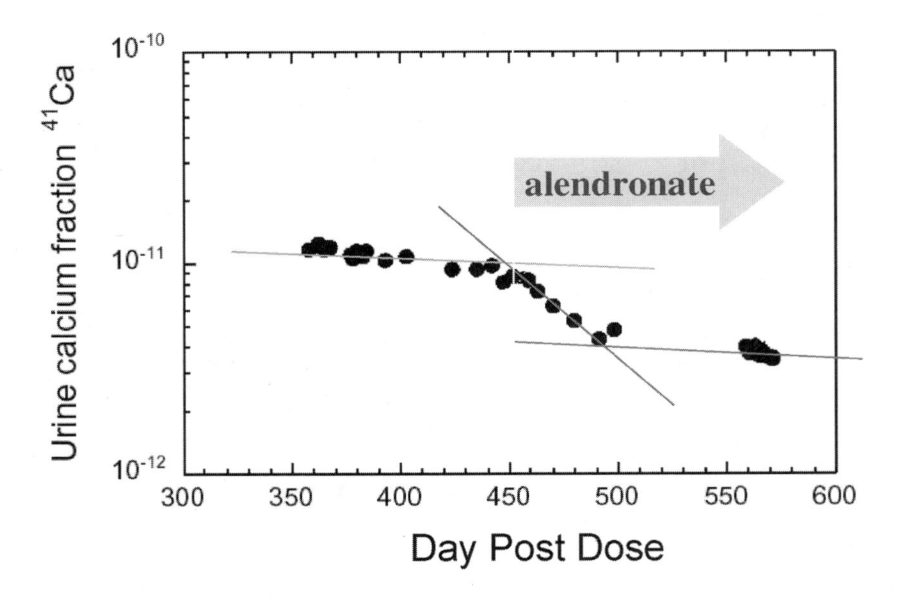

Fig. 5. ^{41}Ca release accurately reflects inhibition in bone loss. The subject in this figure was a postmenopausal women treated with alendronate. (Adapted from ref. *51*.)

of bone turnover *(47)*. Thus, in a small study population, differences in bone formation or resorption rates due to age *(47)* or calcium intake *(8)* are not reflected by differences in biochemical markers of bone turnover. In contrast, calcium supplementation did result in reduced hydroxyproline:creatinine values, a marker of bone resorption, in 14 postmenopausal women in fasting urine samples *(50)*. The variance of biomarkers of bone turnover is reduced in 24-h urine samples compared to fasting urine samples.

Bone resorption rates of change can be monitored sensitively by monitoring urinary ^{41}Ca output from prelabeled skeleton. The suppression of urinary fraction of ^{41}Ca by alendronate is shown clearly and abruptly, as displayed in Fig. 5 *(51)*.

5. CONCLUSIONS

Quantitative nutrition studies such as the balance and kinetic studies described here are seldom undertaken because of the resources required. They typically are limited to rather small sample sizes and can be criticized for lack of generalizability to the larger public. However, when properly conducted, they contribute much information about the quantitative relationship between a nutrient or diet to bone health. Furthermore, they provide insights on the point of metabolism affected by a dietary or nondietary intervention, that is, the gut, kidney, or bone. Rates of changes in calcium metabolism and bone turnover can be determined in rather short studies (days to weeks) compared to changes in bone properties such as bone mineral density (months to years).They provide opportunities for research on the same subject as well as for evaluating intervention therapies.

ACKNOWLEDGMENTS

This work was supported by Public Health Service grants R01AR40553, R01 HD36609, and P50 AT00477.

REFERENCES

1. Beaton GH, Milner BA, Corey P, et al. Source of variance in 24-hour dietary recall data: implications for nutrition study design and interpretation. Am J Clin Nutr 1979; 32:2456–2559.
2. Bauer W, Aub C. Studies of inorganic salt metabolism. I. The ward routine and methods. J Am Dietet Assoc 1927; 3:106–115.
3. Reifenstein EC, Albright F, Wells SL. The accumulation, interpretation, and presentation of data pertaining to metabolic balances, notably those of calcium, phosphorus, and nitrogen. J Clin Endocrinol 1945; 5:367–395.
4. Weaver CM, Martin BR, Plawecki KL, et al. Differences in calcium metabolism between adolescent and adult females. Am J Clin Nutr 1995; 61:577–581.
5. DeSantiago S, Alonso L, Halkali A, Larrea F, Isoard F, Bourges H. Negative calcium balance during lactation in rural Mexican women. Am J Clin Nutr 2002; 76:845–851.
6. Bryant RJ, Wastney ME, Martin BR, et al. Racial differences in bone turnover and calcium metabolism in adolescent females. J Clin Endocrinol Metab 2003; 88:1043–1047.
7. Rambaut PC, Leach CS, Whedon GD. A study of metabolic balance in crewmembers of Skylab W. Acta Astronaut 1979; 6:1313–1322.
8. Wastney ME, Martin BR, Peaock M, et al. Changes in calcium kinetics in adolescent girls induced by high calcium intake. J Clin Endocrinol Metab 2000; 85:4470–4475.
9. Jackman LA, Millane SS, Martin BR, et al. Calcium retention in relation to calcium intake and postmenarcheal age in adolescent females. Am J Clin Nutr 1997; 66:327–333.
10. Palacios C, Wigertz K, Martin BR, Weaver CM. Sweat mineral loss from whole body, patch and arm bag in white and black girls. Nutr Res, 2003; 23:401–411.
11. Charles P, Jensen FT, Mosekilde L, Hanson HH. Calcium metabolism evaluated by ^{47}Ca kinetics estimation of dermal calcium loss. Clin Sci 1983; 65:415–422.
12. Weaver CM, Martin BR, Peacock M. Calcium metabolism in adolescent girls. Challenges of modern medicine. In: Burckhardt P, Heaney RP, eds. Nutritional Aspects of Osteoporosis '94. Ares-Serono Symposium Publications, Rome, Italy 1995; 7:123–128.
13. Dawson-Hughes B, Harris S, Kramich C, DaMal G, Rasmussen HM. Calcium retention and hormonal levels in black and white women on high- and low-calcium diets. J Bone Miner Res 1993; 8:779–787.
14. Malm OJ. Calcium requirement and adaptation in adult men. Scand J Clin Lab Invest 1958; 10(suppl 36):1–280.
15. Isaksson B, Lindholm B, Sjögren B. A critical evaluation of the calcium balance technic. II. Dermal calcium losses. Metabolism 1967; 16:303–313.
16. Wilkinson R. Polyethylene glycol 4000 as a continuously administered non-absorbable fecal marker for metabolic balance studies in human subjects. Gut 1971; 12:654–660.
17. Eastell R, Dewanjee MK, Riggs BL. Comparison of polyethylene glycol and chromium-51 chloride as nonabsorbable stool markers in calcium balance studies. Bone Miner 1989; 6:95–105.
18. Bingham SA, Cummings JH. The use of creatinine ouput as a check on the completeness of 24-hour urine collections. Hum Nutr: Clin Nutr 1985; 39C:343–352.

19. Abrams SA, Yergey Al, Heaney RP. Relationship between balance and dual tracer isotopic measurements of calcium absorption and excretion. J Clin Endocrinol Metab 1994; 79:965–969.
20. Heaney RP. Evaluation and interpretation of calcium-kinetic data in man. Clin Ortho Related Res 1963; 31:153–183.
21. Roth P, Werner E. Interrelations of radiocalcium absorption tests and their clinical relevance. Miner Electrolyte Metab 1985; 11:351–357.
22. Stürup S, Hansen M, Mølgaard C. Measurements of ^{44}Ca:^{43}Ca and ^{42}Ca:^{43}Ca isotopic ratios in urine using high resolution inductively coupled plasma mass spectrometry. J Anal Atomic Spectrom 1997; 12:919–923.
23. Kastenmayer P. Thermal ionization mass spectrometry (TIMS). In: Mellon FA, Sandström B, eds. Stable Isotopes in Human Nutrition. Academic, London, 1996, pp. 81–86.
24. Smith DL. Determination of stable isotopes of calcium in biological fluids by fast atom bombardment mass spectrometry. Anal Chem 1983; 55:2391–2393.
25. Jackson GS, Weaver C, Elmore D. Use of accelerator mass spectrometry for studies in nutrition. Nutr Res Rev 2001; 14:317–334.
26. DeGrazia JA, Ivanovich P, Fellows H, Rich C. A double isotope method for measurement of intestinal absorption of calcium in man. J Lab Clin Med 1965; 66:822–829.
27. Smith DL, Atkin C, Westenfelder C. Stable isotopes of calcium as tracers: methodology. Clin Chim Acta 1985; 146:97.
28. Smith SM, Nyquist LE, Shih C-Y, et al. Calcium kinetics using microgram stable isotope doses and saliva sampling. J Mass Spectrom 1996; 31:1265–1270.
29. Wastney ME, Patterson BH, Linares OA, Greif PC, Boston RC. Investigating Biological Systems Using Modeling: Strategies and Software. Academic Press, San Diego, CA, 1998.
30. Wastney ME, Ng J, Smith D, Martin BR, Peacock M, Weaver CM. Differences in calcium kinetics between adolescent girls and young women. Am J Physiol 1996; 271:R208–R216.
31. Wastney ME, Martin B, Bryant R, Weaver CM. Calcium utilization in young women: new insights from modeling. In: Novotny J, Green MH, Boston RC, eds. Mathematical Modeling in Nutrition and in Health *Sciences*. Kluwer/Plenum, New York, NY 2003, pp. 193–205.
32. Jung A, Bartholdi P, Mermillod B, Reeve J, Neer R. Critical analysis of methods for analyzing human calcium kinetics. J Theor Biol 1978; 73:131–157.
33. Weiss GH, Goans RE, Gitterman M, Abrams SA, Vieira NE, Yergey AL. A non-Markovian model for calcium kinetics in the body. J Pharmacokinet Biopharmaceut 1994; 22(5):367–379.
34. Eastell R, Vieira NE, Yergey AL, Riggs BL. One-day test using stable isotopes to measure true fractional calcium absorption. J Bone Miner Res 1989; 4:463–468.
35. Martin BR, Weaver CM, Heaney RP, Packard PT, Smith DL. Calcium absorption from three salts and $CaSO_4$-fortified bread in premenopausal women. J Agric Food Chem 2002; 50(13):3874–3876.
36. O'Brien KO, Abrams SA. Effects of development on techniques for calcium stable isotope studies in children. Biol Mass Spectrom 1994; 23:357–361.
37. van den Heuvel EG, Mays T, van Dokkum W, Schaafsma G. Oligofructose stimulates calcium absorption in adolescents. Am J Clin Nutr 1999; 69:544–548.
38. Heaney RP, Recker RR. Estimation of true calcium absorption. Ann Intern Med 1985; 103:516–521.
39. Heaney RP, Recker RR. Estimating true fractional calcium absorption. Ann Intern Med 1988; 108:905–906.
40. Beck AB, Bügel S, Stürup S, et al. A novel dual radio- and stable-isotope method for measuring calcium absorption in humans: comparison with the whole-body radioisotope retention method. Am J Clin Nutr 2003; 77:399–405.
41. Weaver CM. Intrinsic Mineral Labeling of Edible Plants: Methods and Uses. CRC Critical Reviews in Food Science and Nutrition 1985; 23:75–101.

42. Weaver CM, Martin BR, Costa NMB, Saleeb FZ, Huth PJ. Absorption of calcium fumarate salts is equivalent to other calcium salts when measured in the rat model. J Agric Food Chem 2002; 50:4974–4975.
43. Weaver CM, Proulx WR, Heaney R. Choices for achieving adequate dietary calcium with a vegetarian diet. Am J Clin Nutr 1999; 70:543S–548S.
44. Heaney RP, Recker RR. Determinants of endogenous fecal calcium in healthy women. J Bone Miner Res 1994; 9:1621–1627.
45. Abrams SA, Sidbury JB, Muenzer J, Esteban NV, Vieira NE, Yergey AL. Stable isotopic measurement of endogenous fecal calcium excretion in children. J Ped Gastroent Nutr 1991; 12:469–473.
46. Neer R, Berman M, Fisher L, Rosenberg LE. Multicompartmental analysis of calcium kinetics in normal adult males. J Clin Invest 1967; 46:1364–1379.
47. Weaver CM, Peacock M, Martin BR, et al. Quantification of biochemical markers of bone turnover by kinetic markers of bone formation and resorption in young healthy females. J Bone Miner Res 1997; 12:1714–1720.
48. Lauffenbarger T, Olah AJ, Dambacher A, Guricaga J, Leutner C, Haas HG. Bone remodeling and calcium kinetic and biochemical study in patients with osteoporosis and Paget's disease. Metabolism 1977; 26:589–605.
49. Charles P, Poser JW, Mosekilde L, Jensen FT. Estimation of bone turnover evaluated by 47Ca-kinetics. J Clin Invest 1985; 76:2254–2258.
50. Horowitz M, Need AG, Philcox JC, Nordin BEC. Effect of calcium supplementation on urinary hydroxyproline in osteoporotic postmenopausal women. Am J Clin Nutr 1984; 39:857–859.
51. Freeman SPHT, Beck B, Bierman J, et al. The study of skeletal Ca metabolism with ^{41}Ca and^{45} Ca. Nuclear Instrum Meth Phys Res 2000; 172B:930–933.

20 Sodium, Potassium, Phosphorus, and Magnesium

Robert P. Heaney

1. INTRODUCTION

Bone health is not a mononutrient issue. Although predominant attention has been given to calcium in recent years (with vitamin D getting honorable mention), other nutrients are also known to affect calcium economy and bone status, even if they are less commonly factored into dietary recommendations for the prevention of osteoporosis or the support of antiosteoporosis therapy. This chapter discusses sodium, potassium, phosphorus, and magnesium, and overlaps with the previous two chapters in matters of the acid/alkaline ash characteristic of the diet. Taken together, these four minerals make up about 6% of the dry, fat-free mass of the human body. Table 1 *(1)* presents their contributions individually, together with that of chloride and calcium—the former because chloride accompanies sodium, both in extracellular fluid and in the diet, and the latter (which is the subject of other chapters in this book) for comparative purposes.

Principal among the ways these nutrients influence bone is the effect that several of them have on obligatory urinary calcium loss. In Nordin's series, urine calcium accounts for approx 40% of the variability in calcium balance *(2),* and in my own, approx 50%. Thus, nutrients affecting urinary calcium loss could in theory have a profound influence on bone maintenance or age-related bone loss. At the same time, it must be noted that the effects of sodium and potassium, for example, on the handling of renal calcium, are much more firmly established than are their putative consequences for bone status. This may be because bony outcomes are harder to study in relation to intakes of these minerals, but it may also be that compensatory mechanisms are, to a greater or lesser extent, offsetting their renal effects and hence preventing or reducing skeletal consequences.

It may be helpful to note at the outset that contemporary intakes of sodium and potassium, particularly, are very different from those to which human physiology is adapted by evolution. Primitive diets are generally recognized to have been very low in sodium (2–30 mmol/d) and relatively very high in potassium (150–250 mmol/d)

From: *Nutrition and Bone Health*
Edited by: M. F. Holick and B. Dawson-Hughes © Humana Press Inc., Totowa, NJ

Table 1
Mineral Composition of the Adult Human Body

	Content[a]	*Percent*[b]
Sodium	80. (1.84)	0.66
Potassium	69. (2.69)	0.96
Phosphorus	387. (12)	4.29
Magnesium	19.6 (0.47)	0.17
Chloride	50. (1.78)	0.63
Calcium	560. (22.4)	8.00

[a] mmol (g)/kg fat-free mass.

[b] Based on dry fat-free mass.

Source: Ref. *1*.

(*see* Chapter 1). By contrast, the contemporary Western diet typically contains 10× the sodium that our hunter–gatherer ancestors would have consumed, and one-third or less the potassium. The potassium:sodium ratio has thus dropped from approx 14:1 to approx 0.5:1. This chapter considers the possible bony consequences of this reversal.

2. SODIUM

The best studied of these four nutrients is sodium, the effects of which are nicely summarized in several review papers *(3–5)*. As long ago as 1937, Aub et al. observed that sodium chloride increased urine calcium *(6)*, and in 1961 Walser showed that sodium and calcium competed for the same reabsorption mechanism in the proximal renal tubule *(7)*. This means that an increase in the filtered load of either sodium or calcium leads to increased clearance of both ions, thereby establishing the mechanistic basis by which a sodium load produces calciuria. A possible role for sodium intake in the pathogenesis of osteoporosis was first emphasized by Goulding, who, in a series of animal and human experiments, showed that sodium intake could affect bone mass in animals, and that the effect required a functioning parathyroid apparatus *(8–12)*.

Numerous studies have found statistically significant positive correlations between 24-h urine sodium excretion and 24-h urine calcium *(6,9–11,13–22)*. Taken together, the available studies indicate that urine calcium rises by from 0.5 to 1.5 mmol (20–60 mg) for every 100 mmol (2300 mg) sodium ingested *(5)*. Most reviewers have used the midpoint of that range (i.e., 1.0 mmol/100 mmol) to characterize the effect.

Given the fact that contemporary sodium intakes generally fall between 100 and 200 mmol/d (i.e., 2300–4600 mg/d), it follows that approx 1.0–2.0 mmol (40–80 mg) of the 24-h total urine calcium excretion is being pulled out of the body by sodium. Ho et al. *(21)* concluded that sodium intake was the principal determinant of urine calcium in Hong Kong Chinese, and Matkovic et al. *(22)* came to a similar conclusion for pubertal girls in the United States. Itoh and Suyama *(13)*, in a study of nearly 900 Japanese adults, in whom sodium intakes tend to be much higher than among Europeans or North Americans, found a positive correlation between sodium intake and urine calcium in both sexes, and across all age groups, even after adjusting for weight and for dietary intakes of protein, phosphorus, and calcium.

Thus, on prevailing diets, sodium intake accounts for most of the obligatory urinary loss of calcium from the body. Clearly, if absorbed calcium is less than the amount needed to offset this loss (in addition to what is needed to cover cutaneous and digestive juice losses), then bone mass must suffer.

It would be expected that increased urinary loss following a sodium load would produce a fall in extracellular fluid calcium ion concentration, and this has been described in some studies *(16,23)*. Such a fall would also produce a rise in parathyroid hormone (PTH), with a consequent increase in synthesis of 1,25-dihydroxyvitamin D [$1,25(OH)_2D_3$], and ultimately in calcium absorption efficiency. Breslau et al. demonstrated that a change in urine calcium evoked the predicted change in PTH, $1,25(OH)_2D_3$, and calcium absorption efficiency, at least in premenopausal women *(16,18)*. However, they failed to observe changes in absorption in a small study involving postmenopausal women. These findings suggested, at least qualitatively, that premenopausal women can handle contemporary sodium intakes with less skeletal impact than postmenopausal women and, by implication, that sodium intakes may be contributing to postmenopausal osteoporosis.

At least two groups of investigators have found that calcium absorption efficiency varies inversely with induced calciuria, whether from a sodium load *(16,23)* or from the calcium-sparing effect of thiazides *(24)*. Breslau et al. *(16)* reported that the change, while occurring in normal subjects, did not occur in two patients with surgical hypoparathyroidism, consistent with Goulding's findings in rats (8). By contrast, Meyer et al. *(23)* did find an increase in calcium absorption efficiency in two patients with hypoparathyroidism following a sodium load, suggesting that the response was mediated by some mechanism other than increased PTH secretion. Although the discrepancy between these studies cannot be resolved with available data, it does appear reasonably certain that, other effects aside, a sodium load leads to increased PTH secretion with all of its usual consequences [increased $1,25(OH)_2D_3$ synthesis, increased calcium absorption, increased bone resorption, and improved renal tubular reabsorption of calcium]. However, additional mechanisms may be operative as well.

Additionally, several groups of investigators have shown that bone remodeling, as measured by various remodeling biomarkers, varied inversely with sodium intake *(9,11,25,26)* and that sodium restriction reduced excretion of resorption biomarkers. This finding is consistent with the effect of sodium loads on PTH secretion. Often, this remodeling effect has been taken to indicate that sodium increases bone loss, although this does not necessarily follow, and there are few reports of a direct connection between sodium intake and subnormal bone status in humans at typical sodium intakes.

There is at least one case report of probably salt-associated osteoporosis *(27)*. A 50-yr-old postmenopausal woman with adequate hormone replacement therapy had high-turnover osteoporosis with vertebral compression fractures and a urine calcium in excess of 7.5 mmol/d (300 mg). She was observed to be using table salt from a paper bag, in quantities so large as to make the food on her plate white, and she reported having done so for the previous 20 yr. Reduction in salt intake reduced her urinary calcium loss to below 2.5 mmol (100 mg)/d.

Whether clinically significant bone loss actually occurs at more typical salt intakes has been the subject of very few studies. Greendale et al. found no association between sodium intake from diet records and bone status 15 yr later *(28)*. Sodium intakes in their subjects averaged about 150 mmol (3450 mg)/d in men and 112 mmol (2576 mg)/d in women. However, estimates of sodium intake from diet records correlate poorly with actual sodium intake, and thus this negative finding cannot absolve sodium intake in this connection. Accurate estimates of sodium intake require measurement of 24-h urine sodium, preferably over a several-day interval. This need probably explains the virtual absence of epidemiological studies showing an association of sodium intake with bone mass.

Dawson-Hughes et al. *(14)*, in a 4-yr prospective study, found a highly significant correlation between sodium intake (as measured by urinary sodium) and urinary calcium excretion in healthy elderly men and women, but no correlation of sodium intake with bone mineral density at any site, in either sex. This failure to find skeletal differences is suggestive of some degree of compensation for the sodium-induced calciuria. Sodium intakes in their study averaged 156 mmol (3600 mg)/d in men, and 118 mmol (2700 mg)/d in women.

The principal human study linking high salt intake to bone loss was by Devine et al. *(29)*, who showed that change in bone mineral density at the total hip site over a 2-yr period was inversely related to sodium intake estimated from urine sodium content. However, they found no such effect at the spine, femoral neck, intertrochanteric region, or radius. From multiple regression models these investigators calculated that halving the sodium intake of their subjects would have obliterated hip bone loss. However, in the same model, doubling of calcium intake (i.e., raising it into the currently recommended range) would have produced approximately the same beneficial effect.

Summarizing the data available up to 2000, much as has been done in this chapter, Burger et al. *(30)* concluded that a sodium–osteoporosis link was still conjectural.

Comment. Based on the multiple regression model of Devine et al. *(29)*, one might conclude that contemporary sodium intakes elevate the calcium requirement—at least for bone status. However, the two strategies suggested by the Devine model are not equivalent. High calcium intakes confer numerous nonskeletal health benefits *(31)*, whereas the benefits of low sodium intakes, although widely touted, are at best problematic *(32)*. Furthermore, from the standpoint of feasibility, higher calcium intakes are much easier to achieve and sustain than are reductions in sodium intake of the magnitude required to offset sodium's effect on obligatory urinary calcium excretion *(33)*. Finally, one must note that, even if sodium's effect on bone mass is normally compensated for by adaptive increases in calcium absorption (or by high calcium intakes), any accompanying increase in bone remodeling may constitute a risk factor for fracture *(34)*. Hence, choosing one or the other of the options offered by the Devine model would seem to be more prudent than doing neither.

For the most part, when the papers cited in this connection speak of "sodium," what is meant is "sodium *chloride,*" i.e., table salt, the form in which about 90% of contemporary sodium intakes are ingested. The accompanying anion is usually ignored. This is probably a mistake. Berkelhammer et al. *(35)* showed clearly, in patients receiving total parenteral nutrition (TPN), that substituting acetate for chloride in TPN solutions reduced urine calcium losses dramatically. In oral feeding studies, Lutz *(36)* showed that substituting sodium bicarbonate for sodium chloride promptly reduced urine calcium. Similarly, sodium bicarbonate loads do not induce an increase in urine calcium, unlike sodium chloride *(36,37)*. Thus, clearly the anion is important, at least for the understanding of what is happening. Nevertheless, it remains true that contemporary diet sodium is overwhelmingly in the form of sodium chloride, and as such is usually hypercalciuric in its effect. Even this statement, however, is not absolute, as the next section will show.

3. POTASSIUM

Potassium is a largely intracellular cation. Bone mineral does not contain appreciable quantities of potassium—only that small amount which is trapped when calcium phosphate is precipitated out of an extracellular fluid phase that happens to contain some potassium. There are no recognized abnormalities of bone or bone cellular function associated with values of serum potassium in the range of concentrations typically encountered.

Probably the principal importance of potassium lies in its effects on the processes that maintain calcium homeostasis, particularly urinary calcium conservation and excretion. Low-potassium diets are known to increase urinary calcium loss, and high-potassium diets to reduce it *(38–41)*. Such observations, by themselves, could mean simply that diet potassium is a marker for other food constituents responsible for the effect. This, in fact, is partly correct (see below). But pure potassium salts—typically the bicarbonate or citrate salts—exhibit the same inverse relationship to urine calcium, pointing to a role specifically for potassium itself.

Perhaps most striking of potassium's effects is the fact that potassium (as the citrate) completely blocks the calciuria of a large sodium chloride load (Fig. 1) *(42)*. It is believed that both the potassium cation and the bicarbonate anion (to which citrate is metabolized) work in the distal renal tubule, facilitating reabsorption of the extra calcium not reclaimed in the proximal tubule because of competition with

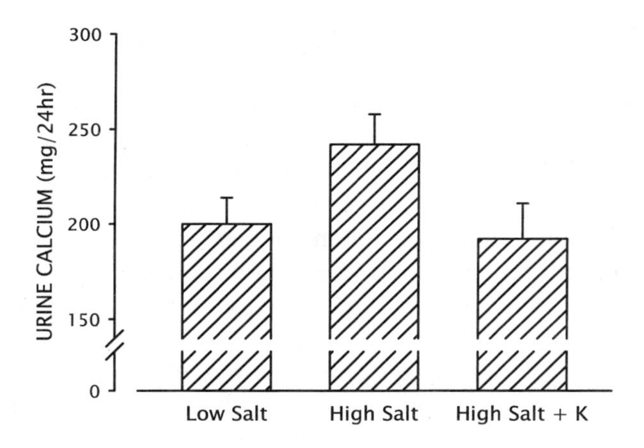

Fig. 1. Effect of a high salt load, with and without supplemental potassium citrate, on 24-h urine calcium excretion in postmenopausal women. The low-salt regimen provided 87 mmol (5 g) salt/d and the high-salt regimen, 225 mmol (13.2 g)/d. The potassium supplement provided 90 mmol (29.2 g) potassium citrate/d. $N = 26$ for each of the treatment groups. The rise in urine calcium on the high-salt regimen was highly statistically significant ($p < 0.005$). Plotted from the data of Sellmeyer et al. *(42)*.

sodium for the transport mechanism. Figure 2 illustrates, schematically, the differing effects on urine calcium of various sodium and potassium salts.

However, just as the undoubted effects of sodium on urine calcium have not yet been unambiguously shown to have corresponding effects on bone (see prior section), so, therefore, amelioration of those effects by potassium has not been clearly shown to confer a skeletal benefit, although, in short-term metabolic experiments, potassium bicarbonate does produce a positive calcium balance shift *(41,43)*.

There are, however, limited data with regard to the association of bone status and dietary potassium intake. New et al. *(44,45),* for example, in observational studies, showed a significant inverse relationship between potassium intake and bone mineral density (BMD) at both hip and spine. Once again, it is not certain that the effect is due solely to the higher potassium content or whether potassium instead is a marker for other food constituents (*see* Chapter 15).

Potassium is ubiquitous in the diet, but is found most abundantly in green and root vegetables, followed closely by fruits, then by legumes and milk (or yogurt). A diet high in potassium will necessarily be high in vegetables. In addition to their potassium content, such foods have an alkaline/ash characteristic and thus, effectively, the anion associated with potassium in such foods will be bicarbonate. By contrast, wheat, rice, corn, and other cereal grains have very low potassium contents and generally exhibit an acid-ash characteristic (because of their high content of sulfur-containing amino acids). Thus, in brief, foods high in

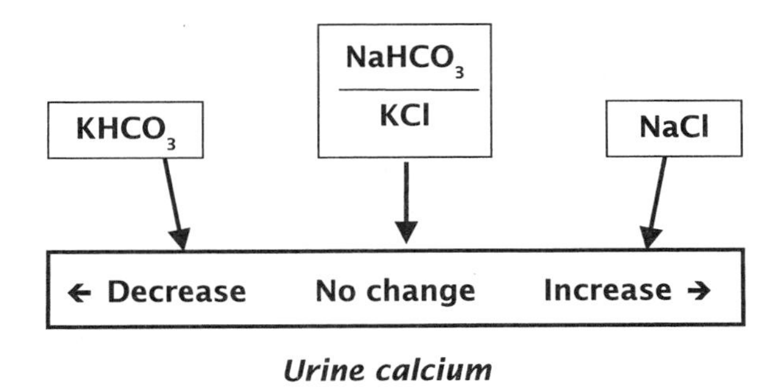

Fig. 2. Effects of various sodium and potassium salts on urine calcium.

potassium generally have an alkaline ash characteristic, and alkaline ash foods will generally be good sources of potassium.

New et al. *(46,47)* have also shown a significant inverse relationship in free-living subjects between net endogenous acid production (NEAP) from ingested foods and lumbar spine bone mineral density. The effect was small (less than 2.5% difference in BMD between upper and lower quartiles of NEAP), but consistent with studies by others showing a calciuric effect of food-based acid production *(38,39)*. New has also shown significantly higher excretion of bone remodeling biomarkers at the highest quartile of NEAP, and she also reports a small but significant difference in NEAP between postmenopausal women with and without fracture.

Given the generally higher potassium content of vegetarian diets, it may be useful to contrast bone status in individuals with vegan and omnivore diets, which represent quasi-extremes of food intake patterns. Studies using contemporary bone assessment technologies *(48–50)* indicate that, in general, not only do the vegans not have denser bones, they tend actually to have somewhat lower BMD, despite the fact that they have higher potassium intakes than do omnivores. Although the vegetable intake of vegans will usually be higher than that of omnivores, their intake of cereal grain products will usually be higher as well. Cereals, as already noted, are very poor sources of potassium and generally produce an acid-ash residue as well (i.e., high NEAP). Hence, any switch between vegan and omnivore diets involves trade-offs. Clearly, the mere substitution of vegetable for animal protein sources does not seem to confer a skeletal advantage, and may, in fact, do the opposite. One may note, in passing, that the primitive human diet was omnivorous.

Comment. Given the abundant mechanistic evidence of beneficial effects on the calcium economy of high-potassium foods and the absence of evidence to support a sometimes presumed dichotomy between animal and vegetable protein sources, the most prudent recommendation for total health (and possibly skeletal health as well) would seem to substitute fruits and green and root vegetables for the high-energy, low-nutrient-density foods (so-called "junk foods") common in the Western diet. This, by itself, would move contemporary intakes of potassium toward the paleolithic norm and would decrease NEAP as well, thereby helping to restore the environmental context to which human physiology is adapted.

4. PHOSPHORUS

Bone mineral is generally characterized as an imperfect hydroxyapatite, i.e., a calcium phosphate salt with a Ca:P molar ratio approximating 1.7:1. The phosphate anion actually makes up more than half the mass of bone mineral. Like the calcium cation, the phosphate of bone mineral is derived from the blood flowing past a bone mineralizing site. Although the formation of the apatite crystal nucleus requires osteoblast work, subsequent crystal growth is purely passive, involving diffusion of the constituent ions down a concentration gradient from the solution phase in blood to the solid phase in bone. At average adult concentrations of calcium and phosphorus, blood is approximately twice saturated with respect to hydroxyapatite, and hence it contains a sufficient quantity of both ions to sustain usual levels of bone formation. However, at the mineralizing site, concentrations in the extracellular fluid around the osteoblast are lower than in the general circulation because calcium and phosphorus are being pulled into bone. In fact, adult serum concentrations may not suffice during growth, when a great deal of mineral is being transferred into the skeleton. In essentially all growth situations—animal and human—serum inorganic phosphorus (P_i) is higher than in adult humans, by as much as two- to threefold. Growth hormone, which raises the renal phosphorus threshold and thereby helps to sustain higher serum P_i concentrations, is a part of the explanation.

Low serum P_i concentrations, on whatever basis, lead to osteomalacia, and are one of the hallmarks of that disorder, whatever may be its ultimate cause. As noted, the bone-forming cell (the osteoblast) is particularly sensitive to ambient P_i concentrations, mainly because the mineralizing process it has induced in the osteoid deposited beneath it depletes its microenvironment of phosphate. This is the reason why essentially all of the osteomalacias exhibit not only impaired mineralization, but impaired osteoblast function as well.

At the other extreme, high serum P_i concentrations increase the risk of extraskeletal calcification. At physiological pH, and without an apatite crystal nucleus, calcium and phosphorus would come out of solution as $CaHPO_4 \cdot H_2O$, and for this solid phase, serum is only half-saturated. Thus the same serum concentrations of calcium and P_i are indefinitely stable in the absence of an apatite crystal nucleus, but provide abundant mineral to support crystal growth at suitably nucleated sites. The risk of extraskeletal calcification occurs when the serum Ca × P product rises to or past saturation. Preventing this result is a principal reason for controlling phosphorus absorption in patients with endstage renal disease.

Also, it is sometimes argued, high phosphorus intakes lead to increased PTH secretion, an outcome presumed to be bad for bone. High phosphorus intakes have even been proposed as contributing to the pathogenesis of osteoporosis *(51)*.

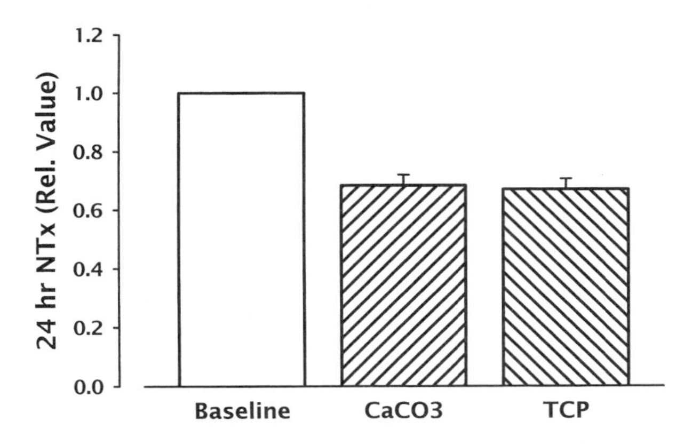

Fig. 3. Twenty-four-hour urine NTx excretion in 28 postmenopausal women under conditions of low calcium intake ("Baseline") and then following 1 wk each of supplementation with calcium at 45 mmol (1800 mg)/d, either as calcium carbonate or as tricalcium phosphate (TCP). The latter provided, in addition to its calcium, 30 mmol (930 mg) supplemental phosphorus daily and resulted in an increase over baseline intake of 150%. All values are expressed relative to each woman's baseline excretion. NTx excretion was decreased by 32% and 33%, respectively ($p < 0.001$ relative to baseline). There was no difference in effect between the two calcium salts.

However, when this issue has been examined directly *(52,53),* it has generally been found that phosphorus supplements decrease bone turnover markers, rather than increasing them, an effect probably due to interference by ambient phosphate with osteoclast response to PTH *(54).* Figure 3 shows the results of one such experiment, in which 24-h urinary *N*-telopeptide excretion after 1 wk of supplementation with tricalcium phosphate was contrasted with that following calcium carbonate. As the figure shows clearly, remodeling suppression was the same for both salts. Moreover, as is generally recognized, phosphorus supplements lower, rather than raise, urine calcium loss, and increased phosphorus intake does not lead to negative calcium balance *(55)*—as would be expected if phosphorus were somehow adversely affecting bone mass. Thus, in the remainder of this section I shall ignore the high end of the distribution of phosphorus intakes, and will focus instead on the possible importance of low phosphorus intakes, particularly in the context of antiosteoporosis therapy.

Phosphorus is widely distributed in natural food sources and, because of its incorporation into the structure and machinery of most cellular tissues, protein and phosphorus tend to go together in the diet. As a result, a diet containing an amount of protein adequate for health will also contain adequate phosphorus.

Unlike calcium, for which net absorption is low (10–15% of intake), net absorption of phosphorus is much more efficient, ranging from 50% to 70% from typical diets. The distinction between net and gross absorption is important because of the fact that there is a considerable amount of both calcium and phosphorus entering the digestive stream from endogenous sources. Much of this is in the form of digestive secretions, but for phosphorus particularly, shed mucosal cells contribute importantly. Fecal phosphorus of endogenous origin has not been frequently measured, but available data suggest that it is of the order of 2–3 mmol (60–90 mg)/d and total digestive juice phosphorus is of the order of 8 mmol (250 mg)/d *(56)*.

For individuals with normal or even moderately impaired renal function, the kidneys are able to handle the relatively large amount of absorbed phosphorus without difficulty, and with only minor elevation of the serum P_i. In Nordin's calculations *(57)* for normal renal functioning, serum P_i rose by less than one-third over a tripling of phosphorus intake. (This is one of the reasons why high phosphorus intakes are probably not much of a problem for individuals with adequate renal function.)

As long as serum P_i concentrations are above 1.0 mmol (3.1 mg/dL)/L, working osteoblasts and bone mineralizing sites in a typical mature adult will "see" adequate quantities of phosphorus to support their activities. (Although the reference lower limit of normal for serum phosphorus extends down to as low as 0.7 mmol [2.2 mg/dL]/L, effects of concentrations at the lower end of the "normal" range on osteoblast and osteoclast function have not been well studied. Hence the *functional* lower limit of normal has yet to be rigorously defined.) However, hypophosphatemia, by whatever definition, is uncommon in older adults; fully two-thirds of patients in our osteoporosis clinic have serum P_i concentrations above 1.2 mmol (3.6 mg/dL)/L. It is important to stress, before the analyses that follow, that a patient who does not experience a fall in serum phosphorus to suboptimal levels is not experiencing effective phosphorus deficiency, no matter what else may be happening in the operation of the calcium and phosphorus economies.

The recommended daily allowance (RDA) in the United States for phosphorus is 700 mg (23 mmol)/d for adults *(58),* and median intakes for adults are above that level at all ages *(59)*. Hence, unlike with calcium, there is no evidence of prevailing phosphorus deficiency. However, the low tail of the distribution of phosphorus intakes reveals a different picture *(59)*. Five percent of all adult women ingest less than 70% of the RDA on any given day; 10% of women over age 60, and 15% of women aged 80 or older also have such low intakes. As already hinted, low phosphorus intakes mean insufficient intakes of protein and calcium as well, and they thus reflect a more global situation of malnutrition. Many of these individuals with low phosphorus intakes will have osteoporosis as well, and will be recipients of current-generation antiosteoporosis therapies, essentially all of which now include recommendations for supplemental calcium. The potential for phosphorus deficiency increases precisely in the context of individuals with low phosphorus intakes, who are taking calcium supplements, and who are receiving antiosteoporosis therapies (Fig. 4).

Although calcium supplements are widely recognized (and employed) in nephrology to bind food phosphorus, the field of osteoporosis has largely

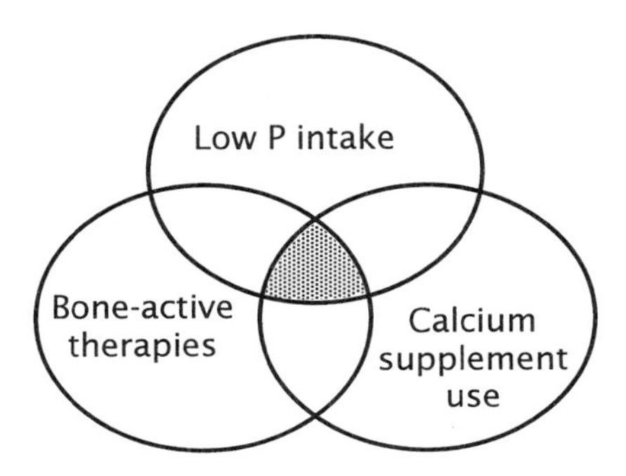

Fig. 4. Venn diagram illustrating the set of patients with osteoporosis most likely to exhibit effective phosphorus deficiency (i.e., the intersection of the three sets). For these purposes, "low P intake" refers to intakes below 70% of the RDA; "calcium supplement use" refers to the carbonate or citrate salts (principally); and "bone active therapies" refers to both antiresorptives and anabolics, but particularly the latter. The sizes of the sets and of their intersections are not intended to be quantitatively accurate.

ignored that fact *(60)*. In individuals with already low intakes, the effect of calcium supplementation will be a reduction in available phosphorus, which could, potentially, induce hypophosphatemia. The extent of absorptive interference will depend on the relative quantities of calcium and phosphorus ingested and on the timing of their ingestion. Heaney and Nordin *(60)* showed that, for mixed food intake, each 12.5 mmol of calcium (500 mg) reduces phosphorus absorption by approx 5.4 mmol (167 mg). As would be expected, given the fact that the mechanism is chemical complexation of the two species, calcium not ingested at the same time as phosphorus will produce less interference with phosphorus absorption *(61)*. The binding observed by Heaney and Nordin was considerably lower than what would be predicted from the chemistry involved, and probably reflects variability in timing of the calcium and phosphorus-containing foods in their subjects' diets.

Using the binding relationship just defined, it can be calculated that calcium supplement intake amounting to 1500 mg (37.5 mmol) Ca/d as the carbonate (or citrate), and ingested with meals (in accordance with usual instruction), will bind approx 16 mmol (~500 mg) phosphorus, i.e., essentially all of the ingested phosphorus in individuals with intakes below 70% of the RDA. Given the ability of the kidney to reabsorb essentially all of the filtered phosphorus at low filtered loads, even this total binding would probably not lead to negative phosphorus balance or to hypophosphatemia in most mature adults, and would likely not adversely affect ordinary bone repair and maintenance. The potential for deficiency arises principally in the context of antiosteoporosis therapy. Antiresorptive agents produce as much as a 5% bone gain in the first year of therapy, followed by slow gain thereafter at a rate of between 0.5 and 1% per year *(62)*. Anabolic agents increase bone mass, at least at the axial skeleton, by as much as 15% per year *(63,64)*. It is not clear that the associated positive phosphorus balance could develop or be sustained if most or all of the diet phosphorus were to be rendered unavailable.

The steady-state slow bone gain of antiresorptive therapy poses the smaller challenge. A gain of 0.5 to 1% per year translates to a positive calcium balance of 0.25–0.5 mmol (10–20 mg)/d, and a positive phosphorus balance of 0.15–0.3 mmol (4.5–9 mg)/d. Given the likelihood that not all food phosphorus would be co-ingested with a calcium supplement, it is unlikely that binding would be complete; thus sufficient phosphorus would probably be absorbed to sustain the amount of bone gain plausible with antiresorptive treatment.

By contrast, the steady-state bone gain produced by the anabolic agents is a full order of magnitude greater. A gain of 15% per year translates to a positive calcium balance of approx 7.7 mmol (308 mg)/d and to a phosphorus balance of approx 5.0 mmol (155 mg)/d. With complete or near complete binding of food phosphorus in individuals on low phosphorus intakes, such a high positive balance would not be possible. Moreover, teriparatide, the only anabolic agent approved for osteoporosis treatment in the United States, directly lowers the renal phosphorus threshold, thus lowering serum P_i concentration in its own right, apart from any effect of net movement of P_i out of serum and into bone.

It must be stressed that this analysis is largely theoretical. The calculations above are for something approaching the largest bone gain possible with each class of treatments—unlikely at a total skeletal level. Furthermore, no data exist showing impairment of treatment response in patients with low phosphorus intakes (as would be predicted from the foregoing analysis). At the same time, it must be noted that a phosphate effect has not actually been looked for. Subjects entered into clinical trials typically exhibit a healthy volunteer effect and are thus unlikely to show the prevalence of low phosphorus intakes found in a true population sample. Furthermore, calcium supplement doses in most published trials have probably been suboptimal (and hence food phosphorus binding correspondingly incomplete). Finally, phosphorus intake would not have been assessed in trials of antiosteoporosis agents. Hence, to the extent that phosphorus deficiency may have been present in their treatment groups, or contributed to the variability of their response, the effect of such deficiency would necessarily have gone unrecognized.

In brief, beyond the certainty of phosphorus binding by nonphosphate calcium supplements (such as the carbonate or citrate salts), there are very few facts on either side of the question. Actually, the question itself probably never would have arisen had it not been for advances in the treatment of osteoporosis, which now make possible bone gains rivaling in magnitude those experienced during the adolescent growth spurt.

Comment. It should be noted that the combination of low food phosphorus intake and large calcium supplement doses constitutes a distinctly unnatural situation.

Typical high-calcium diets from food sources exhibit a Ca:P molar ratio in the range of 0.8:1–1.2:1, whereas the treatment context just analyzed has a Ca:P ratio of about 2.5:1, something essentially never found in nature.

Several courses of action present themselves at this stage of our ignorance. Perhaps most obvious is to use a calcium phosphate supplement, instead of the carbonate or citrate salts, in patients being treated with potent anabolic agents. Another, of course, is to assess total nutritional status in our patients, and to correct insufficiencies. This option, although preferable, is unlikely to occur. Typically, patients with osteoporosis will be placed on pharmacotherapy and given a calcium supplement recommendation without regard to their home diets and nutritional status. Although this is unfortunate, it is also the reality.

5. MAGNESIUM

Although 50–60% of total body magnesium is found in bone, functionally magnesium is largely an intracellular cation, and signs of its deficiency include system-wide cellular dysfunctions. In the field of calcium and bone, severe magnesium deficiency, with hypomagnesemia, results in refractory hypocalcemia, resulting from both impaired parathyroid gland response to hypocalcemia and impaired bone cell response to PTH. In brief, the entire calcium homeostatic apparatus is disabled. However, there is no evidence that this kind of dysfunction occurs at levels of magnesium nutrition associated with normal serum magnesium levels or typical dietary magnesium intakes.

Magnesium is also considered to condition bone mineral crystal solubility, largely by substitution of magnesium for calcium in surface positions of the hydroxyapatite lattice. Although this effect may help optimize the processes that control fluxes of calcium into and out of bone, at the same time, this magnesium effect is largely passive, and will be a function of ambient magnesium levels in the extracellular fluid. Although serum magnesium is generally considered an unreliable indicator of magnesium nutritional status, still it may be the only relevant factor with respect to this effect on bone mineral solubility.

The RDA for magnesium is approx 13 mmol (320 mg)/d in women and 17.5 mmol (420 mg)/d in men *(58)*. More than 70% of the adult population in the United States falls below these recommended intakes *(59)*. It is unclear whether this shortfall has skeletal consequences. Rude *(65)* has recently reviewed the evidence relating magnesium intake to bone status. In general, observational studies of the association of magnesium intake and bone mass or age-related bone loss have produced mixed results—some showing a weak interrelationship, but most showing no apparent effect. Similarly, magnesium supplementation trials have produced mixed results—some showing small skeletal effects; others, nothing at all.

There is a widespread misconception among the general public that magnesium is necessary for calcium absorption, or at least for calcium to exert its proper efficacy. This is the reason given for the popularity of combined calcium–magnesium supplements, such as Dolomite. If there is any factual basis for this belief, it must lie in the effects of magnesium in animals or humans with severe magnesium deficiency in whom, as already noted, calcium regulation is crippled. However, at typical magnesium intakes and prevailing levels of serum magnesium, there is abundant evidence showing that no such relationship exists. Spencer et al. *(66)* more than doubled magnesium intake in normal volunteers and found no effect on calcium absorption, using rigorous absorption methodology.

Even more to the point, the large body of literature showing an effect of calcium on bone accrual during growth, and the reduction of age-related bone loss and fragility fractures in the elderly, consists of studies all of which were done without magnesium supplementation. Both the Chapuy *(67)* and the Dawson-Hughes *(68)* trials produced dramatic fracture reductions (40% and 55%, respectively) using calcium and vitamin D supplementation alone, without extra magnesium. It might be argued that giving supplemental magnesium would have improved these results still further, but that is pure speculation, without a base in evidence (except possibly in patients with celiac sprue, in whom magnesium supplementation may confer a skeletal benefit *[69]*).

Comment. The richest dietary source of magnesium is legumes. Grains and root and green vegetables also have high magnesium densities, but their content of magnesium per serving is lower than in legumes. As a consequence, just as a diet without dairy products will fall short of the calcium intake recommendation, so a diet without legumes will tend to fall short of the magnesium RDA. To the extent that currently recommended magnesium intakes may reflect the paleolithic norm, our hunter–gatherer forebears would have gotten their magnesium principally from roots, greens, and nuts.

6. CONCLUSIONS

The health of bone is dependent on total nutrition, including adequate intakes not only of calcium and vitamin D, but of potassium, phosphorus, magnesium, and trace metals as well. Furthermore, bone health may be compromised by excessive sodium intakes. Bone health is manifested in two principal ways: bone massiveness (expressed as structural strength) and bone cell function (expressed as bone growth and repair). Acute nutrient deficiencies may significantly impair bone cell function without appreciably affecting bone strength, at least immediately. Hence, such deficiencies tend to be silent. Conversely, excess of sodium chloride or deficiencies of potassium and phosphorus, which can impair calcium conservation or bone mineralization, when they have any effect at all, alter bone mass, although slowly. Hence, their effects, if any, are hard to detect and often delayed in onset. Outside of certain special therapeutic or disease situations, the usually encountered variations in intakes of sodium, potassium, phosphorus, and magnesium are without major skeletal consequences.

REFERENCES

1. Documenta Geigy. Scientific Tables. 7th ed. Basle, Switzerland, 1970.
2. Nordin BEC, Morris HA. Osteoporosis and vitamin D. J Cell Biochem 1992; 49:19–25.
3. Nordin BEC, Need AG, Morris HA, Horowitz M. The nature and significance of the relationship between urinary sodium and urinary calcium in women. J Nutr 1993; 123:1615–1622.
4. Nordin BEC, Need AG, Morris HA, Horowitz M. Sodium, calcium and osteoporosis. In: Burckhardt P, Heaney RP, eds. Nutritional Aspects of Osteoporosis. Serono Symposia, Vol. 85. Raven Press, New York, 1991, pp. 279–295.
5. Massey LK, Whiting SJ. Dietary salt, urinary calcium, and bone loss. J Bone Miner Res 1996; 11:731–736.
6. Aub JC, Tibbetts DM, McLean R. The influence of parathyroid hormone, urea, sodium chloride, fat and of intestinal activity upon calcium balance. J Nutr 1937; 113:635–655.
7. Walser M. Calcium clearance as a function of sodium clearance in the dog. Am J Physiol 1961; 200:769–773.
8. Goulding A. Effects of dietary NaCl supplements on parathyroid function, bone turnover and bone composition in rats taking restricted amounts of calcium. Miner Electrolyte Metab 1980; 4:203–208.
9. Goulding A. Fasting urinary sodium/creatinine in relation to calcium/creatinine and hydroxyproline/creatinine in a general population of women. N Z Med J 1981; 93:294–297.
10. Goulding A, Campbell D. Dietary NaCl loads promote calciuria and bone loss in adult oophorectomized rats consuming a low calcium diet. J Nutr 1983; 113:1409–1414.
11. Goulding A, Lim PE. Effects of varying dietary salt intake on the fasting excretion of sodium, calcium and hydroxyproline in young women. N Z Med J 1983; 96:853–854.
12. Goulding A. Osteoporosis: why consuming less sodium chloride helps to conserve bone. N Z Med J 1990; 103:120–122.
13. Itoh R, Suyama Y. Sodium excretion in relation to calcium and hydroxyproline excretion in a healthy Japanese population. Am J Clin Nutr 1996; 63:735–740.
14. Dawson-Hughes B, Fowler SE, Dalsky G, Gallagher C. Sodium excretion influences calcium homeostasis in elderly men and women. J Nutr 1996; 126:2107–2112.
15. Muldowney FP, Freaney R, Moloney MF. Importance of dietary sodium in the hypercalciuria syndrome. Kidney Int 1982; 22:292–296.
16. Breslau NA, McGuire JL, Zerwekh JE, Pak CYC. The role of dietary sodium on renal excretion and intestinal absorption of calcium and on vitamin D metabolism. J Clin Endocrinol Metab 1982; 55:369–373.
17. Castenmiller JJ, Mensink RP, van der Heijden L, et al. The effect of dietary sodium on urinary calcium and potassium excretion in normotensive men with different calcium intakes. Am J Clin Nutr 1985; 41:52–60.
18. Breslau NA, Sakhaee K, Pak CYC. Impaired adaptation to salt-induced urinary calcium losses in postmenopausal osteoporosis. Trans Assoc Am Physicians 1985; 98:107–116.
19. Need AG, Morris HA, Cleghorn DB, De Nichilo D, Horowitz M, Nordin BEC. Effect of salt restriction on urine hydroxyproline excretion in postmenopausal women. Arch Intern Med 1991; 151:757–759.
20. Zemel MB, Gualdoni SM, Walsh MF, et al. Effects of sodium and calcium on calcium metabolism and blood pressure regulation in hypertensive black adults. J Hypertension 1986; 4(suppl 5):S364–S366.
21. Ho SC, Chen YM, Woo JL, Leung SS, Lam TH, Janus ED. Sodium is the leading dietary factor associated with urinary calcium excretion in Hong Kong Chinese adults. Osteoporos Int 2001; 12:723–731.
22. Matkovic V, Ilich JZ, Andon MB, et al. Urinary calcium, sodium, and bone mass of young females. Am J Clin Nutr 1995; 62:417–425.

23. Meyer WJ III, Transbol I, Bartter FC, Delea C. Control of calcium absorption: effect of sodium chloride loading and depletion. Metabolism 1976; 25:989–993.
24. Zerwekh JE, Pak CYC. Selective effects of thiazide therapy on serum 1α,25-dihydroxyvitamin D and intestinal calcium absorption in renal and absorptive hypercalciurias. Metabolism 1980; 29:13–17.
25. McParland BE, Goulding A, Campbell AJ. Dietary salt affects biochemical markers of resorption and formation of bone in elderly women. Br Med J 1989; 299:834–835.
26. Need AG, Morris HA, Cleghorn DB, DeNichilo DD, Horowitz M, Nordin BEC. Effect of salt restriction on urine hydroxyproline excretion in postmenopausal women. Arch Intern Med 1991; 151:757–759.
27. Palmieri GMA. Osteoporosis and hypercalciuria secondary to excessive salt ingestion. J Lab Clin Med 1995; 126:503.
28. Greendale GA, Barrett-Connor E, Edelstein S, Ingles S, Haile R. Dietary sodium and bone mineral density: results of a 16-year follow-up study. J Am Geriatr Soc 1994; 42:1050–1055.
29. Devine A, Criddle RA, Dick IM, Kerr DA, Prince RL. A longitudinal study of the effect of sodium and calcium intakes on regional bone density in postmenopausal women. Am J Clin Nutr 1995; 62:740–745.
30. Burger H, Grobbee DE, Drueke T. Osteoporosis and salt intake. Nutr Metab Cardiovasc Dis 2000; 10:46–53.
31. Heaney RP. Ethnicity, bone status, and the calcium requirement. Nutr Res 2002; 22:153–178.
32. Taubes G. The (political) science of salt. Science 1998; 281:898–907.
33. Hooper L, Bartlett C, Smith GD, Ebrahim S. Systematic review of long term effects of advice to reduce dietary salt in adults. Br Med J 2002; 325:628–636.
34. Heaney RP. Is the paradigm shifting? Bone 2003; 33:457–465.
35. Berkelhammer CH, Wood RJ, Sitrin MD. Acetate and hypercalciuria during total parenteral nutrition. Am J Clin Nutr 1988; 48:1482–1489.
36. Lutz J. Calcium balance and acid-base status of women as affected by increased protein intake and by sodium bicarbonate ingestion. Am J Clin Nutr 1984; 39:281–288.
37. Lemann J Jr, Gray RW, Pleuss JA. Potassium bicarbonate, but not sodium bicarbonate, reduces urinary calcium excretion and improves calcium balance in healthy men. Kidney Int 1989; 35:688–695.
38. Morris RC Jr, Frassetto LA, Schmidlin O, Forman A, Sebastian A. Expression of osteoporosis as determined by diet-disordered electrolyte and acid-base metabolism. In: Burckhardt P, Dawson-Hughes B, Heaney RP, eds. Nutritional Aspects of Osteoporosis. Academic, New York, 2001, pp. 357–378.
39. Buclin T, Cosma M, Appenzeller M, et al. Diet acids and alkalis influence calcium retention in bone. Osteoporos Int 2001; 12:493–499.
40. Lemann J Jr, Pleuss JA, Gray RW, Hoffmann RG. Potassium administration reduces and potassium deprivation increases urinary calcium excretion in healthy adults. Kidney Int 1991; 39:973–983.
41. Lemann J Jr, Pleuss JA, Gray RW. Potassium causes calcium retention in healthy adults. J Nutr 1993; 123:1623–1626.
42. Sellmeyer DE, Schloetter M, Sebastian A. Potassium citrate prevents increased urine calcium excretion and bone resorption induced by a high sodium chloride diet. J Clin Endocrinol Metab 2002; 87:2008–2012.
43. Sebastian A, Harris ST, Ottaway JH, Todd KM, Morris RC Jr. Improved mineral balance and skeletal metabolism in postmenopausal women treated with potassium bicarbonate. N Engl J Med 1994; 330:1776–1781.
44. New SA, Bolton-Smith C, Grubb DA, Reid DM. Nutritional influences on bone mineral density: a cross-sectional study in premenopausal women. Am J Clin Nutr 1997; 65:1831–1839.

45. New SA, Robins SP, Campbell MK, et al. Dietary influences on bone mass and bone metabolism: further evidence of a positive link between fruit and vegetable consumption and bone health. Am J Clin Nutr 2000; 71:142–151.
46. New SA. The role of the skeleton in acid-base homeostasis. Proc Nutr Soc 2002; 61:151–164.
47. New SA. Impact of food clusters on bone. In: Burckhardt P, Dawson-Hughes B, Heaney RP, eds. Nutritional Aspects of Osteoporosis. Academic Press, New York, 2001, pp. 379–397.
48. Barr SI, J. Prior C, Janelle KC, Lentle BC. Spinal bone mineral density in premenopausal vegetarian and nonvegetarian women: cross-sectional and prospective comparisons. J Am Diet Assoc 1998; 98:760–765.
49. Chiu JF, Lan SJ, Yang CY, et al. Long-term vegetarian diet and bone mineral density in postmenopausal Taiwanese women. Calcif Tissue Int 1997; 60:245–249.
50. Lau EMC, Kwok T, Woo J, Ho SC. Bone mineral density in Chinese elderly female vegetarians, vegans, lacto-vegetarians and omnivores. Eur J Clin Nutr 1998; 52:60–64.
51. Calvo MS, Park YK. Changing phosphorus content of the U.S. diet: potential for adverse effects on bone. J Nutr 1996; 1168S–1180S.
52. Silverberg S, Shane E, Clemens TL, et al. The effect of oral phosphate administration on major indices of skeletal metabolism in normal subjects. J Bone Miner Res 1986; 1:383–388.
53. Bizik BK, Ding W, Cerklewski FL. Evidence that bone resorption of young men is not increased by high by high dietary phosphorus obtained from milk and cheese. Nutr Res 1996; 16:1143–1146.
54. Raisz L, Niemann I. Effect of phosphate, calcium and magnesium on bone resorption and hormonal responses in tissue culture. Endocrinology 1969; 85:446–452.
55. Spencer H, Kramer L, Osis D, Norris N. Effect of phosphorus on the absorption of calcium and on the calcium balance in man. J Nutr 1978; 108:447–457.
56. Wilkinson R. Absorption of calcium, phosphorus and magnesium. In: Nordin BEC, ed. Calcium, Phosphate and Magnesium Metabolism. Churchill Livingstone, London, 1976, pp. 36–111.
57. Nordin BEC. Phosphorus. J Food Nutr 1988; 45:62–75.
58. Dietary Reference Intakes for Calcium, Magnesium, Phosphorus, Vitamin D, and Fluoride. Food and Nutrition Board, Institute of Medicine. National Academy Press, Washington, DC, 1997.
59. Alaimo K, McDowell MA, Briefel RR, et al. Dietary intake of vitamins, minerals, and fiber of persons 2 months and over in the United States: Third National Health and Nutrition Examination Survey, Phase 1, 1988–91. Advance data from vital and health statistics; no. 258. National Center for Health Statistics, Hyattsville, MD, 1994.
60. Heaney RP, Nordin BEC. Calcium effects on phosphorus absorption: implications for the prevention and co-therapy of osteoporosis. J Am Coll Nutr 2002; 21:239–244.
61. Schiller LR, Santa Ana CA, Sheikh MS, Emmett M, Fordtran JS. Effect of the time of administration of calcium acetate on phosphorus binding. N Engl J Med 1989; 320:1110–1113.
62. Heaney RP, Yates AJ, Santora AC II. Bisphosphonate effects and the bone remodeling transient. J Bone Miner Res 1997; 12:1143–1151.
63. Neer RM, Arnaud CD, Zanchetta JR, et al. Effect of parathyroid hormone (1-34) on fractures and bone mineral density in postmenopausal women with osteoporosis. N Engl J Med 2001; 344:1434–1441.
64. Arnaud CD, Roe EB, Sanchez MS, Bacchetti P, Black DM, Cann CE. Two years of parathyroid hormone 1–34 and estrogen produce dramatic bone density increases in postmenopausal osteoporotic women that dissipate only slightly during a third year of treatment with estrogen alone: results from a placebo-controlled randomized trial. Bone 2001; 28:S77.
65. Rude RK. Magnesium deficiency: a possible risk factor for osteoporosis. In: Burckhardt P, Dawson-Hughes B, Heaney RP, eds. Nutritional Aspects of Osteoporosis. Academic Press, New York, 2001, pp. 263–271.
66. Spencer H, Fuller H, Norris C, Williams D. Effect of magnesium on the intestinal absorption of calcium in man. J Am Coll Nutr 1994; 13:483–492.

67. Chapuy MC, Arlot ME, Duboeuf F, et al. Vitamin D_3 and calcium to prevent hip fractures in elderly women. N Engl J Med 1992; 327:1637–1642.
68. Dawson-Hughes B, Harris SS, Krall EA, Dallal GE. Effect of calcium and vitamin D supplementation on bone density in men and women 65 years of age or older. N Engl J Med 1997; 337:670–676.
69. Rude RK, Olerich M. Magnesium deficiency: possible role in osteoporosis associated with gluten-sensitive enteropathy. Osteoporos Int 1996; 6:453–461.

21 Fluoride and Bone Health

Johann D. Ringe

1. INTRODUCTION

The halogen fluorine is ubiquitous in nature, counted among the 15 more abundant elements on the earth's surface. Accordingly, it is commonly found in soils, water, plants, and animal tissues *(1)*. The ionic form of fluorine, fluoride, is the most electronegative of the elements of the periodic system. It combines reversibly with hydrogen to form the very aggressive acid "hydrogen fluoride" (HF). Because of its high affinity for calcium, fluoride is attracted by calcified tissues, i.e., bone and teeth.

Fluorine is an indispensable trace element and plays a role in normal development and maintenance of the skeleton and teeth *(2)*. As is true for other essential trace elements, deficiency or excess has clinical consequences: intakes below the recommended daily dose result in growth and development retardation, whereas long-term high intake leads to hyperostosis or even severe skeletal sclerosis (= fluorosis). The latter was described for the first time in the 1940s after observation of cases with severe endemic or industrial fluorosis *(3–5)*.

The same is true for the teeth. Excessive intake of fluoride may produce dental fluorosis, whereas adequate supply reduces the risk of caries. In this latter regard, a highly successful water and topical fluoridation program has dramatically reduced the prevalence and severity of dental caries.

The use of fluoride as a fracture-preventing agent for the skeleton has not yet been as successful. There is no doubt that pharmacological doses of fluoride stimulate osteoblastic bone formation and significantly augment bone mass. Positive effects on bone mineral density (BMD) in human subjects are well documented. Fluoride has the advantages of being highly bone-specific, very inexpensive, and it can be administered orally.

The seemingly narrow therapeutic window for the effects of fluoride on the skeleton, however, has hampered progress in this field. Despite four decades of clinical experience with fluoride, uncertanties still exist about the quality of the newly formed bone tissue and the effectiveness of fluoride treatment in reducing vertebral fracture rate *(6)*. Large studies powered to evaluate definitively vertebral fracture rates are not available, owing, in part, to the low price of fluoride salts and

From: *Nutrition and Bone Health*
Edited by: M. F. Holick and B. Dawson-Hughes © Humana Press Inc., Totowa, NJ

the lack of patent protection of the product. Ringe and Meunier have therefore defined fluoride as an orphan drug *(7)*.

The scientific literature on the ability of fluoride to inhibit the initiation and progression of dental caries and on fluoride as a therapeutic agent in osteoporosis is huge, whereas the data concerning the physiological role of fluoride for bone health are relatively scarce *(8)*.

2. FLUORIDE METABOLISM

2.1. Gastrointestinal Absorption and Skeletal Accretion

In the stomach, fluoride is absorbed by passive nonionic diffusion in the form of HF, the amount of HF formed being dependent on gastric acidity. Absorption is more rapid and more complete in the fasting state, when the gastric contents are more acidic. Fluoride not absorbed in the stomach is readily absorbed from the upper small intestine.

On average, about 50% of orally ingested fluoride is absorbed by the gastrointestinal (GI) tract. The amount absorbed in the individual case, however, is largely influenced by dietary concentrations of calcium and other cations with which fluoride forms insoluble and poorly absorbed salts. The fluoride concentrations of body fluids and tissues are related to the long-term levels of intake. There is no homeostatic regulation *(9)*.

About 99% of total body fluoride content is deposited in the calcified tissues, where it is strongly but not irreversibly bound. A rapidly exchangeable part is located in the hydration shells of bone crystals, whereas the slowly exchangeable pool is the fluoride deposited as fluoroapatite. The latter can only be mobilized by bone resorption during the process of bone remodelling.

The GI absorption of fluoride given in supraphysiological doses as a treatment of osteoporosis differs considerably for different preparations. Whereas intestinal absorption of sodium fluoride (NaF) is as high as 95% when used as a plain preparation, delayed-release and/or enteric-coated preparations will show a significantly reduced bioavailability of the fluoride ions to variable degrees. The major advantage of these preparations, however, lies in their fewer GI side effects.

Disodium-monofluorophosphate (Na_2PO_4F, usually called MFP) has the same high intestinal absorption rate as plain NaF (95–100%), but a significantly lower rate of gastrointestinal side effects *(10)*. This fluoride compound dissociates in a acqueous milieu into Na^+ and $(FPO_3)^{2-}$ ions. The monofluorophosphate ion is absorbed before hydrolysis mainly in the upper small intestine. It is then hydrolyzed by alkaline phosphatases in the intestinal mucosa into fluoride and phosphate ions. The low rate of epigastric side effects of MFP is the result of the fact that no irritating free fluoride ions are released in the GI tract before absorption.

2.2. Renal Excretion

In normal adults, who are not knowingly ingesting fluoride, fasting plasma levels range from 0.5 to 2.3 μM (8.5–45 ng/mL). Concentrations progressively increase with age *(11)*.

After absorption, fluoride is either deposited in bone or excreted by the kidney. Renal excretion of fluoride is positively correlated with serum fluoride levels *(12)*. The renal clearance of fluoride is about 30–40 mL/min in adults *(13)*. It results from almost unrestricted glomerular filtration followed by a variable degree of tubular reabsorption, which is inversely related to tubular fluid pH.

The extrarenal clearance of fluoride (about 40 mL/min) reflects mainly accretion into bone *(12)*. The sum of renal and calcified tissues clearance amounts to approx 75 mL/min in healthy subjects. That means that, in adults, about 50% from the absorbed fluoride is retained in the calcified tissues and about 50% is excreted in the urine. However, in the developing skeleton and teeth of children, up to 80% may be retained because of an increased uptake during growth *(14)*.

For therapeutic use of fluoride it is important to note that the uptake of fluoride by bone tissue is not homogeneous. There is a higher proportion taken up by cancellous sites than by cortical sites. This may account for the heterogeneous response of different skeletal regions observed in clinical studies.

It should be emphasized that, inasmuch as the kidney is the major excretory route for fluoride, patients with impaired renal function should be given fluoride cautiously. Fluoride is not contraindicated, but the dosage must be carefully adjusted depending on the fluoride serum level.

3. INTAKE OF FLUORIDE

3.1. Water and Beverages

The local water fluoride concentration is not suitable for estimating the average daily fluoride intake of a population. In Western countries the consumption of fabricated beverages is increasingly displacing the consumption of tap or well water. The different beverages, however, are prepared with water of quite different degrees of fluoridation.

A study of 225 children aged 2–10 yr showed that more than 50% of the daily fluid intake consisted of soft drinks, juices, tea, and other beverages, altogether ranging from 585 to 756 mL/d *(15)*. The fluoride concentration of these beverages ranged from nondetectable to 6.7 mg/L.

The fluoride concentration in human milk ranges from 0.007 to 0.011 mg/L *(14)*, i.e., with an average intake of 800 mL of human milk per day, the nursing infant is provided with 6–9 μg fluoride. The average intake of fluoride by formula-fed infants depends on whether the product is ready-to-feed or requires the addition of water. When using powdered or liquid-concentrate infant formulas, the fluoride concentration will depend mainly on the water used to reconstitute the respective products. The fluoride contents of ready-to-feed formulas in the United States generally range from 0.1 to 0.2 mg/L, i.e., the fluoride intake is higher compared to breast-fed babies. The average daily fluoride intakes from infancy to early childhood in fluoridated areas are relatively constant at about 0.05 mg/kg body weight per day *(8)*. Ten independent North American studies have shown that dietary fluoride intakes in adults ranged from 1.4 to 3.4 mg/d in areas where the water fluoride concentration

Table 1
Fluoride Concentrations of Selected Foods

Food	*Fluoride concentration (mg/L or kg)*	
	Average	*Range*
Fruits	0.06	0.02–0.08
Meat, fish, poultry	0.22	0.04–0.51
Oils and fats	0.25	0.02–0.44
Dairy products	0.25	0.02–0.82
Leafy vegetables	0.27	0.08–0.70
Sugar and adjunct substances	0.28	0.02–0.78
Root vegetables	0.38	0.27–0.48
Grain and cereal products	0.42	0.08–2.01
Potatoes	0.49	0.21–0.84
Legume vegetables	0.53	0.49–0.57

Source: Adapted from ref. *17.*

was 1.0 mg/L. In areas where the water concentration was less than 0.3 mg/L, the daily intakes ranged from 0.3 to 1.0 mg/d *(16).* On a body-weight basis, adults' fluoride intake is generally lower than it is in children.

Brewed tea contains fluoride at rather high concentrations, ranging from 1 to 6 mg/L. This is due to the ability of tea leaves to accumulate fluoride to amounts exceeding 10 mg/100 g dry weight.

3.2. Intake From Food

Average fluoride concentrations of some foods are shown in Table 1. The data are based on concentrations found in prepared foods that were served to adult hospital patients *(17).* When preparation of foods requires the use of water (e.g., boiling of vegetables), the respective fluoride content will of course influence the fluoride concentration of the food.

The fluoride content of marine fish is often overestimated. It may be relatively higher in small fish that are consumed including bones. The intestinal resorption rate of fluoride from different foods may be reduced by the calcium content of the respective nutrient or simultaneous ingestion with calcium-rich dairy products or calcium supplements.

3.3. Intake From Dental Products

Major contributors to total daily fluoride intake may be toothpastes. The amounts often approach or even exceed intake from the diet, particularly in young children who have poor control of the swallowing reflex. They introduce approx

0.8 mg of fluoride into the mouth with each brushing. The fraction swallowed and absorbed ranges from about 10% to nearly 100%. An average of about 0.30 mg of fluoride is ingested with each brushing by young children *(18)*.

4. EFFECTS ON CALCIFIED TISSUES

4.1. Effects on Bone Cells

Fluoride stimulates bone formation directly, without the need for prior bone resorption. The first in vitro evidence that fluoride acts directly on osteoblasts to stimulate proliferation was reported by Farley et al. in 1983 *(19)*. This study was induced by earlier bone morphometric studies showing that the positive effect of fluoride treatment on bone balance was due mainly to augmented bone formation without increased bone resorption. The increased formation rate was related to an increased number of osteoblasts *(20)*.

The in vitro mitogenic activity of fluoride was confirmed subsequently by several groups studying bone cells of different species, including humans *(21–23)*. It has been demonstrated that fluoride-treated bone tissue is more resistant to resorption than untreated bone *(24)*. This is more likely to be the consequence of a more stable crystal lattice of fluorapatite as compared to hydroxyapatite than any direct inhibitory effect of fluoride on osteoclasts

4.2. Molecular Mechanisms of Action

The precise mechanism(s) of the bone-specific mitogenic potential of fluoride on preosteoblasts is still unknown. Several hypotheses have been proposed, but none explains all observations so far. Fluoride requires the presence of bone cell growth factors, such as insulin-like growth factor (IGF)-1 or transforming growth factor (TGF)-β to stimulate bone cell proliferation *(25)*. Therefore, fluoride may be considered as a mitogen enhancer rather than a bone cell mitogen per se. The mitogenic effect of fluoride is dependent on the phosphate concentration of the medium.

There is evidence that an increased level of tyrosyl phosphorylation of a protein that regulates mitosis (MAP kinase) could be involved in the stimulatory effect on osteoblasts *(26)*. The most likely explanation for the increased tyrosyl phosphorylation of key mitogenic proteins is the ability of fluoride to inhibit a specific osteoblastic phosphotyrosyl protein phosphatase. This growth factor-dependent tyrosine phosphorylation pathway of fluoride action is enhanced further in the presence of aluminum *(27)*.

4.3. Effects on Teeth

The most important effect of fluoride on teeth is the carioprotective action. During tooth development, fluoride is incorporated into the tooth's mineralized structures. The presence of fluoride in the developing dental enamal probably increases later resistance to demineralization when the tooth surface is exposed to organic acids. Posteruptive effects, however, are believed to be more important for the carioprotective properties of fluoride *(28)*. Two main mechanisms contribute to this phenomenon:

1. Effects on the metabolism of bacteria in dental plaques *(29)*
2. Effects on de- and remineralization of enamel during acidogenic challenge

Plaque fluoride concentrations are related to the frequencies of exposure to water, beverages, foods, and dental products and their respective fluoride concentrations. Another important source is the fluoride content of salvia, which depends on the individual's fluoride supply. The halogen inhibits several enzymes of plaque bacteria, which limits the uptake of glucose and thus reduces the amount of acid secreted into the extracellular plaque fluid *(30)*. As a consequence, the pH drop in plaque fluid is reduced and thereby the acidic challenge to the enamel.

In erupted teeth an optimum fluoride content reduces the acid solubility and promotes the remineralization of incipient enamel lesions. These are initiated at the ultrastructural level several times each day according to the frequency of eating or drinking carbohydrate-rich foods metabolizable by plaque bacteria. Accordingly, the erupted teeth of children and adults require frequent exposures to fluoride throughout life to achieve and maintain adequate concentrations of the ion in enamel and dental plaque.

5. FLUORIDE INSUFFICIENCY

5.1. Skeleton

There is a strong affinity toward fluoride in bone, and small amounts of fluoride are found in healthy skeletons of animals and humans. The fluoride content of bone increases with age, suggesting that fluoride is rather firmly sequestered once it is deposited in bone *(31)*. Accordingly, a "normal fluoride content" for bone tissue is not defined, and the amounts differ considerably depending on age, geographic differences in the respective fluoride concentrations of natural drinking water, and individual alimentation habits *(32)*.

Nevertheless, it is a general view that fluoride is essential for normal skeletal development *(2,33)*. Unequivocal scientific data proving that insufficient fluoride supply will lead to skeletal abnormalities and growth retardation have not been published. To our knowledge there are no animal trials that studied the skeletal effects of a low- or fluoride-free diet. According to Bunker, fluoride may have a role in bone mineralization, but no essential function has been proven in humans *(34)*.

5.2. Teeth

There are also no experimental data on the effects of inadequate fluoride intake on teeth. However, in epidemiological studies, sufficient indirect evidence can be found that in communities with low fluoride concentrations in water, caries is significantly more frequent than in regions with "optimal" fluoride contents (range 0.7–1.2 mg/L). The lack of fluoride at any age obviously places an individual at an increased risk for dental caries *(35)*.

6. HIGH FLUORIDE EXPOSURE

6.1. Skeletal Fluorosis

Chronic fluoride intoxication as manifested by diffuse osteosclerosis has been observed in areas where the drinking water contained fluorine in concentrations higher than 4 mg/L. The risk of fluorosis is further increased in regions with a hot and dry climate, i.e., a need for high water consumption. First cases of the so-called *endemic fluorosis* were described in India in the state of Madras as early as 1937 *(5)*. Subsequently, cases of endemic fluorosis have been documented in other parts of India and sporadically from almost all parts of the world. Even earlier, in 1932, a Danish group had described in workers with long-term exposure to cryolithe dust, a "curious whiteness and density of the ribs, claviculae and vertebrae" in X-rays of the lungs *(3)*. After documentation of further cases of *industrial fluorosis,* Roholm *(4)* described three different stages of this disease. Although stage 1 and 2 represent different degrees of hyperostosis or osteosclerosis, in stage 3 additional severe calcifications of tendons and ligaments and sometimes even extraskeletal ossifications are observed. A study from Switzerland on 43 aluminum industry workers reported relations between skeletal X-ray findings, urinary fluoride excretion, and articular pain *(36)*.

The obvious fluoride-related increase in bone density in mild cases of endemic and industrial fluorosis was the basis for the concept of treating osteoporotic patients with fluoride. The first therapeutic trial in humans was published by Rich and Ensick in 1961 *(37)*. With increasing therapeutical use of fluoride, then, as a third form *iatrogenic fluorosis* became a problem *(38)*.

From our own long-term experience with fluoride treatment it must be stated that iatrogenic fluorosis is not a drug side effect as such, but the result of incorrect therapy. We observed two forms. The first type very quickly (often in less than 24 mo) produces a distinct sclerosed bone structure in the lateral roentgenogram or the spine. In such cases it has always been found in retrospect that the diagnosis of osteoporosis was based on a poor X-ray film, i.e., that osteoporosis was probably not present at the start of fluoride treatment. An individually strong response to fluoride therapy might be involved or a potentiation of fluoride effect by aluminum *(27)*. The second type of iatrogenic fluorosis occurs in patients with established osteoporosis only after excessively long-term and/or high-dosage fluoride treatment, i.e., usually after 6–10 yr. While in the first type normally shaped vertebrae show a more or less homogeneous white, ivory-like appearance, for the second type, coarsening of the previously impressed end plates and a remarkable thickening of the trabeculae are characteristic (Fig. 1).

6.2. Dental Fluorosis

In the early part of the 20th century it was observed that residents of certain areas developed brown strains on their teeth and that these stained teeth were highly resistant to decay. Further research proved that the prevalence and severity

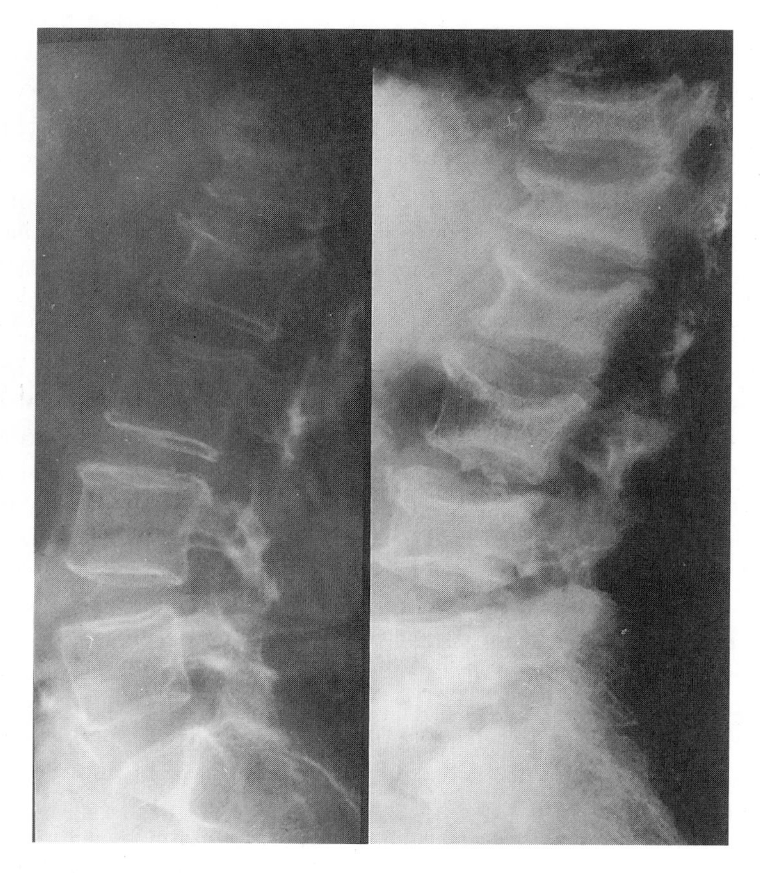

Fig. 1. Lateral X-rays of the lumbar spine of a 68-yr-old woman with established osteoporosis. *Left side:* At onset of fluoride therapy, fractures at L1 and L2. *Right side:* after 8 yr high-dose fluoride therapy (36 mg fluoride ions per day), fluorosis and new fractures at L3 and L4 (iatrogenic fluorosis type II, see text).

of "mottled enamel" (severe dental fluorosis) was directly associated with the concentration of fluoride in the water of the respective region. Three grades of dental changes have been described *(39)*. Teeth affected with mild fluorosis have the appearance of being hard, lustrous, and polished, with minute opaque flecks scattered irregularly. These white opaque spots increase in size until they occupy about half the tooth surface. With increasing degree of fluorosis, pits, grooves, and depressions of irregular area and depth appear, with a typical discoloration varying in shade from light to dark brown. Epidemiological studies proved that fluoride intake at "optimal concentrations" protected against dental caries *(40)*.

7. PREVENTION OF FRACTURES

7.1. Water Fluoridation

The benefit of fluoridation of drinking water in the prevention of dental caries has been overwhelmingly substantiated *(40,41)*, while the effects on BMD and fracture rates have turned out inconsistent *(42,43)*.

The evidence regarding the effect of fluoride in drinking water on fracture incidence is based on comparisons of fracture rates by geographic regions with different concentrations of fluoride in drinking water either naturally or adjusted.

Already in 1955, Leone et al. observed that radiographic signs of osteoporosis were substantially less severe in people living in Bartlett County, Texas, with a water fluoride content of 8 ppm (ppm = mg/kg) than in Framingham, Massachusetts, with a water fluoride content of only 0.09 ppm *(44)*. A subsequent study in two communities in North Dakota with different water fluoride levels showed higher bone densities on X-ray of 300 subjects living in the high-fluoride area as compared to subjects with low fluoride intake. They also noted a significant reduction in osteoporotic fractures in the lumbar vertebrae in the women of the high-fluoride community. However, there was no difference in the prevalence of osteoporotic fractures in the men in the two areas *(45)*. A rather large number of following studies comparing rates of fractures between fluoridated and nonfluoridated communities have variously found that exposure to fluoride by water increases the risk of hip and other nonvertebral fractures *(46,47)*, has no effect on fracture incidence *(48,49)*, or decreases the risk of hip fractures *(50,51)*. In an animal study, femoral bending strength was measured in rats on fluoride intakes that ranged from low levels to levels well above natural high-fluoride drinking water *(52)*. Fluoride had a positive effect on bone strength for lower and a negative for higher intakes, i.e., bone strength followed a biphasic relationship with bone fluoride content. The maximum bone strength occurred at 1.216 ppm of fluoride in the bone ash.

In a study from Germany, data from two towns of the former German Democratic Republic could be compared *(53)*. In the city of Chemnitz, drinking water had been fluoridated with 1ppm (1 mg/L) for a period of 30 yr, while the town of Halle had no fluoridation during the same interval. Calculating age-adjusted incidences, significantly fewer hip fractures occurred in Chemnitz in both men and women (Fig. 2). There were, however, no significant differences in BMD values between the population samples of the two towns, and the authors speculated that with 1 ppm fluoride no anabolic tissue level of fluoride can be reached to stimulate osteoblasts. The lower fracture rate, however, could be explained by a higher content of less soluble fluorapatite in the skeletons of the population of Chemnitz, i.e., a long-lasting lowered rate of bone resorption *(54)*.

Recently, a large prospective multicenter study on 9704 elderly women was published *(55)*. The aim of this study was to determine whether older women with long-term exposure to fluoridated water had different BMD and fracture rates compared with women with no exposure. The women (all ambulatory and without

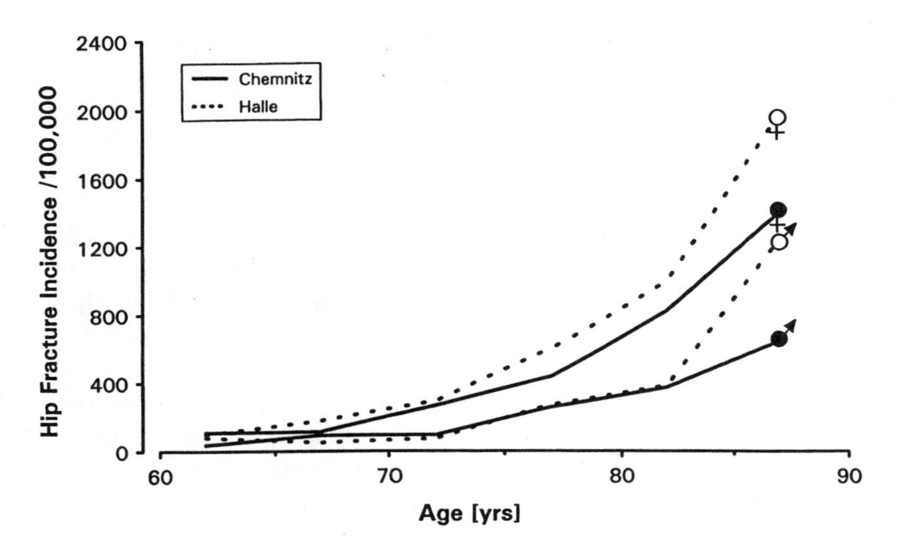

Fig. 2. Annual incidences of hip fractures given in 5-yr groups for men and women in the towns of Chemnitz with water fluoridation, and Halle without fluoridation (solid line, Chemnitz; dotted lines, Halle (according to Lehmann et al., 1998).

Table 2
Bone Mineral Density in g/cm² at the Different Measuring Sites by Exposure to Fluoride in Drinking Water Over 20 yr

Category	*A* *No Exposure*	*B* *Mixed Exposure*	*C* *Continuous Exposure*
Patients *(n)*	2563	1348	3218
Mean age (yr)	74.5	74.2	73.9
Lumbar spine	0.849	0.853	0.871*
Femoral neck	0.647	0.652	0.664*
Distal radius	0.371	0.362	0.364*

* *p*-value A vs C 0.001.

Source: Adapted from ref. *55.*

bilateral hip replacement) had been enrolled during 1986–1988. Of them, 7129 could provide information on exposure to fluoride for each year from 1950 to 1994. The BMD results of this study are given in Table 2. In women with continuous exposure (group C), BMD was 2.5% higher at the lumbar spine, 2.6% at the femoral neck, and 1.9% lower at the distal radius ($p < 0.001$ for all three sites). Women in group C had a significantly reduced risk of hip fractures (RR = 0.69, 95% CI 0.50 to 0.96, $p = 0.028$). The relative risk for vertebral fractures was

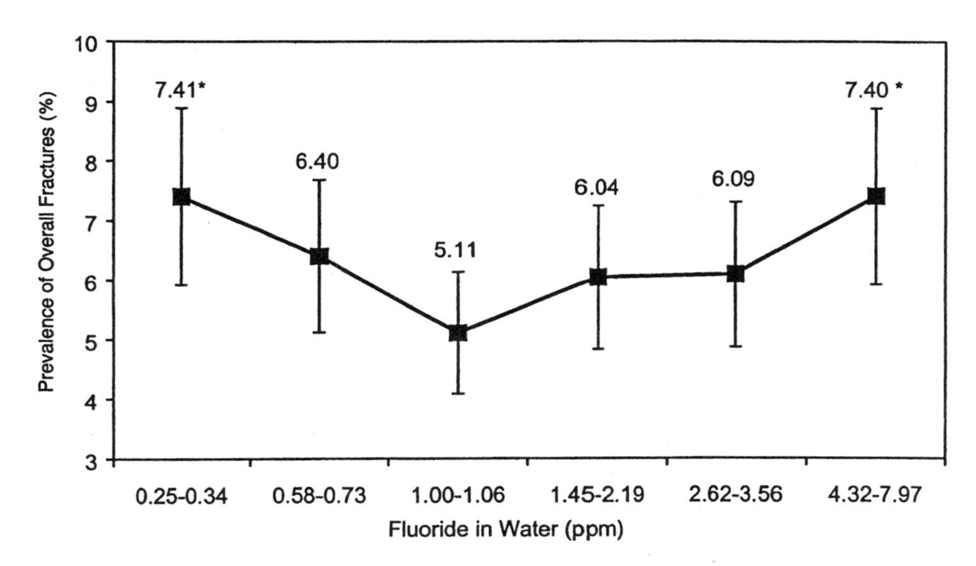

Fig. 3. Prevalence of overall fractures related to fluoride concentrations in drinking water in six Chinese populations (*$p < 0.05$ as compared to the group of 1.00–1.06). (Adapted from Li et al., 2001.)

reduced to 0.73 (CI 0.55–0.97, $p = 0.033$). There was no significant effect on fracture incidence in women with intermittent "mixed exposure."

These findings may have enormous importance for public health. It is the first prospective study with adequate power to examine the risk of specific fractures associated with fluoride on an individual rather than community basis. The study shows that long-term exposure to fluoridated water will lead to a significant reduction in the incidence of hip and vertebral fractures in elderly women. The authors conclude that fluoridation may be one of the most cost-effective methods for reducing the incidence of fractures related to osteoporosis *(55)*.

A positive effect of fluoride on overall fracture incidence was also confirmed by a large study from China on 8266 subjects. The comparison of bone fractures in six subgroups with different drinking water fluoride contents (ranging from 0.25 to 7.97 mg/L, Fig. 3) showed the lowest fracture rate with 1.00–1.06 mg/L *(56)*.

7.2. Fluoride for Prevention of Osteoporosis

As discussed above, prevention of osteoporotic fractures by life-long ingestion of small amounts of fluoride (e.g, by drinking fluoridated water with 1 mg/L) could be a very effective and inexpensive approach, especially for a population-based unselected prophylaxis *(55)*.

A still open question remains whether pharmacological doses of fluoride could reduce the risk of fractures in individual patients with risk factors for osteoporosis or densitometric-proved osteopenia (= preclinical osteoporosis).

Theoretically, bone-forming drugs should be able to increase bone mass above the critical fracture threshold in all stages of osteoporosis. It must be taken into consideration, however, that the lower the initial bone mass, the more advanced is the microarchitectural deterioration of spongy bone, and the goal to restore a more or less normal spongy bone architecture may be difficult to achieve.

On the other hand, an anabolic treatment with fluoride might be extremely effective in healthy subjects with mild osteopenia and rather well-preserved trabecular bone connectivity and still-extended endosteal surfaces *(47,57)*. This interesting concept of an early therapy of postmenopausal spinal osteoporosis has never been tested in sufficiently large trials, and an optimal dosage is not defined.

Only three small fluoride studies have been performed to prevent osteoporosis, or increase BMD, in early postmenopausal women with moderate osteopenia *(58–60)*. Consistent increases in lumbar spine BMD were recorded, but the studies were too small to evaluate fracture events.

Interestingly, combinations of fluoride with hormone replacement therapy (HRT) are more effective than fluoride alone. In a first small trial it could be shown that a dose as low as 10 mg fluoride ions per day given as MFP was associated with a significant additive effect on lumbar spine BMD in early postmenopausal women in comparison with HRT alone *(61)*.

A Danish group studied 100 healthy postmenopausal women treated with four different regimens: HRT, MFP, HRT + MFP, or placebo *(62)*. In the combined-treatment group they found a strong synergistic effect of HRT and MFP and biochemical markers indicating a dissociation of bone formation and resorption.

In an own open-label prospective study we were able to show that MFP/calcium plus HRT was more effective that MFP/calcium alone in protecting from new vertebral and nonvertebral fractures *(63)*.

7.3. Treatment of Established Osteoporosis

There is a large body of radiological and densitometric evidence that long-term fluoride treatment increases bone mass mainly at bone sites with a high proportion of trabecular bone *(57,64–66)*, but sufficiently large randomized controlled trials with new vertebral fractures as primary endpoint are missing.

The first prospective controlled study showing a positive effect of 50 mg/d NaF on vertebral fracture incidence was published by a French group in 1988 *(67)*. In 1990/1991, however, two placebo-controlled studies from the United States reported a discrepancy between gain in BMD and fracture rates *(68,69)*. In the first of these, 75 mg/d of plain NaF had been given over 4 yr. The gain of bone density at the spine was 36% at the end of the study, i.e., an average increase of 9% per year. There was also a small increase at the proximal femur but a significant loss at the radius. The rate of vertebral fractures after 4 yr, however, was not significantly lower in the fluoride–calcium group than in the placebo–calcium group, and the authors reported even a higher incidence of nonvertebral fractures in fluoride-treated patients. The reason was obviously the very high fluoride dose, corresponding to approx 34 mg bioavailable fluoride ions per day, inducing a too-rapid new bone formation of minor quality. Interestingly, an extension and new analysis of the Riggs study, published 4 yr later, showed a reduction of the vertebral fracture rate with lower fluoride doses; i.e., it was conceded that the original dosage of fluoride was too high *(70)*. The possible role of fluoride-containing crystals in this context was discussed controversially. The larger fluorapatite crystals are more resistant to osteoclastic attack. However, if the crystals are excessively large, as in the case of skeletal fluorosis, bones may become brittle and more fragile *(71,72)*.

Several important studies have appeared since 1994 demonstrating positive effects of long-term fluoride therapy on fracture incidence in postmenopausal osteoporosis *(73–75)*. These studies were performed with NaF or MFP at different doses, but for all the dosages of daily bioavailable fluoride ions was between 10 and 20 mg.

The 4-yr study of Reginster et al. was a randomized, double-blind, controlled trial in 200 cases of postmenopausal osteoporosis (initial lumbar spine BMD < –2.5 T-Score) *(75)*. There was a progressive but very moderate increase at the spine, reaching only 10% after 4 yr. The rate of new vertebral fractures after 4 yr was significantly lower for the MFP group, 2.4%, than in the calcium group, 10%. Obviously, a slow and moderate increase of BMD appeared to be better than a rapid and high gain in BMD.

Subsequent studies again found discrepant results. In a randomized controlled French study of 2 yr duration with four different treatment arms, no significant effect of fluoride on fracture rate could be documented in comparison to vitamin D/calcium supplementation *(76)*. A prospective controlled three-arm trial from our group performed on 134 women with established postmenopausal osteoporosis showed a dose-dependent fracture-reducing potency *(77)*. A possible explanation for the positive results was a low-dose intermittent fluoride schedule in the most effective treatment group (114 mg MFP = 15 mg fluoride ions, 3 mo on, 1 mo off). Histomorphometric analysis had previously documented significant advantages in bone quality after intermittent administration of fluoride *(78)*.

Altogether, today many more positive results are documented in the literature than negative experiences. A meta-analysis on the efficacy of fluoride in postmenopausal osteoporosis, however, is difficult because of the differences in the fluoride salts, preparations, and dosages in the different studies.

Bone mineral density results from individual male subjects or small male cohorts in mixed patient groups suggested that the anabolic effect of fluoride is similar in men and women *(47)*. There is only one larger prospective controlled study that included only male patients *(79)*. In that study of 64 men with primary osteoporosis without fractures at onset, we used again the above-mentioned low-dose intermittent fluoride schedule. We found a moderate but highly significant increase of lumbar spine BMD, amounting to 9% after 3 yr, and even small increases at the radius shaft and femoral neck. In comparison to the controls, this treatment was able to reduce significantly the rate of vertebral fractures over 3 yr; i.e., fluoride treatment with the adopted schedule was able to reduce the rate of patients developing established osteoporosis. The moderate increase of 3%/yr at the lumbar spine is consistent with the results of Reginster et al. (1998) in postmenopausal osteoporosis *(75)*.

The therapeutic potential of fluoride salts in glucocorticoid-induced osteoporosis on bone density has been shown in different studies *(80–83)*, but fracture data are not available.

8. CONCLUSION

Fluoride is regarded as an essential trace element, although its physiological role for development and maintenance of bone tissue and teeth is not sufficiently elucidated. There is, however, no doubt about the strong carioprotective effect of fluoride, and the American Dietetic Association *(28)* strongly reaffirms its endorsement to use systemic and topical fluoride, including water fluoridation, at appropriate levels of intake, as an important public health measure. Fluoridation of public water supplies is generally regarded as the most effective dental public health measure in existence. Worldwide, only a few populations receive the maximum benefits possible from community water fluoridation and the use of fluoride products. It is estimated that in the United States a sufficient supply today is guaranteed for about one-half of the population.

A simultaneous positive effect on bone strength is less clearly documented and was discussed mostly controversially in the past. Larger and more recent studies, however, are in favor of a significant fracture-reducing potency of long-term ingestion of fluoridated water with a fluoride content of about 1 mg/L *(53,55,56)*. In an optimistic view, water fluoridation is regarded as the most cost-effective approach for reducing worldwide the huge socioeconomic burden of osteoporotic fractures *(55)*. The question of the risk–benefit relation of using fluoride at pharmacological doses for prevention or treatment of postmenopausal, male, and corticoid-induced osteoporosis is still unanswered after 40 yr of use in different countries *(66)*.

The existing therapeutic data on fluoride treatment of osteoporosis do not reach the level of evidence-based medicine. One important problem of fluoride is the low price and the lack of protection by a patent. That means that very large studies with enough power to assess definitely the therapeutic efficacy will probably never be performed. A meta-analytic approach looking at all randomized controlled studies may come to wrong conclusions due to considerable differences in dosage and bioavailability of fluoride between studies, depending on fluoride salts used and the respective galenic preparations. Accordingly, one recent analysis of 11 trials published between 1980 and 1998 including older studies with high-dose fluoride (35 mg ions per day) and newer trials with lower doses (15–20 mg ions per day) concluded that fluoride increases lumbar spine BMD but does not reduce the incidence of vertebral fractures *(84)*. A meta-analysis including only studies with low fluoride dosage published between 1994 and 2002 showed positive effects on lumbar spine BMD and vertebral fracture rate. Obviously, the selection of papers included is crucial in every meta-analytic approach.

As long as no other potent osteoanabolic substances are available, fluoride should be kept within the choice of therapeutics for osteoporosis. Accepting a certain therapeutic value of a low-dose and well-controlled fluoride therapy, the low price may be an important advantage, especially in countries with limited resources for the health care system. The upcoming bone anabolic treatment with parenteral human recombinant parathyroid hormone will be a very expensive option.

The chance to improve therapeutic results with fluoride by combination with different primarily antiresorptive drugs (e.g., HRT, raloxifen, alfacalcidol, bisphosphonates) has not been sufficiently studied. Impressive additive effects on BMD could be shown *(62,63,85),* but fracture data, histological assessment of bone quality, and micro-computed tomography evaluation of bone architecture are needed.

REFERENCES

1. Navia JM, Aponte-Merced L, Punyasingh K. Fluoride metabolism in humans. In: Prasad AS, ed. Current Topics in Nutrition and Disease. Vol. 18. Lisbon, 1988, pp. 229–250.
2. Mertz W. The essential trace elements. Science 1981; 213:1332–1338.
3. Flemming Moller P, Gudjonsson SV. Massive fluorosis of bones and ligaments. Acta Radiol 1932; 13:267–294.
4. Roholm K. Fluorine Intoxication. A Clinical Hygienic Study with a Review of the Literature and Some Experimental Investigations. Lewis, London, 1937.
5. Shortt HE, McRobert GR, Barnard TW, Nayar ASM. Endemic fluorosis in the Madras Presidency. Indian J Med Res 1937; 25:553–554.
6. Heaney RL. Fluoride and osteoporosis. Ann Intern Med 1994; 120:689–690.
7. Ringe JD, Meunier PJ. What is the future for fluoride in the treatment of osteoporosis? Osteopor Int 1995; 5:71–74.
8. Standing Committee on the Scientific Evaluation of Dietary Reference Intakes. Dietary Reference Intakes for Calcium, Phosphorous, Magnesium, Vitamin D and Fluoride. National Academy Press, Washington, DC, 1997.
9. Guy WS. Inorganic and organic fluorine in human blood. In: Johansen E, Taves DR, Olsen TO, eds. Continuing Evaluation of the Use of Fluorides. AAAS Selected Symposium. Westview, Boulder, CO, 1979.
10. Müller P, Schmid K, Warneke G, Setnikar I. Sodium fluoride-induced gastric mucosal lesions: comparison with sodium monofluorophosphate. Z Gastroenterol 1992; 30:252–254.
11. Husdan H, Vogl R, Oreopoulos D, Gryfe C, Rapoport A. Serum ionic fluoride: normal range and relationship to age and sex. Clin Chem 1976; 22:1884–1888.
12. Ekstrand J, Ehrnebo M, Boreus LO. Fluoride bioavailability after intravenous and oral administration: importance of renal clearance and urine flow. Clin Pharmacol Therapeut 1978; 23:329–337.
13. Schiffl H, Binswanger U. Renal handling of fluoride in healthy man. Renal Physiol 1982; 5:192–196.
14. Ekstrand J, Fomon SJ, Ziegler EE, Nelson SE. Fluoride pharmacokinetics in infancy. Pediatr Res 1994; 35:157–163.
15. Pang DT, Philips CL, Bawden JW. Fluoride intake from beverage consumption in a sample of North Carolina children. J Dent Res 1992; 71:1382–1388.
16. Singer L, Ophaug RH, Harland BF. Dietary fluoride intake of 15-19-year-old male adults residing in the United States. J Dent Res 1985; 64:1302–1305.
17. Taves DR. Dietary intake of fluoride ashed (total intake) v. unashed (inorganic fluoride) analy sis of individual foods. Br J Nutr 1983; 49:295–301.

18. Bruun C, Thylstrup A. Dentifrice usage among Danish children. J Dent Res 1988; 71:1114–1117.
19. Farley JR, Wergedal JE, Baylink DJ. Fluoride directly stimulates proliferation and alkaline phosphatase activity of bone forming cells. Science 1983; 222:330–332.
20. Briancon D, Meunier PJ. Treatment of osteoporosis with fluoride, calcium, and vitamin D. Orthoped Clin N Am 1981; 12:629–668.
21. Bellows CG, Heersche JNM, Aubin JE. The effects of fluoride on osteoblast progenitors in vitro. J Bone Miner Res 1990; 5:S101–S105.
22. Marie PJ, de Vernejoul MC, Lomri A. Fluoride induced stimulation of bone formation is associated with increased DNA synthesis by osteoblastic cells in vitro. J Bone Miner Res 1990; 5:S140.
23. Khokher MA, Dandona P. Fluoride stimulates ^{3}H-thymidine incorporation and alkaline phosphatase production by human osteoblasts. Metabolism 1990; 39:1118–1121.
24. Grynpas MD, Cheng PT. Fluoride reduces the rate of dissolution of bone. Bone Miner 1988; 5:1–9.
25. Libanati C, Lau KHW, Baylink D. Fluoride therapy for osteoporosis. In: Marcus R, Feldman D, Kelsey J, eds. Osteoporosis. Academic, New York, 1996.
26. Lau K, Baylink DJ. Molecular mechanism of action of fluoride on bone cells. J Bone Miner Res 1998; 13:1660–1667.
27. Caverzasio J, Imai T, Amman P, Burgener D, Bonjour JP. Aluminium potentiates the effect of fluoride on tyrosine phosphorylation and osteoblast replication in vitro and bone mass in vivo. J Bone Miner Res 1996; 11:46–55.
28. Position of the American Dietetic Association: the impact of fluoride on health. J Am Dietet Assoc 2001; 101:126–132.
29. Tatevossian A. Fluoride in dental plaque and its effects. J Dent Res 1990; 69:645–652.
30. Marquis RE. Antimicrobial actions of fluoride for oral bacteria. Can J Microbiol 1995; 41:955–964.
31. Guo MK, Nopakun JN, Messer HH, Ophaug R, Singer L. Retention of skeletal fluoride during bone turnover in rats. J Nutr 1988; 118:362–366.
32. Eble D, Deaton TG, Wilson FC, Bawden JW. Fluoride concentrations in human and rat bone. J Public Health Dent 1992; 52:288–292.
33. Sehgal S, Kawatra A. Trace elements in health and disease. J Intern Med India 1999; 2:204–211.
34. Bunker VW. The role of nutrition in osteoporosis. Br J Biomed Sci 1994; 51:228–240.
35. Ripa LW. A half-century of community water fluoridation in the United States: review and commentary. J Public Health Dent 1993; 53:17–44.
36. Boillat MA, Baud CA, Lagier R, Donath A, Dettwiler W, Courvoisier B. Fluorose industrielle. Schweiz Med Wschr 1976; 106:1842–1844.
37. Rich C, Ensinck J. Effect of sodium fluoride on calcium metabolism of human beings. Nature 1961; 191:184–185.
38. Grennan DM, Palmer DG, Maltthus RS, Metangi MF, de Silva RTD. Iatrogenic fluorosis. Austral N Z J Med 1978; 8:528–531.
39. Siddiqui AH. Fluorosis in Nalgonda district, Hyderabad-Deccan. Br Med J 1955; ii:1408–1411.
40. Winston AE, Bhaskar SN. Caries prevention in the 21st century. J Am Dent Assoc 1998; 129:579–587.
41. McDonagh MS, Whiting PF, Wilson PM, et al. Systemic review of water fluoridation. Br Med J 2000; 321:855–859.
42. Ringe JD. Stimulators of bone formation for the treatment of osteoporosis. In: Meunier PJ, ed. Osteoporosis: Diagnosis and Management. Martin Dunitz, London, 1997, pp. 131–148,
43. Ringe JD. What is proven about hip fracture rate and fluoride treatment? Osteologie 1998; 7:151–156.
44. Leone NC, Stevenson CA, Hilbish TF, Sosman MC. A roentgenologic study of a human population exposed to high fluoride domestic water. Am J Roentgenol 1955; 74:874–878.

45. Bernstein DS, Sadowsky N, Hegsted DM, Guri CD, Stare FJ. Prevalence of osteoporosis in high and low fluoride areas in North Dakota. JAMA 1996; 198:499–504.
46. Jacobsen SJ, O'Fallon WM, Melton LJ III. Hip fracture incidence before and after fluoridation of the public water supply: Rochester, Minnesota 1950–1969. Am J Public Health 1993; 83:743–745.
47. Danielson C, Lyon JL, Egger M, Goodenough G. Hip fractures and fluoridation in Utah's elderly population. JAMA 1992; 268:746–748.
48. Madans J, Kleinman JC, Cornoni-Huntley J. The relationsship between hip fracture and water fluoridation: an analysis of national data. Am J Public Health 1983; 73:296–298.
49. Cooper C, Wickham C, Lacey RF, Barker DJP. Water fluoride concentration and fracture of the proximal femur. J Epidemiol Community Health 1990; 44:17–19.
50. Simonen O, Laitinen O. Does fluoridation of drinking water prevent bone fragility and osteoporosis? Lancet 1985; 2:432–433.
51. Jacobsen SJ, Goldberg J, Miles TP. Regional variation on hip fracture: U.S. white women aged 65 years and older. JAMA 1990; 264:500–502.
52. Turner CH, Akhter MP, Heaney RP. The effects of fluoridated water on bone strength. J Orthoped Res 1992; 10:581–587.
53. Lehmann R, Wapniarz M, Hofmann B, Pieper B, Haubitz I, Allolio B. Drinking water fluoridation: bone mineral density and hip fracture incidence. Bone 1998; 22:273–278.
54. Allolio B, Lehmann R. Drinking water fluoridation and bone. Exp Clin Endocrinol Diabetes 1999; 107:12–20.
55. Phipps KR, Orwoll ES, Mason JD, Cauley JA. Community water fluoridation, bone mineral density, and fractures: prospective study of effects in older women. Br Med J 2000; 321:860–864.
56. Li Y, Liang C, Slemenda CW, et al. Effect of long-term exposure to fluoride in drinking water on risks of bone fractures. J Bone Miner Res 2001; 16:932–939.
57. Kleerekoper M, Mendlovic B. Sodium fluoride therapy of postmenopausal osteoporosis. Endocrine Rev 1993; 14:312–323.
58. Pouilles JM, Tremollieres F, Causse E, Louvet JP, Ribot C. Fluoride therapy in postmenopausal osteopenic women: effect on vertebral and femoral bone density and prediction of bone response. Osteopor Int 1991; 1:103–109.
59. Affinito P, Di Carlo C, Primizia M, Petrillo G. A new fluoride preparation for the prevention of postmenopausal osteoporosis: calcium monofluorophosphate. Gynecol Endocrinol 1993; 7:201–205.
60. Sebert JL, Richard P, Mennecier P, Bisset J. Monofluorophosphate increases lumbar bone density in patients with low bone mass but no vertebral fractures. A double-blind randomized study. Osteopror Int 1995; 5:108–114.
61. Gambacciani M, Spinetti A, Cappagli B, et al. Effects of low-dose monofluorophosphate and transdermal oestradiol on postmenopausal vertebral bone loss. Eur Menopause J 1995; 2:16–20.
62. Alexandersen P, Riis BJ, Christiansen C. Monofluorophosphate combined with hormone replacement therapy induces a synergistic effect on bone mass by dissociating bone formation and resorption in postmenopausal women: a randomized study. J Clin Endocriol Metab 1999; 84:3013–3020.
63. Ringe JD, Setnikar I. Monofluorophosphate combined with hormone replacement therapy ion postmenopausal osteoporosis. An open-label pilot efficacy and safety study. Rheumatol Int 2002; 22:27–32.
64. Ringe JD, Kruse HP, Kuhlencordt F. Long-term treatment of primary osteoporosis by sodium fluoride. In: Courvoisier B, Donath A, Baud CA, eds. Fluoride and Bone. Hans Huber, Bern, Switzerland, 1978, pp. 228–232.
65. Eriksen EF, Hodgson SF, Riggs BL. Treatment of osteoporosis with sodium fluoride. In: Riggs BL, Melton LJ III, eds. Osteoporosis: Etiology, Diagnosis and Management. Raven, New York, 1988, pp. 415–432.
66. Ringe JD. Fluoride in osteoporosis. In: Bilezikian JP, Raisz LG, Rodan GA, eds. Principles of Bone Biology. 2nd ed. Academic, San Diego, CA, 2002, pp. 1387–1399.

67. Mamelle N, Dusan R, Martin JL, et al. Risk-benefit ratio of sodium fluoride treatment in primary vertebral osteoporosis. Lancet 1988; 2:361–365.
68. Riggs BL, Hodgson SF, O'Fallow WM, et al. Effect of fluoride treatment on the fracture rate in postmenopausal women with osteoporosis. N Engl J Med 1990; 322:802–809.
69. Kleerekoper M, Peterson EL, Nelson DA, et al. A randomized trial of sodium fluoride as a treatment for postmenopausal osteoporosis. Osteopor Int 1991; 1:155–161.
70. Riggs BL, O'Fallon WM, Lane A, et al. Clinical trial of fluoride therapy in postmenopausal osteoporotic women: extended observations and additional analysis. J Bone Miner Res 1994; 9:265–275.
71. Grynpas MD. Fluoride effects on bone crysdtals. J Bone Miner Res 1990; suppl 1:S169–S175.
72. Ilich JZ, Kerstetter JE. Nutrition in bone health revisited: a story beyond calcium. J Am Coll Nutr 2000; 19:715–737.
73. Pak CYC, Sakhaee K, Piziak V, et al. Slow-release sodium fluoride in the management of postmenopausal osteoporosis. A Randomized controlled trial. Ann Intern Med 1994; 120:625–632.
74. Farrerons J, Rodriguez de la Serna A, Guanabens N, et al. Sodium fluoride treatment is a major protector against vertebral and nonvertebral fractures when compared with other common treatments of osteoporosis: a longitudinal, observtaional study. Calcif Tissue Int 1997; 60:250–254.
75. Reginster JY, Meurmans L, Zegels B, et al. The effect of sodium monofluorophosphate plus calcium on vertebral fracture rate in postmenopausal women with moderate osteoporosis. A randomized, controlled trial. Ann Intern Med 1998; 129:1–8.
76. Meunier PJ, Sebert J-L, Reginster JY, et al., and the FAVOStudy group. Fluoride salts are no better at preventing new vertebral fractures than calcium-vitamin D in postmenopausal osteoporosis: the FAVOStudy. Osteopor Int 1998; 8:4–12.
77. Ringe JD, Kipshoven C, Cöster A, Umbach R. Therapy of established postmenopausal osteoporosis with monofluorophosphate plus calcium: dose-related effects on bone density and fracture rate. Osteopor Int 1999; 9:171–178.
78. Schnitzler CM, Wing JR, Raal FJ, et al. Fewer bone histomorphometric abnormalities with intermittent than with continuous slow-release sodium fluoride therapy. Osteopor Int 1997; 7:376–389.
79. Ringe JD, Dorst A, Kipshoven C, Rovati LC, Setnikar I. Avoidance of vertebral fractures in men with idiopathic osteoporosis by a three year therapy with calcium and low-dose intermittent monofluorophosphate. Osteopor Int 1998; 8:47–52.
80. Meys E, Terraux-Duvert F, Beaume-Six T, Dureau G, Meunier PJ. Bone loss after cardiac transplantation: effects of calcium, calcidiol and monofluorophosphate. Osteopor Int 1993; 3:1–8.
81. Rizzoli R, Chevalley Th, Slosman DO, Bonjour JP. Sodium monofluoro-phosphate increases vertebral bone mineral density in patients with corticoid-induced osteoporosis. Osteopor Int 1995; 5:39–46.
82. Guaydier-Souquieres G, Kotzki PO, Sabatier JP, Basse-Cathalinat B, Loeb G. In corticosteroid-treated respiratory diseases monofluorophosphate increases lumbar bone density: a double-masked randomized study. Osteopor Int 1996; 6:171–177.
83. Lippuner K, Haller B, Casz JP, Montandon A, Jaeger P. Effect of disodium monofluorophosphate, calcium and vitamin D supplementation on bone mineral density in patients chronically treated with glucocorticosteroids: a prospective, randomized, double-blind study. Miner Electrolyte Metab 1996; 22:207–213.
84. Haguenauer D, Welch V, Shea B, Tugwell P, Adachi JD, Wells G. Fluoride for the treamentof postmenopausal osteoporosis. Osteopor Int 2000; 11:727–738.
85. Ringe JD, Rovati LC. Treatment of osteoporosis in men with fluoride alone or in combination with bisphosphonates. Calcif Tissue Int 2001; 69:252–255.

22 Lead Toxicity in the Skeleton and Its Role in Osteoporosis

J. Edward Puzas, James Campbell, Regis J. O'Keefe, and Randy N. Rosier

1. INTRODUCTION

This discussion focuses on the role that lead plays as a risk factor for osteoporosis and as an agent that may compromise skeletal development and fracture healing. The adverse effects of lead in our environment have been demonstrated in many systems. However, its effect on the skeleton have not been widely studied. With the advent of laws removing this toxic agent from our environment, the harmful effects on soft tissues will most certainly begin to decline. Nevertheless, with nearly 95% of the body burden of lead residing in bone and with a lead half-life in the skeleton measured in decades, the effects of this heavy metal on bone cell function could persist for an entire lifetime.

2. IS LEAD A PHYSIOLOGICALLY IMPORTANT ENVIRONMENTAL TOXICANT TODAY?

The answer is a definitive "Yes." And it may pose serious health issues in the skeleton at levels well below the adopted safety threshold in blood of 10 μg/dL. That is, bone and cartilage cells are exposed to lead from two different sources, the extracellular fluid and the mineralized compartment of bone. This is the result of their proximity to and metabolism of mineralized tissues.

Lead toxicity has become a politically controversial topic. Some organizations suggest that the levels of lead in the environment and in humans are not dangerous. It is true that the average blood lead level has been decreasing since the late 1960s and 1970s and that the childhood lead levels are the lowest in decades, averaging around 7 μg/dL. These data have been used by organizations such as the Small Property Owners of America (www.spoa.com) to make the case that lead is no longer a safety issue in our country. They claim that de-leading houses and raising concerns for new sources of lead is an antiquated idea worthy of abandoning.

From: *Nutrition and Bone Health*
Edited by: M. F. Holick and B. Dawson-Hughes © Humana Press Inc., Totowa, NJ

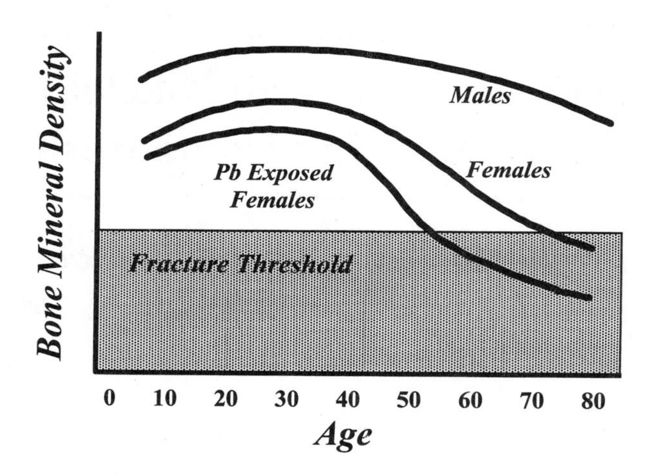

Fig. 1. Diagrammatic representation of bone mineral density vs age for normal males and females and lead-exposed females. *Hypothesis 1:* Because of compromised growth plate activity, lead-exposed females never attain the peak density that normal females attain. *Hypothesis 2:* Lead-exposed individuals (especially females) lose bone at a faster rate than normal individuals. That is, the slope of bone density loss in the "Pb-exposed females" is greater than the slope of normal females. This situation contributes to lead-exposed women crossing the fracture threshold at an earlier age. They will be predisposed to premature osteoporosis.

The Small Property Owners of America also suggests that the risk of lead poisoning in inner-city homes can be brought to near zero with minimal measures such as wiping one's feet upon entering the house.

Contrary to this belief, a careful examination of the literature shows overwhelming evidence in support of the fact that lead remains an environmental problem. In fact, as our ability to perform careful clinical and basic science research becomes more sophisticated, it is apparent that 10 μg/dL is no longer a safe threshold. A recent case in point shows that there is a decrease in IQ in children with each 1.0 μg/dL increase in blood lead levels, and that the effect is largest at levels below 10 μg/dL *(1)*. This toxin, previously thought to be harmless below 10 μg/dL, is not.

Our hypotheses state that lead hinders growth plate function and prevents skeletally immature individuals from attaining an adequate peak bone mass in early adulthood. We also speculate that lead can accelerate the rate of bone loss in aging by virtue of its regulation of bone formation and bone resorption, thus inducing a more negative bone balance. A diagrammatic representation of these hypotheses is presented in Fig. 1.

3. LEAD AND THE SKELETON

Research on the adverse effects of high lead exposure in humans has focused primarily on neurocognitive outcomes among children *(2)*. However, a growing body of literature reports that the effects of childhood lead exposure continue into adolescence and adulthood. These effects include delinquent behavior *(3)*, dental caries *(4)*, cardiovascular disease *(5)*, cardiac arrhythmias *(6)*, and renal dysfunction *(7)*. These delayed effects are probably due to the fact that lead is harbored in bone for years after initial exposure, and thus may be a source for later exposure *(8)*.

The skeleton is the major reservoir of lead in the body and consequently its lead content reflects cumulative lead exposure *(9–12)*. However, little is known specifically about the effects of lead on the skeleton and the mechanisms of action through which it may affect bone growth and remodeling.

3.1. Effects of Lead Exposure in Endochondral Bone Development and the Growth Plate

Lead has been associated with both low birth weight and short stature. Although there has been some discussion as to whether this results from specific skeletal effects or is related to systemic and nutritional factors *(9,13,14)*, the weight of experimental evidence supports a direct effect on the developing skeletal tissues.

It appears that lead alters skeletal maturation, at least in part because of its direct effect on normal growth plate function. This point of view is supported by histomorphometric analyses that have identified thickening of the growth plate and disorganized architecture in both rat and mouse models following lead exposure *(15,16)*. The subcellular mechanisms by which this occurs can be described, in part, by the ability of lead to substitute for biologically essential elements such as Ca^{2+} and Zn^{2+} in the metal-binding domains of kinases and transcription factors *(17)*. Pb^{2+} binds and activates the calcium signaling peptide, calmodulin, at concentrations similar to those necessary for activation by Ca^{2+} in numerous cell types *(18–21)*. The long list of other signaling effects of Pb^{2+} includes potentiation of *ERK/MAPK* activation in PC12 cells *(22)*, activation of *NFκB* and degradation of *IκBα (23)*, and activation/potentiation of *AP-1*, *MEK*, and *JNK* in pheochromocytoma cells *(24)*. Pb^{2+} also perturbs Ca^{2+} transport mechanisms, including voltage-dependent Ca^{2+} channels in adrenal chromaffin cells *(25,26)*. Lead affects intracellular Ca^{2+} release by potentiation of inositol-1,4,5-trisphosphate (IP_3) binding and the binding of other inositol polyphosphates to their respective receptors in cerebellar membrane preparations *(27)*. The net effect includes downstream Ca^{2+}-dependent signaling events. The impact of Pb^{2+} on these ubiquitous signaling mechanisms implies that Pb^{2+} toxicity may cause a broad spectrum of effects in a number of tissues, including bone and cartilage.

Because Pb^{2+} directly affects second messenger signaling, and because skeletal formation is dependent on the proper integration of multiple signaling cascades, it follows that perturbation of signaling events by Pb^{2+} will have deleterious effects on the skeleton. The importance of this issue is underscored by work showing that growth plate chondrocytes display decreased proliferation, altered type X collagen expression, and increased proteoglycan synthesis when exposed to Pb^{2+} in vitro *(13)*. Furthermore, Pb^{2+}-exposed mice and rats have altered growth

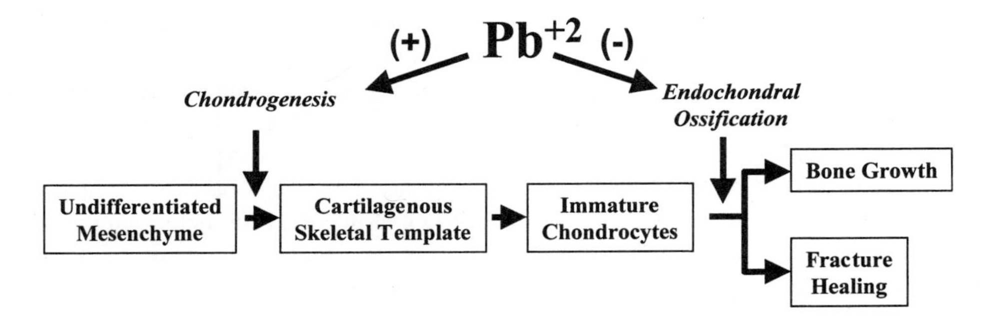

Fig. 2. Effects of lead on growth plate cartilage. From in vitro data, and clinical observations, it appears that lead may influence cartilage differentiation at two stages. Lead appears to enhance chondrogenesis but arrest late-stage differentiation and transitions to bone.

plate morphology and decreased longitudinal bone growth *(15,16,28)*. These results are consistent with effects of Pb^{2+} on children, which include decreased stature due to reduced longitudinal growth *(29)*. Because these effects are not due to systemic toxicity *(13,16)* and occur at low doses, it seems that the effect is due to molecular events associated with altered second messenger signaling.

Our laboratory and others have established that several of the above-mentioned Pb^{2+}-sensitive signaling pathways are critical to the action of factors that regulate the chondrocyte phenotype during development and during the process of endochondral ossification. We have determined that Pb^{2+} does indeed affect the action of parathyroid hormone-related protein and transforming growth factor (TGF)-β as well as the action of *AP-1* and *NF-κB*. These findings not only establish the growth plate as an important target tissue for Pb^{2+} toxicity, but underscore the importance of specific signaling pathways. Other possible signaling pathways that could be affected by Pb^{2+} include *JNK* and *NF-κB*, which are involved in the induction of matrix metalloproteinases in chondrocytes *(30)*, as well as *Erk-1* and *MAPK*, which are involved in chondrogenesis *(31)*.

Evidence in the literature as well as clinical observations hint that lead may influence the endochondral pathway at at least two different steps in differentiation. We have outlined this possibility in Fig. 2. Chondrogenesis is defined as the commitment of undifferentiated mesenchymal cells to a cartilage phenotype. This step is controlled by early signaling pathways and it appears that lead may actually accelerate these steps. However, as the chondrocytes mature, and prepare for the final stages of endochondral ossification, it appears that lead has an inhibitory effect on these processes. Moreover, it follows that since normal growth plate function shares similar mechanisms with fracture healing, this process would also be affected. Thus, lead seems to control long bone growth by prematurely accelerating early stages of differentiation and simultaneously blocking late-stage transitions to bone.

3.2. Effects of Lead on Bone Cell Metabolism

3.2.1. Osteoblasts

There have been a number of reports describing the effect of lead on osteoblasts *(32–36)*. These studies have been performed in transformed cell lines, freshly isolated normal cells, and in in vivo experiments. Uniformly, it has been observed that lead is deleterious to the functioning of these cells. The metal ion has been shown to inhibit secretion of osteonectin in a rat osteogenic sarcoma (ROS) 17/2.8 cells without affecting collagen synthesis *(32)*. A similar specific effect, also in ROS 17/2.8 cells, has been documented with an inhibition of both alkaline phosphatase and type I collagen gene expression and no effect on β-actin or glyceraldehyde-3-phosphate dehydrogenase *(33)*. Although there is a discrepancy between these two manuscripts with regard to collagen synthesis, both reports suggest that the effects of lead are not limited to a common decrease in all mRNA and protein synthesis.

Lead also affects osteoblast cell proliferation. Although some authors claim that in confluent cultures of ROS 17/2.8 cells there is no effect on DNA synthesis *(33)*, the majority of reports indicate that lead affects basal levels as well as stimulated levels of bone cell proliferation. Moreover, lead appears to modulate the effect of growth factors and hormones (such as the insulin-like growth factors and vitamin D) on a number of bone-forming parameters in osteoblasts *(34,35)*.

The kinetics of lead metabolism and its effects on intracellular calcium have been investigated in transformed cells and, in a single study, in freshly prepared normal osteoblasts *(35,36)*. These reports demonstrate that there are a number of exchangeable pools for lead in ROS 17/2.8 and normal cells and that lead may alter the calcium homeostasis by increasing steady-state levels.

3.2.2. Osteoclasts

The harmful effects of lead on osteoclasts have been known for many years *(37,38)*. However, because of the difficulty in preparing isolated osteoclasts, the mechanisms of lead inhibition in these are just now being investigated.

Early work has shown that either acute or chronic exposure to lead leads to morphological changes in osteoclasts. The characteristic most commonly found in severely affected cells is the presence of nuclear and cytoplasmic inclusion bodies *(38–40)*. These bodies are homogeneously dense with a fibrillar periphery. They resemble those found in liver cord cells and renal tubular cells. The presence of inclusion bodies is associated with a decreased osteoclast function. This result, along with impaired osteoblast function, contributes to the genesis of "lead lines" *(39–41)* in young patients in whom the serum level is greater than 70 μg/dL.

In addition to a direct effect of the ion on osteoclast function, it is also possible that indirect effects contribute to inefficient bone resorption. This could be the result of the lead in the bone matrix directly affecting the cells, or it could be due to the presence of an incompetent matrix that resists resorption. The second possibility, although theoretically possible, is probably not a key point of regulation, because it is still possible to substantially upregulate osteoclast number and activity on lead-containing wafers with RANK ligand.

A few reports from a single group have appeared that show that lead may actually stimulate osteoclastic activity in vitro *(42,43)*, but the effects appeared only at specific low doses of the ion. At all concentrations above 5 μM the ion had the expected deleterious effect. As far as we can determine, these are the only reports in the literature describing a stimulatory effect of lead on bone resorption.

4. EFFECTS OF LEAD ON SKELETAL MASS

The results of in vitro laboratory investigations on lead show that the ion adversely influences osteoblasts and osteoclasts. However, because changes in total skeletal mass in humans occur very slowly, and because of a general lack of interest in the bone community to investigate skeletal toxicity's, no controlled clinical studies have been published proving that lead can lead to adult-onset osteoporosis. At the present time, the best in vivo evidence that lead affects the adult skeleton is from numerous case studies and animal models.

In humans, there is evidence for reduced bone density in work-related exposure victims *(10,44)*. This has led to diagnoses of "idiopathic osteoporosis" in these cases.

Another key factor in lead intoxication in humans occurs when elevated blood levels occur during activation of the skeleton in situations such as menopause, paraplegic immobilization, or thyrotoxicosis *(44–46)*. In the case of menopausal elevations in blood lead levels, it has been generally unappreciated that there is a significant elevation of lead concentrations in *all women* at the time of the menopause, whether or not they continue to be exposed to the metal *(47)*. Such a finding, combined with the observation that most of the lead in blood comes from long-term bone stores rather than acute exposure *(8)*, argues that the female population could be at significant risk of lead-induced osteoporosis later in life.

Data such as these do not directly document the effects of lead on bone metabolism, but the blood levels achieved during such episodes (reaching 4–7 μg/dL) are high enough to potently affect osteoblastic and osteoclastic activity. Moreover, because the structural component of bone (i.e, cortical bone) is known to contain the greatest amounts of lead *(48)*, remodeling of this tissue at the menopause could ultimately compromize mechanical support.

5. IN VIVO EFFECTS OF LEAD ON THE SKELETON

Some of the best evidence that lead can lead to bone diseases such as osteoporosis has been generated in animals, i.e., dogs, rabbits, and sheep *(49–52)*. In particular, the work by Anderson et al. *(53)* in dogs demonstrates that osteoporosis, similar to that seen in aging adults, can occur after lead exposure. In their dog model, which is probably the best nonprimate model for human bone metabolism, they showed that both the activation frequency for osteoclast activity and

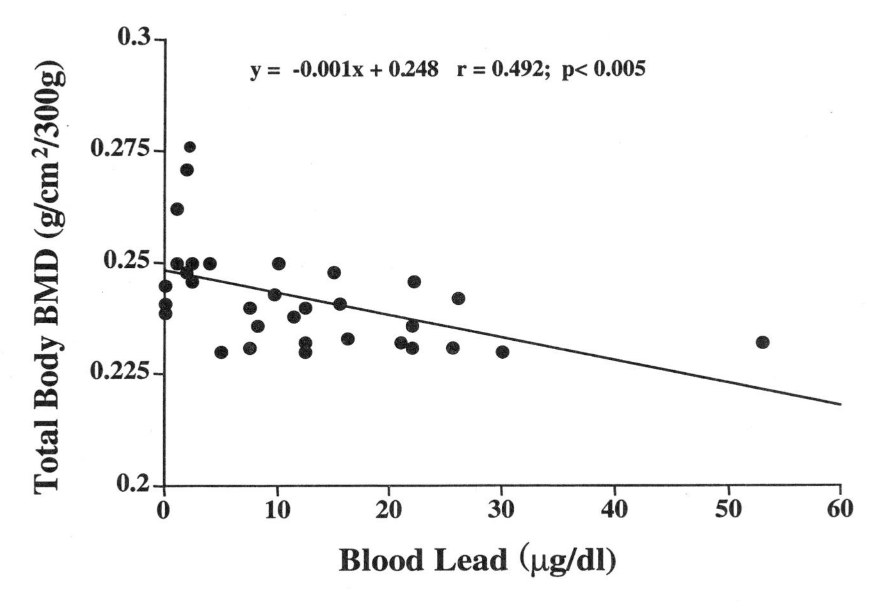

Fig. 3. Lead in the drinking water leads to lower bone density in rats. Male rats exposed to lead for short periods of time show a decrease in skeletal mass in relation to the level of lead in their blood.

bone formation rates were significantly depressed. In fact, the effect on bone formation appeared to be larger than on bone resorption. This would lead to a total decrease in skeletal mass over time. Other strong data demonstrating an effect of lead on bone mass in rats has been presented by Escribano et al. *(54)*. This study showed that rats fed a diet reasonably low in lead developed osteopenia as judged by histomorphometric measurements. They found a significant difference in every parameter measured (i.e., percent bone volume, trabecular number, trabecular thickness, and trabecular space). These findings have been verified by Hamilton et al. *(15)*, who showed a decrease in tail length in lead-treated rats, presumably caused by vertebral body length changes, and by our laboratory (Fig. 3).

The other cited studies in rabbits and sheep *(49,55)* likewise demonstrated osteoporosis in these animal models. The endpoints were not as sophisticated as the Anderson and Escribano studies, but it was clear that these other animal models demonstrated a low bone mass and a tendency to an increase in fracture incidence.

5.1. Effects of Lead Exposure: Clinical Studies

Despite the extensive basic science research on the effects of lead exposure on bone metabolism, human studies on this association are limited. Laraque found that lead-exposed African-American children had a *higher* mean bone mineral content (BMC) than age-matched white norms *(56)*. At age 36–47 mo, the mean BMC of the radius of the lead-exposed subjects and norms were 0.268 gm/cm^2 vs 0.215 gm/cm^2, respectively ($p < 0.005$). The authors attributed the findings to racial differences in bone density. In further analysis, Laraque plotted age against bone mineral density (BMD) among children with low vs high lead exposure (i.e., blood lead level ≤ 29 µg/dL vs ≥30 µg/dL); the difference in BMD was not significant ($p = 0.63$). Laraque concluded that lead exposure is not associated with changes in bone mineralization. However, because the comparison group was made up of children with moderate-level lead exposure (i.e., blood lead level 12–29 µg/dL), such an analysis cannot exclude the possibility that lead exposure has a threshold effect on bone density at lower blood lead levels.

Although recent studies report a dramatic decrease in the prevalence of lead exposure in the United States *(57,58)*, this does not imply that adults have not had exposure in the past. As recently as the late 1970s, 78% of the entire US population—children *and* adults—had blood lead levels of 10 µg/dL or more *(59)*, the current threshold of concern defined by the Centers for Disease Control *(60)*. This means that the majority of adults in the United States had, at some time in the past, an elevated blood lead level, and therefore currently have elevated bone lead levels, given the extremely long half-life of lead in bone ($\tau_{1/2} = 20$ yr) *(11)*. This large number of adults who have elevated bone lead levels, and the morbidity associated with osteoporosis, justifies an aggressive investigation of the role of this heavy metal in skeletal metabolism. This research, however, cannot be limited to adults. Although osteoporosis is an affliction of the elderly, its roots are established during childhood. An individual who does not achieve peak bone mass during childhood may be at risk for osteoporosis in later life *(61)*. Because 90% or more of bone mass is achieved by age 17–20 yr *(62,63)*, the association between lead exposure and bone density should be examined among children and skeletally mature adolescents.

6. LEAD AND DXA MEASUREMENTS FOR BMD

The skeletal data on rats with lead exposure is somewhat controversial. Gruber et al. *(64)* as well as our laboratory have demonstrated that lead causes a decrease in bone mass by dual-energy X-ray absorptiometry (DXA), histomorphometry, and mineral analysis. In a subsequent study, the effect of lead exposure was ameliorated by calcium supplementation *(65)*. Lead-exposed rats have been also shown to have decreased femoral weights and L5 vertebral body heights, as well as histomorphometric evidence of osteopenia *(54)*. However, the bone density in the Escribano study as measured by DXA was higher in the lead-treated animals. This may be explained by an artifact in the DXA measurement recently reported *(66)*. In this work it was shown that bone samples doped with increasing quantities of lead have spuriously high BMD with a direct dose–response effect when measured with pencil-beam DXA instruments from both GE/Lunar and Hologic Inc. The artifact does not appear to be due to the X-ray shielding effects of lead as much as it may be related to the algorithms for the calculation of BMD from the dual X-ray sources. If the artifact occurs in measurements of BMD by DXA in humans, there may be serious implications in the interpretation of clinical DXA data, which may necessitate knowing an individual's bone lead content to correct for this problem. This finding may also explain the discrepancy reported by Escribano et al. *(54)* between the histomorphometric data, which indicated osteoporosis in lead-treated rats, and DXA densitometric data, which showed higher BMD in the same animals.

7. SYSTEMIC TOXICITIES OF SKELETAL LEAD

Because bone has been shown to be the repository for 90–95% of lead in the body *(10,11),* there has been significant concern that release of lead from bone during menopause could lead to elevated blood levels and significant toxicity. A study of blood lead levels in Hispanic women demonstrated higher blood leads in women within the first 4 yr of menopause as compared to women who were more than 4 yr postmenopausal *(47).* Similarly, data from the National Health and Nutrition Examination Survey II (NHANES II, 1976–1980) demonstrated a significant increase in blood leads following menopause *(67).* Further support for this observation was presented in a prior study of antiresorptive therapy with estrogen in lead-exposed women by Webber et al. *(68).* They showed that bone lead was significantly higher in women treated with estrogen than in the untreated group. More potent antiresorptive agents such the bisphosphonates, alendronate, or risedronate, which are now standard therapies for postmenopausal osteoporosis, have not been studied with regard to effects on lead metabolism.

Numerous other conditions of accelerated bone resorption have been shown to increase blood lead levels, including immobilization, thyrotoxicosis, and hyperparathyroidism *(10,44,46,69).* The significance of the accelerated release of lead from bone in postmenopausal women relates to the known toxicities of low levels of lead in other organ systems, such as the central nervous and cardiovascular systems. Lead intoxication at low levels has been associated with hypertension in both children and adults *(70–73),* and has been shown to cause cognitive dysfunction in older adults *(74).* Elevated lead levels have been reported after immobilization for a fracture *(10,44),* with associated renal toxicity in one report *(10).* Local elevation of extracellular fluid lead levels at a fracture site secondary to increased resorption from the combined inflammatory response and immobilization could impair or delay the healing of osteoporotic fractures. The data on systemic toxicities of bone-derived lead support a new wave of clinical trials in lead-exposed menopausal women to determine if a potent antiresorptive can prevent the rise of blood lead during accelerated bone resorption.

8. MEASUREMENT OF BONE LEAD

The method of K-shell X-ray fluorescence (KXRF) has been employed to measure bone lead in occupationally lead-exposed and unexposed individuals *(75).* In vivo bone lead measurements were first developed by Ahlgren and

Mattsson at Lund in Sweden. They used X-rays from a ^{57}Co source to excite lead K-shell electrons. Measurement was made in a finger bone, and this system has been in operation since 1972 *(76)*. Subsequently, an improved system was developed by Chettle, Scott, Laird, and Somervaille in Birmingham, England. This system used X-rays from a ^{109}Cd source to excite the lead K electrons to produce γ-rays (i.e., X-ray fluorescence), and proved to have three particular advantages over the original approach: measurements were more precise; bone lead content could be related directly to bone mineral; and measurements could usefully be made in any superficial bone *(77)*. The first human measurements were made in 1983, and initially the tibia lead concentration was studied. Since then, the calcaneus has frequently been selected as a trabecular bone site. The patella has also been used in this way, and measurements have also been reported of radius, sternum, and skull. This measurement approach has been adopted by a number of laboratories around the world. In 1991, Chettle and Webber at McMaster University developed a new system, with improved precision, based on the same principles *(78)*.

Two particular features of the relationship between bone lead and lead exposure have consistently emerged from studies in which K X-ray fluorescence technology has been employed. First, bone lead concentration reflects cumulative lead exposure. Cumulative exposure can be represented by the time-weighted integral of blood lead, monitored regularly in lead-exposed workers. Second, release of lead from bone contributes to circulating lead in blood, thus constituting an endogenous exposure. This relationship is particularly clear when industrial exposure has ceased. For such people, endogenous exposure can often be the dominant contributor to current blood lead.

9. CONCLUSION

Given the potent effects of low levels of lead on bone and cartilage cells and the widespread exposure of the adult population of the United States, lead may be an unrecognized contributing factor in the current and expanding epidemic of osteoporosis in this country. Skeletal effects of lead during childhood could predispose one to osteoporosis by diminishing achieved peak bone mass through acceleration of bony maturation. The release of skeletal lead during menopause could also contribute to neurocognitive, cardiovascular, or other systemic toxicities. Finally, the documentation of a lead-related artifact in DXA measurements could dramatically alter our interpretation of DXA results, and mandate knowledge of an individual's lead exposure or bone lead content in order to ensure accurate use of this highly prevalent testing and screening modality.

ACKNOWLEDGMENT

This work was supported in part by NIH P30 ES01247 (JEP) and NIH P01 ES11854 (JEP, RJO).

REFERENCES

1. Lanphear BP, Dietrich KN, Berger O. Prevention of lead toxicity in US children. Ambul Pediatr 2003; 3:27–36.
2. National Research Council. Measuring Lead Exposure in Infants, Children, and Other Sensitive Populations. National Academy Press, Washington, DC, 1993.
3. Needleman HL, McFarland C, Ness R, Tobin M, Greenhouse J. Bone lead levels in adjuducated delinquents: a case-control study. Pediatr Res 2000; 47(4):155A.
4. Moss ME, Lanphear BP, Auinger P. Association of dental caries and blood lead levels. JAMA 1999; 281:2294–2298.
5. Schwartz J. Lead, blood pressure, and cardiovascular disease in men and women. Environ Health Perspec 1991; 91:71–75.
6. Cheng Y, Schwartz J, Vokonas PS, Weiss ST, Aro A, Hu H. Electrocardiographic conduction disturbances in association with low-level lead exposure (the Normative Aging Study). Am J Cardiol 1998; 82:594–599.
7. Kim R, Rotnitzky A, Sparrow D, Weiss ST, Wagner C, Hu H. A longitudinal study of low-level lead exposure and impairment of renal function—The Normative Aging Study. JAMA 1996; 275:1177–1181.
8. Gulson BL, Mahaffey KR, Mizon KJ, Korsch MJ, Cameron MA, Vimpani G. Contribution of tissue lead to blood lead in adult female subjects based on stable lead isotope methods. J Lab Clin Med 1995; 125:703–712.
9. Pounds JG, Long GJ, Rosen JF. Cellular and molecular toxicity of lead in bone. Env Health Perspect 1991; 91:17–32.
10. Berlin K, Gerhardsson L, Borjesson J, et al. Lead intoxication caused by skeletal disease. Scand J Work Environ Health 1995; 21:296–300.
11. Wedeen RP. Removing lead from bone: clinical implications of bone lead stores. Neurotoxicology 1992; 13:843–852.
12. Hu H, Payton M, Korrick S, et al. Determinants of bone and blood lead levels among community-exposed middle-aged to elderly men. The Normative Aging Study. Am J Epidemiol 1996; 144:749–759.
13. Hicks DG, O'Keefe RJ, Reynolds KJ, et al. Effects of lead on growth plate chondrocyte phenotype. Toxicol Appl Pharmacol 1996; 140:164–172.
14. Hammond PB, Chernausek SD, Succop PA, Shukla R, Bornschein RL. Mechanisms by which lead depresses linear and ponderal growth in weanling rats. Toxicol Appl Pharmacol 1989; 99:474–486.
15. Hamilton JD, O'Flaherty EJ. Effects of lead exposure on skeletal development in rats. Fundam Appl Toxicol 1994; 22:594–604.
16. Gonzalez-Riola J, Hernandez ER, Escribano A, Revilla M, Villa LF, Rico H. Effect of lead on bone and cartilage in sexually mature rats: a morphometric and histomorphometry study. Environ Res 1997; 74:91–93.
17. Bouton CM, Pevsner J. Effects of lead on gene expression. Neurotoxicololgy 2000; 21:1045–1055.
18. Goldstein GW, Ar D. Lead activities calmodulin sensitive processes. Life Sci 1983; 33:1001–1006.
19. Habermann E, Crowell K, Janicki P. Lead and other metals can substitute for calcium in calmodulin. Arch Toxicol 1983; 54:61–70.
20. Ozawa T, Sasaki K, Umezawa Y. Metal ion selectivity for formation of the calmodulin-metal-target peptide ternary complex studied by surface plasmon resonance spectroscopy. Biochem Biophys Acta 1999; 1434:211–220.
21. Kern M, Wisniewsski M, Cabell L, Audesirk G. Inorganic lead and calcium interact positively in activation of calmodulin. Neurotoxicology 2000; 21:353–363.
22. Williams TM, Ndifor AM, Near JT, Reams-Brown RR. Lead enhances NGF-induced neurite outgrowth in PC12 cells by potentiating ERK/MAPK activation. Neurotoxicology 2000; 21:1081–1089.

23. Kudrin AV. Trace elements in regulation of NF-KB activity. J Trace Elem Med Biol 2000; 14:129–142.
24. Ramesh GT, Manna SK, Aggarwal BB, Jadhav AL. Lead activates nuclear transcription factor-kB, activator protein-1 and amino-terminal C-Jun kinase in pheochromocytoma cells. Toxicol Appl Pharmacol 1999; 155:280–286.
25. Sun LR, Suszkiw JB. Extracellular inhibition and extracellular enhancement of calcium currents by lead in bovine adrenal chromaffin cells. J Neurophysiol 1995; 74:574–581.
26. Pokorski PL, McCabe MJ, Pounds JG. Lead inhibits meso-2,3-dimercaptosuccinic acid induced calcium transients in cultured rhesus monkey kidney cells. Toxicology 1999; 134:19–26.
27. Vig PJ, Pentyqala SN, Chetty CS, Rajanna B, Desaiah D. Lead alters inositol polyphosphate receptor activities: protection by ATP. Pharmacol Toxicol 1994; 75:17–22.
28. Bushnell PJ, Jaeger RJ. Hazards to health from environmental lead exposure: a review of recent literature. Vet Hum Toxicol 1986; 26:255–261.
29. Shukla R, Bornschein RL, Dietrich KN, et al. Fetal and infant lead exposure: effects of growth in stature. Pediatrics 1989; 84:604–612.
30. Mengshol JA, Vincenti MP, Coon CI, Barchowski A, Brinckerhoff CE. Interleukin-1 induction of collagenase 3 (matrix metalloproteinase 13) gene expression in chondrocytes requires P38, C-Jun N-terminal kinase, and nuclear factor kappa B: differential regulation of collangenase 1 and collangenase 3. Arthritis Rheum 2000; 43:801–811.
31. Yoon YM, Oh CD, Kim DY, et al. Epidermal growth factor negatively regulates chondrogenesis of mesenchymal cells by modulating the protein kinase C-a, Erk-1, and P38 signaling pathways. J Biol Chem 2000; 275:12353–12359.
32. Sauk JJ, Smith T, Silbergeld EK, Fowler BA, Somerman MJ. Lead inhibits secretion of osteonectin/SPARC without significantly altering collagen or Hsp47 production in osteoblast-like ROS 17/2.8 cells. Toxicol Appl Pharmacol 1992; 116:240–247.
33. Klein RF, Wiren KM. Regulation of osteoblast gene expression by lead. Endocrinology 1993; 132:2531–2537.
34. Angle CR, Thomas DJ, Swanson SA. Lead inhibits the basal and stimulated responses of a rat osteoblast-like cell line ROS 17/2.8 to 1,25 dihydroxyvitamin D3 and IGF-1. Toxicol Appl Pharmacol 1990; 103:281–287.
35. Schane FAX, Gupta RK, Rosen JF. Lead inhibits 1,25dihydroxyvitamin D3 regulation of calcium metabolism in osteoblastic osteosacoma cells, ROS 17/2.8. Biochem Biophys Acta 1992; 1180:187–194.
36. Long GJ, Rosen JF, Pounds JG. Cellular toxicity and metabolism in primary and clonal osteoblastic bone cells. Toxicol Appl Pharmacol 1900; 102:346–361.
37. vanMullem PJ, Stadhouders AM. Bone marking and lead intoxication; early pathological changes in osteoclasts. Virchows Arch Abt B Zellpathol 1974; 15:345–350.
38. Hsu FS, Krook L, Shivel JN, Duncan JR, Pond WG. Lead inclusion bodies in osteoclasts. Science 1973; 181:447–448.
39. Eisenstein R, Kawanoue S. The lead line in bone-a lesion apparently due to chondroclastic indigestion. Am J Pathol 1975; 80:309–316.
40. Milgram JW. Chemical toxic agents affecting bone growth and structure, lead. In: Radiologic and Histologic Pathology of Nontumorous Diseases of Bones and Joints. Northbrook, Chicago, IL, 1990, pp. 859–860.
41. Park EA. X-ray shadows in growing bones produced by lead, their characteristics, causes, anatomical counterpart in the bone and differentiation. J Pediatr 1933; 3:265–298.
42. Miyahara T, Komiyama H, Miyanishi A, et al. Stimulative effects of lead on bone resorption in organ culture. Toxicology 1995; 97:191–197.
43. Miyahara T, Komiyama H, Miyanishi A, et al. Effects of lead on osteoclast-like cell formation in mouse bone marrow cell cultures. Calcif Tissue Int 1994; 54:165–169.

44. Shannon M, Lindy H, Anast C, Graef J. Recurrent lead poisoning in a child with immobilization osteoporosis. Vet Human Toxicol 1988; 30:586–588.
45. Silbergeld EK, Schwartz J, Mahaffey K. Lead and osteoporosis: mobilization of lead from bone in postmenopausal women. Environ Res 1988; 47:79–94.
46. Goldman RH, White R, Kales SN, Hu H. Lead poisoning from mobilization of bone stores during thyrotoxicosis. Am J Ind Med 1994; 25:417–424.
47. Symanski E, Hertz-Picciotto I. Blood lead levels in relation to menopause, smoking and pregnancy history. Am J Epidemiol 1995; 41:1047–1058.
48. Wittmers LE, Aufderheide AC, Wallgreen J, Rapp G, Alich A. Lead in bone. IV. Distribution of lead in the human skeleton. Arch Environ Health 1988; 43:381–391.
49. Hass GM, Brown DVL, Eisenstein R, Hemmens A. Relations between lead poisoning in rabbit and man. Am J Pathol 1964; 45:691–715.
50. Massie HR, Aiello VR. Lead accumulation in the bones of aging male mice. Gerontology 1992; 38:13–17.
51. Barry PSI. A comparison of concentrations of lead in human tissue. Br J Ind Med 1975; 32:119–139.
52. Whiting SJ. Safety of some calcium supplements questioned. Nutr Rev 1994; 52:95–97.
53. Anderson C, Danylchuk KD. The effect of chronic low level lead intoxication on the haversian remodelling system in dogs. Lab Invest 1977; 37:466–469.
54. Escribano A, Revilla M, Hernandez HR, et al. Effect of lead on bone development and bone mass: a morphometric, densitometric and histomorphometric study in growing rats. Calcif Tissue Int 1997; 60:200–203.
55. Clegg FC, Rylands JM. Osteoporosis and hydronephrosis in young lambs following ingestion of lead. J Comp Pathol 1966; 76:15–22.
56. Laraque D, Arena L, Karp J, Gruskay D. Bone mineral content in black pre-schoolers: normative data using single photon absorptiometry. Pediatr Radiol 1990; 20:461–463.
57. Brody DJ, Pirkle JL, Kramer RA, et al. Blood lead levels in the US population: Phase 1 of the Third National Health and Nutrition Examination Survey (NHANES III, 1988 to 1991). JAMA 1994; 272:277–283.
58. Pirkle JL, Kaufmann RB, Brody DJ, Hickman T, Gunter EW, Paschal DC. Exposure of the U.S. population to lead, 1991–1994. Environ Health Perspect 1998; 106:745–750.
59. Mahaffey KR, Annest JL, Roberts J, Murphy RS. National estimates of blood lead levels: United States, 1976–1980. N Engl J Med 1982; 307:573–579.
60. Centers for Disease Control. Preventing Lead Poisoning in Young Children. U.S. Department of Health and Human Services, Atlanta, GA, 1991.
61. National Institutes of Health. Osteoporosis prevention, diagnosis and therapy. NIH Consensus Statement, 2000 March 27–29; 17(1):1–45.
62. Magarey AM, Boulton TJ, Chatterton BE, Schultz C, Nordin BE, Cockington RA. Bone growth from 11 to 17 years: relationship to growth, gender and changes with pubertal status including timing of manarche. Acta Paediatr 1999; 88:139–146.
63. Lloyd T, Chinchilli VM, Eggli DF, Rollings N, Kulin HE. Body composition development of adolescent white females: the Penn State Young Women's Health Study. Arch Pediatr Adolesc Med 1998; 152:998–1002.
64. Gruber HE, Gonick HC, Khalil-Manesh F, et al. Osteopenia induced by long-term, low- and high-level exposure of the adult rat to lead. Miner Electrolyte Metab 1997; 23(2):65–73.
65. Gruber HE, Ding Y, Stasky AA, et al. Adequate dietary calcium mitigates osteopenia induced by chronic lead exposure in adult rats. Miner Electrolyte Metab 1999; 25:143–146.
66. Puzas JE, Campbell J, O'Keefe R, Schwarz E, Rosier R. Lead in the skeleton interferes with bone mineral measurements. J Bone Miner Res 2002; 17(S1):S314.
67. Silbergeld EK, Schwartz J, Mahaffey K. Lead and osteoporosis: mobilization of lead from bone in postmenopausal women. Environ Res 1988; 47:79–94.

68. Webber CE, Chettle DR, Bowins RJ, et al. Hormone replacement therapy may reduce the return of endogenous lead from bone to the circulation. Environ Health Perspect 1995; 193:1150–1153.
69. Osterloh JD. Observations on the effect of parathyroid hormone on environmental blood lead concentrations in humans. Environ Res 1991; 54:8–16.
70. Hu H, Aro A, Payton M, et al. The relationship of bone and blood lead to hypertension. The Normative Aging Study. JAMA 1996; 276:1037–1038.
71. deCastro FJ, Medley J. Lead in bone and hypertension. Maternal Child Health J 1997; 1:199–200.
72. Korrick SA, Hunter DJ, Rotnitzky A, Hu H, Speizer FE. Lead and hypertension in a sample of middle-aged women. Am J Public Health 1999; 89:330–335.
73. Vig EK, Hu H. Lead toxicity in older adults. J Am Geriatr Soc 2000; 48:1501–1506.
74. Payton M, Riggs KM, Spiro A, Weiss ST, Hu H. Relations of bone and blood lead to cognitive function: the VA Normative Aging Study. Neurotoxicol Teratol 1998; 20:19–27.
75. Aro A, Amarasiriwardena C, Lee ML, Kim R, Hu H. Validation of K x-ray fluorescence bone-lead measurements by inductively coupled plasma mass spectrometry in cadaver legs. Med Phys 2000; 27:119–123.
76. Ahlgren L, Mattsson S. An X-ray fluorescence technique for in vivo determination of lead concentration in a bone matrix. Phys Med Biol 1979; 24(1):136–145.
77. Somervaille LJ, Chettle DR, Scott MC, et al. In vivo tibia lead measurments as an index of cumulative lead exposure in occupationally exposed subjects. Br J Ind Med 1988; 45:174–181.
78. Gordon CL, Chettle DR, Webber CE. An upgraded 109Cd K X-ray fluorescence bone Pb measurement. Basic Life Sci 1993; 60:285–288.

23 Microminerals and Bone Health

Steven A. Abrams and Ian J. Griffin

1. INTRODUCTION

Although the macrominerals calcium, phosphorus, and magnesium are of primary concern in bone health, other minerals, including trace minerals can also play an important role. In this chapter, the role of some of these will be considered. In general, the data supporting and defining the role of the trace minerals in bone health is much less well developed than for the macrominerals. There are early suggestions of the importance of a number of trace elements, but it has been difficult to isolate their individual effects or to study them systematically, especially in humans. Many studies have used animal models, which are difficult to extrapolate to humans. There are very few well-designed intervention studies in humans that address the importance of trace and ultratrace minerals in human bone metabolism. Those that have been carried out often use surrogate markers of bone turnover rather than outcomes such as bone mineral density or fracture risk. Other epidemiological studies have compared dietary intakes of certain trace elements and fracture risk or the risk of osteoporosis. These are confounded by the other lifestyle and socioeconomic factors that may cause such differences in dietary intakes. In addition, low-quality diets may be deficient in more than one nutrient, making it extremely difficult to ascribe the change to any single nutrient.

2. COPPER

Copper is an essential element in human nutrition and is required by many enzymes, including lysyl oxidase, which is responsible for cross-linking of collagen and elastin *(1)*. The prototypical disease of copper deficiency is Menkes' kinky hair syndrome.

2.1. Copper Deficiency

Menkes' kinky hair syndrome is a congenital cause of copper deficiency resulting from impaired copper absorption, which can present with skeletal changes resembling scurvy, fractures, or delayed bone age. Acquired copper deficiency has

From: *Nutrition and Bone Health*
Edited by: M. F. Holick and B. Dawson-Hughes © Humana Press Inc., Totowa, NJ

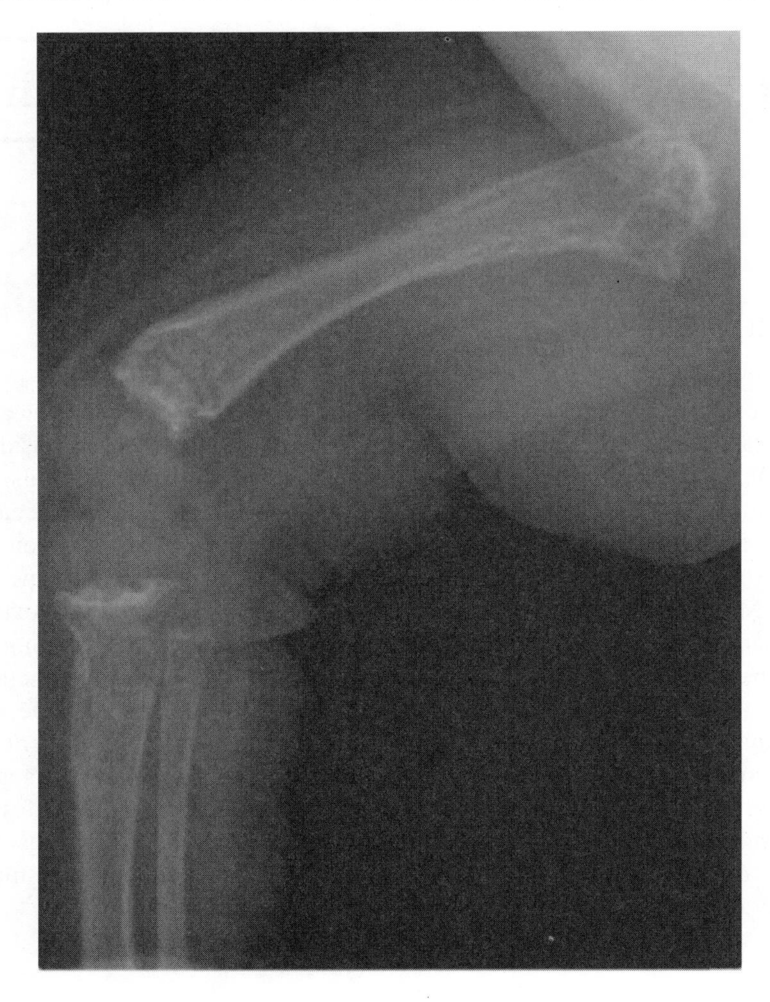

Fig. 1. Radiograph of the lower limb of a former preterm infant demonstrating changes of severe copper deficiency.

also been reported in humans, most commonly in premature or low birth-weight (LBW) infants who had low enteral or parenterally copper intake *(2,3)* or children on prolonged copper-free parenteral nutrition *(4)*. In premature infants the symptoms of copper deficiency may include osteopenia, fractures, or other bony changes, as well as anemia, apnea, and neutropenia *(3,5)*.

Figure 1 shows an extreme example of copper deficiency-induced bone disease in a former preterm infant who had been on prolonged copper-free total parenteral nutrition (TPN). This infant had been born extremely prematurely and developed complications including necrotizing enterocolitis—a severe, potentially fatal gastrointestinal infection. This led to the development of widespread gut necrosis requiring multiple surgeries, and ultimately to severe short-gut syndrome. Because of this, establishment of enteral feeds was extremely difficult, and he required prolonged TPN, and developed cholestatic liver disease ("TPN cholestasis"). Copper is excreted in the bile, and therefore may not always be used in infants with cholestasis resulting from the possibility that it may worsen liver failure. In this infant, copper had been completely removed from his TPN for a prolonged period of time. After several months on copper-free TPN and minimal enteral copper intake, a routine chest X-ray revealed radiological changes consistent with copper deficiency. A low serum copper level and low ceruloplasmin concentration and characteristic changes in long bone films confirmed the diagnosis of acquired copper deficiency. Radiological features suggestive of copper deficiency include osteopenia, metaphyseal cupping and flaring, spurs and fractures, and retarded bone age *(6)*.

2.2. Animal Studies

Several studies have shown that although calcium content may not change, copper-deficient experimental animals have decreased bone strength *(7,8)*, The cause of this is believed to be the reduced activity of lysyl oxidase, the copper metalloenzyme that is responsible for formation of collagen cross-links. Copper deficiency has been shown to cause decreased collagen cross-linking and this is accompanied by decreased bone strength in chicks *(9)*. Furthermore, the reduction in bone strength was reversed by chemical induction of cross-links in vitro, suggesting that the decrease in bone strength results from decreased cross-linking *(9)*. Long-term copper deficiency may also reduce ostrogenesis and reduce osteoclast activity *(10)*.

In ovarectomized rats, copper deficiency increases bone loss *(11)*, whereas copper supplementation may reduce it *(12)*. However, a similar effect is seen with manganese, with no additional benefit coming from copper supplementation *(13)*.

2.3. Human Studies

Although frank copper deficiency clearly has adverse effects of bone health, the importance of mild deficiency or poor copper intakes is much less clear. It has been hypothesized that suboptimal copper nutrition may be a major cause of osteoporosis in Western societies *(14)*, although good evidence for this is lacking. One epidemiological study has described a relationship between copper intake (and indeed iron and zinc intake) and forearm bone mineral content in premenopausal women *(15)*.

Two small, randomized studies have examined the effect of copper intake on markers on bone health. In one study, 11 males aged 20–59 yr were studied sequentially on diets containing low (0.7 mg/d), medium (1.6 mg/d), and high (6.0 mg/d) copper intakes for 8 wk each. When these subjects were switched from the medium to low copper intake there was a significant increase in urinary markers of bone resorption, and a significant decrease when they were switched from the low to high

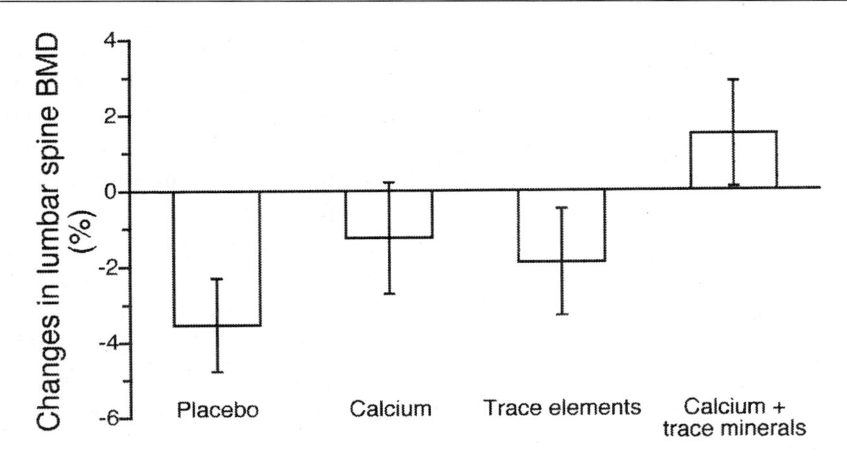

Fig. 2. Changes in percent bone mineral density (BMD) from the study of Strause et al. (data extracted from ref. *19*). Mean changes over the 2-yr study period are shown for the four treatment groups. Error bars represent ±1 standard error of the mean.

copper intake *(16)*. A further study by the same investigators *(17)* considered 24 adults, 22–46 yr, who were studied three times—following 6 wk of treatment with 3 mg/d copper sulfate, 3 mg/d copper-glycine chelate, and 6 mg/d copper-glycine chelator. There were no differences in serum osteocalcin (a marker of bone formation), or the urinary pyridinoline:creatinine ratio or the urinary deoxypyridinoline:creatine ratio (markers of bone resorption). It is worth noting, however, that markers of copper status also did not change among the three treatments *(17)*. A similar study by Cashman et al. *(18)* compared changes in markers of bone turnover and resorption after 4 wk treatment with 3 mg/d copper, 6 mg/d copper, and placebo in 16 healthy females, 20–28 yr old. Copper treatment significantly increased the markers of copper status, serum copper, and erythrocyte superoxide dismutase. However, no differences were noted in markers of bone formation (serum osteocalcin) or bone resorption (urinary pyridinoline:creatinine ratio or urinary deoxypyridinoline:creatinine ratio).

Strause and coworkers reported the only long-term interventional study. They studied 59 postmenopausal women randomized to one of four treatments for 2 yr *(19)*. They were given either a placebo, calcium (1000 mg/d), trace elements alone (15 mg/d zinc, 5 mg/d manganese, and 2.5 mg/d copper), or both calcium and trace elements. After 2 yr of treatment lumbar spine bone mineral density fell in all the groups except those receiving calcium and trace elements (Fig. 2). The only significant difference was between the placebo group and the group receiving both calcium and trace elements. The calcium-only and the trace element-only groups were intermediate between the placebo and calcium-plus-trace element group (Fig. 2).

2.4. Conclusions

Although overt copper deficiency has serious effects of the skeleton, the role of milder forms of copper deficiency remains unclear. Although it has been hypothesized that suboptimal copper intake may be an etiological factor in human osteoporosis, direct evidence for this is lacking. Indeed, the few interventional trials that have looked at the effect of copper supplementation on bone health have shown variable results. These studies have, however, been small and of generally short duration. Furthermore, with the exception of one study *(19),* they have assessed proxy markers of bone formation and resorption rather than bone density. That study, however, does suggest a benefit to addition of trace elements to calcium to reduce postmenopausal bone loss. Whether this benefit is attributable to copper, zinc, or manganese is unclear. Likewise, one study by Baker et al. showed increased bone resorption as copper intake fell, although a second study was unable to confirm this *(16,17).* Clearly, there is much to learn about the role of copper in bone health, especially in populations without clinically apparent copper deficiency. What is urgently needed are large-scale, long-term, randomized studies of copper supplementation on bone density or fracture risk. In the absence of such studies the role of copper as an etiological factor in osteoporosis will remain unclear.

3. ZINC

Zinc is a component of more than 200 enzymes, and overt zinc deficiency is well characterized. The principal clinical features are diarrhea, dermatitis, alopecia, delayed sexual maturation, and decreased taste acuity *(20).* In contrast to copper deficiency, bony changes do not typically feature as symptoms of zinc deficiency; however, even the earliest reports of human zinc deficiency recognized that short stature was a relatively consistent feature *(21).*

3.1. Observational Human Studies

The possible role of zinc as a cause of osteoporosis has increased as several studies have shown that humans with osteoporosis have reduced plasma zinc concentrations *(22,23)* and increased urinary zinc excretion *(24,25).* The latter, however, may be a result of increased bone loss, as approximately one-third of total bone zinc is found in bone *(20).* However, a number of studies have suggested that lower zinc intakes are associated with lower bone mineral content *(15),* lower bone mineral density *(26),* and may be a risk factor for subsequent fractures *(27).*

3.2. Animal Studies

In rats, experimental zinc deficiency can lead to low-turnover osteopenia *(28)* and worsen experimental diabetic osteoporosis *(29).* Conversely, zinc supplementation may ameliorate the bone loss that accompanies skeletal unloading in rats *(30),* and increase serum and bone alkaline phosphatase content in mice *(31)* and rats *(32).* However, studies are contradictory, with some showing increased bone calcium, DNA, and alkaline phosphatase contents following short periods of zinc supplementation *(32),* but decreases after longer periods *(33).* In rats, increasing zinc intake can lead to dose-dependent increases in bone strength *(34).* However, in rats on a low-calcium diet it had the opposite effect, worsening bone strength and elasticity *(35).* One confounding factor may be the anorexia that accompanies zinc deficiency. Zinc-deficient rats have lower femur weights than pair-fed or *ad lib*-fed controls, but bone volume was similarly reduced in zinc-deficient and pair-fed controls, compared to *ad lib*-fed controls *(36).*

There is also an increasing body of evidence from cell and tissue culture models suggesting that zinc is required for protein synthesis *(37,38);* and may increase bone protein content *(39),* DNA content *(39),* and insulin-like growth factor-1 and transforming growth factor-β production *(40,41),* which may be important for fracture healing.

In higher animals, zinc deficiency leads to poor bone growth in pigs *(42).* In rhesus monkeys made marginally zinc deficient from conception to 3 yr of age, bone maturation was delayed. Although bone mineralization was reduced at 6 mo of age, by 3 yr of age it had largely returned to normal *(43).*

3.3. Human Studies

Given the confusion of the animal and basic sciences literature, what is needed is well-designed, large-scale intervention studies in humans. However, there is a paucity of these. In a study of calcium supplementation, Freudenheim et al. *(44)* showed that subjects with higher dietary zinc intake have reduced losses in radial bone mineral density. However, the benefit was seen only in those subjects in the placebo limb of the trial, not in those who received calcium supplementation.

The study of Stause et al. *(19)* is discussed above, although the design of that study does not allow an assessment of whether copper, zinc, or manganese was responsible for the benefits observed. Peretz et al. *(45)* examined the effect of 12 wk of zinc supplementation in healthy men. They demonstrated a significant increase in serum alkaline phosphatase in the zinc-treated group, but not in controls. There was no effect of urinary and C-terminal collagen peptide (a measure of bone resorption).

3.4. Conclusions

Profound zinc deficiency appears to lead to reduced bone growth and maturation, probably through an effect on protein synthesis. However, there is little convincing evidence that increasing zinc intake, in a population with a low incidence of overt deficiency, would have any impact on the incidence of osteoporosis or fractures.

4. BORON

The role of boron in human nutrition remains uncertain. Boron has been hypothesized to enhance bone mineral balance, although its mechanism of action is uncertain. It has been suggested that this occurs by increasing hormonal levels, including estradiol and testosterone levels.

4.1. Animal Data

A recent study in ovariectomized rats showed that a combination of boron and estrogen (17-β-estradiol) increased apparent absorption of calcium, phosphorus, and magnesium. This effect was not seen for boron alone or for estrogen alone. No benefit was seen for boron in combination with parathyroid hormone *(46)*. This study is consistent with previous data in several animal species indicating an increase in mineral balance with supplemental boron *(47)*.

A study in rats has demonstrated increase in bone mass in rats that were subjected to exercise when they were also provided with boron compared to rats who were exercised without boron *(48)*.

4.2. Human Studies

Few human studies have evaluated the role of boron in bone mineral metabolism. In one, providing boron to 12 postmenopausal women, who had been maintained on a low-boron diet for about 4 mo, decreased the urinary excretion of calcium and magnesium. A lowered urinary phosphorous excretion was seen in those with a low-magnesium diet. In that study and in one in adults, it has been suggested that boron may act by increasing serum 17-β-estradiol *(47)*.

4.3. Conclusion

Although some data support a role for boron on bone health, especially in postmenopausal women, substantial further research, including controlled trials, is needed to clarify the role for this nutrient as well as its physiological mechanisms of action.

5. STRONTIUM

The metal strontium (atomic weight 87.6) has been proposed as effective in enhancing bone health. Stable strontium has been widely used as a marker for assessing calcium absorption, as it appears to be absorbed via similar pathways and share physical properties. This similarity includes having its absorption stimulated by vitamin D *(49)*.

5.1. Animal Studies

Several studies have evaluated the effects of strontium in the form of strontium ranelate on bone formation and resorption. These studies have demonstrated a positive effect on bone formation in growing rats as well as prevention of bone resorption in ovariectomized rats *(50)*. The mechanism of action is unknown, but the similarities of calcium and strontium suggest that it may be directly implicated in physically strengthening bone as well as having hormonal effects. Of significant interest is that bone strontium levels are closely correlated with plasma strontium, a relationship not seen with calcium *(51)*. A recent study in mice confirmed a significant increase in trabecular bone mass strontium with long-term strontium ranelate treatment *(52)*.

5.2. Human Studies

Two recent studies in Europe have evaluated the effects of strontium ranelate on bone mineralization in adults. In one of these, a 2-yr randomized placebo control trial of osteoporotic women demonstrated that 2 g/d of this material increased lumbar bone mineral density by 3% relative to placebo. This change was accompanied by biochemical markers suggesting increased bone formation and decreased bone resorption *(53)*. In a different trial, it was shown that strontium ranelate of 1 g/d for 2 yr increased bone density by 2.4% relative to placebo in early-postmenopausal women *(54)*.

5.3. Conclusions

These animal and human studies suggest that relatively large doses of strontium may be beneficial in decreasing bone resorption and enhancing bone mineralization. Larger population studies and human studies evaluating the interaction of strontium therapy with other accepted options for prevention and treatment of osteoporosis are needed. No toxicity or significant adverse effects have been reported with this therapy, although safety and acceptability will need to be evaluated in larger groups as well.

6. SILICON

Silicon has been suggested as an important trace mineral necessary for bone development, but few specific data are available. Rico and coworkers found that ovariectomized rats that were provided silicon had a lower rate of bone loss *(49)*. The mechanism may be related to specific enhancement of bone formation and decrease in bone resorption. There are no prospective studies available in humans. A very small retrospective study suggested a benefit to silicon in bone density in osteoporotic adults *(55)*. No conclusions can be drawn at this time, however, about the role of silicon in bone health in humans.

7. OTHER TRACE AND ULTRATRACE MINERALS

Numerous other minerals have been proposed to have either an enhancing or harmful effect on bone (e.g., aluminum). Among these is manganese, although evidence for an effect is very minimal *(56)*. Because these are uncommonly deficient in diets and are difficult to assess in isolation from other minerals, it has been difficult to obtain solid information regarding their role, and therapeutic use should be considered only in the context of controlled trials.

ACKNOWLEDGMENTS

This work is a publication of the U.S. Department of Agriculture (USDA)/Agricultural Research Service (ARS) Children's Nutrition Research Center, Department of Pediatrics, Baylor College of Medicine and Texas Children's Hospital, Houston, TX. This project has been funded in part with federal funds from the USDA/ARS under Cooperative Agreement number 58-6250-6-001. Contents of this publication do not necessarily reflect the views or policies of the USDA, nor does mention of trade names, commercial products, or organizations imply endorsement by the U.S. Government.

REFERENCES

1. Turnland JR. Copper. In: Shils ME, Olson JA, Shike M, Ross AC, eds. Modern Nutrition in Health and Disease. 9th ed. Williams & Wilkins, Baltimore, 1999.
2. al-Rashid RA, Spangler J. Neonatal copper deficiency. N Engl J Med 1971; 285:841–843.
3. Blumenthal I, Lealman GT, Franklyn PP. Fracture of the femur, fish odour, and copper deficiency in a preterm infant. Arch Dis Child 1980; 55:229–231.
4. Karpel JT, Peden VH. Copper deficiency in long-term parenteral nutrition. J Pediatr 1972; 80:32–36.
5. Sutton A M, Harvie A, Cockburn F. Farquharson J, Logan RW. Copper deficiency in the preterm infant of very low birthweight. Four cases and a reference range for plasma copper. Arch Dis Child 1985; 60:644–651.
6. Grünebaum M, Horodiceanu C, Steinherz R. The radiographic manifestations of bone changes in copper deficiency. Pediatr Radiol 1980; 9:101–104.
7. Jonas J, Burns J, Abel EW, Cresswell MJ, Strain JJ, Paterson CR. Impaired mechanical strength of bone in experimental copper deficiency. Ann Nutr Metab 1993; 37:245–252.
8. Opsahl W, Zeronian H, Ellison M, Lewis D, Rucker RB, Riggins RS. Role of copper in collagen cross-linking and its influence on selected mechanical properties of chick bone and tendon. J Nutr 1982; 112:708–716.
9. Rucker RB, Riggins RS, Laughlin R, Chan MM, Chen M, Tom K. Effects of nutritional copper deficiency on the biomechanical properties of bone and arterial elastin metabolism in the chick. J Nutr 1975; 105:1062–1070.
10. Strause L, Saltman P, Glowacki J. The effect of deficiencies of manganese and copper on osteoinduction and on resorption of bone particles in rats. Calcif Tissue Int 1987; 41:145–150.
11. Yee CD, Kubena KS, Walker M, Champney TH, Sampson HW. The relationship of nutritional copper to the development of postmenopausal osteoporosis in rats. Biol Trace Elem Res 1995; 48:1–11.
12. Rico H, Roca-Botran C, Hernandez ER, et al. The effect of supplemental copper on osteopenia induced by ovariectomy in rats. Menopause 2000; 7:413–416.
13. Rico H, Gomez-Raso N, Revilla M, et al. Effects on bone loss of manganese alone or with copper supplement in ovariectomized rats. A morphometric and densitomeric study. Eur J Obstet Gynecol Reprod Biol 2000; 90:97–101.
14. Strain JJ. A reassessment of diet and osteoporosis—possible role for copper. Med Hypotheses 1988; 27:333–338.
15. Angus RM, Sambrook PN, Pocock NA, Eisman JA. Dietary intake and bone mineral density. Bone Miner 1988; 4:265–277.
16. Baker A, Harvey L, Majask-Newman G, Fairweather-Tait S, Flynn A, Cashman K. Effect of dietary copper intakes on biochemical markers of bone metabolism in healthy adult males. Eur J Clin Nutr 1999; 53:408–412.

17. Baker A, Turley E, Bonham MP, et al. No effect of copper supplementation on biochemical markers of bone metabolism in healthy adults. Br J Nutr 1999; 82:283–290.
18. Cashman KD, Baker A, Ginty F, et al. No effect of copper supplementation on biochemical markers of bone metabolism in healthy young adult females despite apparently improved copper status. Eur J Clin Nutr 2001; 55:525–531.
19. Strause L, Saltman P, Smith KT, Bracker M, Andon MB. Spinal bone loss in postmenopausal women supplemented with calcium and trace minerals. J Nutr 1994; 124:1060–1064.
20. King JC, Keen CL. Zinc. In: Shils ME, Olson JA, Shike M, Ross AC, eds. Modern Nutrition in Health and Disease. 9th ed. Williams & Wilkins, Baltimore, 1999.
21. Prasad A, Miale A, Farid Z, Sandstead HH, Shculet AR. Zinc metabolism in patients with the syndrome of iron deficiency anemia, hepatosplenomegaly, dwarfism and hypogonadism. J Lab Clin Med 1962; 61:537–549.
22. Atik OS. Zinc and senile osteoporosis. J Am Geriatr Soc 1983; 31:790–791.
23. Gür A, Çolpan L, Nas K, et al. The role of trace minerals in the pathogenesis of postmenopausal osteoporosis and a new effect of calcitonin. J Bone Miner Metab 2002; 20:39–43.
24. Herzberg M, Foldes J, Steinberg R, Menczel, J. Zinc excretion in osteoporotic women. J Bone Miner Res 1990; 5:251–257.
25. Relea P, Revilla M, Ripoll E, Arribas I, Villa LF, Rico H. Zinc, biochemical markers of nutrition, and type I osteoporosis. Age Ageing 1995; 24:303–307.
26. New SA, Bolton-Smith C, Grubb DA, Reid DM. Nutritional influences on bone mineral density: a cross-sectional study in premenopausal women. Am J Clin Nutr 1997; 65:1831–1839.
27. Elmståhl S, Gullberg B, Janzon L, Johnell O, Elmståhl B. Increased incidence of fractures in middle-aged and elderly men with low intakes of phosphorus and zinc. Osteoporos Int 1998; 8:333–340.
28. Eberle J, Schmidmayer S, Erben RG, Stangassinger M, Roth HP. Skeletal effects of zinc deficiency in growing rats. J Trace Elem Med Biol 1999; 13:21–26.
29. Fushimi H, Inoue T, Yamada Y, et al. Zinc deficiency exaggerates diabetic osteoporosis. Diabetes Res Clin Pract 1993; 20:191–196.
30. Yamaguchi M, Ehara Y. Zinc decrease and bone metabolism in the femoral-metaphyseal tissues of rats with skeletal unloading. Calcif Tissue Int 1995; 57:218–223.
31. Dimai HP, Hall SL, Stilt-Coffing B, Farley JR. Skeletal response to dietary zinc in adult female mice. Calcif Tissue Int 1998; 62:309–315.
32. Yamaguchi M, Yamaguchi R. Action of zinc on bone metabolism in rats. Increases in alkaline phosphatase activity and DNA content. Biochem Pharmacol 1986; 35:773–777.
33. Yamaguchi M, Mochizuki A, Okada S. Stimulatory effect of zinc on bone growth in weanling rats. J Pharmacobiodyn 1982; 5:619–626.
34. Ovesen J, Moller-Madsen B, Thomsen JS, Danscher G, Mosekilde L. The positive effects of zinc on skeletal strength in growing rats. Bone 2001; 29:565–570.
35. Kenney MA, McCoy H. Adding zinc reduces bone strength of rats fed a low-calcium diet. Biol Trace Elem Res 1997; 58:35–41.
36. Rossi L, Migliaccio S, Corsi A, et al. Reduced growth and skeletal changes in zinc-deficient growing rats are due to impaired growth plate activity and inanition. J Nutr 2001; 131:1142–1146.
37. Ehara Y, Yamaguchi M. Zinc stimulates protein synthesis in the femoral-metaphyseal tissues of normal and skeletally unloaded rats. Res Exp Med (Berl) 1997; 196:363–372.
38. Yamaguchi M, Matsui R. Effect of dipicolinate, a chelator of zinc, on bone protein synthesis in tissue culture. The essential role of zinc. Biochem Pharmacol 1989; 38:4485–4489.
39. Ma ZJ, Yamaguchi M. Stimulatory effect of zinc on deoxyribonucleic acid synthesis in bone growth of newborn rats: enhancement with zinc and insulin-like growth factor-I. Calcif Tissue Int 2001; 69:158–163.
40. Igarashi A, Yamaguchi M. Increase in bone growth factors with healing rat fractures: the enhancing effect of zinc. Int J Mol Med 2001; 8:433–438.

41. Ma ZJ, Misawa H, Yamaguchi M. Stimulatory effect of zinc on insulin-like growth factor-I and transforming growth factor-beta1 production with bone growth of newborn rats. Int J Mol Med 2001; 8:623–628.
42. Nordin RW, Krook L, Pond WG, Walker EF. Experimental zinc deficiency in weanling pigs on high and low calcium diets. Cornell Vet 1973; 63:264–290.
43. Leek JC, Keen CL, Vogler JB, et al. Long-term marginal zinc deprivation in rhesus monkeys. IV. Effects on skeletal growth and mineralization. Am J Clin Nutr 1988; 47:889–895.
44. Freudenheim JL, Johnson NE, Smith EL. Relationships between usual nutrient intake and bone-mineral content of women 35–65 years of age: longitudinal and cross-sectional analysis. Am J Clin Nutr 1986; 44:863–876.
45. Peretz A, Papadopoulos T, Willems D, et al. Zinc supplementation increases bone alkaline phosphatase in healthy men. J Trace Elem Med Biol 2001; 15:175–178.
46. Sheng M, Taper LJ, Veit H, Thomas EA, Ritchey SJ, Lau KH. Dietary boron supplementation enhances the effects of estrogen on bone mineral balance in ovariectomized rats. Biol Trace Miner Res 2001; 81:29–45.
47. Nielson FH. The justification for providing dietary guidance for the nutritional intake of boron. Biol Trace Elem Res 1998; 66:319–330.
48. Rico H, Crispo E, Hernandez ER, Seco C, Crespo R. Influence of boron supplementation on vertebral and femoral bone mass in rats on strenuous treadmill exercise. J Clin Densitom 2002; 5:187–192.
49. Dijkgraaf-Ten Bolscher M, Netelonbos JC, Barto R, van Der Vijgh WJ. Strontium as a marker for intestinal calcium absorption: the stimulatory effect of calcitriol. Clin Chem 2000; 46:248–251.
50. Marie PJ, Hott M, Modrowski D, et al. An uncoupling agent containing strontium prevents bone loss by depressing bone resorption and maintaining bone formation in estrogen-deficient rats. J Bone Miner Res 1993; 8:607–615.
51. Dahl SG, Allain P, Marie PJ, et al. Incorporation and distribution of strontium in bone. Bone 2001; 28:446–453.
52. Delannoy P, Bazot D, Marie PJ. Long-term treatment with strontium ranelate increases vertebral bone mass without deleterious effect in mice. Metabolism 2002; 51:906–911.
53. Meunier PJ, Slosman DO, Delmas PD, et al. Strontium ranelate: dose-dependent effects in established postmenopausal vertebral osteoporosis—A 2-year randomized placebo controlled trial. J Clin Endocrinol Metab 2002; 87:2060–2066.
54. Reginster JY, Deroisy R, Dougados M, Jupsin I, Coette J, Roux C. Prevention of early postmenopausal bone loss by strontium ranelate: the randomized, two-year, double-masked, dose-ranging, placebo-controlled PREVOS trial. Osteoporos Int 2002; 13:925–931.
55. Eisinger J, Clairet D. Effects of silicon, fluoride, etidronate and magnesium on bone mineral density: a retrospective study. Magnes Res 1993; 6:247–249.
56. Rico H, Gallego-Lago JL, Hernandez ER, et al. Effect of silicon supplement on osteopenia induced by ovariectomy in rats. Calcif Tissue Int 2000; 66:53–55.

V

Fat-Soluble Vitamins/ Micronutrients

24 Vitamin A and Bone Health

Peter Burckhardt

1. INTRODUCTION

Vitamin A is an unsaturated 20-carbon cyclic alcohol. Its main sources are preformed vitamin A (retinol or retinyl esters) from animal food, especially liver and fish oil, in a lesser degree from dairy products and fortified foods, and precursors as β-carotene and other vitamin A-forming carotenoids from green plants, carrots, and some fruits *(1)*.

Carotene is cleaved in the intestine to retinol, which is transported by chylomicrons to the liver, where it is stored mainly as retinyl palmitate. It is transported and distributed as retinol in the plasma and in the cells, bound to a specific binding protein, the retinol-binding protein. Intracellular metabolites are retinoic acid and retinal.

Vitamin A deficiency is characterized by xerophtalmia, night blindness, cessation of growth, and increased susceptibility to infections. It is a public health problem in developing countries, vitamin A sufficiency has been linked to decreased cancer incidence; however, the data are not conclusive, although β-carotene is a dietary antioxydant *(2)*, having a favorable influence on the balance between pro- and antioxydants. There is evidence that β-carotene intake decreases lung cancer, mainly in alcohol abusers and heavy smokers *(3)*. In addition, vitamin A supplementation in neonates decreased mortality in several studies *(2)*. On the other side, very high vitamin A intake might affect, among others, bone and bone metabolism, as has been long known *(4)*.

2. THE POTENTIAL PROBLEM OF VERY HIGH VITAMIN A INTAKE

Vitamins are substances that are presumably not produced by the human organism although they are vital for its health. They were discovered by their efficacy in healing the detrimental consequences of deficiency states. This differs from drugs, which are discovered by their efficacy in preventing or healing disease states. Vitamins are taken as supplements and in the form of fish oil preparations

From: *Nutrition and Bone Health*
Edited by: M. F. Holick and B. Dawson-Hughes © Humana Press Inc., Totowa, NJ

by large proportions of the population, especially the elderly, and are also used as drugs, i.e., in pharmacological doses, for their presumed health effects beyond those known for the correction of a given deficiency. Another explanation for their extended use is the uncertain possibility of recognizing latent states of vitamin deficiencies. For these reasons, their use differs from that of drugs by the lack of previous dose-finding studies and the insufficient prevalent knowledge of toxicity. The often unconsidered overuse of vitamins led to the discovery of hazardous adverse effects, published as case reports, which had to be analyzed retrospectively and documented by appropriate studies. As a consequence, the recommendations for an adequate vitamin intake moved from the "Recommended Daily Allowances" (RDA), which prevent deficiency states in most of the population, to the definition of safety limits, and to "Estimated Safe and Adequate Daily Dietary Intakes," rather confusing terms for the consumer *(3)*. This chapter tries to untangle the conflicting data on the eventual impact of vitamin A supplementation and treatment on bone health.

3. VITAMIN A INTOXICATION

Vitamin A and its precursors are quantified in international units (IU) and retinol equivalents (RE), 3 μg RE corresponding to 10 IU. Recommended intakes go as high as 1500 μg RE per day (5000 IU). Recently, 700 μg RE (2330 IU) for women and 900 μg (3000 IU) for men were proposed as Adequate Intakes (AIs) *(5,6)*. It was also observed that not just 6 μg, but 21 μg of β-carotene are required to provide 1 μg of retinol or 1 RE, which would mean that the intakes more often fall below the AI than has been assumed so far *(7)*. In developed countries, however, intakes often exceed these recommendations *(5,8)*. In Norway an average intake of vitamin A of 1500–2000 μg RE was reported *(9)*, in the United States an intake of about 2300 μg *(10)*, with reported intakes up to of, e.g., 17,000 IU/d *(11)*. Therefore, in some countries large proportions of the population exceed the recommended AI by severalfold.

This overuse is enhanced by the frequent use of vitamin supplements. For instance, more than one-third of women physicians in the United States and about 20–50% of adults take supplements containing vitamin A *(10,12,13,17,70)*. As a consequence, many subjects take a daily dose of even 25,000 IU, resp. 7500 μg *(5)*. Intakes of 25,000–50,000 IU (15,000 μg) per day taken over several months *(14)* or an acute ingestion of more than 100,000 IU *(15)* have been associated with adverse effects, such as birth defects, various bone abnormalities, and liver toxicity, which are rarely observed at lower doses *(3)*.

Vitamin A can be toxic because of its long half-life, which is partially due to the large storage capacity of the liver. For these reasons, the US Food and Nutrition Board set the safety limit at 3000 μg/d retinol (10,000 IU/d) *(16,17)*. However, this limit is not based on fracture risk and may still not provide an adequate margin of safety *(14)*. In general, skeleton-related adverse effects of vitamin A are observed when supplement intakes reach 25,000 IU, but sometimes less *(3)*.

No vitamin A intoxication was observed with β-carotene *(18),* because its absorption or its transformation into vitamin A seems to be sufficiently controlled. For the same reason, supplementation with β-carotenes does not increase serum retinol concentrations *(19).*

4. FISH OIL

An analysis of 22 fish oil preparations commercialized in Germany showed too high vitamin A contents in a few brands *(20),* the content depending mainly on the fish species: in three samples the content of retinyl esters was above 200 μg per capsule, whereas nine had less than 1 μg. In none were the levels so high that an intake of three to four times over the recommended dose would be dangerous. Therefore, the uncontrolled consumption of fish oil preparations is unlikely to lead to intakes beyond the safety limits, but added to relatively vitamin A-rich food and vitamin supplements, it can be a crucial contribution to the total intake of vitamin A. Regarding the high content of *n*-3 polyunsaturated fatty acids, no beneficial effect on bone has been shown so far in men *(21),* their beneficial action being essentially the prevention of coronary heart disease and sudden death *(22).* In growing rats, however, dietary fish oil decreased osteoclastic activity *(23),* whereas in ovariectomoized rats, eicosapentaenoic acid (EPA) neutralized the negative effect of a low-calcium diet on the strength of bones *(24),* and *n*-3 fatty acids stimulated bone formation and decreased bone resorption *(25).* The mechanisms by which *n*-3 fatty acids alter biochemical and molecular processes involved in bone metabolism are unknown, but these observations led to the conclusion that consuming diets rich in *n*-3 fatty acids will help to maintain a healthy skeleton in humans *(21).* However, this extrapolation to men remains hypothetical, comparable to the extrapolation of the antiresorptive effect of onions in rats to human nutrition.

5. ASSESSMENT OF VITAMIN A INTAKE

Vitamin A can be evaluated either by assessing intake, which is difficult, or by measuring plasma levels. Intake can be evaluated by dietary records, food-frequency questionnaires, and records of supplement intake, and has to include either the total vitamin A intake, or that of retinol and of provitamin A carotenoids separately, mainly β-carotene. However, this more accurate approach is rarely applied to larger studies. In fact, most studies calculated the total vitamin A intake in IU, while any accurate evaluation would need a separate quantification of the intake of preformed vitamin A, β-carotene, and other vitamin A precursors. In addition, the daily variation of vitamin A intake renders the use of a 24-h recall unfit for evaluating the average intake over a longer period *(5).* It has even been stated that more than 365 d are required to estimate the long-term intake of vitamin A because of the irregular intake resulting from the unequal distribution of vitamin A in food *(26).* In some recent reports this statistical requirement was compensated for by the large number of individuals included in the study. On the other hand, the average vitamin A intake seems to be quite constant because in a 5-yr study the annual average serum retinol level of the cohort varied by less than 10% *(72)*.

Measurements of plasma levels can include retinol, which correlates with the consumption of meat, fish, oil, and alcohol, or β-carotene, which reflects the intake of green vegetables *(27)*. The normal range of plasma retinol is very wide, which almost excludes the interpretation of an individual value in search of an insufficient or a high intake *(28)*. In several studies, however, serum retinol has been significantly associated with vitamin A intake *(73,74)*, although not in all *(75)*. In a recent study of 2322 men, individual serum levels could be related to fracture risk *(71)*, whereas the relation with individual vitamin A intakes reached only a borderline significance, although this difference might be the result of the smaller number (1138 men). The serum level of β-carotene was not associated with the fracture risk. The results of this study are discussed later.

Retinyl esters are markers of excess intake *(28,29)*. Their levels increase when intake exceeds either the capacity of liver storage or that of retinol-binding protein production. An increased ratio of retinyl esters over the sum of retinyl esters and protein-bound retinol is a marker of excess intake. Vitamin A status can also be assessed in humans by measuring the dilution after injection of a labeled vitamin A, a technique that is not used in relation to bone health *(30)*. The measurement of retinol-binding protein evaluates the nutritional state of an individual and not specifically the vitamin A intake, and for this reason only it was found to be decreased in osteoporosis *(31)*.

6. ANIMAL DATA

In general, vitamin A toxicity was reported to lead to accelerated bone resorption and fractures *(14)*; hypercalcemia and various bone abnormalities were also observed *(32)*. The various effects of vitamin A on bone were studied in several laboratory animals and cell cultures, with the additional advantage that this allows interventional trials, which are lacking in human investigation. The former results could recently been explained by the observation that both osteoblasts and osteoclasts express nuclear receptors for retinoic acid *(33,34)*.

Histomorphometry in rats showed that high doses of vitamin A increased osteoclast numbers and reduced osteoid surfaces, therefore stimulating bone resorption and inhibiting bone formation *(35)*. Studies in vitro showed, in bone cells and calvaria cultures, that vitamin A, and in some experiments also retinoic acid, inhibited bone collagen synthesis *(36–38)*, by stimulating collagenase synthesis in osteoblasts *(39)*, and stimulated formation of osteoclasts and osteoclastic bone resorption *(37,38,40–42)*. Several of these studies applied very high doses, 10,000–75,000 IU/d in rats, which would be beyond safety limits even for humans. Probably for this reason, in vivo studies including direct measurements of bone density were not conclusive. Vitamin A-intoxicated rats were already reported in the 1940s to have spontaneous fractures. On the other hand, high vitamin A intake in rats resulted in increased bone mineral density (BMD), although trabecular area decreased *(43)*. But high vitamin A intake in dogs, 3 times the recommended amount, had no influence on BMD measured by quantitative computed tomography *(44)*.

A new light was shed on the bone effect of vitamin A with the investigation of the interaction with vitamin D. In aged rats, a model closer to human osteoporosis than young and growing animals, a high vitamin A intake had negative effects, although only in conjunction with a low calcium intake *(45)*. Later investigations pointed to an interference with the utilization of vitamin D and its hydroxylated metabolites. Vitamin A antagonized the ability of vitamin D to maintain normal plasma calcium levels in the rat *(46)*, resulting in higher vitamin D requirements. On the other hand, it has a protective effect against hypervitaminosis D *(47,48)*, specifically against the bone resorption induced by $1,25(OH)_2D_3$ in vitro *(42)*. There are nuclear receptors of vitamin A in the bone of rats, and vitamin A influences gene expression *(49)* in conjunction with vitamin D, the receptors of both vitamins binding to the target genes as heterodimers, with $1,25(OH)_2D_3$ activating the vitamin D receptor–retinoid X receptor complex *(50)*. This interaction has recently been demonstrated also in humans by the observation that retinyl palmitate decreased plasma calcium and diminished the calcium response to $1,25(OH)_2D_3$ *(51)*.

7. HUMAN DATA (SEE TABLE 1)

The first reports on chronic adverse effects of vitamin A were isolated case reports of children and adults published from the 1960s into the 1990s *(52)*. They included effects on the skeleton and on bone metabolism, such as hypercalcemia, retardation or arrest of growth, bone pain, hyperostosis, accelerated bone loss, etc. *(5)*. These observations were usually made with doses of 100,000 IU/d or more. Acute adverse effects of vitamin A supplements given to neonates, typically bulging of the anterior fontanella, are also bone-related, but they are not lasting *(2)*. The question arises whether long-term high intake of vitamin A in adults has a remaining negative effect on bone health.

A series of *cross-sectional studies* that correlated vitamin A intake with bone mineral content (BMC) showed no such negative association *(13,53,54)*. These studies used single-photon absorptiometry (SPA) to measure bone density (i.e., BMC) on the forearm, which involves mostly cortical bone of the radius shaft. Trabecular bone, the bone tissue that is most sensitive to changes in bone metabolism, is present mainly in the ultradistal part of the radius and in vertebral bodies, but these sites were rarely included in the studies. Another reason why these studies might have missed an effect of vitamin A on bone is the relatively small number of subjects included. The study in which plasma levels of vitamin A were measured, and no correlation with the fracture history was found, was again not powered to detect an eventual association *(13)*. Furthermore, the studies were not focused on elderly subjects, in whom the impact of vitamin A would probably have been easier to demonstrate. In addition, the pitfalls in evaluating the nutritional intake of vitamin A, as discussed above, also add to the difficulties in performing conclusive studies.

Table 1
Studies of the Negative Effect of High Vitamin A Intake on Bone Health in Humans

Author, Year	*Ref. Population*	*Measurements*	*Bone Assessments*[a]	*Negative Association*
Cross-sectional studies				
Yano et al., 1985 *(53)*	1208 men >60 yr and 912 women >50 yr	Vitamin A intake[b]	BMC radius, ulna . . .	None
Sowers et al., 1985 *(54)*	324 postmenopausal women	Vitamin A intake[b]	BMC radius	Almost signif.
Sowers and Wallace, 1990 *(13)*	246 women, 55–80 yr	Vitamin A intake[b] and serum retinol	BMC mid-radius	None
			Fracture history	None
Melhus et al., 1998 *(56)*	175 women, 40–76 yr	Retinol intake[c]	BMD LS, prox. fem.	Signif.
Melhus et al., 1998 *(56)*	247 hip-fx, 873 contr.	Retinol intake[c]	Hip-fx incidence	Signif.
Sigurdson et al., 2001 *(55)*	232 women 70 yr	Vitamin A intake[d]	BMD LS, prox. fem.	None
Ballew et al., 2001 *(52)*	5790 adults, 20–>80 yr	Serum retinyl esters	BMD prox. fem.	None
Longitudinal studies				
Houtkooper et al., 1995 *(8)*	66 women, 28–39 yr	Vitamin A intake[e]	TB-BMD loss/time	Signif.
		Carotene intake[e]	TB-BMD loss/time . . . when combined with fat mass	Signif.
Freudenheim et al., 1986 *(11)*	99 women, 35–65 yr	Vitamin A intake[f]	BMC ulna/4 yr	Signif.
Feskanich et al., 2002 *(10)*	72,337 women, 34–77 yr	Vitamin A intake and	Hip-fx incidence	Signif.
		β-carotene intake[g]	Hip-fx incidence	None
Promislov et al., 2002 *(70)*	570 women + 388 men 55–92 yr	Retinol intake[g]	BMD loss/time	Signif.
Michaëlsson et al., 2003 *(71)*	2322 men ± 50 + 24 yr	Serum retinol	Fracture incidence	Signif.
Interventional studies		*Intervention*		
Kawahara et al., 2002 *(61)*	80 men, 18–58 yr	Retinol palmitate	Bone markers	None
Johansson and Melhus, 2001 *(51)*	9 men, 24–41 yr	Retinyl palmitate	Response to 1,25$(OH)_2$VitD_3	Interaction

[a] LS = lumbar spine.

In addition to registration of supplement intake: [b] 24-h recall; [c] 1-wk dietary record, 4 times; [d] FFQ including retinol, β-carotene; [e] 12 randomly assigned 24-h recalls; [f] structured 24-h record, 72 times/3 yr; [g] FFQ, type of fat/oil.

More relevant are the studies that used dual X-ray absorptiometry (DXA) and that included large numbers of subjects. One such very large survey in the third National Health and Nutrition Examination Survey (NHANES III) used plasma levels of retinyl ester, a marker of excessive retinol intake, and DXA for the measurement of BMD. Despite that, it found no association between these two parameters *(52)*. A smaller study using DXA also could not show a correlation between vitamin A intake and bone density *(55)*. However, a Scandinavian cross-sectional study on 175 women did relate high vitamin A intake to lowered BMD, and a nested case-control study of 247 hip fractures and 873 controls linked a retinol intake of more than 1500 μg (5000 IU)/d to an elevated risk of hip fractures, with an odds ratio of 1.54 when calcium intake was included in the model *(56)*. This elevated fracture risk was linked to a dose of vitamin A that exceeds the recommended limit, and probably concerns only a small subgroup of the population *(57)*. However, since older adults have a diminshed capacity to clear high levels of ingested retinol, it was hypothetized that excess retinol intake may explain the high incidence of osteoporosis in northern Europe *(58)*. An indirect argument in favor of the negative effect of vitamin A on bone health in cross-sectional studies is its effect as a confounding factor: The odds ratio for hip fracture of the highest quartile of calcium intake approached 1.0 when it was adjusted for retinol intake. And its inclusion in the multivariate model gave a significantly elevated odds ratio for hip fracture for the highest intake of iron, vitamin C, and magnesium, which were close to already reported relationships *(59,60)*.

Follow-up studies, which by design are of higher statistical power than cross-sectional studies, also revealed significant correlations between the loss of total body BMD and vitamin A intake *(8)*. Even with SPA on the ulna, but not on the radius, some significant correlations could be found with vitamin A intake in a longitudinal study *(11)*. The largest longitudinal study, the Nurses Health Study, assessed the vitamin A intake of 72,337 postmenopausal women and found a significant correlation with hip fracture incidence, the highest quintile of vitamin A intake being significantly linked to an increased fracture risk, except in women on hormone replacement therapy (HRT). The relative risk (RR) for hip fracture was 1.69 in women with increased retinol intake from food, whereas β-carotene showed no contribution *(10)*. Women who consumed liver once or more per week already had an almost significantly increase by 69% in relative hip fracture risk. The impact of vitamin A supplements on hip fracture risk was also important, but not significant *(10)*. However, considering that the results were at the edge of significance (RR 1.40; CL 0.99–1.99), and that the population was relatively young for hip fractures (34–77 yr), it is probable that a new evaluation of the same population about 10 yr later will yield a significant relationship with the consumption of vitamin supplements.

A 4-yr follow-up study of 958 men and women revealed that retinol intake was associated with BMD at baseline and 4 yr later, as well as with BMD change *(70)*. Interestingly, however, the association was positive up to intakes of 2000–2800 IU, reminding one of the importance of also avoiding vitamin deficiencies for bone health. The association became negative when the intake exceeded these amounts, which were close to the recommended daily allowances of 2330–3000 IU, in accordance with other studies. These high intakes were reached mostly by supplement users.

A Swedish population-based follow-up study of 2322 men followed for ±24 yr provided not only similar but also highly significant results *(71)*. The comparison of baseline serum levels of retinol with the clinical follow-up showed, that for every increase of 1 SD in serum retinol, the risk of fractures increased by 26%. The association was not linear. The multivariate rate ratio for hip fractures was 2.47 in the highest quintile of retinol levels as compared with the third quintile. The risk increase became substantial mostly in the highest 5th percentiles, and was exponential within this category.

Interventional studies are rare, and none measured bone density changes over a longer period. The already-mentioned acute trial in nine healthy men confirmed the animal data of an interaction with vitamin D *(51)*. And a controlled 6-wk trial, in which 25,000 IU (7576 μg) of retinol palmitate were given to adult men, could not detect any effect on bone markers *(61)*.

Taken together, these studies seem to be contradictory. However, when only those with high statistical power and optimal methods of bone measurements are taken into account, there remains some evidence that high intake of vitamin A is linked to an increased risk of osteoporosis and hip fractures.

A cross-sectional study in Chinese children and adolescents described a positive correlation of plasma levels of vitamin A with body mass index (BMI) and indices of growth, such as body weight. Noncorrected for energy intake, this result reflects the importance of the nutritional state of these children in general, and cannot be taken as evidence for a specific effect of vitamin A on bone development *(50)*.

8. ADVERSE EFFECTS ON BONE OF TREATMENT WITH SYNTHETIC RETINOIDS

The pharmacological use of synthetic retinoids, such as isotretinoin against acne, and etetrinate against psoriasis, are often accompanied by adverse side effects, which illustrates the relatively narrow therapeutic windows of these substances. Observational reports link various rheumatological complications to the use of retinoids *(62)*. These include bone pain, premature closure of epiphyses, development of osteophytes, calcification of ligaments, etc. *(63)*. The main question, whether long-term treatments lead to decreased BMD and to osteoporosis, is not yet answered. Short-term treatments with isotretinoin did not affect BMD *(64)*. Long-term treatments with etetrinate had no negative effect on bone scans *(65)*, but this is not a technique appropriate for detecting loss of bone mass. When BMD was measured, etetrinate caused bone loss, whereas isotretinoin had no effect *(66,67)*. Indeed, isotretinoin reduced bone markers only for 1–2 wk *(68)*. In another study, isotretinoin also reduced BMD, but the decrease was significant only at the Wards triangle, whereas the measurements of more relevant areas of interest did not reveal any significant changes *(69)*.

These observations, and several others, illustrate the presence of adverse effects of synthetic retinoids on the musculoskeletal system, but they cannot serve as proofs of a negative effect of vitamin A on bone mass; they just point to a probable impact of vitamin A on bone metabolism and bone health.

9. CONCLUSION

Human hypervitaminosis A clearly involves the skeleton, as other reviews have already revealed *(14)*. High vitamin A intake seems to accelerate bone loss. Despite this conclusion, the association of a high retinol intake with an increased risk of hip fractures was long discussed and considered uncertain. As pointed out by Hathcock, "the issue for vitamin A and bone health is not whether mechanisms exist, however, but instead whether the effect occurs at the usual levels of retinol intake experienced by most persons" *(16)*. More recent studies of considerable dimensions provided evidence for a relationship between a high intake of vitamin A and adverse effects on bone. High intakes and high serum retinol levels could be related to an increased fracture risk, especially the risk of hip fracture. The usual levels of vitamin A or retinol intake are often close to or beyond the approximate safety limit, especially due to the widespread use of vitamin supplements. This points to a narrow window between the amount necessary for health maintenance and the supraoptimal doses linked to increased fracture risks.

REFERENCES

1. Sklan D. Vitamin A in human nutrition. Prog Food and Nutr Sci 1987; 11:39–55.
2. Humphrey JH, Agoestina T, Juliana A, et al. Neonatal vitamin D supplementation: effect on development and growth at 3 y of age. Am J Clin Nutr 1998; 68:109–117.
3. Hathcock JN. Vitamins and minerals: efficacy and safety. Am J Clin Nutr 1997; 66:427–437.
4. Moore T, Wang Y. Hypervitaminosis A. Biochem J 1945; 39:222–228.
5. Hathcock JN, Hattan DG, Jenkins MY, McDonald JT, Sundaresan PR, Wilkening VL. Evaluation of vitamin A toxicity. Am J Clin Nutr 1990; 52:183–202.
6. Feskanich D, Willett WC, Colditz GA. Letter. JAMA 2002; 287:1397.
7. West CE. Meeting requirements for vitamin A. Nutr Rev 2000; 50:341–345.
8. Houtkooper LB, Ritenbaugh C, Aickin M, et al. Nutrients, body composition and exercise are related to change in bone mineral density in premenopausal women. J Nutr 1995; 125:1229–1237.
9. Johansson L, Solvoll K, Bjorneboe GE, Drevon C. Dietary habits among Norwegian men and women. Scand J Nutr 1997; 41:63–70.
10. Feskanich D, Singh V, Willett WC, Colditz GA. Vitamin A intake and hip fractures among postmenopausal women. JAMA 2002; 287:47–54.
11. Freudenheim JL, Johnson NE, Smith EL. Relationships between usual nutrient intake and bone-mineral content of women 35–36 years of age: longitudinal and cross-sectional analysis. Am J Clin Nutr 1986; 44:863–876.
12. Frank E, Bendich A, Denniston M. Use of vitamin-mineral supplements by female physicians in the United States. Am J Clin Nutr 2000; 72:969–975.

13. Sowers MFR, Wallace RB. Retinol, supplemental vitamin A and bone status. J Clin Epidemiol 1990; 43:693–699.
14. Binkley N, Krueger D. Hypervitaminosis A and bone. Nutr Rev 2000; 58:138–144.
15. Bendich A, Langseth L. Safety of vitamin A. Am J Clin Nutr 1989; 49:358–371.
16. Hathcock JN. Does high intake of vitamin A pose a risk for osteoporotic fracture? Letter to the editor. JAMA 2002; 287:1396–1397.
17. Panel on Micronutrients, Subcommittees on Upper Reference Levels of Nutrients and of Interpretation and Use of Dietary Reference Intakes, Standing Committee on the Scientific Evaluation of Dietary Reference Intakes, Food and Nutrition Board. Vitamin A. In: Dietary Reference Intakes for Vitamin A, Vitamin K, Arsenic, Boron, Chromium, Copper, Iodine, Iron, Manganese, Molybdenum, Nickel, Silicon, Vanadium, and Zinc. National Academy Press, Washington, DC, 2001, pp. 65–126.
18. Heywood R, Palmer AK,Gregson RI, Hummler II. The toxicity of beta-carotene. Toxicology 1985; 36:91–100.
19. Nierenberg DW, Dain BJ, Mott LA, Baron JA, Greenberg ER. Effect of 4 y of oral supplementation with beta-carotene on serum concentrations of retinol, tocopherol, and five carotenoids. Am J Clin Nutr 1997; 66:315–319.
20. Koller H, Luley C, Klein B, Baum H, Biesalski HK. Contaminating substances in 22 over-the-counter fish oil and cod liver oil preparations: cholesterol, heavy metals and vitamin A (in German). Z Ernährungswiss 1989; 28:76–83.
21. Watkins BA, Li Y, Lippman HE, Seifert MF. Omega-3 polyunsaturated fatty acids and skeletal health. Exp Biol Med 2001; 226:485–497.
22. Connor WE. *n*-3 Fatty acids from fish and fish oil: panacea or nostrum. Am J Clin Nutr 2001; 74:415–416.
23. Iwami-Morimoto Y, Yamaguchi K, Tanne K. Influence of dietary *n*-3 polyunsaturated fatty acids on experimental tooth movement in rats. Angle Orthod 1999; 69:365–371.
24. Sakaguchi K, Morita I, Murota S. Eicosapentaenoic acid inhibits bone loss due to ovaricetomy in rats. Prostaglandins Leukot Essent Fatty Acids 1994; 50:81–84.
25. Schlemmer CK, Coetzer H, Claassen N, Kruger MC. Oestrogen and fatty acid supplementation corrects bone loss due to ovariectomy in the female Sprague Dawley rat. Prostaglandins Leukot Essent Fatty Acids 1999; 61:381–390.
26. Basiotis PB, Welsh SO, Cronin FJ, Kelsay JL, Mertz W. Number of days of food intake recors required to estimate individual and group nutrient intakes with defined confidence. J Nutr 1987; 117:1638–1641.
27. Wang G, Brun TA, Geissler CA, et al. Vitamin A and carotenoid status in rural China. Br J Nutr 1996; 76:809–820.
28. Goodman DS. Vitamin A transport and delivery and the mechanism of vitamin A toxicity. In: Orfanos CE, Braun-Falco O, Farber EM, Grupper C, Polano MK, Schuppli R, eds. Retinoids: Advances in Basic Research and Therapy. Springer Verlag, New York, 1981, pp. 31–39.
29. Krasinski SD, Russell RM, Otradovec CL, et al. Relationship of vitamin A and vitamin E intake to fasting plasma retinol, retinol-binding protein, retinyl esters, carotene, α-tocopherol, and cholesterol among elderly people and young adults: increased plasma retinyl esters among vitamin A-supplement users. Am J Clin Nutr 1989; 49:112–120.
30. Tang G, Qin J, Hao L, Yin S, Russell RM. Use of short-term isotope-dilution method for determining the vitamin A status of children. Am J Clin Nutr 2002; 76:413–418.
31. Rico H, Relea P, Crespo R, et al. Biochemical markers of nutrition in type-I and type-II osteoporosis. J Bone Joint Surg 1995; 77:148–151.
32. Frame B, Jackson CE, Reynolds WA, Umphrey JE. Hypercalcemia and skeletal effects in chronic hypervitaminosis A. Ann Intern Med 1974; 80:44–48.
33. Kindmark A, Törmä H, Johansson A, Ljunghall S, Melhus H. Reverse transcritpion polymerase chaine reaction assay demonstrates that the 9-cis retinoic acid receptor alpha is expressed in human osteoblasts. Biochem Biophys Res Commun 1993; 192:1367–1372.

34. Saneshige S, Mano H, Tezka K, et al. Retinoic acid directly stimulates osteoclastic bone resorption and gene expression of acthepsin K/OC-2. Biochem J 1995; 309:721–724.
35. Frankel TL, Seshadri MS, McDowall DB, Cornish CJ. Hypervitaminosis A and calcium-regulating hormones in the rat. J Nutr 1986; 116:578–587.
36. Dickson IR, Walls J, Webb S. Vitamin A and bone formation. Different responses to retinol and retinoic acid of chick bone cells in organ culture. Biochim Biophys Acta 1989; 1013:254–258.
37. Togari A, Kondo M, Arai M, Matsumoto S. Effects of retinoic acid on bone formation and resorption in cultured mouse calvaria. Gen Pharmacol 1991; 22:287–292.
38. Hough S, Avioli LV, Muir G, et al. Effects of hypervitaminosis A on the bone and mineral metabolism of the rat. Endocrinology 1988; 122:2933–2939.
39. Heath JK, Reynolds JJ, Meikle MC. Osteopetrotic (grey-lethal) bone produces collagenase and TIMP in organ culture: regulation by vitamin A. Biochem Biophys Res Commun 1990; 168:1171–1176.
40. Oreffo RO, Teti A, Triffitt JT, Francis MJ, Carano A, Zallone AZ. Effect of vitamin A on bone resorption: evidence for direct stimulation of isolated chicken osteoclasts by retinal and retinoic acid. J Bone Miner Res 1988; 3:203–210.
41. Scheven BA, Hamilton NJ. Retinoid acid and 1,25 dihydroxy vitamin D_3 stimulate osteoclast formation by different mechanisms. Bone 1990; 11:53–59.
42. Kindmark A, Melhus H, Ljunghall S, Ljunggren O. Inhibitory effects of 9-cis and all-transretinoic acid on 1.25 (O)2 vitamin D3-induced bone resorption. Calcif Tissue Int 1995; 57:242–244.
43. Lind PM, Larsson S, Johansson S, et al. Bone tissue composition, dimensions and strength in female rats given an increased dietary level on vitamin A or exposed to 3,3′,4,4′,5-pentachlorobiphenyl (PCB126) alone or in combination with vitamin C. Toxicology 2000; 151:11–23.
44. Cline JL, Czarnecki-Maulden GL, Losonsky JM, Sipe CR, Easter RA. Effect of increasing dietary vitamin A on bone density in adult dogs. J Anim Sci 1997; 75:2980–2985.
45. Li XF, Dawson-Hughes B, Hopkins R, et al. The effects of chronic vitamin A excess on bone remodeling in aged rats. Proc Soc Exp Biol Med 1989; 191:103–107.
46. Rohde CM, Manatt M, Clagett-Dame M, de Luca HF. Vitamin A antagonises the action of vitamin D in the rat. J Nutr 1999; 129:2246–2250.
47. Aburto A, Edwards HM Jr, Britton WM. The influence of vitamin A on the utilization and amelioration of toxicity of choleclacifero, 25-hydroxycholecalciferol, and 1,25 dihydroxycholecalciferol in young broiler chickens. Poultry Sci 1998; 77:585–593.
48. Metz AL, Walser MM, Olson WG. The interaction of dietary vitamin A and vitamin D related to skeletal development in the turkey poult. J Nutr 1985; 115:929–935.
49. Harada H, Miki R, Masushige S, Kato S. Gene expression of retinoic acid receptors, retinoid-X receptors, and cellular retinol-binding protein I in bone and its regulation by vitamin A. Endocrinology 1995; 136:5329–5335.
50. Haussler MR, Haussler CA, Jurutka PW, et al. The vitamin D hormone and its receptor: molecular actions and disease states. J Endocrinol 1997; 154 (suppl):S57–S73.
51. Johansson S, Melhus H. Vitamin A antagonizes calcium response to vitamin D in man. J Bone Miner Res 2001; 16:1899–1905.
52. Ballew C, Galuska D, Gillespie C. High serum retinyl esters are not associated with reduced bone mineral density in the Third National Health and Nutrition Examination Survey, 1988–1994. J Bone Miner Res 2001; 16:2306–2312.
53. Yano K, Heilbrun LK, Wasnich RD, Hankin JH, Vogel JM. The relationship between diet and bone mineral content of multiple skeletal sites in elderly Japanese-American men and women living in Hawaii. Am J Clin Nutr 1985; 42:877–888.
54. Sowers MR, Wallace RB, Lemke JH. Correlates of mid-radius bone density among postmenopausal women: a community study. Am J Clin Nutr 1985; 41:1045–1053.
55. Sigurdsson G, Franzson L, Thorgeirsdottir H, Steingrimsdottir L. A lack of association between excessive dietary intake of vitamin A and bone mineral density in seventy-year-old icelandic

women. In: Burckhardt P, Dawson-Hughes B, Heaney R, eds. Nutritional Aspects of Osteoporosis. Academic Press, San Diego, CA, 2001, pp. 295–302.

56. Melhus H, Michaëlsson K, Kindmark A, et al. Excessive dietary intake of vitamin A is associated with reduced bone mineral density and increased risk for hip fracture. Ann Intern Med 1998; 129:770–778.
57. Sigurdsson G. Dietary vitamin A intake and risk for hip fracture (Letter). Ann Intern Med 1999; 131:392.
58. Whiting SJ, Lemke B. Excess retinol intake may explain the high incidence of osteoporosis in northern Europe. Nutr Rev 1999; 57:192–195.
59. Michaelsson K, Holmberg L, Mallmin H, et al. Diet and hip fracture risk: a case control study. Int J Epidemiol 1995; 24:771–782.
60. Michaelsson K. The complicated research field of nutrients and osteoporosis (corresp). Nutr Rev 2000; 58:249–250.
61. Kawahara TN, Krueger DC, Engelke JA, Harke JM, Binkley NC. Short-term vitamin A supplementation does not affect bone turnover in men. J Nutr 2002; 132:1169–1172.
62. Nesher G, Zuckner J. Rheumatologic complications of vitamin A and retinoids (Review). Semin Arthritis Rheum 1995; 24:291–296.
63. Guire J, Lawson JO. Skeletal changes associated with chronic isotretinoin and etetrinate administration. Dermatologica 1987; 175(suppl 1):169–181.
64. Margolis DJ, Attie M, Leyden JJ. Effects of isotretinoin on bone mineralization during routine therapy with isotretinoin for acne vulgaris. Arch Dermatol 1996; 132:769–774.
65. Glover MT, Peters AM, Atherton. Surveillance for skeletal toxicity of children treated with etretinate. Br J Dermatol 1987; 116:609–614.
66. DiGiovanna JJ, Sollitto RB, Abangan DL, Steinberg SM, Geynolds JC. Osteoporosis is a toxic effect of long-term etetrinate therapy. Arch Dermatol 1995; 131:1263–1267.
67. Okada N, Nomura M, Morimoto S, Ogihara T, Yoshikawa K. Bone mineral density of the lumbar spine in psoriatic patients with long-term etretinate therapy. J Dermatol 1994; 21:308–311.
68. Kindmark A, Rollman O, Mallmin H, Petrén-Mallmin M, Ljunghall S, Melhus H. Oral isotretinoin therapy in severe acne induces transient suppression of biochemical markers of bone turnover and calcium homeostasis. Acta Dermatol Venereol (Stockh) 1998; 78:266–269.
69. Leachman SA, Insogna KI, Katz L, Ellison A, Milstone LM. Bone densities in patients receiving isotretinoin for cystic acne. Arch Dermatol 1999; 135:961–965.
70. Promislow JHE, Goodman-Gruen D, Slymen DJ, Barrett-Connor E. Retinol intake and bone mineral density in the elderly: the Rancho Bernardo Study. J Bone Miner Res 2002; 17:1349–1358.
71. Michaëlsson K, Lithell H, Vessby B, Melhus H. Serum retinol levels and the risk of fracture. N Engl J Med 2003; 348:287–294.
72. Stauber PM, Sherry B, VanderJagt DJ, Bhagavan HN, Garry PJ. A longitudinal study of the relationship between vitamin A supplementation and plasma retinol, retinyl esters, and lever enzyme activities in a healthy elderly population. Am J Clin Nutr 1991; 54:878–883.
73. Garry PJ, Huut WL, Brandrofchak JL, VanderJagt D, Goodwin JS. Vitamin D intake and plasma retinol levels in healthy elderly men and women. Am J Clin Nutr 1987; 46:989–994.
74. Neuhouser ML, Rock CL, Eldridge AL, et al. Serum concentrations of retinol, alfatocopherol and the carotenoids are influenced by diet, race and obesity in a sample of healthy adolescents. J Nutr 2001; 131:2184–2191.
75. Johnson EJ, Krall EA, Dawson-Hughes B, Dallal GE, Russell RM. Lack of an effect of multivitamins containing vitamin A on serum retinyl esters and liver function tests in healthy women. J Am Coll Nutr 1992; 11:682–686.

25 Vitamin D

Michael F. Holick

1. EVOLUTION OF VITAMIN D

Although it is not certain when vitamin D became critically important for calcium metabolism and bone health for our early ancestors, there is evidence that some of the earliest phytoplankton life forms were photosynthesizing vitamin D more than 750 million years ago *(1–3)*. Life evolved in a fertile soup that contained all of the organic and inorganic compounds necessary for life to evolve. One of the key elements that early life forms used was calcium for regulation of many metabolic processes. As invertebrates and vertebrates evolved, they took advantage of the high calcium content of their ocean environment (approx 400 mmol) and used it as a major component for their exo- and endoskeletons, respectively. When vertebrate life forms ventured onto land, the calcium on which they became dependent was plentiful in the soils, but they had no mechanism to extract it. Plants, however, extracted the precious calcium out of the soils and distributed it throughout their structures. Thus, calcium was harvested by vertebrates from the soil indirectly by the ingestion of these plants. To utilize the dietary calcium there was a need for a mechanism to recognize the calcium status of the organism and to regulate the efficiency of intestinal calcium absorption depending on the organism's calcium needs. It is likely that vitamin D played a crucial role in early vertebrate development by regulating intestinal calcium absorption and calcium metabolism *(1–3)*.

2. VITAMIN D METABOLISM AND ACTION ON THE INTESTINE

Once vitamin D is made in the skin, it enters the circulation. Vitamin D (vitamin D represents either vitamin D_2 or vitamin D_3) from the diet is incorporated in chylomicrons and absorbed into the lymphatic system, where it eventually is deposited into the venous circulation. Both dietary and skin sources of vitamin D are bound in the circulation to a vitamin D-binding protein (DBP) *(4)*. Some of the lipophylic vitamin D in the circulation is deposited in the body fat, while most of it is directed to the liver *(5–7)*. Once it enters hepatocytes, it is metabolized by the vitamin D-25-hydroxylase (CYP27A) and transformed to 25-hydroxyvitamin D

From: *Nutrition and Bone Health*
Edited by: M. F. Holick and B. Dawson-Hughes © Humana Press Inc., Totowa, NJ

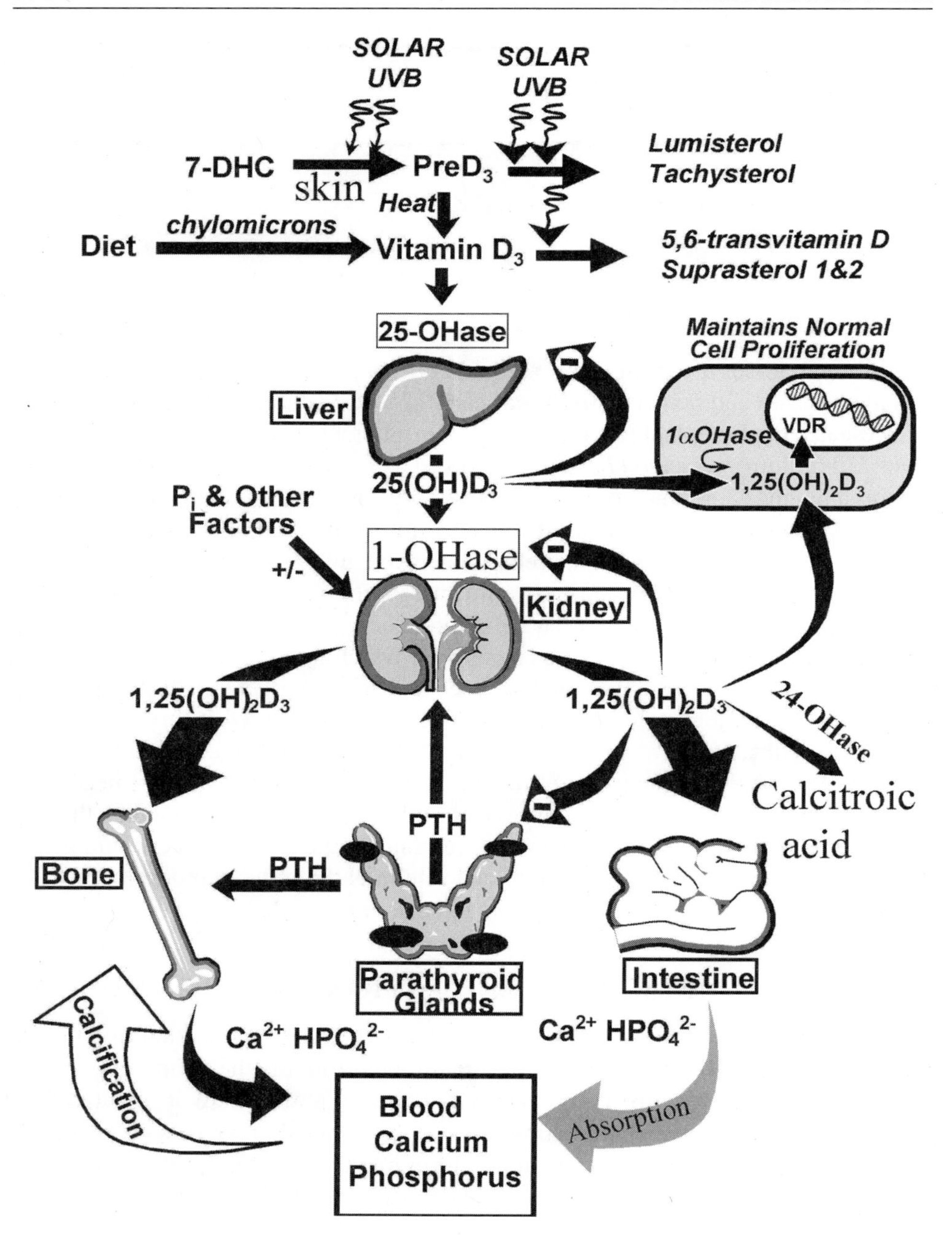

[25(OH)D] *(6,7)*. 25(OH)D leaves the hepatocyte and enters the circulation and is once again bound to the DBP. 25(OH)D is the major circulating form of vitamin D and, as a result, is used to determine the vitamin D status of both children and adults. The 25(OH)D–DBP complex is recognized by megalin that is located in the plasma membrane of the renal tubular cells. Megalin facilitates the endocytic trans-

Fig. 1. Schematic representation for cutaneous production of vitamin D and its metabolism and regulation for calcium homeostasis and cellular growth. During exposure to sunlight, 7-dehydrocholesterol (7-DHC) in the skin absorbs solar ultraviolet (UVB) radiation and is converted to previtamin D_3 (preD_3). Once formed, D_3 undergoes thermally induced transformation to vitamin D_3. Further exposure to sunlight converts preD_3 and vitamin D_3 to biologically inert photoproducts. Vitamin D coming from the diet or from the skin enters the circulation and is metabolized in the liver by the vitamin D-25-hydroxylase (25-OHase) to 25-hydroxyvitamin D_3 [25(OH)D_3]. 25(OH)D_3 reenters the circulation and is converted in the kidney by the 25-hydroxyvitamin D_3-1α-hydroxylase (1-OHase) to 1,25-dihydroxyvitamin D_3 [1,25$(OH)_2D_3$]. A variety of factors, including serum phosphorus (P_i) and parathyroid hormone (PTH), regulate the renal production of 1,25$(OH)_2$D. 1,25$(OH)_2$D regulates calcium metabolism through its interaction with its major target tissues, bone and the intestine. 1,25$(OH)_2D_3$ also induces its own destruction by enhancing the expression of the 25-hydroxyvitamin D-24-hydroxylase (24-OHase). 25(OH)D is metabolized in other tissues for the purpose of regulation of cellular growth.

port of the 25(OH)D–DBP complex into the renal cell *(8)*. 25(OH)D is then released and enters the mitochondria, where the cytochrome P450-25-hydroxyvitamin D-1-hydroxylase (CYP27B1; 1-OHase) converts it to 1,25-dihydroxyvitamin D [1,25$(OH)_2$D] *(3,6,7)* (Fig. 1). The renal 1-OHase is upregulated by hypocalcemia and hypophosphatemia. Parathyroid hormone (PTH) is also a potent stimulator of the renal 1-OHase. During pregnancy and lactation, estrogen and prolactin are also thought to play a role in upregulating the 1-OHase *(6,7)*.

1,25$(OH)_2$D is considered to be the biologically active form of vitamin D. It binds to its specific nuclear vitamin D receptor (VDR), which in turn binds with the retinoic acid X receptor (RXR) to form a herterodimeric complex. This complex interacts with specific sequences in the promoter region of vitamin D-responsive genes, known as vitamin D-responsive element (VDRE) *(3,7,9,10)*. The binding of the VDR-1,25$(OH)_2$D-RXR complex to the VDRE initiates the binding of several transcriptional factors that ultimately results in either an increased or decreased expression of vitamin D-responsive genes *(9–11)*.

1,25$(OH)_2$D is recognized by the VDR in the small intestine, resulting in an increase in the expression of the epithelial calcium channel on the mucosal surface of the intestinal absorptive cell *(2,3,7)*. In addition, there is an increase in the expression of the calcium-binding protein$_{9K}$ (calbindin), calcium-dependent ATPase, and several other brush border proteins *(2,3,7,10,12)*. The ultimate result is that 1,25$(OH)_2$D enhances the efficiency of intestinal calcium absorption from a baseline of approx 10–15% to 30–40%. Most of the dietary calcium is absorbed in the duodenum and to a lesser extent in the jejunum and ileum.

Once 1,25$(OH)_2$D carries out its function in the small intestine, it then induces the expression of the 25-hydroxyvitamin D-24-hydroxylase (CYP-24). This results in the initiation of a cascade of metabolic steps that culminates in the cleavage of the side chain between carbons 23 and 24 to yield the water-soluble, biologically inactive excretory product, calcitroic acid *(3,7,10)*.

3. VITAMIN D ACTION ON BONE CALCIUM MOBILIZATION

Although vitamin D is associated with bone health, the principal physiological function of vitamin D is to support the serum calcium within a physiologically acceptable range in order to maintain neuromuscular and cardiac function and a multitude of other metabolic activities *(2)*. Thus, when dietary calcium is inadequate to satisfy the body's requirement for calcium, this results in vitamin D becoming a catabolic hormone that mobilizes calcium stores from the skeleton.

$1,25(OH)_2D$ increases the removal of calcium from the skeleton by increasing osteoclastic activity. It was originally believed that $1,25(OH)_2D$ interacted with specific nuclear receptors in preosteoclasts to initiate the formation of mature osteoclasts. We now recognize that $1,25(OH)_2D$ initiates the mobilization of preosteoclasts through its interaction with its VDR in osteoblasts. The osteoblast serves as the master cell for regulating bone metabolism. $1,25(OH)_2D$ interacts with the VDR in mature osteoblasts and induces the expression of RANKL (receptor for RANKL) on its plasma membrane surface *(3,7,13–15)*. The precursor monocytic osteoclasts have a membrane receptor for RANKL, known as RANK (receptor activator NFκB). It is the intimate interaction of the preosteoclast's RANK with the osteoblast's RANKL that ultimately signals the preosteoclast to become a mature bone-resorbing multinucleated osteoclast (Fig. 2). Thus, in calcium-deficient states $1,25(OH)_2D$ production is enhanced and in turn mobilizes an army of osteoclasts that resorb bone-releasing precious calcium stores into the circulation to maintain ionized calcium levels in the normal range.

4. VITAMIN D AND BONE MINERALIZATION

$1,25(OH)_2D$ interacts with osteoblasts, not only to increase the expression of RANKL, but also to enhance the expression of osteocalcin, alkaline phosphatase, and osteopontin *(6,7,10,16,17)*. Despite all of these biological functions in the osteoblast, there is no evidence that $1,25(OH)_2D$ is essential for the ossification process of the collagen matrix *(18–20)*. This is based on the observation that severely vitamin D-deficient rats that either received a high-calcium and high-phosphorus-with-lactose diet or received calcium intravenously had bones that had no evidence of rickets or other pathology (Fig. 3; *20a*). This has also been confirmed in rachitic patients with a VDR defect known as $1,25(OH)_2D$-resistant rickets (vitamin D-dependent rickets type 2) and who received an infusion of calcium, resulting in the healing of their rickets *(20)*.

5. DIETARY SOURCES OF VITAMIN D

There are very few foods that naturally contain vitamin D. These foods include oily fish including makerel, eel, and salmon, cod liver oil, sun-exposed mushrooms, and egg yolks (Table 1).

Steenbock *(21)* recognized the importance of promoting antirachitic activity in foods by irradiating them with ultraviolet radiation. He suggested irradiation of

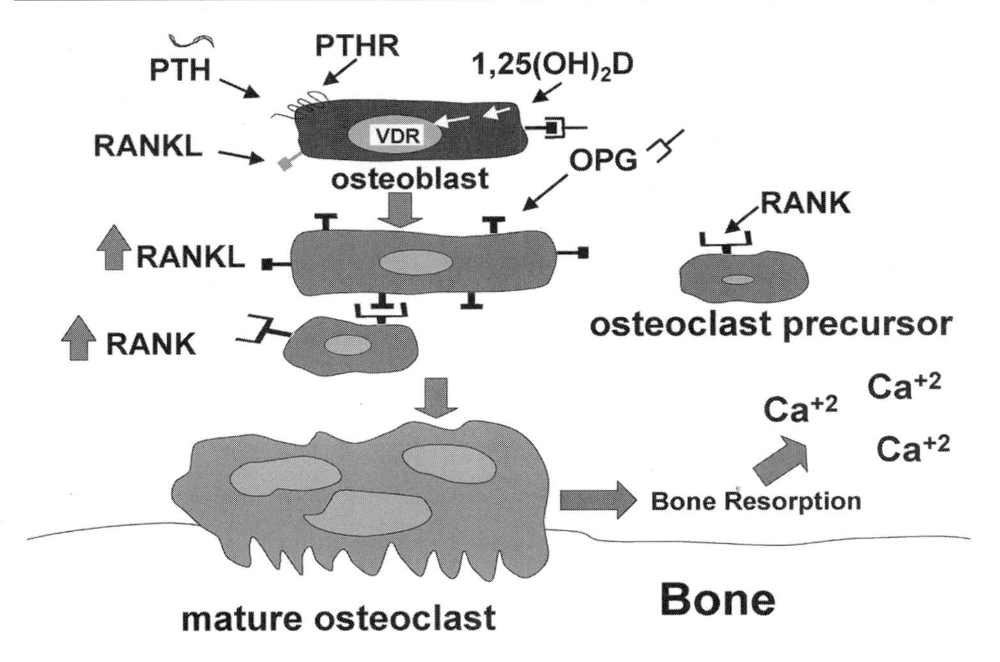

Fig. 2. Schematic representation of the mechanism by which 1,25-dihydroxyvitamin D [1,25$(OH)_2$D] enhances bone calcium mobilization. 1,25$(OH)_2$D interacts with its receptor (VDR) in mature osteoblasts. This increases the expression of receptor activator of NFκB ligand (RANKL) on the osteoblast's plasma membrane. The osteoclast precursor, which has the receptor for RANKL (known as RANK), interacts with RANKL, sending a signal to induce the premature osteoclast to become a mature multinucleated bone-resorbing osteoclast. Osteoprotegerin (OPG) acts as a decoy RANK receptor. It binds to RANKL and decreases the interaction between osteoclast precursors and mature osteoblasts. Parathyroid hormone (PTH) also stimulates osteoclastic activity in a similar manner by binding to its receptor PTHR.

milk that was fortified with ergosterol (provitamin D_2) as a mechanism to provide children with their vitamin D requirement. This recommendation was embraced by the United States, Canada, and Europe, and this simple food fortification program essentially eradicated rickets by 1940.

In the 1930s, the fortification of milk with vitamin D was a novelty and many companies became interested in fortifying their products with vitamin D. This included, among others, Bond bread, Rickter's hot dogs, and Twang soda. Schlitz Brewery cleverly marketed their beer as containing the sunshine vitamin D (Fig. 4). In Europe, custards, milk, and other foods were fortified with vitamin D *(22)*.

In the late 1930s, the US Food and Drug Administration forbade any nutritional claims for alcoholic beverages, and vitamin D fortification of beer was halted. In Europe in the 1950s there were several outbreaks of vitamin D intoxication, that is, hypercalcemia in children, which caused great alarm *(23)*. This resulted in most European countries forbidding the fortification of any food product with vitamin D.

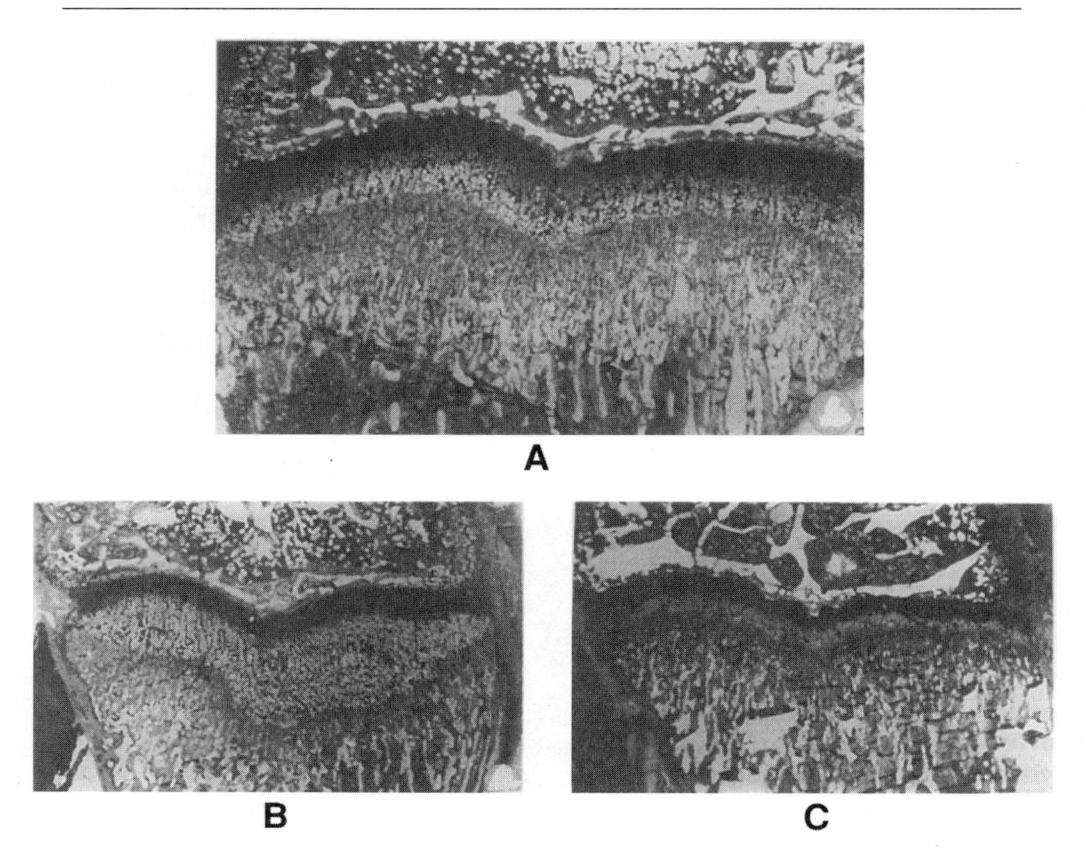

Fig. 3. Epiphyseal plates of tibias from rats that were fed **(A)** a vitamin D-deficient diet and supplemented with 125 ng (5 IU) of vitamin D_3 orally five times a week, **(B)** a vitamin D-deficient diet containing 3% calcium and 0.65% phosphorus, and **(C)** a vitamin D-deficient diet with 20% lactose, 4% calcium, and 1% phosphorus. Note the wide and disorganized hypertrophic zone in the vitamin D-deficient rat's tibial epiphyseal **(B)** fed high calcium and normal phosphorus diet compared with normal tibial epiphyseal plates from the rats that were either vitamin D repleted **(A)** or maintained on normal serum calcium and phosphorus by being on a high-calcium lactose, high-phosphorus diet **(C)**. (Reproduced with permission from ref. *20a.*)

In the United States and Canada, milk, some breads, cereals, and yogurts are fortified with vitamin D. There is 100 IU (2.5 µg) of vitamin D in 8 oz of milk. In most European countries, margarine and some cereals are fortified with vitamin D.

The reason milk was the vehicle for the vitamin D supplementation program was that children drank milk and they were at risk for developing rickets. However, with the awareness that vitamin D deficiency is an epidemic in both young, middle-aged, and older adults, there is a need for other dietary sources of vitamin D other than milk. Tangpricha et al. *(24)* observed that the fat content in milk does not influence vitamin D bioavailability. They also demonstrated that vitamin D added to orange

Table 1
Vitamin D Content in Foods and Supplements

1 tbsp cod liver oil	1360 IU
Salmon (3.5 oz)	360
Mackerel (3.5 oz)	345
Sardines (3.5 oz)	270
Eel (3.5 oz)	200
Dry cereal* 1/2 cup milk	99
Milk, 1 cup	100
Cereal bar*	50
Beef liver	30
Egg yolk	25
Multivitamin	400 IU
Vitamin D supplement	400 or 1000 IU

* These vary widely; check the label.

juice was bioavailable for young and middle-aged adults (Fig. 5). Thus, the recent introduction of vitamin D-fortified orange juice and other juice products heralds a new era in the vitamin D fortification process and should have a significant impact on vitamin D status of children and adults who consume these products.

6. VITAMIN D FROM SUNLIGHT EXPOSURE

Because very few foods contain vitamin D, most children and adults receive their vitamin D requirement from exposure to sunlight. During sunlight exposure, the solar ultraviolet B photons (UVB; with energies 290–315 nm) penetrate into the epidermis and are absorbed by 7-dehydrocholesterol (provitamin D_3) that resides in the plasma membrane of the epidermal cells *(3,15,25)*. This absorption results in a rearrangement of the double bonds that causes the B ring to open to form previtamin D_3 (Fig. 1). Previtamin D_3 exists in two conformeric forms, the s-*cis*, s-*cis* (czc) and its more thermodynamically stable counterpart the s-*trans*, s-*cis* (tzc) conformer (Fig. 6). It is only the czc conformer that can undergo rearrangement of its double bonds to form vitamin D_3. In order for the skin to efficiently convert previtamin D_3 to vitamin D_3, the previtamin D_3 is made in the plasma membrane and is locked into the czc conformation, which then can rapidly isomerize to vitamin D_3 *(26,27)*. Once formed, this molecule no longer is sterically compatible to reside in the cell's plasma membrane and is released into the extracellular space, where it is picked up in the dermal capillary bed and bound to the DBP (Fig. 1).

Unlike vitamin D that is absorbed in the small intestine into the chylomicron fraction, where no more than two-thirds of it is bound to DBP, essentially 100% of the vitamin D_3 that comes from the skin and enters into the venous circulation is bound to the DBP *(28)*. This gives the cutaneous vitamin D_3 a more prolonged

Fig. 4. In 1932–1936, Schlitz fortified their beer with vitamin D to market it as a unique nutrient-enriched product. However, in 1937 the FDA forbid any nutrient claims for alcoholic beverages and vitamin D was removed from beer.

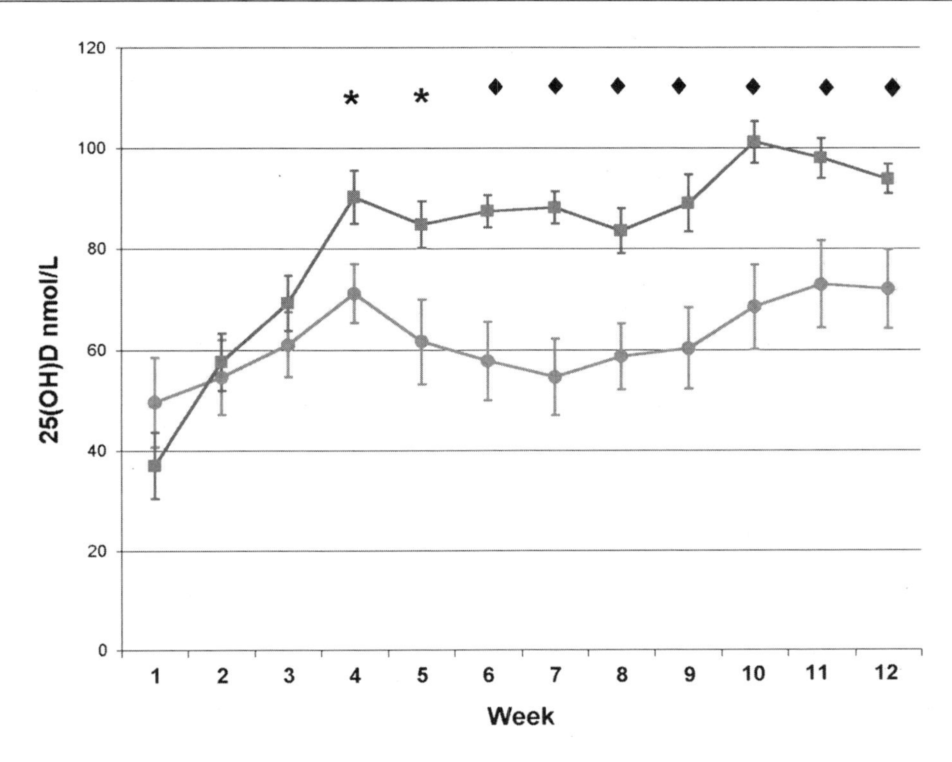

Fig. 5. Weekly 25-hydroxyvitamin D (25(OH)D) levels in healthy adults ingesting vitamin D (1000 IU/8 oz/d) fortified (—■—) and unfortified orange juice (—●—). Error bars represent standard error of the means. $p < 0.05$, ◆$p \leq 0.01$. (Reproduced with permission from ref. *24*.)

half-life in the circulation and thus provides an advantage for obtaining vitamin D from exposure of the skin to the sun.

7. FACTORS THAT INFLUENCE THE CUTANEOUS PRODUCTION OF VITAMIN D_3

Since the vitamin D_3 synthetic process is dependent on the number of UVB photons that enters into the epidermis, anything that interferes with the number of photons reaching the Earth's surface and ultimately penetrating into the viable epidermis results in an alteration in the production of vitamin D_3 in the skin.

During exposure to sunlight, the UVB photons enter into the skin and initiate the photochemistry necessary for producing previtamin D_3. The UVB photons also signal melanocytes to increase the production of melanin. Melanin acts as a natural sunscreen and is efficiently packaged into melanosomes that migrate upward to the upper layers of the epidermis, where they efficiently absorb UVB and ultraviolet A (321–400 nm) radiation. An increase in skin pigmentation is inversely

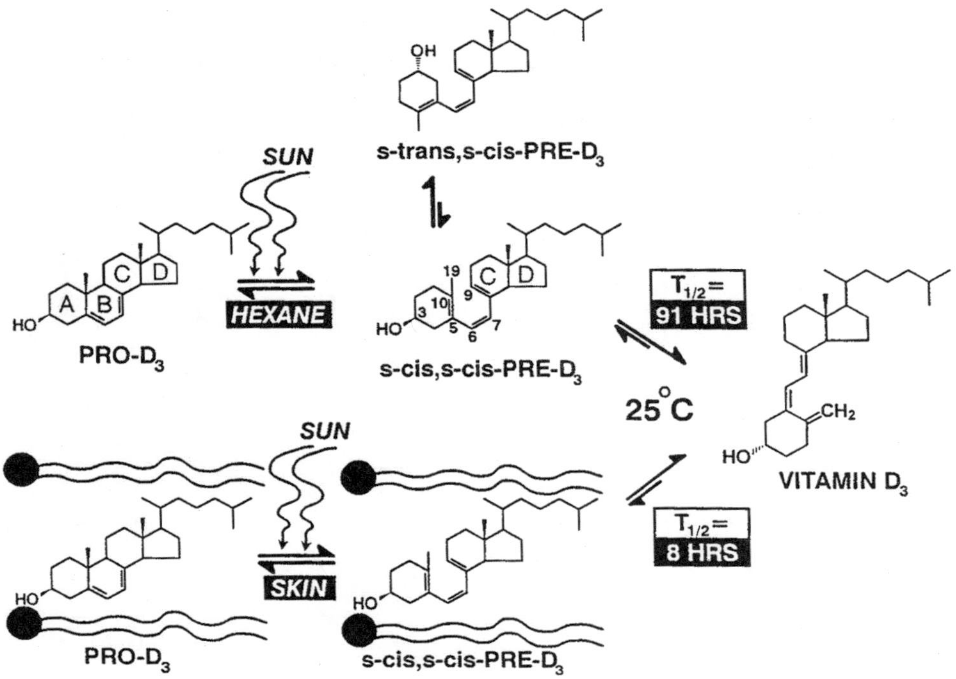

Fig. 6. Photolysis of provitamin D_3 (pro-D_3) into previtamin D_3 (pre-D_3) and its thermal isomerization of vitamin D_3 in hexane and in lizard skin. In hexane pro-D_3 is photolyzed to s-*cis*,s-*cis*-pre-D_3. Once formed, this energetically unstable conformation undergoes a conformational change to the s-*trans*,s-*cis*-pre-D_3. Only the s-*cis*,s-*cis*-pre-D_3 can undergo thermal isomerization to vitamin D_3. The s-*cis*,s-*cis* conformer of pre-D_3 is stabilized in the phospholipid bilayer by hydrophilic interactions between the 3β-hydroxyl group and the polar head of the lipids, as well as by the van der Waals interactions between the steroid ring and side-chain structure and the hydrophobic tail of the lipids. These interactions significantly decrease the conversion of the s-*cis*,s-*cis* conformer to the s-*trans*,s-*cis* conformer, thereby facilitating the thermal isomerization of s-*cis*,s-*cis*-pre-D_3 to vitamin D_3. (Reproduced with permission from ref. *26*.)

related to the number of UVB photons that can penetrate into the epidermis and dermis. Thus, the efficiency in utilizing UVB photons to produce vitamin D_3 in the skin is inversely related to the amount of skin pigmentation. This effect can be quite dramatic. A person with deep skin pigmentation of African origin (skin type 5), who is exposed to the same amount of sunlight as a person with minimum skin pigmentation of Celtic or Scandinavian origin (skin type 2), will produce no more than 5–10% of that produced in the lighter-skinned individual *(3,15,29)* (Fig. 7).

Sunscreens are heavily promoted for the prevention of skin cancer and wrinkles. Sunscreens, like melanin, efficiently absorb UVB radiation when applied topically to the skin. As a result, there is a marked diminishment in the penetration of UVB

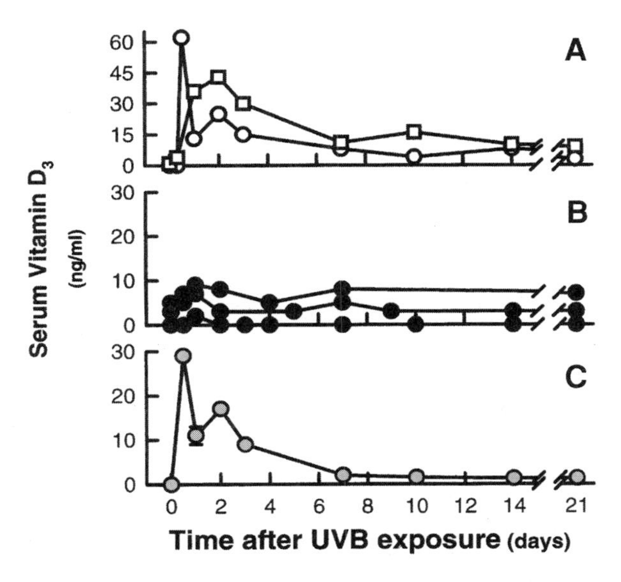

Fig. 7. Change in serum concentrations of vitamin D in two lightly pigmented white (skin type 2) **(A)** and three heavily pigmented black subjects (skin type 5) **(B)** after total-body exposure to 54 mJ/cm² of UVB radiation. **(C)** Serial change in circulation vitamin D after reexposure of one black subject in **B** to a 320-mJ/cm² dose of UVB radiation. (Reproduced with permission from ref. *29*.)

photons into the epidermis. The proper use of a sunscreen (2 mg sunscreen/cm² skin surface, i.e., about 1 oz or 25% of a 4-oz bottle applied to all sun exposed skin of a person wearing a bathing suit) with an SPF of 8 reduces the production of pre-vitamin D_3 by more than 95% *(30)* (Fig. 8A). Clothing absorbs 100% of the incident UVB radiation, and thus no vitamin D_3 is made in the skin covered by clothing *(31)*. This is the reason why women who wear veils and cover all sun-exposed skin with clothing when outside are often vitamin D deficient *(32,33)*. Glass also absorbs all UVB photons. Therefore, exposure of the skin from sunlight that has passed through glass will not promote vitamin D_3 synthesis in the skin *(34)*.

Aging causes a decrease in the amount of 7-dehydrocholesterol in the epidermis *(34,35)*. Elders exposed to the same amount of sunlight as a young adult will produce approx 25% of the amount of previtamin D_3, compared to a young adult *(34)* (Fig. 8B).

The angle by which sunlight penetrates the Earth's atmosphere also dramatically influences the production of previtamin D_3 in the skin. This angle, known as the zenith angle, is related to season, time of day, and latitude. There is a direct relationship with increase in latitude and in the zenith angle of the sun. The higher the zenith angle, the longer is the path length that solar UVB photons have to travel through the ozone layer, which efficiently absorbs most of these vitamin

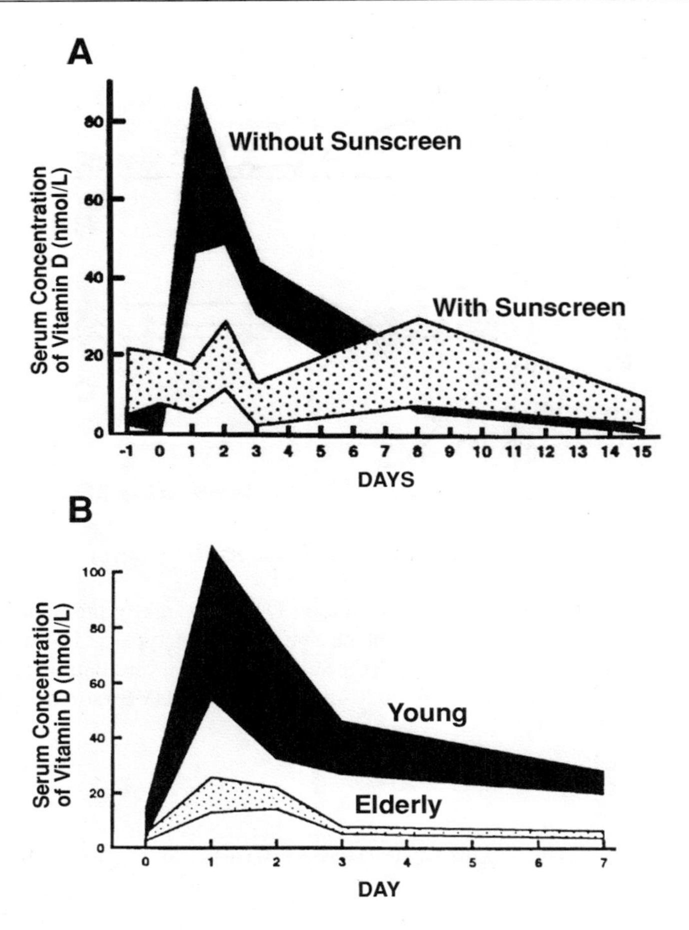

Fig. 8. **(A)** Circulating concentrations of vitamin D after a single exposure to 1 minimal erythemal dose of simulated sunlight with either a sunscreen, with a sun protection factor of (SPF-8) 8, or a topical placebo cream. **(B)** Circulating concentrations of vitamin D in response to a whole-body exposure to 1 minimal erythemal dose in healthy young and elderly subjects. (Reproduced with permission from ref. *34*.)

D_3-producing photons. Typically, in the summer no more than about 0.1% of the solar UVB photons that hit the outer stratosphere reach the Earth's surface. The lowest zenith angle, which permits more UVB photons to penetrate to the Earth's surface, occurs at around noontime and in the middle of the summer at the Equator.

During the winter (i.e., November–February) above and below 35° latitude, the zenith angle is so oblique that essentially all of the UVB photons are absorbed by the stratospheric ozone layer. As a result, very little, if any, previtamin D_3 can be produced in human skin. At very high latitudes, such as Bergen, Norway, and Edmonton, Canada, little, if any, previtamin D_3 is produced between the months

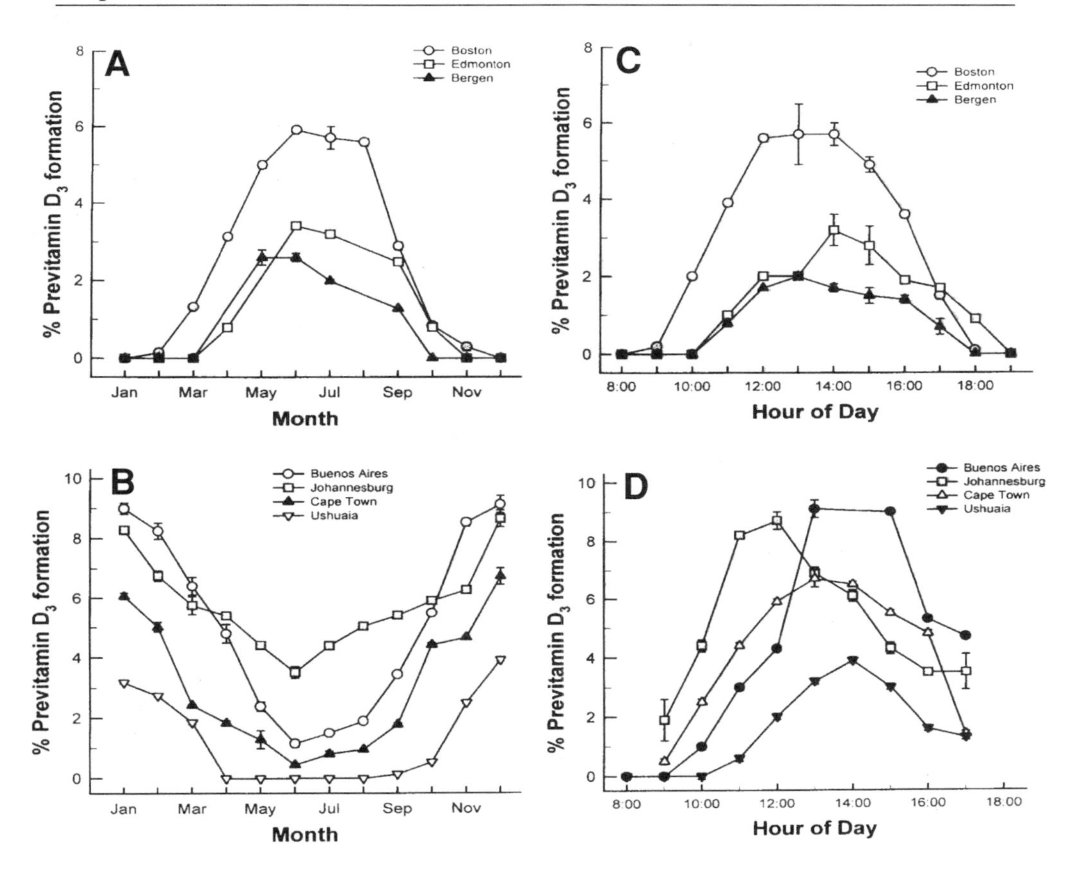

Fig. 9. Influence of season, time of day, and latitude on the synthesis of previtamin D_3 in Northern (**A** and **C**) and Southern Hemispheres (**B** and **D**). The hour indicated in **C** and **D** is the end of the 1-h exposure time. (Reproduced with permission from ref. *34a*.)

of October and March. Figure 9 shows how latitude, season, and time of day dramatically influence the production of previtamin D_3 in the skin *(15)*.

8. CONSEQUENCES OF VITAMIN D DEFICIENCY ON MUSCULOSKELETAL HEALTH

Chronic vitamin deficiency in infants and young children causes the bone-deforming disease commonly known as rickets. Vitamin D deficiency disrupts chondrocyte maturation and inhibits the normal mineralization of the growth plates. This causes a widening of the epiphyseal plates that is commonly seen at the ends of the long bones in rachitic children, as well as bulging of the costo-

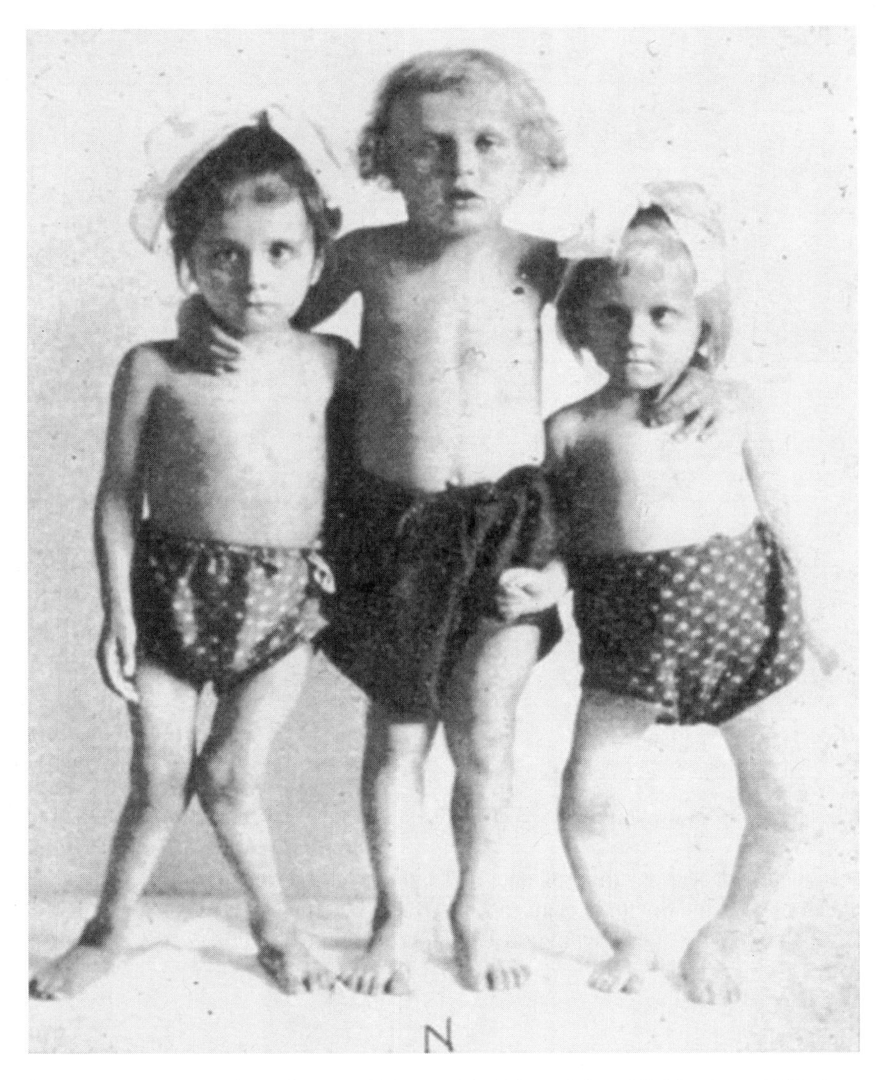

Fig. 10. Typical presentation of two children with rickets. The child in the middle is normal; the children on either side have severe muscle weakness and bone deformities including bowed legs (right) or knock knees (left).

chondral junctions that results in what is known as the rachitic rosary. The skeleton is also poorly mineralized, due to the low calcium × phosphate product. This poor mineralization makes the skeleton less rigid, and when the rachitic child begins to stand, gravity causes either inward or outward bowing of the long bones in the lower extremities, resulting in bowed legs or knocked knees, respectively (Fig. 10).

In adults after the epiphyseal plates have been fused, the skeletal abnormalities resulting from vitamin D deficiency are more subtle. Vitamin D deficiency results in a decrease in efficiency of intestinal calcium absorption. This causes a decrease in the serum ionized calcium, which is immediately recognized by the calcium sensor in the parathyroid glands *(36)*. This results in an increase in the expression and production of PTH. PTH, in turn, has three options to maintain serum calcium levels within a physiologically acceptable range. It can increase the efficiency of the renal tubules, especially the distal convoluted tubules, to increase the reabsorption of calcium from the ultrafiltrate. It also stimulates the kidney to produce more $1,25(OH)_2D$, which in turn increases intestinal calcium absorption (Fig. 1). If these actions are not adequate to maintain the serum calcium levels, then PTH will stimulate the expression of RANKL in osteoblasts to mobilize preosteoclasts to become mature bone-resorbing osteoclasts by a mechanism similar to $1,25(OH)_2D$ *(13,37)* (Fig. 2). Thus, an increase in osteoclastic activity results in the destruction of the matrix and release of calcium into the extracellular space. The net effect is to increase the porosity of the skeleton, thereby causing a decrease in bone mineral density and precipitating or exacerbating osteoporosis.

A more subtle, but important, effect of PTH on skeletal health is its effect on phosphorus metabolism in the kidney. PTH causes an increase in the urinary excretion of phosphorus. Although subtle in nature, the low-normal or low serum phosphorus is inadequate to maintain a supersaturated level of calcium × phosphorus product, resulting in a mineralization defect of the newly laid-down osteoid by osteoblasts. Histologically this appears as widened osteoid seams (Fig. 11) and is known as osteomalacia. Because osteoid has no mineral component, it provides little, if any, structural support to the skeleton and increases risk of fracture *(38–41)*. In addition, the lack of calcium hydroxyapatite deposition in newly laid-down osteoid results in no increase in bone mineral density. It is not possible to detect either by standard X-rays or bone densitometry the difference between osteoporosis, that is, holes in the skeleton, vs osteomalacia, which is simply a collagen matrix without mineral *(42,43)*.

Unlike osteoporosis, which is a silent disease until a fracture occurs, osteomalacia is often associated with bone discomfort. Patients often complain of an aching in their skeleton that is unexplained. This can be detected on physical exam by palpating the sternum with minimum pressure of the thumb or forefinger on the sternum or on the anterior tibia. The patient often complains of discomfort with minimum to moderate applied pressure. Although the exact cause for this pain is not known, its possible that the collagen-rich osteoid that is laid down on the periosteal surface of the skeleton becomes hydrated similar to gelatin in Jell-O and causes an outward pressure on the periostial covering that is innervated with sensory pain receptors *(44)*.

Patients with osteomalacia often complain of muscle aches and muscle weakness. There is mounting evidence that vitamin D deficiency results in muscle weakness and increases sway, which can result in increase in falling, thereby increasing risk of skeletal fractures *(33,45,46)*.

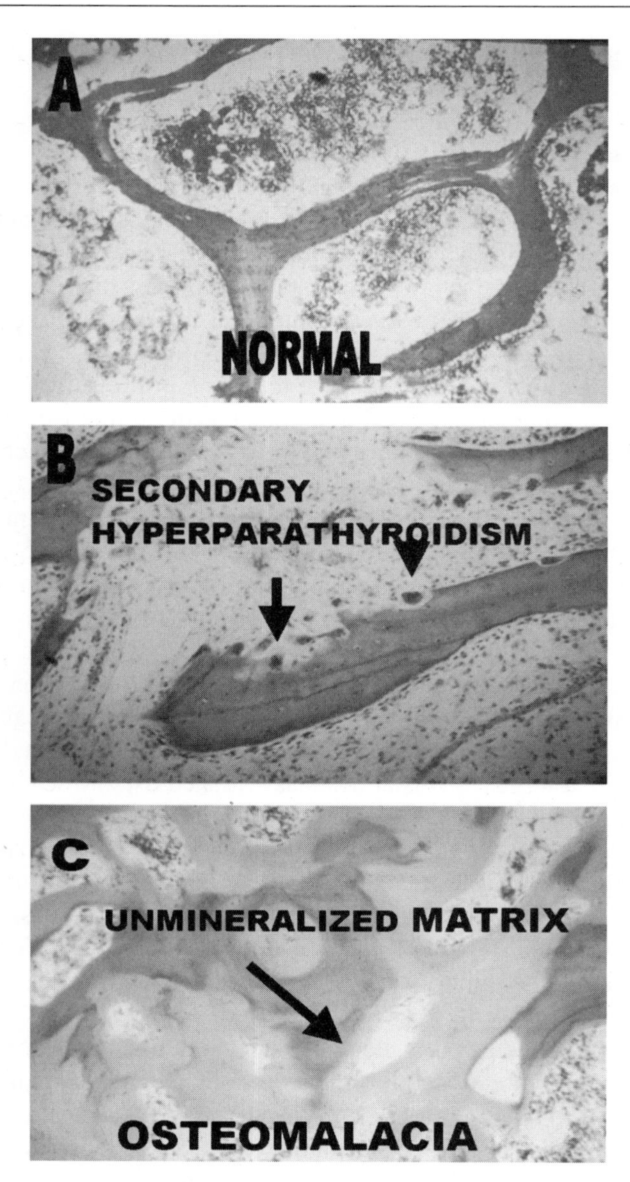

Fig. 11. Bone histology demonstrating **(A)** normal mineralized trabecular bone, **(B)** increased osteoclastic bone resorption due to secondary hyperparathyroidism, and **(C)** osteomalacia with widened unmineralized osteoid light grey areas. (Reproduced with permission from ref. *15*.)

Patients often complain to their physicians about nonspecific bone aches, muscle aches, and discomfort. Often after a thorough workup, including a sedimentation rate, rheumatoid factor, and even a bone scan, the physician will inform the patient that no specific cause has been found and often these patients are given the diagnosis of fibromyalgia. It has been estimated that upwards of 40–80% of patients complaining of nonspecific bone pain and muscle aches and weakness are suffering not from fibromyalgia, but from chronic vitamin D deficiency *(33,44)*.

9. PREVALENCE OF VITAMIN D DEFICIENCY IN CHILDREN AND ADULTS

It is both surprising and alarming that vitamin D deficiency continues to plague both children and adults *(38–55)*. Infants who receive their total nutrition from breast feeding are at high risk of vitamin D deficiency because human milk contains very little, if any, vitamin D to satisfy their requirement *(53)*. This is especially true for infants of color, because their mothers are often vitamin D deficient as well and provide no vitamin D nutrition in breast milk *(52,54)*. Even in Caucasian and African American women who had a mean intake of 457 IU/d, the concentrations of vitamin D and 25(OH)D in their milk was 12.6 IU/L and 37.6 IU/L, respectively *(53)*. It has been estimated that human milk contains no more than about 15 IU of vitamin D in 8 oz.

Children who are active and outdoors are at little risk of vitamin D deficiency as long as there is a short period of time when they wear no sun protection, such as clothing or sunscreen, on face, arms, and legs.

It has been recognized for more than three decades that elders are at high risk of developing vitamin D deficiency *(47–50)*. Vitamin D deficiency is extremely common in older adults in Europe because essentially no foods are fortified with vitamin D. In the United States and Canada, vitamin D deficiency is also more common than expected. Gloth et al. *(47)* reported 54% of community dwellers and 38% of nursing home residents in the Baltimore area were severely vitamin D deficient [25(OH)D < 10 ng/mL]. Numerous studies have reported that between 25% and more than 60% of adults aged 50+ years were vitamin D deficient. In Boston, we observed in independently living elders (83 ± 8 yr; 50 white, 14 Hispanic, and 5 African American subjects) in August of 1997 30%, 43%, and 84% of white, Hispanic, and black elders were vitamin D deficient (Fig. 12) *(15)*. Inpatients are especially at high risk of vitamin D deficiency *(55)*. It was reported that 57% of middle-aged and older adults were vitamin D deficient. Sixty percent of the patients consumed less than the recommended adequate intake of vitamin D, and 37% who had intakes above the recommended daily allowance were found to be vitamin D deficient.

It would be expected that young and middle-aged active adults would not be at risk of vitamin D deficiency. However, they have several risk factors for vitamin D deficiency, including long hours of work indoors with little exposure to sunlight, and they are also more likely to wear sun protection on all sun-exposed areas because of their worry about increased risk of skin cancer and wrinkles. As a result, when exposed to sunlight they make little vitamin D_3 in their skin. In Boston, we observed 32% of medical students and young doctors, aged 18–29 yr, were vitamin D deficient *(51)*.

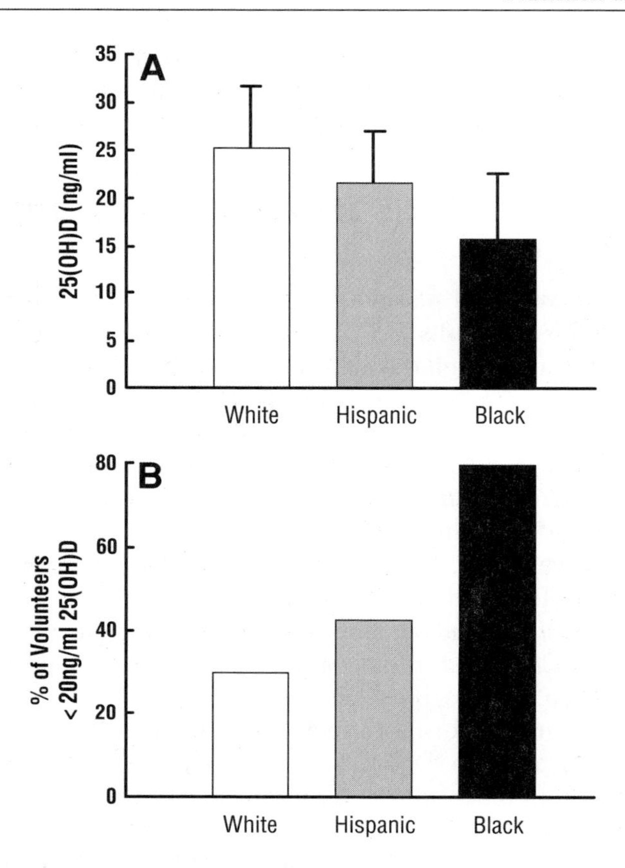

Fig. 12. **(A)** Serum 25(OH)D levels in free-living senior citizens in August in Boston. Mean ± SEM. **(B)** Percentage of free-living senior citizens who were vitamin D insufficient in August in Boston. (Reproduced with permission from ref. *15.*)

Fifteen percent had secondary hyperparathyroidism, and 4% of the students and residents remained vitamin D deficient at the end of the summer.

10. CAUSES OF VITAMIN D DEFICIENCY

The major cause of vitamin D deficiency is that it is not appreciated that very few foods naturally contain vitamin D and that most (80–100%) of our vitamin D requirement comes from casual exposure to sunlight. Even though oily fish contain vitamin D, it is highly variable depending on what season they were caught and whether they were farm raised and what their vitamin D intake was from their diet. Furthermore, it would require that a person eat oily fish at least two to three times a week. To satisfy the vitamin D requirement by drinking milk, would require ingesting two, four, and six glasses a day for children and adults up to the age of 50, and adults aged 51–70, 70+ yr, respectively *(53)*. Because the vitamin D content in milk is highly variable and often contains less than 50% of what is stated on the label, it may not provide an adequate amount of vitamin D *(56)*.

Intestinal malabsorption syndromes, especially of the small intestine where vitamin D is absorbed, can lead to severe vitamin D deficiency *(57,58)* (Fig. 13). Patients with end-stage hepatic failure not only are unable to produce an adequate amount of 25(OH)D, but often suffer from fat malabsorption and are unable to absorb dietary vitamin D. Patients who are on total parenteral nutrition often suffer from a severe metabolic bone disease that is characteristic of vitamin D deficiency osteomalacia. However, the inclusion of 400 IU of vitamin D in the total parenteral nutrition solution does not protect the patient from vitamin D deficiency bone disease *(59,60)*.

The principal cause of vitamin D deficiency is lack of adequate exposure to sunlight. The skin has a large capacity to produce vitamin D_3. Exposure of an adult in a bathing suit to simulated sunlight that mimicked the amount of time that would be one minimal erythemal dose (1 MED), that is, cause a minimum pinkness to the skin, resulted in an increase in blood levels of vitamin D_3 comparable to ingesting between 10,000 and 25,000 IU of vitamin D *(15)* (Fig. 14). Although aging substantially reduces the amount of 7-dehydrocholesterol in the skin, it still has an adequate capacity to make vitamin D *(15,61,62,62a)* (Fig. 15).

Patients with obesity often complain of bone aches, muscle aches, and weakness, which exacerbates their inability to be active and their obesity. It is recognized that obesity is associated with vitamin D deficiency *(63)*. This is due to the fact that body fat acts as a sink for vitamin D. Thus, whether vitamin D is produced in the skin or ingested in the diet, a majority of it is deposited in an almost irreversible manner into the body fat and is not bioavailable to the body (Fig. 16).

11. DIAGNOSIS OF VITAMIN D DEFICIENCY

Often physicians assume that the most sensitive indicator to detect vitamin D deficiency is to observe a below-normal serum calcium value. Unfortunately, as explained previously, the body is vigilant to maintain the serum calcium within the normal range in order to maintain most bodily functions. As a result, a person with vitamin D deficiency develops secondary hyperparathyroidism and maintains serum calcium within the normal range until most available calcium is depleted from the skeleton. The secondary hyperparathyroidism results in mild to moderate hypophosphatemia. However, this is also difficult to detect, especially if the patient's blood is taken in a nonfasting state. Serum phosphorus levels are influenced by dietary phosphorus intake, sugar intake, and by acidosis and alkalosis *(6)*.

With the exception of observing widened epiphyseal plates and Looser's pseudo-fractures in the long bones, it is not possible to detect vitamin D deficiency by X-rays.

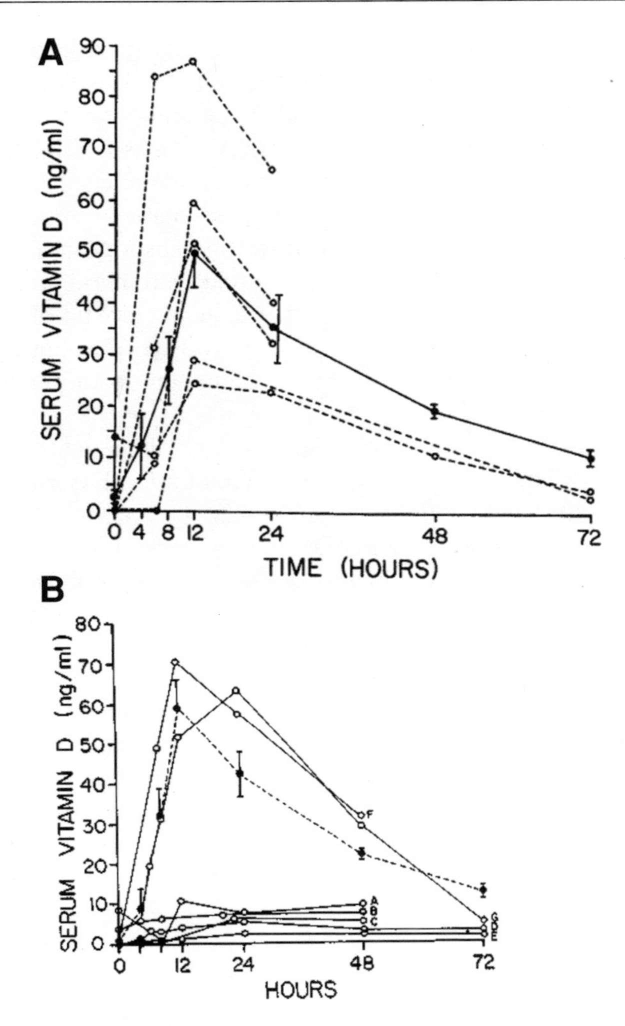

Fig. 13. (A) Serum vitamin D concentrations in 7 patients with intestinal fat malabsorption syndromes after a single oral dose of 50,000 IU (1.25 mg) of vitamin D_2. For comparison, the means and standard errors of vitamin D concentrations measured in seven normal control subjects after a similar dose are indicated by the filled circles and dotted lines (—●—). Note that two patients, one with Crohn's ileocolitis (patient F) and one with ulcerative colitis (patient G), had essentially normal absorption curves. Five patients, however, absorbed very little, if any, vitamin D_2. **(B)** Vitamin D absorption in young (filled circles) and elderly (open circles) adults. Each subject received an oral dose of 50,000 IU of vitamin D_2 and at various times blood determinations were made for circulating concentrations of vitamin D. (Reproduced with permission from ref. *57*.)

The only method to determine vitamin D deficiency is to measure the blood level of the major circulating form of vitamin D, 25(OH)D. Although $1,25(OH)_2D$ is the biologically active form of vitamin D and would appear to be the ideal marker for vitamin D deficiency, it is not. There are several reasons for

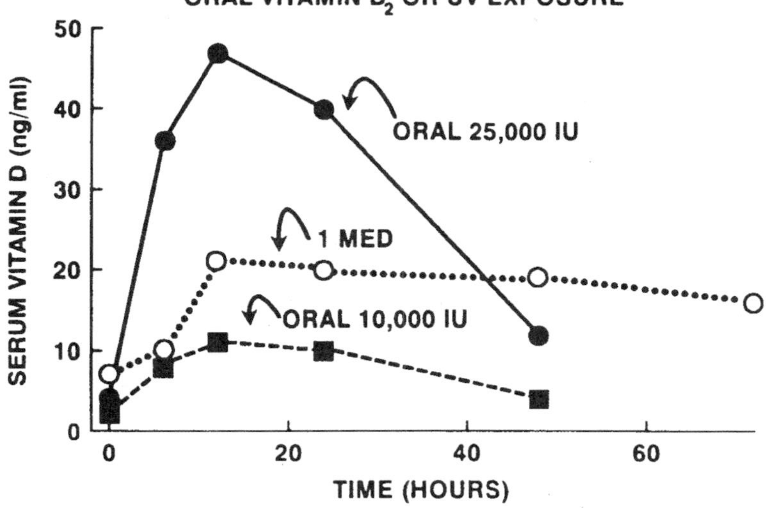

Fig. 14. Comparison of serum vitamin D levels after a whole-body exposure to 1 MED (minimal erythemal dose) of simulated sunlight compared with a single oral dose of either 10,000 or 25,000 IU of vitamin D_2. (Reproduced with permission from ref. *15*.)

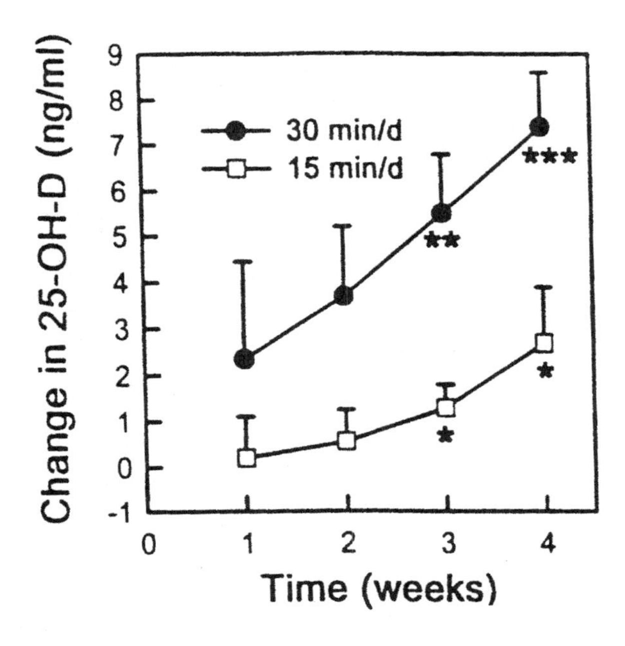

Fig. 15. Change in serum 25(OH)D levels from baseline in elderly rest home residents in Auckland, New Zealand (37°C) spending 15 or 30 min/d outdoors in the spring, who exposed their heads, necks, forearms, and lower legs to sunlight. $N = 5$ each group; $*p < 0.06$, $**p < 0.02$, and $***p < 0.005$. (Reproduced with permission from ref. *62a*.) Lund B, Sorensen OH. Scand J Clin Lab Invest 1979; 39:23–30.)

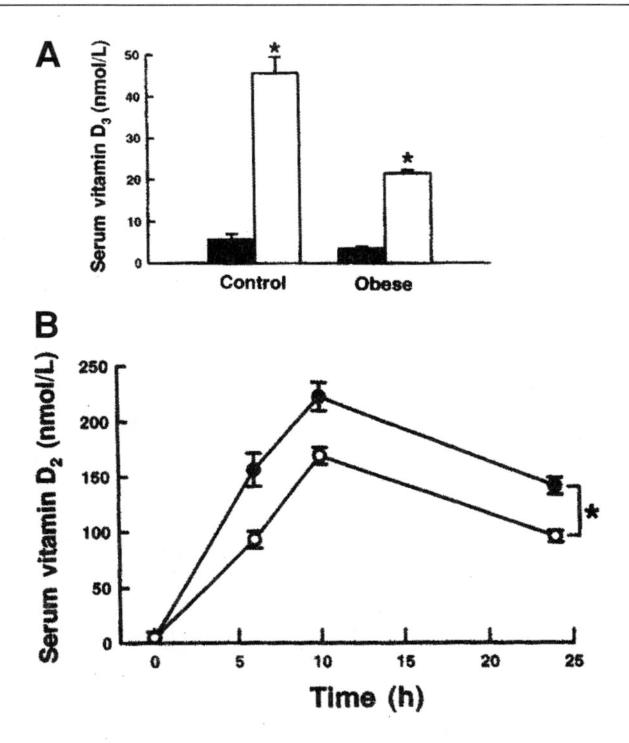

Fig. 16. (**A**) Mean (± SEM) serum vitamin D_3, concentrations before (■) and 24 h after (□) whole-body irradiation (27 mJ/cm^2) with UVB radiation. The response of the obese subjects was attenuated when compared with that of the control group. There was a significant time-by-group interaction, $p = 0.003$. *Significantly different from before values ($p < 0.05$). (**B**) Mean (± SEM) serum vitamin D_2 concentrations in the control (●) and obese (○) groups before and after 25 h after oral intake of vitamin D_2 (50,000 IU, 1.25 mg). Vitamin D_2 rose rapidly until ~ 10 h after intake and then declined slightly thereafter. *Significant time and group effects by ANOVA ($p < 0.05$) but no significant time-by-group interaction. The difference in peak concentrations between the obese and nonobese control subjects was not significant. (Reproduced with permission from ref. *5*.)

this. The circulating concentration of $1,25(OH)_2D$ is 1000th the concentration of 25(OH)D (pg vs ng/mL). The half-life for $1,25(OH)_2D$ is only 4–6 h, compared to 2 wk for 25(OH)D *(15)*. Finally, as a person becomes vitamin D deficient and develops secondary hyperparathyroidism, the kidney's 1-OHase produces more $1,25(OH)_2D$ *(3,6,7,15)*. Thus, when a patient is vitamin D insufficient there is often a normal or even elevated blood level of $1,25(OH)_2D$ *(64)*. The measurement of $1,25(OH)_2D$ as a gauge of vitamin D status is not only useless, but often misleads physicians into thinking their patient is vitamin D sufficient since the $1,25(OH)_2D$ levels can be normal.

Table 2
Adequate Intake (AI), Reasonable Daily Allowance, Tolerable Upper Limit (UL), and Reasonable Safe Upper Limit for Vitamin D

Age	*AI* [IU (μg)/d]	*Reasonable daily allowance* (IU/d)	*UL* [IU (μg)/d]	*Reasonable safe upper limit* [IU (μg)/d]
0–12 mo	200 (5)	200–400	1000 (25)	2000 (50)
1–18 yr	200 (5)	400–1000	2000 (50)	5000 (125)
19–50 yr	200 (5)	400–1000	2000 (50)	5000 (125)
51–70 yr	400 (10)	800–2000	2000 (50)	5000 (125)
71+ yr	600 (15)	800–2000	2000 (50)	5000 (125)
Pregnancy	200 (5)	400–1000	2000 (50)	5000 (125)
Lactation	200 (5)	400–1000	2000 (50)	5000 (125)

12. VITAMIN D REQUIREMENT: ADEQUATE INTAKE VS HEALTHY INTAKE

In 1997, the Institute of Medicine announced the new recommended Adequate Intakes (AI) of vitamin D (1 μg = 40 IU) for children and adults aged 0–50, 51–70, and 71+ yr to be 200, 400, and 600 IU/d, respectively (Table 2) *(3,6,15,53)*. These recommendations were based on literature published before 1996 that evaluated the effect of vitamin D intake on calcium metabolism and bone health. Since 1996, several investigators have reported on the effect of vitamin D intake on circulating concentrations of 25(OH)D. Vieth et al. *(65,66)* gave healthy adults (41 ± 9 yr) 4000 IU of vitamin D a day for 2–5 mo and did not observe any untoward toxicity. Their 25(OH)D levels during the winter increased from 10.2 ± 4 to 24.1 ± 4 ng/mL. Barger-Lux et al. *(67)* evaluated a dose response of vitamin D and 25(OH)D intake in healthy males for 4 and 8 wk, respectively. The groups of adults treated with 1000, 10,000, or 50,000 IU of vitamin D_3/d for 8 wk demonstrated increases in their serum vitamin D_3 levels of 5.0, 52.6, and 300.2 ng/mL, respectively. In the same groups, the 25(OH)D increased by 11.6, 58.4, and 257.2 ng/mL, respectively. Male adults who received 10, 20, or 50 μg of 25(OH)D_3/d for 4 wk demonstrated increases of 25(OH)D by 11.6, 58.4, and 257.2 ng/mL, respectively. None of the men demonstrated any significant change in either their calcium or 1,25(OH)$_2$D levels. In a follow-up study, Heaney et al. *(68)* gave 67 men who were in general good health either 0, 25, 125, or 250 μg of vitamin D_3 for approx 20 wk during the winter. They observed serum 25(OH)D levels increased in direct proportion to dose with a slope of approximately 0.28 ng/mL for each additional 1 μg of vitamin D_3 ingested. The calculated oral input required to sustain serum 25(OH)D concentrations present in the men during autumn was 12.5 μg/d. The total amount from all sources (supplement, food, tissue stores) needed to sustain the starting 25(OH)D

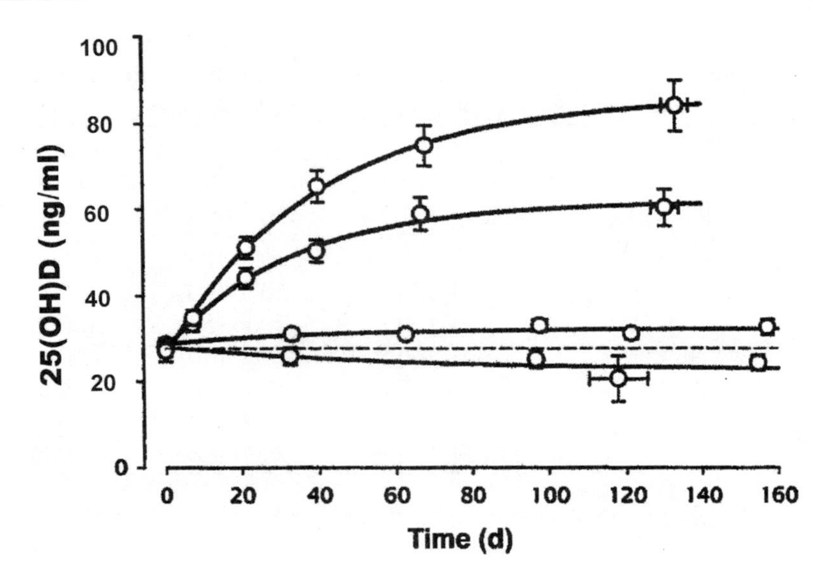

Fig. 17. Time course of serum 25-hydroxyvitamin D_3 [25(OH)D] concentration for the four dose groups. The points represent the mean values, and error bars are 1 SEM. The curves are the plot of the Equation *1*, fitted to the mean 25(OH)D_3 values for each dosage group. The curves, from the lowest upward, are for 0.25, 125, and 250 μg vitamin D_3 (labeled dose)/d. The horizontal dashed line reflects zero change from baseline. (Reproduced with permission from ref. *68*.)

level was estimated at 96 μg (approx 3800 IU/d). They concluded that healthy men used between 3000 and 5000 IU vitamin D_3/d to meet greater than 80% of their winter vitamin D requirement that was provided by cutaneous production of vitamin D_3 during the previous spring, summer, and fall (Fig. 17). Tangpricha et al. observed that healthy young and middle-aged female and male adults who ingested 1000 IU of vitamin D/d for 3 mo increased their blood levels of 25(OH)D from 15 ± 3 to 38 ± 8 ng/mL after 2 mo. Continued intake of 1000 IU of vitamin D/d did not increase blood levels of 25(OH)D above 40 ng/mL (Fig. 5).

13. INTERPRETING SERUM 25(OH)D LEVELS

The normal blood level of 25(OH)D varies from different laboratories, but generally is in the range of 10–55 ng/mL *(69)*. This normal range is determined by collecting blood from hundreds of healthy volunteers, and the mean ± 2 SD is considered to be the normal range. However, in light of the fact that many adults are vitamin D insufficient, it is likely that some of the so-called normal population from which the normal range was determined were vitamin D deficient. This would result in a lower-than-expected normal range. This was substantiated by Malabanan et al. *(48),* who gave 39 healthy middle-aged and older adults, who had blood levels that were considered to be in the low/normal range of between 11 and 25 ng/mL,

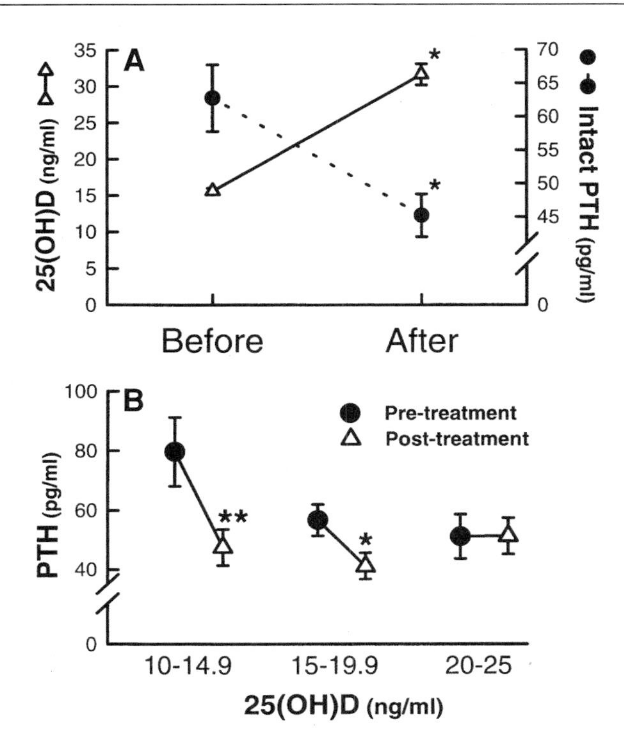

Fig. 18. **(A)** Serum levles of 25(OH)D (—△—) and PTH (—●—) before and after therapy with 50,000 IU of vitamin D_2 and calcium supplementation once a week for 8 wk. **(B)** Serum levels of PTH levels in patients who had serum 25(OH)D levels of between 10 and 25 ng/mL and who were stratified in increments of 5 ng/mL before and after receiving 50,000 IU of vitamin D_2 and calcium supplementation for 8 wk. (Reproduced with permission from ref. *48*.)

50,000 IU of vitamin D_2 once a week for 8 wk. As can be seen in Fig. 18, 25(OH)D levels increased by more than 100% and on average the PTH values decreased by 22%. Those who had blood levels of 25(OH)D of between 11 and 15 ng/mL on average had a 55% decrease in their PTH values, and those with a 25(OH)D of 16–20 ng/mL had a 35% decline. There was no significant decrease in PTH levels in those adults who had 25(OH)D levels of at least 20 ng/mL. Thus, at a minimum, a 25(OH)D should be at least 20 ng/mL. Chapuy et al. *(50,70)* plotted 25(OH)D levels with PTH levels and concluded that a 25(OH)D of 28 ng/mL was required for no further decline in PTH values.

The upper range of normal by most assays is 55–60 ng/mL. However, this upper normal range again is simply based on +2 standard deviations above the mean from the normal population. This does not provide any useful information about what the blood level of 25(OH)D needs to be to cause toxicity. Indeed, lifeguards routinely have blood levels of 25(OH)D of 100 ng/mL with no untoward consequences. Heaney et al. *(68)* observed blood levels of $25(OH)D_3$ of 100 ng/mL without any

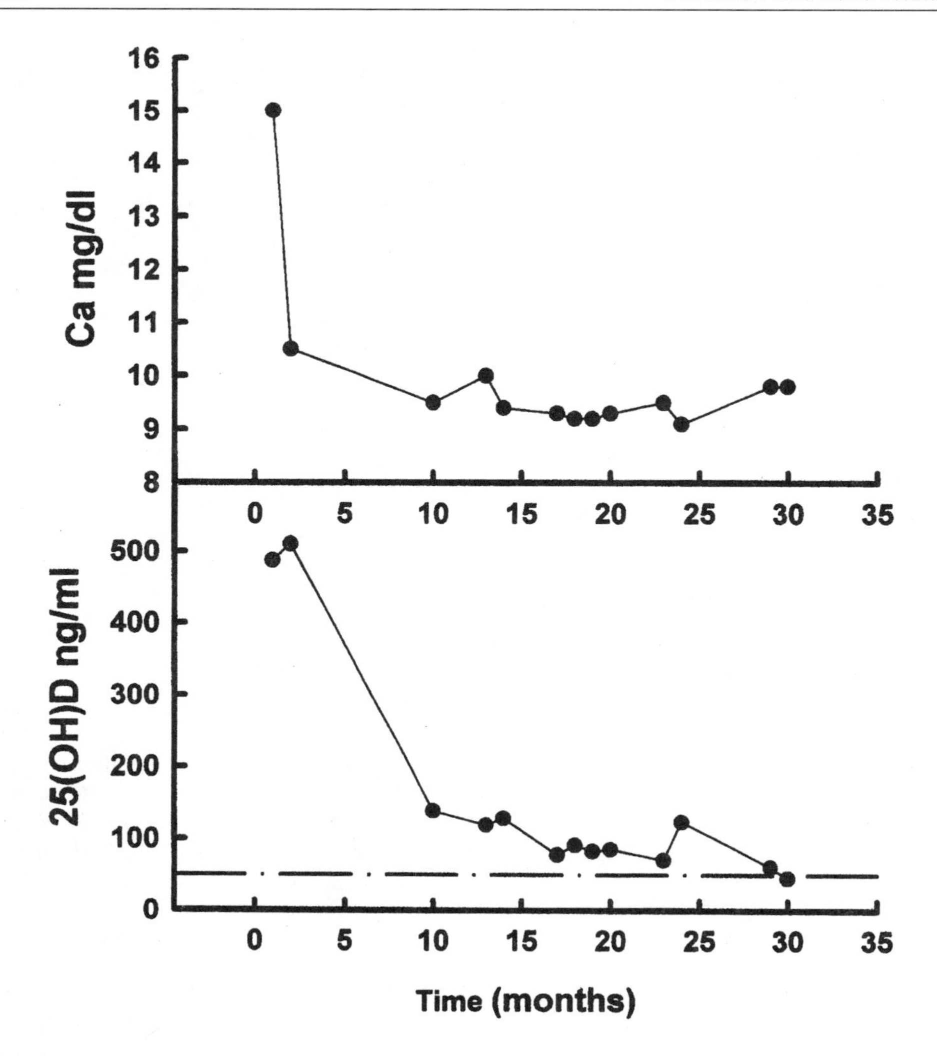

Fig. 19. Serum calcium level (upper panel) and 25(OH)D level (lower panel) in a patient who had vitamin D intoxication after ingestion of an over-the-counter vitamin D supplement that contained as much as 1 million units of vitamin D_3 in a teaspoon. The patient stopped all vitamin D intake and wore sunscreen before going outside after his hospitalization (mo 0). The dotted line (lower panel) represents the upper limit for the 25(OH)D assay that was 46.7 ng/mL. (Reproduced with permission from ref. *71.*)

untoward side effects or hypercalcemia. Based on reports of vitamin D intoxication, that is, associated with hypercalcemia, and suppressed PTH levels, 25(OH)D need to be at least 150 ng/mL (Fig. 19) *(71–74).*

Thus, based on the available literature today, it has been suggested that in the absence of any exposure to sunlight, the vitamin D requirement for children and adults is at least 1000 IU of vitamin D/d (Table 2). Furthermore, a 25(OH)D level of between 30 and 50 ng/mL should be considered as a healthy range for 25(OH)D.

14. TREATMENT FOR VITAMIN D DEFICIENCY

The best method to treat vitamin D deficiency is to give pharmacological doses of vitamin D. This can be accomplished by giving an oral dose of 50,000 IU of vitamin D once a week for 8 wk *(48)* (Fig. 18). Alternatively, intramuscular injection of up to 500,000 IU of vitamin D has been demonstrated to prevent vitamin D deficiency in elderly nursing home residents when given twice a year *(75)*. However, the intramuscular preparation available to us has been ineffective in raising blood levels of 25(OH)D when given intramuscularly. This may be a bioavailability problem. In addition, a relatively large volume of oil in which the vitamin D is dissolved when given intramuscularly can be quite uncomfortable, which again is a good reason to give a pharmacological doses of vitamin D orally to correct vitamin D deficiency. Aging does not alter vitamin D absorption *(15)*. It is worthwhile to check 25(OH)D after the 8-wk therapy. In some cases, the vitamin D deficiency can be so severe that the blood levels of 25(OH)D do not increase substantially. Another 8-wk course of 50,000 IU of vitamin D once a week is reasonable.

An alternative and an inexpensive method to treat vitamin D deficiency is to encourage patients to be exposed to some sunlight. The amount depends on the person's skin sensitivity, time of day, season of the year, and latitude. For example, for an adult in Boston with a skin type 2, who would get a sunburn after being outside for 30 min at noontime in July, the recommendation is exposure to approx 20–30% of that time or 6–9 min of face, arms, and hands, two to three times a week. For those concerned about increased risk of wrinkles or skin damage to the face, exposure of arms and legs or back and legs would be adequate. No sunscreen or sun protection should be used for this brief period of time. However, if the person wishes to stay outdoors for a longer period of time, then use of a sunscreen with an SPF of at least 15 and sun protection with clothing is recommended. For those with marked increased skin pigmentation, the time outside could be as much as 30–60 min, again depending on the person's skin sensitivity, time of day, season of the year, and latitude *(3,15,76)*.

Patients with severe intestinal malabsorption syndrome and who are on total parenteral nutrition can obtain their vitamin D requirement from sun exposure. However, if they cannot go outside or the season will not permit them to make any vitamin D in their skin, then the use of a UVB radiation source, either a home device or a tanning bed at a tanning salon, would be appropriate. In one patient who had only 2 ft of small intestine left, Koutia et al. *(77)* reported that exposure to 0.75 MED of tanning bed radiation three times a week markedly increased blood levels of 25(OH)D by 700% and decreased PTH values into the normal range (Fig. 20). In addition, the patient, who suffered from severe bone pain and muscle aches and weakness, had complete relief of her symptoms.

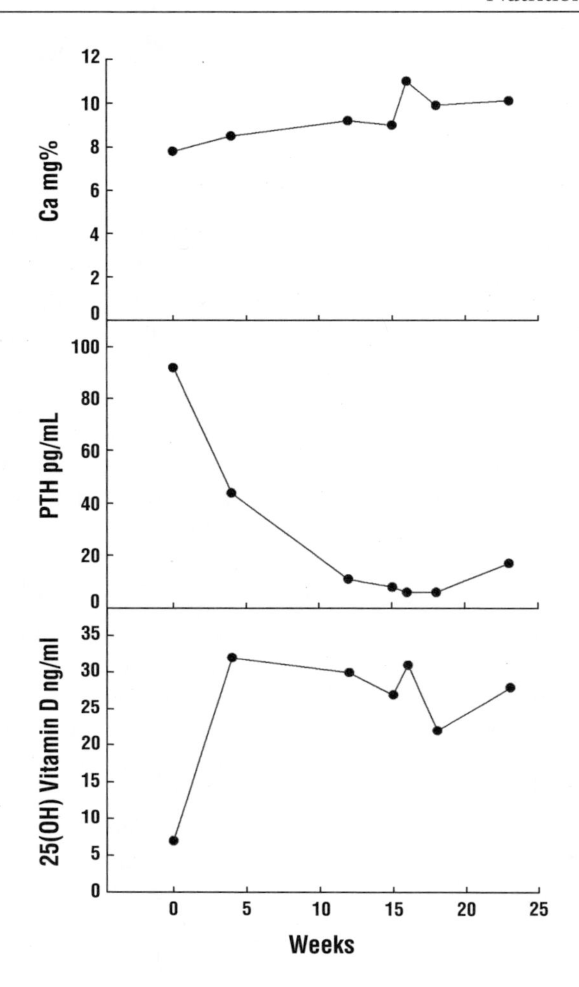

Fig. 20. Serum 25(OH)D, PTH, and calcium levels in a patient with Crohn's disease who had whole-body UVB exposure for 10 min, 3 times in a week for 6 mo. (Reproduced with permission from ref. *77.*)

Chuck et al. *(62)* also have demonstrated that the use of subliminal UVB lighting in an activity room in a nursing home was the most effective means to sustain 25(OH)D levels within the normal range and was far superior to taking a multivitamin that contained 400 IU of vitamin D a day (Fig. 21).

15. NONSKELETAL CONSEQUENCES OF VITAMIN D DEFICIENCY

As early as 1941 it was reported that people who lived in higher latitudes were at higher risk of dying of cancer *(78)*. A multitude of epidemiological studies have

Fig. 21. Exposure of nursing home residents to ultraviolet-B lamps that were installed near the ceiling in the day room. This was found to be the most effective method of maintaining serum 25(OH)D levels in these residents. (Reproduced with permission from ref. *62.*)

now confirmed this early observation *(82–90)*. There is firm evidence that people living at higher latitudes are at higher risk of developing and dying of breast, colon, ovarian, and prostate cancers *(76,79–85)*. Indeed, mortality rates in both men and women are related to their exposure to sunlight *(85)* (Fig. 22).

There is also a latitudinal association with increased risk of developing hypertension and multiple sclerosis *(86,87)*.

It is now recognized that most tissues and cells possess a VDR. The exact function of $1,25(OH)_2D$ in tissues, such as the brain, breast, prostate, skin, β-islet cells in the pancreas, monocytes, and activated T- and B-lymphocytes, is not fully understood. However, it is known that $1,25(OH)_2D$ is extremely effective in downregulating cellular growth in cells that possess a VDR. Indeed, the potent antiproliferative activity of $1,25(OH)_2D$ has been taken advantage of by the development of activated vitamin D analogs for the treatment of the hyperproliferative disorder psoriasis *(88)*.

It is recognized that the β-islet cells have a VDR and that $1,25(OH)_2D$ modulates insulin production and secretion *(3,7,89)*. $1,25(OH)_2D$ also modulates the immune system by regulating the activity of both activated T- and B-lymphocytes and

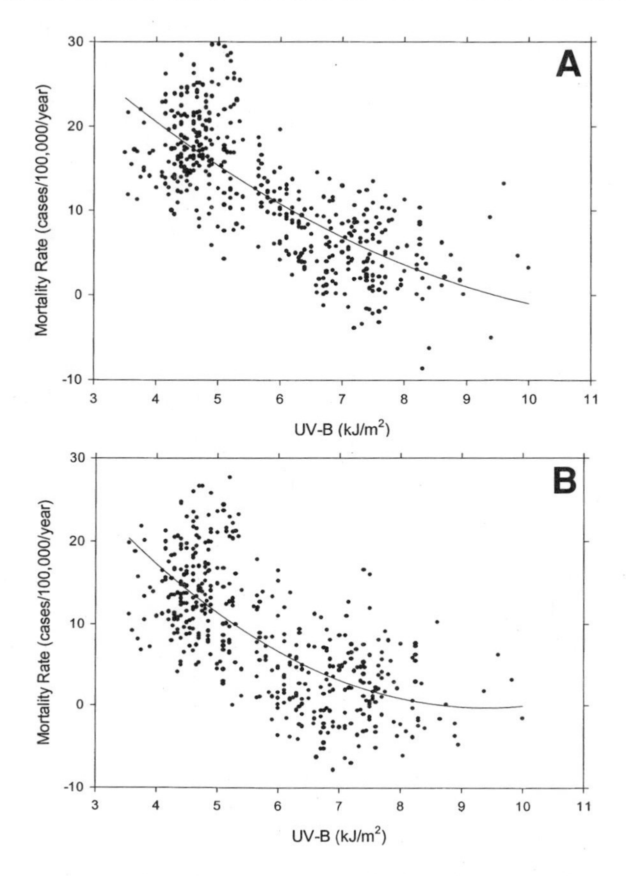

Fig. 22. **(A)** Premature mortality due to cancer, white females, vs total ozone mapping spectrometer (TOMS), July 1992, DNA-weighed UV-B. **(B)** Premature mortality due to cancer with insufficient UV-B, white males, U.S., 1970–1994, vs July 1992 DNA-weighted UV-B radiation. (Reproduced with permission from ref. *85*.)

activated macrophages *(3,7,90–92)*. This may be the explanation for why Hyponnen et al. *(93)* observed that children treated with at least 2000 IU of vitamin D a day reduced their risk of developing type 1 diabetes by 80%. This was similar to what was observed when NOD mice, which invariably develop type 1 diabetes, received $1,25(OH)_2D_3$ *(91,92)*: they showed an 80% reduction in developing the disease.

The kidney is an endocrine organ for producing $1,25(OH)_2D$ for regulating calcium metabolism. Recently, it was recognized that $1,25(OH)_2D$ also downregulates the production of renin in the kidney *(94)*. This may be the explanation for why vitamin D deficiency is associated with hypertension and increased risk of coronary artery disease and congestive heart failure *(44,95–99)*. Krause et al. *(97)* reported that exposure of hypertensive adults to a tanning bed that emitted UVB radiation raised the blood levels of 25(OH)D by more than 100% and controlled their hypertension. A

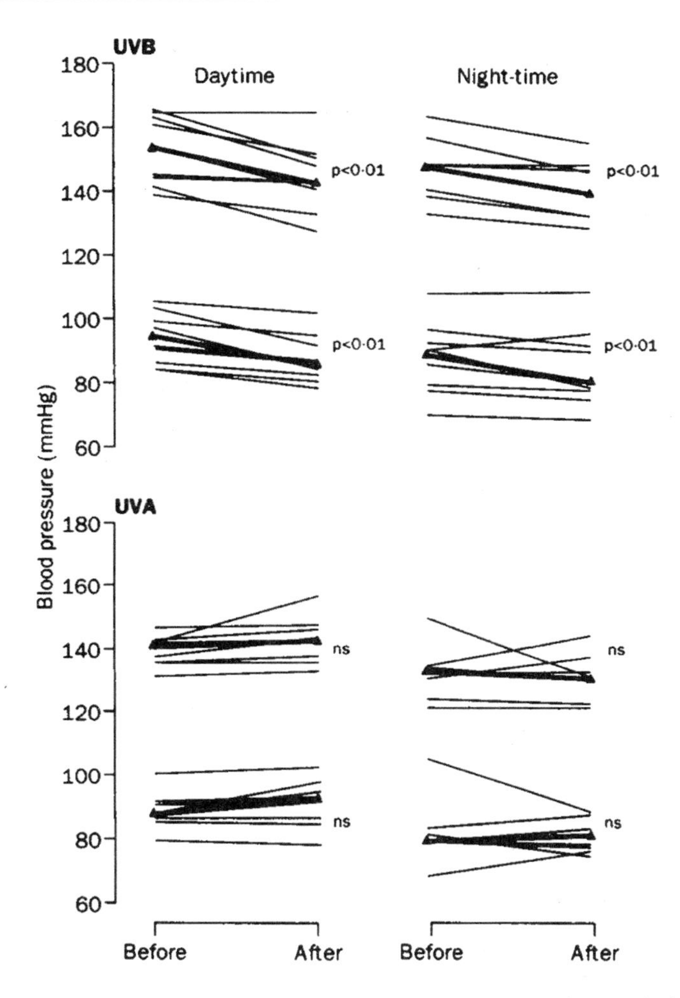

Fig. 23. Effect of UV-B and UV-A irradiation on ambulatory daytime and night-time blood pressure in hypertensive adults. ns = nonsignificant. Thick line = mean. (Reproduced with permission from ref. 97.)

similar group of hypertensive adults exposed to a similar tanning bed for 3 mo that emitted UVA but no UVB radiation not only did not increase their blood levels of 25(OH)D, but also had no effect on their hypertension (Fig. 23).

16. VITAMIN D AND THE CANCER CONNECTION

Although in the 1990s there were several reports that some of the most common cancers occurred in people living at higher latitudes and that colon cancer and prostate cancer rates were significantly reduced in individuals with higher circulating levels of 25(OH)D, it was difficult to understand how increased exposure to sunlight could impact on decreasing risk of common cancers.

The reason for this is that it was well known that any significant increase in vitamin D intake or exposure to sunlight did not raise blood levels of 1,25$(OH)_2$D. Thus, it was difficult to understand how increasing one's 25(OH)D levels would be able to regulate cellular growth and prevent some cancers, since circulating levels of 1,25$(OH)_2$D, the antiproliferative hormone, were not increased. The mystery was solved when it was observed that prostate cells and prostate cancer cells expressed a functional 1-OHase similar to what was observed in the skin *(3,15,99–101)*. Since this initial observation, it is now recognized that normal colon tissue and colon cancer, breast and breast cancer cells, as well as variety of other cell types have the enzymatic machinery to convert 25(OH)D directly to 1,25$(OH)_2$D *(3,15,16,102–104)*. Thus, it appears that when 25(OH)D levels are adequate, probably above 30 ng/mL, it acts a substrate for the extra renal 1-OHase in these tissues. The local production of 1,25$(OH)_2$D may be necessary to maintain and regulate genes responsible for cellular growth and to prevent the cells from becoming autonomous, that is, carcinogenic. It has been suggested that once it carries out its function, it induces the 25-hydroxyvitamin D-24-hydroxylase, which in turn catabolizes 1,25$(OH)_2$D to the inactive water-soluble calcitropic acid (Fig. 24).

17. CONCLUSION

Vitamin D deficiency is extremely common and needs to be recognized. Vitamin D deficiency in children and teenagers can result in poor bone health and the inability to attain the genetically predetermined peak bone mass. In young, middle-aged, and older adults, vitamin D deficiency causes osteomalacia and can precipitate and exacerbate osteoporosis. In addition, many of the symptoms associated with vitamin D deficiency mimic fibromyalgia, and as a result, many patients go undiagnosed.

Vitamin D deficiency, however, may have extremely important health consequences that heretofore have not been fully appreciated. Maintenance of an adequate 25(OH)D level of at least 20 ng/mL and preferably 30–50 ng/mL throughout life may help reduce the risk of developing many chronic diseases, including type 1 diabetes, hypertension, multiple sclerosis, and cancers of the breast, prostate, colon, and ovary (Fig. 25). Thus, there needs to be a reawakening about the appreciation of maintaining a healthy vitamin D status throughout life. The best method to determine vitamin D adequacy is to measure 25(OH)D. Similar to evaluating patients for their blood pressure and blood lipid profile on their yearly exam, they should also be evaluated with a 25(OH)D to measure their vitamin D status. This will ensure vitamin D health and mitigate the consequences of vitamin D deficiency.

ACKNOWLEDGMENTS

This work was supported in party by National Institutes of Health Grants MO100533 and AR 36963.

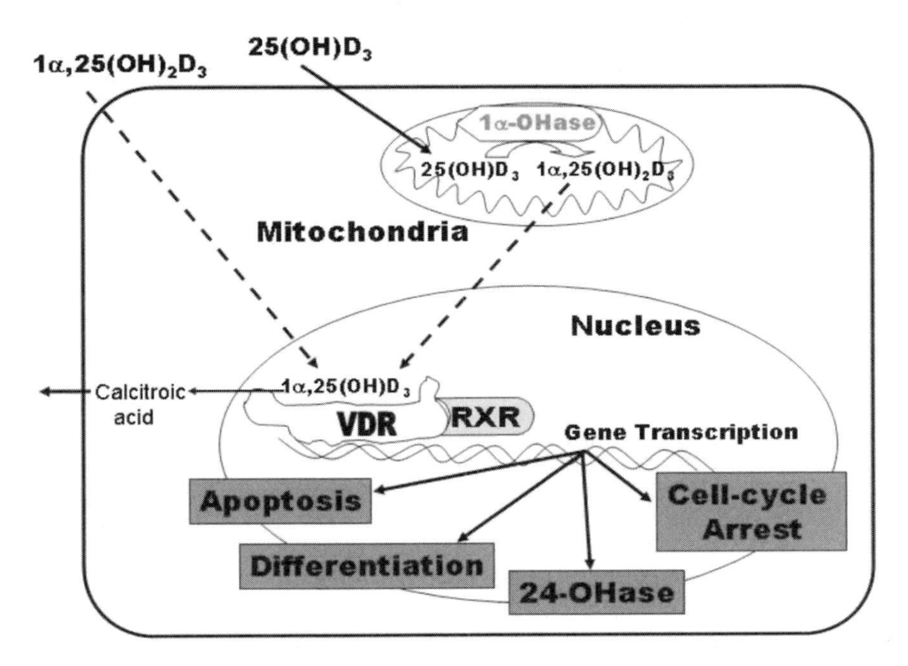

Fig. 24. Schematic representation of the metabolism of 25-hydroxyvitamin D [25(OH)D] to 1,25-dihydroxyvitamin D [$1,25(OH)_2D$] in nonrenal tissues and the role of 1,25(OH)D in regulating gene expression of genes responsible for cellular proliferation, differentiation, and death and for the catabolism of $1,25(OH)_2D$.

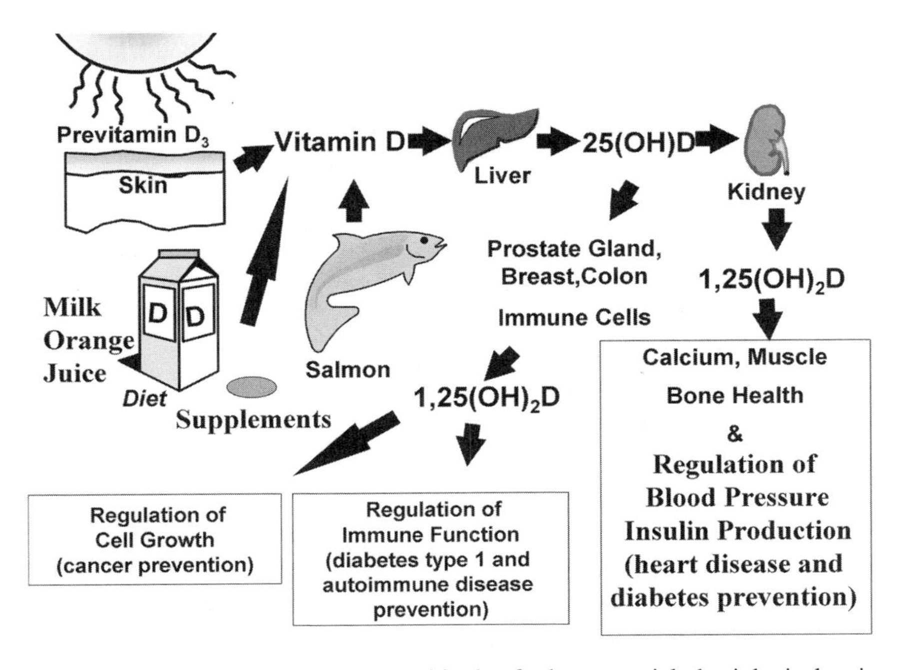

Fig. 25. Schematic representation of the multitude of other potential physiological actions of vitamin D for cardiovascular health, cancer prevention, regulation of immune function, and decreased risk of autoimmune diseases.

REFERENCES

1. Holick MF. Phylogenetic and evolutionary aspects of vitamin D from phytoplankton to humans. In: Pang PKT, Schreibman MP, eds. Vertebrate Endocrinology: Fundamentals and Biomedical Implications, Vol. 3. Academic, Orlando, FL, 1989, pp. 7–43.
2. Holick MF. Calcium and vitamin D in human health. Annales Nestlé, in press.
3. Holick MF. Vitamin D: a millennium perspective. J Cell Biochem 2003; 88:296–307.
4. Cooke NE, David EV. Serum vitamin D-binding protein is a third member of the albumin and alpha fetoprotein gene family. J Clin Invest 1985; 76:2420–2424.
5. Wortsman J, Matsuoka LY, Chen TC, Lu Z, Holick MF. Decreased bioavailability of vitamin D in obesity. Am J Clin Nutr 2000; 72:690–693.
6. Holick MF. Evaluation and treatment of disorders in calcium, phosphorus, and magnesium metabolism. In: Noble J, ed. Textbook of Primary Care Medicine, 3rd ed. Mosby, St. Louis, 2001; 100:886–898.
7. Bouillon R. Vitamin D: from photosynthesis, metabolism, and action to clinical applications. In: DeGroot LJ, Jameson JL, eds. Endocrinology. Saunders, Philadelphia, 2001; pp. 1009–1028.
8. Nykjaer A, Dragun D, Walther D, et al. An endocytic pathway essential for renal uptake and activation of the steroid 25-(OH) vitamin D_3. Cell, 1999; 96:507–515.
9. MacDonald PN. Molecular biology of the vitamin D receptor. In: Holick MF, ed. Vitamin D: Physiology, Molecular Biology and Clinical Applications. Humana, Totowa, NJ, 1999, pp. 109–128.
10. Holick, MF. Vitamin D: photobiology, metabolism, mechanism of action, and clinical applications. In: Favus MJ, ed. Primer on the Metabolic Bone Diseases and Disorders of Mineral Metabolism, 4th ed. Lippincott-Raven, Philadelphia, 1999, pp. 92–98.
11. Freedman LP. Multimeric coactivator complexes for steroid/nuclear receptors. Trends in Endocrinol Metab 1999; 10:403–407.
12. Raval-Pandya M, Porta AR, Christakos S. Mechanism of action of 1,25-dihydroxyvitamin D_3 on intestinal calcium absorption and renal calcium transport. In: Holick MF, ed. Vitamin D: Physiology, Molecular Biology and Clinical Applications. Humana, Totowa, NJ, 1999, pp. 163–173.
13. Khosla S. The OPG/RANKL/RANK system. Endocrinology 2001; 142:5050–5055.
14. Jimi E, Nakamura I, Amano H, et al. Osteoclast function is activated by osteoblastic cells through a mechanism involving cell-to-cell contact. Endocrinology 1996; 137:2187–2190.
15. Holick MF. Vitamin D: the underappreciated D-lightful hormone that is important for skeletal and cellular health. Curr Opin Endocrinol Diabetes 2002; 9:87–98.
16. Zerwekh JE, Sakhaee K, Pak CYC. Short-term 1,25-dihydroxyvitamin D_3 administration raises serum osteocalcin in patients with postmenopausal osteoporosis. J Clin Endocrinol Metab 1985; 60:615–617.
17. Lian JB, Staal A, van Wijnen JA., Stein JL, Stein GS. Biologic and molecular effects of vitamin D on bone. In: Vitamin D: Physiology, Molecular Biology, and Clinical Applications. Holick MF, ed. Humana Totowa, NJ, 1999; 175–193.
18. Underwood JL, DeLuca HF. Vitamin D is not directly necessary for bone growth and mineralization. Am J Physiol 1984; 246:E493–E498.
19. Holtrop ME, Cox KA, Carnes DL, Holick MF. Effects of serum calcium and phosphorus on skeletal mineralization in vitamin D-deficient rats. Am J Physiol 1986; 251(2 pt 1):E234–E240.
20. Balsan S, Garabedian M, Larchet M, et al. Long-term nocturnal calcium infusions can cure rickets and promote normal mineralization in hereditary resistance to 1,25-dihydroxyvitamin D. J Clin Invest 1986; 77:1661–1667.

20a. Holick MF. Evolution, biologic functions, and recommended dietary allowances for vitamin D. In: Vitamin D: Physiology, Molecular Biology, and Clinical Applications. Holick MF, ed. Humana Press, Totowa, NJ, 1999:1–16.

21. Steenbock H. The induction of growth-prompting and calcifying properties in a ration exposed to light. Science 1924; 60:224–225.
22. Holick MF. Vitamin D: importance for bone health, cellular health and cancer prevention. In: Holick MF, ed. Biologic Effects of Light Kluwer, Boston, 2001, pp. 155–173.
23. Oppé TE. Infantile hypercalcemia, nutritional rickets, and infantile survey in Great Britain. Br Med J 1964; 1:1659–1661.
24. Tangpricha V, Koutkia P, Rieke SM, Chen TC, Perez AA, Holick MF. Fortification of orange juice with vitamin D: a novel approach to enhance vitamin D nutritional health. Am J Clin Nutr, 2003; 77:1478–1483.
25. MacLaughlin JA, Anderson RR, Holick MF. Spectral character of sunlight modulates the photosynthesis of previtamin D_3 and its photo isomers in human skin. Science 1982; 1001–1003.
26. Holick MF, Tian XQ, Allen M. Evolutionary importance for the membrane enhancement of the production of vitamin D_3 in the skin of poikilothermic animals. Proc Natl Acad Sci USA 1995; 92(8):3124–3126.
27. Tian XQ, Chen TC, Matsuoka LY, Wortsman J, Holick MF. Kinetic and thermodynamic studies of the conversion of previtamin D_3 to vitamin D_3 in human skin. J Biol Chem 1993; 268(20):14888–14892.
28. Haddad JG, Matsuoka LY, Hollis BW, Hu YZ, Wortsman J. Human plasma transport of vitamin D after its endogenous synthesis. J Clin Invest 1993; 91:2552–2555.
29. Clemens TL, Henderson SL, Adams JS, Holick MF. Increased skin pigment reduces the capacity of skin to synthesize vitamin D_3. Lancet 1982; 74–76.
30. Matsuoka LY, Ide L, Wortsman J, MacLaughlin J, Holick MF. Sunscreens suppress cutaneous vitamin D_3 synthesis. J Clin Endocrinol Metab 1987; 64:1165–1168.
31. Matsuoka LY, Wortsman J, Dannenberg MJ, Hollis BW, Lu Z, Holick MF. Clothing prevents ultraviolet-B radiation-dependent photosynthesis of vitamin D_3. J Clin Endocrinol Metab 1992; 75(4):1099–1103.
32. Taha SA, Dost SM, Sedrani SH. 25-Hydroxyvitamin D and total calcium: Extraordinarily low plasma concentrations in Saudi mothers and their neonates. Pediatr Res 1984; 18:739–741.
33. Glerup H, Mikkelsen K, Poulsen L, et al. Hypovitaminosis D myopathy without osteomalacic bone involvement. Calcif Tissue Int 2000; 66(6):419–424.
34. Holick, MF. McCollum Award Lecture, 1994: Vitamin D: new horizons for the 21st century. Am J Clin Nutr 1994; 60:619–630.
34a. Chen TC. Photobiology of vitamin D. In: Vitamin D: Physiology, Molecular Biology, and Clinical Applications. Holick, MF, ed. Humana Press, Totowa, NJ, 1999: 17–37.
35. MacLaughlin J, Holick MF. Aging decreases the capacity of human skin to produce vitamin D_3. J Clin Invest 1985; 76:1536–1538.
36. Brown EM, Pollak M, Seidman CE, et al. Calcium-ion-sensing cell-surface receptors. N Engl J Med 1995; 333:234–240.
37. Jüppner H, Brown EM, Kronenberg HM. Parathyroid hormone. In: Favus MJ, ed. Primer on the Metabolic Bone Diseases and Disorders of Mineral Metabolism, 4th ed. Lippincott-Raven, Philadelphia, 1999; pp. 80–87.
38. Hordon LD, Peacock M. Osteomalacia and osteoporosis in femoral neck fracture. Bone Miner 1990; 11:247–259.
39. Chapuy MC, Arlot M, Duboeuf F, et al. Vitamin D_3 and calcium to prevent hip fractures in elderly women. N Engl J Med 1992; 327:1627–1642.
40. Dawson-Hughes B, Harris SS, Krall EA, Dallal GE. Effect of calcium and vitamin D supplementation on bone density in men and women 65 years of age or older. N Engl J Med 1997; 337:670–676.
41. Schnitzler CM, Solomon L. Osteomalacia in elderly White South African women with fractures of the femoral neck. S Afr Med J 1983; 64:527–530.

42. Al-Ali H, Fuleihan GEH. Nutritional osteomalacia: substantial clinial improvement and gain in bone density post-therapy. J Clin Densitometr 2000; 3:97–101.
43. Malabanan AO, Turner AK, Holick MF. Severe generalized bone pain and osteoporosis in a premenopausal black female: effect of vitamin D replacement. J Clin Densitometr 1998; 1:201–204.
44. Holick MF. Sunlight and vitamin D, both good for cardiovascular health (editorial). J Gen Intern Med 2002; 17:733–735.
45. Rimaniol J, Authier F, Chariot P. Muscle weakness in intensive care patients: initial manifestation of vitamin D deficiency. Intensive Care Med 1994; 20:591–592.
46. Bischoff HA, Stähelin HB, Dick W, et al. Effect of vitamin D and calcium supplementation on falls: a randomized controlled study. J Bone Miner Res 2003; 18(2):343–351.
47. Gloth FM, Gundberg CM, Hollis BW, Haddad HG, Tobin JD. Vitamin D deficiency in homebound elderly persons. JAMA 1995; 274:1683–1686.
48. Malabanan A, Veronikis IE, Holick MF. Redefining vitamin D insufficiency. Lancet 1998; 351:805–806.
49. Lips P. Vitamin D deficiency and secondary hyperparathyroidism in the elderly: consequences for bone loss and fractures and therapeutic implications. Endocrine Rev 2001; 22(4):477–501.
50. Chapuy MC, Preziosi P, Maaner M, et al. Prevalence of vitamin D insufficiency in an adult normal population. Osteopor Int 1997; 7:439–443.
51. Tangpricha V, Pearce EN, Chen TC, Holick MF. Vitamin D insufficiency among free-living healthy young adults. Am J Med 2002; 112:659–662.
52. Kreiter SR, Schwartz RP, Kirkman HN, Charlton PA, Calikoglu AS, Davenport M. Nutritional rickets in African American breast-fed infants. J Pediatr 2000; 137:2–6.
53. Standing Committee on the Scientific Evaluation of Dietary Reference Intakes. Vitamin D. In: Institute of Medicine. Dietary Reference Intakes for Calcium, Phosphorus, Magnesium, Vitamin D, and Fluoride. National Academy Press, Washington, DC, 1997, p. 263.
54. Welch TR. Vitamin D deficient rickets: the reemergence of a once-conquered disease. J Pediatr 2000; 137(2):143–145.
55. Thomas MK, Lloyd-Jones DM, Thadhani RI, et al. Hypovitaminosis D in medical inpatients. N Engl J Med 1998; 338:777–783.
56. Holick MF, Shao Q, Liu WW, Chen TC. The vitamin D content of fortified milk and infant formula. N Engl J Med 1992; 326(18):1178–1181.
57. Lo CW, Paris PW, Clemens TL, Nolan J, Holick MF. Vitamin D absorption in healthy subjects and in patients with intestinal malabsorption syndromes. Am J Clin Nutr 1985; 42:644–649.
58. Shane E, Silverberg SJ, Donovan D, et al. Osteoporosis in lung transplantation candidates with end-stage pulmonary disease. Am J Med 1996; 101:262–269.
59. Shike M, Harrison J, Sturtridge C, et al. Metabolic bone disease in patients receiving long-term parenteral nutrition. Ann Intern Med 1994; 92:343–350.
60. Shike M, Shils ME, Heller A, et al. Bone disease in prolonged parenteral nutrition: osteopenia without mineralization defect. Am J Clin Nutr 1986; 44:89–98.
61. Chel VGM, Ooms ME, Popp-Snijders C, et al. Ultraviolet irradiation corrects vitamin D deficiency and suppresses secondary hyperparathyroidism in the elderly. J Bone Miner Res 1998; 13:1238–1242.
62. Chuck A, Todd J, Diffey B. Subliminal ultraviolet-B irradiation for the prevention of vitamin D deficiency in the elderly: a feasibility study. Photochem Photoimmun Photomed 2001; 17(4):168–171.
62a. Lund B, Sorensen OH. Measurement of 25-hydroxyvitamin D in serum and its relation to sunshine, age and Vitamin D intake in the Danish population. Scand J Clin Lab Invest 1979; 39:23–30.
63. Bell NH, Epstein S, Greene A, Shary J, Oexmann MJ, Shaw S. Evidence for alteration of the vitamin D-endocrine system in obese subjects. J Clin Invest 1985; 76:370–373.

64. Eastwood JB, De Wardener HE, Gray RW, Lemann JL, Jr. Normal plasma 1,25-$(OH)_2$-vitamin D concentrations in nutritional osteomalacia. Lancet 1979; 1377–1378.
65. Vieth R. Vitamin D supplementation, 25-hydroxyvitamin D concentrations, and safety. Am J Clin Nutr 1999; 69:842–856.
66. Vieth R, Chan PC, MacFarlane GD. Efficacy and safety of vitamin D_3 intake exceeding the lowest observed adverse effect level[1–3]. Am J Clin Nutr 2001; 73:288–294.
67. Barger-Lux MJ, Heaney RP, Dowell S, Chen TC, Holick MF. Vitamin D and its major metabolites: serum levels after graded oral dosing in healthy men. Osteopor Int 1998; 8:222–230.
68. Heaney RP, Davies KM, Chen TC, Holick MF, Barger-Lux MJ. Human serum 25-hydroxycholecalciferol response to extended oral dosing with cholecalciferol. Am J Clin Nutr 2003; 77:204–210.
69. Lips P, Chapuy MC, Dawson-Hughes B, Pols HA, Holick MF. An international comparison of serum 25-hydroxyvitamin D measurements. Osteoporos Int 1999; 9(5):394–397.
70. Chapuy MC, Schott AM, Garnero P, et al. Healthy elderly French women living at home have secondary hyperparathyroidism and high bone turnover in winter. J Clin Endocrinol Metab 1996; 81:1129–1133.
71. Koutkia P, Chen TC, Holick MF. Vitamin D intoxication associated with an over-the-counter supplement. N Engl J Med 2001; 345(1):66–67.
72. Bauer JM, Freyberg RH. Vitamin D intoxication and metastatic calcification. JAMA 1946; 1208–1215.
73. Adams JS, Lee G. Gains in bone mineral density with resolution of vitamin D intoxication. Ann Intern Med 1997; 127:203–206.
74. Jacobus CH, Holick MF, Shao Q, et al, Hypervitaminosis D associated with drinking milk. N Engl J Med 1992; 326:1173–1177.
75. Heikinheimo RJ, Ubjivaaram JA, Jantti PO, Maki-Jokela PL, Rajala SA, Sievanen H. Intermittant parenteral vitamin D supplementation in the elderly in nutritional aspects of osteoporosis. In: Burckhard P, Heaney RP, eds. Challenges of Modern Medicine. Ares-Serono, Geneva, Switzerland, 1994, pp. 335–340.
76. Holick MF, Jenkins M. The UV Advantage. In press.
77. Koutkia P, Lu Z, Chen TC, Holick MF. Treatment of vitamin D deficiency due to Crohn's disease with tanning bed ultraviolet B radiation. Gastroenterology 2001; 121(6):1485–1488.
78. Apperly FL. The relation of solar radiation to cancer mortality in North America. Cancer Res 1941; 1:191–195.
79. Garland CF, Garland FC, Shaw EK, Comstock GW, Helsing KJ, Gorham ED. Serum 25-hydroxyvitamin D and colon cancer: eight-year prospective study. Lancet 1989; 18:1176–1178.
80. Garland C, Shekelle RB, Barrett-Connor E, Criqui MH, Rossof AH, Oglesby P. Dietary vitamin D and calcium and risk of colorectal cancer: a 19-year prospective study in men. Lancet 1985; 9:307–309.
81. Garland CF, Garland FC, Gorham ED, Raffa J. Sunlight, vitamin D, and mortality from breast and colorectal cancer in Italy. In: Biologic Effects of Light. Holick MF, Kligman A. eds. Walter de Gruyter, New York, 1992, pp. 39–43.
82. Garland FC, Garland CF, Gorham ED, Young JF. Geographic variation in breast cancer mortality in the United States: a hypothesis involving exposure to solar radiation. Prevent Med 1990; 19:614–622.
83. Hanchette CL, Schwartz GG. Geographic patterns of prostate cancer mortality. Cancer 1992; 70:2861–2869.
84. Ahonen MH, Tenkanen L, Teppo L, Hakama M, Tuohimaa P. Prostate cancer risk and prediagnostic serum 25-hydroxyvitamin D levels (Finland). Cancer Causes Control 2000; 11:847–852.
85. Grant WB. An ecologic study of dietary and solar ultraviolet-B links to breast carcinoma mortality rates. Am Cancer Soc 2002; 94:272–281.

86. Rostand SG. Ultraviolet light may contribute to geographic and racial blood pressure differences. Hypertension 1979; 30:150–156.
87. Hernan MA, Olek MJ, Ascherio A. Geographic variation of MS incidence in two prospective studies of US women. Neurology 1999; 51:1711–1718.
88. Holick MF. Clinical efficacy of 1,25-dihydroxyvitamin D_3 and its analogues in the treatment of psoriasis. Retinoids 1998; 14(1):12–17.
89. Wongsurawat N, Armbrecht HJ, Zenser TV, Davis BB, Thomas ML, Forte LR. 1,25-Dihydroxyvitamin $D_3$and 24,25-dihydroxyvitamin D_3 production by isolated renal slices is modulated by diabetes and insulin in the rat. Diabetes 1983; 32:302–306.
90. Adorini L. 1,25-Dihydroxyvitamin D3 analogs as potential therapies in transplantation. Curr Opin Invest Drugs 2002; 3(10):1458–1463.
91. Gregori S, Giarratana N, Smiroldo S, Uskokovic M, Adorini L. A 1a,25-dihydroxyvitamin D_3 analog enhances regulatory T-cells and arrests autoimmune diabetes in NOD mice. Diabetes 2002; 51:1367–1374.
92. Mathieu C, Waer M, Laureys J, Rutgeerts O, Bouillon R. Prevention of autoimmune diabetes in NOD mice by 1,25 dihydroxyvitamin D_3. Diabetologia 1994; 37:552–558.
93. Hypponen E, Laara E, Jarvelin M-R, Virtanen SM. Intake of vitamin D and risk of type 1 diabetes: a birth-cohort study. Lancet 2001; 358:1500–1503.
94. Li YC, Kong J, Wei M, Chen ZF, Liu SQ, Cao LP. 1,25-Dihydroxyvitamin D_3 is a negative endocrine regulator of the renin-angiotensin system. J Clin Invest 2002; 110:229–238.
95. Scragg R, Jackson R, Holdaway IM, Lim T, Beaglehole R. Myocaardial infarction is inversely associated with plasma 25-hydroxyvitamin D3 levels: a community-based study. Int J Epidemiol 1990; 19:559–563.
96. Zitterman A, Schulze Schleithoff S, Tenderich C, Berthold H, Koefer R, Stehle P. Low vitamin D status: a contributing factor in the pathogenesis of congestive heart failure? J Am Coll Cardiol 2003; 41(1):105–112.
97. Krause R, Buhring M, Hopfenmuller W, Holick MF, Sharma AM. Ultraviolet B and blood pressure. Lancet 1998; 352(9129):709–710.
98. Scragg R. Seasonality of cardiovascular disease mortality and the possible protective effect of utlra-violet radiation. Int J Epidemiol 1981; 10:337–341.
99. Schwartz GG, Whitlatch LW, Chen TC, Lokeshwar BL, Holick MF. Human prostate cells synthesize 1,25-dihydroxyvitamin D_3 from 25-hydroxyvitamin D_3. Cancer Epidemiol Biomarkers Prev 1998; 7(5):391–395.
100. Bikle DD, Nemanic MK, Gee E, Elias P. 1,25-Dihydroxyvitamin D_3 production by human keratinocytes: kinetics and regulation. J Clin Invest 1986; 78:557–566.
101. Lehmann B, Knuschke P, Meurer M. UVB-induced conversion of 7-dehydrocholesterol to 1α,25-dihydroxyvitamin D_3 (calcitriol) in the human keratinocyte line HaCaT. Photochem Photobiol 2000; 72(6):803–809.
102. Cross HS, Bareis P, Hofer H, Bischof MG, Bajna E, Kriwanek S. 25-Hydroxyvitamin D_3-1α-hydroxylase and vitamin D receptor gene expression in human colonic mucosa is elevated during early cancerogenesis. Steroids 2001; 66:287–292.
103. Tangpricha V, Flanagan JN, Whitlatch LW, et al. 25-hydroxyvitamin D-1a-hydroxylase in normal and malignant colon tissue. Lancet 2001; 357(9269):1673–1674.
104. Holick MF. Sunlight "D"ilemma: risk of skin cancer or bone disease and muscle weakness. Lancet 2001; 357:4–6.

26 Vitamin D Utilization in Subhuman Primates

Lessons Learned at the Los Angeles Zoo

John S. Adams, Rene F. Chun, Shaoxing Wu, Songyang Ren, Mercedes A. Gacad, and Hong Chen

Man with all his noble qualities . . . with his god-like intellect which has penetrated into the movements and constitution of the solar system . . . still bears in his bodily frame, the indelible stamp of his lowly origin

Charles Robert Darwin, in the Descent of Man

1. EARLY PRIMATE EVOLUTION

In the Eocene period, 50–100 million yr ago, the great southern hemispheric landmass, Pangea, broke apart. These tectonic events resulted in the American land mass and Madagascar moving away from Africa. This continental separation occurred early in the process of primate evolution, trapping primordial primates in South America, Africa, and Madagascar, respectively. As a consequence, the three major primate infraorders, platyrrhines or New World primates, catarrhines or Old World primates and lemurs, evolved independently of one another *(1)* (Fig. 1A). Unlike Old World primates, including our own species, which have populated virtually every land mass on our planet, New World primates have remained confined to Central and South America for the last 50 million yr. Compared to Old World primates, especially some of the terrestrial species such as gorillas, New World primates are smaller in stature. This is a characteristic well suited to their lifestyle as plant-eating, arboreal sunbathers, residing in the canopy of the periequatorial rain forests of the Americas.

From: *Nutrition and Bone Health*
Edited by: M. F. Holick and B. Dawson-Hughes © Humana Press Inc., Totowa, NJ

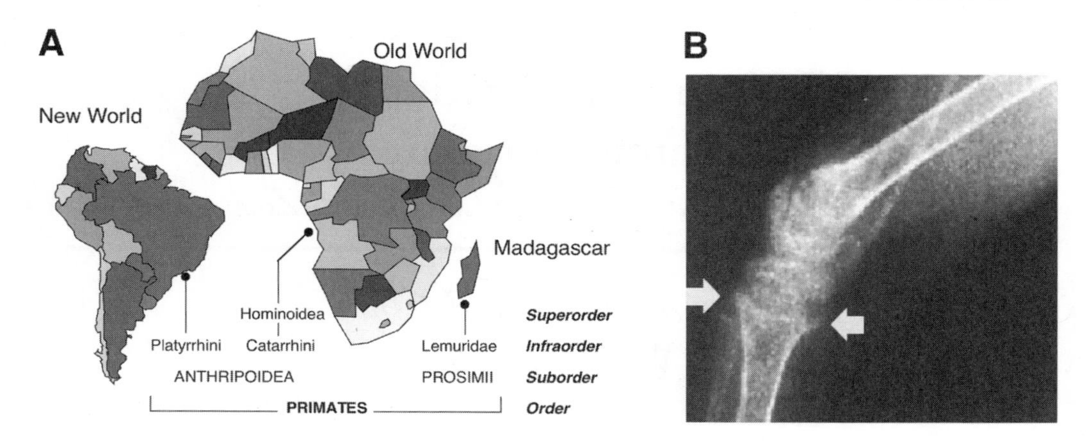

Fig. 1. New World primate evolution and rachitic bone lesion. **A** describes in geographic terms the independent evolution of the three primate suborders, Platyrrhini, Catarrhini, and Lemuridae, in South America (the New World), Africa (the Old World), and Madagascar, respectively. **B** displays the characteristic "cupping" and "fraying" of the tibial metaphysis (arrows) in a New World primate resident of the Los Angeles Zoo with rickets.

2. SIMIAN BONE DISEASE

The appearance of generalized metabolic bone disease in captive primates has been recognized for the last 150 yr *(2)*. The disease, which has not been well studied from a histopathological standpoint, carries the clinical and radiological stigmata of rickets and osteomalacia *(3)* (Fig. 1B). Compared to Old World primates reared in captivity, New World primates or platyrrhines are particularly susceptible to the disease. The disorder affects primarily young, growing animals and results in muscle weakness, skeletal fragility, and in many instances death of the affected individual. Rachitic bone disease of this sort has long presented a problem to veterinarians caring for captive platyrrhines, particularly in North American and European zoos *(4)*, because death of preadolescent and adolescent primates prior to sexual maturity severely limits on-site breeding programs.

Because the disease was reported to be ameliorated by either the oral administration of vitamin D_3 in large doses or by ultraviolet B irradiation of affected primates, it was presumed to be caused by vitamin D deficiency *(4)*. The frequent occurrence of rickets and osteomalacia in New World primates was also ascribed to the relative inability of platyrrhines, compared to Old World primates including man, to effectively employ vitamin D_2 in their diet *(5)*; a similar observation had been made for chickens *(6)*. Using an assay technology that does not discriminate between 25-hydroxylated vitamin D_2 and vitamin D_3 metabolites, Marx and colleagues *(7)* determined that 25-hydroxyvitamin D [25(OH)D] levels were 2–3-fold higher when platyrrhines were dosed with supplemental vitamin D_3 than with vitamin D_2. These data suggested that 25-hydroxylation of vitamin D substrate in New

World primates was much more effective when vitamin D_3 was employed as substrate. However, in the same study, two species of Old World primates demonstrated similar discrimination against vitamin D_2, in favor of vitamin D_3, to promote significantly more 25(OH)D produced. In summary, these results seemed to indicate that all subhuman primates, whether Old or New World, were relatively resistant to vitamin D_2 in terms of its ability to engender an increase in serum levels of 25(OH)D. Finally, Hay and colleagues *(8)* suggested that New World primates may transport 25(OH)D in the serum by means and proteins that are dissimilar from those encountered in Old World primate species. This hypothesis was disproven by Bouillon et al. *(9),* who showed that the vitamin D-binding protein was the major carrier of 25(OH)D in ther serum of both New and Old World primates.

3. VITAMIN D AND STEROID HORMONE RESISTANCE IN NEW WORLD PRIMATES

The question of why platyrrhines were more susceptible to vitamin D deficiency than were catarrhines began to be answered with the detection of extraordinarily high circulating levels of the active vitamin D metabolite, 1,25-dihydroxyvitamin D [1,25$(OH)_2$D] in New World primates *(10,11).* These data pointed to the fact that New World primates were resistant to the active vitamin D hormone. The concept of generalized steroid hormone resistance in New World primates was first revealed by Brown et al. in 1970 *(12).* These investigators discovered greatly elevated serum cortisol levels in platyrrhini compared to catarrhini. Despite biochemical evidence of resistance to glucocorticoids, platyrrhines affected with high cortisol levels showed no sign of glucocorticoid deficiency or toxicity at the level of the target organ; glucose homeostasis, electrolyte balance, blood pressure, and life expectancy were all similar to that observed in Old World primates *(13).*

These data indicated that glucocorticoid resistance in New World primates was physiologically compensated by increased synthesis of the hormone. Increased production of the hormone was achieved by lack of feedback inhibition of pituitary adrenocorticotropic hormone (ACTH) production *(14),* adrenal (zona fasiculata) hypertrophy *(15),* and increased enzymatic synthesis and decreased catabolism of cortisol *(16,17).* A relative increase in the availability of glucocorticoid to target tissues was also proposed to occur in New World primates *(18)* and participate in the response to cortisol resistance. What is the proximate cause of glucocorticoid resistance in New World primates? Early studies from the group of Lipsett and Loriaux *(13,19,20)* suggested that resistance was caused by expression of a glucocorticoid receptor (GR) in New World primates with a lower affinity for cortisol. Most recently, relative overexpression of FKBP51, the FK506-binding immunophilin that normally interacts with the heat shock protein 90 (hsp90)-GR complex, was postulated to be the cause of lowered affinity of the New World primate GR for its cognate ligand *(21).* It remains to be determined whether constitutive overexpression of the New World primate immunophillin in Old World primate cells will squelch GR-directed transactivation.

New World primates are also resistant to steroid hormones produced by the ovary *(22)*. Until recently, it was considered that estrogen and progesterone resistance resulted from a diminishment of the estrogen receptor (ER) and progesterone receptor (PR) population in target tissues. A similar mechanism was proposed for vitamin D resistance *(23)*. As will be discussed below, it now appears that the steroid hormone and vitamin D hormone receptor complement of New World primates is not functionally distinct from that of their Old World primate counterparts. What is different is the relative overexpression in New World primate cells of at least two distinct families of intracellular proteins, the heterogeneous nuclear ribonucleoproteins and the heat-shock proteins 70 (hsp70), which conspire to legislate, by receptor-independent means, the degree of steroid/sterol hormone resistance among the various New World primate species.

4. OUTBREAK OF RICKETS IN THE NEW WORLD PRIMATE COLONIES OF THE LOS ANGELES ZOO

Beginning in 1985, our research team was asked to investigate an outbreak of rickets among New World primate species at the Los Angeles Zoo. The index case in our original studies was a preadolescent New World primate of the Emperor tamarin species. When investigated radiographically (Fig. 1B), this tamarin and those like him displayed classical rickets complete with growth retardation, metaphyseal cupping, and fraying characteristic of rickets. In order to investigate this rachitic syndrome, blood and urine was collected from involved monkeys as well as from control, nonrachitic New and Old World primates. Compared to Old World primates and as shown in Fig. 2A, that comparison yielded a biochemical phenotype that was most remarkable for an elevated serum $1,25(OH)_2D$ level in rachitic New World primates *(11)*. In fact, with the exception of nocturnal primates in the genus *Aotus,* New World primates in all other genera had vitamin D hormone levels ranging up to two orders of magnitude higher than that observed in Old World primates, including man *(24–26)*.

In the initial analysis, New World primates affected with rickets were those with the lowest $1,25(OH)_2D$ levels, while their healthy counterparts were those with the highest serum $1,25(OH)_2D$ levels. These data were interpreted to mean that most New World primate genera were naturally resistant to the vitamin D hormone, and that the resistant state could be compensated by maintenance of high $1,25(OH)_2D$ levels. If this was true, then an increase in the serum $1,25(OH)_2D$ concentration in affected primates should result in biochemical compensation for the resistant state and resolution of their rachitic bone disease. When rachitic New World primates were exposed to 6 mo of artificial sunlight in their enclosures, both substrate serum 25(OH)D and product $1,25(OH)_2D$ levels rose dramatically, resulting in cure of rickets *(27)* (Fig. 2B).

In summary, New World primates are periequitorial sunbathers for a reason. As depicted by the oversized arrows in a simplified scheme of vitamin D synthesis and metabolism (Fig. 3), New World primates require a lot of cutaneous vitamin D syn-

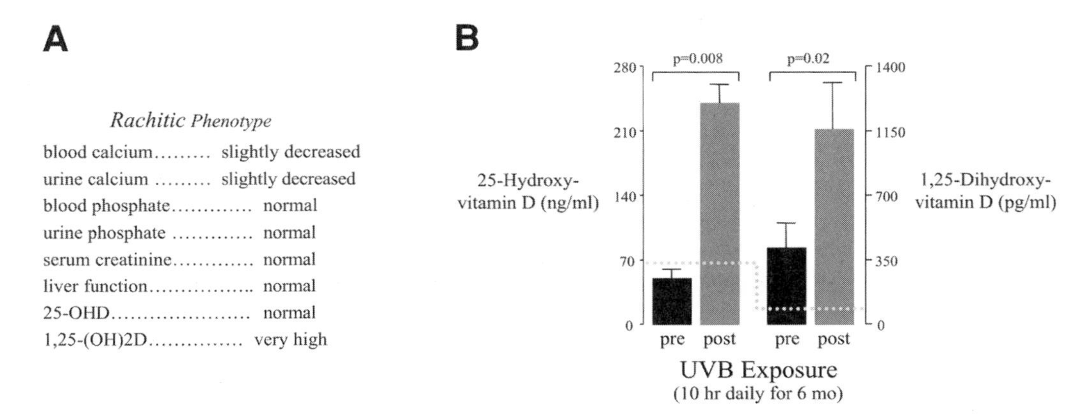

Fig. 2. Biochemical phenotype of rachitic New World primates. **A** demonstrates biochemical indices of bone health in New World primate suffering from rickets compared to developmental age- and sex-matched nonrachitic Old World primates. The outstanding characteristic is a 1,25-dihydroxyvitamin D [$1,25(OH)_2D$] level two-to-three orders of magnitude greater than that observed in Old World primates, including man. **B** shows the mean 25-hydroxyvitamin D (left) and 1,25-dihydroxyvitamin D levels (right) in 7 different rachitic New World primates before (pre) and after (post) exposure to 6 mo of artificial sunlight in their enclosures. The upper limits of the normal human Old World primate range is described by the dotted line. Both substrate and product rose significantly with light therapy and resulted in cure of rickets. (Data from ref. *27.*)

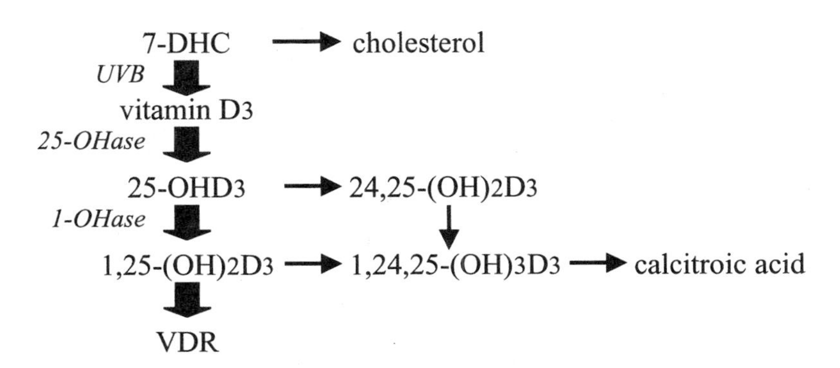

Fig. 3. Scheme of vitamin D synthesis and metabolism in New World primates. The **bold arrows** describe the means by which these high 1,25-dihydroxyvitamin D_3 [$1,25(OH)_2D_3$] levels are achieved and maintained. Ultraviolet B photon *(UVB)* exposure is increased in the natural habitat of New World primates, the canopy of the equatorial rain forests of Central and South America. Increased cutaneous vitamin D_3 synthesis results in increased production, via the hepatic vitamin D-25-hydroxylase *(25-OHase),* of 25-hydroxyvitamin D_3 [$25(OH)D_3$]. Elevated $1,25(OH)_2D_3$ levels are achieved by increased synthesis of the hormone via the 25-hydroxyvitamin D-1-hydroxylase *(1-OHase)* as well as by diminished catabolism to scheme $24,25(OH)_2D_3$. In this way $1,25(OH)_2D_3$ becomes available to the VDR in relatively large quantities to compensate for the hormone-resistant state characteristic of New World primates.

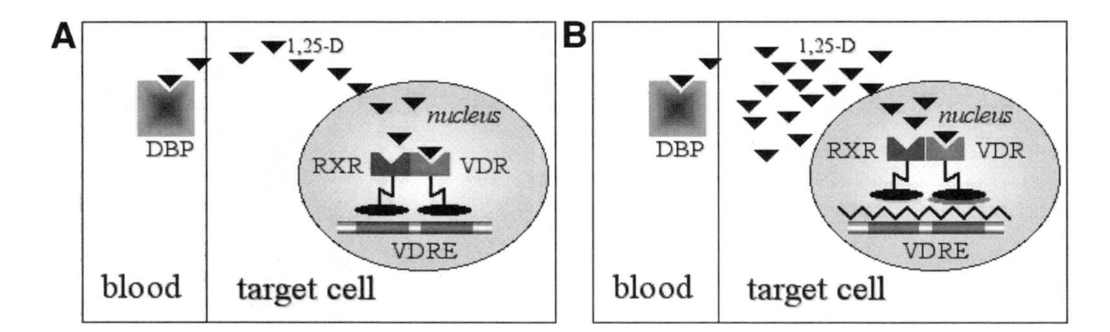

Fig. 4. Pathway of hormone 1,25-dihydroxyvitamin D (1,25-D) from the blood to nucleus of the target cell in the vitamin D-resistant New World primate. In Old World primate cells **(A)** the vitamin D hormone (dark triangles) moves normally from the circulating vitamin D-binding protein (DBP), through the cell membrane and cytoplasm, and onto the vitamin D receptor (VDR) paired with the retinoid X receptor (RXR) forming a heterodimeric complex in the cell nucleus. This heterodimeric complex then interacts with its cognate cis element, the vitamin D response element (VDRE), and initiates transcription of vitamin D-regulated genes. **(B)** demonstrates similar events as they occur in New World primate cells. The jagged line at the VDRE represents a relative inability of the heterodimeric receptor complex to engage the cis element. This and the accumulation of hormone in the cell cytoplasm represent salient disparities in hormone handling and action in the New World and Old World primate cells.

thesis in order to push their 25(OH)D and $1,25(OH)_2D$ levels high enough to interact effectively with the vitamin D receptor (VDR). The question remained as to why these primates are resistant to all but the highest levels of the vitamin D hormone.

5. INVESTIGATING THE BIOCHEMICAL NATURE OF VITAMIN D RESISTANCE IN NEW WORLD PRIMATES

In order to answer the above question, cultured fibroblasts and immortalized cell lines from both resistant and hormone-responsive New and Old World primates were used to track, step by step, the path taken by the vitamin D hormone from the serum vitamin D-binding protein (DBP) in the blood in route to the nucleus and transactivation of hormone-responsive genes *(24–33)* (Fig. 4A). It was determined that the movement of hormone from DBP across the cell membrane and through the cell cytoplasm and nuclear membrane in New World primate cells was indistinguishable from that observed in Old World primate cells. It was also determined that the ability of the New World primate VDR to bind to $1,25(OH)_2D_3$ or $1,25(OH)_2D_2$ and induce receptor dimerization with the retinoid X receptor (RXR) was normal. In fact, when removed from the intranuclear environment and in distinction to previous reports *(23),* the VDR in New World primates was similar to the Old World primate VDR in all biochemical and functional

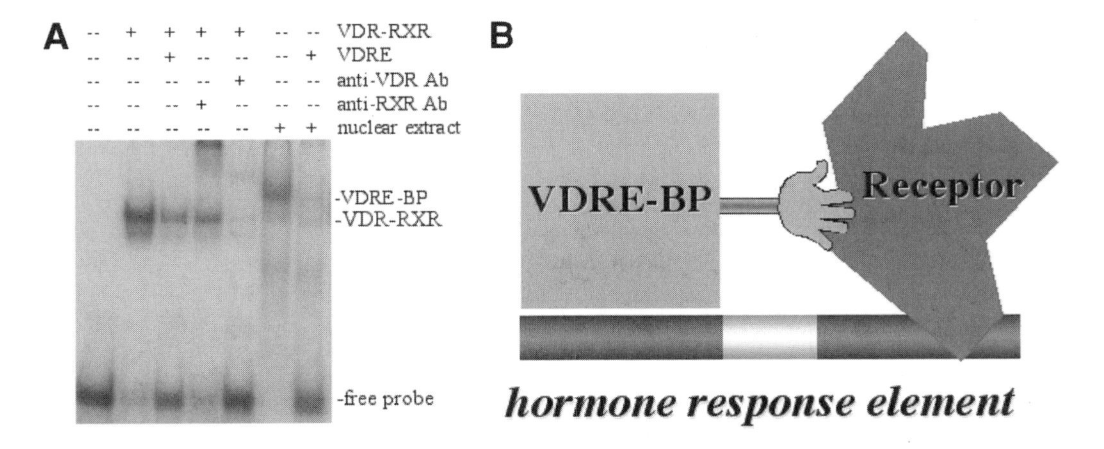

Fig. 5. Evidence for the dominant-negative action of the New World primate vitamin D response element-binding protein (VDRE-BP). **A** is an electromobilty shift assay using consensus vitamin D response element as probe, showing the presence of a second trans-binding protein, in addition to the vitamin D receptor (VDR)–retinoid X receptor (RXR), in nuclear extracts of vitamin D-resistant New World primate cells. Addition of excess unlabeled VDRE is shown in lanes 3 and 7. Addition of anti-RXR*alpha* antibody (lane 4) supershifts, while addition of either anti-VDR antibody (lane 5) or New World primate nuclear extract containing the VDRE-BP (lane 6) competes away probe–VDR–RXR binding; these data are reprinted with permission of the authors from ref. *34.* **B** is a cartoon representing the proposed competition for binding to the VDRE between the VDRE–BP and VDR (receptor).

respects *(31)*. What was not the same in New World primate cells (*see* Fig. 4B) was the reduced ability of VDR–RXR complex to bind to its cognate cis element and transact. In addition to this failure was the apparent buildup of hormone in the cytoplasm of the New World primate cell.

In order to elucidate nuclear receptor events in New World primate cells, the nuclei of New World primate cells were isolated and extracted. In addition to the VDR–RXR, it was determined that these extracts contained a second protein that was bound by the VDRE. This protein was coined the vitamin D response element-binding protein (VDRE-BP) *(34)*. In electromobility shift assay (EMSA) using the VDRE as probe (Fig. 5A), Old World primate cell extract contained only the VDR–RXR bound to the VDRE probe, while the New World primate extract contained two probe-reactive bands, one compatible with the VDR–RXR and a second, more pronounced VDRE–BP–VDRE band. This VDRE–BP–VDRE binding reaction was specific, as the VDRE–BP was competed away from VDRE probe by the addition of excess unlabeled VDRE. These data suggested that VDRE–BP might function as a dominant-negative inhibitor of receptor-response element binding by competing *in trans* with receptor, “knocking it off” the VDRE (Fig. 5B). That was the case. When recombinant human VDR and RXR were per-

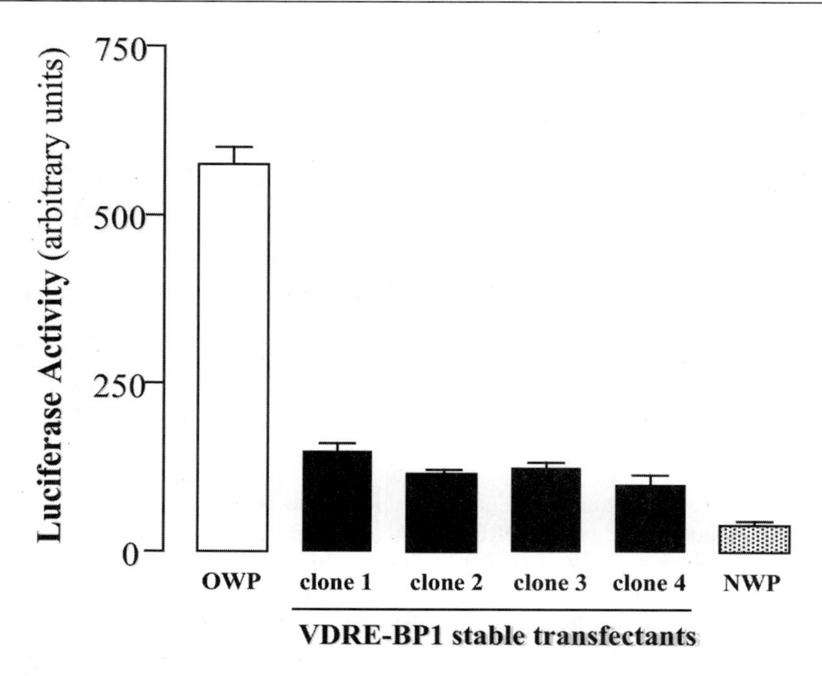

Fig. 6. Vitamin D response element-binding protein-1 (VDRE-BP1) squelched transactivation. Shown is significant squelching ($p < 0.001$) of VDR–RXR-directed, VDRE–reporter-driven transactivation in four different clones of Old World primate (OWP) cells after stable transfection with the New World primate VDRE–BP-1; reporter activity in untransfected New World primate (NWP) cells is shown for comparison. (Data from ref. *34*.)

mitted to interact in EMSA with increasing amounts of nuclear extract from either vitamin D-resistant cells containing a VDRE–BP or from normal vitamin D-responsive cells, the addition of more control extract acted only to amplify the VDR–RXR–retarded probe on the gel. By contrast, increasing amounts of the hormone-resistant extract competed away VDR–RXR–probe binding in favor VDRE–BP–probe binding.

6. VITAMIN D RESPONSE ELEMENT-BINDING PROTEINS

To date three distinct VDRE–BPs, two in New World primates and one in humans, have been identified. All are members of the heterogeneous nuclear ribonucleoprotein A family *(34),* previously considered to be only single-strand mRNA-binding proteins *(35).* However, as pointed out (*see* Fig. 5A), VDRE–BPs can also bind specifically to double-strand DNA. In fact, it is by virtue of their ability to bind DNA that they can be distinguished from traditional co-repressor proteins *(36).* When overexpressed, they can effectively squelch VDR-directed transactivation. This ability to squelch transactivation is shown in Fig. 6. Depicted is VDRE-directed reporter activity in four different subclones of wild-type Old World primate cells stably overexpressing the New World primate VDRE–BP-1 as well as New World primate cells that are naturally hormone-resistant. In all instances, stable overexpression of VDRE–BP-1 squelched VDRE-directed luciferase activity substantially compared to the untransfected, wild-type host cell to levels observed in hormone-resistant New World primate cells that naturally overexpress the protein; VDRE–BP-2 overexpression reduced, but not significantly, VDRE-directed transactivation. These data provide strong confirmatory evidence that when naturally overexpressed in vivo, VDRE–BP-1 is the cause of vitamin D resistance in these monkeys.

7. INTRACELLULAR VITAMIN D-BINDING PROTEINS

On the way to the discovery of the VDRE–BPs in New World primate cells, it was also observed that these cells were extraordinarily efficient at accumulating 25-hydroxylated vitamin D metabolites in the cytoplasmic space (*see* Fig. 4B). Accumulation here was the result of expression of a second set of resistance-associated proteins. These intracellular vitamin D-binding proteins *(37,38),* or IDBPs as they have come to be called, exhibit both high capacity and high affinity for 25-hydroxylated vitamin D metabolites. In fact, among all of the vitamin D metabolites that have been tested, IDBP purified from vitamin D-resistant New World primate cells binds $25(OH)D_3$ and $25(OH)D_2$ best *(26,38);* in a competitive displacement assay using radioinert $25(OH)D_3$ as competitor and $[^3H]25(OH)D_3$ as labeled ligand, the concentration of metabolite required to achieve half-maximal displacement of labeled hormone (EC_{50}) < 1 nM. Although normally present in Old World primate, including human cells, these proteins can be overexpressed some 50-fold in New World primate cells. They are highly homologous to proteins in the heat-shock protein-70 family *(39).* The first three members of this family cloned and characterized by our laboratory, IDBP-1, -2, and -3, bear a high degree of sequence identity with constitutively expressed human heat-shock protein-70, heat-shock-inducible heat-shock protein-70, and mitochondrial-targeted grp-75, respectively. The general domain structure of the IDBPs *(39)* is shown in Fig. 7A. They all contain an ATP-binding-ATPase domain ahead of a protein–protein interaction domain. Some, such as IDBP-3, also harbor an N-terminal organelle-targeting domain. Preliminary studies indicate that the vitamin D ligand-binding domain is in the middle of the molecule *(40).*

What are these IDBPs doing inside the hormone-resistant New World primate cell? Two countervailing hypotheses were considered to explain the function of these proteins (Fig. 7B). One hypothesis held that these IDBPs were "sink" molecules that worked in cooperation with the VDRE–BP in the nucleus to exert vitamin D resistance by disallowing access of the hormone to the VDR and the nucleus of the cell. The opposing hypothesis held that these were "swim" molecules that actually promoted the delivery of ligand to the vitamin D receptor, improving the ability of the VDR to dimerize and bind to DNA, antagonizing the actions of the VDRE–BP that was overexpressed in New World primate cells. In order to determine which of these hypotheses was correct, we

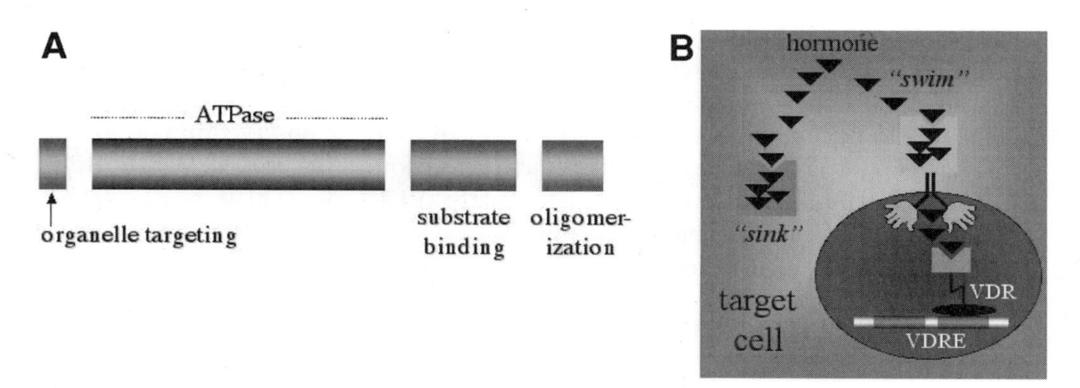

Fig. 7. The hsp-70-related intracellular vitamin D-binding proteins and their proposed function in vitamin D-resistant New World primate cells. **A** shows the general domain structure of these hsp-70-like proteins. They all contain an ATP-binding ATPase domain ahead of two protein–protein interaction domains. Some also harbor an N-terminal organelle-targeting domain. Two countervailing hypotheses for the function of these proteins were considered **(B).** One hypothesis held that these IDBPs were "sink" molecules that worked in cooperation with the response element-binding proteins to exert vitamin D resistance by disallowing access of the hormone to the vitamin receptor (VDR) and the nucleus of the cell. The opposing hypothesis held that these were "swim" molecules, promoting the delivery of ligand to the vitamin D receptor, improving the ability of the VDR to dimerize and bind to the vitamin D response element (VDRE–BP), antagonizing the actions of the vitamin D response element-binding protein that is overexpressed in New World primate cells.

stably overexpressed the most abundant of these IDBPs, IDBP-1, in wild-type Old World primate cells and demonstrated that IDBP-1 imparted protransactivating potential *(40);* the endogenous transcriptional activity of three different 1,25$(OH)_2$D-responsive genes, the vitamin D-24-hydroxylase, osteopontin, and osteocalcin genes, in Old World primate wild-type cells was markedly enhanced (Fig. 8A). It was concluded from these studies, at least for the function of transactivation, that IDBP-1 was a "swim" molecule for the vitamin D hormone, promoting delivery of ligand to the VDR.

Considering the facts that New World primates are required to maintain very high serum levels of 1,25$(OH)_2$D in order to avert rickets (*see* Figs. 2 and 3), it was also hypothesized that the IDBPs, which are known to bind 25(OH)D even better than 1,25$(OH)_2$D, will also promote the synthesis of the active vitamin D metabolite via promotion of the 25(OH)D-1-hydroxylase. Evidence that this is the case is provided in Fig. 8B. When human kidney cells expressing the 25(OH)D-1-hydroxylase gene were stably transfected with IDBP-1 and incubated with substrate 25$(OH)D_3$, 1,25$(OH)_2D_3$ production went up 4–8-fold compared to untransfected wild-type cells *(41).* This increase in specific 25(OH)D-1-hydroxylase activity occurred independent of a change in expression of the 25(OH)D-1-hydroxylase gene *(41).* In fact, current data (R. Chun, unpublished) now strongly

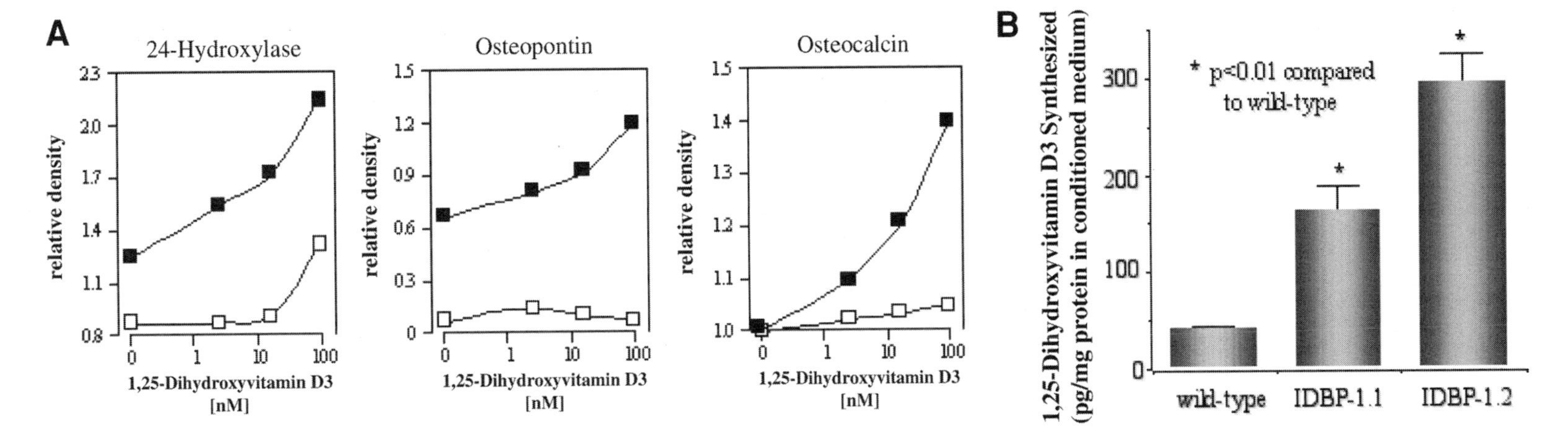

Fig. 8. Consequences of stable overexpression of members of the New World primate intracellular vitamin D-binding protein (IDBP) family in vitamin D-responsive Old World primate cells. **A** depicts the 1,25-dihydroxyvitamin D concentration-dependent relative endogenous expression level, by Northern blot analysis, of three hormone-responsive genes in Old World primate cells before (open squares) and after (filled squares) stable overexpression of IDBP-1. **B** demonstrates the 1,25-dihydroxyvitamin D synthetic capacity of Old World primate (wild-type) before and after and stable overexpression of IDBP-1; IDBP-1.1 and -1.2 represent different clones of cells stably transfected with IDBP-1. (Data from refs. *40* and *41*.)

indicate that this increase in hormone production is the result of the ability of IDBPs to promote the delivery of substrate 25(OH)D to the inner mitochondrial membrane and the 25(OH)D-1-hydroxylase stabled there.

8. A NEW MODEL FOR INTRACELLULAR VITAMIN D TRAFFICKING

Dogma has held that sterol/steroid hormones such as vitamin D, by nature of their lipid solubility, move through the plasma membrane of the target cell and "ping-pong" around the cell interior until they encounter another specific binding protein such as the 25(OH)D-1-hydroxylase or the VDR with which to bind. Our most recent results, developed from a compendium of confocal imaging studies with fluorescently labeled IDBPs and vitamin D metabolites as well as with gst-pull-down, co-immunoprecipitation, and yeast 2-hybrid binding experiments (R. Chun, unpublished), indicate that the hormone does not haphazardly "ping-pong" around the cell interior. Rather, the hormone enters the cell and is distributed to specific intracellular destinations by a series of protein–protein interactions that involve the hsp family of intracellular vitamin D-binding proteins.

For example, it is now know from the work of Willnow and co-workers *(42,43)* that vitamin Ds can enter target cells via internalized vesicles (Fig. 9). The vitamin D stays bound to the serum vitamin D-binding protein (DBP), which is in turn bound by megalin and cubulin, members of the LDL superfamily of proteins. Once inside the cell there is interaction between the C-terminal domain of megalin, which protrudes into the cytoplasm, and the N-terminal domain of at least two different IDBPs, IDBP-1 and -3 (R. Chun, unpublished). If one overexpresses either IDBP-1, the hsc-70 homolog, or IDBP-3, a mitochondrially targeted hsp, and incubates transfected IDBP-overexpressing cells with a fluorescently labeled 25-hydroxylated vitamin D metabolite, one will observe a significant increase in the uptake of the labeled hormone. Moreover, if the protein–protein interaction is between megalin and IDBP-1 and the ligand is $1,25(OH)_2D_3$, then the ultimate destination for that hormone and its chaperone is the unliganded VDR (J. Barsony and R. Chun, unpublished), residing in the perinuclear region of the cell. If, on the other hand, megalin interacts with IDBP-3, which contains an N-terminal targeting sequence for the inner mitochondrial membrane, then the ultimate destination for the hormone is the mitochondria. Confirmation of a protein–protein interaction between a substrate-carrying IDBP molecule and a target enzyme, the 25(OH)D-1-hydroxylase, has been accomplished with gst "pull-down" assays using the carboxy-terminal domain of the 25(OH)D-1-hydroxylase as bait (R. Chun, unpublished); this is the portion of the 25(OH)D-1-hydroxylase that is exposed to the intermembrane space of the mitochondria. Employing this substrate-accessible part of the enzyme to capture 25(OH)D-1-hydroxylase-interacting proteins, it has recently been shown that the grp-75-like IDBP-3, but not the hsc-like IDBP-1, interacts with the 25(OH)D-1-hydroxylase (R. Chun, unpublished).

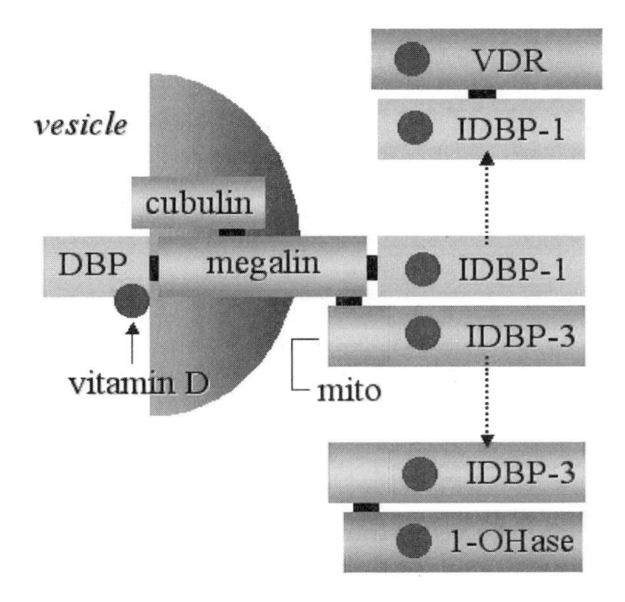

Fig. 9. Proposed roles for the intracellular vitamin D-binding proteins (IDBP-s) in the intracellular trafficking of vitamin D metabolites. Vitamin Ds enter target cells via internalized vesicles. The vitamin D remains bound to the serum vitamin D-binding protein (DBP), which is, in turn, bound by members of the LDL superfamily of proteins, such as megalin and cubulin. Once inside the cell there is the potential for interaction between the C-terminal domain of megalin, which protrudes into the cytoplasm, and the N-terminal domain of at least two different IDBPs, IDBP-1 and -3. It is proposed that if the interaction is with IDBP-1, then there is transfer of the vitamin D to IDBP-1. IDBP-1 can undergo protein–protein interaction with the vitamin D receptor (VDR), resulting in delivery of its vitamin D cargo to the receptor and promotion of transactivation. On the other hand, if the megalin interaction is with IDBP-3, which possesses an amino-terminal mitochondrial targeting sequence (mito), then there exists the potential for the IDBP-3 and its cargo to enter mitochondrial. A protein–protein interaction between the vitamin D-1-hydroxylase and IDBP-3 can result in transfer of the vitamin D cargo to the enzyme for metabolism.

9. CONCLUSION

In summary, it is proposed that these hsp-mediated-chaperoning events are normally active in man but overly active in hormone-resistant New World primates, where they function to compensate for the vitamin D-resistant state by augmenting receptor function and increasing vitamin D hormone synthesis. We anticipate that further analysis of these events will more clearly define the protein-mediated vitamin D trafficking circuits that contribute to a determination of the action and metabolism of vitamin D metabolites in target cells harboring the VDR and/or vitamin D hydroxlases.

ACKNOWLEDGMENTS

This work was supported by National Institutes of Health grants AR37399 and DK58891 to John S. Adams. The authors would like to acknowledge the useful discussions and critiques of this work provided over the years by Dr. Thomas Clemens and the late Dr. Bayard "Skip" Catherwood.

REFERENCES

1. Pilbeam D. The descent of hominoids and hominoids and hominoids. Sci Am 1984; 250:84–96.
2. Bland Sutton JB. Observation on rickets etc. in wild animals. J Anat 1884; 18:363–397.
3. Krook L, Barrett RB. Simian bone disease—a secondary hyperparathyroidism. Cornell Vet 1962; 52:459–492.
4. Hershkovitz, P. Living New World Monkeys (Platyrrhini): With an Introduction to Primates. University of Chicago Press, Chicago, 1977.
5. Hunt RD, Garcia FG, Hegsted DM. A comparison of vitamin D2 and D3 in New World primates. I. Production and regression of osteodystrophia fibrosa. Lab Anim Care 1967; 17:222–234.
6. Steenbock H, Kletzein SWF, Halpin JG. Reaction of chicken irradiated ergosterol and irradiated yeast as contrasted with natural vitamin D of fish liver oils. J Biol Chem 1932; 97:249–266.
7. Marx SJ, Jones G, Weinstein RS, Chrousos GP, Renquist DM. Differences in mineral metabolism among nonhuman primates receiving diets with only vitamin D3 or only vitamin D2. J Clin Endocrinol Metab 1989; 69:1282–1290.
8. Hay AW. The transport of 25-hydroxycholecalciferol in a New World monkey. Biochem J 1975; 151:193–196.
9. Bouillon R, Van Baelen H, De Moor P. The transport of vitamin D in the serum of primates. Biochem J 1976; 159:463–466.
10. Shinki T, Shiina N, Takahashi Y, Tamoika H, Koizumi H, Suda T. Extremely high circulating levels of 1,25-dihydroxy vitamin D3 in the marmoset, a New World Monkey. Biochem Biophys Res Commun 1983; 114:452–457.
11. Adams JS, Gacad MA, Baker AJ, Gonzales B, Rude RK. Serum concentrations of 1,25-dihydroxyvitamin D in Platyrrhini and Catarrhini: a phylogenetic appraisal. Am J Primatol 1985; 9:219–224.
12. Brown GM, Grota LJ, Penney DP, Reichlin S. Pituitary-adrenal function in the squirrel monkey. Endocrinology 1970; 86:519–529.
13. Brandon DD, Markwick AJ, Chrousos GP, Loriaux DL. Glucocoorticoid resistance in humans and nonhuman primates. Cancer Res 1989; 49:2203–2213.
14. Chrousos GP, Brandon D, Renquist DM, et al. Uterine estrogen and progesterone receptors in estrogen and progesterone-resistant primates. J Clin Endocrinol Metab 1984; 58:516–520.
15. Albertson BD, Maronian NC, Frederick KL, et al. The effect of ketoconazole on steroidogenesis. II. Adrenocortical enzyme activity in vitro. Res Commun Chem Pathol Pharmacol 1988; 61:27–34.
16. Albertson BD, Frederick KL, Maronian NC, et al. The effect of ketoconazole on steroidogenesis: I. Leydig cell enzyme activity in vitro. Res Commun Chem Pathol Pharmacol 1988; 61:17–26.
17. Moore CC, Mellon SH, Murai J, Siiteri PK, Miller WL. Structure and function of the hepatic form of 11 beta-hydroxysteroid dehydrogenase in the squirrel monkey, an animal model of glucocorticoid resistance. Endocrinology 1993; 133:368–375.
18. Klosterman LL, Murai JT, Siiteri PK. Cortisol levels, binding, and properties of corticosteroid-binding globulin in the serum of primates. Endocrinology 1986; 118:424–434.
19. Chrousos GP, Renquist DM, Brandon D, Fowler D, Loriaux DL, Lipsett MB. The squirrel monkey: receptor-mediated end-organ resistant to progesterone? J Clin Endocrinol Metab 1982; 55:364–368.

20. Brandon DD, Markwick AJ, Flores M, Dixon K, Albertson BD, Loriaux DL. Genetic variation of the glucocorticoid receptor from a steroid-resistant primate. Mol Endocrinol 1991; 7:89–96.
21. Scammell JG, Denny WB, Valentine DL, Smith DF. Overexpression of the FK506-binding immunophilin FKBP51 is the common cause of glucocorticoid resistance in three New World primates. Gen Comp Endocrinol 2001; 124:152–165.
22. Chrousos GP, Brandon D, Renquist DM, et al. Uterine estrogen and progesterone receptors in an estrogen- and progesterone-"resistant" primate. J Clin Endocrinol Metab 1984; 58:516–520.
23. Liberman UA, de Grange D, Marx SJ. Low affinity of the receptor for 1 alpha,25-dihydroxyvitamin D3 in the marmoset, a New World monkey. FEBS Lett 1985; 182:385–388.
24. Adams JS, Gacad MA, Rude RK, Endres DB, Mallette LE. Serum concentrations of immunoreactive parathyroid hormone in Platyrrhini and Catarrhini: a comparative analysis with three different antisera. Am J Primatol 1987; 13:425–433.
25. Adams JS, Gacad MA, Baker AJ, Keuhn G, Rude RK. Diminished internalization and action of 1,25-dihydroxyvitamin D in dermal fibroblasts cultured from New World primates. Endocrinology 1985; 116:2523–2527.
26. Gacad MA, Adams JS. Evidence for endogenous blockage of cellular 1,25-dihydroxyvitamin D-receptor binding in New World primates. J Clin Invest 1991; 87:996–1001.
27. Gacad MA, Adams JS. Influence of ultraviolet B radiation on vitamin D3 metabolism in vitamin D3-resistant New World primates. Am J Primatol 1992; 28:263–270.
28. Adams JS, Gacad MA. Phenotypic diversity of the cellular 1,25-dihydroxyvitamin D-receptor interaction among different genera of New World primates. J Clin Endocrinol Metab 1988; 66:224–229.
29. Gacad MA, Adams JS. Specificity of steroid binding in New World primate cells with a vitamin D-resistant phenotype. Endocrinology 1992; 131:2581–2587.
30. Gacad MA, Adams JS. Identification of a competitive binding component in vitamin D-resistant New World primate cells with a low affinity but high capacity for 1,25-dihydroxyvitmain D3. J Bone Miner Res 1993; 8:27–35.
31. Chun RF, Chen H, Boldrick L, Sweet C, Adams JS. Cloning, sequencing and functional characterization of the vitamin D receptor in vitamin D-resistant New World primates. Am J Primates 2001; 54:107–118.
32. Arbelle JE, Chen H, Gacad MA, Allegretto EA, Pike JW, Adams JS. Inhibition of vitamin D receptor-retinoid X receptor-vitamin D response element complex formation by nuclear extracts of vitamin D-resistant New World primate cells. Endocrinology 1996; 137:786–789.
33. Chen H, Arbelle JE, Gacad MA, Allegretto EA, Adams JS. Vitamin D and gonadal steroid-resistant New World primate cells express an intracellular protein which competes with the estrogen receptor for binding to the estrogen response element. J Clin Invest 1997; 99:769–775.
34. Chen H, Hu B, Allegretto EA, Adams JS. The vitamin D response element binding proteins: novel dominant-negative-acting regulators of vitamin D-directed transactivation. J Biol Chem 2000; 275:35557–35564.
35. Dreyfuss G, Matunis MJ, Pinol-Roma S, Burd CG. hnRNP proteins and the biogenesis of mRNA. Annu Rev Biochem 1993; 62:289–321.
36. Horwitz KB, Jackson TA, Bain DL, Richer JK, Takamoto GS, Tung L. Nuclear receptors coactivators and corepressors. Mol Endocrinol 1996; 10:1167–1177.
37. Gacad MA, LeBon TR, Chen H, Arbelle JE, Adams JS. Functional characterization and purification of an intracellular vitamin D binding protein in vitamin D resistant New World primate cells: amino acid sequence homology with proteins in the hsp-70 family. J Biol Chem 1997; 272:8433–8440.
38. Gacad MA, Adams JS. Proteins in the heat shock-70 family specifically bind 25-hydroxylated vitamin D metabolites and 17β-estradiol. J Clin Endocrinol Metab 1998; 83:1264–1267.
39. Hartl FU. Molecular chaperones in cellular protein folding. Nature 1996; 3381:571–579.

40. Wu S, Ren S-Y, Gacad MA, Adams JS. Intracellular vitamin D binding proteins: novel facilitators of vitamin D-directed transactivation. Mol Endocrinol 2000; 14:1387–1397.
41. Wu S, Chun R, Ren S, Chen H, Adams JS. Regulation of 1,25-dihydroxyvitamin D synthesis by intracellular vitamin D binding protein-1. Endocrinology 2002; 143:4135–4138.
42. Christensen EI, Willnow TE. Essential role of megalin in renal proximal tubule for vitamin homeostasis. J Am Soc Nephrol 1999; 10:2224–2236.
43. Nykjaer A, Dragun D, Walther D, et al. An endocytic pathway essential for renal uptake and activation of the steroid 25(OH) vitamin D3. Cell 1999; 96:507–515.

27 Vitamin K, Oral Anticoagulants, and Bone Health

Sarah L. Booth and Anne M. Charette

1. SOURCES OF VITAMIN K

Vitamin K is a fat-soluble vitamin that may have a protective role against age-related bone loss. Vitamin K refers to a family of compounds with a common chemical structure, 2-methyl-1,4-napthoquinone (Fig. 1). Phylloquinone, or vitamin K_1, is present in foods of plant origin. Green, leafy vegetables contain the highest content of phylloquinone, and contribute up to 60% of total phylloquinone intake *(1,2)*. Certain plant oils, margarine, spreads, and salad dressings, derived from plant oils, are also important dietary sources of phylloquinone, whereas animal fat sources, such as butter, are not *(3,4)*. Recently, phylloquinone has been added in varying amounts to some dietary supplements. During the process of hydrogenation of certain phylloquinone-rich vegetable oils, phylloquinone is converted to 2′,3′-dihydrophylloquinone, which differs from the parent form by saturation of the 2′,3′ double bond of the phytyl side chain *(5)* (Fig. 1). Dihydrophylloquinone is found exclusively in hydrogenated phylloquinone-rich vegetable oils, which are widely used by industry in food preparation because of their physical characteristics and oxidative stability.

Bacterial and other forms of vitamin K, referred to as the menaquinones or vitamin K_2, differ in structure from phylloquinone in their 3-substituted lipophilic side chain. The major menaquinones contain 4–10 repeating isoprenoid units, indicated by MK-4 to MK-10; forms of up to 13 isoprenoid groups have been identified (Fig. 1). Of the limited data available, MK-6 to MK-9 are found in significant amounts in cow livers, some animal meats, and in foods whose preparation involves bacterial fermentation, such as cheese and fermented soybean products *(6–8)*. Very little is known about the contribution of dietary menaquinones to overall vitamin K nutrition, and although it is a generally held belief that approx 50% of the daily requirement for vitamin K is supplied by the gut flora, there is insufficient experimental evidence to support this conviction *(9)*. Menaquinone-4 (MK-4) is not a major constituent of bacterial production; instead it is alkylated

From: *Nutrition and Bone Health*
Edited by: M. F. Holick and B. Dawson-Hughes © Humana Press Inc., Totowa, NJ

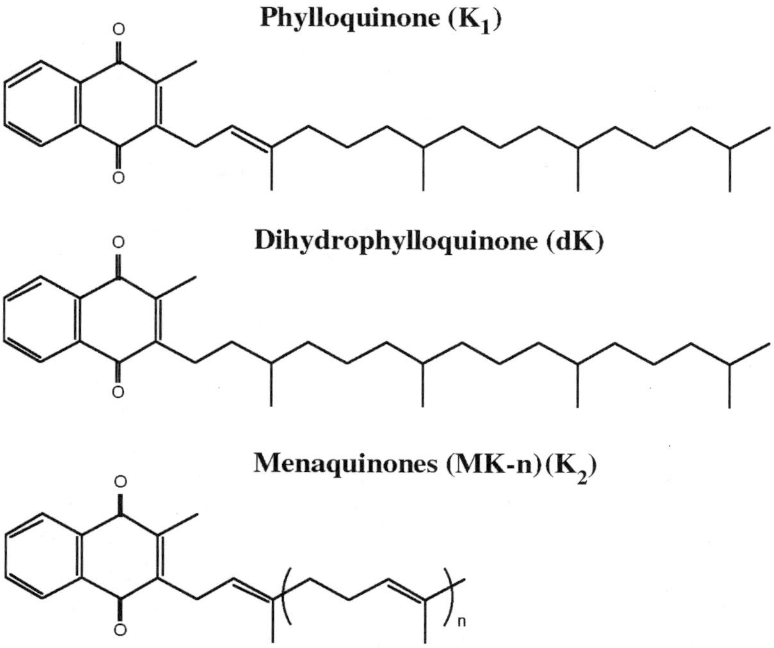

Fig. 1. Structures of Vitamin K.

from menadione present in animal feeds or is the product of tissue-specific conversion directly from dietary phylloquinone *(10)*. Poultry products are the primary dietary sources of MK-4 in the US food supply (unpublished data); MK-4 is also available in therapeutic doses for treatment of osteoporosis in Japan.

Different forms of vitamin K have a tissue-specific distribution. The liver, which is the major storage site for vitamin K, contains long-chain menaquinones (MK-7 through MK-13), with limited capacity for storing phylloquinone *(11)*. In plasma, serum, and bone, the major forms of vitamin K are phylloquinone, followed by short-chain menaquinones, such as MK-4 through MK-8 *(12)*. Vitamin K may be delivered to osteoblasts by triglyceride-rich lipoproteins, although the precise mechanisms are still not well understood *(13)*.

Based on representative dietary intake data, the Adequate Intake (AI) for vitamin K was set at 120 and 90 μg/d for men and women, respectively *(14)*. Recent surveys of dietary phylloquinone intakes in North America, Europe, and Asia indicate that dietary intakes of phylloquinone are lower than previously assumed *(15)*. In the United States and the Netherlands, older adults report higher average phylloquinone intakes (80–210 μg/d) compared to younger adults. However, reported mean phylloquinone intakes in a study among a national sample of British elderly men and women were below the current AI (70 and 61 μg/d, respectively) *(1)*.

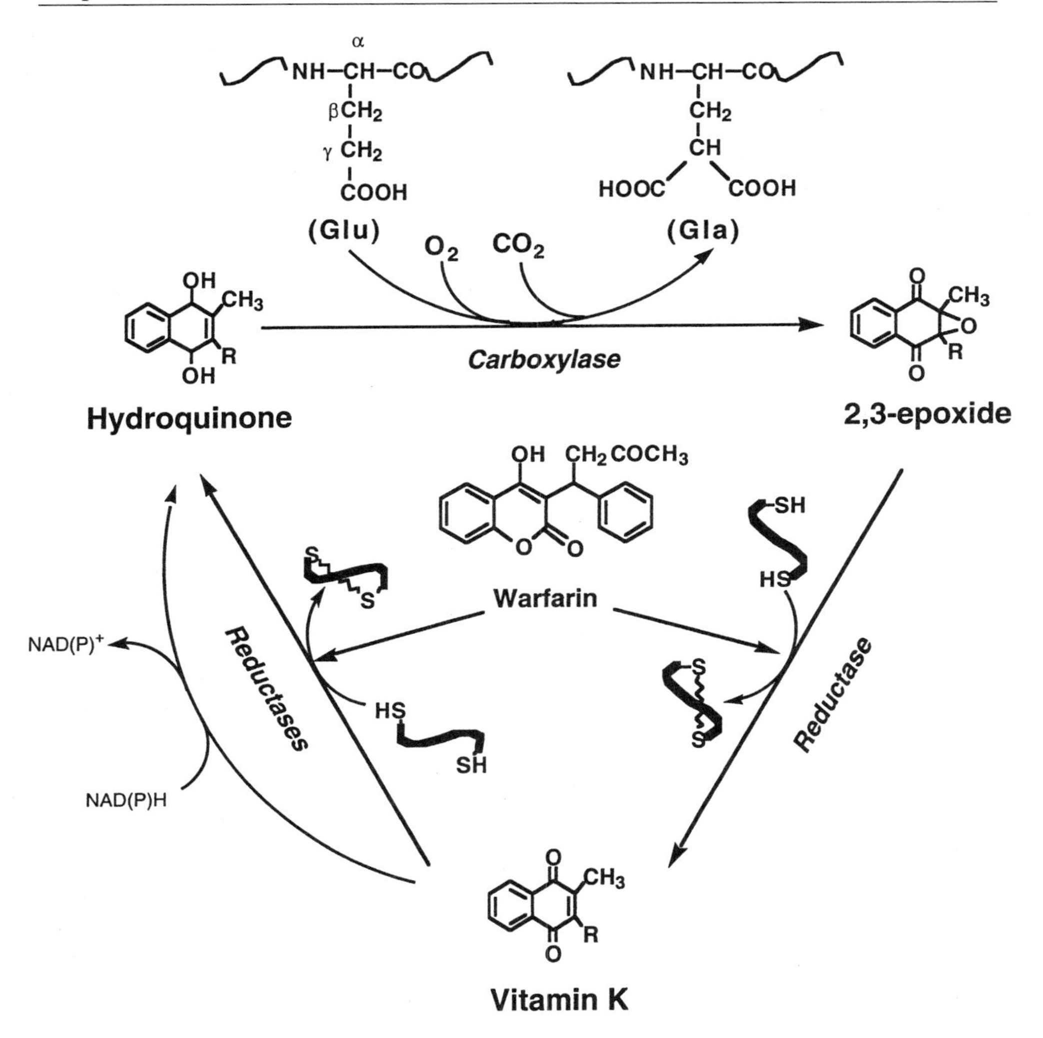

Fig. 2. Vitamin K cycle.

2. POTENTIAL MECHANISMS

The observations that vitamin K and vitamin K-dependent proteins are present in the skeleton have been used to support the hypothesis that vitamin K has a role in bone biology. Vitamin K is a cofactor specific to the formation of γ-carboxyglutamyl (Gla) residues from specific glutamate residues in certain proteins *(16)*. The Gla residues in these vitamin K-dependent proteins, also termed Gla-containing proteins, confer mineral-binding properties *(17)*.

Much of our understanding of the vitamin K cycle, as shown in Fig. 2, is still hypothetical. The naturally occurring forms of vitamin K are quinones (i.e., phylloquinone and menaquinones), so vitamin K must be reduced to the vitamin K hydroquinone by various reductases prior to catalyzing the γ-carboxylation reaction *(16)*. Once formed, vitamin K epoxide can, after enzymatic reduction, be recycled as a cofactor for the vitamin K-dependent γ-glutamyl carboxylase. The reductase that operates at normal tissue concentrations of vitamin K is the dithiol-dependent vitamin K epoxide reductase *(18)*. This warfarin-sensitive enzyme reduces the epoxide as well as the quinone, and is essential to the recycling of vitamin K. An unrelated vitamin K-reducing enzyme, the NAD(P)H dehydrogenase, DT-diaphorase, is relatively insensitive to warfarin and other coumarin-based compounds. This enzyme reduces the quinone form of vitamin K to the hydroquinone when vitamin K is present in high tissue concentrations, but is unable to reduce the vitamin K epoxide and therefore has limited capacity to participate in the recycling of vitamin K.

Because the only known function of vitamin K is as a posttranslational cofactor for the vitamin K-dependent γ-carboxylation reaction, it is currently assumed that any putative role of vitamin K in age-related bone loss is mediated through the carboxylation of vitamin K-dependent proteins in bone. At least three vitamin K-dependent proteins are present in bone and cartilage: osteocalcin, matrix Gla protein (MGP), and protein S *(19)*.

Osteocalcin is a small 49-amino acid protein produced by osteoblasts during bone matrix formation, and is one of the most abundant noncollagenous proteins in bone *(20)*. Osteocalcin synthesis is regulated by, but not absolutely dependent on, 1,25-dihydroxyvitamin D, and its hydroxyapatite-binding capacity is associated with the vitamin K-dependent γ-carboxylation of its three glutamate residues (residues 17, 21, and 24). Osteocalcin is thought to act as a regulator of bone mineral, although its precise physiological function has yet to be elucidated *(21)*. The degree of carboxylation of Glu residues in osteocalcin obtained from human adult bone has been shown to be incomplete, particularly at the first potential Gla residue at position 17 *(22)*. The Gla residue at position 17 is essential for the conformation that facilitates the selective binding of osteocalcin to hydroxyapatite *(23)*, such that partially carboxylated osteocalcin may have reduced binding to the hydroxyapatite.

MGP is present in cartilage and bone *(24)*. Although there is some sequence homology between MGP and osteocalcin, and the synthesis of both are stimulated by 1,25-dihydroxyvitamin D, these proteins differ by 80% of their amino acid sequence, and their solubility in water. MGP also appears earlier in calcification than osteocalcin and binds to both the organic and mineral components of bone, whereas osteocalcin is anchored to bone exclusively by its binding to hydroxyapatite. Whereas osteocalcin can only be detected in bone, other tissues, such as kidney, lung, and spleen, synthesize MGP. In a knockout mouse model, MGP-deficient mice have abnormal calcification, leading to osteopenia, fractures, and premature death owing to arterial calcification *(25)*.

Protein S, a protein cofactor for the anticoagulant activities of protein C, is also synthesized by bone. Osteopenia has been reported among children with inherited protein S deficiency *(26)*. Protein S has significant homology to a vitamin

K-dependent growth arrest-specific gene product (gas6), which has been identified in chondrocytes and shown to regulate osteoclast activity *(27)*. Although multiple vitamin K-dependent proteins have been identified in the skeleton, analytical developments in biological markers of vitamin K status have been limited to osteocalcin.

Vitamin K may also have a role in bone health as mediated through a mechanism other than carboxylation of vitamin K-dependent proteins, as proposed by investigators based on in vitro studies using MK-4 *(28–30)*. Whereas the active site for the carboxylation reaction is on the napthoquinone ring, which is identical in phylloquinone and MK-4 (Fig. 1), MK-4 differs structurally from phylloquinone in its side-chain configuration. In vitro studies indicate that MK-4 has greater capacity to enhance in vitro mineralization *(30)* and decrease bone resorption *(28,29)*, compared to equivalent doses of phylloquinone. Simultaneous exposure to the vitamin K antagonist, warfarin, does not effect MK-4 capacity to reduce bone resorption, consistent with the hypothesis that there is a vitamin K effect on bone that is independent of carboxylation. Takeuchi et al. *(31)* demonstrated that MK-4 inhibits osteoclast formation via its direct effects on the ODF/RANKL expression in bone marrow cells, thereby suppressing bone resorption. In contrast, phylloquinone had no effect in the same model system, lending further support to the hypothesis that the side chain of MK-4 has an effect on bone, independent of γ-carboxylation.

3. EVIDENCE OF ROLE IN BONE HEALTH

3.1. Animal Models

Evidence from animal studies to support a role of vitamin K in bone health is equivocal, in part due to the appropriateness of the animal models used. The use of the rat model to study the effects of vitamin K deficiency on bone health has been problematic, as reviewed elsewhere *(32)*. Feces contain significant amounts of menaquinones such that coprophagy has to be eliminated in order to create a true vitamin K-deficient state, either by use of anal cups or use of germ-free rats. Prevention of coprophagy coupled with consumption of a very low vitamin K-containing diet results in severe vitamin K deficiency, as indicated by abnormal coagulation. To overcome the challenge of conserving normal coagulation while depleting vitamin K in bone, Price and colleagues developed a model in which rats are simultaneously maintained on high doses of vitamin K and the vitamin K antagonist, warfarin *(33)*. This model is based on the premise that vitamin K has a differential ability to counteract the effect of warfarin, depending on the tissue. Whereas vitamin K has a well-established ability to counteract the effect of warfarin in the synthesis of blood coagulation factors by the liver, vitamin K does not have equivalent ability to counteract the effect of warfarin on the synthesis of vitamin K-dependent proteins in extrahepatic tissues, such as bone. Although the mechanism underlying this observation is controversial *(34,35)*, this joint administration of warfarin and vitamin K continues to be refined as a useful model to study the effects of carboxylation of extrahepatic vitamin K-dependent proteins *(36)*.

Use of this warfarin and vitamin K model has yielded inconsistent results in terms of bone health, as reviewed elsewhere *(37)*. Warfarin treatment has been reported to have no effect on fracture healing, bone strength, or overall bone and tooth morphology in rats by some investigators *(24,38,39)*. In contrast, warfarin treatment has been reported to decrease bone volume and femoral neck strength in rats *(40)* and lambs *(41)*. Some of these discrepancies may reflect subtle differences in study design, including the degree to which coprophagy was limited, the doses of warfarin and vitamin K used to create a deficiency state, the age of the animal studied, and the animal model used. The effect of warfarin on osteocalcin also appears to be species-specific *(37)*. Normal concentrations of osteocalcin in humans are less than 5% of rats *(24)*, such that in the presence of warfarin, which results in a decrease in femur osteocalcin in rats, the circulating osteocalcin concentrations are still greater than those of healthy humans not on warfarin *(39)*. One study reported that short-term vitamin K deficiency and subsequent repletion did not affect the percentage of circulating undercarboxylated osteocalcin in rats *(42)*, which is in contrast to the response in humans *(15)*, but is consistent with an earlier rat study in which there was no change in the carboxylation of osteocalcin in fracture calluses, despite being fed a vitamin K-deficient diet *(38)*.

A more recent line of study in animals has been the combined effect of vitamins D and MK-4 on bone. A combined intake of vitamins D, calcium, and MK-4 has been shown to significantly increase peak bone mass in juvenile rats compared to intake of each of the individual nutrients *(43)*. Among ovariectomized rats, menaquinone-4 alone *(44)* and menaquinones-4 plus vitamin D *(45)* suppressed ovariectomy-induced bone loss. Whether these results reflect an additive effect of vitamins D and MK-4 through independent actions or a synergistic effect of vitamin D on vitamin K-dependent γ-carboxylation and/or modulation of osteocalcin synthesis remains unknown *(12)*.

3.2. Observational Studies

In humans, most of the evidence in support of a role for dietary vitamin K and bone health has been based on associations between dietary intakes or biochemical markers of vitamin K status and age-related bone loss, as measured by bone mineral density (BMD) or fracture rate (Table 1).

Undercarboxylated osteocalcin (ucOC) has been used in several epidemiological studies to demonstrate a positive association between vitamin K status and BMD *(46–48)*. In prospective and nested case-control studies, an association between baseline ucOC and risk of hip fracture has been reported among elderly institutionalized *(49,50)* and free-living women *(47)*, respectively. Luukinen et al. *(51)* reported an inverse association between a low ratio of carboxylated osteocalcin:total osteocalcin (consistent with a high undercarboxylated osteocalcin) and hip fracture risk in men and women.

In the study by Szulc et al. *(49)*, there was an inverse association between ucOC and serum 25-hydroxyvitamin D [25(OH)D] concentrations, and a subsequent

Table 1
Summary of Selected Observational Studies of Vitamin K and Bone Health

Author (Reference)	*Population*	*Design*	*Findings*
Undercarboxylated osteocalcin (ucOC) and hip fracture (hip fx) risk:			
Szulc et al., 1993 *(49)*	195 institutionalized women (70–101 yr)	18-mo follow-up: ucOC and hip fx	RR for elevated ucOC and hip fx: 5.9, 99.9% CI: 1.5, 22.7
Szulc et al., 1996 *(50)*	183 institutionalized women (70–97 yr)	3-yr follow-up: ucOC and hip fx	OR for elevated ucOC and hip fx: 2.6, 95% CI:1.05, 6.4
Vergnaud et al., 1997 *(47)*	359 women (mean age: 82 yr)	Nested case-control: ucOC & hip fx	OR for elevated ucOC and hip fx: 1.8, 95% CI: 1.0–3.0
Luukinen et al., 2000 *(51)*	792 men and women (mean age: 75 yr)	5-yr follow-up: ucOC[a] & hip fx	HR for elevated ucOC and hip fx: 5.32, 95% CI: 3.26–8.68
Dietary phylloquinone (VK) and hip fx risk:			
Feskanich et al., 1999 *(52)*	72, 327 women (38–63 yr)	10-yr follow-up: VK intake and hip fx	RR for high VK intake and hip fx: 0.70, 95% CI: 0.53, 0.93
Booth et al., 2000 *(54)*	900 men and women (mean age: 75 yr)	7-yr follow-up: VK intake and hip fx	RR for high VK intake and hip fx: 0.35, 95% CI: 0.13, 0.94
UcOC, plasma VK and MK-7, and bone mineral density (BMD):			
Szulc et al., 1994 *(46)*	98 institutionalized women (mean age: 81 yr)	Cross-sectional: ucOC and BMD	Partial correlation between hip BMD & ucOC = –0.38
Knapen et al., 1998 *(48)*	212 women (20–90 yr)	Cross-sectional: ucOC and BMD	Correlation coefficient between low hip BMD & ucOC = 0.5–0.7
Kanai et al., 1997 *(81)*	71 postmenopausal women (mean age: 54 yr)	Cross-sectional: plasma VK, MK-7, & BMD	Plasma VK & MK-7 significantly lower among women with low BMD
Dietary VK and BMD:			
Booth et al., 2000 *(54)*	888 men and women (mean age: 75 yr)	Cross-sectional: VK intake and BMD	No association
Booth et al., 2003 *(56)*	2591 men and women (mean age: 59 yr)	Cross-sectional: VK intake and BMD	In men, no association; in women, high dietary VK intakes are associated with high BMD

[a] Authors report ucOC as a low-carboxylated osteocalcin:total osteocalcin ratio (COC:TOC); however, for the purposes of consistency, a low COC:TOC is reported as a high ucOC in this table.

RR, relative risk; OR, odds ratio; HR, hazards ratio; CI, confidence interval.

reduction of ucOC by vitamin D supplementation. These findings cannot be explained by our current understanding of the biochemical role of vitamin K, but suggest that vitamin D may influence the ucOC level. In light of the animal data that describe a combined effect of vitamins D and K on bone health in juvenile *(43)* and ovariectomized rats *(45)*, these observational data in humans highlight the need to consider vitamin D status when examining the associations between vitamin K and age-related bone loss.

There are several reports of statistically significant associations between low dietary phylloquinone intakes and hip fracture risk, notably among women *(52–54)* (Table 1). Low dietary intakes of MK-7 have also been associated with a higher risk for hip fracture among postmenopausal Japanese women *(55)*. In contrast, the associations between dietary phylloquinone and BMD are equivocal. Among elderly men and women, low dietary phylloquinone intakes were not associated with low BMD (either cross-sectionally or prospectively), even though there was a positive association with hip fracture risk *(54)*. However, among a younger cohort of men and women (mean age: 59 yr), women in the lowest quartile of phylloquinone intake had significantly lower BMD at the femoral neck and spine compared to those in the highest quartile of phylloquinone intake *(56)*. These associations remained after controlling for potential confounders, such as dietary potassium, and stratification by age or supplement use. However, no significant association was found between dietary vitamin K intake and BMD among men.

Gender-specific differences in response to vitamin K depletion, supplementation, or antagonism have not been reported in metabolic studies *(57–59)*. Furthermore, at least two studies report an inverse association between vitamin K status and hip fracture risk in men and women *(51,54)*. When stratified by estrogen replacement use, there was a suggestion that phylloquinone intake had a protective effect on hip fracture risk only among postmenopausal women reporting no estrogen use *(52)*. This observation is consistent with the observation that MK-4 suppressed ovariectomy-induced bone loss in rats *(44)*. However, it is not known from the limited data available if there are gender-specific and/or hormonal influences on any putative role of vitamin K in the prevention of age-related bone loss.

While suggestive of a role of poor vitamin K status in age-related bone loss, one consistent criticism of the observational data has been the potential confounding effect of overall poor nutrition. A high phylloquinone intake may simply be a marker for an overall healthy diet that includes high fruit and vegetable consumption. In addition, high intakes of alkaline-producing foods, specifically fruits and vegetables, and their associated minerals, potassium and magnesium, have been associated with greater BMD *(60,61)*. Likewise, soybean is a rich dietary source of MK-7; it is also a rich source of isoflavones, also associated with reduced bone loss *(62)*. By the nature of their study design, observational studies preclude the ability to isolate the effects of a single nutrient on bone health from those of the dietary patterns associated with high intakes of food(s) rich in that nutrient. Therefore, the associations between low phylloquinone or MK-7 intakes and low

BMD or increased risk of hip fracture may be suggestive of causality or simply consistent with an overall healthy diet.

3.3. Clinical Trials

There is very little information available concerning the long-term effect of vitamin K supplementation on age-related bone loss and fracture rate. Age-related bone loss at the spine was reduced among postmenopausal women in response to MK-4 supplementation compared to controls *(63)*. In a 24-mo randomized, open-label study among 241 postmenopausal osteoporotic women, MK-4 supplementation increased lumbar bone mineral density and reduced occurrence of new fractures compared to a control group *(64)*. When combined with vitamin D_3, administration of MK-4 partially reduced bone loss in postmenopausal women *(65)* and women with induced estrogen deficiency as treatment for estrogen-dependent diseases *(66)*. While consistent in the observation that MK-4 can reduce bone loss, which is enhanced with concurrent vitamin D supplementation, these studies need to be interpreted with caution when extrapolating to the potential nutritional role of vitamin K in bone health. These investigators used doses of 45 mg of MK-4/d, in contrast to the current AI of 90–120 μg of phylloquinone/d *(14)*. Pharmacological doses of MK-4 may have an antiresorptive effect on bone that is mediated through a modulation of cellular differentiation by its geranylgeranyl side chain, and independent of its effect on γ-carboxylation of vitamin K-dependent proteins *(31)*. That ucOC decreased in response to MK-4 supplementation in the study by Shiraki and colleagues *(64)* suggests that there is a change in the γ-carboxylation of vitamin K-dependent proteins in response to MK-4 supplementation. However, the extent to which the γ-carboxylation of those proteins in bone contributed to the observed reduction in age-related bone loss cannot be determined based on the current published studies using MK-4 as a supplement.

At the time of this writing, the results of only one long-term clinical trial investigating phylloquinone supplementation on bone mineral density were available. Braam et al. *(67,68)* reported that 3-yr intake of a dietary supplement containing calcium, vitamin D, and 1 mg/d of phylloquinone significantly reduced bone loss at the femoral neck in postmenopausal women, aged 50–60 yr, compared to a placebo or a supplement containing calcium and vitamin D. In contrast, no beneficial effect of phylloquinone was observed in lumbar BMD. While encouraging, more clinical trials need to be completed before conclusions can be made regarding the efficacy of phylloquinone supplementation in reducing age-related bone loss.

4. WHAT IS THE OPTIMAL DIETARY INTAKE OF VITAMIN K FOR BONE HEALTH?

4.1. Phylloquinone Intakes

An AI for vitamin K was set at 120 and 90 μg/d for men and women, respectively, because there were insufficient data in which to establish an average requirement *(14)*. The only defined clinical indication of vitamin K deficiency is vitamin K-responsive hypoprothrombinemia, as indicated by prolonged coagulation times. Therefore, determination of optimally nutritional intakes of vitamin K has been difficult to establish, particularly in the context of its putative role in bone health. Adding further complexity is the possibility that requirements for vitamin K in bone may be higher than those in liver *(69,70)*.

In an attempt to establish dietary vitamin K requirements, there have been numerous human studies that have evaluated the short-term effects of dietary manipulation of phylloquinone on measures of vitamin K status, and more recently, bone turnover. A selection of these studies is summarized in Table 2. Several measures of vitamin K status have not been included in this table, including coagulation times, for which there is no evidence of change in response to mild vitamin K deficiency or antagonism among healthy adults.

Plasma and serum phylloquinone concentrations consistently decrease at the population level in response to dietary intakes below 100 μg/d and increase at intakes above 200 μg/d (Table 2). Liver phylloquinone concentrations also appear to be rapidly depleted in response to short-term phylloquinone restriction *(71)*. However, there are large intra- and interindividual variations in dietary phylloquinone intakes, with concomitant large variations at the individual level in response of plasma and serum phylloquinone concentrations to various levels of intakes *(72)*. Some of this variation in circulating concentrations is also attributable to nondietary factors, such as season of plasma and serum collection, plasma and serum triglyceride concentrations, and smoking status *(2)*. Therefore, it is difficult to estimate the optimal dose of phylloquinone, in response to which plasma and serum phylloquinone concentrations are stable.

There are widely divergent reports of the absolute percentage of osteocalcin that is undercarboxylated. How the degree of carboxylation is assessed is dependent on the amount of osteocalcin in the sample and the amount of hydroxyapatite used for binding *(73)*. The current methodology precludes exact determination of undercarboxylated osteocalcin; instead it is a relative term, with the direction change being critical and not the absolute percentages. As summarized in Table 2, there is a consistent increase in the percentage of undercarboxylated osteocalcin (%ucOC) in response to dietary intakes of phylloquinone approx 10 μg/d *(74)*, and a consistent decrease in response to vitamin K supplementation *(57,59,74–79)*. There was one report of an age-dependent effect, with no change in %ucOC in response to vitamin K supplementation among younger women *(80)*; however, this may be a spurious finding because it has not been substantiated in other studies. In one dose-response study among healthy young adults *(79)*, a daily intake of 1000 μg/d of phylloquinone was required to maximally γ-carboxylate osteocalcin, which is approx 10× the current AI for vitamin K *(14)*. Although it is assumed that maximal carboxylation is optimal for bone health, the physiological relevance of maximal carboxylation of osteocalcin is not known. Therefore, it is premature to extrapolate the results of this single dose–response study to the determination of

Table 2
Response of Biomarkers of Vitamin K Status and Bone Turnover to Phylloquinone (VK) Restriction and Supplementation

			Biomarkers[a]				
Author (Reference)	*Subjects*	*Intervention*	*VK*	*% ucOC*	*Total OC*	*Bone Resorption*[b]	*Urinary Calcium*
Knapen et al., 1989 *(80)*	100 women 25–40 yr; 55–75 yr	14 d @ 1mg/d	n/a[c]	↓ (55–75 yr) no Δ (24–40 yr)	↑ (55–75 yr) no Δ (24–40 yr)	n/a	no Δ[d]
Knapen et al., 1993 *(75)*	145 women 20–85 yr	14 d @ 1 mg/d[e]	n/a	↓	↑	n/a	[d]
Douglas et al., 1995 *(76)*	10 women 52–73 yr	(a) 14 d @ 1 mg/d[e]	n/a	↓	no Δ	no Δ	no Δ
		(b) 14 d @ 1 mg/d + 400 IU vit D[e]	n/a	↓	↑	no Δ	no Δ
Cracuin et al., 1998 *(77)*	8 women mean age: 35 yr	30 d @ 10 mg/d[e]	n/a	↓	n/a	↓	↓
Booth et al., 1999 *(59)*	36 men and women 20–40 yr; 60–80 yr	5 d @ 400 μg/d[e]	↑	↓	no Δ	n/a	n/a
Binkley et al., 2000 *(57)*	112 men and women 18–30 yr; ≥65 yr	14 d @ 1000 μg/d[e]	↑	↓	↓	no Δ	n/a
Booth et al., 2001 *(74)*	15 men and women 20–40 yr	(a) 15 d @ 10 μg/d[e]	↓	↑	↑	↑	no Δ
		(b) 10 d @ 200 μg/d[f]	↑	↓	↓	↓	no Δ
Binkley et al., 2002 *(79)*	10 men and women 19–36 yr	7 d @ 500–2000 μg/d[e]	n/a	↓	no Δ	n/a	n/a
	100 men and women mean age: 26 yr	14 d @ 250 μg/d[g]	↑	↓	no Δ	n/a	n/a
		14 d @ 375–1000 μg/d[g]	↑	↓	no Δ	n/a	n/a

[a] No attempt has been made to standardize the specific assays used for each biomarker.

[b] May include urinary NTx, CTx, or DPD.

[c] n/a = not available.

[d] Difficult to interpret, as subjects were divided into two groups based on urinary calcium losses prior to statistical analysis.

[e] Compared to baseline.

[f] Compared to depletion period.

[g] Compared with a placebo group.

dietary vitamin K requirements for bone health. However, a comparable study among elderly would shed light on any age-specific differences in the response of %ucOC to vitamin K supplementation.

There is a current lack of consensus regarding vitamin K supplementation and its effect on total osteocalcin. Because vitamin K is a posttranslational cofactor for the γ-carboxylation of vitamin K-dependent proteins, total osteocalcin concentrations are not markers of carboxylation status. In vitro studies confirm that vitamin K has no effect on osteocalcin synthesis, although it may increase osteocalcin accumulation within the osteoblasts *(30)*. Instead, total osteocalcin is a common measure of bone formation. As summarized in Table 2, total osteocalcin has been reported to increase *(75)*, decrease *(57,74)*, and remain unchanged *(79)* in response to vitamin K supplementation. In one study, total osteocalcin increased in response to a combination of vitamins D and K, but did not change in response to vitamin K supplementation alone *(76)*. Others have suggested an age effect, with a response of total osteocalcin to vitamin K supplementation limited to postmenopausal women *(80)*. Because different antibodies are used for measurement of total osteocalcin among the different studies, the discrepancy in results may also be an artifact of the variation in affinities of different antibodies for the carboxylated form of osteocalcin *(82)*.

More recently, investigators have examined the effects of manipulating dietary vitamin K on bone resorption markers (Table 2). To date, there is a lack of consensus on the response of bone resorption markers on vitamin K restriction or supplementation. One study reported a modest response of both total OC and NTx to short-term vitamin K depletion and repletion among young adults *(74)*. Another study reported a response in bone turnover to vitamin K supplementation among female athletes *(83)*, whereas others have reported no change *(57,76)*. Significant changes in markers of bone turnover in response to some bone-specific pharmacological agents are only observed after 1 mo of treatment *(84)*, which may explain the lack of response to short-term phylloquinone supplementation. Braam *(68)* reported that bone turnover, as measured by total osteocalcin and bone-specific alkaline phosphatase, did show a transient increase in the first 12 mo among a vitamin D-, calcium-, and vitamin K-supplemented group, but returned to baseline levels by 36 mo. Conversely, there is considerable interlaboratory variation in bone turnover marker values, and the current lack of standardization may have considerable impact on the ability to detect changes in bone marker values in response to treatments *(85)*.

There has been some suggestion that vitamin K intake influences urinary calcium excretion in humans *(86)*, consistent with rat studies in which urinary calcium excretion increased in response to vitamin K deficiency *(87)*. Among female athletes, there was a significant decrease in urinary calcium in response to short-term vitamin K supplementation *(83)*. Other studies have reported a reduction in calcium excretion in response to vitamin K supplementation among women classified as "fast calcium losers," based on their baseline urinary calcium:creatinine values *(75,80)*. That an effect of vitamin K on urinary calcium loss may be limited to this subset of study subjects, and not applicable to the study group as a whole *(80)*, is supported by the lack of association reported among studies that did not classify subjects based on baseline urinary calcium: creatinine (Table 2). Likewise, in the only phylloquinone-supplementation trial published to date, there were no changes in urinary calcium:creatinine excretion in response to vitamin K supplementation *(68)*.

4.2. Other Dietary Forms of Vitamin K

Vitamin K-dependent reactions are related to both the length and the isomeric configuration of the side chain *(88)*. As previously discussed under Subheading 2., the side chain of MK-4 also appears to confer a unique bone resorptive effect when administered in pharmacological doses, independent of γ-carboxylation. Metabolic studies comparing the absorption and biological activity of phylloquinone with other dietary forms of vitamin K, such as MK-4 and MK-6, at intakes that can be achieved in the diet have yet to be conducted. Comparison of plasma concentrations of phylloquinone and dihydrophylloquinone following dietary repletion suggest that dihydrophylloquinone is not as well absorbed as phylloquinone *(74)*. The results of this single study also suggest that carboxylation of hepatic proteins was partially conserved, whereas carboxylation in the extrahepatic vitamin K-dependent proteins, such as osteocalcin, was not conserved when dihydrophylloquinone was the exclusive form of vitamin K consumed following short-term phylloquinone restriction. By implication, the presence of dihydrophylloquinone in margarines and processed foods containing hydrogenated phylloquinone-rich plant oils may reduce the contribution of these food sources to overall dietary vitamin K available for γ-carboxylation of vitamin K-dependent proteins in bone. However, more metabolic studies are required examining the role of dihydrophylloquinone in bone health.

5. ROLE OF ORAL ANTICOAGULANTS IN BONE HEALTH

Sodium warfarin, a 4-hydroxycoumarin compound, belongs to the coumarin class of anticoagulants. Warfarin prevents vitamin K from acting as a cofactor in the synthesis of factors II, VII, IX, and X and proteins C and S. Warfarin has been used since the 1940s to interfere with blood clotting in patients who develop potentially life-threatening blood clots or who are at risk for developing those clots. Patients who develop deep-vein thrombosis, acute myocardial infarction, atrial fibrillation, and those with artificial heart valves are candidates for warfarin treatment. Many of the conditions for which warfarin is indicated occur more often in elderly patients. Length of treatment is variable but can last for several years, depending on the indication and the individual patient profile. The question of whether adequate levels of vitamin K are important for bone health becomes one of greater concern for patients taking warfarin if the hypothesis is accurate that by antagonizing vitamin K, bone density is likely to diminish. Low chronic intakes of vitamin K typically found in US diets combined with treatment with a vitamin K antagonist has the potential for significant effects on bone density *(89)*. In addition to oral anticoagulants disrupting vitamin K function, many health care providers counsel patients who take those medications to avoid vitamin K-rich foods or to maintain a stable vitamin K intake, which frequently results in low vitamin K intake *(89)*. As a corollary, warfarin use may parallel the effects of vitamin K dietary restriction, thereby providing a model for studying the effects of inadequate γ-carboxylation of vitamin K-dependent proteins.

Several studies have attempted to evaluate the effect of warfarin on bone density and fracture risk, but those studies have not reached a consensus, as summarized in Table 3. Carabello et al. *(90)* conducted a cohort study of 572 women, ages 35 yr and older (mean 63.9 ± 15.8 yr), who were treated with an oral anticoagulant for their first thromboembolic event between 1966 and 1990. They assessed the risk of vertebral and rib fractures in those treated women compared with the number of fractures expected in sex- and age-specific fracture incidence rates for the general population; 99% of study subjects were Caucasian, 86% were postmenopausal, and 66% had a history of one or more conditions associated with an increase fracture risk. Of those fractures that occurred in the study subjects, 78% were related to minimal or moderate trauma; vertebral and rib fractures were the only fractures associated with a greater exposure to anticoagulants. After adjusting for other risk factors, these authors found that treatment of 12 mo or more was associated with a twofold increase in vertebral fractures and a 2.1-fold increase in rib fractures. No measurements of vitamin K status were evaluated in this study.

Another study, conducted by Philip et al. *(91)*, also found an association between treatment with oral anticoagulants and bone loss. This team studied 40 Caucasian men who were treated with warfarin for 1 to 302 mo (median 40.5 mo). Study subjects were matched to 40 male controls for age (±5 yr), similar condition (but not treated with an anticoagulant), concomitant diseases, treatment with other medications, and general medical history. All study participants were living at home and had normal activity levels. Disease states and medications known to affect bone density were excluded. The measurement of adequate anticoagulation used was the International Normalized Ratio (INR), which in the treatment group ranged from 1.3 to 3.9 (median 2.65). The INR measurements were the result of averaging two measurements, one taken before and one after the time of bone density measurement. These authors found a statistically significant reduction in bone mineral density in cancellous-rich bone, such as the distal radius (9% reduction) and lumbar spine (10.4 % reduction). No relationship was found between duration of therapy, warfarin dose, or level of anticoagulation and bone mineral density.

Sato et al. *(92)* examined chronic oral anticoagulation effects on both second metacarpals in patients following ischemic stroke. Sixty-four patients with histories of nonrheumatic atrial fibrillation treated with 12 mo or more of warfarin (mean 5.1 yr) were matched with 63 stroke patients who were not treated; all

Table 3
Summary of Selected Studies of Patients on Long-Term Warfarin and Bone Health

Author (Reference)	*Population*	*Design*	*Findings*
Effect of warfarin on BMD:			
Philip et al., 1995 *(91)*	40 men on warfarin *and* 40 men not on warfarin	Case control	↓ BMD at all sites[a]: 9% ↓ at distal radius, 10.4% ↓ at lumber spine
Rosen et al., 1993 *(94)*	50 men and women on warfarin *and* 50 men and women not on warfarin	Case control	No association between warfarin and BMD
Sato et al., 1997 *(92)*	64 stroke patients on warfarin *and* 63 stroke patients not on warfarin *and* 39 control subjects	Case control	BMD is lower in both sides[b] ↓ serum phylloquinone ↓ OC ↓ 25(OH)D[a]
Jamal et al., 1998 *(93)*	6201 elderly postmenopausal women	3.5-yr follow-up	No association between warfarin and BMD
Warfarin and fracture (fx) risk:			
Jamal et al., 1998 *(93)*	6201 elderly postmenopausal women	3.5-yr follow-up	No association between warfarin and hip fx risk
Caraballo et al., 1999 *(90)*	572 women ≥ 35 yr warfarin use for venous thromboembolism	Population-based retrospective cohort study	↑ vertebral and rib fx with warfarin exposure >12 mo[c]

[a] Compared to control group.

[b] Compared to stroke patients not on warfarin and control group.

[c] Compared to non-warfarin users and/or 12 mo or less use of warfarin.

stroke patients were hemiplegic. Exclusions included rheumatic atrial fibrillation, patients who had received any drug known to alter bone metabolism, history of multiple strokes, total disability, or patients with severe renal or hepatic insufficiency. Non-warfarin-treated stroke patients were matched for age, sex, illness duration, severity of hemiplegia, and Barthel index (functional evaluation). In addition, there were 39 healthy controls for comparison. Sera were assayed for phylloquinone, MK-4, osteocalcin, and 25(OH)D. There was a significant decrease in BMD in both the hemiplegic and the intact side among patients treated with warfarin, compared to the untreated stroke patients. Serum measurements of osteocalcin were lower in treated patients than in untreated patients, and serum phylloquinone, but not MK-4 measurements, were lower in the treated group than in untreated patients or controls. Both warfarin-treated and -untreated patients had a lower dietary intake of vitamin K than controls. Measurements of 25(OH)D were lower in both patient groups, which supports the hypothesis that significant differences in BMD between the two groups of patients were attributed to differences in vitamin K status, consistent with a significant association between serum phylloquinone and BMD in the warfarin-treated patients but not in the untreated patients.

In contrast to those findings, Jamal et al. *(93)* found no association between warfarin use and either bone loss or an increase in fracture risk. They conducted a prospective, observational study of 6201 postmenopausal women at four US centers to evaluate warfarin effect on hip and heel bone density. Participants were 65 yr of age and older and were ambulatory; there were 149 warfarin users and 6052 nonusers. Study subjects took warfarin for at least 2 yr and BMD measurements were taken over a 2-yr period. Study participants were also followed for fractures for 3.5 yr. All fractures were confirmed by review of a radiographic report. Self-reported vertebral fractures and fractures from severe trauma were excluded from the analysis. Confounders that were adjusted for in the analysis included age, weight, estrogen use, clinic site, health status, involuntary weight loss, nonthiazide diuretics, frailty, and inability to rise from a chair independently. These authors reported nontraumatic, nonvertebral fractures in 10% of warfarin users and in 9.3% of nonusers. There were no measurements of vitamin K status, such as %ucOc conducted as part of this study.

Similarly, Rosen et al. *(94)* found no association between warfarin treatment and changes in BMD in the hip or spine. These authors compared 50 patients, ages 41–85 yr old (20 men and 30 women) treated with warfarin for more than 1 yr, with 50 age-, sex-, and race-matched untreated controls. Subjects were excluded if they had any medical conditions or were taking any medications that would adversely affect bone density. Potential confounders such as calcium intake, duration of use of warfarin, activity levels, smoking history, and use of thyroid hormone, diuretics, or estrogen were adjusted for in the analysis. In a second phase of this study, measurements were taken from 113 nonanticoagulated adults over the age of 65 yr (35 men and 78 women). Body mass index, BMD, physical activity scores, calcium intake, smoking history, and use of estrogens, diuretics, and thyroid hormone were adjusted for in the analysis. Vitamin K status was determined by measuring plasma phylloquinone and PIVKA-II (Protein Induced by Vitamin K Absence—Factor II or prothrombin). BMD in the hip and spine of treated and untreated patients were similar; furthermore, BMD did not correlate with measurements of vitamin K status among normal subjects. These authors concluded that vitamin K status was not a major factor in determining adult skeletal health. An alternative interpretation of the data might be that because the vitamin K indices were measurements of recent vitamin K consumption, whereas nutrient influences on bone mineral density are long-term, there may have been misclassification of the long-term vitamin K status of subjects in the analysis.

In a meta-analysis of 11 published studies that evaluated the association between bone mineral density and long-term warfarin use, it was concluded that oral anticoagulation was associated with a very modest reduction in bone density in the radius, but no overall change in bone density in the hip and spine *(95)*. If vitamin K mediates its putative effect on bone exclusively through the γ-carboxylation of vitamin K-dependent proteins in bone, then one would predict that warfarin would disrupt the carboxylation reaction and diminish the function of these proteins. Alternatively, if vitamin K has an effect on bone that is independent of γ-carboxylation of vitamin K-dependent proteins, as suggested by in vitro studies *(31)*, chronic warfarin use would not be expected to increase age-related bone loss or fracture risk. Until the mechanisms by which this nutrient can affect bone loss are determined, interpretation of the results from warfarin studies will continue to be controversial.

6. CONCLUSIONS

Vitamin K is a fat-soluble vitamin that may have a protective role against age-related bone loss. It is assumed that any putative role of vitamin K in bone health is mediated through the γ-carboxylation of the vitamin K-dependent proteins present in bone, although other mechanisms of action have been implicated based on in vitro studies. Furthermore, a potential synergistic effect of vitamins D and K on bone has been suggested based on animal and human studies. Much of the evidence in humans is derived from observational studies, the results of which may be confounded by overall poor nutrition. Dietary vitamin K intakes are lower than previously assumed, but in the absence of sufficient data in which to establish an average requirement, optimal intakes of vitamin K have yet to be defined. Evaluation of studies among patients on long-term use of oral anticoagulants has not resulted in a consensus regarding the role of vitamin K antagonism and fracture risk, although to date no study has reported a statistically significant association between chronic warfarin use and risk of hip fracture. The majority of clinical trials reporting a positive effect of vitamin K supplementation in reducing age-related bone loss used pharmacological doses of a form of vitamin K not widely consumed in North America or Europe. Although a role of vitamin K in bone health is biologically plausible, multiple clinical trials are required that use vitamin K in widely available forms and doses attainable in the diet to isolate any putative effects of vitamin K in the prevention of age-related bone loss.

ACKNOWLEDGMENTS

This material is based on work supported by the U.S. Department of Agriculture, under agreement 58-1950-9-001. Any opinions, findings, conclusion, or recommendations expressed in this publication are those of the authors and do not necessarily reflect the view of the U.S. Department of Agriculture.

REFERENCES

1. Thane C, Paul A, Bates C, Bolton-Smith C, Prentice A, Shearer M. Intake and sources of phylloquinone (vitamin K-1): variation with socio-demographic and lifestyle factors in a national sample of British elderly people. Br J Nutr 2002; 87:605–613.
2. McKeown NM, Jacques PF, Gundberg CM, et al. Dietary and non-dietary determinants of vitamin K biochemical measures in men and women. J Nutr 2002; 132:1329–1334.
3. Piironen V, Koivu T, Tammisalo O, Mattila P. Determination of phylloquinone in oils, margarines and butter by high-performance liquid chromatography with electrochemical detection. Food Chem 1997; 59:473–480.
4. Peterson JW, Muzzey KL, Haytowitz D, Exler J, Lemar L, Booth SL. Phylloquinone (vitamin K-1) and dihdyrophylloquinone content of fats and oils. J Assoc Offic Anal Chem 2002; 79:641–646.
5. Davidson K, Booth S, Dolnikowski G, Sadowski J. Conversion of vitamin K-1 to 2′,3′-dihydrovitamin K1 during the hydrogenation of vegetable oils. J Agric Food Chem 1996; 44:980–983.
6. Shearer MJ, Bach A, Kohlmeier M. Chemistry, nutritional sources, tissue distribution and metabolism of vitamin K with special reference to bone health. J Nutr 1996; 126:1181S–1186S.
7. Schurgers L, Geleijnse J, Grobbee D, et al. Nutritional intake of vitamins K1 (phylloquinone) and K2 (menaquinone) in the Netherlands. J Nutr Environ Med 1999; 9:115–122.
8. Sakano T, Notsumoto S, Nagaoka T, et al. Measurement of K vitamins in food by high-performance liquid chromatography with fluorometric detection. Vitamins (Japan) 1988; 62:393–398.
9. Suttie JW. The importance of menaquinones in human nutrition. Annu Rev Nutr 1995; 15:399–417.
10. Davidson RT, Foley AL, Engelke JA, Suttie JW. Conversion of dietary phylloquinone to tissue menaquinone-4 in rats is not dependent on gut bacteria. J Nutr 1998; 128:220–223.
11. Newman P, Shearer MJ. Vitamin K metabolism. Subcell Biochem 1998; 30:455–88.
12. Shearer MJ. The roles of vitamins D and K in bone health and osteoporosis prevention. Proc Nutr Soc 1997; 56:915–937.
13. Newman P, Bonello F, Wierzbicki AS, Lumb P, Savidge GF, Shearer MJ. The uptake of lipoprotein-borne phylloquinone (vitamin K1) by osteoblasts and osteoblast-like cells: role of heparan sulfate proteoglycans and apolipoprotein E. J Bone Miner Res 2002; 17:426–433.
14. Institute of Medicine. Dietary Reference Intakes for Vitamin A, Vitamin K, Arsenic Boron, Chromium, Copper, Iodine, Iron, Manganese, Molybdenum, Nickel, Silicon, Vanadium, and Zinc. National Academy Press, Washington, DC, 2001.
15. Booth SL, Suttie JW. Dietary intake and adequacy of vitamin K. J Nutr 1998; 128:785–788.
16. Furie B, Bouchard BA, Furie BC. Vitamin K-dependent biosynthesis of gamma-carboxyglutamic acid. Blood 1999; 93:1798–1808.
17. Furie BC, Furie B. Structure and mechanism of action of the vitamin K-dependent gamma-glutamyl carboxylase: recent advances from mutagenesis studies. Thromb Haemost 1997; 78:595–598.

18. Cain D, Hutson SM, Wallin R. Assembly of the warfarin-sensitive vitamin K 2,3-epoxide reductase enzyme complex in the endoplasmic reticulum membrane. J Biol Chem 1997; 272:29068–29075.
19. Ferland G. The vitamin K-dependent proteins: an update. Nutr Rev 1998; 56:223–230.
20. Hauschka PV, Lian JB, Cole DE, Gundberg CM. Osteocalcin and matrix Gla protein: vitamin K-dependent proteins in bone. Physiol Rev 1989; 69:990–1047.
21. Ducy P, Desbois C, Boyce B, et al. Increased bone formation in osteocalcin-deficient mice. Nature 1996; 382:448–452.
22. Cairns JR, Price PA. Direct demonstration that the vitamin K-dependent bone Gla protein is incompletely gamma-carboxylated in humans. J Bone Miner Res 1994; 9:1989–1997.
23. Nakao M, Nishiuchi Y, Nakata M, Kimura T, Sakakibara S. Synthesis of human osteocalcins: gamma-carboxyglutamic acid at position 17 is essential for a calcium-dependent conformational transition. Peptide Res 1994; 7:171–174.
24. Price PA. Gla-containing proteins of bone. Connect Tissue Res 1989; 21:51–57.
25. Luo G, Ducy P, McKee MD, et al. Spontaneous calcification of arteries and cartilage in mice lacking matrix GLA protein. Nature 1997; 386:78–81.
26. Pan EY, Gomperts ED, Millen R, Gilsanz V. Bone mineral density and its association with inherited protein S deficiency. Thromb Res 1990; 58:221–231.
27. Nakamura YS, Hakeda Y, Takakura N, et al. Tyro 3 receptor tyrosine kinase and its ligand, Gas6, stimulate the function of osteoclasts. Stem Cells 1998; 16:229–238.
28. Kameda T, Miyazawa K, Mori Y, et al. Vitamin K2 inhibits osteoclastic bone resorption by inducing osteoclast apoptosis. Biochem Biophys Res Commun 1996; 220:515–519.
29. Hara K, Akiyama Y, Nakamura T, Murota S, Morita I. The inhibitory effect of vitamin K2 (menatetrenone) on bone resorption may be related to its side chain. Bone 1995; 16:179–184.
30. Koshihara Y, Hoshi K, Ishibashi H, Shiraki M. Vitamin K2 promotes 1alpha,25(OH)2 vitamin D3-induced mineralization in human periosteal osteoblasts. Calcif Tissue Int 1996; 59:466–473.
31. Takeuchi Y, Suzawa M, Fukumoto S, Fujita T. Vitamin K(2) inhibits adipogenesis, osteoclastogenesis, and ODF/RANK ligand expression in murine bone marrow cell cultures. Bone 2000; 27:769–776.
32. Vermeer C, Jie KS, Knapen MH. Role of vitamin K in bone metabolism. Annu Rev Nutr 1995; 15:1–22.
33. Price PA, Williamson MK. Effects of warfarin on bone. Studies on the vitamin K-dependent protein of rat bone. J Biol Chem 1981; 256:12754–12759.
34. Wallin R, Rossi F, Loeser R, Key LL Jr. The vitamin K-dependent carboxylation system in human osteosarcoma U2-OS cells. Antidotal effect of vitamin K1 and a novel mechanism for the action of warfarin. Biochem J 1990; 269:459–464.
35. Price PA, Kaneda Y. Vitamin K counteracts the effect of warfarin in liver but not in bone. Thromb Res 1987; 46:121–131.
36. Price P, Faus S, Williamson M. Warfarin causes rapid calcification of the elasetic lamellae in rat arteries and heart valves. Arterioscler Thromb Vasc Biol 1998; 18:1400–1407.
37. Binkley NC, Suttie JW. Vitamin K nutrition and osteoporosis. J Nutr 1995; 125:1812–1821.
38. Einhorn TA, Gundberg CM, Devlin VJ, Warman J. Fracture healing and osteocalcin metabolism in vitamin K deficiency. Clin Orthoped 1988:219–225.
39. Haffa A, Krueger D, Bruner J, et al. Diet- or warfarin-induced vitamin K insufficiency elevates circulating undercarboxylated osteocalcin without altering skeletal status in growing female rats. J Bone Miner Res 2000; 15:872–878.
40. Simon RR, Beaudin SM, Johnston M, Walton KJ, Shaughnessy SG. Long-term treatment with sodium warfarin results in decreased femoral bone strength and cancellous bone volume in rats. Thromb Res 2002; 105:353–358.
41. Pastoureau P, Vergnaud P, Meunier PJ, Delmas PD. Osteopenia and bone-remodeling abnormalities in warfarin-treated lambs. J Bone Miner Res 1993; 8:1417–1426.

42. Sato T, Ohtani Y, Yamada Y, Saitoh S, Harada H. Difference in the metabolism of vitamin K between liver and bone in vitamin K-deficient rats. Br J Nutr 2002; 87:307–314.
43. Hirano J, Ishii Y. Effects of vitamin K2, vitamin D, and calcium on the bone metabolism of rats in the growth phase. J Orthoped Sci 2002; 7:364–369.
44. Akiyama Y, Hara K, Kobayashi M, Tomiuga T, Nakamura T. Inhibitory effect of vitamin K2 (menatetrenone) on bone resorption in ovariectomized rats: a histomorphometric and dual energy X-ray absorptiometric study. Jpn J Pharmacol 1999; 80:67–74.
45. Matsunaga S, Ito H, Sakou T. The effect of vitamin K and D supplementation on ovariectomy-induced bone loss. Calcif Tissue Int 1999; 65:285–289.
46. Szulc P, Arlot M, Chapuy MC, Duboeuf F, Meunier PJ, Delmas PD. Serum undercarboxylated osteocalcin correlates with hip bone mineral density in elderly women. J Bone Miner Res 1994; 9:1591–1595.
47. Vergnaud P, Garnero P, Meunier PJ, Breart G, Kamihagi K, Delmas PD. Undercarboxylated osteocalcin measured with a specific immunoassay predicts hip fracture in elderly women: the EPIDOS Study. J Clin Endocrinol Metab 1997; 82:719–724.
48. Knapen MH, Nieuwenhuijzen-Kruseman AC, Wouters RSME, Vermeer C. Correlation of serum osteocalcin fractions with bone mineral density in women during the first 10 years after menopause. Calcif Tissue Int 1998; 63:375–379.
49. Szulc P, Chapuy MC, Meunier PJ, Delmas PD. Serum undercarboxylated osteocalcin is a marker of the risk of hip fracture in elderly women. J Clin Invest 1993; 91:1769–1774.
50. Szulc P, Chapuy MC, Meunier PJ, Delmas PD. Serum undercarboxylated osteocalcin is a marker of the risk of hip fracture: a three year follow-up study. Bone 1996; 18:487–488.
51. Luukinen H, Kakonen SM, Pettersson K, et al. Strong prediction of fractures among older adults by the ratio of carboxylated to total serum osteocalcin. J Bone Miner Res 2000; 15:2473–2478.
52. Feskanich D, Weber P, Willett WC, Rockett H, Booth SL, Colditz GA. Vitamin K intake and hip fractures in women: a prospective study. Am J Clin Nutr 1999; 69:74–79.
53. Stone K, Duong T, Sellmeyer D, Cauley J, Wolfe R, Cummings S. Broccoli may be good for bones: dietary vitamin K-1, rates of bone loss and risk of hip fracture in a prospective study of elderly women. J Bone Miner Res 1999; 14:S263.
54. Booth SL, Tucker KL, Chen H, et al. Dietary vitamin K intakes are associated with hip fracture but not with bone mineral density in elderly men and women. Am J Clin Nutr 2000; 71:1201–1208.
55. Kaneki M, Hedges S, Hosoi T, et al. Japanese fermented soybean food as the major determinant of the large geographic difference in circulating levels of vitamin K2—possible implications for hip-fracture risk. Nutrition 2001; 17:315–321.
56. Booth S, Broe K, Gagnon D, et al. Vitamin K intakes and bone mineral density in women and men. Am J Clin Nutr, 2003; 77:512–516.
57. Binkley NC, Krueger DC, Engelke JA, Foley AL, Suttie JW. Vitamin K supplementation reduces serum concentrations of under-gamma-carboxylated osteocalcin in healthy young and elderly adults. Am J Clin Nutr 2000; 72:1523–1528.
58. Bach AU, Anderson SA, Foley AL, Williams EC, Suttie JW. Assessment of vitamin K status in human subjects administered "minidose" warfarin. Am J Clin Nutr 1996; 64:894–902.
59. Booth SL, O'Brien-Morse ME, Dallal GE, Davidson KW, Gundberg CM. Response of vitamin K status to different intakes and sources of phylloquinone-rich foods: comparison of younger and older adults. Am J Clin Nutr 1999; 70:368–377.
60. New SA, Bolton-Smith C, Grubb DA, Reid DM. Nutritional influences on bone mineral density: a cross-sectional study in premenopausal women. Am J Clin Nutr 1997; 65:1831–1839.
61. Tucker KL, Hannan MT, Chen H, Cupples LA, Wilson PW, Kiel DP. Potassium, magnesium, and fruit and vegetable intakes are associated with greater bone mineral density in elderly men and women. Am J Clin Nutr 1999; 69:727–736.

62. Arjmandi BH, Getlinger MJ, Goyal NV, et al. Role of soy protein with normal or reduced isoflavone content in reversing bone loss induced by ovarian hormone deficiency in rats. Am J Clin Nutr 1998; 68:1358S–1363S.
63. Iwamoto I, Kosha S, Noguchi S, et al. A longitudinal study of the effect of vitamin K2 on bone mineral density in postmenopausal women a comparative study with vitamin D3 and estrogen-progestin therapy. Maturitas 1999; 31:161–164.
64. Shiraki M, Shiraki Y, Aoki C, Miura M. Vitamin K2 (menatetrenone) effectively prevents fractures and sustains lumbar bone mineral density in osteoporosis. J Bone Miner Res 2000; 15:515–521.
65. Ushiroyama T, Ikeda A, Ueki M. Effect of continuous combined therapy with vitamin K(2) and vitamin D(3) on bone mineral density and coagulofibrinolysis function in postmenopausal women. Maturitas 2002; 41:211–221.
66. Somekawa Y, Chigughi M, Harada M, Ishibashi T. Use of vitamin K2 (menatetrenone) and 1,25-dihydroxyvitamin D3 in the prevention of bone loss induced by leuprolide. J Clin Endocrinol Metab 1999; 84:2700–2704.
67. Braam L, Knapen M, Geusens P, et al. Vitamin K1 supplementation retards bone loss in postmenopausal women between 50 and 60 yr of age. Calcif Tissue Int 2003; 73:21–26.
68. Braam L. Effects of high vitamin K intake on bone and vascular health. Department of Biochemistry. Unviersity of Maastricht, Maastricht, 2002:139.
69. Vermeer C, Schurgers LJ. A comprehensive review of vitamin K and vitamin K antagonists. Hematol Oncol Clin N Am 2000; 14:339–353.
70. Kohlmeier M, Salomon A, Saupe J, Shearer MJ. Transport of vitamin K to bone in humans. J Nutr 1996; 126:1192S–1196S.
71. Usui Y, Tanimura H, Nishimura N, Kobayashi N, Okanoue T, Ozawa K. Vitamin K concentrations in the plasma and liver of surgical patients. Am J Clin Nutr 1990; 51:846–852.
72. Booth SL, Tucker KL, McKeown NM, Davidson KW, Dallal GE, Sadowski JA. Relationships between dietary intakes and fasting plasma concentrations of fat-soluble vitamins in humans. J Nutr 1997; 127:587–592.
73. Gundberg CM, Nieman SD, Abrams S, Rosen H. Vitamin K status and bone health: an analysis of methods for determination of undercarboxylated osteocalcin. J Clin Endocrinol Metab 1998; 83:3258–3266.
74. Booth SL, Lichtenstein AH, O'Brien-Morse M, et al. Effects of a hydrogenated form of vitamin K on bone formation and resorption. Am J Clin Nutr 2001; 74:783–790.
75. Knapen MH, Jie KS, Hamulyak K, Vermeer C. Vitamin K-induced changes in markers for osteoblast activity and urinary calcium loss. Calcif Tissue Int 1993; 53:81–85.
76. Douglas AS, Robins SP, Hutchison JD, Porter RW, Stewart A, Reid DM. Carboxylation of osteocalcin in post-menopausal osteoporotic women following vitamin K and D supplementation. Bone 1995; 17:15–20.
77. Craciun AM, Groenen-van Dooren MM, Thijssen HH, Vermeer C. Induction of prothrombin synthesis by K-vitamins compared in vitamin K- deficient and in brodifacoum-treated rats. Biochim Biophys Acta 1998; 1380:75–81.
78. Binkley N, Krueger D, Todd H, Foley A, Engelke J, Suttie J. Serum undercarboxylated osteocalcin concentration is reduced by vitamin K supplementation. FASEB J 1999; 13:A238.
79. Binkley NC, Krueger DC, Kawahara TN, Engelke JA, Chappell RJ, Suttie JW. A high phylloquinone intake is required to achieve maximal osteocalcin gamma-carboxylation. Am J Clin Nutr 2002; 76:1055–1060.
80. Knapen MH, Hamulyak K, Vermeer C. The effect of vitamin K supplementation on circulating osteocalcin (bone Gla protein) and urinary calcium excretion. Ann Intern Med 1989; 111:1001–1005.
81. Kanai T, Takagi T, Masuhiro K, et al. Serum vitamin K level and bone mineral density in postmenopausal women. Intl J Gyn Ob 1997; 56:25–30.

82. Gundberg C. Biology, physiology, and clinical chemistry of osteocalcin. J Clin Ligand Assay 1998; 21:128–138.
83. Craciun AM, Wolf J, Knapen MH, Brouns F, Vermeer C. Improved bone metabolism in female elite athletes after vitamin K supplementation. Int J Sports Med 1998; 19:479–484.
84. Hannon R, Blumsohn A, Naylor K, Eastell R. Response of biochemical markers of bone turnover to hormone replacement therapy: impact of biological variability. J Bone Miner Res 1998; 13:1124–1133.
85. Seibel MJ, Lang M, Geilenkeuser WJ. Interlaboratory variation of biochemical markers of bone turnover. Clin Chem 2001; 47:1443–1450.
86. Vermeer C, Gijsbers BL, Craciun AM, Groenen-van Dooren MM, Knapen MH. Effects of vitamin K on bone mass and bone metabolism. J Nutr 1996; 126:1187S–1191S.
87. Robert D, Jorgetti V, Lacour B, et al. Hypercalciuria during experimental vitamin K deficiency in the rat. Calcif Tissue Int 1985; 37:143–147.
88. Suttie J, Grossman C, Benton M. Specificity of the vitamin K and glutamyl binding sites of the liver microsomal gamma glutamyl carboxylase. Proceedings of the First International Congress on Vitamins and Biofactors in Life Science in Kobe, 1991, 1992, pp. 405–408.
89. Booth SL, Centurelli MA. Vitamin K: a practical guide to the dietary management of patients on warfarin. Nutr Rev 1999; 57:288–296.
90. Caraballo PJ, Heit JA, Atkinson EJ, et al. Long-term use of oral anticoagulants and the risk of fracture. Arch Intern Med 1999; 159:1750–1756.
91. Philip WJ, Martin JC, Richardson JM, Reid DM, Webster J, Douglas AS. Decreased axial and peripheral bone density in patients taking long-term warfarin. Q J Med 1995; 88:635–640.
92. Sato Y, Honda Y, Kunoh H, Oizumi K. Long-term oral anticoagulation reduces bone mass in patients with previous hemispheric infarction and nonrheumatic atrial fibrillation. Stroke 1997; 28:2390–2394.
93. Jamal SA, Browner WS, Bauer DC, Cummings SR. Warfarin use and risk for osteoporosis in elderly women. Study of Osteoporotic Fractures Research Group. Ann Intern Med 1998; 128:829–832.
94. Rosen HN, Maitland LA, Suttie JW, Manning WJ, Glynn RJ, Greenspan SL. Vitamin K and maintenance of skeletal integrity in adults. Am J Med 1993; 94:62–68.
95. Caraballo PJ, Gabriel SE, Castro MR, Atkinson EJ, Melton LJ 3rd. Changes in bone density after exposure to oral anticoagulants: a meta- analysis. Osteoporos Int 1999; 9:441–448.

VI

Lifestyle Effects/ Supplements

28 Smoking, Alcohol, and Bone Health

Douglas P. Kiel

1. INTRODUCTION

Smoking and alcohol consumption are two lifestyle factors that have important contributions to skeletal health. Deleterious effects of smoking on the skeleton have been recognized for several decades. The 2000 Surgeon General's Report on Women and Smoking *(1)* recognized smoking as a significant contributor to bone health and fracture risk. That report concluded that smoking adversely affects bone density and increases hip fracture risk in postmenopausal women. However, as male osteoporosis has been recognized as a considerable disease burden, the role of smoking in male bone health deserves equal consideration.

The role of alcohol on skeletal health has not been as well studied as that of smoking, and results from these studies suggest both beneficial as well as deleterious effects on the skeleton.

2. SMOKING AND BONE HEALTH

2.1. Smoking Effects on the Skeleton

There are potential direct and indirect effects of smoking on skeletal health and fracture risk. Direct toxic effects of smoking on bone cells may be related to nicotine effects *(2,3)* or possibly to cadmium *(4)*. Indirect effects of smoking on bone may result from decreased intestinal calcium absorption *(5)*, alterations in vitamin D status *(6)*, or alterations in metabolism of adrenal cortical and gonadal hormones *(7–9)* of smokers. These effects may account for the generally observed decrease in markers of bone formation, such as osteocalcin, in smokers *(6,10)*. Smoking may also indirectly influence bone density and the risk of fractures indirectly through reductions in body weight. Body weight tends to be lower for smokers than for nonsmokers, and this weight difference may itself lead to lower bone density and increased risk for fracture *(11,12)*. Finally, smokers may be less physically active, which itself may reduce bone density and increase fracture risk *(13)*.

From: *Nutrition and Bone Health*
Edited by: M. F. Holick and B. Dawson-Hughes © Humana Press Inc., Totowa, NJ

In several analyses involving women, weight explains part of the increased risk of low bone mineral density (BMD) associated with smoking *(14);* however, there are differences in BMD and fracture between smokers and nonsmokers, even after adjusting for weight differences *(10,15,16).* The lower weight in smokers compared to nonsmokers may increase the risk of fractures, such as hip fractures, through several mechanisms: reduced soft tissue mass overlying the trochanter, resulting in less energy absorption from a fall on the hip; decreased skeletal loading; or even reduced conversion of adrenal steroids into sex steroids in the adipose tissue. The antiestrogenic effect of smoking may also contribute to osteoporosis in women *(17,18).* Interestingly, although estrogen appears to be a critical hormone for male skeletal health *(19–21),* smoking does not appear to attenuate the association between estradiol levels and bone density in men *(22).* Finally, smoking may increase the risk of fracture through a reduction in physical performance capacity, which itself may increase the risk of falls *(23).*

2.2. Smoking and Bone Density

2.2.1. Skeletal Change Over the Lifespan

In adults, bone mass is dependent on the amount achieved at the peak, and on losses due to aging and other factors. The pace of skeletal growth is rapid in infancy, slower during childhood, and accelerated during puberty, such that by age 20–30 yr the peak skeletal mass is attained *(24,25).* Gains in BMD continue into the third decade, after bone growth has ceased *(26).* After menopause, bone loss is accelerated over the rates during the premenopausal period. These rates continue or actually increase with aging *(27)* and also occur in men *(28,29).* Because of these age-related patterns, smoking influences on bone density may be observed in the attainment of peak bone mass, in premenopausal and postmenopausal women, and in men.

2.2.2. Smoking and Attainment of Peak Bone Mass

The contribution of smoking to the attained level of peak bone mass is uncertain because there are limited data on the skeletal effects of smoking around the time of puberty. Furthermore, it is possible that relatively short times of exposure in this age group would have little effect on bone density measurements. One prospective cohort study of children and adolescents (ages 9–18 yr) in Finland repeatedly ascertained lifestyle factors and followed participants for 11 yr, at which time they underwent bone density testing. In males, but not in females, smoking was associated with lower BMD of the hip and spine after adjustment for covariates *(30).* A cross-sectional study of 15-yr-old Swedish adolescents did not find any association between smoking and total body bone mineral content *(31).*

Because so few data exist on the role of smoking in the attainment of peak bone mass, additional information can be gleaned from studies of premenopausal women. In a recent metaanalysis of cigarette smoking, BMD, and risk of hip fracture, 10 cross-sectional studies of premenopausal women were identified *(30,32–40).* The difference between the average BMD of current smokers and nonsmokers in each of the studies included was recorded as a proportion of one standard deviation, since absolute bone density units varied between studies according to the bone and the measurement technique. Each bone density difference was weighted by the inverse of its variance, and was age-adjusted only. As shown in Table 1, the mean ages of the study samples ranged from 22 to 52 yr. Bone densities were reported for current smokers vs nonsmokers in most studies, but in current vs former and never smokers combined in a few studies. There was no evidence of a significant difference in BMD between smokers and nonsmokers in this age group (*see* Fig. 1).

A second meta-analysis of the effects of cigarette smoking on BMD was reported in which studies were included regardless of bone site *(41)*. In this meta-analysis, a "combined bone site" was created for studies that measured multiple bone sites by averaging the sites. Associations between smoking and BMD were also examined at the four most frequently reported sites (lumbar spine, os calcis, hip, and forearm). In addition, this study accounted for whether smoking effects were adjusted for body weight differences, menopausal status, estrogen replacement status, use of oral contraceptives, calcium intake, height, use of medications that affect bone metabolism, physical activity, coffee intake, and alcohol intake; however, only calcium intake, and physical activity were included as covariates in a sufficient number of studies to analyze. When the effects were examined stratified by menopausal status, no significant effects were observed for premenopausal women. There were no results for premenopausal women according to their body mass index (BMI) or weight. The consistency in results for premenopausal women in both meta-analyses suggests that bone density does not differ between smokers and nonsmokers up to the time of menopause in women. No significant interaction effects were observed between age and gender, indicating that age effects were similar for women and men. This implies that bone density would not likely differ between young male smokers and nonsmokers.

2.2.3. Smoking and Bone Density in Mid and Late Life

In contrast to the results for younger persons, bone density studies performed in populations well beyond the years of peak bone mass demonstrate significant differences between smokers and nonsmokers. As demonstrated in Fig. 1, based on the meta-analysis by Law and colleagues *(42)*, bone density was lower in smokers than in nonsmokers, and the difference increased linearly with age. For every 10-yr increase in age, the bone density of smokers fell below that of nonsmokers by approx 2% of the average bone density at the time of the menopause, and this was true regardless of the skeletal site that was measured. As shown in Table 2, studies reported since the Law meta-analysis have been conflicting. In the meta-analysis of Ward, postmenopausal current smokers had significantly reduced bone mass compared with nonsmokers, at combined bone sites (standardized difference = –0.13 [95% CI –0.17, –0.09]), the forearm (standardized difference = –0.07 [95% CI –0.13, –0.01]), and the hip (standardized difference = –0.22 [95% CI

Table 1

Cross-Sectional Studies of Bone Density According to Smoking Status in Women (See Fig. 1 for Results)

	Mean (Range) Age (yr)	*Smoking Status*	*Site of Bone Density Measurement*
Studies in premenopausal women			
Fehily et al., 1992 *(32)*	22 (20–23)	104 current smokers	Radius
		78 never/former smokers	
Valimaki et al., 1994 *(30)*	24 (20–29)	9 current smokers	Femur
		47 never smokers	
McCulloch et al., 1990 *(90)*	28 (20–35)	25 current smokers	Calcaneus
		76 never/former smokers	
Ortego-Centeno et al., 1997 *(34)*	28 (SD = 10)	47 current smokers	Femur
		51 never/former smokers	
Daniel et al., 1992 *(35)*	29 (20–35)	25 current smokers	Femur
		27 never/former smokers	
Mazess and Barden, 1991 *(36)*	30 (20–39)	23 current smokers	Femur
		195 never/former smokers	
Sowers et al., 1992 *(37)*	36 (22–54)	31 current smokers	Radius
		77 never/former smokers	
Law et al., 1997 *(38)*	37 (35–39)	28 current smokers	Radius
		72 never smokers	
	42 (40–44)	63 current smokers	Radius
		115 never smokers	
	47 (45–49)	50 current smokers	Radius
		107 never smokers	
	52 (50–54)	14 current smokers	Radius
		79 nonsmokers	
Hopper and Seeman, 1994 *(39)*	42 (27–49)	9 current smokers	Femur
		9 never smokers	

Johnell and Nilsson, 1984 *(40)*	49 (49)	186 current smokers 185 never/former smokers	Radius
Law et al., 1997 *(38)*	45 (39–49)	24 current smokers 56 never smokers	Radius
	52 (50–54)	31 current smokers 83 never smokers	Radius
	57 (55–59)	32 current smokers 135 never smokers	Radius
	62 (60–64)	27 current smokers 65 never smokers	Radius
Jensen and Christiansen, 1988 *(18)*	50 (44–53)	56 current smokers 54 never/former smokers	Radius
Jensen et al., 1985 *(17)*	51 (44–56)	67 current smokers 69 never/former smokers	Radius
Slemenda et al., 1989 *(91)*	51 (45–57)	21 current smokers 63 never/former smokers	Radius
McDermott and Witte, 1988 *(92)*	53 (SD = 10)	24 current smokers 24 never smokers	Radius
Guthrie et al., 1996 *(93)*	54 (48–57)	7 current smokers 39 never/former smokers	Femur
Cheng et al., 1991 *(94)*	54 (50–60)	25 current smokers 82 never/former smokers	Calcaneus
Krall and Dawson Hughes, 1991 *(95)*	59 (40–70)	35 current smokers 267 never/former smokers	Femur
Hopper and Seeman, 1994 *(39)*	62 (50–73)	7 current smokers 7 nonsmokers	Femur
Sowers et al., 1985 *(96)*	62 (55–80)	119 current smokers 278 never smokers	Radius

(table continues)

Table 1
(continued)

	Mean (Range) Age (yr)	*Smoking Status*	*Site of Bone Density Measurement*
Hansen et al., 1991 *(75)*	63 (59–67)	61 current smokers 60 never/former smokers	Femur
Egger et al., 1996 *(97)*	66 (63–68)	23 current smokers 99 never smokers	Femur
Hollo et al., 1979 *(98)*	68 (61–75)	41 current smokers 125 never smokers	Radius
Nguyen et al., 1994 *(88)*	70 (>60)	102 current smokers 765 never smokers	Femur
Jensen, 1986 *(99)*	70 (70)	77 current smokers 103 never smokers	Radius
Johansson et al., 1992 *(100)*	70 (70)	38 current smokers 200 never smokers	Calcaneus
Rundgren and Mellstrom, 1984 *(101)*	70 (70)	43 current smokers 243 never smokers	Calcaneus
	75 (75)	49 current smokers 364 never smokers	Calcaneus
	79 (79)	19 current smokers 218 never smokers	Calcaneus
Bauer et al., 1993 *(14)*	71 (65–84)	485 current smokers 4367 never smokers	Radius
Kiel et al., 1996 *(16)*	74 (68–98)	77 current smokers 340 never smokers	Femur
Cheng et al., 1993 *(102)*	75 (75)	10 current smokers 161 never smokers	Calcaneus
Hollenbach et al., 1993 *(103)*	76 (60–89)	42 current smokers 320 never smokers	Femur

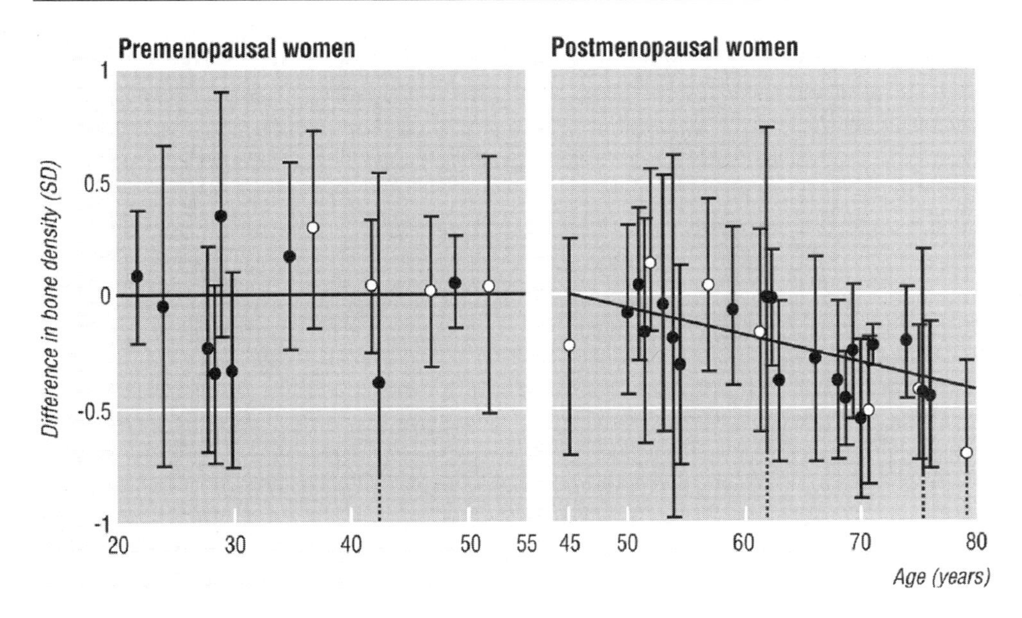

Fig. 1. Differences (95% confidence intervals), as a proportion of 1 SD, in bone mineral density between female smokers and nonsmokers according to age and menopuasal status. Fitted regression lines are shown. The 11 open circles refer to two studies *(38,101)*, solid circles refer to the other studies in the order listed in Table 1 (Fehily through Johnell and Law through Hollenbach). (From ref. *42*, with permission.)

–0.28, –0.16]) *(41)*. In this same meta-analysis, absolute effect sizes for bone density differences with smoking were greater in men than in women at combined bone sites and the lumbar spine. Similar trends were observed at the forearm and the os calcis, although the differences were not statistically significant (*see* Fig. 2).

More recently, data have begun to emerge that implicate smoking as a significant risk factor for loss of bone in longitudinal studies of older men and women (Table 3). Of the six longitudinal studies reporting smoking data (three involving women and men, two women only, and one men only) *(28,29,37,43–45)*, three reported significantly higher bone loss in female smokers *(37,43,44)*, and three reported higher rates of loss among male smokers *(29,44,45)*. Studies that failed to find a statistically significant relation tended to have relatively lower percentages of smokers.

Smoking may also interfere with the treatment of osteoporosis with estrogen replacement therapy (ERT), as levels of estradiol are lower in smokers taking estrogen than in nonsmokers taking estrogen *(46)*, and bone density values in women taking estrogen are lower in smokers than in nonsmokers *(15)*. Recently, a study investigating the influence of smoking on the antiosteoporotic efficacy of raloxifene demonstrated no influence on BMD differences or vertebral fracture risk after 4 yr of treatment *(47)*.

Table 2

Studies of Bone Density According to Smoking Status in Women and Men Since the 1997 Meta-Analysis by Law and Colleagues *(48)*

Study	*Sample, Age (yr)*	*Smoking Status*	*Measurement/Site*	*Principal Finding*
Women				
Kim et al., 2000 *(104)*	238 Korean women; mean age 24.2 ± 2.5, scanned only as reference population 552 postmenopausal Korean women mean age 62.5 ± 8.2	Not specified	BUA calcaneus	No association between history of smoking and low quantitative ultrasound values after controlling for age and duration following menopause
Cheng et al., 1999 *(85)*	200 Caucasian women; aged 20–79	38% had history of tobacco use (avg. 8.2 packs/yr); 7% current smokers	BUA calcaneus	Smoking not associated with BUA ($p > 0.05$)
Gregg et al., 1999 *(86)*	393 women (7.4% non-Caucasian; 12.2% peri- or postmenopausal); aged 45–53	9.2% current smokers	BUA, SOS calcaneus; BMD spine, hip	Smoking not significantly associated with calcaneal BUA or SOS
Varenna et al., 1999 *(105)*	6160 postmenopausal Italian women; mean age 54.5 ± 6.4 yr	74.9% never smokers; 5.0% ex-smokers; 20.1% current smokers	BMD spine	Smoking not associated with BMD, osteoporosis risk
Jones and Scott, 1999 *(106)*	263 premenopausal women; mean age 33 ± 4.5 yr	45% current smokers	BMD spine, hip, and whole body	Current smoking associated with significantly lower BMD at the hip and trends at the spine and whole body
Grainge et al., 1998 *(83)*	580 postmenopausal women; aged 45–59 yr	25.7% current smokers, 74.3% nonsmokers at time of scan	BMD spine, hip, radius/ulna, whole body	BMD more strongly related to number of months spent smoking than to pack-years at all five sites ($p < 0.05$ at all sites except femoral neck)

Smeets-Goevaers et al., 1998 *(82)*	5896 perimenopausal, white Dutch women; aged 46–54	Never smokers, past or current smokers said to be identified, but no data given	BMD spine	Increased risk for low BMD (osteopenia and osteoporosis) associated with smoking (OR 1.25, 95% CI 1.08–1.44)
Brot et al., 1997 *(107)*	433 perimenopausal Danish women; aged 45–58. Of these, 87 were followed for 2 yr	49% current smokers; 39% never smokers; 12% former smokers	BMC of whole body measured at enrollment and after 1 and 2 yr	Smoking (pack-years) a significant and independent predictor of total BMC ($p < 0.001$)
Takada et al., 1997 *(108)*	3867 Japanese women; aged 37–69	Dichotomous category current smoking (yes/no), but no data given	BMD at distal radius 1/3	The combined variable of no drinking (i.e., consumption of alcohol 3 or fewer days per week) and being a current smoker has a statistically significant negative effect on radial BMD among older (56–69) women ($p < 0.05$)
Men				
Hagiwara and Tsumura, 1999 *(87)*	1736 Japanese men; aged 20–64	35.5% nonsmokers; 15.7% ex-smokers; 48.8% current smokers	BMD of the calcaneus	Men in the highest BMD quintile were younger and had a higher BMI and lower mean pack-years history of smoking than men in the lowest quintile
Vogel et al., 1997 *(45)*	1303 men of Japanese descent living in Hawaii; aged 61–82	35% never smoked 45% past smokers 20% current smokers	BMD calcaneus, distal and proximal radius	Current and to some degree past smokers had 1.8–4.8% lower BMD in the calcaneus and distal radius

BUA, broadband ultrasound attenuation; SOS, speed of sound; BMD, bone mineral density; OR, odds ratio; CI, confidence interval; BMC, bone mineral content; BMI, body mass index.

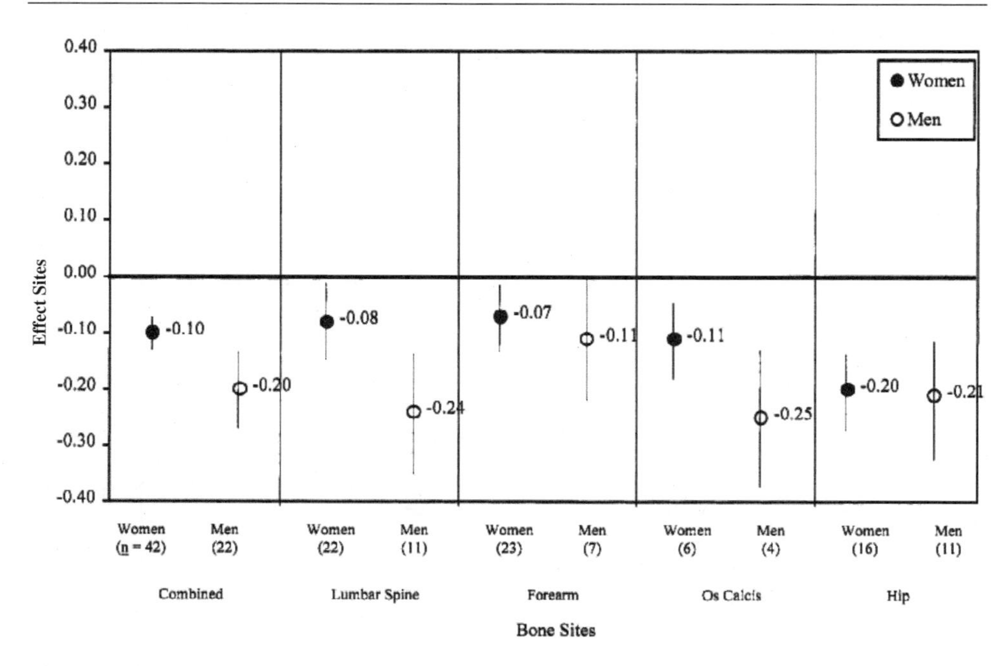

Fig. 2. Effect size comparing current smokers to nonsmokers (standardized mean differences with 95% confidence intervals) according to gender. (From ref. *41,* with permission.)

Given these data, in postmenopausal women there is sufficient evidence to infer a causal relationship between smoking and low bone density. There may be an even greater effect size in older men. Longitudinal studies suggest there may be a causal relationship between smoking and bone loss in older women and men. Data comparing current smokers and former smokers to never smokers have tended to show greater differences in bone density between current and never smokers than between former and never smokers. These findings indirectly suggest that smoking cessation may slow, or partially reverse, the accelerated bone loss caused by years of smoking.

These effects on bone density are significant in mid and late life, since for every 10-yr increase in age, the bone density of smokers falls below that of nonsmokers by about 0.14 SD, or 2% of the average bone density at the time of the menopause. Because a 1.0-SD decrese in bone density doubles the risk of fracture, and because fracture incidence increases with age, the proportion of all fractures attributable to smoking would be expected to increase as smokers continue smoking into old age. Attempts to decrease smoking as early in life as possible are likely to reduce fractures that occur in old age among smokers.

Because bone loss is relatively small over short periods of time, studies with longer duration of follow-up and minimal avoidable losses to follow-up could add important information to the understanding of the contributions of smoking to

Table 3
Studies of Bone Loss According to Smoking Status in Women and Men

Study	*Sample, Age (yr)*	*Smoking Status*	*Measurement/Site*	*Principal Finding*
Sowers et al., 1992 *(37)*	217 women aged 22–54	Mean lifetime packs of cigarettes = 2447	BMD distal radius	In postmenopausal, but not premenopausal women, smoking at baseline was associated with lower follow-up BMD.
Jones et al., 1994 *(28)*	626 (385 women, 241 men); average follow-up 2.5 yr	Women had median of 9 pack-yr smoking. Men had median of 31 pack-years smoking	BMD hip and spine	No difference in rates of loss between current smokers and nonsmokers.
Vogel et al., 1997 *(45)*	1303 Japanese-American men (aged 51–82); average follow-up 5 yr	20% current smokers; 45% past smokers; 35% never smokers	BMD distal and proximal radius and calcaneus	Compared to never smokers, current smokers had significantly greater rates of bone loss from the calcaneus (29.4% greater) and distal radius (33.8% greater) both ($p < 0.05 = 0.001$) and 4.8% ($p < 0.0001$, respectively). Analyses adjusted for age, height, weight, physical activity, and alcohol and thiazide use.
Guthrie et al., 1998 *(43)*	224 women (74 pre-, 90 peri-, and 60 post-menopausal); follow-up 2 yr	Premenopausal women 14% current smokers. Early perimenopausal women 14% current smokers. Late perimenopausal women 25% current smokers. Postmenopausal women 15% current smokers	BMD hip and spine	In the women who became postmenopausal during the study, 6 were current smokers and their mean annual change in spine BMD was slightly greater (–3.3%) than the 36 nonsmokers (–2.3%); $p = 0.10$.
Burger et al., 1998 *(44)*	1856 Dutch men (mean age, 66.7), 2452 Dutch women (mean age, 67.2); average follow-up 2 yr	Current smokers: men (23%) women (19%)	BMD hip	Smoking accompanied by a significantly higher rate of bone loss in both men and women (men $p = 0.02$; women $p = 0.01$). Association stronger when not adjusting for BMI.
Hannan et al., 2000 *(29)*	468 women 273 men (mean age 74.5); average follow-up 4 yr	Current smokers: women (10%); men (8%)	BMD hip, spine, and radius	Compared to women who had never smoked, bone loss in current smokers was not increased. In men, In men, current smokers had greater bone loss (4–5%) than never smokers.

BMD, bone mineral density; BMI, bone mass index.

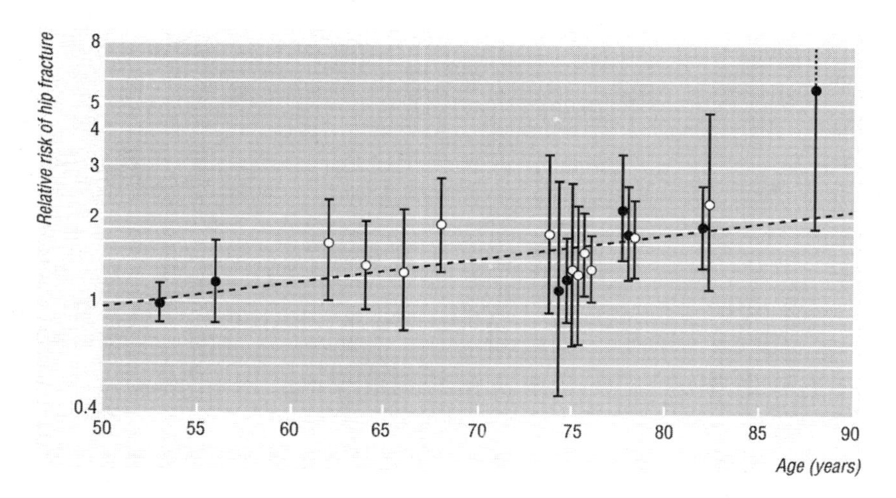

Fig. 3. Relative risk (95% confidence intervals) of hip fracture in smokers compared with nonsmokers in postmenopausal women according to age, in cohort studies (solid circles) and case-control studies (open circles), each in the same order as in Table 2. Fitted regression line is shown.

bone loss. Additional information is likely to come from studies of biochemical markers of bone turnover, which might yield an underlying explanation of the mechanism whereby smoking accelerates bone loss.

2.3. Smoking and Fracture Risk

Hip fractures, the most frequently studied of fractures in relation to smoking, account for a significant proportion of the morbidity and mortality attributed to osteoporosis. In the meta-analysis by Law and colleagues *(48),* 19 cohort and case-control studies of the risk of hip fracture in postmenopausal women according to smoking habits were reviewed (*see* Table 4). The studies differed with regard to the ages of the study participants, duration of follow-up, and whether former smokers were included in the smoking group or the nonsmoking group. For the cohort studies, the duration of follow-up ranged from 3 yr *(49)* to 26 yr *(15).* Figure 3 shows the risk of hip fracture in smokers relative to nonsmokers according to age. The risk of hip fracture in smokers compared to nonsmokers increased with increasing age.

Data on the association between smoking and fractures at other sites are more limited. Studies from the 1980s and early 1990s that examined fractures other than those of the hip rarely found associations with smoking, although more recent studies have demonstrated positive associations between smoking and vertebral fractures *(50,51),* ankle fracture *(52),* and the general category of non-hip fractures *(53).* Also, since the publication of the meta-analysis by Law and colleagues, many *(54–57),* but not all, subsequent studies of hip fracture *(58–60)* have continued to support the association between smoking and an increased risk of hip fracture (*see* Table 5).

Table 4

Studies of Smoking and Relative Risk of Hip Fractures Used in Meta-Analysis by Law and Colleagues *(48)*

Type of Fracture/Study	*Age on Entry*	*Mean Age at Fracture (yr)*	*No. of Subjects (% Smokers)*	
			With Fracture	*Without Fracture*
Hip-Cohort Studies				
Hemenway et al., 1988 *(109)*	34–59	53	662	68,056 (28)
Meyer et al., 1993 *(110)*	35–49	56	124	20,881 (37)
Holbrook et al., 1988 *(111)*	50–79	75	33	924
Kiel et al., 1992 *(15)*	28–62	75	167 (22)	2,243 (37)
Cummings et al., 1995 *(11)*	≥ 65	78	192	9,324 (10)
Forsen et al., 1994 *(49)*	≥ 50	78	220 (16)	14,598 (20)
Paganini-Hill et al., 1991 *(112)*	Any	82	242 (13)	5,558 (13)
Wickham et al., 1989 *(113)*	≥ 65	88	44	1375
Hip-Case Control Studies				
La Vecchia et al., 1991 *(114)*	29–74	62	158 (11)	1,096 (6)
Williams et al., 1982 *(115)*	50–74	64	160 (60)	567 (53)
Kreiger et al., 1982 *(116,117)*	45–74	66	98	801
Michaelsson et al., 1999 *(89)*	40–75	68	205 (18)	765 (10)
Kreiger et al., 1992 *(118)*	50–84	74	102 (29)	277 (17)
Grisso et al., 1994 *(119)*	≥ 45	75	109 (29)	169 (15)
Paganini-Hill et al., 1981 *(120)*	< 80	75	83 (35)	166 (30)
Jaglal et al., 1993 *(121)*	55–84	75	381 (22)	1,138 (16)
Lau et al., 1988 *(122)*	Any	76	400	800
Cooper et al., 1988 *(123)*	≥ 50	78	300 (48)	600 (37)
Cumming and Klineberg, 1994 *(124)*	≥ 65	82	209	207

Table 5
Studies of Smoking and Relative Risk of Fractures of the Hip and Other Sites

Type of Fracture/Study	*Study Design*	*Sample*	*Results*
Hip fracture			
Fujiwara et al., 1997 *(60)*	Cohort study	1,586 Japanese men, 2,987 Japanese women; mean age, 58.5 ± 12.2 yr) During up to 14-yr follow-up, 55 incident hip fractures not due to traffic accidents identified	Smoking not related to hip fracture risk.
Clark et al., 1998 *(58)*	Case-control	45 Mexican men and 107 Mexican women with hip fracture; aged 45 and older (mean, 70.2 for men, 73.5 for women) 143 healthy controls (37 men, 106 women) without hip fracture; mean age, 68.9 for men, 71.1 for women	Smoking not associated with risk of hip fracture.
Forsen et al., 1998 *(57)*	Cohort study	14,428 Norwegian men, 15,364 Norwegian women; aged 50–75+ During 3-yr follow-up, 421 new cases of incident hip fracture identified	Among subjects ≤ age 75, RR of hip fracture elevated for current smokers (men: RR 5.0, 95%CI 1.5, 16.9; women: RR 1.9; 95%CI 1.2, 3.1). For ex-smokers, including those who had quit smoking >5 yr previously (men: RR 4.4; 95% CI 1.2, 15.3; women: RR 1.3; 95% CI 0.6, 3.0).
Mussolino et al., 1998 *(59)*	Cohort study	2,879 white US men; aged 45–74; during up to 22-yr follow-up, 71 cases of hip fracture identified	Smoking not significantly associated with hip fracture.
Turner et al., 1998 *(125)*	Cross-sectional	2325 women ages 50+ from NHANES III queried about history of wrist or hip fracture	Bivariate analysis showed that the percentage of former smokers in the wrist or hip fracture group were greater than in the nonfracture group. Smoking not associated with fracture in multivariate analyses.
Burger et al., 1999 *(56)*	Cohort study	2,193 Dutch men, 3,015 Dutch women; aged 55 and over; during 4-yr follow-up, 47 persons (14 men) experienced first hip fracture	When adjusted for age and gender, current smoking was a statistically significant indicator of hip fracture risk (OR 2.6; 95% CI 1.4–5.1).

Cornuz et al., 1999 *(126)*	Cohort study	116,229 female nurses (98% Caucasian); aged 34–59; during 12-yr follow-up, 377 hip fractures occurred due to low or moderate trauma	Current smokers experienced higher rates of hip fracture than women who never smoked; risk increased with the number of cigarettes smoked daily. Age-adjusted relative risk of hip fracture was 1.3 (95% CI 1.0–1.7) among all cigarette smokers and 1.6 (95% CI 1.1–2.3) among those who smoked 25+ cigarettes per day ($p = 0.09$ for trend). After 10 yr of quitting, risk of fracture no longer significant.
Hoidrup et al., 1999 *(127)*	Cohort study	6,159 postmenopausal Danish women; during 15- to 17-yr follow-up, 363 hip fractures identified and validated	Use of HRT was associated with a lower risk of hip fracture in former (RR 0.55; 95% CI 0.22, 1.37) and current (RR 0.61; 95% CI 0.38, 0.99) smokers but not in never smokers (RR 1.10; 95% CI 0.60, 2.03).
Kanis et al., 1999 *(55)*	Case-control	730 Southern European men with hip fracture; age 50 and older (mean, 73.9) 1,132 age-stratified controls	A long history of smoking (>49 yr) associated with a significant increase in risk of hip fracture (RR = 1.44; 95% CI 1.10–1.89; $p < 0.01$).
Melhus et al., 1999 *(54)*	Case-control	247 Swedish women with hip fracture and 873 controls from a cohort study of 66,651 Swedish women; aged 40–76	OR for hip fracture among current smokers was 2.1 (1.3–3.2). OR for hip fracture among current smokers with low intake of vitamin E was 3.0 (95% CI 1.6–5.4) and of vitamin C (1.6–5.6). OR decreased to 1.1 (95% CI 0.5–2.4) and 1.4 (95% CI 0.7–3.0) with high intakes of vitamins E and C respectively. In current smokers with a low intake of vitamins E and C, OR increased to 4.9 (95% CI 2.2–11.0).

(table continues)

Table 5
(Continued)

Type of Fracture/Study	*Study Design*	*Sample*	*Results*
Vertebral fracture			
Aloia et al., 1985 *(128)*	Age-matched, case-control study	58 cases 58 controls Volunteer women Mean age 64 yr United States	Percentage of smokers ($p < 0.01$) Cases: 59% Controls: 30%
Kleerekoper et al., 1989 *(129)*	Case-control	266 cases 263 controls Postmenopausal women screened for osteoporosis trial Aged 45–75 yr United States	Percentage of current smokers ($p > 0.05$) Cases: 27% Controls: 20%
Cooper et al., 1991 *(130)*	Survey of general practice patients	1,012 women 79 fractures Aged 48–81 yr United Kingdom	Smoking > 10 cigarettes/d for > 10 yr not related to fracture risk.
Santavirta et al., 1992 *(131)*	Population-based survey	27,278 girls and women 105 fractures Aged ≥ 15 yr Finland	RR = 1.1 (95% CI, 0.6–2.0), for current smokers adjusted for age, history of trauma, tuberculosis, peptic ulcer, BMI, occupation.
Scane et al., 1999 *(51)*	Case-control	91 UK men with vertebral fractures; aged 27–79 (median, 64) 91 age-matched controls	Current smoking associated with a significantly increased risk of vertebral fracture (OR = 2.8; 95% CI 1.2–6.7).
Lau et al., 2000 *(50)*	Cross-sectional	396 community-dwelling Chinese men aged 70–79	Heavy smoking a significant risk factor for vertebral deformity (OR 6.5; 95% CI 1.3–32.7).

Distal forearm fracture			
Williams et al., 1982 *(115)*	Population-based, case-control study	184 cases 567 controls Aged 50–74 yr United States	Higher fracture risk in women smokers using estrogen.
Kelsey et al., 1992 *(132)*	Cohort study	9,704 women 171 fractures over 2.2 yr (mean) Aged ≥ 65 yr United States	RR = 1.0 (95% CI 0.96–1.0), for current smokers (10 cigarettes/d) vs never smoked.
Kreiger et al., 1992 *(118)*	Hospital case-control study	54 fractures Aged 50–84 yr Canada	RR = 1.5 (95% CI, 0.9–2.6), for current smokers vs former smokers or never smoked adjusted for age, BMI.
Mallmin et al., 1994 *(133)*	Population-based, case-control study	385 cases 385 controls Aged 40–80 yr Sweden	RR = 0.9 (95% CI, 0.5–1.6), for current smokers adjusted for multiple factors, including age, BMI, physical activity, hormone use.
Honkanen et al., 1998 *(52)*	Retrospective survey	12,192 women 345 fractures Aged 47–56 yr Finland	Current smoking RR = 0.9 (0.6–1.4) Any smoking RR = 0.6 (0.3–1.1), 1–10 cig/d RR = 1.4 (0.9–2.3), >10 cig/d Adjusted for age, BMI, menopausal status, chronic health disorders.
Proximal humerus fracture			
Kelsey et al., 1992 *(132)*	Cohort study	9,704 women 79 fractures over 2.2 yr Aged ≥ 65 years United States	RR = 1.2 (95% CI, 0.9–1.6), for current smokers (10 cig/d)

(table continues)

Table 5
(Continued)

Type of Fracture/Study	*Study Design*	*Sample*	*Results*
Ankle fracture			
Honkanen et al., 1998 *(52)*	Retrospective survey	12,192 women 210 fractures Aged 47–56 yr Finland	Current smoking RR = 2.2 (95%, CI, 1.6–3.2) Any smoking RR = 1.6 (95% CI, 0.9–2.8), for 1–10 cig/d RR = 3.0 (95% CI, 1.9–4.6), for > 10 cig/d Adjusted for age, BMI, menopausal, status, chronic health disorders
Seeley et al., 1996 *(134)*	Cohort study	9704 women aged ≥ 65 yr 191 fractures over 5.9 yr (mean)	Current smokers No association
Foot fracture			
Seeley et al., 1996 *(134)*	Cohort study	9704 women aged ≥ 65 yr 204 fractures over 5.9 yr (mean)	Current smokers No association
Non-hip fracture			
Jacqmin-Gadda et al., 1998 *(53)*	Cohort study	3216 French men and women; aged 65 and older (mean, 74.8 years); during 5-yr follow-up, 265 persons (8.2%) reported one fracture, 19 (0.6%) reported two fractures, and one (0.03%) reported three fractures	Current smoking status associated with a higher risk of non-hip fractures (OR 1.68; 95% CI 1.08–2.60), but not hip fractures (OR 0.73, 95% CI 0.24–2.20).

RR, relative risk, CI, confidence interval, OR, odds ratio; HRT, hormone replacement therapy; BMI, body mass index.

Based on these data, smoking appears to increase the risk of hip fracture; however, there are fewer studies of smoking and fracture risk at other skeletal sites. Because the risk of hip fractures in smokers increases with age, and hip fracture incidence also increases with age, the proportion of hip fractures attributable to smoking increases with age. Interventions aimed at helping smokers quit are likely to result in a significantly reduced number of hip fractures. Although hip fractures carry the greatest risk of mortality, morbidity, and cost, other fractures also contribute significantly to these outcomes. Further research is necessary to quantify the risk of these fractures in smokers.

3. ALCOHOL AND BONE HEALTH

3.1. Alcohol Effects on the Skeleton

The direct effects of alcohol on bone and mineral metabolism have been described in both rats and in men. Studies of chronic alcohol consumption in growing male and female rats have indicated that bone growth is suppressed, leading to a failure to acquire a normal peak bone mass *(61)*. Bone loss in adult rats fed ad libitum a liquid diet containing increasing concentrations of ethanol until receiving the appropriate percentage of total caloric intake, resulted in a dose-dependent decrease in trabecular thickness, bone turnover, and bone formation rate *(62)*. When equated to humans, the doses used in the adult rat experiments ranged from the low end of moderate (3% of caloric intake) to alcoholic levels (35% of caloric intake). These findings in rats suggest that even moderate levels of alcoholic beverage consumption in humans may have the potential to reduce bone turnover and possibly to have deleterious effects on the skeleton. In rats fed ethanol over long periods, Peng et al. reported a greater risk of tibial fractures and a decrease in trabecular bone volume and bone strength *(63)*, which is consistent with Saville's observation of lower density of tibial bone in rats fed with alcohol *(64)*.

In humans, alcoholics have been shown to have low bone mineral density that is due to an inhibition of bone remodeling by a mechanism independent of the calciotropic hormones *(65,66)*. Others have compared alcoholics to controls and found that serum concentrations of 25-hydroxyvitamin D3 and 1,25-dihydroxyvitamin D3 were significantly reduced among the alcoholics as compared to the controls *(67,68)*. These low levels have been suggested to be the result of a deficient diet, reduced exposure to sunlight, malabsorption of vitamin D, increased biliary excretion of 25-hydroxyvitamin D metabolites, or to be the result of reduced reserves of vitamin D owing to a reduction of adipose and muscle tissue in alcoholics *(69)*. Laitenen and colleagues demonstrated that the low serum levels of vitamin D metabolites in noncirrhotic alcoholics were not because of nutritional deficiency, and hypothesized that there was increased degradation of vitamin D metabolites in the liver. However, they showed that high calcium intake could counteract the vitamin D abnormalities *(70)*. Alcohol may also have deleterious effects on bone homeostasis through increased excretion of calcium and magnesium *(71)*. Consistent with the observations of reduced bone formation in rats fed alcohol, reductions in osteoblastic activity have been observed in acute alcohol intoxication and in moderate use over 3 wk time in adults *(72,73)*.

3.2. Alcohol and Bone Density

The above suggestions that alcohol has deleterious effects on bone and mineral metabolism have not been substantiated in a series of larger cohort studies of BMD. Laitinen et al. first reported that in postmenopausal Finnish women, those who consumed any alcohol had higher bone mineral density at the femur and spine sites than nondrinking women. This same relation was not seen in alcohol-drinking premenopausal women, who tended to have lower BMDs than nondrinkers *(74)*. Hansen and colleagues examined the relation between alcohol consumption and radial bone loss over 12 yr of follow-up in 121 healthy postmenopausal women and found that after the first 2 yr of menopause, the rate of postmenopausal bone loss was significantly reduced in the group of women with a regular intake of alcohol *(75)*. In a group of 182 men and 267 women aged 45 and over being studied for hyperlipidemia, alcohol intake, whether estimated from average self-reported weekly consumption or from a 24-h recall, was positively associated with BMD at the femur in men and with BMD at the spine in women. These results were adjusted for potential confounders such as age, BMI, exercise, smoking, and ERT in women *(76)*. Results from the Framingham Study revealed that women who drank at least 7 oz of alcohol per week had higher BMDs at most skeletal sites (4.2–13%) than women consuming the least amount of alcohol (<1 oz) after adjusting for age, weight, height, smoking, and estrogen use. In men the differences were less than in women. This degree of alcohol consumption could be achieved by drinking two drinks of hard liquor per day, almost three beers per day, or three glasses of wine per day. In general, moderate intake of alcohol did not affect bone density *(77)*. Finally, in 1999, Feskanich et al. found that in a sample of 188 white postmenopausal women ages 50–74 from the Nurses' Health Study, alcohol intake assessed from food-frequency questionnaires was associated with BMD at the spine and hip. There was a linear increase in spinal bone density over increasing categories of alcohol intake, suggesting that even modest alcohol intakes might be of benefit *(78)*. Several other studies of men and women also demonstrated significant protective effects of alcohol consumption on BMD both cross-sectionally *(74,76–82)* and longitudinally *(44,75)*. The remaining studies examining the relation between alcohol consumption and BMD found no significant associations *(14,32,83–88)* (*see* Table 6).

3.3. Alcohol and Fracture Risk

Despite the suggestion from some of the above studies that alcohol may have beneficial effects on the skeleton, the vast majority of studies examining alcohol consumption and risk of fractures have either shown no significant association, or in some cases, an increased risk of fracture among those men and women with high intakes of alcohol (*see* Table 7). There is only a single case-control study

Table 6

Studies of Bone Density According to Alcohol Use

Study	*Sample, Age (yr)*	*Alcohol Status*	*Measurement/Site*	*Principal Finding*
Tanaka et al., 2001 *(84)*	325 male volunteers ≥ 50 yr old	Total intake per day without indication of beverage type	Femoral neck	No association between alcohol consumption and BMD.
	355 men > age 60 yr	Current alcohol intake without indication of beverage type	Spine and hip	After age and weight adjustment, alcohol intake was positively associated with BMD at all sites.
Forsmo et al., 2001 *(80)*	1652 peri- and postmenopausal women age 50–59 yr	Units per 2 wk, although definition of "units" not specified	Distal and ultradistal radius	Alcohol consumption was positively associated with BMD more strongly at the ultradistal radius than the distal site.
Orwoll et al., 2000 *(79)*	355 men > 60 yr	Current alcohol intake without indication of beverage type	Spine and hip	After age and weight adjustment, alcohol intake was positively associated with BMD at all sites.
Hannan et al., 2000 *(29)*	468 women 273 men (mean age 74.5); average follow-up 4 yr	Grams of alcohol per week based on published formula for calculating alcohol from beer, wine, and liquor.	BMD hip, spine, and radius	No association between alcohol consumption and changes in BMD at any site.
Huuskonen et al., 2000 *(81)*	140 Finnish men aged 54–63 yr	Grams per day of absolute ethanol based on 4-wk alcohol records	Spine and hip	Alcohol consumption positively associated with spine BMD.
Feskanich et al., 1999 *(78)*	188 postmenopausal women from the Nurses Health Study ages 50–70	Food frequency questionnaires listing beer, wine, and liquor. Used formula from USDA data on grams of alcohol in each beverage type	Spine and hip	Alcohol was positively associated with spine BMD (0.005 g/cm^2 increased BMD for an increase of 10 g of alcohol per week). No association with hip BMD.
Gregg et al., 1999 *(86)*	393 women aged 45–53	Grams of alcohol per day but not stated as to how ascertained or calculated	Quantitative ultrasound of calcaneus and BMD spine and hipand hip	No association between alcohol consumption and either calcaneus ultrasound or BMD.

(table continues)

Table 6
Studies of Bone Density According to Alcohol Use

Study	*Sample, Age (yr)*	*Alcohol Status*	*Measurement/Site*	*Principal Finding*
Cheng et al., 1999 *(85)*	200 Caucasian women; aged 20–79	Total drinks per week of beer, wine, and spirits	Quantitative ultrasound of calcaneus	No association between alcohol consumption and calcaneus ultrasound.
Hagiwara and Tsumura, 1999 *(87)*	1,736 Japanese men; aged 20–64	Ethanol per week using a "standard conversion table"	Calcaneal bone mineral density	No association between alcohol consumption and calcaneal BMD.
Burger et al., 1998 *(44)*	4333 (1856 men and 2452 women) ≥ 55 yr old from Rotterdam	Total alcohol intake calculated on basis of 1 unit of alcoholic beverage = 10 g of alcohol	Femur measured at baseline and after 1.9 yr median follow-up	In men, as intake increased from 0 to 0.1–10 g/d to 10–20 g/d, bone loss decreased. When intake was 20+ g/d, bone loss was not different from 0 g/d. Overall trend not significant; in women, no association.
Grainge et al., 1998 *(83)*	580 postmenopausal British women 45–59 yr from general practices of primary physicians	Alcohol intake recorded as number of $^1/_2$-pints beer or cider, glasses of wine and spirits per week for each year of life since age 10 yr	Spine and hip	Neither lifetime alcohol nor current alcohol consumption was associated with BMD at any site.
Smeets-Goevaers et al., 1998 *(82)*	5896 Dutch women aged 46–54 yr	Current consumption without indication of beverage type	Spine	Current users had lower risk of having low bone density (T-score < –1) RR = 0.71, 95% CI 0.61–0.83).
Felson et al., 1995 *(77)*	1154 men and women from the Framingham Cohort, ages 68–96	Used published formula for calculating alcohol from beer, wine, and liquor. Consumption divided into <1 oz/wk, 1–3 oz/wk, 3–7 oz/wk, and ≥ 7 oz/wk	Spine, hip, radius	Women who drank ≥ 7 oz/wk had higher BMD at most sites (4.2–13% range) than women in the lowest intake category.

Nguyen et al., 1994 *(88)*	709 elderly men and 1,080 women over age 60 yr	Total intake per day in grams without indication of beverage type	Spine and hip	No association between alcohol consumption and BMD.
Bauer et al., 1993 *(14)*	9704 postmenopausal women ≥ 65 from four US cities	Total weekly and lifetime ounces intake without indication of beverage type	Radius and calcaneus	No association between alcohol consumption and BMD.
Holbrook et al., 1993 *(76)*	182 men and 267 women aged ≥ 45 yr	Weekly intake assessed using interviewer and a 24-h recall using dietician	Spine and hip	In men hip BMD increased significantly with increasing alcohol intake, while in women spine BMD increased with increasing intake. Similar trends in both sexes at other bone sites but not statistically significant.
Fehily et al., 1992 *(32)*	371 subjects aged 20–23 who were part of a previous milk supplement intervention study	Grams of alcohol per day but not stated as to how ascertained or calculated	Radius	In men only at proximal radius site, inverse association between alcohol intake and BMD but not in multivariate models.
Laitenen et al., 1991 *(74)*	351 healthy Finnish women aged 20–76	Total intake without indication of beverage type	Spine and hip	Premenopausal women who used alcohol had slightly decreased Ward's area BMD.
Hansen et al., 1991 *(75)*	121 healthy postmenopausal women	Regular intake defined as the weekly consumption of one or more units of alcohol for at least 1 yr	Spine and hip	Rate of postmenopausal bone loss was significantly reduced in the group of women with regular intake of alcohol.

BMD, bone mineral density.

Table 7
Studies of Alcohol and Relative Risk of Fractures of the Hip and Other Sites

Type of Fracture/Study	*Study Design*	*Sample*	*Results*
Hip fracture			
Hoidrup et al., 1999 *(135)*	Cohort study	Three population studies conducted in 1964–1992 in Denmark involving 17,868 men and 13,917 women; 500 hip fractures in women and 307 in men	A low to moderate weekly alcohol intake (1–27 drinks for men and 1–13 drinks for women) was not associated with hip fracture. Among men, the RR of hip fracture increased in those who drank 28 drinks or more per week (RR = 1.75; 95% CI 1.06–2.89 for 28–41 drinks and RR = 5.28; 95% CI 2.60–10.70) for 70 or more drinks.
Hoidrup et al., 1999 *(127)*	Cohort study	6159 postmenopausal women followed from 1976 to 1993; 363 hip fractures occurred	Alcohol not significantly associated with hip fracture.
Melhus et al., 1999 *(54)*	Case-control study	247 Swedish women with hip fracture and 873 controls from a cohort study of 66,651 Swedish women; aged 40–76	Alcohol not significantly associated with hip fracture.
Michaelsson et al., 1999 *(89)*	Case-control study	1610 cases of hip fracture and 2–4 population controls per case in each 5-yr age group and county of residence	Ever-use of alcohol associated with a 19–25% significant reduction in hip fracture risk.
Clark et al., 1998 *(58)*	Case-control study	45 Mexican men and 107 Mexican women with hip fracture; aged 45 and older (mean, 70.2 for men, 73.5 for women) 143 healthy controls (37 men, 106 women) without hip fracture; mean age, 68.9 for men, 71.1 for women	Any alcohol intake increased the risk of hip fracture (OR = 1.73; 95% CI 1.04–2.90) for total group and for women (OR = 2.78; 95% CI 1.25–6.14).
Jacqmin-Gadda et al., 1998 *(53)*	Cohort study	3216 men and women age 65 yr and older followed for 5 yr; 63 hip fractures	Daily use of alcohol increased the risk of hip fractures (RR = 5.41; 95% CI 1.78–16.4).
Mussolino et al., 1998 *(59)*	Cohort study	2879 white US men; aged 45–74; during up to 22 yr follow-up, 71 cases of hip fracture identified	Alcohol not significantly associated with hip fracture.
Turner et al., 1998 *(125)*	Cross-sectional study	2325 women ages 50 yr and older who were part of NHANES III	Alcohol not significantly associated with hip fracture.
Fujiwara et al., 1997 *(60)*	Cohort study	1586 Japanese men, 2987 Japanese women; mean age, 58.5 ± 12.2 yr); during up to 14-yr follow-up, 55 incident hip fractures not due to traffic accidents identified	Regular alcohol intake doubled the risk of hip fracture (RR = 1.91; 95% CI 1.07–3.42).

Johnell et al., 1995 *(136)*	Case-control	2086 cases from 14 centers in six countries in Southern Europe; 3532 nonhospital controls from population registers or from neighborhoods to cases	Alcohol not significantly associated with hip fracture.
Cummings et al., 1995 *(11)*	Cohort study	9516 Caucasian women at least 65 yr of age recruited from four centers in the US; during 4.1 yr average follow up there were 192 hip fractures	Alcohol was associated with lower hip fracture risk (RR = 0.7; 95% CI 0.5–0.9) in age-adjusted models but not in multivariate-adjusted models (RR = 0.8; 95% CI 0.6–1.1).
Cumming and Klineberg, 1994 *(124)*	Case-control study	209 cases of hip fracture admitted to 12 hospitals in Western suburbs of Sydney Australia and 207 controls living in private homes in same area selected with an area probability sampling method	Alcohol not significantly associated with hip fracture.
Grisso et al., 1994 *(119)*	Case-control study	144 cases offirst hospital admissions for hip fracture in black women compared to 181 hospitalized black women controls matched to age and hospital	$\geq$ 7 drinks/wk increased risk of hip fracture (OR = 3.1; 95% CI 1.2–8.1).
Hemenway et al., 1994 *(137)*	Cohort study	51,529 male health professionals followed for 6 yr; 67 hip fractures	Alcohol not significantly associated with hip fracture.
Kreiger et al., 1992 *(118)*	Case-control study	565 cases (ages 50–84) admitted to hospital for either hip or wrist fractures, and 490 controls stratified on age, hospital and inpatient or outpatient admission	Nonsignificant increase in fracture risk among alcohol users.
LaVecchia et al., 1991 *(114)*	Case-control study	209 cases of hip fracture ages 29–74 and 1449 controls admitted to same hospitals as the cases	Alcohol not significantly associated with hip fracture.
Paganini-Hill et al., 1991 *(112)*	Cohort study	13,987 residents of retirement community followed from 1981 to 1988; 332 women and 86 men sustained first hip fractures	Alcohol not significantly associated with hip fracture.
Holbrook et al., 1988 *(111)*	Cohort study	957 adults aged 50–79 residing in a geographically defined upper-middle-class white community followed for 14 yr; 33 hip fractures	Alcohol not significantly associated with hip fracture.

(table continues)

Table 7
(Continued)

Type of Fracture/Study	*Study Design*	*Sample*	*Results*
Hemenway et al., 1988 *(109)*	Cohort study	96,508 nurses followed for 4 yr; 925 hip or forearm fractures	Among women consuming 15 g or more of alcohol per day RR = 1.24; 95% CI 1.04–1.48, compared to abstainers and light drinkers
Paganini-Hill et al., 1981 *(120)*	Case-control study	91 hip fracture cases occurring during a 5-yr period in female residents under age 80 living in a retirement community compared to age- and race-matched community controls	8+ shots of liquor per week associated with increased risk of hip fracture but not statistically significant.
Vertebral fracture			
Lau et al., 2000 *(50)*	Cross-sectional study	396 community-dwelling Chinese men aged 70–79	Heavy drinking (≥ 3 drinks/d) associated with definite vertebral deformity (RR = 2.9; 95% CI 1.0–8.4)
Scane et al., 1999 *(51)*	Case-control study	91 UK men with vertebral fractures; aged 27–79 (median, 64) 91 age-matched controls	Consumption of more than 250 g/wk associated with increased risk of vertebral fractures (OR = 3.8; 95% CI 1.7–8.7)
Cooper et al., 1991 *(130)*	Cross-sectional study	1012 peri- and postmenopausal women aged 48–81 yr seeing their practitioner; 79 (7.8%) had one or more vertebral fractures	Alcohol not significantly associated with vertebral fracture.
Kleerekoper et al., 1989 *(129)*	Case-control study	266 postmenopausal white females with nontraumatic vertebral fractures, 134 nonfractured women from a medical clinic and 263 nonfractured women presenting for osteoporosis screening	Alcohol not significantly associated with vertebral fracture.
Forearm fracture			
Seeley et al., 1996 *(134)*	Cohort study	9704 women aged 65 yr and older from 4 U.S. cities with 5.9 yr of follow-up	Alcohol not significantly associated with fractures.
Kelsey et al., 1992 *(132)*	Cohort study	9704 women aged 65 yr and older from four US cities with 2.2 yr of follow-up; 171 forearm fractures and 79 humerus fractures	Alcohol not significantly associated with fractures.

RR, relative risk; CI, confidence interval; OR, odds ratio;

demonstrating a reduction in fracture risk among alcohol users *(89)*. Thus, given some suggestion of a beneficial effect of alcohol on BMD, yet an increased risk of fracture in some studies for men and women in the higher intake categories, it is conceivable that the fracture risk is mediated through an increased risk for falls or other types of trauma. At the current time, it would be reasonable to conclude that modest alcohol consumption does not pose significant risk for osteoporosis and related fractures. Higher intakes may predispose to trauma-associated fracture outcomes. Finally, alcoholics appear to have low bone density and metabolic abnormalities that threaten bone health.

4. CONCLUSION

Cigarette smoking, which interferes with calcium absorption, alters estrogen metabolism, and which may expose the skeleton to direct toxic effects, is clearly a risk factor for low bone density, bone loss, and hip fracture, especially with advancing age. On the other hand, the evidence that alcohol use is deleterious for the skeleton appears to be limited to situations of more extreme intake, which has been associated with impaired osteoblast function and greater risk of trauma. In fact, there is some evidence that modest intakes may result in higher bone density, although fracture risk does not appear to be lowered by modest intake.

REFERENCES

1. U.S. Department of Health and Human Services. Reducing tobacco use: a report of the Surgeon General. U.S. Department of Health and Human Services, Centers for Disease Control and Prevention, National Center for Chronic Disease Prevention and Health Promotion, Office on Smoking and Health, Atlanta, GA, 2000.
2. Riebel GD, Boden SD, Whitesides TE, Hutton WC. The effect of nicotine on incorporation of cancellous bone graft in an animal model. Spine 1995; 20:2198–2202.
3. Fang MA, Frost PJ, Iida-Klein A, Hahn TJ. Effects of nicotine on cellular function in UMR 106-01 osteoblast-like cells. Bone 1991; 12:283–286.
4. Bhattacharyya MH, Whelton BD, Stern PH, Peterson DP. Cadmium accelerates bone loss in overiectomized mice and fetal rat limb bones in culture. Proc Natl Acad Sci USA 1988; 85:8761–8765.
5. Krall EA, Dawson-Hughes B. Smoking increases bone loss and decreases intestinal calcium absorption. J Bone Miner Res 1999; 14:215–220.
6. Brot C, Jorgensen NR, Sorensen OH. The influence of smoking on vitamin D status and calcium metabolism. Eur J Clin Nutr 1999; 53:920–926.
7. Baron JA, Comi RJ, Cryns V, Brinck-Johnsen T, Mercer NG. The effect of cigarette smoking on adrenal cortical hormones. J Pharmacol Exp Ther 1995; 272:151–155.
8. Khaw KT, Tazuke S, Barrett-Connor E. Cigarette smoking and levels of adrenal androgens in postmenopausal women. N Engl J Med 1988; 318:1705–1709.
9. Michnovicz JJ, Hershcopf RJ, Naganuma H, Bradlow HL, Fishman J. Increased 2-hydroxylation of estradiol as a possible mechanixm for the anti-estrogenic effect of cigarette smoking. N Engl J Med 1986; 315:1305–1309.
10. Bjarnason NH, Christiansen C. The influence of thinness and smoking on bone loss and response to hormone replacement therapy in early postmenopausal women. J Clin Endocrinol Metab 2000; 85:590–596.

11. Cummings SR, Nevitt MC, Browner WS, et al. Risk factors for hip fracture in white women. N Engl J Med 1995; 332:767–773.
12. Kiel DP, Felson DT, Anderson JJ, Wilson PW, Moskowitz MA. Hip fracture and the use of estrogens in postmenopausal women. The Framingham Study. N Engl J Med 1987; 317:1169–1174.
13. Gregg EW, Cauley JA, Seeley DG, Ensrud KE, Bauer DC. Physical activity and osteoporotic fracture risk in older women. Study of Osteoporotic Fractures Research Group (see comments). Ann Intern Med 1998; 129:81–88.
14. Bauer DC, Browner WS, Cauley JA, et al. Factors associated with appendicular bone mass in older women. The Study of Osteoporotic Fractures Research Group (see comments). Ann Intern Med 1993; 118:657–665.
15. Kiel DP, Baron JA, Anderson JJ, Hannan MT, Felson DT. Smoking eliminates the protective effect of oral estrogens on the risk for hip fracture among women. Ann Intern Med 1992; 116:716–721.
16. Kiel DP, Zhang Y, Hannan MT, Anderson JJ, Baron JA, Felson DT. The effect of smoking at different life stages on bone mineral density in elderly men and women. Osteopor Int 1996; 6:240–248.
17. Jensen J, Christiansen C, Rodbro P. Cigarette smoking, serum estrogens, and bone loss during hormone-replacement therapy early after menopause. N Engl J Med 1985; 313:973–975.
18. Jensen J, Christiansen C. Effects of smoking on serum lipoproteins and bone mineral content during postmenopausal hormone replacement therapy. Am J Obstet Gynecol 1988; 159:820–825.
19. Slemenda CW, Longcope C, Zhou L, Hui SL, Peacock M, Johnston CC. Sex steroids and bone mass in older men positive associations with serum estrogens and negative associations with androgens. J Clin Invest 1997; 100:1755–1759.
20. Khosla S, Melton LJ 3rd, Atkinson EJ, O'Fallon WM, Klee GG, Riggs BL. Relationship of serum sex steroid levels and bone turnover markers with bone mineral density in men and women: a key role for bioavailable estrogen. J Clin Endocrinol Metab 1998; 83:2266–2274.
21. Amin S, Zhang Y, Sawin CT, et al. Association of hypogonadism and estradiol levels with bone mineral density in elderly men from the Framingham study. Ann Intern Med 2000; 133:951–963.
22. Amin S, LaValley MP, Zhang Y, et al. Is the effect of smoking on bone mineral density (BMD) in elderly men mediated through estradiol (E2)? The Framingham Osteoporosis Study. J Bone Miner Res 1999; 14(suppl 1):S147.
23. Nelson HD, Nevitt MC, Scott JC, Stone KL, Cummings SR. Smoking, alcohol, and neuromuscular and physical function of older women. Study of Osteoporotic Fractures Research Group (see comments). JAMA 1994; 272:1825–1831.
24. Kroger H, Kotaniemi A, Vainio P, Alhava E. Bone densitometry of the spine and femur in children by dual-energy x-ray absorptiometry [published erratum appears in Bone Miner 1992 Jun; 17(3):429]. Bone Miner 1992; 17:75–85.
25. Lu PW, Cowell CT, SA LL-J, Briody JN, Howman-Giles R. Volumetric bone mineral density in normal subjects, aged 5–27 years. J Clin Endocrinol Metab 1996; 81:1586–1590.
26. Recker RR, Davies KM, Hinders SM, Heaney RP, Stegman MR, Kimmel DB. Bone gain in young adult women (see comments). JAMA 1992; 268:2403–2408.
27. Ensrud KE, Palermo L, Black DM, et al. Hip and calcaneal bone loss increase with advancing age: longitudinal results from the study of osteoporotic fractures. J Bone Miner Res 1995; 10:1778–1787.
28. Jones G, Nguyen T, Sambrook P, Kelly PJ, Eisman JA. Progressive loss of bone in the femoral neck in elderly people: longitudinal findings from the Dubbo osteoporosis epidemiology study. Br Med J 1994; 309:691–695.
29. Hannan MT, Felson DT, Dawson-Hughes B, et al. Risk factors for longitudinal bone loss in elderly men and women: the Framingham Osteoporosis Study. J Bone Miner Res 2000; 15:710–720.

30. Valimaki MJ, Karkkainen M, Lamberg-Allardt C, et al. Exercise, smoking, and calcium intake during adolescence and early adulthood as determinants of peak bone mass. Cardiovascular Risk in Young Finns Study Group. Br Med J 1994; 309:230–235.
31. Lotborn M, Bratteby LE, Samuelson G, Ljunghall S, Sjostrom L. Whole-body bone mineral measurements in 15-year-old Swedish adolescents. Osteopor Int 1999; 9:106–114.
32. Fehily AM, Coles RJ, Evans WD, Elwood PC. Factors affecting bone density in young adults. Am J Clin Nutr 1992; 56:579–586.
33. McCulloch RG, Whiting SJ, Bailey DA, Houston CS. The effect of cigarette smoking on trabecular bone density in premenopausal women, aged 20–35 years. Can J Public Health 1991; 82:434–435.
34. Ortego-Centeno N, Munoz-Torres M, Jodar E, Hernandez-Quero J, Jurado-Duce A. Effect of tobacco consumption on bone mineral density in healthy young males. Calcif Tissue Int 1997; 60:496–500.
35. Daniel M, Martin AD, Drinkwater DT. Cigarette smoking, steroid hormones, and bone mineral density in young women. Calcif Tissue Int 1992; 50:300–305.
36. Mazess RB, Barden HS. Bone density in premenopausal women: effects of age, dietary intake, physical activity, smoking, and birth-control pills. Am J Clin Nutr 1991; 53:132–142.
37. Sowers MR, Clark MK, Hollis B, Wallace RB, Jannausch M. Radial bone mineral density in pre- and perimenopausal women: a prospective study of rates and risk factors for loss. J Bone Miner Res 1992; 7:647–657.
38. Law MR, Cheng R, Hackshaw AK, Allaway S, Hale AK. Cigarette smoking, sex hormones and bone density in women. Eur J Epidemiol 1997; 13:553–558.
39. Hopper JL, Seeman E. The bone density of female twins discordant for tabacco use. N Engl J Med 1994; 330:387–431.
40. Johnell O, Nilsson BE. Life-style and bone mineral mass in perimenopausal women. Calcif Tissue Int 1984; 36:354–356.
41. Ward KD, Klesges RC. A meta-analysis of the effects of cigarette smoking on bone mineral density. Calcif Tissue Int 2001; 68:259–270.
42. Law MR, Hackshaw AK. A meta-analysis of cigarette smoking, bone mineral density and risk of hip fracture: recognition of a major effect. Br Med J 1997; 315:841–846.
43. Guthrie JR, Ebeling PR, Hopper J, et al. A prospective study of bone loss in menopausal Australian-born women. Osteopor Int 1998; 8:282–890.
44. Burger H, de Laet CE, van Daele PL, et al. Risk factors for increased bone loss in an elderly population: the Rotterdam Study. Am J Epidemiol 1998; 147:871–879.
45. Vogel JM, Davis JW, Nomura A, Wasnich RD, Ross PD. The effects of smoking on bone mass and the rates of bone loss among elderly Japanese-American men. J Bone Miner Res 1997; 12:1495–1501.
46. Komulainen M, Kroger H, Tuppurainen MT, Heikkinen AM, Honkanen R, Saarikoski S. Identification of early postmenopausal women with no bone response to HRT: results of a five-year clinical trial. Osteopor Int 2000; 11:211–218.
47. Chapurlat RD, Ewing SK, Bauer DC, Cummings SR. Influence of smoking on the antiosteoporotic efficacy of raloxifene. J Clin Endocrinol Metabol 2001; 86:4178–4182.
48. Law M. Smoking and osteoporosis. In: Wald N, Baron J, eds. Smoking and Hormone-Related Disorders. Oxford University Press, New York, 1990, pp. 83–92.
49. Forsen L, Bjorndal A, Bjartveit K, et al. Interaction between current smoking, leanness, and physical inactivity in the prediction of hip fracture. J Bone Miner Res 1994; 9:1671–1678.
50. Lau EM, Chan YH, Chan M, et al. Vertebral deformity in chinese men: prevalence, risk factors, bone mineral density, and body composition measurements. Calcif Tissue Int 2000; 66:47–52.
51. Scane AC, Francis RM, Sutcliffe AM, Francis MJ, Rawlings DJ, Chapple CL. Case-control study of the pathogenesis and sequelae of symptomatic vertebral fractures in men. Osteopor Int 1999; 9:91–97.

52. Honkanen R, Tuppurainen M, Kroger H, Alhava E, Saarikoski S. Relationships between risk factors and fractures differ by type of fracture: a population-based study of 12,192 perimenopausal women. Osteopor Int 1998; 8:25–31.
53. Jacqmin-Gadda H, Fourrier A, Commenges D, Dartigues JF. Risk factors for fractures in the elderly. Epidemiology 1998; 9:417–423.
54. Melhus H, Michaelsson K, Holmberg L, Wolk A, Ljunghall S. Smoking, antioxidant vitamins, and the risk of hip fracture. J Bone Miner Res 1999; 14:129–135.
55. Kanis J, Johnell O, Gullberg B, et al. Risk factors for hip fracture in men from southern Europe: the MEDOS study. Mediterranean Osteoporosis Study. Osteopor Int 1999; 9:45–54.
56. Burger H, de Laet CE, Weel AE, Hofman A, Pols HA. Added value of bone mineral density in hip fracture risk scores. Bone 1999; 25:369–374.
57. Forsen L, Bjartveit K, Bjorndal A, Edna TH, Meyer HE, Schei B. Ex-smokers and risk of hip fracture. Am J Public Health 1998; 88:1481–1483.
58. Clark P, de la Pena F, Gomez Garcia F, Orozco JA, Tugwell P. Risk factors for osteoporotic hip fractures in Mexicans. Arch Med Res 1998; 29:253–257.
59. Mussolino ME, Looker AC, Madans JH, Langlois JA, Orwoll ES. Risk factors for hip fracture in white men: the NHANES I Epidemiologic Follow-up Study. J Bone Miner Res 1998; 13:918–924.
60. Fujiwara S, Kasagi F, Yamada M, Kodama K. Risk factors for hip fracture in a Japanese cohort. J Bone Miner Res 1997; 12:998–1004.
61. Turner RT, Greene VS, Bell NH. Demonstration that ethanol inhibits bone matrix synthesis and mineralization in the rat. J Bone Miner Res 1987; 2:61–66.
62. Turner RT, Kidder LS, Kennedy A, Evans GL, Sibonga JD. Moderate alcohol consumption suppresses bone turnover in adult female rats. J Bone Miner Res 2001; 16:589–594.
63. Peng TC, Kusy RP, Hirsch PF, Hagaman JR. Ethanol-induced changes in morphology and strength of femurs of rats. Alcohol Clin Exp Res 1988; 12:655–659.
64. Saville PD, Lieber CS. Effect of alcohol on growth, bone density and muscle magnesium in the rat. J Nutr 1965; 87:477–484.
65. Bikle DD, Genant HK, Cann C, Recker RR, Halloran BP, Strewler GJ. Bone disease in alcohol abuse. Ann Intern Med 1985; 103:42–48.
66. Crilly RG, Anderson C, Hogan D, Delaquerriere-Richardson L. Bone histomorphometry, bone mass, and related parameters in alcoholic males. Calcif Tissue Int 1988; 43:269–276.
67. Bjorneboe GE, Bjorneboe A, Johnsen J, et al. Calcium status and calcium-regulating hormones in alcoholics. Alcohol Clin Exp Res 1988; 12:229–232.
68. Feitelberg S, Epstein S, Ismail F, D'Amanda C. Deranged bone mineral metabolism in chronic alcoholism. Metabolism 1987; 36:322–326.
69. Laitinen K, Valimaki M. Alcohol and bone. Calcif Tissue Int 1991; 49(suppl):S70–S73.
70. Laitinen K, Valimaki M, Lamberg-Allardt C, et al. Deranged vitamin D metabolism but normal bone mineral density in Finnish noncirrhotic male alcoholics. Alcohol Clin Exp Res 1990; 14:551–556.
71. Kalbfleisch JM, Lindemann RD, Ginn HE, Smith WD. Effects of ethanol administration on urinary excretion of magnesium and other electrolytes in alcoholic and normal subjects. J Clin Invest 1963; 42:1471.
72. Laitinen K, Lamberg-Allardt C, Tunninen R, Karonen SL, Ylikahri R, Valimaki M. Effects of 3 weeks' moderate alcohol intake on bone and mineral metabolism in normal men. Bone Miner 1991; 13:139–151.
73. Garcia-Sanchez A, Gonzalez-Calvin JL, Diez-Ruiz A, Casals JL, Gallego-Rojo F, Salvatierra D. Effect of acute alcohol ingestion on mineral metabolism and osteoblastic function. Alcohol Alcohol 1995; 30:449–453.
74. Laitinen K, Valimaki M, Keto P. Bone mineral density measured by dual-energy X-ray absorptiometry in healthy Finnish women. Calcif Tissue Int 1991; 48:224–231.

75. Hansen MA, Overgaard K, Riis BJ, Christiansen C. Potential risk factors for development of postmenopausal osteoporosis—examined over a 12-year period. Osteopor Int 1991; 1:95–102.
76. Holbrook TL, Barrett-Connor E. A prospective study of alcohol consumption and bone mineral density. Br Med J 1993; 306:1506–1509.
77. Felson DT, Zhang Y, Hannan MT, Kannel WB, Kiel DP. Alcohol intake and bone mineral density in elderly men and women The Framingham Study. Am J Epidemiol 1995; 142:485–492.
78. Feskanich D, Korrick SA, Greenspan SL, Rosen HN, Colditz GA. Moderate alcohol consumption and bone density among postmenopausal women. J Womens Health 1999; 8:65–73.
79. Orwoll ES, Bevan L, Phipps KR. Determinants of bone mineral density in older men. Osteopor Int 2000; 11:815–821.
80. Forsmo S, Schei B, Langhammer A, Forsen L. How do reproductive and lifestyle factors influence bone density in distal and ultradistal radius of early postmenopausal women? The Nord-Trondelag Health Survey, Norway. Osteopor Int 2001; 12:222–229.
81. Huuskonen J, Vaisanen SB, Kroger H, et al. Determinants of bone mineral density in middle aged men: a population-based study. Osteopor Int 2000; 11:702–708.
82. Smeets-Goevaers CG, Lesusink GL, Papapoulos SE, et al. The prevalence of low bone mineral density in Dutch perimenopausal women: the Eindhoven Perimenopausal Osteoporosis Study. Osteopor Int 1998; 8:404–409.
83. Grainge MJ, Coupland CA, Cliffe SJ, Chilvers CE, Hosking DJ. Cigarette smoking, alcohol and caffeine consumption, and bone mineral density in postmenopausal women. The Nottingham EPIC Study Group. Osteopor Int 1998; 8:355–363.
84. Tanaka T, Latorre MR, Jaime PC, Florindo AA, Pippa MG, Zerbini CA. Risk factors for proximal femur osteoporosis in men aged 50 years or older. Osteopor Int 2001; 12:942–949.
85. Cheng S, Fan B, Wang L, et al. Factors affecting broadband ultrasound attenuation results of the calcaneus using a gel-coupled quantitative ultrasound scanning system. Osteopor Int 1999; 10:495–504.
86. Gregg EW, Kriska AM, Salamone LM, et al. Correlates of quantitative ultrasound in the Women's Healthy Lifestyle Project. Osteopor Int 1999; 10:416–424.
87. Hagiwara S, Tsumura K. Smoking as a risk factor for bone mineral density in the heel of Japanese men. J Clin Densitom 1999; 2:219–222.
88. Nguyen TV, Kelly PJ, Sambrook PN, Gilbert C, Pocock NA, Eisman JA. Lifestyle factors and bone density in the elderly: implications for osteoporosis prevention. J Bone Miner Res 1994; 9:1339–1346.
89. Michaelsson K, Weiderpass E, Farahmand BY, et al. Differences in risk factor patterns between cervical and trochanteric hip fractures. Swedish Hip Fracture Study Group. Osteopor Int 1999; 10:487–494.
90. McCulloch RG, Bailey DA, Houston S, Dodd BL. Effects of physical activity, dietary calcium intake and selected lifestyle factors on bone density in young women. Can Med Assoc J 1990; 142:221–227.
91. Slemenda CW, Hui SL, Longcope C, Johnston CC. Cigarette smoking, obesity, and bone mass. J Bone Miner Res 1989; 4:737–741.
92. McDermott MT, Witte MC. Bone mineral content in smokers. So Med J 1988; 81:477–480.
93. Guthrie JR, Ebeling PR, Hopper JL, Dennerstein L, Wark JD, Burger HG. Bone mineral density and hormone levels in menopausal Australian women. Gynecol Endocrinol 1996; 10:199–205.
94. Cheng S, Suominen H, Rantanen T, Parkatti T, Heikkinen E. Bone mineral density and physical activity in 50-60-year-old women. Bone Miner 1991; 12:123–132.
95. Krall EA, Dawson-Hughes B. Smoking and bone loss among postmenopausal women. J Bone Miner Res 1991; 6:331–338.
96. Sowers M, Wallace RB, Lemke JH. Correlates of mid-radius bone density among postmenopausal women: a community study 1–3. Am J Clin Nutr 1985; 41:1045–1053.

97. Egger P, Duggleby S, Hobbs R, Fall C, Cooper C. Cigarette smoking and bone mineral density in the elderly. J Epidemiol Community Health 1996; 50:47–50.
98. Hollo I, Gergely I, Boross M. Influence of heavy smoking upon the bone mineral content of the radius of the aged and effect of tobacco smoke on the sensitivity to calcitonin of rats. Aktuelle Gerontologic 1979; 9:365–368.
99. Jensen GF. Osteoporosis of the slender smoker revisited by epidemiologic approach. Eur J Clin Invest 1986; 16:239–242.
100. Johansson C, Mellstrom D, Lerner U, Osterberg T. Coffee drinking: a minor risk factor for bone loss and fractures. Age Ageing 1992; 21:20–26.
101. Rundgren A, Mellstrom D. The effect of tobacco smoking on the bone mineral content of the ageing skeleton. Mech Ageing Devel 1984; 28:273–277.
102. Cheng S, Suominen H, Heikkinen E. Bone mineral density in relation to anthropometric properties, physical activity and smoking in 75-year-old men and women. Aging Clin Exp Res 1993; 5:55–62.
103. Hollenbach KA, Barrett-Connor E, Elelstein SL, Holbrook T. Cigarette smoking and bone mineral density in older men and women. Am J Public Health 1993; 83:1265–1270.
104. Kim CH, Kim YI, Choi CS, et al. Prevalence and risk factors of low quantitative ultrasound values of calcaneus in Korean elderly women. Ultrasound Med Biol 2000; 26:35–40.
105. Varenna M, Binelli L, Zucchi F, Ghiringhelli D, Gallazzi M, Sinigaglia L. Prevalence of osteoporosis by educational level in a cohort of postmenopausal women. Osteopor Int 1999; 9:236–2341.
106. Jones G, Scott FS. A cross-sectional study of smoking and bone mineral density in premenopausal parous women: effect of body mass index, breastfeeding, and sports participation. J Bone Miner Res 1999; 14:1628–1633.
107. Brot C, Jensen LB, Sorensen OH. Bone mass and risk factors for bone loss in perimenopausal Danish women. J Intern Med 1997; 242:505–511.
108. Takada H, Washino K, Iwata H. Risk factors for low bone mineral density among females: the effect of lean body mass. Prevent Med 1997; 26:633–638.
109. Hemenway D, Colditz GA, Willett WC, Stampfer MJ, Speizer FE. Fractures and lifestyle: effect of cigarette smoking, alcohol intake, and relative weight on the risk of hip and forearm fractures in middle-aged women. Am J Public Health 1988; 78:1554–1558.
110. Meyer HE, Tverdal A, Falch JA. Risk factors for hip fracture in middle-aged Norwegian women and men. Am J Epidemiol 1993; 137:1203–1211.
111. Holbrook TL, Barrett-Connor E, Wingard DL. Dietary calcium and risk of hip fracture: 14-year prospective population study. Lancet 1988; 2:1046–1049.
112. Paganini-Hill A, Chao AR, Ross RK, Henderson B. Exercise and other factors in the prevention of hip fracture: the Leisure World Study. Epidemiology 1991; 2:16–25.
113. Wickham CA, Walsh K, Cooper C, et al. Dietary calcium, physical activity, and risk of hip fracture: a prospective study (see comments). Br Med J 1989; 299:889–892.
114. La Vecchia C, Negri E, Levi F, Baron JA. Cigarette smoking, body mass and other risk factors for fractures of the hip in women. Int J Epidemiol 1991; 20:671–675.
115. Williams AR, Weiss NS, Ure CL, Ballard J, Daling JR. Effect of weight, smoking, and estrogen use on the risk of hip and forearm fractures in postmenopausal women. Obstet Gynecol 1982; 60:695–699.
116. Kreiger N, Kelsey JL, Holford TR, O'Connor T. An epidemiologic study of hip fracture in postmenopausal women. Am J Epidemiol 1982; 116:141–148.
117. Kreiger N, Hilditch S. Cigarette smoking and estrogen-dependent diseases. In Letters to the Editor. Am J Epidemiol 1986; 123:200.
118. Kreiger N, Gross A, Hunter G. Dietary factors and fracture in postmenopausal women: a case-control study. Int J Epidemiol 1992; 21:953–958.

119. Grisso JA, Kelsey JL, Strom BL, et al. Risk factors for hip fracture in black women. N Engl J Med 1994; 330:1555–1559.
120. Paganini-Hill A, Ross RK, Gerkins VR, Henderson BE, Arthur M, Mack TM. Menopausal estrogen therapy and hip fractures. Ann Intern Med 1981; 95:28–31.
121. Jaglal SB, Kreiger N, Darlington G. Past and recent physical activity and risk of hip fracture. Am J Epidemiol 1993; 138:107–118.
122. Lau E, Donnan S, Barker DJP, Cooper C. Physical activity and calcium intake in fracture of the proximal femur in Hong Kong. Br Med J 1988; 297:1441–1443.
123. Cooper C, Barker DJ, Wickham C. Physical activity, muscle strength, and calcium intake in fracture of the proximal femur in Britain. Br Med J 1988; 297:1443–1446.
124. Cumming RG, Klineberg RJ. Case-control study of risk factors for hip fractures in the elderly. Am J Epidemiol 1994; 139:493–503.
125. Turner LW, Fu Q, Taylor JE, Wang MQ. Osteoporotic fracture among older U.S. women: risk factors quantified. J Aging Health 1998; 10:372–391.
126. Cornuz J, Feskanich D, Willett WC, Colditz GA. Smoking, smoking cessation, and risk of hip fracture in women. Am J Med 1999; 106:311–314.
127. Hoidrup S, Gronbaek M, Pedersen AT, Lauritzen JB, Gottschau A, Schroll M. Hormone replacement therapy and hip fracture risk: effect modification by tobacco smoking, alcohol intake, physical activity, and body mass index. Am J Epidemiol 1999; 150:1085–1093.
128. Aloia JF, Cohn SH, Vaswani A, Yeh JK, Yuen K, Ellis K. Risk factors for postmenopausal osteoporosis. Am J Med 1985; 78:95–100.
129. Kleerekoper M, Peterson E, Nelson D, et al. Identification of women at risk for developing postmenopausal osteoporosis with vertebral fractures: role of history and single photon absorptiometry. Bone Miner 1989; 7:171–186.
130. Cooper C, Shah S, Hand DJ, et al. Screening for vertebral osteoporosis using individual risk factors. The Multicentre Vertebral Fracture Study Group. Osteopor Int 1991; 2:48–53.
131. Santavirta S, Konttinen YT, Heliovaara M, Knekt P, Luthje P, Aromaa A. Determinants of osteoporotic thoracic vertebral fracture. Screening of 57,000 Finnish women and men. Acta Orthoped Scand 1992; 63:198–202.
132. Kelsey JL, Browner WS, Seeley DG, Nevitt MC, Cummings SR. Risk factors for fractures of the distal forearm and proximal humerus. The Study of Osteoporotic Fractures Research Group (published erratum appears in Am J Epidemiol 1992 May 15; 135[10]:1183). Am J Epidemiol 1992; 135:477–489.
133. Mallmin H, Ljunghall S, Persson I, Bergstrom R. Risk factors for fractures of the distal forearm: a population-based case-control study. Osteopor Int 1994; 4:298–304.
134. Seeley DG, Kelsey J, Jergas M, Nevitt MC. Predictors of ankle and foot fractures in older women. The Study of Osteoporotic Fractures Research Group. J Bone Miner Res 1996; 11:1347–1355.
135. Hoidrup S, Gronbaek M, Gottschau A, Lauritzen JB, Schroll M. Alcohol intake, beverage preference, and risk of hip fracture in men and women. Copenhagen Centre for Prospective Population Studies. Am J Epidemiol 1999; 149:993–1001.
136. Johnell O, Gullberg B, Kanis JA, et al. Risk factors for hip fracture in European women: the MEDOS Study. Mediterranean Osteoporosis Study. J Bone Miner Res 1995; 10:1802–1815.
137. Hemenway D, Azrael DR, Rimm EB, Feskanich D, Willett WC. Risk factors for hip fracture in US men aged 40 through 75 years. Am J Public Health 1994; 84:1843–1845.

29 Exercise and Bone Health

Maria A. Fiatarone Singh

1. OVERVIEW

Bone mass begins to decrease well before the menopause in women (as early as the 20s in the femur of sedentary women), and accelerates in the perimenopausal years, with continued declines into late old age *(1)*. Similar patterns are seen in men, without the acceleration related to loss of ovarian function seen in women *(2)*. As with losses of muscle tissue [sarcopenia *(3)*], many genetic, lifestyle, nutritional, and disease and medication-related factors enter into the prediction of bone density at a given age *(4–12)*. A wealth of animal and human data provide evidence for a relationship between physical activity and bone health at all ages. Mechanical loading of the skeleton generally leads to favorable site-specific changes in bone density, morphology, or strength *(13–18)*, whereas unloading (in the form of bed rest, immobilization, casting, spinal cord injury, or space travel) produces rapid and sometimes dramatic resorption of bone, increased biochemical markers of bone turnover, changes in morphology such as increased osteoclast surfaces, and increased susceptibility to fracture *(11,19–22)*. Among these models, spinal cord injury results in the most profound loss of skeletal mass (up to 45% at the pelvis), limited to the weight-bearing bones of the lower extremities and lumbar spine *(22)*.

Less extreme variations in mechanical loading patterns seen within normal populations are also associated with differences in bone morphology and strength. Comparative studies of athletic and nonathletic populations usually demonstrate significantly higher bone density in the active cohorts, ranging from 5% to 30% higher, depending on the type, intensity, and duration of exercise training undertaken, and the characteristics of the athletes studied *(23–25)*. Exceptions occur with non-weight-bearing activities such as swimming, or amenorrheic or competitive distance runners *(26)*, who appear similar to controls. Similarly, on a smaller scale, differences are often observed between habitually active and sedentary nonathletic individuals *(6,8,26–28)*. Consistent with such bone density findings, hip fracture incidence has been observed to be as much as 30–50% lower in older adults with a history of higher levels of physical activity in daily life, compared to age-matched, less active individuals *(29–35)*.

From: *Nutrition and Bone Health*
Edited by: M. F. Holick and B. Dawson-Hughes © Humana Press Inc., Totowa, NJ

Given this epidemiological and experimental background, it is important for health care professionals to understand the rationale and current recommendations for the use of exercise in the prevention and treatment of osteoporosis and osteoporotic fracture, and to place it in context with the other available strategies for this syndrome. *The optimal use of exercise in this syndrome is dependent on the practice of a sustained, adequate dose of the correct modality of exercise/physical activity in the target populations, while minimizing the risk of side effects.* The approach is thus similar to the pharmacological or nutritional management of osteoporosis. A "blanket" exercise prescription for all ages and all health conditions is not possible, as the relative importance of skeletal and nonskeletal contributants to osteoporotic fracture shifts over the course of the lifespan and in relation to other health conditions. Although there are still many unanswered questions with regard to the optimal role of exercise and physical activity in bone health, and in particular its ultimate efficacy for fracture prevention, we have learned a great deal from studies carried out over the past few decades. Although physical activity cannot be expected to rebuild extremely osteopenic bone to normal architecture and strength, existing literature supports the role of early, sustained, and appropriate patterns of loading in beneficial adaptations in bone size, shape, cortical wall thickness, trabecular architecture, and ultimately resistance to fracture. A summary of the current state of knowledge in this field, as well as the author's recommendations for effective and safe implementation of physical activity in various settings, is reviewed in the sections that follow.

2. EPIDEMIOLOGICAL ASSOCIATIONS BETWEEN PHYSICAL ACTIVITY PATTERNS AND BONE HEALTH

A large number of studies have attempted to define the role of physical activity and exercise patterns across the lifespan on bone density. Most of these studies are cross-sectional comparisons of healthy pre- or postmenopausal women. Due to the fact that higher physical activity level is often correlated with better health and nutritional status, the well-designed studies have attempted to control for these other risk factors for osteopenia in their analyses, as well as to identify specific periods of life during which physical activity confers benefit. For example, Krall *(28)* studied 239 healthy postmenopausal women prior to their entry into a trial of vitamin D supplementation, and found that those who reported walking more than 7.5 miles per week in the previous month had higher bone density (legs, trunk, and whole body) than those who reported walking less than 1 mile per week. Current walking was correlated with walking behavior earlier in life, suggesting that the bone density observed was a reflection of life-long physical activity patterns. Furthermore, in a prospective 1-yr follow-up in these women, the rate of loss of bone mineral density (BMD) in the legs (but not other sites) was inversely related to the current number of miles walked per week, pointing to the site-specific nature of the physical activity effect. Conflicting or nonsignificant results from other studies of walking may be due to differences in the measurement of physical activity, differences in calcium intake, as well as small sample sizes in most studies *(36)*. In another analysis of lifetime physical activity habits and bone density in old age, Vuillemin *(37)* found that sporting activity in youth was associated with lumbar spine BMD in older men and women, whereas more recent physical activity helped to preserve femoral BMD.

Fewer data are available for men, but they are generally consistent with the findings in women. Need *(38)* found that energy expenditure in physical activity was associated with age-corrected bone density at the femoral neck in men aged 20–83 in a dose-dependent fashion, and was associated with lumbar spine and well as all femoral sites in men under 50 yr of age. When men over 50 were evaluated separately, no relationship was found, the authors suggesting that exercise may have its major role in peak bone density in men. However, other studies of older men, such as Kenny's study of 83 men (average age 75) with low testosterone levels *(39)*, and larger studies of healthy community-dwelling cohorts *(40,41)*, do support an independent role for physical activity in femoral and total-body BMD among older men.

Physical activity has been linked to reduced osteoporotic fracture prevalence or incidence as well as higher bone density in older adults. In the prospective study of Paganini-Hill, men who were active for 1 h or more per day had a 49% reduction in the risk of hip fracture compared to less active men. In the Study of Osteoporotic Fractures *(34)*, women who reported walking for exercise had a significant 30% reduction in hip fracture risk compared to women who did not walk for exercise. In more detailed analyses of this cohort *(42)*, including types and intensities of all recreational and household activities, 9704 women over the age of 65 were followed for 7.6 yr for fracture incidence. Incidence of hip fracture was 42% lower in the most active compared to the least active quintile. A dose–response effect was seen for both volume and intensity, with low-intensity activity conferring a 27% risk reduction, compared to a 45% reduction associated with moderate-vigorous activity. Multivariate adjustments indicated that the relationship of physical activity to hip fracture in this cohort was minimally confounded by health status, functional status, history of falling, smoking, calcium and alcohol intake, estrogen use, or body weight, resulting in a significant 36% reduction in risk after adjustment. The protective effect of exercise on hip fracture was only marginally reduced by adding bone density, muscle strength, and falls to the model, suggesting that its mechanism of action is multifactorial and not completely understood. Additionally, adjusted vertebral fracture risk was reduced by 33% in women who reported moderate-vigorous exercise, whereas wrist fracture was unrelated to physical activity volume or intensity. In another large cohort, the prospective EPIDOS study of 6901 white women over the age of 75 followed for 3.6 yr, investigators found that a low level of physical activity increased the risk for proximal humerus fracture by more than twofold *(43)*. The relative risk (RR) of fracture in sedentary women (RR = 2.2) was greater than that attributable to low bone density (RR = 1.4), maternal history of hip fracture (RR = 1.8), or impaired balance (RR = 1.8). The interaction of these risk factors is indicated by the fracture rate, which rose from about 5 per 1000 women years in individuals with either bone fragility or high fall risk to 12 per 1000 woman-years for women with both types of risk factors. Such data suggests the great potential utility of multifactorial prevention programs for osteoporotic fracture that can address *both* bone density and fall risk (sedentary behavior, sarcopenia, muscle weakness, poor balance, polypharmacy, etc.) simultaneously.

In summary, cross-sectional and prospective data support a relationship between lifetime physical activity patterns and preservation of bone density into old age, as well as a protective effect for hip, humerus, and vertebral fractures. These reduced risks for fracture remain after adjustment for most major known risk factors for osteoporosis, and are not simply accounted for by alterations in bone density, muscle strength, or fall rates. Experimental evidence in animal models as well as some human data suggest that other changes in bone strength apart from density changes may contribute to the overall benefits of mechanical loading for skeletal integrity (e.g., increased bone volume or altered trabecular morphology) *(44–46)*, so that evaluating bone density changes alone may underestimate the skeletal effects of loading. This foundation of epidemiological research has laid the groundwork for the many experimental trials of exercise and bone health that have been carried out in healthy and clinical populations, as reviewed in the sections that follow.

3. PHYSICAL ACTIVITY AND BONE HEALTH DURING CHILDHOOD AND ADOLESCENCE

The goal of physical activity in relation to bone health in youth is to maximize peak bone mass, which is attained at various sites by age 16–26 yr in most studies *(47)*, in order to potentially reduce the burden and delay the onset of osteoporotic fracture in adults. Thus, any attempts to influence peak bone mass must occur very early in life, and have residual benefits that span 50 yr or more. It is thought that approx 20–50% of the variation in bone mass is due to modifiable factors such as hormonal status, physical activity patterns, and nutrition, while the remainder is explained by sex, race, hereditary, and familial factors.

Available data comes from cross-sectional studies of children and adolescents involved to various degrees in sport and recreational play, as well as specific exercise intervention trials.

3.1. Cross-Sectional Studies of Young Athletes

Studies of athletes are most definitive with regard to the relationship between BMD and physical activity. Loaded bones have 5–30% higher bone density than that measured in unloaded limbs or nonathletic control subjects *(48–52)*. Although genetic factors could partially explain the differences between athletes and nonathletes, the contralateral limb studies demonstrate that the side-to-side differences are primarily the result of specific effects of mechanical loading during training, as the BMD of nondominant arms in athletes are not different than in nonathletic controls *(53,54)*. The kinds of activities that appear to be most robust in their effect on the skeleton are those that include:

1. High-impact, rapid, forceful loading (jumping, running, gymnastics, volleyball)
2. Changing, diverse, or novel loading angles and magnitudes of forces over time (ball sports, gymnastics)
3. Weight-bearing, high forces (dancing, weight lifting, not swimming, water polo, cycling, cross-country skiing)
4. Activities that impact directly on the bone of interest (dominant arm of tennis players)

3.2. Cross-Sectional Studies of Nonathletic Children and Adolescents

As important as these studies of athletes are for understanding the potential of bone to respond to large volumes of forces of high magnitude, they do not adequately address the issue of the effectiveness of normal activities and play in the vast majority of children and adolescents who will never engage in competitive athletics. This kind of data has been gathered in a number of cross-sectional studies, which generally show that the difference in BMD between low and high activity or fitness categories in normal populations of children varies between 5% and 15% *(47)*. The effect of physical activity translates to slightly less than 1 SD, or an average of 7–8% higher BMD compared to sedentary children and adolescents. An increase in peak bone mass of this magnitude, if it were to be sustained until old age, would theoretically substantially lower the risk of osteoporotic fractures, as has been observed in some studies of lifelong physical activity patterns *(55)*. For example, each standard deviation (10%) decrease in femoral neck BMD may be associated with a 2.6-fold increase in the risk of fracture at this site *(3)*.

Some studies suggest that exercise effects on peak bone mass are particularly potent when the activity is begun before the onset of puberty, and is sustained throughout the young adult years. For example, the benefit of playing tennis or squash for BMD is approximately two to four times as great if female players started at or before menarche, as opposed to after the onset of puberty *(56)*. Mature bone appears to be far less responsive to even vigorous and extended training than developing bone *(57)*, suggesting that major public health impacts of physical activity recommendations for osteoporosis prevention should be directed at young children for optimal efficacy. Detraining results in a return of BMD toward normal values of sedentary individuals *(58)*, so that any activity undertaken should be ideally feasible and sustainable in some fashion throughout adult life.

3.3. Exercise Intervention Trials in Children

Experimental trials in children and adolescents have sought to corroborate the cross-sectional observations described above, as well as define the optimum modality, dose, duration, and intensity of mechanical loading required for robust skeletal adaptations. At the present time, this experimental evidence is not as strong as the observational data, and there is no consensus as to the optimal nature of the stimulus that should be applied to result in clinically meaningful alterations in peak bone mass and susceptibility to fracture in later life. Randomized controlled trials in children have all been short-term (2 yr or less), and generally have resulted in relatively small improvements in BMD of 5% or less *(59–62),* as compared to the larger differences observed in cross-sectional studies of athletic children and sedentary peers. For example, Witzke *(63)* conducted a controlled 9-mo study of plyometric jump training and exercises using weighted vests in 53 girls of average age 15 yr. There were no significant differences between groups in bone mineral content changes over time (perhaps because of the short duration of the intervention), strength, or balance, although exercisers tended to have larger improvements, particularly at the greater trochanter, as well as better balance performance. However, a recent systematic review of controlled trials of weight-bearing activity in children under 18 yr found that activity consistently resulted in positive effects on the total body, lumbar spine, and femoral neck *(64).* Exercise regimens used in successful trials have included jumping, hopping, ball sports, weight lifting, running, dancing, and gymnastic activities, often in the setting of modified school physical education classes.

3.4. Physical Activity Recommendations for Optimizing Peak Bone Mass

While acknowledging the need for a more complete experimental basis for a precise exercise prescription for skeletal health in children and young adults, some recommendations can be tentatively offered at this time. Beginning in childhood (before puberty), individuals who are able should be encouraged to engage in regular weight-bearing exercise, via a combination of lifestyle choices (such as walking to school or errands), structured sports and school-based physical education, and unstructured games (outdoor play). Participation in competitive sports, although associated with even more robust effects on bone, may be hindered by many barriers, including skill level, self-efficacy, gender bias, financial burden, lack of parental encouragement, time commitments, and travel requirements, and is not attractive to all children. Therefore, emphasizing the replacement of sedentary activities (TV, video games) with active outdoor play, physical education in school, and less reliance on mechanical modes of transportation (cars, elevators, escalators) is likely to have a far greater impact on public health and sustained behavioral patterns in adult life than simply encouraging more competitive team sports. If sports are chosen, those involving jumping from different angles, fast rates of loading, running, and lifting appear to have the greatest effect on bone. Games that can be chosen or modified to emphasize high-impact, rapid loading (such as jumping rope, hopping, skipping, jumping over objects or steps, etc.) should be promoted if otherwise safe to do so. It is critically important for growing children and adolescents to maintain adequate energy, protein, and calcium intake, and other health habits, as bone health is compromised in the presence of eating disorders, excess phosphate from carbonated drinks, smoking, and hormonal disturbances associated with very low levels of body fat, sometimes seen in very active young women with menstrual disturbances and/or eating disorders. The benefits of mechanical loading may be completely undone by concomitant amenorrhea in young females in the very important period from puberty to the attainment of peak bone mass in the mid-20s *(65,66)*.

4. PHYSICAL ACTIVITY AND BONE HEALTH IN PREMENOPAUSAL WOMEN

4.1. Overview of Experimental Trials

Although the majority of randomized controlled trials of exercise and bone health have been conducted in postmenopausal women, there have been a number of similar trials in premenopausal women conducted over the past decade, as reviewed in three recent meta-analyses *(67–69)*, and summarized in Table 1. Although the majority of these trials have lacked statistical power on their own to demonstrate significant treatment effects due to small sample sizes, the meta-analyses all concur that exercise has positive effects on BMD at the lumbar spine in young women. Aerobic training, resistance training, and combined aerobic and resistance programs all increase lumbar spine BMD by about 1% per year on average, relative to sedentary controls. The magnitude of the exercise effect is approximately equivalent to that seen with calcium supplementation, and is not substantially different at the lumbar spine than that observed in postmenopausal women (see below and Table 1). Although this treatment effect may appear small relative to the efficacy of new pharmacological agents, it does approximate the rate of loss of bone with age (1% per year), outside of the accelerated losses in the early postmenopausal years. Thus, it may be enough to counteract typical age-related losses of bone. Changes at the femoral neck or greater trochanter have been assessed less frequently in these studies of young women, but femoral neck changes have been seen to be significant in programs that combine aerobic and strength training *(68)* as well as in high-impact aerobic jumping/stepping exercise *(70)*, and trochanteric changes have been significant after high-impact exercise including jumping and skipping *(71)*, 50 jumps/6 d/wk *(72)* and jumping/lower extremity resistance training with a weighted vest *(73)*.

The studies in premenopausal women have ranged from 6 to 36 mo in duration, enrolling women with mean ages ranging from 16 to 44 yr. Unfortunately, the dropout rate has been relatively high, ranging from 0 to 68%, with five of the eight trials reported by Wallace *(69)* having dropout rates greater than 25%, for example. Such high dropout rates raise the issue of generalizability and feasibility of exercise programs for osteoporosis, which would presumably have to be sustained for decades to exert a meaningful influence on fracture rates in this cohort. Withdrawal of training likely results in rapid reversal of skeletal adaptation. For example, in a controlled trial of unilateral limb training, Vuori *(74)* trained young women 5 d/wk for 12 mo at high intensity on a leg-press machine, and found that

Table 1
Meta-Analyses of Physical Activity and Bone Density

Reference	*Population*	*Studies included*	*Total number of trials, subjects*	*Type of Exercise*	*Study treatment effect*[a]	*Significance Level*
Berard, 1997 *(157)*	Healthy women > 50 yr of age without osteoporosis	Prospective controlled intervention trials	18 trials 1966–1996	Walking, running, physical conditioning, aerobics	Lumbar spine effect size = 0.8745	Lumbar spine < 0.05 Forearm = ns Femoral neck = ns
Kelley, 1998 *(90)*	Postmenopausal women	Prospective controlled intervention trials	10 trials 1975–1994; 330 subjects	Aerobic activity	Lumbar spine +2.83%	Lumbar spine < 0.05
Kelley, 1998 *(91)*	Postmenopausal women	Prospective controlled intervention trials	6 studies, 1978–1995	Aerobic exercise	Hip +2.42%	Hip < 0.05
Kelley, 1998 *(155)*	Postmenopausal women	Randomized controlled trials	11 studies 1975–1995; 719 subjects	Aerobic or strength training	Any exercise +0.27% Aerobic +1.62% Strength +0.65%	All sites < 0.05
Wolff, 1999 *(68)*	Pre- and post-menopausal women	Prospective controlled intervention trials	25 studies, 1966–1996	Aerobic, high-impact, or strength training at least 16 wk duration	*Premenopausal* *Aerobic + strength* Lumbar spine +0.90% Femoral neck +0.90% *Postmenopausal* *Aerobic* Lumbar spine +0.96% Femoral neck +0.90%	All sites and modalities significant (< 0.05) except for strength training in post-menopausal women

					Strength training Lumbar spine +0.44% Femoral neck 0.86% *Aerobic + strength* Lumbar spine 0.79% Femoral neck 0.89%	
Wallace, 2000 *(69)*	Pre- and post-menopausal women	Randomized controlled trials	32 studies, 1966–1998	Impact (aerobic or heel drops) and strength training	*Premenopausal* *Impact* Lumbar spine +1.5% Femoral neck +0.90% *Strength training* Lumbar spine 1.2% Femoral neck insufficient data *Postmenopausal* *Impact* Lumbar spine +1.6% Femoral neck 0.9% *Strength training* Lumbar spine 1% Femoral neck 1.4%	All sites and modalities significant (< 0.05) except for femoral neck in pre-menopausal women
Kelly, 2001 *(67)*	Pre- and Post-menopausal women	Prospective controlled trials	29 studies 1966–1998, 1123 women	Resistance training	Femur +0.38% Lumbar spine +1.26% Radius +2.17%	Femur nonsignificant Lumbar spine <0.05 Radius <0.05

[a] Study treatment effect is the difference between the percentage changes per year in bone mass (bone density or bone mineral content) in the training group minus the control group. A positive figure indicates a protective effect of exercise. Results are annualized for studies less than 12 mo duration, assuming a linear rate of change in bone mass.

BMD increased and then returned to baseline values within 3 mo of detraining. A similar return to baseline values was seen after 6 mo of detraining in 30- to 45-year-old women who had undergone 1 yr of jumping/lower-extremity resistance training *(73)*. Thus, skeletal adaptations to training is rapidly eroded, even after relatively long-term exposure to high-intensity loading paradigms.

There are several other limitations to the current literature on exercise training in premenopausal women. There is significant heterogeneity in the exercise interventions used in terms of exercise modality, frequency, intensity, impact loading, duration, and specific muscles and joints targeted. This makes it extremely difficult to compare results across studies, and crude categorization into "impact" and "nonimpact" or "endurance" vs "strength" training, as is done in some meta-analyses, does not sufficiently describe the type or dose of exercise received, or explain potential differences between studies. For example, some lower-intensity strength-training regimens may be insufficiently robust to be grouped together with high-intensity, progressive-resistance training programs, resulting in a weakening of the aggregate effect size *(67)*. Even studies of high-impact exercises are not uniform, as investigators have created impact in various ways (jumping, stepping, skipping, plyometrics, weighted vests), and have prescribed these movements alone *(71)*, as part of an aerobic exercise routine *(70)*, or as part of a resistance-training routine *(73)*, making it very difficult to compare relative efficacy or offer firm recommendations. This heterogeneity has led some to question the validity or appropriateness of combining these studies analytically via meta-analysis, which may serve to hide important distinctions between exercise trials that could lead to improved preventive and treatment strategies *(75)*. This would be analogous to performing a meta-analysis of "drug treatment for osteoporosis," merging the data on calcium, vitamin D, estrogen, bisphosphonates, and selective estrogen receptor modulators into one overall effect size. Unlike some other health outcomes related to exercise, the osteogenic response to mechanical loading appears to be extremely exacting in its requirements. Even apparently minor variations, such as waiting a few more seconds between loading cycles, results in large differences in bone cell response *(76)*.

With the above caveats in mind, some general statements may be inferred from the existing data. Resistance and aerobic training regimens appear to have relatively equal efficacy on bone, although they differentially affect aerobic capacity and strength, as would be expected *(77)*. It is also not clear whether multimodal exercise programs are superior to single-exercise modalities, which have the potential advantage of greater simplicity and adherence. For example, in one of the largest and best-designed studies conducted in this cohort, a combination of aerobic and resistance training over 2 yr significantly improved BMD at the spine, femoral neck, trochanter, and calcaneal sites, as well as improving maximal aerobic capacity and muscle strength, compared to stretching exercise in 127 women aged 20–35 yr *(78)*. However, the dropout rate was 50%, and the compliance rate in the remaining 50% averaged only 61%. Significant bone changes were not observed until the second year of treatment, at a time when attendance was at its low point of 54% in the remaining subjects. A final limitation is that very few of these studies have enrolled women at high risk for osteoporotic fracture or followed them long enough to assess the impact of the exercise training on fall rates or fracture incidence. It is possible that the results seen in healthy women would not be replicable in high-risk individuals. For example, Hakkinen *(79)* applied a program of moderate-intensity resistance training for 2 yr in patients with rheumatoid arthritis, who are at elevated risk for osteopenia due to their disease, inactivity, and corticosteroid medications. Although muscle strength and disease activity improved with exercise, only small differences in femoral neck BMD (0.51%) were noted. Whether this modest adaptation in bone was related to the clinical status of the subjects or the less intense nature of the strength-training intervention (home-based, using elastic bands) is not clear. Thus, unlike analysis of the dose-effect in drug trials, it is difficult or impossible to define a single strategy that is supported by a strong evidence base or known to be *optimally* effective for bone at this time, as well as feasible over the long term in representative populations of young women.

4.2. Physical Activity Recommendations for Young Women

The recommendations offered for this age group are based on the available literature reviewed above, and represent the author's interpretation of current evidence, in the areas of modality, intensity, frequency, and dose of exercise to prescribe.

4.2.1. Modality

After the age of 30, the physical activity prescription for the maintenance of bone health should be viewed more comprehensively than that offered to children, as both skeletal and nonskeletal risks for osteoporotic fracture need to be considered. Peak levels of muscle and bone have already been attained, and femoral bone mass may have already started to decline. Chronic health conditions that may influence bone density or temper the exercise prescription may have started to emerge. Recreational and occupational physical activity levels have simultaneously started to fall in many adults. Although weight-bearing aerobic exercise, high-impact training, and resistance training have all been shown to maintain or augment bone density in this stage of life, resistance training has the added benefit of increasing muscle mass and strength, as well as balance. In the meta-analysis by Kelley *(67),* for example, resistance training in premenopausal women resulted in significant changes in lean mass (+2 kg) and muscle strength (+40%) and losses of body fat (–2%), compared to minimal changes in the control groups. This combination of effects on body composition and muscle function is a direct antidote to age-associated changes in these domains, and offers potential benefit for many health conditions in addition to osteoporosis. Aerobic exercise does not increase muscle mass and strength, and does not improve balance, and is therefore less comprehensive in its effects on the multiple risk factors for osteoporotic fracture (*see* Table 2). Additionally, there is no evidence in young women to support the isolated use of

Table 2
Multifactorial Risk Factor Targeting for Bone Health

Risk Factor for Osteoporotic Fractures	*Preventive or Therapeutic Options*
Osteopenia	Bisphosphonates, SERMS, HRT, PTH Vitamin D Calcium **Resistance or aerobic training, high-impact training**
Sedentary behavior	**Exercise or physical activity prescription**
Falls	**Resistance training** **Balance training** Multifactorial risk factor interventions (polypharmacy, vision, lower extremity dysfunction, footwear, assistive devices, home safety, etc.) Hip protectors Evaluate and treat postural hypotension
Muscle weakness	**Resistance training** Vitamin D
Impaired balance	**Balance training** Hip protectors
Depression, antidepressant medications	**Substitute aerobic or resistance training for antidepressant medication if possible**
Protein and calorie undernutrition, weight loss	Nutritional counseling and support **Resistance training to increase nitrogen retention and appetite**
Polypharmacy	Drug review and modification as appropriate
Visual impairment	Ophthalmologic evaluation and treatment as appropriate
Smoking and excess alcohol intake	Reduce or eliminate

SERMS, selective estrogen receptor modulators; HRT, hormone replacement therapy; PTH, parathyroid hormone.

aerobic training that does not involve high-impact forces as a means to maintain or augment femoral bone density, whereas programs that include resistance training and/or high-impact training have been shown to benefit the skeleton at this clinically vital site. Therefore, the most economical prescription with the broadest benefits for body composition and bone health as well as neuromuscular function would be resistance training as the exercise modality. Adding high-impact forces/movements may further enhance benefits for the femoral neck or trochanter, lower-extremity muscle power, and dynamic balance *(70),* but no direct comparisons of these two modalities are available. Rest periods between sets of weight-lifting exercise may be used to complete 10–20 jumps if feasible (depending on the presence or absence of previous injuries or osteoarthritis of the knees and hips). Such a routine incorporates resistance training and high-impact loading in one session without extending the time required, an economical prescription for busy adults.

4.2.2. Intensity

The physiological response in bone and muscle is proportional to the magnitude and rate of strain imposed *(76)*, and successful programs have utilized intensities at the higher ranges in general. Therefore, moderate to high-intensity progressive resistance training and/or high-impact training is recommended as the primary intensity of planned exercise in this age group. It should be noted that high-impact programs have successfully increased trochanteric BMD by 3–4% in young women via jumps approx 8 cm off the ground. This kind of jump produces ground reaction forces that are three to four times body weight (thus high impact), but are feasible for nonathletic women, are infrequently associated with injuries, and able to be completed in approx 2 min per day *(72)*.

4.2.3. Volume

Two or three days of weight lifting, aerobic exercise, or high-impact programs per week have been shown to augment bone density significantly compared to sedentary controls if continued for at least 1–2 yr. This volume of weight-lifting training is also sufficient for the other body composition changes and improvements in muscle strength, power, and balance as well (*see* Table 3). The total number of loading cycles, or repetitions, needed is not known, but animal studies do not show benefits of very high volumes compared to small numbers of loading cycles. For example, classical early studies showed that 36 loading cycles in a turkey ulna is not different than 1800 loading applications of the same load *(44)*, and more recent evidence in rat models confirms this finding: after 50–100 cycles of loading in a given bout, additional repetitions are largely ineffective for stimulating further osteogenic response *(80)*. Bassey showed that 50 jumps of 8.5 cm height, 6 d/wk over 6 mo, was associated with a 2.8% increase in trochanteric BMD compared to controls *(72)*. It is possible that the accelerated adaptation in bone in this study compared to other studies of high-impact exercise in young women was related to the greater number of sessions (six per week compared to three per week) typically prescribed. Overall, the clinical trials literature would support a recommendation of approx 40–50 jumps or repetitions of a given weight-lifting exercise per training day, and this is consistent with our current understanding of osteogenic stimuli from animal studies.

4.2.4. Frequency

Animal data on osteogenic adaptation also suggest that the capacity of bone cells to respond to mechanical signals is quickly saturated by repetitive loading cycles without rest periods *(76)*. Although comparable human data are not available and

Table 3
Comparison of Exercise Recommendations for Specific Body Composition Outcomes in Older Adults

Exercise Recommendations	*Adipose Tissue Mass and Visceral Deposition*	*Muscle Mass and Strength*	*Bone Mass and Density; Fracture Risk*
Modality	Aerobic or resistance training	Resistance training	Resistance training or aerobic training[a] High-impact activities (jumping using weighted vest during exercise) if tolerated by joints Balance training
Frequency	3–7 d/wk	3 d/wk	3 d/wk Balance training: up to 7 d/wk
Dose	30–50 min/session	Two to three sets of 8–10 repetitions of six to eight muscle groups	Two to three sets of 8–10 repetitions of six to eight muscle groups 50 jumps per session for high impact Two to three repetitions of 5–10 different static and dynamic balance postures
Intensity	60–75% of maximal exercise capacity (VO_2 max or maximal heart rate) or 13–14 on the Borg Scale of perceived exertion	70–80% of maximal strength (one repetition maximum)	70–80% of maximal capacity (one repetition maximum) as load 5–10% of body weight in vest during jumps; jumps or steps of progressive height Practice most difficult balance posture not yet mastered

[a] Aerobic exercise should be weight-bearing modalities of exercise with high ground-reaction forces (e.g., walking, jogging, running, stepping, rather than swimming or cycling).

are needed to confirm these findings, turkey and rat models strongly suggest that optimal recovery periods are 10–14 s between loading cycles (repetitions), and 8 h between bouts of loading (training) *(76)*. Such long rest intervals between repetitions are longer than currently prescribed by most trainers, who wait only 1–2 s between repetitions, but are certainly not detrimental to the muscle function outcomes, and are likely to enhance excellent adherence to form and thus minimize injury. Understanding of muscle adaptation to mechanical loading also indicates that recovery periods between bouts are necessary to maximize hypertrophy and prevent injury *(81)*, and are even longer than those recommended for bone (usually 1 d off between sessions). As noted earlier, it is possible to do isolated jumping exercise as frequently as 6 d/wk and produce significant bone density changes in the femur, although this regimen did not significantly improve leg power or balance in healthy young women *(72)*.

Thus, recommending exercise no more frequently than every other day (approx 3 d/wk) satisfies both muscle and bone requirements, and is not overly burdensome to most individuals.

5. PHYSICAL ACTIVITY IN POSTMENOPAUSAL WOMEN AND OLDER MEN

5.1. Overview of Experimental Trials

Many of the observations made in young women in the previous section are also relevant to older adults, and therefore similarities, differences, and special considerations in the older population will be highlighted in the sections that follow. As indicated in the meta-analyses including postmenopausal women in Table 1, significant changes in the femur, lumbar spine, and radius have been seen following aerobic training, resistance training, and combined programs of aerobic and resistive exercise. Although not enough data are yet available for meta-analysis, high-impact training has not been found to be effective in several studies in postmenopausal women. Although earlier studies suffered from small subject numbers, nonrandomized designs, short intervention periods, and other methodological flaws, more recent trials have enrolled larger populations, targeted women with osteoporosis *(82)* or previous osteoporotic fractures *(83)*, and continued observations for 2–5 yr *(84,85)* in some cases. High dropout rates (30–50%) have continued to be problematic in this cohort, as in younger women, raising the issue of generalizability and sustainability of the outcomes observed. This is particularly relevant to bony adaptation, as several studies have shown complete or partial reversal of gains in BMD after the cessation of training *(82)*.

5.2. Modality of Exercise for Bone Health in Older Women

At this stage in the lifecycle, a combination of decreased anabolic hormones (estrogen, testosterone, growth hormone), increased catabolic milieu (higher leptin and cortisol associated with visceral adipose tissue) *(86–89)*, the emergence of musculoskeletal and other diseases, retirement, and reduced recreational activities have a major negative impact on bone as well as muscle tissue. The majority of studies demonstrating the efficacy of aerobic or resistive exercise on bone density have been conducted in women between 50 and 70 yr of age, and it is not yet known if efficacy would be similar in older women with multiple co-morbidities, who have usually been excluded from such trials. Both types of exercise have approximately equivalent effects on bone health in postmenopausal women of about 1–1.5% per year between exercisers and nonexercisers in meta-analyses of well-designed trials *(67–69,90,91)*. Meta-analysis may not distinguish sufficiently between exercise modalities, however, since intensity and adequacy of training techniques are rarely considered in such analyses, and nonrobust "resistance training" interventions using body weight or elastic bands may dilute the effectiveness of more appropriate physiological stimuli in high-intensity weight-lifting programs, for example.

5.2.1. Aerobic Exercise vs Resistance Training in Older Women

In general, the older the individual, the more favorable resistance training appears, due to its broader benefits on muscle, bone, balance, and fall risk, relative to aerobic training (*see* Tables 2 and 3). If aerobic training is chosen, however, activities that are weight bearing and high impact have a greater efficacy than non-weight-bearing or low-impact aerobic activities. For example, women who walked 4 d/wk for 50 min at 75–80% of maximum heart rate while wearing a 3.1-kg leaded belt maintained lumbar spine trabecular bone mineral density (+0.5%) compared to a 7% loss in sedentary controls after 1 yr *(92)*. Despite this vigorous level of aerobic exercise involving the legs, no changes were seen in the femur. In contrast, McMurdo *(93)* randomized 118 older women to aerobic exercise classes 3 d/wk for 2 yr and found no changes at the lumbar spine and a modest effect at the distal radius. It is possible that the lower intensity of this exercise regimen limited the skeletal response. Even longer intervention periods may be required for positive results at the femur. In a 3-yr study of self-paced brisk walking vs upper-extremity exercise 3 d/wk in 165 women recruited from an emergency department after upper-extremity fracture, Ebrahms found that there was a trend for less decline in femoral neck BMD in walkers (–0.25% vs –2.8%, $p < 0.056$), and no significant difference at the lumbar spine between groups. Notably, fall rates were increased in the walkers, although fracture rates were similar, pointing to the possible risks of prescribing isolated aerobic exercise without other fall-prevention measures such as concurrent balance or strength training in women with a history of fall-related fractures.

Effective resistance training regimens have usually involved high-intensity (70–80% of peak capacity as the training load) training that is progressed continually over the course of the intervention *(67,94,95)*. However, using a similar protocol (2 d/wk at 80% of the one-repetition maximum), McCartney reported no bone changes in 142 older men and women after 2 yr of high-intensity weight-lifting exercise, compared to *increases* in whole-body and lumbar spine BMD and content in sedentary controls. There is no obvious explanation for these results. In some resistance-training studies, although significant increases in bone density have not been seen, changes in muscle strength correlate with local changes in BMD *(96),* suggesting that more robust adaptations in muscle parallel osteogenic responsiveness.

The relative efficacy of aerobic vs resistive exercise regimens for postmenopausal women may be perhaps best assessed via recent studies that have directly compared various intensities of these two exercise modalities in randomized subjects. Kohrt *(97)* found that both aerobic activities with high ground-reaction forces (walking, jogging, stair climbing) and exercises with high joint-reaction forces (weight lifting, rowing) significantly increased BMD of the whole body, lumbar spine, and Ward's triangle, whereas only the ground-reaction group increased BMD at the femoral neck *(97).* The weight-lifting group preserved femoral neck BMD relative to controls, as has been seen in other resistance-training studies *(84,95).* However, lean mass and muscle strength increased only in the weight-lifting group, leaving overall benefits of these two types of exercise for ultimate fall and fracture prevention still unresolved. Heinonen *(98)* compared 18 mo of moderate–high-intensity (55–75% maximal aerobic capacity) weight-bearing aerobic training to low-intensity (body weight plus 1–2 kg) strengthening calisthenics or a stretching control group. The femoral neck BMD was preserved only in the aerobic group, relative to losses over time in the other two groups ($p = 0.043$), with no changes at the lumbar spine. The low intensity of the strengthening regimen employed here probably explains the results observed. By contrast, Humphries *(99)* compared high-intensity strength training to low-intensity walking over 2 yr in 64 postmenopausal women. In this study, walking was associated with decreased lumbar spine BMD of 1.3% at 6 mo, compared to minimal changes in the weight lifters ($p = 0.06$ for group effect). In contrast to the high ground-reaction forces employed in the brisk walking, jogging, and stair climbing regimens of Kohrt *(97)* and Heinonen *(98)* described above, the low ground-reaction forces associated with self-paced group walking in Humphries' study were insufficient to preserve or augment BMD losses of aging. In a well-designed comparative study *(84),* Kerr randomized 126 postmenopausal women to 2 yr of high-intensity weight-lifting exercise, moderate-intensity aerobic training (circuit training and stationary cycling), or sedentary control condition. Total hip and intertrochanteric BMD was improved only by strength training, and was significantly different than aerobic training or control groups (+3.2% at 2 yr). Thus, it is important to consider not only the optimal modality of exercise, but also the relative intensity, as the skeletal adaptation is critically linked to the *intensity* of the loading (whether due to increased amount of weight lifted during resistance training, or higher ground-reaction forces during aerobic/jumping activities). As most comparative studies other than Kohrt's *(97)* and Kerr's *(84)* have not sought to optimize both modalities, it is still not possible to definitively choose one best modality for all bone sites. A consideration of nonskeletal risk factors for osteoporotic fracture (muscle weakness, poor balance, sarcopenia), however, would clearly favor high-intensity resistance training over high-intensity aerobic training *(67,92,95,97).* The relative importance of these risk factors in specific individuals, as well as co-morbidities that may affect their tolerance of specific exercises (see below), may guide the prescription of exercise therefore.

5.2.2. Intensity of Resistance Training

Within the resistance training and bone literature, there is a great deal of heterogeneity in intervention techniques utilized, as well as in skeletal adaptations observed. The predominant training factor that appears to influence effectiveness is the intensity and novelty of the load, rather than the number of repetitions, sets, or days per week, or even total duration of the program. This observation is also true for animal models of mechanical loading, in which bone is most sensitive to short periods of loading characterized by unusual strain distribution, high strain magnitudes, and rapid rate of loading *(76)*. For example, Sinaki *(100)* prescribed back extension exercises at 30% of the 1RM and shoulder girdle exercises at low to moderate intensity and found no significant improvements in spine or femoral BMD in 96 women even after 3 yr. In a well-designed randomized trial comparing two different intensities of weight-lifting exercise in postmenopausal women, Kerr *(101)* found that 1 yr of strength training at high intensity (3 sets of 8 repetitions) significantly increased BMD at the femoral trochanter, intertrochanteric site, and Ward's triangle, as well as the ultra-distal forearm, compared to low-intensity training (three sets of 20 repetitions), which produced no significant changes in BMD at any site except the mid-forearm. Changes in muscle strength were correlated with changes in BMD only in the high-intensity group. Interesting results have been reported more recently by Cusslet et al. *(102)*, in a randomized trial of 140 postmenopausal women participating in a multimodal exercise program (high-intensity resistance training, and a weight-bearing circuit of moderate-impact activities including walking/jogging, skipping, hopping, stair climbing/stepping with weighted vests). Bone density improvements at the femoral trochanter were significantly and linearly related to total weight lifted during the 12 mo, as well as total weight lifted in leg press, squats, and military press exercises, but not to volume or quality of the nonresistance training components of the program. High-intensity resistance training is also more beneficial than low-intensity training for muscle strength gains and muscle hypertrophy, as well as associated gait disorders, functional impairments, and disability, making it ideal as a multiple-risk-factor intervention strategy for injurious falls in osteopenic women *(95,103–108)*.

5.2.3. High-Impact Exercise in Older Women

The theoretical utility of high-impact exercise for bone health is not matched by a wealth of experimental data in older women. This is likely due to the difficulty in implementing such exercise regimens in a cohort who are more likely to have osteoarthritis and other underlying joint abnormalities predisposing to injury. However, a small number of studies have been conducted. Bassey randomized postmenopausal women to heel drops (1.5 times body weight) or control conditions, and found no difference in BMD after 12 mo, perhaps due to the smaller impact of this regimen compared to jumping *(109)*. Subsequently, Bassey and colleagues also reported that the same jumping intervention (50 jumps, 6 d/wk) successfully utilized in premenopausal women did not significantly improve BMD in 123 postmenopausal women exercising for 12 mo, nor in a smaller subset of 38 women after 18 mo. It is not clear why the jumping was ineffective, as the height of the jumps was just as great in the older women, and the rate of loading and magnitude (four times body weight) of the ground-reaction forces were even higher than in the premenopausal women, who gained 2.8% in femoral BMD after only 6 mo *(72)*. Snow *(110)* conducted a randomized trial of lower-extremity resistance training and jumping while wearing a weighted vest in 40 postmenopausal women, which improved neuromuscular function but not bone density at 9 mo. However, in a nonrandomized 5-yr follow-up of a subset of 18 women from the original cohort who agreed to continued testing or training, BMD was increased or maintained at all hip sites (including the femoral neck) relative to losses of 3.4–4.4% in controls who were active, but not using weighted vests, jumping, or doing resistance exercise *(85)*. Thus, to date there is very little evidence that prescribed high-impact activities that appear optimally effective in adolescents and premenopausal women are osteogenic in older women. It is possible that alterations in the loading frequency or recovery period are needed, or that insufficient muscle contractile forces are generated by the older women due to underlying sarcopenia or muscle weakness at baseline. Additional studies are clearly warranted to determine what is limiting bone adaptation to impact loading in this cohort.

5.3. Exercise in Older Men

Very few intervention trials have been conducted in older men, despite the growing importance of osteoporosis in this cohort as well *(2)*. Specific subgroups of older men at increased risk include:

1. Men with habitually low lifetime levels of physical activity *(6,38)*
2. Men with a history of alcohol or tobacco dependence *(4,111)*
3. Hypogonadal men *(39)*
4. Men on chronic corticosteroid therapy for chronic lung disease, organ transplant, immunosuppression, etc. *(112,113)*
5. Men with chronic renal failure
6. Men with spinal cord injury or other neurological disease associated with mobility impairment
7. Men with protein-calorie malnutrition

Osteopenia associated with corticosteroid usage appears to be completely eliminated by concurrent progressive resistance training, and should be recommended for all such patients *(113)*. An excellent target group for such health promotion efforts is older men with steroid-dependent chronic lung disease *(114,115)*, in whom pulmonary cachexia,malnutrition, tobacco use, and steroid myopathy and osteoporosis combine to produce profound wasting, osteoporotic fracture, and impaired exercise tolerance *(116,117)*. Aerobic training will improve functional status in this clinical cohort, but is insufficient to address the musculoskeletal wasting *(118,119)*.

In healthy older men, high-intensity resistance training has been shown to increase BMD at the lumbar spine and greater trochanter compared to controls *(120)*, similar to results in older women. In one of the few studies of older men and women with physical frailty, Kohrt *(121)* compared low-intensity home-based physical therapy to supervised high-intensity resistance training over 9 mo. The high-intensity weight-lifting group had significantly better BMD at the whole body and Ward's triangle compared to the low-intensity exercise group at the end of the study, again demonstrating the superior efficacy of more intensive exercise, as has been shown in most studies of healthy pre- and postmenopausal women.

6. REHABILITATION AFTER OSTEOPOROTIC FRACTURE

Although not the focus of this review, in older men and women who have already sustained an osteoporotic fracture, exercise is still extremely important to assist in recovery of function as well as prevent recurrent injurious falls *(122)*. Progressive resistance training has been shown to be superior to standard physical therapy during the recovery from hip fracture in elderly patients. In addition, resistance training has been shown to be a potent treatment for depression in the elderly, and may thus be able to substitute for antidepressant medications, which are known to increase the risk of falls and hip fracture *(123,124)*. A combination of resistance training and balance training may offer the best approach to rehabilitation in this setting, as it optimally targets the remediable physiological risk factors for falls, fractures, and disability in this cohort. Additional studies are needed to define the effects of training in this clinical setting on bone density and strength itself, as well as the optimal timing and duration of such interventions in the post-fracture recovery period.

7. EXERCISE, MEDICATION, AND NUTRIENT INTERACTIONS AND BONE HEALTH

Medication–exercise interactions need to be considered in the overall context of bone health. Some drugs, such as corticosteroids (see above) and thyroid hormone, increase the risk of osteopenia and fracture, and should always be accompanied by an appropriately intensive physical activity prescription. In the case of corticosteroids, this should be resistance training, due to the need to oppose both the myopathy and osteopenia of corticosteroid administration *(113,125)*. By contrast, aerobic exercise has not been shown to significantly improve BMD in the setting of chronic prednisone therapy *(126)*. Both oral and inhaled corticosteroid treatment carry this risk of osteopenia and therefore the need for prophylaxis. There is some evidence that excess alcohol consumption may adversely effect bone mineral density or increase risk of osteoporotic fracture *(12,111)*, and therefore moderation of alcohol intake is recommended. The adverse consequences of excess alcohol intake on testicular atrophy, myopathy, peripheral and central neurological function, gait, balance, coordination, and judgment, in addition to impairment of bone metabolism, may combine to produce a high risk profile for injurious falls and fractures in the elderly. Finally, as is the case with muscle mass, weight loss, and inadequate intake of protein and energy will lead to losses of bone and increase hip fracture risk *(31)*, and are associated with poor recovery and excess mortality after hip fracture. Thus, prevention and treatment of energy or protein malnutrition and early assessment of unintentional weight loss is part of the management plan for optimal bone mass and health.

Oral contraceptive (OC) use in young women has been associated with reduced bone turnover and lower BMD in this cohort. Exercise in the setting of oral contraceptives appears to have a complex and site-specific effect on bone. The negative effect at the femoral neck was only partially counteracted by a 24-mo program of combined cycling, jumping rope, and high-intensity resistance training in 123 women aged 18–31 *(127)*. However, in a subsequent report of this same trial, the investigators noted that the exercise resulted in an increase of total body BMC, a decline in femoral neck BMD, while preventing increases in BMD and content seen in the spine of the OC–nonexercise group *(128)*. The authors suggest that suppression of bone turnover and resorption due to OC use modifies the skeletal response to exercise, and that inadequate calcium intake may worsen this interaction. Effective countermeasures for this pharmacological effect are warranted, given the prevalence of OC use in young women during the years when peak spinal BMC is usually attained.

On the other hand, some investigators have sought to maximize bony adaptation via the combination of exercise, nutritional, and pharmacological treatments for osteoporosis. Most studies of exercise have supplemented subjects with calcium, and in some cases vitamin D, to equalize baseline status of these nutrients. It makes sense to ensure adequate intake in these nutrients so as to optimize the skeletal milieu available for osteogenesis before exercise is begun. Specker has reported that exercise resulted in increased spinal BMD only in postmenopausal women consuming more than 1 g of calcium per day in an analysis of 16 trials *(129)*. It appears that adequate mineral must be present for bone remodeling to occur under the stimulus of mechanical loading, Given the prevalence of calcium and vitamin D deficiencies in the general population, the recommendation to ensure adequate dietary intake of calcium in all exercising men and women, and adequate vitamin D status in older men and women, is supported by current knowledge of skeletal physiology and nutrient requirements *(130)*.

The combination of exercise and postmenopausal estrogen replacement therapy for bone accretion has been studied by several groups of investigators. For example, Kohrt *(131)* reported a trial of 32 postmenopausal women assigned to groups by matching for body weight. The combination of exercise plus hormone replacement therapy (HRT) resulted in increased BMD at all sites except the wrist, with effects being additive for the lumbar spine and Ward's triangle and synergistic for the total body. Based on reductions in serum osteocalcin levels, these authors attributed increases in BMD in response to HRT and exercise plus HRT to decreased bone turnover (decreased bone resorption) and not increased formation, although increases in osteoblast-mediated bone formation have been linked to exercise benefits by other investigators *(16,132)*. Additive or synergistic effects of estrogens and various forms of exercise, including resistance training, have been reported by others in both animal *(133)* and human studies *(94,134)*. Thus, it appears that for women on HRT, additional skeletal benefits are likely when combined with either resistance training or aerobic exercise, along with the spectrum of nonskeletal health benefits of increased physical activity.

Bisphosphonates are now the mainstay of pharmacological osteoporosis prevention and treatment for many postmenopausal women, and it is reasonable to consider combining this antiresorptive treatment with the potentially anabolic effect of exercise on bone. Five studies have been identified that address this topic. Tamaki found that the bisphosphonate etidronate decreased osteoclast number while exercise increased osteoblast number at the proximal femur in ovariarectomized rats, thus providing complementary pathways to the increased bone density observed with combined treatments *(135)*. In the earliest human report by Vico, young men were randomized to etidronate, exercise, both treatments, or placebo during 120 d of bedrest *(136)*. The two groups receiving bisphosphonates had decreased markers of bone resorption on iliac crest biopsy, while exercise alone resulted in increased bone resorption. In a subsequent study, Grigoriev *(137)* reported that exercise and etidronate improved calcium balance more than exercise alone in young men during 1 yr of bed rest. Using a similar randomized factorial design to that of Vico, comparing etidronate, resistance training, both treatments, and placebo in 48 postmenopausal women for 1 yr, Chilibeck *(138)* found that etidronate significantly increased whole-body and lumbar spine BMC, whereas exercise had no direct or interactive effect on bone. However, only the resistance training improved muscle strength and lean body mass and decreased fat mass. The only human study to report a beneficial interaction of bisphosphonates and exercise was recently reported by Uusi-Rasi, in a 12-mo trial of jumping and alendronate in postmenopausal women *(139)*. Alendronate plus jumping preserved bone mass and exercise increased bone area in the distal tibia compared to controls. It is not clear whether differences in drug, type of exercise, site of bone measurement, or assessment of bone geometry rather than simply mass or density explain discrepancies between these studies, but clearly more research is warranted. It is important to note, however, that even if exercise does not add to bisphosphonate effects on bone, the combination of effects on bone, muscle, fat, and function noted in Chilibeck's study *(138)* provide a comprehensive package of health-related benefits.

8. PRACTICAL IMPLEMENTATION OF EXERCISE PROGRAMS FOR BONE HEALTH

Habitual exercise has a relatively potent effect on BMD in epidemiological and cross-sectional investigations *(8,140,141)*, and both weight-bearing aerobic exercise *(131,142–148)* as well as resistive exercises *(95,113,149–152)* have positive effects in experimental trials. The weight of the evidence suggests that while aerobic exercise may be effective as a prevention strategy in younger adults, it is likely that significant shifts in bone mineral compartment in older adults are more robust with weight-lifting exercise. In addition, it has been shown that high-impact forces to bone (jumping, using weighted vest with stepping, jumping, and resistive activities) are likely to have greater effects than low-impact or low-loading activities *(85)*. By contrast, wearing a weighted vest without the additional prescription of specific exercises has no impact on bone density or functional status *(153)*. The difficulty comes in the attempt to prescribe high-impact activities (such as jumping while wearing a weighted vest) in older adults with both osteoarthritis of the hips and knees as well as risk of osteoporotic fracture and falling. It is doubtful that high-impact activities such as jumping would be feasible in such a patient profile, and could result in exacerbation of arthritis as well as fall-related injuries. In such cases, therefore, a *low-impact* but *high-loading* form of exercise (such as seated and standing weight lifting with machines or free weights) would be both effective and tolerable. In general, because the effects of muscle contraction on bone appear to be primarily regional (electromagnetic field stimulation of osteoblast function) rather than systemic, it is advised that muscle groups connected to bones of relevance to osteoporotic fracture be emphasized in such a program (e.g., spinal extensor muscles, hip abductors, hip extensors, knee knee flexors) as well as those related to gait and balance (ankle plantar flexors and dorsiflexors). Specific exercise recommendations and suggested modifications for common co-morbid diseases of older men and women are presented in Table 4.

In addition to the above considerations, activity recommendations for the older age group should include avoidance of forward flexion of the spine, particularly while carrying an object (bowling, bending over to pick up something from the floor, sit ups with straight legs, etc.). Such actions increase the risk of anterior compression fractures of thoracic vertebrae in the presence of osteopenia. Similarly high-risk activities or hazardous environments that may lead to falls in those with poor balance are best avoided. Potential risks of exercise in individuals and suggested means to avoid such complications are presented in Table 5. It should be noted that the literature to date documents very few adverse events attributable to exercise, including resistance training and impact loading in older women, attesting to the relative safety of such prescriptions in supervised and unsupervised settings. However, few trials have included subjects with significant medical conditions that might increase the risk of exercise-related injury, and such trials are clearly needed.

Table 4
Specific Exercises for Bone Health and Modifications for Physical Limitations

Exercise Modality	*Standard or Optimal Mode*	*Modification for Arthritis*	*Modification for Frailty/ Neuromuscular Impairment*	*Modification for Cardiovascular/ Pulmonary Disease*
Progressive resistance training	8–10 exercises for major muscle groups, including muscles attaching to greater trochanter and vertebral bodies, as well as gait and balance[a] Include novel planes of movement, free weights, standing postures if possible High intensity (approximately 80% of peak capacity, progressed continuously)	Provide exacting attention to form to prevent injuries May need to limit range to pain-free motion, provide good back support, adjust machines or free weights to accommodate joint deformities or restrictions Intensity may need to be individualized for some exercises May need to medicate for pain prior to exercise	Usually little modification needed May need to alter certain exercises for neurological impairment Supervision usually needs to be more intensive for safety and progression	Usually no modification needed If angina or ischemia is provoked by exercise, keep intensity below the level at which this occurs May need to perform exercises in seated rather than standing positions due to fatigue or poor balance Avoid breath holding, Valsalva maneuver, sustained isometric contractions, or tight handgrip during weightlifting
Aerobic training	Moderate to high intensity Weight-bearing High ground-reaction forces (jogging, stepping, jump rope, etc.)	May need to reduce or eliminate weight-bearing or high-impact component: substitute brisk walking, stair climbing for jogging, step aerobics	May need to substitute seated exercises if weakness or poor balance prevents standing postures May need to begin with low–moderate intensity level and short sessions until improved	Keep training intensity below the level that causes ischemia or severe dyspnea Walk or exercise beyond the onset of claudication if possible (1–2 min); then rest and repeat Avoid breath holding, Valsalva maneuver, sustained isometric contractions, or tight handgrip during

High-impact exercise	Jumping, stepping off boxes, jump rope Progressively increase height of jumps or boxes, hop on one leg	May need to reduce or eliminate high ground-reaction forces (heel drops instead of jumps) Substitute power training (rapid concentric muscle contraction against moderate to high load on weight-lifting machine) to produce rapid onset of high muscle contraction forces as in take off of jump, but with no impact	Start with heel drops instead of jumps Perform exercises under supervision and while holding onto a support rail initially Gradually reduce hand support as tolerated	Keep training intensity below the level that causes ischemia or severe dyspnea
Balance training	Combine progressively more difficult static and dynamic postures Reduce base of support Perturb center of mass Withdrawn vision Increase compliance of standing surface (decrease proprioception) by using pads, mattress, pillows to stand on Incorporate postures from yoga and t'ai chi which emphasize the above principles	May not be able to place full body weight on osteoarthritic joints—use less painful leg to perform one-legged postures, assist weight bearing with use of cane Keep sessions short to avoid pain from prolonged weight bearing Reduce angle of flexion at knee during t'ai chi movements	Perform exercises under supervision and while holding onto a support rail initially Gradually reduce hand support as tolerated	Usually none

[a] Most important exercises include leg press, squats, knee extension, hip abduction, hip flexion, dorsiflexion, military press, lat pull down, back extension, abdominal muscles

[b] One-legged standing, tandem walking, crossover walking, turning, stepping over objects, leaning to limits of sway, etc.

Table 5
Risks of Exercise in Osteoporosis

Potential Risk	*Preventive Strategy*
Injurious fall	Prescribe balance training prior to aerobic training if gait and balance are impaired
	Prescribe progressive resistance training for sarcopenia and muscle weakness
	Optimize lighting, visual aids, safety of exercise environment, climate conditions, footwear, judgment
	Review medications for agents which may cause falls, postural hypotension, or altered central nervous system function
Spinal compression fractures	Avoid forward flexion with loading of the spine
	Avoid twisting movements of the spine
	Emphasis good sitting and standing posture
	Avoid or modify sports/activities involving spinal flexion (bowling, biking, golf, gardening, vacuuming)
	Bend knees rather than spine to pick up or reach low objects
Dislocation of total hip prosthesis	Avoid internal rotation and flexion of the hip
Pain from osteoarthritis	Use low-impact, high-intensity exercises (such as weight lifting) rather than high-impact exercises (jumping, stepping, jogging)
	Emphasize brief, novel loading of bones with adequate rest periods rather than prolonged, repetitive loading bouts
Pain from hip fracture, spinal osteoporosis, or old compression fractures	Rule out new fractures or dislocation of surgical prostheses
	Brace or support spine during exercise if needed
	Use analgesia or local pain relieving techniques (heating, massage, etc.)

The evidence presented in this review is consistent in the finding that the volume of exercise required for bony adaptation is small [only 12 min per week of jumping in one study *(72)*], whereas the need for high-impact or intensity of loading which is progressed over time is the critical factor. There is a great need to improve on behavioral strategies to provide adequate instruction, supervision, and compliance with such an exercise prescription, as most trials have suffered from high dropout rates and low compliance, even when fully supervised. The skeletal adaptations are sustained only when the loading is continued, so that any impact on future fracture risk is dependent on long-term adherence to exercise prescriptions. Given the very short time (several minutes per day) that is needed for impact or resistive loading, finding ways to incorporate such episodes into daily activities may be more successful than planning structured exercise classes away from home. For example, inserting a few jumps during television commercials or hopping rather than walking up a flight of stairs may provide an effective stimulus if such habits can be effectively behaviorally reinforced. Such integrated lifestyle intervention programs should be based on the physiology of bone mechano-sensors, as currently known, since the requirements for bone remodeling are quite stringent compared to other health outcomes achievable through general physical activity recommendations. In addition, it is important to consider both skeletal and nonskeletal risk factors for osteoporotic fracture, and assess the responsiveness of each of these to targeted intervention programs (*see* Table 2). Given the multifactorial nature of this syndrome, it is likely that successful prevention programs will need to be multifaceted as well.

9. CONCLUSION AND FUTURE DIRECTIONS

At all ages, an exercise prescription is important for the prevention and treatment of osteoporosis. A combination of lifestyle choices, organized sports, unstructured play, and household and occupational tasks can all contribute to a desirable exposure to physical activity that will be lifelong and robust enough to counteract age and disease-related losses of bone. An initial emphasis on weight-bearing aerobic and high-impact activities in youth, shifting toward resistive loading and balance-enhancing exercises in old age, appears to address optimally the needs and capacities of the musculoskeletal system throughout the lifespan.

Evidence has been presented that a stabilization or increase in bone mass in pre- and postmenopausal women is achievable by either resistive *(67,85,94,95,154)* weight-bearing aerobic exercise *(27,28,39,131,142,143)* or high-impact loading *(71,85)*. Such effects on bone density (differences of 1–2% per year associated with exercise) may be important for both prevention and treatment of osteoporosis and related fractures and disability, as reviewed in several recent meta-analyses *(68,69,90,155,156)*. Additional data are needed in men, as is refinement of the exercise prescription for bone health in terms of the optimal modality, dose, frequency, and intensity of activity recommended. Even if exercise alone is an insufficient stimulus to maintain bone density at youthful levels, the combination of exercise effects on bone strength, muscle mass, muscle strength, and balance should lower the risk of injurious falls substantially in physically active individuals. However, large, long-term, randomized, controlled trials of any exercise modality with osteoporotic fracture itself as a primary outcome remain to be conducted, and are a priority for advances in this field.

REFERENCES

1. Mazess R. On aging bone loss. Clin Orthoped 1982; 165:239–252.
2. Glynn NW, Meilahn EN, Charron M, Anderson SJ, Kuller LH, Cauley JA. Determinants of bone mineral density in older men. J Bone Miner Res 1995; 10:1769–1777.

3. Matkovic V, Jelic T, Wardlaw G. TIming of peak bone mass in Caucasian females and its implication for the prevention of osteoporosis. Inference from a cross-sectional model. J Clin Invest 1994; 93:799–808.
4. Nguyen TV, Kelly PJ, Sambrook PN, Gilbert C, Pocock NA, Eisman JA. Lifestyle factors and bone density in the elderly: implications for osteoporosis prevention. J Bone Miner Res 1994; 9:1339–1346.
5. Ward JA, Lord SR, Williams P, Anstey K, Zivanovic E. Physiologic, health and lifestyle factors associated with femoral neck bone density in older women. Bone 1995; 16:373S–378S.
6. Snow-Harter C, Whalen R, Myburgh K, Arnaud S, Marcus R. Bone mineral density, muscle strength, and recreational exercise in men. J Bone Miner Res 1992; 7:1291–1296.
7. Tajima O, Ashizawa N, Ishii T, et al. Interaction of the effects between vitamin D receptor polymorphism and exercise training on bone metabolism. J Appl Physiol 2000; 88:1271–1276.
8. Ulrich CM, Georgiou CC, Snow-Harter CM, Gillis DE. Bone mineral density in mother-daughter pairs: relations to lifetime exercise, lifetime milk consumption, and calcium supplements. Am J Clin Nutr 1996; 63:72–79.
9. Young D, Hopper JL, Nowson CA, et al. Determinants of bone mass in 10- to 26-year-old females: a twin study. J Bone Miner Res 1995; 10:558–567.
10. Pocock N, Eisman J, Gwinn T, et al. Muscle strength, physical fitness and weight but not age predict femoral neck bone mass. J Bone Miner Res 1989; 4:441–448.
11. Mazess R, Whedon G. Immobilization and bone. Calcif Tissue Int 1983; 35:265–267.
12. May H, Murphy S, Khaw KT. Alcohol consumption and bone mineral density in older men. Gerontology 1995; 41:152–158.
13. Barengolts EI, Curry DJ, Bapna MS, Kukreja SC. Effects of endurance exercise on bone mass and mechanical properties in intact and ovariectomized rats. J Bone Miner Res 1993; 8:937–942.
14. Barengoits E, Curry D, Bapna S, Kukreja S. Effects of two non-endurance protocols on established bone loss in ovariectomized adult rats. Calcif Tissue Int 1993; 52:239–243.
15. Notomi T, Lee S, Okimoto N, et al. Effects of resistance exercise training on mass, strength, and turnover of bone in growing rats. Eur J Appl Physiol 2000; 82:268–274.
16. Rubin C, Turner A, Mallinckrodt C, Jerome C, Mcleod K, Bain S. Mechanical strain, induced noninvasively in the high-frequency domain, is anabolic to cancellous bone, but not cortical bone. Bone 2002; 30:445–452.
17. Frost H. Skeletal structural adaptations to mechanical usage. Anat Rec 1990; 26:403–413.
18. Smith E, Gilligan C. Physical activity effects on bone metabolism. Calcif Tissue Int 1991; 49S:S50–S54.
19. Greenleaf J, Kuzlowski S. Physiological consequences of reduced physical activity during bedrest. Exer Sci Sports Rev 1982; 10:84–119.
20. Tallarida G, Peruzzi G, Castrucci F, et al. Dynamic and static exercises in the countermeasure programmes for musculo-skeletal and cardiovascular deconditioning in space. Physiologist 1991; 34(1 suppl):S114–S117.
21. Donaldson C, Hulley SB, Vogl JM, et al. Effects of prolonged bed rest on bone mineral. Metabolism 1970; 19:1071–1084.
22. Giangregorio L, Blimskie C. Skeletal adaptations to alterations in weight-bearing activity. Sports Med 2002; 32:459–476.
23. Huddleston AL, Rockwell D, Kulund DN. Bone mass in lifetime tennis players. JAMA 1980; 244:1107–1109.
24. Talmage R, Stinnett S, Landwehr J, Vincent L, McCartney W. Age-related loss of bone mineral density in nonathletic and athletic women. Bone Miner 1986; 1:15–25.
25. Chilibeck P, Sale D, Webber C. Exercise and bone mineral density. Sports Med 1995; 19:103–122.
26. Hagberg J, Zmuda J, McCole S, et al. Moderate physical activity is associated with higher bone mineral density in postmenopausal women. J Am Geriatr Soc 2001; 49:1411–1417.

27. Aloia J, Vaswani A, Yeh J, Cohn S. Premenopausal bone mass is related to physical activity. Arch Intern Med 1988; 148:121–123.
28. Krall EA, Dawson-Hughes B. Walking is related to bone density and rates of bone loss. Am J Med 1994; 96:20–26.
29. Lauritzen JB, McNair PA, Lund B. Risk factors for hip fractures. A review. Danish Med Bull 1993; 40:479–485.
30. White C, Farmer M, Brody J. Who is at risk? Hip fracture epidemiology report. J Gereontological Nursing 1984; 10:26–30.
31. Farmer ME, Harris T, Madans JH, Wallace RB, Carnoni-Huntley J, White LH. Anthropometric indicators and hip fracture: the NHANES I epidemiologic follow-up study. J Am Geriatr Soc 1989; 37:9–16.
32. Grisso JA, Kelsey JL, Strom BL, et al. Risk factors for falls as a cause of hip fracture in women. N Engl J Med 1991; 324:1326–1331.
33. Lau EM, Donnan SP. Falls and hip fracture in Hong Kong Chinese. Public Health 1990; 104:117–121.
34. Cummings SR, Nevitt MC, Browner WS, et al. Risk factor for hip fracture in white women. N Engl J Med 1995; 332:767–773.
35. Coupland C, Wood D, Cooper C. Physical inactivity is an independent risk factor for hip fracture in the elderly. J Epidemiol Community Health 1993; 47:441–443.
36. Cavanaugh D, Ce C. Brisk walking does not stop bone loss in postmenopausal women. Bone 1988; 9:201–204.
37. Vuillemin A, Guillemin F, Jouanny P, Denis G, Jeandel C. DIfferential influence of physical activity on lumbar spine and femoral neck bone mineral density in the elderly population. J Gerontol (Biol Sci Med Sci) 2001; 56A:B248–B253.
38. Need A, Wishart J, Scopacasa F, Horowitz M, Morris H, Nordin B. Effect of physical activity on femoral bone density in men. Br Med J 1995; 310:1501–1502.
39. Kenny A, Prestwood K, Marcello K, Raisz L. Determinants of bone density in healthy older men with low testosterone levels. J Gerontol (Med Sci) 2000; 55A:M492–M497.
40. Greendale G, Barrett-Connor E, Edelstein S, Ingles S, Haile R. LIfetime leisure exercise and osteoporosis. J Epidemiol 1995; 141:951–959.
41. Paganini-HIll A, Chao A, Ross R, Henderson B. Exercise and other risk factors in the prevention of hip fracture. Epidemiology 1991; 2:16–25.
42. Gregg E, Cauley J, Seeley D, Ensrud K, Bauer D. Physical activity and osteoporotic fracture risk in older women. Ann Intern Med 1998; 129:81–88.
43. Lee S, Dargent-Molina P, Breart G. Risk factors for fractures of the proximal humerus: results from the EPIDOS prospective study. J Bone Miner Res 2002; 17:817–825.
44. Rubin C, Lawton L. Regulation of bone formation by applied dynamic loads. J Bone Joint Surg 1984; 66-A:397–402.
45. Qin Y-X, Rubin C, McLeod K. Nonlinear dependence of loading intensity and cycle number in the maintenance of bone mass and morphology. J Orthoped Res 1998; 16:482–487.
46. Jarvinen T, Kannus P, Sievanen H, Jolma P, Heinonen A, Jarvinen M. Randomized controlled study of effects of sudden impact loading on rat femur. J Bone Miner Res 1998; 13:1475–1482.
47. Vuori I. Peak bone mass and physical activity: a short review. Nutr Rev 1996; 54:S11–S14.
48. Kirchner E, Lewis R, O'Connor P. Bone mineral density and dietary intake of female college gymnasts. Med Sci Sports Exerc 1995; 27:543–549.
49. Heinonen A, Oja P, Kannus P. Bone density of female athletes in different sports. Bone Miner 1993; 23:1–14.
50. Snow-Harter C. Bone health and prevention of osteopososis in active and athletic women. Clin Sports Med 1994; 13:389–404.
51. Taaffe D, Snow-Harter C, Connolly D. 10. J Bone Miner Res 1995; 586–593.

52. LIma F, Falco VD, Baima J, Carazzato J, Pereira R. Effect of impact load and active load on bone metabolism and body composition of adolescent athletes. Med Sci Sports Exer 2001; 33:1318–1323.
53. Haapasalo H, Kannus P, Sievanen H. Long-term unilateral loading and bone mineral density and content in female squash players. Calcif Tissue Int 1994; 54:249–255.
54. Kannus P, Haapasalo H, Sievanen H. The site-specific effects of long-term unilateral actiivity on bone mineral density and content. Bone 1994; 15:279–284.
55. Gregg E, Ma P, Caspersen C. Physical activity, falls, and fractures among older adults: a review of the epidemiologic evidence. J Am Geriatr Soc 2000; 48:883–893.
56. Kannus P, Haapasalo H, Sankelo M, et al. Effect of starting age of phsysical activity on bone mass in the dominant arm of tennis and squash players. Ann Intern Med 1995; 123:27–31.
57. Boussein M, Marcus R. Overview of exercise and bone mass. Rheum Dis Clin N Am 1994; 20:787–802.
58. Sievanen H, Kannus P, Heinonen A. Bone mineral density and muscle strength of lower extremities after long-term strength training, subsequent knee ligament injury and rehabilitation: a unique 2-year follow-up of a 26 year-old-female student. Bone 1994; 15:85–90.
59. Karlsson M, Bass S, Seeman E. The evidence that exercise during growth or adultnood reduces the risk of fragility fractures is weak. Best Pract Res Clin Rheumatol 2001; 15:429–450.
60. Fuchs R, Bauer J, Snow C. Jumping improves hip and lumbar spine bone mass in prepubescent children: a randomized controlled trial. J Bone Miner Res 2001; 116:148–156.
61. McKay H, Petit M, Schutz R, Prior J, Barr S, Kahn K. Augmented trochanteric bone mineral density after modified physical education classes: a randomized school-based exercise intervention study in prepubescent and early pubescent children. J Pediatr 2000; 136:156–162.
62. Bradney M, Pearce G, Naughton G, et al. Moderate exercise during growtn in prepubertal boys: changes in bone mass, size, volumetric density, and bone strength: a controlled prospective study. J Bone Miner Res 1998; 13:1814–1821.
63. Witzke K, Snow C. Effects of plyometric jump training on bone mass in adolescent girls. Med Sci Sports Exer 2000; 32:1051–1057.
64. French S, Fulkerson J, Story M. Increasing weight-bearing physical activity and calcium intake for bone mass growth in children and adolescents: a review of intervention trials. Prevent Med 2000; 3:22–31.
65. Okano H, Mizunuma H, Soda M, et al. Effects of exercise and amenorrhea on bone mineral density in teenage runners. Endocrine J 1995; 42:271–276.
66. Hetland M, Haarbo J, Christiansen C. Running induces menstrual disturbances but bone mass is unaffected, except in amenorrheic women. Am J Med 1993; 95:53–60.
67. Kelley G, Kelley D, Kristi S, Tran Z. Resistance training and bone mineral density in women: a meta-analysis of controlled trials. Am J Phys Med Rehab 2001; 80:65–77.
68. Wolff I, Croonenborg J, Kemper H, Kostense P, Twisk J. The effect of exercise training programs on bone mass: a meta-analysis of published controlled trials in pre- and postmenopausal women. Osteopor Int 1999; 9:1–12.
69. Wallace M, Cumming R. Systematic review of randomized trials of the effect of exercise on bone mass in pre- and postmenopausal women. Calcif Tissue Int 2000; 67:10–18.
70. Heinonen A, Kannus P, Sievanen H, et al. Randomised controlled trial of effect of high-impact exercise on selected risk factors for osteoporotic fractures. Lancet 1996; 348:1343–1347.
71. Bassey E, Ramsdale S. Increase in femoral bone density in young women following high-impact exercise. Osteopor Int 1994; 4:72–75.
72. Bassey E, Rothwell M, Littlewood J, Pye D. Pre- and postmenopausal women have different bone mineral density responses to the same high-impact exercise. J Bone Miner Res 1998; 13:1805–1813.
73. Winters K, Snow C. Detraining reverses positive effects of exercise on the musculskeletal system in premenopausal women. J Bone Miner Res 2000; 15:2495–2503.

74. Vuori I, Heinonen A, Sievanen H, Kannus P, Pasanen P, Oja P. Effects of unilateral strength training and detraining on bone mineral density and content in young women: a study of mechanical loading and deloading on human bones. Calcif Tissue Int 1994; 55:59–67.
75. Ernst E. Exercise for female osteoporosis: a systematic review of randomised clinical trials. Sports Med 1998; 25:359–368.
76. Burr D, Robling A, Turner C. Effects of biomechanical stress on bones in animals. Bone 2002; 30:781–786.
77. Snow-Harter C, Bouxsein M, Lewis B, Carter D, Marcus R. Effects of resistance and endurance exercise on bone mineral status of young women: a randomized exercise intervention trial. J Bone Miner Res 1992; 7:761–769.
78. Friedlander A, Genant H, Sadowsky S, Byl N, Gluer C-C. A two-year program of aerobics ad weight training enhances bone mineral density of young women. J Bone Miner Res 1995; 10:574–585.
79. Hakkinen A, Sokka T, Kotaniemi A, Hannonen P. A randomized two-year study of the effects of dynamic strength training on muscle strength, disease activity, functional capacity, and bone mineral density in early rheumatoid arthritis. Arthr Rhem 2001; 44:515–522.
80. Turner C, Forwood M, Otter M. Mechanotransduction in bone: do bone cells act as sensors of fluid flow? FASEB J 1994; 8:875–878.
81. ACSM. ACSM's Guidelines for Exercise Testing and Prescription. In: Kenner W, ed. American College of Sports Medicine. ACSM's Guidelines for Exercise Testing and Prescription, 5th ed. Williams & Wilkins, Philadelphia, 1997, p. 373.
82. Iwamoto J, Takesda T, Ichimura S. Effect of exercise training and detraining on bone mineral density in postmenopasual women with osteoporosis. J Orthoped Sci 2001; 6:128–132.
83. Ebrahim S, Thompson P, Baskaran V, Evans K. Randomized placebo-controlled trial of brisk walking in the prevention of postmenopausal osteoporosis. Age Ageing 1997; 26:253–260.
84. Kerr D, Ackland T, Maslen B, Morton A, Prince R. Resistance training over 2 years increases bone mass in calcium-replete postmenopausal women. J Bone Miner Res 2001; 16:175–181.
85. Snow C, Shaw J, Winters K, Witzke K. Long-term exercise using weighted vests prevents hip bone loss in postmenopausal women. J Gerontol (Med Sci) 2000; 55A:M489–M491.
86. Brochu M, Starling R, Tchernof A, Matthews D, Garcia-Rubi E, Poehlman E. Visceral adipose tissue is an independent correlate of glucose disposal in older obese postmenopausal women. J Clin Endocrinol Metab 2000; 85:2378–2384.
87. Lemieux I, Pascot A, Couillard C, et al. Hypertriglyceridemic waist: a marker of the atherogenic metabolic triad (hyperinsulinemia; hyperapolipoprotein B; small, dense LDL) in men? Circulation 2000; 102:179–184.
88. Taylor S, Barr V, Reitman M. Does leptin contribute to diabetes caused by obesity? Science 1996; 274:1151–1152.
89. Chumlea W, Guo S, Glaser R, Vellas B. Sarcopenia, function and health. Nutr Health Aging 1997; 1:7–12.
90. Kelley G. Aerobic exercise and lumbar spine bone mineral density in postmenopausal women: a meta-analysis. J Am Geriatr Soc 1998; 46:143–152.
91. Kelley G. Aerobic exercise and bone density at the hip in postmenopausal women: a meta-analysis. Prevent Med 1998; 27:798–807.
92. Nelson ME, Fisher E, Dilmaniam F, et al. 1-year walking program and increased dietary calcium in post menopausal women. Effect on bone. Am J Clin Nutr 1991; 53:1304–1311.
93. McMurdo M, Mole P, Paterson C. Controlled trial of weight bearing exercise in older women in relation to bone density and falls. Br Med J 1997; 314:569.
94. Notelovitz M, Martin D, Tesar R, et al. Estrogen and variable-resistance weight training increase bone mineral in surgically menopausal women. J Bone Miner Res 1991; 6:583–590.
95. Nelson M, Fiatarone M, Morganti C, Trice I, Greenberg R, Evans W. Effects of high-intensity strength training on multiple risk factors for osteoporotic fractures. JAMA 1994; 272:1909–1914.

96. Rhodes E, Martin A, Taunton J, Donnelly M, Warren J, Elliot J. Effects of one year of resistance training on the relation between muscular strength and bone density in elderly women. Br J Sports Med 2000; 34:18–22.
97. Kohrt W, Ehsani A, Birge S. Effects of exercise involving predominantly either joint-reaction or ground-reaction forces on bone mineral density in older women. J Bone Miner Res 1997; 12:1253–1261.
98. Heinonen A, Oja P, Sievanen H, Pasanen P, Vuori I. Effect of two training regimens on bone mineral density in healthy perimenopausal women. J Bone Miner Res 1998; 13:483–490.
99. Humphries B, Newton R, Bronks R, et al. Effect of exercise intensity on bone density, strength, and calcium turnover in older women. Med Sci Sports Exer 2000; 32:1043–1050.
100. Sinaki M, Wahner H, Bergstralh E, et al. Three-year controlled, randomized trial of the effect of dose-specified loading and strengthening exercises on bone mineral density of spine and femur in nonathletic, physically active women. Bone 1996; 19:233–244.
101. Kerr D, Morton A, Dick I, Prince R. Exercise effects on bone mass in postmenopausal women are site-specific and load-dependent. J Bone Miner Res 1996; 11:218–225.
102. Cussler E, Lohman T, Going S, et al. Weight lifted in strength training predicts bone change in postmenopausal women. Med Sci Sports Exer 2003; 35:10–17.
103. Fiatarone MA, O'Neill EF, Ryan ND. The Boston FICSIT study:the effects of resistance training and nutritional supplementation on physical frailty in the oldest old. J Am Geriatr Soc 1993; 41:333–337.
104. Mazzeo R, Cavanaugh P, Evans W, et al. Exercise and physical activity for older adults. Med Sci Sports Exer 1998; 30:992–1008.
105. Foldvari M, Clark M, Laviolette L, et al. Assocation of muscle power with functional status in community dwelling elderly women. J Gerontol (Med Sci) 2000; 55A:M192–M199.
106. Morris J, Fiatarone M, Kiely D, et al. Nursing rehabilitation and exercise strategies in the nursing home. J Gerontol 1999; 54A:M494–M500.
107. Hausdorff J, Nelson M, Kaliton D, et al. The etiology and plasticity of gait instability in older adults: a randomized controlled trial of exercise. J Appl Physiol 2001; 90:2117–2129.
108. Fiatarone Singh M. Exercise comes of age: rationale and recommendations for a geriatric exercise prescription. J Gerontol (Med Sci) 2002; 57:M262–M282.
109. Bassey E, Ramsdale S. Weight-bearing exercise and ground reaction forces: a 12-month randomized controlled trial of effects on bone mineral density in healthy postmenopausal women. Bone 1995; 16:469–476.
110. Shaw J, Snow C. Weighted vest exercise improves indices of fall risk in older women. J Gerontol (Biol Sci Med Sci) 1998; 53:M53–M58.
111. Dalen N, Lamke B. Bone mineral losses in alcoholics. Acta Orthoped Scand 1976; 47:469–471.
112. Adinoff A, Hollister J. Steroid-induced fractures and bone loss in patients with asthma. N Engl J Med 1983; 309:265–268.
113. Braith R, Mills R, Welsch M, Keller J, Pollock M. Resistance exercise training restores bone mineral density in heart transplant recipients. J Am Coll Cardiol 1996; 28:1471–1477.
114. Storer T. Exercise in chronic pulmonary disease: resistance exercise prescription. Med Sci Sports Exer 2001; 33:S680–S686.
115. Simpson K, Killian K, McCartney N, Stubbing D, Jones N. Randomised controlled trial of weightlifting exercise in patients with chronic airflow limitation. Thorax 1996; 47:70–75.
116. Casaburi R. Skeletal muscle dysfunction in chronic obstructive pulmonary disease. Med Sci Sports Exer 2001; 33:S662–S670.
117. Nishimura Y, Nakata H, Matsubara M, Maeda H, Yokoyama H. [Bone mineral loss in patients with chronic obstructive pulmonary disease]. Nippon Kyobu Shikkan Gakkai Zasshi—Jpn J Thoracic Dis 1993; 31:1548–1552.

118. Cambach W, Wagenaar R, Koelman T, van Keimpema T. The long-term effects of pulonary rehabilitation in patients with asthma and chronic obstructive pulmonary disease: a research synthesis. Arch Phys Med Rehab 1999; 80:103–111.
119. Olopade C, Beck K, Viggiano R, Staats B. Exercise Limitation and pulmonary rehabilitation in chronic obstructive pulmonary disease. Mayo Clin Proc 1992; 67:144–157.
120. Maddalozzo G, Snow C. High intensity resistance training: effects on bone in older men and women. Calcif Tissue Int 2000; 66:399–404.
121. Kohrt W, Yarasheski K, Hollosy J. Effects of exercise training on bone mass in elderly women and men with physical frailty. Bone 1998; 23:S499.
122. Hauer K, Rost B, Rutschle K, et al. Exercise training for rehabilitation and secondary prevention of falls in geriatric patients with a history of injurious falls. J Am Geriatr Soc 2001; 49:10–20.
123. Singh N, Stavrinos T, Scarbeck Y, Galambos G, Fiatarone Singh M. The effectiveness and appropriate intensity of progressive resistance training required to treat clinical depression in the elderly. Austral N Z J Med 2000; 30:304.
124. Singh N, Fiatarone Singh M. Exercise and depression in the older adult. Nutr Clin Care 2000; 3:197–208.
125. Braith R, Welsch M, Mills R, Keller J, Pollock M. Resistance exercise prevents glucocorticoid-induced myopathy in heart transplant recipients. Med Sci Sports Exer 1998; 30.
126. Westby M, Wade J, Rangno K, Berkowitz J. A randomized controlled trial to evaluate the effectiveness of an exercise program in women with rheumatoid arthritis taking low dose prednisone. J Rheumatol 2000; 27:1674–1680.
127. Burr D, Yoshikawa T, Teegarden D, et al. Exercise and oral contraceptive use suppress the normal age-related increase in bone mass and strength of the femoral neck in women 18–31 years of age. Bone 2000; 27:855–863.
128. Weaver C, Teegarden D, Lyle R, et al. Impact of exercise on bone health and contraindication of oral contraceptive use in young women. Med Sci Sports Exer 2001; 33:873–880.
129. Specker B. Evidence for an interaction between calcium intake and physical activity on changes in bone mineral density. J Bone Miner Res 1996; 11:1539–1544.
130. Committee on the Scientific Evaluation of Dietary Reference Intakes. Dietary reference intakes for calcium, phosphorus, magnesium, vitamin D, and fluoride. National Academy Press, Food and Nutrition Board, Institute of Medicine, Washington, DC, 1997.
131. Kohrt WM, Snead DB, Slatopolsky E, Birge SJ Jr. Additive effects of weight-bearing exercise and estrogen on bone mineral density in older women. J Bone Miner Res 1995; 10:1303–1311.
132. Ryan A, Treuth M, Rubin M, et al. Effects of strength training on bone mineral density: hormonal and bone turnover relationships. J Appl Physiol 1994; 77:1678–1684.
133. Yeh JK, Aloia JF, Barilla ML. Effects of 17 beta-estradiol replacement and treadmill exercise on vertebral and femoral bones of the ovariectomized rat. Bone Miner 1993; 24:223–234.
134. Kohrt W, Ehsani A, Birge S. HRT preserves increases in bone mineral density and reductions in body fat after a supervised exercise program. J Appl Physiol 1998; 84:1506–1512.
135. Tamaki H, Akamine T, Goshi N, Kurata H, Sakou T. Effects of exercise training and etidronate treatment on bone mineral density and trabecular bone in ovariectomized rats. Bone 1998; 23:147–151.
136. Vico L, Chappard D, Alexandre C, et al. Effects of a 120 day period of bed-rest on bone mass and bone cell activities in man: attempts at countermeasure. Bone Miner 1987; 2:383–394.
137. Grigoriev A, Morskov B, Oganov V, Rakhmanov A, Buravkova L. Effect of exercise and bisphosphonate on mineral balance and bonde density during 360-day antiorthostatic hypokinesia. J Bone Miner Res 1992; 7:S449–S455.
138. Chilibeck P, Davison K, Whiting S, Suzuki Y, Janzen C, Peloso P. The effect of strength training combined with bisphosphonate (etidronate) therapy on bone mineral, lean tissue, and fat mass in postmenopausal women. Can J Physiol Pharmacol 2002; 80:941–950.

139. Uusi-Rasi K, Sievanen H, Kannus P, et al. Effects of alendronate and exercise on tibia and physical performance in early postmenopausal women. J Bone Miner Res 2001; 16:S409.
140. Drinkwater B, Grimson S, Cullen-Raab D, Harter-Snow C. ACSM position stand on osteoporosis and exercise. Med Sci Sports Exer 1995; 27:i–vii.
141. Hamdy R, Anderson J, Whalen K, Harvill L. Regional differences in bone density of young men involved in different exercises. Med Sci Sports Exer 1994; 26:884–888.
142. Aloia J, Cohn S, Ostuni J, Cane R, Ellis K. Prevention of involutional bone loss by exercise. Ann Intern Med 1978; 89:356–358.
143. Dalsky G, Stocke K, Ehsani A, Slatopolsky E, Lee W, Birge S. Weight-bearing exercise training and lumbar bone mineral content in postmenopausal women. Ann Intern Med 1988; 108:824–828.
144. Hughes-D B. Nutrition, exercise, and lifestyle factors that affect bone health. In: Kotsonis F, Mackey, MA, eds. Nutrition in the 90's: Current Controversies and Analysis. Vol. 2. Marcel Dekker, New York, 1994, p. 170.
145. Chow R, Harrison JE, Notarius C. Effect of two randomised exercise programes on bone mass of healthy postmenopausal women. Br Med J 1987; 295:1441–1444.
146. Krolner B, Toft B, Nielsen SP, Tondevold E. Physical exercise as prophylaxix against involutional vertebral bone loss: a controlled trial. Clin Sci 1983; 64:541–546.
147. White MK, Mathin RB, Yeater RA, Butcher RL, Rodink EL. The effects of exercise on the bones of postmenopausal women. Int Orthopaed 1984; 7:209–214.
148. Nelson M, Fisher E, Dilmanian F, Dallal G, Evans W. A 1-y walking program and increased dietary calcium in postmenopausal women: effects on bone. Am J Clin Nutr 1991; 53:1394–1311.
149. McCartney N, Hicks A, Martin J, Webber C. Long-term resistance training in the elderly: effects on dynamic strength, exercise capacity, muscle, and bone. J Gerontol 1995; 50A:B97–B104.
150. Simkin A, Ayalon J, Leichter I. Increased trabecular bone density due to bone-loading exercises in postmenopausal osteoporotic women. Calcif Tissue Int 1987; 40:59–63.
151. Snow-Harter C, Bouxsein M, Lewis B, Carter D, Marcus R. Effects of resistance and endurance exercise on bone mineral status of young women: a randomized exercise interventionm trial. J Bone Miner 1992; 7:761–769.
152. Ayalon J, Simkin A, Leichter I, Raifman S. Dynamic bone loading exercises for postmenopausal women: effect on the density of the distal radius. Arch Phys Med Rehabil 1987; 68:280–282.
153. Greendale G, Salem G, Young J, et al. A randomized trial of weighted vest use in ambulatory older adults: strength, performance, and quality of life outcomes. J Am Geriatr Soc 2000; 48:305–311.
154. Menkes A, Mazel R, Redmond R, et al. Strength training increases regional bone mineral density and bone remodeling in middle-aged and older men. J Appl Physiol 1993; 74:2478–2484.
155. Kelley G. Exercise and regaional bone mineral density in postmenopausal women: a meta-analytic review of randomized trials. Am J Phys Med Rehabil 1998; 77:76–87.
156. Vuori I. Dose-response of physical activity and low pack pain, osteoarthritis, and osteoporosis. Med Sci Sports Exer 2001; 33:S551–S586.
157. Berard A, Bravo G, Gauthier P. Meta-analysis of the effectiveness of physical activity for the prevention of bone loss in postmenopausal women. Osteopor Int 1997; 7:331–337.

30 Body Weight/Composition and Weight Change

Effects on Bone Health

Sue A. Shapses and Mariana Cifuentes

A low body weight (LBW) in older individuals is a major risk factor for fracture *(1,2)*, and maintenance of body weight can prevent bone loss *(3–6)*. The importance of LBW has been highlighted by the National Osteoporosis Foundation *(7)*, suggesting that it is one of the top four major risk factors for osteoporotic fractures. The relationship between body weight and bone is discussed here with particular reference to obesity and adipose tissue. Weight reduction, depending on whether it is involuntary or voluntary, will affect bone differently. Although mechanisms regulating bone loss are uncertain, it is clear that the method to achieve voluntary weight reduction (through different diets, medication, or increasing levels of activity) will determine the bone response. Finally, alterations in bone quality and strength parameters due to weight reduction and regain are discussed and future directions are proposed.

1. BODY WEIGHT, BONE MASS, AND FRACTURE RISK

1.1. Obesity

The prevalence of obesity is increasing worldwide. Importantly, it is predicted that the current generation of children will grow into the highest number of obese adults in US history *(8)*. Obesity associated with visceral adiposity (upper body obesity or a high waist-to-hip ratio) increases the risk of co-morbid conditions such as type II diabetes, hypertension, and dyslipidemia. Overconsumption of energy and the lack of physical activity are two important factors contributing to the rising prevalence of obesity *(9)*. Poor diet and sedentary behavior are also factors believed to contribute to the rising prevalence of osteoporosis. Both osteoporosis and obesity are diseases of Western cultures, yet there is typically a reduced incidence of osteoporosis in obese populations. The relationship between obesity and osteoporosis and mechanisms involved are discussed here.

From: *Nutrition and Bone Health*
Edited by: M. F. Holick and B. Dawson-Hughes © Humana Press Inc., Totowa, NJ

1.1.1. Body Mass Index

Body mass index (BMI; kg/m^2) is a widely used tool to assess body weight adequacy and a good estimate of body fat in most subjects. In addition, it is a means to predict co-morbid conditions associated with excess body weight. In the National Institutes of Health (NIH) obesity guidelines established in 1998, a new overweight category was defined (BMI 25–29.9 kg/m^2), associated with co-morbid conditions similar to obese individuals (BMI > 30) *(10)*. Therefore, many more individuals are no longer considered within the normal range and weight reduction is recommended. In most older individuals or nonathletes, it can reasonably be assumed that excess weight corresponds to excess fat. However, the accuracy of BMI in measuring body fatness in older populations needs to be questioned because of the height loss experienced by some osteopenic and osteoporotic patients due to vertebral crush fractures. It may be more accurate to ask an older patient his or her height as a young adult and to gain the patient's weight history to determine osteoporosis risk due to height loss. Recommendations for weight reduction may not be desirable for older, slightly overweight individuals who have become shorter as a result of atraumatic vertebral fracture.

1.1.2. Obesity and Bone

Studies show a higher bone mass with high body weight *(11–14)*, but the accuracy of bone mass measurements should be considered (*see* Subheadings 1.2.1 and 3.3). This is attributed to either increased mechanical loading of body weight (including a greater lean body mass) and/or linked to a greater fat mass and greater conversion of adrenal androgens to estrogens in adipose tissue, as well as other potential mechanisms *(15)*. Because hormone profiles vary by fat depot (visceral vs subcutaneous) and gender *(16,17)*, the benefit of excess adipose tissue to bone may not be uniform in all heavier individuals. For example, multiple endocrine perturbations are found with visceral fat accumulation. These include elevated cortisol, androgens, and growth hormone in women, as well as low testosterone secretion in men *(16)*, all of which are known to regulate bone mass. There are other hormonal differences in the obese that may increase bone mass, such as higher serum levels of calcitonin *(18,19)*. Some have found no differences in serum parathyroid hormone (PTH) between lean and obese subjects *(18,20)*, whereas others have found slightly increased PTH levels *(19,21,22)* and lower 25-hydroxyvitamin D_3 [25(OH)D] *(22,23)* in the obese. Decreased bioavailability of 25(OH)D in obese subjects may be the result of its deposition into adipose stores *(23)*. It is possible that the decreased levels of 25(OH)D found in physically inactive compared to active individuals *(24)* is because of their greater adipose stores. It is not clear whether these effects are detrimental or beneficial to bone. Serum leptin levels are elevated in the obese, and its effect on bone is discussed here.

In children, there is a higher bone mass in overweight and obese compared to lean children for a given chronological age *(25,26)*, but their growth spurt during puberty is less than that of lean children *(26)*. It is interesting that in overweight and obese boys and girls there is a mismatch between body weight and bone development during growth, such that their bone mass and bone area are low for their body weight *(27)*. One study found that adiposity in boys is associated with lower bone mass, and also an increased risk of distal forearm fracture *(28)*. These data are disturbing in light of the rising incidence of childhood obesity.

1.2. Low Body Weight

Evidence that the lean population is at greater risk of osteoporosis is strong. It is believed that genetics contribute to most of the variance in bone mass, with 20–25% being under environmental influences *(29)*. Therefore, it is especially important that individuals with LBW use other preventive behaviors (i.e., prevention of further weight loss, high dairy intake, physical activity) to reduce their risk of osteoporosis.

1.2.1. Lower Bone Density in Lean Than Obese Individuals

There is a large body of evidence that bone density is lower in lean than in obese women *(30,31)*. The extent to which bone mass is greater at weight-bearing sites is important to help determine the mechanism increasing bone density in the heavier population. Although studies show that overweight and obese women have a higher vertebral bone density than the normal reference population *(32)*, more recent studies suggest that this may be owing to a higher prevalence of spinal osteoarthritis in obese postmenopausal women *(33)*, which can artificially increase bone mineral density (BMD). Hence, it may be more accurate to compare bone mass between obese and nonobese populations by evaluating sites with a lower prevalence of artifact error due to arthritis, such as the hip and forearm. By measuring these sites in the obese, one may be able to differentiate the effect of weight-bearing compared to endogenous hormones (i.e., estrogen) on bone. Even in lean perimenopausal women, there is discordance of BMD between skeletal sites *(34)*, in which the forearm measurements differ from the axial sites (total body, spine, and hip). One study suggested that there were greater differences of BMD at weight-bearing sites between lean and obese, but a limited number of sites was measured *(35)*.

Hormone Replacement Therapy. In postmenopausal women, hormone replacement therapy (HRT) is especially effective for maintaining bone mass in normal weight, but not in overweight or obese women *(14)*. This is important in light of findings showing a greater reduction in coronary heart disease mortality with estrogen use in thinner compared to obese women *(36)*. Finally, because thinner women typically have a reduced risk of breast cancer risk than obese women *(37)*, the benefits of HRT should not be underestimated in the lean population. This is despite our concerns about the risks associated with HRT based on data from the Women's Health Initiative *(38)*.

1.2.2. Increased Rate of Bone Turnover and Loss

In lean women, there is an increased annual rate of bone loss compared to obese women *(29,39)*. A higher rate of bone loss in leaner postmenopausal women is supported by their higher rate of bone turnover compared to heavier women

(14,40). In contrast, among overweight and obese women, the relationship between high bone turnover and lower body weight is not observed, suggesting that bone mass is regulated differently in heavier women *(14)*. It is still debated whether increased weight bearing *(41)* or excess fat tissue is more protective against bone loss. In one interesting comparison, the amount of bone loss was compared between two obese groups, postmenopausal and premenopausal women *(42)*. These researchers found that adiposity determined bone loss only in the obese postmenopausal women *(42)*. This might suggest that excess fat tissue is a more important determinant of bone loss than weight bearing in older, but not younger populations. Hormonal differences other than estrogen levels between lean and obese individuals may also contribute to the higher rate of bone remodeling in the lean, and should be addressed in future studies.

1.2.3. Increased Fracture Risk

LBW increases the risk of developing osteoporotic fractures compared to obese individuals (relative risk of 2–2.4) *(31,43)*. However, in the obese who fall, the fracture severity could be greater *(44)*. In addition, there is a higher risk of falling in obese (particularly those with an abnormal distribution of body fat in the abdominal area) than lightweight individuals *(45)*. On the other hand, the obese may be protected against fractures because of a cushioning effect of the fat surrounding crucial areas such as the hip, particularly in those with lower body obesity. Consistent with this, fracture risk is consistently higher in lean than obese populations, and therefore prevention and treatment of low BMD is important. For both lean and obese individuals, the prevention of fractures and their severity should include prevention of falls.

1.3. Summary and Recommendations

A low BMI (<20 or possibly higher) increases the risk of fracture, whereas a BMI over 25 (overweight or obese) confers some protection. The recommendation that all individuals achieve a normal body weight may be inappropriate for those individuals at high risk for osteoporosis. To reduce fracture risk, it is possible that a higher body weight (within high normal or slightly overweight) is beneficial for certain postmenopausal women or older men, but not for younger individuals. This recommendation coincides with data showing that normal-weight individuals have a greater life expectancy than thin older individuals (BMI <20) *(46)*. It is suggested that weight recommendations should be addressed by considering both fracture risk and the co-morbid conditions associated with excess body weight.

2. BODY COMPOSITION

2.1. Cellular Level: Osteoblasts and Adipocytes

Factors that signal mesenchymal cells to differentiate into their mature cell types will eventually constitute the balance between bone, adipose tissue, and muscle mass.

Marrow stromal cells or stem cells can be differentiated in culture into osteoblasts, chondrocytes, adipocytes, and even myoblasts. For example, cultured myoblasts will differentiate into osteocytes or adipocytes following treatment with bone morphogenetic proteins (BMPs) or adipogenic inducers, respectively *(47)*. Gene array technology has been applied to identify new genes that are activated or repressed during the process of osteoblast, adipocyte, and myoblast differentiation. For example, there is selective downregulation of human urea transporter during adipogenesis that may be a marker for the switch from an osteoblast to an adipocyte phenotype *(48)*, whereas others have shown that there are as many as 27,000 genes involved in in vitro differentiation of primary osteoblasts isolated from mouse calvaria *(49)*.

There are different regulators of mesenchymal cell differentiation. Estrogen increases osteoblastogenesis *(50,51)*, whereas increases in intracellular cAMP (i.e., 3-isobutyl-1-methylxanthine plus dexamethasone) will increase adipocyte production *(52)*. In addition, skeletal unloading increases adipocyte differentiation and inhibits osteoblast differentiation. These abnormalities are prevented and reversed by transforming growth factor (TGF)-$\beta 2$ *(53)*. Interestingly, low-density lipoprotein (LDL) oxidation products promote osteoporotic loss of bone by directing progenitor marrow stromal cells to undergo adipogenic instead of osteogenic differentiation *(54)*. It is possible that aging-related bone loss and increased marrow adipose tissue is the result of a switch in differentiation of stromal cells from the osteoblastic to the adipocytic lineage *(55)*. If this "lipid hypothesis of osteoporosis" is true, then obese individuals with upper-body obesity and dyslipidemia may be at greater risk of bone loss than those with lower-body obesity. This is supported by a cross-sectional study showing that fat distribution is associated with an altered mineral metabolism more than BMI *(20)*. A gender-specific study (to avoid the influence of sex steroids) would help evaluate this hypothesis.

Overall, potential drugs that inhibit marrow adipogenesis with the parallel enhancement of osteoblastogenesis would be a goal in the prevention of osteoporosis. However, such agents may impact organs or tissues other than bone or fat. Consequently, therapeutic agents with potential stromal cell receptor targets must have tissue specificity. The plasticity between adipose and osteoblast cells is an interesting area showing the relationship between different components of body composition that could further our understanding of obesity and osteoporosis.

2.2. Leptin and Body Composition

Leptin is a 16-kDa adipocyte-derived hormone that circulates in the serum in the free and bound form. Serum levels of leptin are highly correlated with body fat mass in adults, children, and newborns. Obese individuals have significantly higher circulating leptin than lean subjects. In addition, females have higher serum leptin than males with equivalent fat mass *(56)*. Although leptin correlates with fat mass, circulating concentrations are altered by extremes in energy intake, such as fasting and overfeeding. In addition, serum leptin regulates food intake, with higher levels reducing the desire to eat and lower levels (which occur with weight loss) resulting in hyperphagia.

The role of leptin on bone regulation is conflicting. Leptin-deficient *(ob/ob)* and leptin receptor deficient *(db/db)* mice that are obese and hypogonadic have a high bone mass phenotype, whereas in humans, a genetically based leptin deficiency presents with morbid obesity but not with high bone mass *(57)*. Investigators have suggested that leptin has a detrimental effect by inhibiting bone formation through a central mechanism *(58)*. However, others have shown that leptin administration increases bone mass by increasing osteopotegrin mRNA levels *(59,60)* and may redirect the differentiation of stem cells from adipocytes to osteoblasts *(61)*. Clinical studies show a positive effect of serum leptin levels on bone *(62–64),* or none at all *(65,66)*. The positive relationship between leptin and bone may be present only in nonobese individuals *(62,67)* who are not leptin resistant. In contrast, others have shown an inverse relationship between serum leptin and bone mass (corrected for BMI) in younger men participating in NHANES III *(68)*. Overall, the relationship between leptin and bone appears to be dependent on a number of factors (i.e., obesity, gender, age, leptin resistance), whereas the determinants influencing whether central hypothalamic control or the local anabolic effects of leptin predominate are not known.

2.3. Relationship Between Bone, Lean Soft Tissue, and Fat Tissue Mass

The extent to which lean or fat tissue mass influence bone mass is not uniformly reported in the literature. An important determinant of bone mass could be the amount of lean tissue mass, because it may reflect weight-bearing activity *(69,70)* or fat mass, because this is known to influence bone density as a source of estrogen *(71)*. Low muscle mass is a risk factor for low BMD in young adult women, whereas higher fat is protective only when it is associated with substantial muscle mass *(72)*. Consistent with this, premenopausal and perimenopausal women show a beneficial effect of increased body weight on bone mass, but only when it is comprised primarily of lean mass *(73)*. The strong association between lean mass (rather than fat mass) and BMD in younger women may be attributed to exercise, lifestyle factors, estrogen levels, or a combination of these factors. In postmenopausal women, Chen and co-workers *(74)* observed that body weight was a better estimate of bone mass than either lean or fat tissue alone. However, the annual changes in bone mass were better predicted by changes in fat than lean mass in these postmenopausal women *(74)*. It has been suggested that even though most studies indicate that lean mass and strength are the main determinants of bone mass *(75,76),* the influence of fat mass increases with aging *(42,74,77)* and is more important in women than men *(78–80)*. In postmenopausal years, this phenomenon may be explained through the influence of fat tissue on serum estrogen levels.

The study of the influence of lean vs fat tissue on bone mass is complicated by the fact that this effect varies depending on the bone site being evaluated *(75,81,82)*. Differences in trabecular content of bone, as well as weight bearing of the specific site, may confound the observations. Establishing the relative importance of fat and lean mass in different populations is important in that it may lead to measures to prevent bone loss in certain physiological states.

2.3.1. Calcium Intake and Body Composition

Calcium supplementation decreases bone resorption, age-related increases in PTH, and bone loss, particularly when initial dietary calcium is low *(83,84)*. More recently, there has also been evidence pointing toward an influence of calcium intake on other parameters of body composition *(85,86)*. In an analysis including observational and controlled studies, Heaney *(87)* concluded that there is a consistent effect of calcium intake on body fat and weight. Animal studies support these observations that high dietary calcium accelerates weight and fat loss when compared to a low-calcium, isoenergetic diet *(88)*. It has been proposed that a low calcium intake stimulates 25(OH)D and PTH, and that these calcitropic circulating substances in turn stimulate adipocyte calcium uptake. Elevated intracellular calcium promotes fatty acid synthase transcription and activity and inhibits lipolysis *(89,90)*. In addition, dietary patterns characterized by increased dairy consumption have a strong inverse association with insulin-resistance syndrome among overweight adults and may reduce risk of type II diabetes and cardiovascular disease *(91)*. Although promising, the studies available are limited and a retrospective examination of women in our weight-reduction studies at two levels of calcium intake found only a weak association between calcium intake and body fat (unpublished data). Therefore, prospective, adequately powered, controlled clinical trials are needed to confirm the relationship between calcium, fat, and body weight.

2.4. Genetic Markers

It is known that genetics contribute to the majority of variance in bone mass *(29)*. A few genetic markers of bone mass have been proposed, including several polymorphisms of the vitamin D receptor (VDR), the LDL receptor-related protein 5, estrogen receptor, interleukin-6, TGF-β, apolipoprotein E, and the Sp 1-binding site of the collagen type I α 1 gene *(92–96)*. The polymorphisms of VDR have been associated with body weight and more strongly with lean body mass and serum concentrations of insulin-like growth factor (IGF)-I *(97)*, which in turn is known to influence bone and lean body mass directly *(98)*. Genetic markers of bone that also reflect fat mass could help determine the relationship between body composition and disease states such as osteoporosis and obesity.

3. WEIGHT REDUCTION

3.1. Involuntary Weight Reduction

The difference between voluntary weight reduction (i.e., a conscious effort to achieve a negative energy balance by restricting caloric intake and/or increasing energy expenditure) and involuntary weight loss is a very important factor when considering the effect on inducing bone loss. Involuntary weight loss is likely associated with an underlying illness or disorder that may have an independent effect on bone, therefore confounding the direct effect of caloric restriction. For example, weight loss is typically observed in patients with depression, cancer or gastroin-

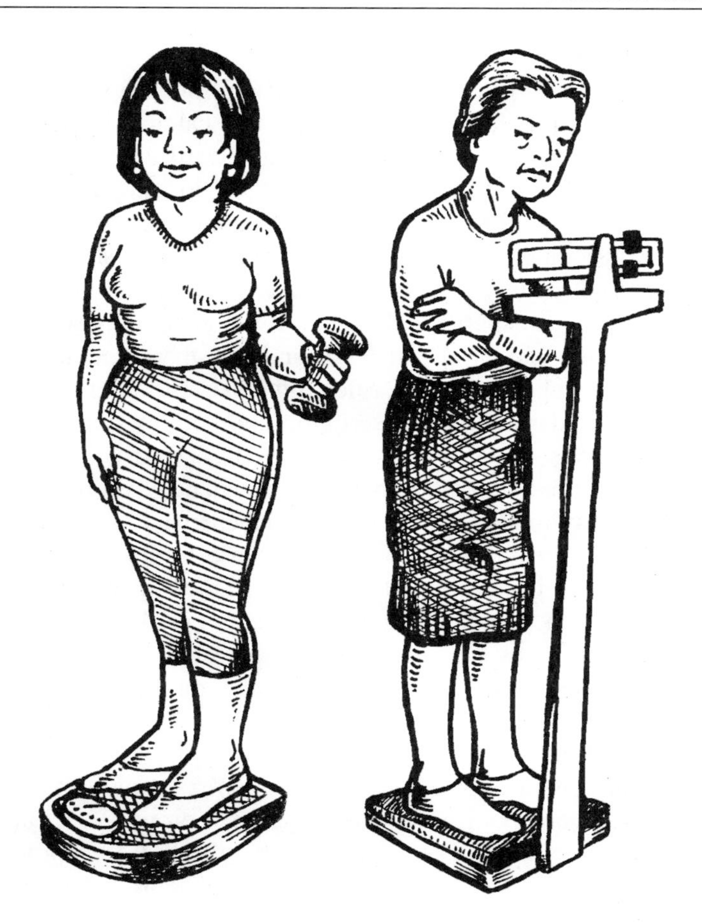

Illustration 1. Voluntary (left) compared to involuntary (right) weight reduction may have a different effect on bone mass.

testinal disorders, all of which have been shown to affect bone negatively *(99–102)*. In addition, studies of involuntary weight loss show bone loss *(3,4),* and an increased risk of hip fracture in older women *(103,104)* and men *(105,106),* particularly when there is low initial body weight. Aging, sedentarism, and antigravity may also cause unintentional loss of body weight and muscle mass (sarcopenia) *(107).* In summary, for the same degree of weight reduction, the negative impact on bone may be greater than expected in cases of involuntary weight loss when compared to the same weight lost intentionally in a healthy subject (Illustration 1).

3.2. Voluntary Weight Reduction

In overweight or obese individuals, weight reduction of approx 10% is recommended because researchers have found that this is achievable and reduces

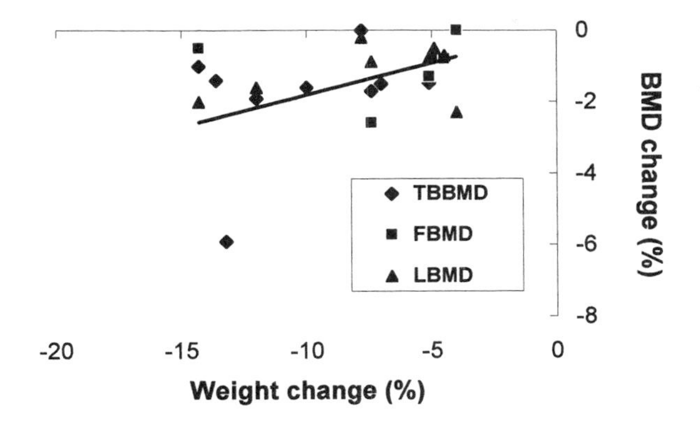

Fig. 1. Weight loss and bone mineral density change for total body (TBBMD, ◆), femoral neck (FBMD, ■) and lumbar spine (LBMD, ▲) with voluntary weight reduction trials of 3–18 months duration *(109–120)*. trend line given for TBBMD; $r = 0.40$.

co-morbid risk factors *(108)*. Although regional differences in bone loss might be expected, with greater loss from more trabecular than cortical regions (or differences owing to weight bearing), the data currently suggest that bone changes of about 1–2% occur at all sites with a 10% weight reduction. Most studies measured total body bone density, with less consistency in measuring other sites. Together these studies in different populations show an association between the amount of weight reduction and total body bone loss ($r = 0.40$) (Fig. 1) *(109–120)*. The association between weight reduction and bone loss may be stronger, but is limited due to imaging limitations of dual-energy X-ray absorptiometry (DXA) *(121,122)*, and the mix of different populations (men, pre- and postmenopausal women, different ages, different initial body weights) between studies. Our data suggest that hormonal changes (specifically estrogens) are important regulators of bone loss as a result of weight reduction, which would vary significantly between different protocols and study populations *(14,112,113)*. Others suggest that reduced weight bearing is the more important regulator of bone loss *(42,114,117)*, but thus far studies have not been able to discern differences in bone loss between sites with different weight-bearing stimulus (i.e., forearm compared to femoral neck).

The amount of bone loss resulting from weight reduction does seem to vary between populations. For example, studies examining a population of postmenopausal women (without other populations) during weight reduction have found a 1–2.5% loss of bone (total or lumbar spine) compared to a control, weight-stable group *(112,119)*. Lean or overweight pre- or perimenopausal women *(118)* respond to moderate weight loss (~5%) in a similar manner as described for postmenopausal women, and show a loss of 0.8% bone loss at the hip. Evidence of bone loss in younger obese premenopausal women (<45 yr) is not clear *(111,113,116,119,123)*, suggesting that weight reduction may not pose a risk

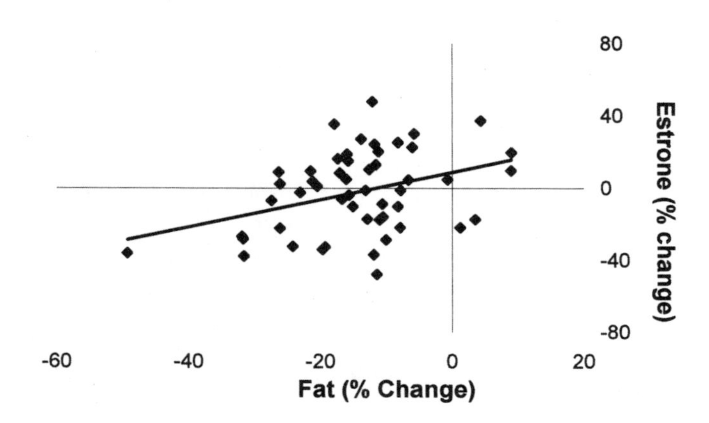

Fig. 2. Association between the loss of body fat tissue and changes in serum estrone after 6 mo of weight reduction (r = 0.36) in overweight and obese postmenopausal women (*112* and unpublished data).

factor for bone loss in estrogen-replete women. In a study of overweight middle-aged men, moderate weight reduction (7%) resulted in a 1% bone loss (total body), but this is the only study examining a homogenous population of men *(115)*.

In general, the studies suggest that there is little or no bone loss in obese younger individuals who lose a moderate amount of weight, and therefore weight reduction (in light of fracture risk) can be recommended with confidence for these individuals. Weight reduction in leaner women may result in more bone loss *(4)*, and is the focus of our ongoing studies. It is suggested that all individuals interested in losing a moderate amount of weight will benefit from additional calcium *(124,125)* or exercise *(117,118)*.

3.2.1. Calcium Absorption and Endocrine Regulation

Calcium absorption is modified by different components of the diet (e.g., calcium, protein, fat, oxalate, lactose, citrate) and physiological states (e.g., aging, pregnancy, growth, lactation, menopause). Recent evidence indicates that energy restriction also influences calcium absorption by reducing its efficiency *(126,127)*, and this could be one of the mechanisms by which weight reduction induces bone loss. Importantly, this may elevate the requirements for calcium during weight reduction *(124,125)*.

Several endocrine changes occurring during weight reduction may be involved in the decrease in calcium absorption. For example, energy restriction and/or weight loss are associated with reduced estrogen levels *(112,128–130)* (Fig. 2), and this may have a direct detrimental effect on calcium absorption *(127,131–133)*. Weight reduction may also induce an increase in glucocorticoids *(127,134,135)* and thus negatively influence calcium absorption *(136)*. Consistent with these observations, a rise in serum PTH has been associated with weight reduction *(112,124)*

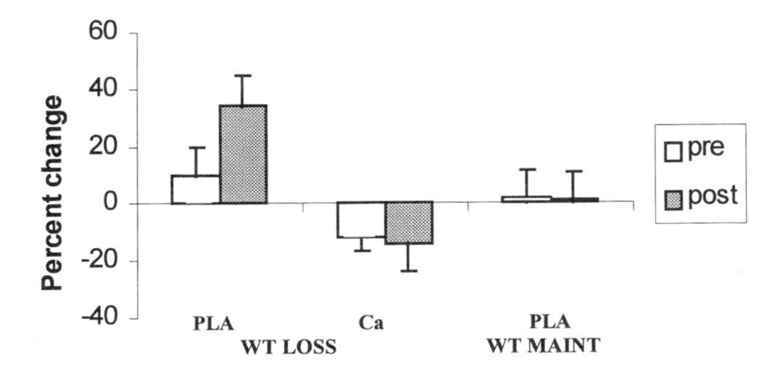

Fig. 3. The change in serum intact parathyroid hormone (PTH) in pre(□) and post(■)-menopausal women after 6 mo of moderate weight loss (Wt Loss) randomized to either placebo (PLA) or 1 g/d Ca supplement (Ca) or weight maintenace (Wt Maint) without Ca supplementation *(112,113,124)*. Bars represent the change mean ±SEM.

(Fig. 3), which may be explained by an increased calcium–PTH axis due to decreased calcium absorption *(126)*.

In addition to an altered endocrine regulation of calcium absorption during energy restriction, direct effects of hormones on bone can be expected. A reduction in estrogen, as occurs with weight reduction (Fig. 2), will affect osteoblasts and the differentiation and activity of osteoclasts to increase bone resorption and loss. A decrease in IGF-I *(137)* or leptin *(112)* resulting from weight reduction could inhibit bone formation, whereas a rise in glucocorticoids *(127)* will increase the rate of bone resorption and decrease bone formation *(138)*. A change in endocrine profile owing to energy restriction can affect bone both directly and indirectly. Identifying the primary regulator(s) and/or organ(s) that responds to energy restriction would be important in minimizing bone loss.

3.2.2. Fast Weight Reduction

A faster compared to slower rate of weight loss results in a more negative nitrogen balance and loss of skeletal muscle mass *(139)*. It is therefore possible that bone is also susceptible to the detrimental effects of rapid compared to more moderate weight reduction. However, short-term studies using a very-low-calorie diet (VLCD), which results in a more rapid rate of weight loss, do not show a greater amount bone loss, but do show a greater variability in response to the diet *(111,120,123,140,141)*. Because VLCDs are typically used by obese or morbidly obese individuals, this may have affected the DXA results. Measurements of bone density using DXA are less sensitive in obese individuals and may not be able to detect small changes in bone mass with weight reduction due to changes in the soft tissue surrounding bone. Nevertheless, there is evidence suggesting that a more rapid weight loss will be more detrimental to bone due to an activation of the calcium–PTH axis in women who lose weight slightly faster (0.7 kg/wk) than others (0.3 kg/wk) *(126)*.

3.2.3. Special Diets and Nontraditional Methods of Weight Reduction

This chapter, thus far, has focused on moderate or severe weight-reduction studies that achieve weight reduction using a reduced energy intake with a proportional reduction in all macronutrients. Nevertheless, there are other popular diets that result in successful weight reduction. For example, there are very-low-carbohydrate weight-loss diets that produce ketosis (also known as the Atkins Diet), and other mildly low-carbohydrate Diets (i.e., Sugar Busters Diet or The Zone Diet) that do not typically produce ketosis. It has been speculated that, during starvation, acidosis and ketone body formation may play a role in calcium mobilization *(141–144)*. There are at least two studies examining bone parameters during weight loss on a low-carbohydrate ketogenic diet *(145,146)*. The findings in the carefully designed study by Reddy et al. *(145)* showed that 6 wk of a ketogenic diet in healthy subjects results in a marked increase in acid load to the kidney. Urinary calcium levels increased (~60%) without a commensurate increase in intestinal fractional calcium absorption, resulting in a decrease in calcium balance. Markers of bone formation decreased, whereas resorption markers showed no change. Other investigators have shown that a low-carbohydrate, ketogenic diet in the treatment of morbidly obese adolescents results in a 15-kg weight loss over 8 wk, but also increases calcium excretion and reduces total body bone mineral content *(146)*. On the other hand, there are a number of studies showing that an alkaline diet in weight-stable conditions is beneficial to bone *(144,147–150)*. Overall, an acidic diet in conjunction with caloric restriction may exacerbate the usual potential side effects of a weight-loss diet, thereby increasing the risk of bone loss and kidney stone formation. However, prospective studies that would contribute valid evidence for the response of bone mass to a very-low-carbohydrate diet have not been completed, yet are currently underway in a number of laboratories.

Other diets that could potentially affect bone are those aided with medication. The pancreatic lipase inhibitor orlistat (OLS) induces an increase in bone resorption relative to formation *(151)* that is more dramatic than the change in the control group without OLS. Both groups showed a decrease in vitamin D status, but only the OLS group showed a significant increase in serum PTH. However, no changes in bone mass or density were seen after 1 yr of OLS treatment with adequate calcium and vitamin D intake, apart from those explained by the weight loss itself. A vitamin D and calcium supplement should be taken during the treatment with orlistat for weight reduction *(151)*. In one preliminary study using sibutramine (a satiety medication) to promote weight reduction, bone loss was no greater than in a group losing weight without sibutramine *(152)*.

Severely obese individuals (BMI > 35 kg/m^2) with co-morbid conditions may undergo gastric reduction surgery when less invasive methods of weight loss have failed *(10)*. With an average weight loss that ranges from 50 to 100 kg in 1 yr, the method has gained popularity because of the rising incidence of obesity and because it is the only known successful method to reduce weight and mortality in this population *(153,154)*. Despite the high bone density in these obese individuals before weight loss, osteoporosis and fracture may be a long-term consequence of weight reduction due to gastric reduction surgery *(155)*. Our studies in obese women show that about 3 yr after gastric reduction surgery (with 30% weight loss), the prevalence of osteopenia is about 33% and 16% at the hip and spine, respectively *(156)*. The rate of bone loss may be greater for these women *(156)*, due to an increased activity of the calcium–PTH axis *(21,156)*. We currently recommend that individuals who have undergone gastric bypass surgery consume at least 1500 mg of calcium citrate and adequate vitamin D after gastric reduction surgery, to avoid excessive bone loss.

3.2.4. Calcium and Other Micronutrients During Weight Reduction

Calcium intake is an important determinant of bone loss during weight-stable conditions and also during weight reduction. During moderately low-calorie intake, women consume less calcium than during weight-stable conditions, yet both intakes are lower than recommended levels *(116,124)*. In obese postmenopausal women, our double-blind, placebo-controlled trial showed that 1 g of supplemental Ca/d prevented bone mobilization associated with a 10% weight reduction *(124)*. In support of these findings, Jensen et al. *(125)* found that a 1-g Ca supplement/d in a group of pre- and postmenopausal women who lost 5.5% of their body weight prevented bone loss at the femoral neck. Others have found that postmenopausal women losing weight sustain significant bone loss from the spine (but not forearm or total body) despite a total calcium intake of 1–1.2 g/d *(119)*. Consistent with this, our preliminary data in overweight postmenopausal women *(126)* show that 1 g calcium/d does not result in adequate total calcium absorbed and induces an increase in the calcium–PTH axis. Importantly, energy restriction studies in the rat show both significant loss of bone and strength properties despite adequate dietary intakes of calcium and other nutrients *(129)*. This supports the hypothesis that usual recommended intakes of calcium may not be able to suppress bone loss during weight reduction and that higher levels (1600–1800 mg Ca/d) should be recommended *(124,125)*. Importantly, we have found no side effects of consuming this level of calcium intake. In addition, there is an absence of side effects due to dietary calcium, even in individuals with a history of recurrent kidney stones due to hypercalciuria *(157)*.

The level of vitamin D intake and its role with calcium during weight reduction is an important consideration. Vitamin D supplementation alone results in no change in bone resorption, whereas the addition of calcium suppresses resorption in weight-stable individuals. During weight reduction, our preliminary data suggest that a 1-g/d calcium supplement (6 mo) and short-term (8 wk) vitamin D supplementation (200 IU/d) during moderate caloric restriction prevented the weight loss-associated rise in bone resorption and bone loss at the hip *(158)*. Seasonal fluctuations in serum 25(OH)D result in lower levels in the winter months, which are also associated with a rise in PTH, bone turnover, and lower bone mass *(159,160)*. It is possible that weight reduction in the early spring months (which is a popular time for dieting) is associated with a greater bone loss than a summer diet *(113)*. However, additional vitamin D and calcium intake during the winter months may prevent the changes to calcitropic hormones and bone loss *(161)*. Finally, other nutrients (such as magnesium, zinc, vitamin K, etc.) may be limited during caloric restriction, yet their influence on bone mass has not been examined.

3.2.5. Physical Activity and Weight Reduction

A few studies have compared whether adding exercise to caloric restriction prevents the loss of bone mass as a result of weight reduction *(111,117–119)*. Under the assumption that a decrease in weight bearing is an important cause for weight-loss-associated bone loss, maintenance of muscle mass particularly through an increase in resistance exercise would be likely to preserve bone mass. In post-menopausal women, aerobic exercise added to a caloric-restriction program was able to prevent loss of bone density at specific sites such as the hip, which may be relevant for fracture risk *(117)*. In a lifestyle intervention study *(118)*, it was found that those women losing weight (>8%) who were more physically active (primarily aerobic) lost less bone than the more sedentary women. Others found no effect of adding exercise (compared to diet alone) on bone mass, however still recommend it for numerous other health benefits *(111,119)*. In summary, it is unclear whether weight reduction due to increased physical activity differs from energy restriction alone, and it is likely that the impact differs largely with the type of exercise program and the population studied.

3.3. DXA Measurement Error

In bone studies, it is most practical and appropriate to measure bone and other components of body composition with DXA. The sensitivity of DXA, however, may be reduced when measuring an individual before and after weight reduction, due to changes in the soft tissue surrounding the bone *(114,121,122)*. In some DXA studies, lard has been placed on top of the individual to determine the degree of error in bone measurements attributable to the change in the overlying fat tissue *(114,162)*. Because large quantities of fat (lard) were used in these studies, it is likely that with moderate weight reduction, the measurement error is much less or possibly insignificant. Nevertheless, there are other potential sources of error in using DXA in the obese population. For example, spinal osteoarthritis is prevalent in the older overweight and obese population, and osteophytes can artificially elevate the bone density measurement. In addition, nonhomogeneous fat distribution or extremes in body size may reduce the sensitivity of DXA measurements *(109)*. Finally, positioning an obese person on the DXA bed can be a challenge, and we have seen errors by untrained technicians (Illustration 2). As necessary, we use straps around the arms and waist to reduce the spread of patients and to keep excess tissue within the scanning area.

In addition to the limitations of DXA in obese patients and during weight changes, there are intrinsic problems with DXA, such as measuring bone mass for

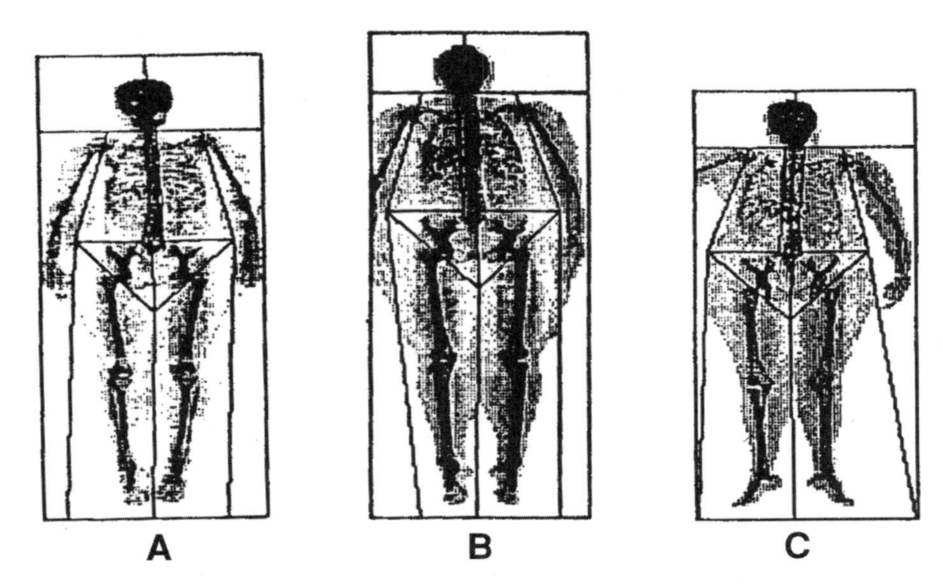

Illustration 2. Total body image using dual-energy X-ray absorptiometry in obese subjects. **(A).** Good image. **(B, C)** Technician error.

a projected bone area (g/cm^2) and content (g) rather than measuring true volumetric BMD (g/cm^3). A three-dimensional result would provide more accurate data and allow for the distinction between cortical and trabecular bone *(163,164)*. Instruments such as computerized tomography or magnetic resonance imaging may be able to overcome some of DXA's shortcomings. Such distinctions are important to provide further insight into bone quality and hence fracture risk.

3.4. Bone Quality After Energy Restriction

Fracture risk is increased with weight reduction or LBW, as evidenced by numerous observational and longitudinal studies. In rodent studies, the relationship between body weight, bone density, and strength properties has also been demonstrated (Fig. 4). In addition, energy restriction in older compared to younger rats is more detrimental to bone density and strength parameters *(129)*. These data are consistent with observations of bone density and strength in human cadavers *(165)* and with epidemiological studies showing that older women who lose weight have an increased risk of fracture *(1,2,105)*. To our knowledge there are no clinical data confirming the hypothesis that a decrease in bone density in young adult bone corresponds to the same rise in fracture risk as would be expected in older bone. Further studies examining changes in bone quality due to energy restriction are important to improve our understanding of how changes in the ultrastructural properties or in the trabecular:cortical ratio of bone influence bone strength and risk of osteoporosis.

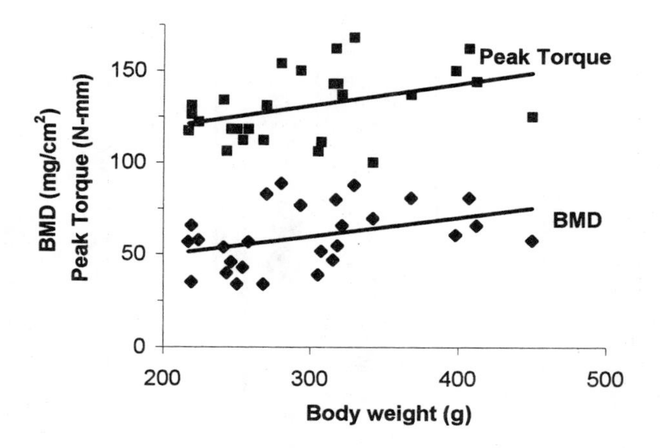

Fig. 4. The relationship between body weight and femoral bone mineral density (BMD; ◆) ($r = 0.39$) and peak torque (■) ($r = 0.40$) in normal and semistarved mature rats.

3.5. Weight Regain

The fact that weight regain after weight loss occurs in most cases raises the question of whether bone that is lost is also regained. In addition, it would be important to evaluate not only if net bone mass is regained, but also the ultra-structural properties and quality of the recovered bone. In postmenopausal women, it was observed that weight regain did not reverse the weight loss-induced reduction in lumbar and spine bone density *(109)*. Similarly, a history of repeated weight loss and regain (i.e., weight cycling), which is common among overweight and obese individuals, may cause cumulative damage on bone and greatly increase the risk for fracture *(166)*. Indeed, lower bone density has been observed in pre-menopausal women with a weight cycling history *(167)*, and rodent studies have confirmed these observations, with reductions in bone quality and strength after regaining weight previously lost through energy restriction *(130)*.

4. CONCLUSIONS AND FUTURE DIRECTIONS

LBW in older individuals is a major risk factor for fracture, and may be the result of low peak bone mass and/or increased rate of bone loss. The rising prevalence of both osteoporosis and obesity is attributed to a poor diet and sedentary behavior, yet there is typically an inverse relationship between the incidence of osteoporosis and obesity. Hormonal differences and/or increased weight bearing in the obese may increase bone mass. Lower levels of estrogen may be a reason that women of normal compared to high body weight are especially responsive to HRT. Fat mass may be a more important predictor of bone mass in older women, whereas lean tissue and physical strength are important determinants of bone in younger populations. Because increased marrow adipose tissue is associated with aging-related bone loss, further knowledge about the differentiation of stromal cells from the osteoblastic to the adipocytic lineage is indicated.

Involuntary weight loss and the method to achieve voluntary weight reduction (through different diets, medication, or increasing levels of activity) will affect bone mass differently. Caloric restriction alters bone-regulating hormones that have both direct and indirect effects on bone. A faster rate of weight loss is more catabolic to skeletal muscle mass, and may also be more detrimental to bone mass. During moderate weight reduction, we currently recommend at least 1500 mg calcium/d through dietary sources, if possible, and adequate vitamin D to reduce the risk of bone loss and ultimate fracture risk.

Future studies of bone health need to examine the impact of current treatments for obese individuals on bone-regulating hormones and the absorption of nutrients that support adequate bone health. Studies of the specific nutrient requirements during weight loss are important for optimizing nutritional recommendations to reduce osteoporosis risk. The addition of exercise to caloric restriction in order to prevent loss of bone mass needs to be explored further, as does the impact of weight regain on bone. Improvements in imaging techniques to measure bone mass and its quality is a promising direction of research that may improve the reliability of current methods when applied to obese populations and/or those undergoing weight loss or gain.

ACKNOWLEDGMENTS

The authors would like to thank N. von Thun and Dr. Y. Schlussel for their help in preparing this manuscript. This work was supported by NIH-AG12161.

REFERENCES

1. Margolis KL, Ensrud KE, Schreiner PJ, Tabor HK. Body size and risk for clinical fractures in older women. Study of Osteoporotic Fractures Research Group. Ann Intern Med 2000; 133:123–127.
2. Ensrud KE, Lipschutz RC, Cauley JA, et al. Body size and hip fracture risk in older women: a prospective study. Study of Osteoporotic Fractures Research Group. Am J Med 1997; 103:274–280.
3. Hannan MT, Felson DT, Dawson-Hughes B, et al. Risk factors for longitudinal bone loss in elderly men and women: the Framingham Osteoporosis Study. J Bone Miner Res 2000; 15:710–720.
4. Nguyen TV, Sambrook PN, Eisman JA. Bone loss, physical activity, and weight change in elderly women: the Dubbo Osteoporosis Epidemiology Study. J Bone Miner Res 1998; 13:1458–1467.
5. Uusi-Rasi K, Sievanen H, Pasanen M, Oja P, Vuori I. Maintenance of body weight, physical activity and calcium intake helps preserve bone mass in elderly women. Osteopor Int 2001; 12:373–379.
6. Wu F, Ames R, Clearwater J, Evans MC, Gamble G, Reid IR. Prospective 10-year study of the determinants of bone density and bone loss in normal postmenopausal women, including the effect of hormone replacement therapy. Clin Endocrinol (Oxf) 2002; 56:703–711.
7. National Osteoporosis Foundation. Physician's Guide To Prevention and Treatment of Osteoporosis. National Osteoporosis Foundation, Washington, DC, 1998.

8. Must A, Jacques PF, Dallal GE, Bajema CJ, Dietz WH. Long-term morbidity and mortality of overweight adolescents. A follow-up of the Harvard Growth Study of 1922 to 1935. N Engl J Med 1992; 327:1350–1355.
9. Jequier E. Pathways to obesity. Int J Obes Relat Metab Disord 2002; 26:S12–S17.
10. National Institutes of Health. Clinical guidelines on the identification, evaluation, and treatment of overweight and obesity in adults—the evidence report. Obes Res 1998; 6:51S–209S.
11. Edelstein SL, Barrett-Connor E. Relation between body size and bone mineral density in elderly men and women. Am J Epidemiol 1993; 138:160–169.
12. Felson DT, Zhang Y, Hannan MT, Anderson JJ. Effects of weight and body mass index on bone mineral density in men and women: the Framingham study. J Bone Miner Res 1993; 8:567–573.
13. Albala C, Yanez M, Devoto E, Sostin C, Zeballos L, Santos JL. Obesity as a protective factor for postmenopausal osteoporosis. Int J Obes Relat Metab Disord 1996; 20:1027–1032.
14. Cifuentes M, Shapses SA, Johnson MA, Lewis R, Chowdhury HA, Modlesky C. Body weight reflects bone resorption in lean, but not overweight or obese postmenopausal women. Osteopor Int 2003; 14:116–122.
15. Reid IR, Ames R, Evans MC, et al. Determinants of total body and regional bone mineral density in normal postmenopausal women—a key role for fat mass. J Clin Endocrinol Metab 1992; 75:45–51.
16. Bjorntorp P. The regulation of adipose tissue distribution in humans. Int J Obes Relat Metab Disord 1996; 20:291–302.
17. Pedersen SB, Hansen PS, Lund S, Andersen PH, Odgaard A, Richelsen B. Identification of oestrogen receptors and oestrogen receptor mRNA in human adipose tissue. Eur J Clin Invest 1996; 26:262–269.
18. Shiraki M, Ito H, Fujimaki H, Higuchi T. Relation between body size and bone mineral density with special reference to sex hormones and calcium regulating hormones in elderly females. Endocrinol Jpn 1991; 38:343–349.
19. Zamboni G, Soffiati M, Giavarina D, Tato L. Mineral metabolism in obese children. Acta Paediatr Scand 1988; 77:741–746.
20. Lind L, Lithell H, Hvarfner A, Pollare T, Ljunghall S. On the relationships between mineral metabolism, obesity and fat distribution. Eur J Clin Invest 1993; 23:307–310.
21. Andersen T, McNair P, Fogh-Andersen N, Nielsen TT, Hyldstrup L, Transbol I. Increased parathyroid hormone as a consequence of changed complex binding of plasma calcium in morbid obesity. Metabolism 1986; 35:147–151.
22. Bell NH, Epstein S, Greene A, Shary J, Oexmann MJ, Shaw S. Evidence for alteration of the vitamin D-endocrine system in obese subjects. J Clin Invest 1985; 76:370–373.
23. Wortsman J, Matsuoka LY, Chen TC, Lu Z, Holick MF. Decreased bioavailability of vitamin D in obesity. Am J Clin Nutr 2000; 72:690–693.
24. Scragg R, Holdaway I, Singh V, Metcalf P, Baker J, Dryson E. Serum 25-hydroxyvitamin D3 is related to physical activity and ethnicity but not obesity in a multicultural workforce. Austral N Z J Med 1995; 25:218–223.
25. Klein KO, Larmore KA, de Lancey E, Brown JM, Considine RV, Hassink SG. Effect of obesity on estradiol level, and its relationship to leptin, bone maturation, and bone mineral density in children. J Clin Endocrinol Metab 1998; 83:3469–3475.
26. De Simone M, Farello G, Palumbo M, et al. Growth charts, growth velocity and bone development in childhood obesity. Int J Obes Relat Metab Disord 1995; 19:851–857.
27. Goulding A, Taylor RW, Jones IE, McAuley KA, Manning PJ, Williams SM. Overweight and obese children have low bone mass and area for their weight. Int J Obes Relat Metab Disord 2000; 24:627–632.
28. Goulding A, Jones IE, Taylor RW, Williams SM, Manning PJ. Bone mineral density and body composition in boys with distal forearm fractures: a dual-energy X-ray absorptiometry study. J Pediatr 2001; 139:509–515.

29. Nguyen TV, Howard GM, Kelly PJ, Eisman JA. Bone mass, lean mass, and fat mass: same genes or same environments? Am J Epidemiol 1998; 147:3–16.
30. Dawson-Hughes B, Shipp C, Sadowski L, Dallal G. Bone density of the radius, spine, and hip in relation to percent of ideal body weight in postmenopausal women. Calcif Tissue Int 1987; 40:310–314.
31. van der Voort DJ, Geusens PP, Dinant GJ. Risk factors for osteoporosis related to their outcome: fractures. Osteopor Int 2001; 12:630–638.
32. Ribot C, Tremollieres F, Pouilles JM, Bonneu M, Germain F, Louvet JP. Obesity and postmenopausal bone loss: the influence of obesity on vertebral density and bone turnover in postmenopausal women. Bone 1987; 8:327–331.
33. Liu G, Peacock M, Eilam O, Dorulla G, Braunstein E, Johnston CC. Effect of osteoarthritis in the lumbar spine and hip on bone mineral density and diagnosis of osteoporosis in elderly men and women. Osteopor Int 1997; 7:564–569.
34. Abrahamsen B, Stilgren LS, Hermann AP, et al. Discordance between changes in bone mineral density measured at different skeletal sites in perimenopausal women—implications for assessment of bone loss and response to therapy: The Danish Osteoporosis Prevention Study. J Bone Miner Res 2001; 16:1212–1219.
35. Takata S, Ikata T, Yonezu H. Characteristics of bone mineral density and soft tissue composition of obese Japanese women: application of dual-energy X-ray absorptiometry. J Bone Miner Metab 1999; 17:206–210.
36. Rodriguez C, Calle EE, Patel AV, Tatham LM, Jacobs EJ, Thun MJ. Effect of body mass on the association between estrogen replacement therapy and mortality among elderly US women. Am J Epidemiol 2001; 153:145–152.
37. Maehle BO, Tretli S, Skjaerven R, Thorsen T. Premorbid body weight and its relations to primary tumour diameter in breast cancer patients; its dependence on estrogen and progesteron receptor status. Breast Cancer Res Treat 2001; 68:159–169.
38. Nelson HD. Assessing benefits and harms of hormone replacement therapy: clinical applications. JAMA 2002; 288:882–884.
39. Zhang HC, Kushida K, Atsumi K, Kin K, Nagano A. Effects of age and menopause on spinal bone mineral density in Japanese women: a ten-year prospective study. Calcif Tissue Int 2002; 70:153–157.
40. Ravn P, Cizza G, Bjarnason NH, et al. Low body mass index is an important risk factor for low bone mass and increased bone loss in early postmenopausal women. Early Postmenopausal Intervention Cohort (EPIC) study group. J Bone Miner Res 1999; 14:1622–1627.
41. Takata S, Yonezu H, Yasui N. Intergenerational comparison of total and regional bone mineral density and soft tissue composition in Japanese women without vertebral fractures. J Med Invest 2002; 49:142–146.
42. Douchi T, Yamamoto S, Oki T, et al. Difference in the effect of adiposity on bone density between pre- and postmenopausal women. Maturitas 2000; 34:261–266.
43. Espallargues M, Sampietro-Colom L, Estrada MD, et al. Identifying bone-mass-related risk factors for fracture to guide bone densitometry measurements: a systematic review of the literature. Osteopor Int 2001; 12:811–822.
44. Spaine LA, Bollen SR. 'The bigger they come…': the relationship between body mass index and severity of ankle fractures. Injury 1996; 27:687–689.
45. Corbeil P, Simoneau M, Rancourt D, Tremblay A, Teasdale N. Increased risk for falling associated with obesity: mathematical modeling of postural control. IEEE Trans Neural Syst Rehab Eng 2001; 9:126–136.
46. Grabowski DC, Ellis JE. High body mass index does not predict mortality in older people: analysis of the Longitudinal Study of Aging. J Am Geriatr Soc 2001; 49:968–979.
47. Asakura A, Komaki M, Rudnicki M. Muscle satellite cells are multipotential stem cells that exhibit myogenic, osteogenic, and adipogenic differentiation. Differentiation 2001; 68:245–253.

48. Prichett WP, Patton AJ, Field JA, et al. Identification and cloning of a human urea transporter HUT11, which is downregulated during adipogenesis of explant cultures of human bone. J Cell Biochem 2000; 76:639–650.
49. Garcia T, Roman-Roman S, Jackson A, et al. Behavior of osteoblast, adipocyte, and myoblast markers in genome-wide expression analysis of mouse calvaria primary osteoblasts in vitro. Bone 2002; 31:205–211.
50. Okazaki R, Inoue D, Shibata M, et al. Estrogen promotes early osteoblast differentiation and inhibits adipocyte differentiation in mouse bone marrow stromal cell lines that express estrogen receptor (ER) alpha or beta. Endocrinology 2002; 143:2349–2356.
51. Dang ZC, van Bezooijen RL, Karperien M, Papapoulos SE, Lowik CW. Exposure of KS483 cells to estrogen enhances osteogenesis and inhibits adipogenesis. J Bone Miner Res 2002; 17:394–405.
52. Nuttall ME, Gimble JM. Is there a therapeutic opportunity to either prevent or treat osteopenic disorders by inhibiting marrow adipogenesis? Bone 2000; 27:177–184.
53. Ahdjoudj S, Lasmoles F, Holy X, Zerath E, Marie PJ. Transforming growth factor beta2 inhibits adipocyte differentiation induced by skeletal unloading in rat bone marrow stroma. J Bone Miner Res 2002; 17:668–677.
54. Parhami F, Jackson SM, Tintut Y, et al. Atherogenic diet and minimally oxidized low density lipoprotein inhibit osteogenic and promote adipogenic differentiation of marrow stromal cells. J Bone Miner Res 1999; 14:2067–2078.
55. Verma S, Rajaratnam JH, Denton J, Hoyland JA, Byers RJ. Adipocytic proportion of bone marrow is inversely related to bone formation in osteoporosis. J Clin Pathol 2002; 55:693–698.
56. Rosenbaum M, Pietrobelli A, Vasselli JR, Heymsfield SB, Leibel RL. Sexual dimorphism in circulating leptin concentrations is not accounted for by differences in adipose tissue distribution. Int J Obes Relat Metab Disord 2001; 25:1365–1371.
57. Ozata M, Ozdemir IC, Licinio J. Human leptin deficiency caused by a missense mutation: multiple endocrine defects, decreased sympathetic tone, and immune system dysfunction indicate new targets for leptin action, greater central than peripheral resistance to the effects of leptin, and spontaneousn correction of leptin-mediated defects. J Clin Endocrinol Metab 1999; 84:3686–3695.
58. Karsenty G. The central regulation of bone remodeling. Trends Endocrinol Metab 2000; 11:437–439.
59. Steppan CM, Crawford DT, Chidsey-Frink KL, Ke H, Swick AG. Leptin is a potent stimulator of bone growth in ob/ob mice. Regul Pept 2000; 92:73–78.
60. Burguera B, Hofbauer LC, Thomas T, et al. Leptin reduces ovariectomy-induced bone loss in rats. Endocrinology 2001; 142:3546–3553.
61. Thomas T, Gori F, Khosla S, Jensen MD, Burguera B, Riggs BL. Leptin acts on human marrow stromal cells to enhance differentiation to osteoblasts and to inhibit differentiation to adipocytes. Endocrinology 1999; 140:1630–1638.
62. Pasco JA, Henry MJ, Kotowicz MA, et al. Serum leptin levels are associated with bone mass in nonobese women. J Clin Endocrinol Metab 2001; 86:1884–1887.
63. Blain H, Vuillemin A, Guillemin F, et al. Serum leptin level is a predictor of bone mineral density in postmenopausal women. J Clin Endocrinol Metab 2002; 87:1030–1035.
64. Yamauchi M, Sugimoto T, Yamaguchi T, et al. Plasma leptin concentrations are associated with bone mineral density and the presence of vertebral fractures in postmenopausal women. Clin Endocrinol (Oxf) 2001; 55:341–347.
65. Sato M, Takeda N, Sarui H, et al. Association between serum leptin concentrations and bone mineral density, andbiochemical markers of bone turnover in adult men. J Clin Endocrinol Metab 2001; 86:5273–5276.
66. Martini G, Valenti R, Giovani S, Franci B, Campagna S, Nuti R. Influence of insulin-like growth factor-1 and leptin on bone mass in healthy postmenopausal women. Bone 2001; 28:113–117.

67. Lee M, Zmuda JM, Wisniewski S, Krishnaswami S, Evans RW, Cauley JA. Serum leptin concentrations and bone mass: differential association among obese and non-obese men. J Bone Mineral Res 2002; 17:S463.
68. Ruhl CE, Everhart JE. Relationship of serum leptin concentration with bone mineral density in the United States population. J Bone Miner Res 2002; 17:1896–1903.
69. Thorsen K, Nordstrom P, Lorentzon R, Dahlen GH. The relation between bone mineral density, insulin-like growth factor I, lipoprotein (a), body composition, and muscle strength in adolescent males. J Clin Endocrinol Metab 1999; 84:3025–3029.
70. Layne JE, Nelson ME. The effects of progressive resistance training on bone density: a review. Med Sci Sports Exerc 1999; Jan; 31:25–30.
71. Simpson E, Rubin G, Clyne C, et al. Local estrogen biosynthesis in males and females. Endocr Relat Cancer 1999; 6:131–137.
72. Sowers MF, Kshirsagar A, Crutchfield MM, Updike S. Joint influence of fat and lean body composition compartments on femoral bone mineral density in premenopausal women. Am J Epidemiol 1992; 136:257–265.
73. Salamone LM, Glynn N, Black D, et al. Body composition and bone mineral density in premenopausal and early perimenopausal women. J Bone Miner Res 1995; 10:1762–1768.
74. Chen Z, Lohman TG, Stini WA, Ritenbaugh C, Aickin M. Fat or lean tissue mass: which one is the major determinant of bone mineral mass in healthy postmenopausal women? J Bone Miner Res 1997; 12:144–151.
75. Blain H, Vuillemin A, Teissier A, Hanesse B, Guillemin F, Jeandel C. Influence of muscle strength and body weight and composition on regional bone mineral density in healthy women aged 60 years and over. Gerontology 2001; 47:207–212.
76. Taaffe DR, Cauley JA, Danielson M, et al. Race and sex effects on the association between muscle strength, soft tissue,and bone mineral density in healthy elders: the Health, Aging, and Body Composition Study. J Bone Miner Res 2001; 16:1343–1352.
77. Lindsay R, Cosman F, Herrington BS, Himmelstein S. Bone mass and body composition in normal women. J Bone Miner Res 1992 Jan; 7:55–63.
78. Kirchengast S, Peterson B, Hauser G, Knogler W. Body composition characteristics are associated with the bone density of the proximal femur end in middle- and old-aged women and men. Maturitas 2001; 39:133–145.
79. Reid IR, Plank LD, Evans MC. Fat mass is an important determinant of whole body bone density in premenopausal women but not in men. J Clin Endocrinol Metab 1992; 75:779–782.
80. Coin A, Sergi G, Beninca P, et al. Bone mineral density and body composition in underweight and normal elderly subjects. Osteopor Int 2000; 11:1043–1050.
81. Hla MM, Davis JW, Ross PD, et al. A multicenter study of the influence of fat and lean mass on bone mineral content: evidence for differences in their relative influence at major fracture sites. Early Postmenopausal Intervention Cohort (EPIC) Study Group. Am J Clin Nutr 1996; 64:354–360.
82. Takata S, Ikata T, Yonezu H. Characteristics of bone mineral density and soft tissue composition of obese Japanese women: application of dual-energy X-ray absorptiometry. J Bone Miner Metab 1999; 17:206–210.
83. Fardellone P, Brazier M, Kamel S, et al. Biochemical effects of calcium supplementation in postmenopausal women: influence of dietary calcium intake. Am J Clin Nutr 1998; 67:1273–1278.
84. Riggs BL, O'Fallon WM, Muhs J, O'Connor MK, Kumar R, Melton LJ 3rd. Long-term effects of calcium supplementation on serum parathyroid hormone level, bone turnover, and bone loss in elderly women. J Bone Miner Res 1998; 13:168–174.
85. Zemel MB, Shi H, Greer B, Dirienzo D, Zemel PC. Regulation of adiposity by dietary calcium. FASEB J 2000 Jun; 14:1132–1138.
86. Carruth BR, Skinner JD. The role of dietary calcium and other nutrients in moderating body fat in preschool children. Int J Obes Relat Metab Disord 2001 Apr; 25:559–566.

87. Heaney RP, Davies KM, Barger-Lux MJ. Calcium and weight: clinical studies. J Am Coll Nutr 2002; 21:152S–155S.
88. Shi H, Dirienzo D, Zemel MB. Effects of dietary calcium on adipocyte lipid metabolism and body weight regulation in energy-restricted aP2-agouti transgenic mice. FASEB J 2001; 15:291–293.
89. Jones BH, Kim JH, Zemel MB, et al. Upregulation of adipocyte metabolism by agouti protein: possible paracrine actions in yellow mouse obesity. Am J Physiol 1996; 270:E192–E196.
90. Xue B, Moustaid-N, Wilkison WO, Zemel MB. The agouti gene product inhibits lipolysis in human adipocytes via a Ca^{2+}-dependent mechanism. FASEB J 1998; 12:1391–1396.
91. Pereira MA, Jacobs DR Jr, Van Horn L, Slattery ML, Kartashov AI, Ludwig DS. Dairy consumption, obesity, and the insulin resistance syndrome in young adults: the CARDIA Study. JAMA 2002; 287:2081–2089.
92. Boyden LM, Mao J, Belsky J, et al. High bone density due to a mutation in LDL-receptor-related protein 5. N Engl J Med 2002; 346:1513–1521.
93. Grant SF, Reid DM, Blake G, Herd R, Fogelman I, Ralston SH. Reduced bone density and osteoporosis associated with a polymorphic Sp1 binding site in the collagen type I alpha 1 gene. Nat Genet 1996; 14:203–205.
94. Uitterlinden AG, Burger H, Huang Q, et al. Relation of alleles of the collagen type Ialpha1 gene to bone density and the risk of osteoporotic fractures in postmenopausal women. N Engl J Med 1998; 338:1016–1021.
95. Cauley JA, Zmuda JM, Yaffe K, et al. Apolipoprotein E polymorphism: A new genetic marker of hip fracture risk—The Study of Osteoporotic Fractures. J Bone Miner Res 1999; 14:1175–1181.
96. Yamada Y, Miyauchi A, Takagi Y, Tanaka M, Mizuno M, Harada A. Association of the C-509→T polymorphism, alone of in combination with the T869→C polymorphism, of the transforming growth factor-beta1 gene with bone mineral density and genetic susceptibility to osteoporosis in Japanese women. J Mol Med 2001; 79:149–156.
97. Matkovic V. Nutrition, genetics and skeletal development. J Am Coll Nutr 1996; 15:556–569.
98. Hedstrom M. Hip fracture patients, a group of frail elderly people with low bone mineral density, muscle mass and IGF-I levels. Acta Physiol Scand 1999; 167:347–350.
99. Valdimarsson T, Lofman O, Toss G, Strom M. Reversal of osteopenia with diet in adult coeliac disease. Gut 1996; 38:322–327.
100. Robbins J, Hirsch C, Whitmer R, Cauley J, Harris T. The association of bone mineral density and depression in an older population. J Am Geriatr Soc 2001; 49:732–736.
101. Michelson D, Stratakis C, Hill L, et al. Bone mineral density in women with depression. N Engl J Med 1996; 335:1176–1181.
102. Mundy GR. Metastasis to bone: causes, consequences and therapeutic opportunities. Nat Rev Cancer 2002; 2:584–593.
103. Langlois JA, Mussolino ME, Visser M, Looker AC, Harris T, Madans J. Weight loss from maximum body weight among middle-aged and older white women and the risk of hip fracture: the NHANES I epidemiologic follow-up study. Osteopor Int 2001; 12:763–768.
104. Ensrud KE, Cauley J, Lipschutz R, Cummings SR. Weight change and fractures in older women. Study of Osteoporotic Fractures Research Group. Arch Intern Med 1997 Apr 28; 157:857–863.
105. Langlois JA, Visser M, Davidovic LS, Maggi S, Li G, Harris TB. Hip fracture risk in older white men is associated with change in body weight from age 50 years to old age. Arch Intern Med 1998 May 11; 158:990–996.
106. Mussolino ME, Looker AC, Madans JH, Langlois JA, Orwoll ES. Risk factors for hip fracture in white men: the NHANES I epidemiologic follow-up study. J Bone Miner Res 1998; 13:918–924.
107. Bales CW, Ritchie CS. Sarcopenia, weight loss, and nutritional frailty in the elderly. Annu Rev Nutr 2002; 22:309–323.
108. Wing RR, Hill JO. Successful weight loss maintenance. Annu Rev Nutr 2001; 21:323–341.

109. Avenell A, Richmond PR, Lean ME, Reid DM. Bone loss associated with a high fibre weight reduction diet in postmenopausal women. Eur J Clin Nutr 1994; 48:561–566.
110. Chao D, Espeland MA, Farmer D, et al. Effect of voluntary weight loss on bone mineral density in older overweight women. J Am Geriatr Soc 2000; 48:753–759.
111. Fogelholm GM, Sievanen HT, Kukkonen-Harjula TK, Pasanen ME. Bone mineral density during reduction, maintenance and regain of body weight in premenopausal, obese women. Osteopor Int 2001; 12:199–206.
112. Ricci TA, Heymsfield SB, Pierson RN Jr, Stahl T, Chowdhury HA, Shapses SA. Moderate energy restriction increases bone resorption in obese postmenopausal women. Am J Clin Nutr 2001; 73:347–352.
113. Shapses SA, Von Thun NL, Heymsfield SB, et al. Bone turnover and density in obese premenopausal women during moderate weight loss and calcium supplementation. J Bone Miner Res 2001; 16:1329–1336.
114. Jensen LB, Quaade F, Sorensen OH. Bone loss accompanying voluntary weight loss in obese humans. J Bone Miner Res 1994; 9:459–463.
115. Pritchard JE, Nowson CA, Wark JD. Bone loss accompanying diet-induced or exercise-induced weight loss: a randomised controlled study. Int J Obes Relat Metab Disord 1996; 20:513–520.
116. Ramsdale SJ, Bassey EJ. Changes in bone mineral density associated with dietary-induced loss of body mass in young women. Clin Sci (Colch) 1994; 87:343–348.
117. Ryan AS, Nicklas BJ, Dennis KE. Aerobic exercise maintains regional bone mineral density during weight loss in postmenopausal women. J Appl Physiol 1998; 84:1305–1310.
118. Salamone LM, Cauley JA, Black DM, et al. Effect of a lifestyle intervention on bone mineral density in premenopausal women: a randomized trial Am J Clin Nutr 1999; 70:97–103.
119. Svendsen OL, Hassager C, Christiansen C. Effect of an energy-restrictive diet, with or without exercise, on lean tissue mass, resting metabolic rate, cardiovascular risk factors, and bone in overweight postmenopausal women. Am J Med 1993; 95:131–140.
120. Van Loan MD, Keim NL. Influence of cognitive eating restraint on total-body measurements of bone mineral density and bone mineral content in premenopausal women aged 18–45 y: a cross-sectional study. Am J Clin Nutr 2000; 72:837–843.
121. Bolotin HH, Sievanen H. Inaccuracies inherent in dual-energy X-ray absorptiometry in vivo bone mineral density can seriously mislead diagnostic/prognostic interpretations of patient-specific bone fragility. J Bone Miner Res 2001; 16:799–805.
122. Tothill P, Avenell A. Errors in dual-energy X-ray absorptiometry of the lumbar spine owing to fat distribution and soft tissue thickness during weight change. Br J Radiol 1994; 67:71–75.
123. Vestergaard P, Borglum J, Heickendorff L, Mosekilde L, Richelsen B. Artifact in bone mineral measurements during a very low calorie diet: short-term effects of growth hormone. J Clin Densitom 2000; 3:63–71.
124. Ricci TA, Chowdhury HA, Heymsfield SB, Stahl T, Pierson RN Jr, Shapses SA. Calcium supplementation suppresses bone turnover during weight reduction in postmenopausal women. J Bone Miner Res 1998; 13:1045–1050.
125. Jensen LB, Kollerup G, Quaade F, Sorensen OH. Bone minerals changes in obese women during a moderate weight loss with and without calcium supplementation. J Bone Miner Res 2001, 16:141–147.
126. Shapses SA, Cifuentes M, Sherrell R, Reidt C. Rate of weight loss influences calcium absorption. J Bone Min Res 2002; 17:S471.
127. Cifuentes M, Morano AB, Chowdhury HA, Shapses SA. Energy restriction reduces fractional calcium absorption in mature obese and lean rats. J Nutr 2002; 132:2660–2666.
128. O'Dea JP, Wieland RG, Hallberg MC, Llerena LA, Zorn EM, Genuth SM. Effect of dietery weight loss on sex steroid binding sex steroids, and gonadotropins in obese postmenopausal women. J Lab Clin Med 1979; 93:1004–1008.

129. Talbott SM, Cifuentes M, Dunn MG, Shapses SA. Energy restriction reduces bone density and biomechanical properties in aged female rats. J Nutr 2001; 131:2382–2387.
130. Wang C, Zhang Y, Xiong Y, Lee CJ. Bone composition and strength of female rats subjected to different rates of weight reduction. Nutr Res 2000; 20:1613–1622.
131. Heaney RP, Recker RR, Saville PD. Menopausal changes in calcium balance performance. J Lab Clin Med 1978; 92:953–963.
132. O'Loughlin PD, Morris HA. Oestrogen deficiency impairs intestinal calcium absorption in the rat. J Physiol 1998; 511:313–322.
133. Kalu DN, Orhii PB. Calcium absorption and bone loss in ovariectomized rats fed varying levels of dietary calcium. Calcif Tissue Int 1999; 65:73–77.
134. Yu BP, Chung HY. Stress resistance by caloric restriction for longevity. Ann N Y Acad Sci 2001; 928:39–47.
135. Harris SR, Brix AE, Broderson JR, Bunce OR. Chronic energy restriction versus energy cycling and mammary tumor promotion. Proc Soc Exp Biol Med 1995; 209:231–236.
136. Arnaud S, Navidi M, Deftos L, et al. The calcium endocrine system of adolescent rhesus monkeys and controls before and after spaceflight. Am J Physiol Endocrinol Metab 2002; 282:E524–E521.
137. Grinspoon SK, Baum HB, Kim V, Coggins C, Klibanski A. Decreased bone formation and increased mineral dissolution during acute fasting in young women. J Clin Endocrinol Metab 1995; 80:3628–3633.
138. Canalis E, Delany AM. Mechanisms of glucocorticoid action in bone. Ann N Y Acad Sci 2002; 966:73–81.
139. Goldstein SA, Elwyn DH. The effects of injury and sepsis on fuel utilization, Annu Rev Nutr 1989; 9:445–473.
140. Gossain VV, Rao DS, Carella MJ, Divine G, Rovner DR. Bone mineral density (BMD) in obesity effect of weight loss. J Med 1999; 30:367–376.
141. Compston JE, Laskey MA, Croucher PI, Coxon A, Kreitzman S. Effect of diet-induced weight loss on total body bone mass. Clin Sci (Lond) 1992; 82(4):429–432.
142. Stein F, Kolanowski J. Bemelmans S. Sesmecht P. Renal handling of calcium in fasting subjects: relation to ketosis and plasma ionized calcium level. Scand J Clin Lab Invest 1983; 43(suppl. 65):99–100.
143. Nishizawa Y, Koyama H, Shoji T, et al. Altered calcium homeostasis accompanying changes of regional bone mineral during a very-low-calorie diet. Am J Clin Nutr 1992; 56:265S–267S.
144. Grinspoon SK, Baum HB, Kim V, Coggins C, Klibanski A. Decreased bone formation and increased mineral dissolution during acute fasting in young women. J Clin Endocrinol Metab 1995; 80:3628–3633.
145. Reddy ST, Wang CY, Sakhaee K, Brinkley L, Pak CY. Effect of low- carbohydrate high-protein diets on acid-base balance, stone-forming propensity, and calcium metabolism. Am J Kidney Dis 2002; 40:265–274.
146. Willi SM, Oexmann MJ, Wright NM, Collop NA, Key LL Jr. The effects of a high-protein, low-fat, ketogenic diet on adolescents with morbid obesity: body composition, blood chemistries, and sleep abnormalities. Pediatrics 1998; 101:61–67.
147. New SA, New SA. The role of the skeleton in acid-base homeostasis. Proc Nutr Soc 2002; 61:151–164.
148. Sebastian A, Harris ST, Ottaway JH, Todd KM, Morris RC Jr. Improved mineral balance and skeletal metabolism in postmenopausal women treated with potassium bicarbonate. N Engl J Med 1994; 330:1776–1781.
149. Maurer M, Riesen W, Muser J, Hulter HN, Krapf R. Neutralization of the acidogenic Western diet inhibits bone resorption independent of K-intake and reduces cortisol secretion in humans. Am J Physiol Renal Physiol 2003; 284:F32–F40.

150. Bushinsky DA. Acid-base imbalance and the skeleton. Eur J Nutr 2001; 40:238–244.
151. Gotfredsen A, Westergren Hendel H, Andersen T. Influence of orlistat on bone turnover and body composition. Int J Obes Relat Metab Disord 2001; 25:1154–1160.
152. Raatz SK, Reck KP, Kwong CA, Swanson JE, Redmon JB, Bantle JP. Bone mineral density of individuals with type 2 diabetes mellitus after combination weight loss therapy using appetite suppressions and meal replacements. J Am Dietet Assoc 2002; 102:A10.
153. Brolin RE. Gastric bypass. Surg Clin N Am 2001; 81(5):1077–1095.
154. MacDonald KG, Long SD, Swanson MS. The gastic bypass operation decreases progression and mortality of non-inulsin diabetes mellitus. J Gastroint Surg 1997; 1:213–220.
155. Kral JG. Surgical treatment of obesity. In: Bray GA, Bouchard C, James WPT, eds. Handbook of Obesity. Marcel Dekker, New york, 1998, pp. 977–993.
156. Goode L, Brolin R, Chowdhury H, Shapses SA. Bone and gastric bypass surgery: effects of dietary calcium and vitamin D. Obesity Res 2003.
157. Borghi L, Schianchi T, Meschi T, et al. Comparison of two diets for the prevention of recurrent stones in idiopathic hypercalciuria. N Engl J Med 2002; 346:77–84.
158. Ricci TA, Heymsfield SB, Shapses SA. Calcium plus vitamin D supplementation prevents bone loss during weight loss. FASEB J 1999; 13:A869.
159. Rosen CJ, Morrison A, Zhou H, et al. Elderly women in northern New England exhibit seasonal changes in bone mineral density and calciotropic hormones. Bone Miner 1994; 25:83–92.
160. Rapuri PB, Kinyamu HK, Gallagher JC, Haynatzka V. Seasonal changes in calciotropic hormones, bone markers, and bone mineral density in elderly women. J Clin Endocrinol Metab 2002; 87:2024–2032.
161. Storm D, Eslin R, Porter ES, et al. Calcium supplementation prevents seasonal bone loss and changes in biochemical markers of bone turnover in elderly New England women: a randomized placebo-controlled trial. J Clin Endocrinol Metab 1998; 83:3817–3825.
162. Milliken LA, Going SB, Lohman TG. Effects of variations in regional composition on soft tissue measurements by dual-energy X-ray absorptiometry. Int J Obes Relat Metab Disord 1996; 20:677–682.
163. Grampp S, Henk CB, Imhof H. CT and MR assessment of osteoporosis. Semin Ultrasound CT MR 1999; 20:2–9.
164. Fujita T. Volumetric and projective bone mineral density. J Muscoloskel Neuron Interact 2002; 2:302–305.
165. Cheng XG, Lowet G, Boonen S, Nicholson PH, Van der Perre G, Dequeker J. Prediction of vertebral and femoral strength in vitro by bone mineral density measured at different skeletal sites. J Bone Miner Res 1998 Sep; 13:1439–1443.
166. Meyer HE, Tverdal A, Selmer R. Weight variability, weight change and the incidence of hip fracture: a prospective study of 39,000 middle-aged Norwegians. Osteopor Int 1998; 8:373–378.
167. Fogelholm M, Sievanen H, Heinonen A, et al. Association between weight cycling history and bone mineral density in premenopausal women. Osteopor Int 1997; 7:354–358.

31 Attenuation of Osteoporosis by *n*-3 Lipids and Soy Protein

Gabriel Fernandes

1. OSTEOPOROSIS

Osteoporosis has become a major health and economic issue in our fast-growing elderly population *(1–3)*. After attaining peak bone mass between the ages of 20 and 30, both men and women start losing bone at a rate of about 0.5–1% per year *(4)*. In the United States alone, nearly $14 billion is spent each year for treatment of complications of osteoporosis *(5)*. As life expectancy increases, there will be an increase in the financial burden on society. It is estimated that by the year 2050, the cost of osteoporosis-related treatments will increase to $131 billion *(6)*. Osteoporosis is characterized by decreased bone mineral density resulting in susceptibility to fractures with minor to moderate trauma. Osteoporosis has been divided into two types *(7)*. Type I osteoporosis, also designated postmenopausal osteoporosis, is the result of estrogen deficiency, whereas type II osteoporosis occurs in the entire aging population of both men and women.

2. BONE METABOLISM

Bone is a dynamic multifunctional organ that is comprised of a structural framework of calcified matrix with many different cell types, including chondrocytes, osteoblasts, osteocytes, osteoclasts, endothelial cells, monocytes, macrophages, lymphocytes, and hematopoeitic cells. Bone tissue is constantly being remodeled throughout life in response to hormonal signals, paracrine and autocrine factors, and physical stresses. The structure of bone is formed, maintained, and re-formed by the collective action of cells that produce and mineralize bone matrix and those that break it down. These cells are termed *osteoblasts* and *osteoclasts (8,9)*.

Osteoblasts are mononucleated cells originating from mesenchymal stem cells. At bone formation sites they produce and actively secrete a material called osteoid consisting mainly of type I collagen and other proteins. This material forms the bone matrix, and under normal conditions it is mineralized rapidly with hydroxyapatite to form the structural framework of bone. Osteoblasts maintain a high

From: *Nutrition and Bone Health*
Edited by: M. F. Holick and B. Dawson-Hughes © Humana Press Inc., Totowa, NJ

alkaline phosphatase activity and produce a number of regulatory effectors including prostaglandins, cytokines, and growth factors that stimulate either bone formation or resorption *(10–13)*.

Osteoclasts, the cells responsible for bone resorption, are large multinucleated cells formed by the fusion of mononuclear hematopoietic precursors brought to the bone through the vasculature. Active osteoclasts reside on the mineralized surface of the bone, where they produce and release lysosomal enzymes, protons, and free radicals into the space in contact with the bone. These products dissolve the mineral and degrade the bone matrix *(14)*. Osteoclastogensis is regulated by several cytokines, growth factors, and eicosanoids.

An accelerated phase of bone loss occurs in women during and immediately following menopause, which results in a 20–30% loss of cancellous bone and a 5–10% loss of cortical bone. This accelerated bone loss occurs also in oophorectomized women, 4–6 yr after surgery *(15)*, and women who have hypothalamic amenorrhea *(16)* have low bone mineral density. Estrogen deficiency is the primary factor in these conditions *(16)*, and estrogen appears to control the production of cytokines that stimulate bone resorption and other factors *(17)*. In addition, availability of adequate dietary calcium and exercise are major determinants in preventing bone loss in the elderly *(18–21)*.

Although estrogen replacement therapy is known to reduce bone loss, its side effects, including increased risk of developing cancer and cardiovascular disease, are of major concern *(22,23)* and have prompted the discontinuation of recent clinical trials. Dietary calcium supplementation has been used for many years, but response to this therapy has been variable. Therapies designed to modulate osteoclast activity, such as bisphophonates and calcitonin, as well as to block signaling of the osteoprotegerin (OPG)/receptor activator of nuclear factor-kB ligand (RANKL)/RANK pathway, are also being investigated *(24–28)*.

3. ROLE OF DIET IN OSTEOPOROSIS

3.1. Omega-3 Fatty Acids and Immunology

In recent years, considerable evidence has accumulated for a role of diet and nutrition in the development and maintenance of the immune system throughout life *(29,30)*. The composition of the dietary lipids modifies the phospholipid fatty acid composition of plasma membranes *(31–33)* and their physicochemical characteristics, which in turn alters cell–cell interactions, intracellular communication, cell activation, membrane lipid–protein interactions, and the expression of various receptors for growth factors and cytokines *(34)* in both lymphoid and bone-forming cells.

Dietary essential fatty acids (EFAs) are derived mainly from either plant or marine (fish) sources. These fatty acids can be divided into two categories depending on the position of the double bonds in the carbon chain *(35)*. The designations omega-3 (ω-3 or *n*-3) and omega-6 (ω-6 or *n*-6) indicate that the third or sixth carbon, respectively, from the methyl terminus is unsaturated. Dietary sources of long-chain ω-3 and ω-6 fatty acids are essential for maintaining optimum health because mammals cannot synthesize fatty acids with a double bond past the Δ-9 position. Longer-chain *n*-3 and *n*-6 fatty acids can be derived from the 18-carbon precursors, α-linolenic acid and linoleic acid, respectively, by chain elongation and desaturation.

The association of cholesterol and saturated fat with an increased risk of cardiovascular disease led to dietary recommendations to reduce the intake of animal fat and to increase the intake of plant oils. The resulting dramatic change in food formulations has led to a greater intake of *n*-6-containing plant oils such as corn, safflower, and soybean oil, which are high in linoleic acid, and an elevated ratio of *n*-6/*n*-3 fatty acids over the last 35 yr *(36)*.

Eicosanoids (prostaglandins, leukotrienes, and related compounds) are formed from long-chain polyunsaturated fatty acids released by the action of lipases from membrane phospholipids. The amount of different types of eicosanoids produced reflects the fatty acids available in the membrane lipids. Inflammatory eicosanoids such as prostaglandin E_2 (PGE_2), which stimulate bone resorption, are produced from arachidonic acid (AA 20:4*n*-6), whereas eicosapentaenoic acid (EPA 20:5*n*-3) competitively inhibits PGE_2 formation and serves as a precursor in the formation of the less inflammatory PGE_3. Thus dietary intake of EFAs has far-reaching consequences on membrane composition in all cells in the body, and influences the production of various cytokines by lymphoid cells. Furthermore, *n*-3 fatty acid deficiency is linked to changes in gene expression and cAMP-dependent protein kinase A and C (PKA, PKC) activities *(37)*.

It has been shown that ω-3 supplements have anti-inflammatory activity against arthritis, lupus, and immunoglobulin A nephropathy in both human and animal studies *(38)*. We, and others, have described the role of *n*-3 fatty acids with or without calorie restriction (CR) in protecting against autoimmune renal disease in murine lupus nephritis models *(39–51)*. We have further observed that inhibition of renal disease in N2B × N2W F (B/W) mice was associated with decreased platelet-derived growth factor (PDGF)-A and thrombin receptor expression and decreased immune complex deposition in the kidney *(48)* as well as decreased expression of proinflammatory growth factors (transforming growth factor β) and cytokines (interleukin [IL]-1β, IL-6, and tumor necrosis factor [TNF]-α) *(52–54)*. Also we found increased antioxidant enzyme activity and decreased C-Myc and c-Ha-Ras oncogene expression in kidney tissue from *n*-3 fatty acid-fed mice. Because IL-1a, IL-6, and TNF-α are closely linked to the activation of osteoclasts *(55–58)* and PGE_2 has also been found to accelerate bone loss *(59,60)*, we have initiated studies on osteoporosis in ovariectomized (OVX) mice fed *n*-6 compared to *n*-3 lipids.

3.2. Omega-3 Fatty Acids and Bone Health

Recently, there has been increasing evidence that, in addition to inadequate dietary calcium and/or vitamin D, deficiency of certain fatty acids may also contribute to bone loss *(61–63)*. In one study, essential fatty acid-deficient animals developed severe osteoporosis coupled with renal and arterial calcification *(62)*. An influence of fatty acids on bone metabolism and increase periosteal bone formation *(64,65)* have also been reported in animal studies. In addition, dietary supplementation of calcium along with γ-linolenic acid (GLA, *n*-6) and eicosapentaenoic acid (EPA, *n*-3) in humans has been reported to decrease bone turnover and increase bone mineral density *(66–68)*. Despite these reports of protection against bone loss, the mechanism of the prevention of bone loss by *n*-3 fatty acids has not been determined.

One of the postulated mechanisms by which fish oil exerts its protective action against bone loss is a decrease in urinary calcium loss *(68–70)*. Also, diets high in saturated fat have been reported to interfere with the absorption of dietary calcium. Wohl et al. *(71)* reported that cancellous bone strength in animals maintained on a low-fat diet consistently exceeded that in animals fed a high-fat diet. There is a great deal of literature indicating that in animal and in vitro cell culture experiments a decrease in the *n*-6/*n*-3 fatty acid ratio can protect against bone mineral loss and production of PGE_2.

Initially, Sakaguchi *(72)* reported that OVX and a low-calcium diet caused a significant decrease in bone weight and bone strength and that EPA prevented this loss of bone weight and bone strength, but it did not result in an increase in bone weight and strength in the normal calcium group. Later, Kruger et al. *(73)* also used OVX rats to study the relationship among EFAs, bone turnover, and bone calcium. The rats were fed a semipurified diet containing different ratios of GLA:EPA+DHA (9:1, 3:1, 1:3, 1:9) from age 12–18 wk. LA:ALA (3:1) was used as a control in a sham-operated and an OVX group. DGLA, DHA, and EPA were negatively correlated with deoxypyridinoline (Dpyd), a marker of bone degradation, but only the DGLA correlation was significant ($r = 0.54$; $p = 0.002$). Based on its positive correlation with bone calcium and negative correlation with Dpyd, DGLA may have an anabolic effect on bone. Furthermore, in a study performed in elderly women whose diet was low in calcium *(68)*, supplementation of calcium and GLA + EPA produced a significant fall in osteocalcin and Dpyd, indicative of a decrease in bone turnover. After 18 mo, the patients on GLA + EPA treatment showed an increase of 3.1% in lumbar spine density and 4.7% in femoral bone mineral density. Calcium transport seems to be enhanced by incorporation of PUFA in the intestinal epithelial cell membranes *(74)*, resulting in enhanced bone mineral density and prevention of osteoporosis. A possible role of EFAs in osteoporosis was discussed by Das *(61)*, with particular reference to their modulating effects on cytokines that may be involved in the process of osteoporosis *(55,75–77)*. Estrogen can suppress IL-1, TNF, and IL-6 synthesis. EFAs are known to inhibit the production of a range of cytokines *(78–81)* and could modify the development of osteoporosis by modifying cytokine secretion.

3.3. Dietary Protein Source and Bone Health

The endogenous acid hypothesis proposed by Wachman and Bernstein in 1968 *(82)* has led to a great deal of controversy regarding the role of dietary protein in osteoporosis. This hypothesis speculates that diets rich in protein will increase bone loss due to the acid produced by protein catabolism and the consequent need for calcium resorption from the skeleton to buffer the elevated acidity of the endogenous medium. Although this hypothesis does not distinguish between animal and vegetable sources of protein, Sebastian et al. have advanced a related hypothesis that a high dietary ratio of animal to vegetable foods, quantified by protein content, increases bone loss and the risk of fracture *(83)*. They postulate that this is due to the higher acid load from animal vs vegetable foods. They present data from a large cohort of elderly women showing no effect of the ratio of animal-to-vegetable protein intake on initial bone density but a higher rate of loss in women consuming a higher ratio of animal-to-vegetable protein after an average of 3.6 yr. While it has generally been found that urinary calcium excretion increases with increasing protein consumption *(84–87)*, there have been reports to the contrary *(88,89)*. Cross-cultural studies implicating the high consumption of animal proteins in Western cultures have often been cited in support of the endogenous acid hypothesis *(90,91)*. In contrast, studies of the effect of protein supplementation in hip fracture patients have suggested a beneficial effect *(92–95)*, but results from population-based cohort studies have been mixed *(96–102)*. The relative effects of animal vs vegetable sources of protein are confounded by other components of the plant sources that may have potent effects on bone metabolism unrelated to protein intake, particularly in cross-cultural studies in which traditional foods may include plants not consumed in quantity by other cultures. This is exemplified by the phytoestrogens found in soy protein sources.

The effects of soy protein on bone metabolism have been attributed largely to their content of phytoestrogens, principally in the form of the isoflavones genistein and daidzein *(103–107)* (*see* chapter on phytoestrogens). These are found mostly as β-glycosides in soy products but are readily hydrolyzed to the aglycone form by intestinal bacteria. Daidzein is further metabolized by gut bacteria to the more active equol, which is thought to be responsible for most of the estrogenic activity *(108,109)*, but this conversion is not so readily achieved and up to 30% of the human population apparently lack the bacteria responsible for this conversion. This has been suggested as an explanation for the variable results in clinical trials of soy products. In future studies of soy products in humans, these "equol nonproducers" should be accounted for. If these nonproducers can be converted to producers, incorporation of more soy protein sources in the diet may hold some promise for a relatively potent and innocuous stimulator of bone formation. We found significant bone loss (20%) in the distal left femur of casein + CO-fed OVX mice that correlated with high RANKL expression ($p < 0.05$) in T-cells, whereas soy protein + CO-fed mice lost 13% (Fig. 1). Casein and soy protein, FO-fed mice lost 10% and 4% bone mineral density (BMD) compared to the corresponding sham-operated mice (Fig. 1), and RANKL remained low (Fig. 2). BMD was significantly higher in casein than in soy protein-fed sham-operated mice but BMD loss and RANKL were less in soy protein-fed OVX mice *(110)*. Thus, it appears that

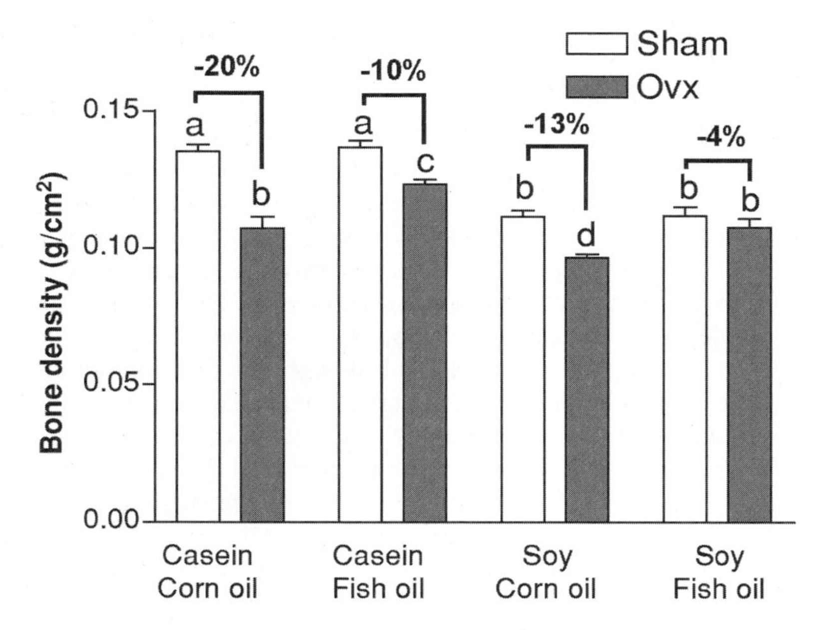

Fig. 1. Effect of casein and soy as dietary protein source and corn and fish oil as dietary fat source on bone mineral density changes with OVX measured by dual-energy X-ray absorptiometry in distal left femur of Balb/C mice. Means of bars not labeled with the same superscript letter are significantly different at $p < 0.05$ by Newman Kuehl's test. (From ref. *110a.*)

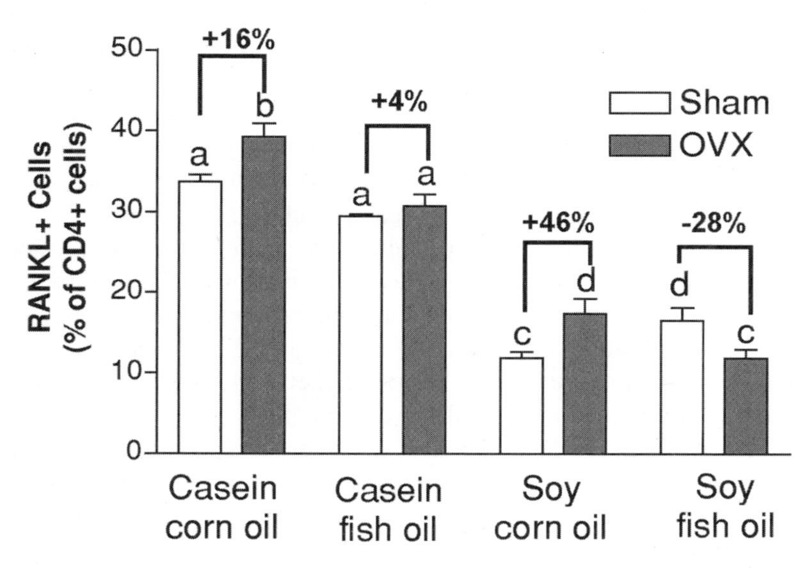

Fig. 2. Effect of casein and soy as dietary protein source and corn and fish oil as dietary fat source on changes with OVX in % of $CD4^+$ cells from Balb/C mice that express RANKL. Means of bars not labeled with the same superscript letter are significantly different at $p < 0.05$ by Newman Kuehl's test. (From ref. *110a.*)

soy protein and fish oil may act synergistically to decrease bone loss during the postmenopausal period. However, new studies are underway with larger numbers of mice to confirm these pilot studies.

As has been pointed out by Promislow et al. *(99)* and others, protein is essential for bone formation because it is a structural component of bone accounting for approximately half of bone volume and one-fourth of bone mass, including the skeletal matrix. The majority of the evidence suggests that, given adequate intakes of other nutrients, excess protein intake from either animal or vegetable sources has no adverse effect on bone density and most probably is beneficial. We are reminded by Heaney *(111)* that any strategy to increase bone density must be accompanied by an adequate intake of the necessary components for bone formation, including calcium and phosphorus, to be effective.

3.4. Diet Effects on OPGL/RANK/RANKL Signaling

One of the mechanisms by which bone loss is increased in estrogen deficiency is release of inhibition of osteoclast differentiation *(56,57)*. This occurs through stimulation of osteoclast progenitors by macrophage colony-stimulating factor (M-CSF) and RANKL simultaneously *(112,113)*. Under normal resting conditions, the differentiation of osteoclast progenitors into mature osteoclasts in bone marrow depends on the balance between OPGL-RANK signaling and the levels of biologically active OPG produced by stromal cells and osteoblasts *(114)*. During stimulation, however, other bone marrow cells participate in regulating osteoclast formation by stimulatory and inhibitory cytokine production. It has been reported recently that activated T-cells play an essential causal role, not only in inflammation-induced bone loss, but also in the bone wasting induced by estrogen deficiency. In fact, T-cells express estrogen receptors, and estrogen directly regulates T-cell function and cytokine production *(115–118)*. Activated T-cells modulate osteoclast formation by expression of RANKL *(119–121)*, osteoprotegerin *(122)*, and IFN-γ *(123)*. During inflammation and autoimmune arthritis, production of RANKL by activated T-cells promotes bone resorption and bone loss, whereas IFN-γ limits T-cell-induced bone wasting.

We therefore measured bone density in ovariectomized (OVX) mice fed diets containing either 5% corn oil or 5% fish oil. Significantly increased bone loss (20% in distal left femur and 22.6% in lumbar vertebrae) was observed in OVX mice fed corn oil (Figs. 1 and 3). In contrast, fish oil-fed mice showed much less loss (10% in distal left femur and no change in lumbar vertebrae). A protective role of *n*-3 fatty acids, with or without calorie restriction, on autoimmune renal diseases in animal models has been reported *(40,42,47,48)*. Inhibition of renal disease in B/W mice was accompanied by decreased expression of proinflammatory cytokines (IL-1, IL-6, and TNF-α) *(52,124,125)*. The *n*-3 fatty acids also suppress lymphocyte proliferation *(126)*. Since IL-1, IL-6, and TNF-α are associated with osteoclastogenesis and T-cells play a critical role in the pathogenesis of osteoporosis *(121)*, dietary *n*-3 fatty acids may prevent bone loss induced by ovariectomy by regulating *T*-cell function and cytokine production, thereby inhibiting

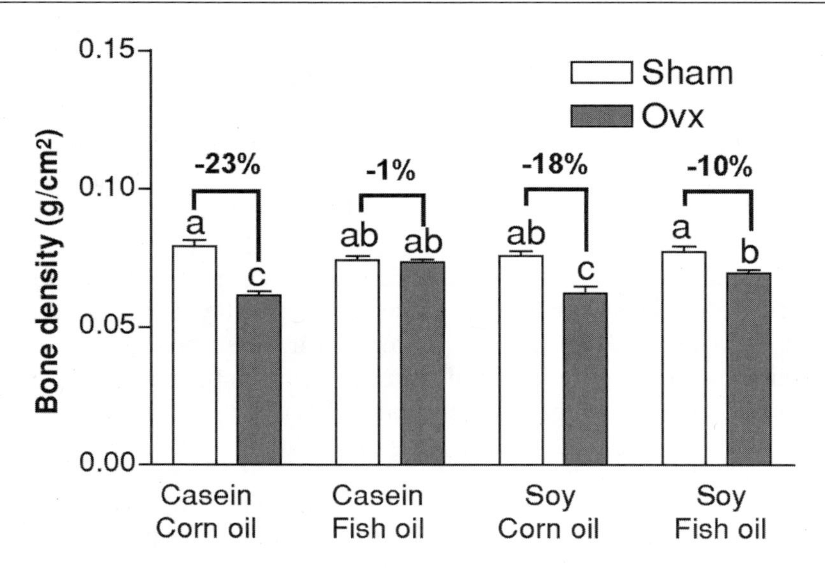

Fig. 3. Effect of casein and soy as dietary protein source and corn and fish oil as dietary fat source on bone mineral density changes with OVX in lumbar vertebrae of Balb/C mice. Means of bars not labeled with the same superscript letter are significantly different at $p < 0.05$ by Newman Kuehl's test. (From ref. *110a.*)

activation and maturation of osteoclasts. The role of fatty acids in bone metabolism was confirmed by the observation that EPA-deficient animals developed severe osteoporosis coupled with increased renal and arterial calcification *(62).* This situation is somewhat similar to the clinical picture seen in elderly humans, in whom osteoporosis is associated with calcification of the arteries and kidneys *(62).* It was reported that the deposition of calcium in both kidneys and aorta was reduced by fish oil or EPA in male Sprague-Dawley rats *(127).* A well-controlled human study also showed that polyunsaturated fatty acids (PUFAs) such as GLA and EPA have beneficial effects on bone density in elderly people *(128).* In this study, PUFA given orally were reported to be incorporated into the intestinal membrane vesicles resulting in enhanced calcium transport *(74).* This appears to be one of the possible mechanisms by which GLA and EPA are able to prevent osteoporosis.

Loss of BMD was correlated with increased receptor activator of NF-κB ligand (RANKL) expression in activated $CD4^+$ T-cells from corn oil-fed OVX mice, whereas no change in RANKL expression was observed in cells from fish oil-fed mice (Fig. 4). It has been demonstrated that the decline in estrogen production in postmenopausal women results in increased production of several osteoclastogenic cytokines such as IL-1, IL-6, and TNF-α *(57,129).* These cytokines induce the expression of COX-II in osteoblastic and stromal cells, resulting in an increase in the production of PGE_2 *(130).* TNF-α and PGE_2 increase the expression of RANKL, not only in stromal cells *(131),* but also in pre-B-cells, which then cooperatively

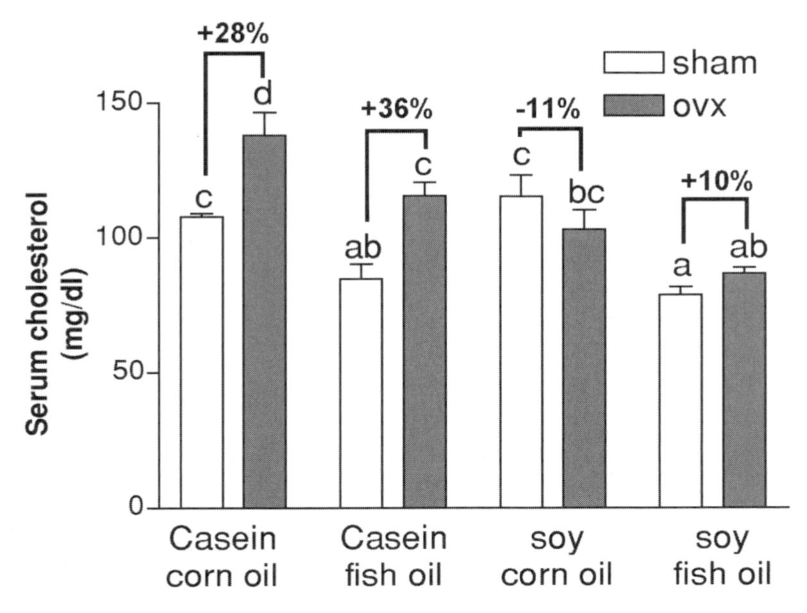

Fig. 4. Effect of casein and soy as dietary protein source and corn and fish oil as dietary fat source on serum cholesterol changes with OVX in Balb/C mice. Means of bars not labeled with the same superscript letter are significantly different at $p < 0.05$ by Newman Kuehl's test. (From ref. *110a.*)

stimulate osteoclast progenitors through RANK–RANKL interaction and promote the differentiation of these cells into mature osteoclasts. This process ultimately leads to osteoporosis *(55)*. We have found that fish oil downregulates TNF-α and IL-6 *(132)*. Fish oil has been found to decrease levels of mRNA coding for proinflammatory cytokines (IL-1β, IL-6, and TNF-α) in immune cells and kidney tissue in vivo *(39–41,43,45,47–51,133)* and also in osteoblasts in vitro *(134)*. In addition, *n*-3 fatty acids can decrease the production of PGE_2 and may downregulate cyclooxygenase-II (COX-II) activity *(135)* in some tissues. NO is postulated to play an important role in bone metabolism, and it is also known that both DHA and EPA enhance NO formation *(136,137)*. This may be another mechanism of the antiosteoporotic effects of *n*-3 fatty acids.

NF-κB has also been reported to play a role in the pathogenesis of osteoporosis. Mice null for NF-κB1 and NF-κB2 developed osteopetrosis and produced very few osteoclasts compared with normal controls *(138)*. This confirms the essential role of the NF-κB signal pathway in osteoclast generation and activation *(139–141)*. We have found that EPA and DHA alone and in combination inhibited RANKL-induced NF-κB activation in bone marrow-derived macrophages, whereas the *n*-6 fatty acids LA and AA had no effect. Camandola et al. *(142)* also showed that fish oil can inhibit LPS-induced NF-κB activation in a macrophage

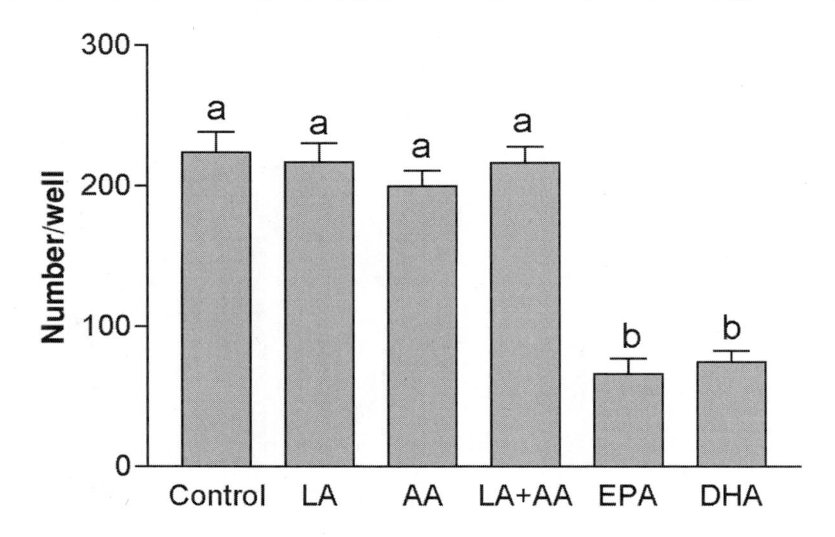

Fig. 5. Effect of fatty acids on osteoclastogenesis in vitro in bone marrow-derived macrophages. Means of bars not labeled with the same superscript letter are significantly different at $p < 0.05$ by Newman Kuehl's test. (From ref. *132*.)

cell line. We measured the effects of *n*-3 (docosahexaenoic acid, DHA, and eicosapentaenoic acid, EPA) and *n*-6 (linoleic acid, LA, and arachidonic acid, AA) fatty acids on the tartrate-resistant acid phosphatase (TRAP) activity and TRAP-positive multinuclear cell formation associated with osteoclastogenesis in bone marrow cells cultured with $1,25(OH)_2D_3$. Consistent with these results, we also found that the *n*-3 fatty acids DHA and EPA, alone or in combination, inhibited TRAP activity in primary bone marrow cells cultured with vitamin D_3 for 7 d, and significantly inhibited osteoclastogenesis in vitro as assessed by TRAP staining *(132)*. No inhibition was observed with LA and AA (Fig. 5). Furthermore, DHA and EPA, alone or combination, were able to inhibit activation of macrophage NF-κB production induced by RANKL in vitro. One of the mechanisms by which *n*-3 fatty acids inhibit bone loss could be inhibition of RANKL expression by T-cells.

When lymphocyte subset proportions were analyzed in ovariectomized mice fed corn or fish oil, a significant increase in the number of splenic B-cells and decrease in the number of $CD4^+$ T-cells was found in the corn oil-fed OVX group. However, there was no difference found between sham-operated and OVX mice in the fish oil-fed group. It has been shown that OVX increases B-cell numbers, particularly in the bone marrow, and may also promote osteoclastogenesis in the presence of T-cells *(143,144)*, and that OVX does not cause bone loss in nude mice in the absence of T-cells *(145)*. Steroids, particularly estrogens, are powerful regulators of bone metabolism, and they are involved in the differentiation of the B-lymphocyte hematopoietic lineage. B-lymphocyte precursors declined dramatically in the bone marrow of pregnant or estrogen-treated mice, and the same populations increased in hypogonadal, ovariectomized mice *(144,146)*. A role of B-cells in the pathogenesis of OVX-induced bone loss has been postulated *(147,148)*. Onoe et al. *(149)* postulated an effect of different cytokines and RANKL in bone loss caused by increased B-lymphocyte numbers. Other investigators showed that osteoclasts and B-lymphocytes have a common precursor *(150)*, and osteoclasts can form from a highly purified $B220^+$ bone marrow cell population when stimulated with M-CSF and RANKL. This indicates that the direct effect of estrogen withdrawal on osteoclast and B-lymphocyte precursors is an important event in the pathogenesis of bone loss after ovariectomy *(60)*. Therefore, the effect of fish oil on B-lymphocyte lineage development and on osteoclastogenesis through the B-cell lineage pathway could be one important mechanism by which it decreases osteoporosis.

4. CONCLUSION

In conclusion, there is strong support for the contention that dietary fish oil can decrease OVX-induced bone loss. Fish oil downregulates expression of RANKL on activated T-cells and inhibits NF-κB activation in osteoclast progenitors, thereby inhibiting osteoclastogenesis. This may be one of the primary mechanisms of the protective effects of fish oil on bone density. Animal protein is necessary for adolescent bone formation and remodeling. However, dietary animal protein has been shown in some studies to increase calcium loss, but others have not found the same results. Other components of dietary protein sources such as isoflavones found in soy protein may account for protection by soy protein compared to animal proteins. However, additional studies in both animals and humans are needed to determine the effects of both *n*-3 fatty acids and soy protein in protecting against perimenopausal bone loss and fractures in the elderly.

ACKNOWLEDGMENTS

This work was supported in part by National Institutes of Health grants AG14541 and AG20239.

REFERENCES

1. Nevitt MC. Epidemiology of osteoporosis. Rheum Dis Clin N Am 1994; 20:535–559.
2. Melton LJ 3rd. How many women have osteoporosis now? J Bone Miner Res 1995; 10:175–177.
3. Mundy G, Garrett R, Harris S, et al. Stimulation of bone formation in vitro and in rodents by statins. Science 1999; 286:1946–1949.
4. McGarry KA, Kiel DP. Postmenopausal osteoporosis. Strategies for preventing bone loss, avoiding fracture. Postgrad Med 2000; 108:79–82, 85–78, 91.
5. Ray NF, Chan JK, Thamer M, Melton LJ 3rd. Medical expenditures for the treatment of osteoporotic fractures in the United States in 1995: report from the National Osteoporosis Foundation. J Bone Miner Res 1997; 12:24–35.

6. Johnell O. The socioeconomic burden of fractures: today and in the 21st century. Am J Med 1997; 103:20S–25S; discussion 25S–26S.
7. Kassem M, Melton LI, Riggs B. The type I and type II model for involutional osteoporosis. In: Marcus R, Feldman D, Kelsey J, eds. Osteoporosis. Academic, New York, 1996, pp. 691–702.
8. Ducy P, Schinke T, Karsenty G. The osteoblast: a sophisticated fibroblast under central surveillance. Science 2000; 289:1501–1504.
9. Teitelbaum SL. Bone resorption by osteoclasts. Science 2000; 289:1504–1508.
10. Baylink DJ, Finkelman RD, Mohan S. Growth factors to stimulate bone formation. J Bone Miner Res 1993; 8(suppl 2):S565–S572.
11. Mundy GR. Cytokines and growth factors in the regulation of bone remodeling. J Bone Miner Res 1993; 8(suppl 2):S505–S510.
12. Raisz LG. Bone cell biology: new approaches and unanswered questions. J Bone Miner Res 1993; 8(suppl 2):S457–S465.
13. Canalis E. Regulation of bone remodeling. In: Favus M, ed. Pimer on the Metabolic Bone Diseases and Disorders of mineral Metabolism. Raven, New York, 1993, pp. 33–37.
14. Blair HC, Schlesinger PH, Ross FP, Teitelbaum SL. Recent advances toward understanding osteoclast physiology. Clin Orthoped 1993; 294:7–22.
15. Lindsay R. Sex steroids in the pathogenesis and prevention of osteoporosis. In: Riggs B, Melton LI, eds. Osteoporosis: Etiology, Diagnosis and Management. Raven, New York, 1988.
16. Sanborn CF, Martin BJ, Wagner WW Jr. Is athletic amenorrhea specific to runners? Am J Obstet Gynecol 1982; 143:859–861.
17. Bilezikian JP. Estrogens and postmenopausal osteoporosis: was Albright right after all? J Bone Miner Res 1998; 13:774–776.
18. Weaver CM. The growing years and prevention of osteoporosis in later life. Proc Nutr Soc 2000; 59:303–306.
19. Weaver CM, Heaney RP. Dairy consumption and bone health. Am J Clin Nutr 2001; 73:660–661.
20. Weaver CM, Teegarden D, Lyle RM, et al. Impact of exercise on bone health and contraindication of oral contraceptive use in young women. Med Sci Sports Exerc 2001; 33:873–880.
21. Weaver CM. Calcium requirements of physically active people. Am J Clin Nutr 2000; 72:579S–584S.
22. Lacey JV Jr, Mink PJ, Lubin JH, et al. Menopausal hormone replacement therapy and risk of ovarian cancer. JAMA 2002; 288:334–341.
23. Risks and benefits of estrogen plus progestin in healthy postmenopausal women: principal results From the Women's Health Initiative randomized controlled trial. JAMA 2002; 288:321–333.
24. Karpf DB, Shapiro DR, Seeman E, et al. Prevention of nonvertebral fractures by alendronate. A meta-analysis. Alendronate Osteoporosis Treatment Study Groups. JAMA 1997; 277:1159–1164.
25. Stevenson JC, Abeyasekera G, Hillyard CJ, et al. Calcitonin and the calcium-regulating hormones in postmenopausal women: effect of oestrogens. Lancet 1981; 1:693–695.
26. Tiegs RD, Body JJ, Wahner HW, Barta J, Riggs BL, Heath H 3rd. Calcitonin secretion in postmenopausal osteoporosis. N Engl J Med 1985; 312:1097–1100.
27. Riggs BL, Melton LJ 3rd. Evidence for two distinct syndromes of involutional osteoporosis. Am J Med 1983; 75:899–901.
28. Horowitz MC, Xi Y, Wilson K, Kacena MA. Control of osteoclastogenesis and bone resorption by members of the TNF family of receptors and ligands. Cytokine Growth Factor Rev 2001; 12:9–18.
29. Calder PC. Dietary fatty acids and the immune system. Lipids 1999; 34:S137–S140.
30. Fernandes G. Nutrition and immunity. In: Dulbecco R, ed. Encyclopedia of Human Biology. Vol. 5. Academic, New York, 1991, pp. 503–516.

31. Lewis RA, Austen KF, Soberman RJ. Leukotrienes and other products of the 5-lipoxygenase pathway. Biochemistry and relation to pathobiology in human diseases. N Engl J Med 1990; 323:645–655.
32. Endres S, Ghorbani R, Kelley VE, et al. The effect of dietary supplementation with *n*-3 PUFA on the synthesis of IL-1 and TNF by mononuclear cells. N Engl J Med 1989; 320:265–271.
33. Wang H, Chen X, Fisher EA. N-3 fatty acids stimulate intracellular degradation of apoprotein B in rat hepatocytes. J Clin Invest 1993; 91:1380–1389.
34. Hwang D, Rhee SH. Receptor-mediated signaling pathways: potential targets of modulation by dietary fatty acids. Am J Clin Nutr 1999; 70:545–556.
35. Speizer LA, Watson MJ, Brunton LL. Differential effects of omega-3 fish oils on protein kinase activities in vitro. Am J Physiol 1991; 261:E109–E114.
36. Simopoulos AP. Evolutionary aspects of omega-3 fatty acids in the food supply. Prostaglandins Leukot Essent Fatty Acids 1999; 60:421–429.
37. Mirnikjoo B, Brown SE, Kim HF, Marangell LB, Sweatt JD, Weeber EJ. Protein kinase inhibition by omega-3 fatty acids. J Biol Chem 2001; 276:10888–10896.
38. Kehn P, Fernandes G. The importance of omega-3 fatty acids in the attenuation of immune-mediated diseases. J Clin Immunol 2001; 21:99–101.
39. Calder PC. Immunomodulatory and anti-inflammatory effects of omega-3 polyunsaturated fatty acids. Proc Nutr Soc 1996; 55:737–774.
40. Robinson DR, Knoell CT, Urakaze M, et al. Suppression of autoimmune disease by omega-3 fatty acids. Biochem Soc Trans 1995; 23:287–291.
41. Robinson DR, Kremer JM. Rheumatoid arthritis and inflammatory mediators. World Rev Nutr Diet 1991; 66:44–47.
42. Prickett JD, Robinson DR, Steinberg AD. Dietary enrichment with the PUFA eicosapentaenoic prevents proteinuria and prolongs survival in NZB×NZW$_{F1}$ mice. J Clin Invest 1981; 658:556–559.
43. Kelley VE, Ferretti A, Izui S, Strom TB. A fish oil diet rich in eicosapentanoic acid reduces cyclooxygenese metabolites and suppresses lupus in MRL/lpr mice. J Immunol 1985; 134:1914–1919.
44. Fernandes G. Effect of dietary fish oil supplement on autoimmune disease: changes in lymphoid cell subsets, oncogene mRNA expression and neuroendocrine hormones. In: Chandra R, ed. Health Effects of Fish and Fish Oils. ARTS Biomedical Publications, St. John's, Newfoundland, Canada, 1989, pp. 409–433.
45. Jeng KC, Fernandes G. Effect of fish oil diet on immune response and proteinurea in mice. Proc Natl Sci Council ROC 1991; 15(2):105–110.
46. Fernandes G, Venkatraman J. Role of w-3 fatty acids in health and disease. Nutr Res 1993; 13:S19–S45.
47. Venkatraman JT, Chandrasekar B, Kim JD, Fernandes G. Effects of *n*-3 and *n*-6 fatty acids on the activities and expression of hepatic antioxidant enzymes in autoimmune-prone NZB×NZW F1 mice. Lipids 1994; 29:561–568.
48. Fernandes G, Bysani C, Venkatraman JT, Tomar V, Zhao W. Increased TGF-beta and decreased oncogene expression by omega-3 fatty acids in the spleen delays onset of autoimmune disease in B/W mice. J Immunol 1994; 152:5979–5987.
49. Fernandes G. Dietary lipids and risk of autoimmune disease. Clin Immunol Immunopathol 1994; 72:193–197.
50. Fernandes G. Effect of calorie restriction and omega-3 fatty acids on autoimmunity and aging. Nutr Rev 1995; 53:S72–S79.
51. Fernandes G, Chandrasekar B, Luan X, Troyer DA. Modulation of antioxidant enzymes and programmed cell death by *n*-3 fatty acids. Lipids 1996; 31(suppl):S91–S96.
52. Chandrasekar B, Fernandes G. Decreased pro-inflammatory cytokines and increased antioxidant enzyme gene expression by ω-3 lipids in murine lupus nephritis. Biochem Biophys Res Commun 1994; 200:893–898.

53. Chandrasekar B, Troyer DA, Venkatraman JT, Fernandes G. Dietary omega-3 lipids delay the onset and progression of autoimmune lupus nephritis by inhibiting transforming growth factor beta mRNA and protein expression. J Autoimmun 1995; 8:381–393.
54. Chandrasekar B, McGuff HS, Aufdermorte TB, Troyer DA, Talal N, Fernandes G. Effects of calorie restriction on transforming growth factor beta 1 and proinflammatory cytokines in murine Sjogren's syndrome. Clin Immunol Immunopathol 1995; 76:291–296.
55. Horowitz MC. Cytokines and estrogen in bone: anti-osteoporotic effects. Science 1993; 260:626–627.
56. Manolagas SC, Jilka RL. Bone marrow, cytokines, and bone remodeling. Emerging insights into the pathophysiology of osteoporosis. N Engl J Med 1995; 332:305–311.
57. Pacifici R. Estrogen, cytokines, and pathogenesis of postmenopausal osteoporosis. J Bone Miner Res 1996; 11:1043–1051.
58. Lorenzo JA, Naprta A, Rao Y, et al. Mice lacking the type I interleukin-1 receptor do not lose bone mass after ovariectomy. Endocrinology 1998; 139:3022–3025.
59. Cenci S, Weitzmann MN, Roggia C, et al. Estrogen deficiency induces bone loss by enhancing T-cell production of TNF-alpha. J Clin Invest 2000; 106:1229–1237.
60. Kanematsu M, Sato T, Takai H, Watanabe K, Ikeda K, Yamada Y. Prostaglandin E2 induces expression of receptor activator of nuclear factor-kappa B ligand/osteoprotegrin ligand on pre-B cells: implications for accelerated osteoclastogenesis in estrogen deficiency. J Bone Miner Res 2000; 15:1321–1329.
61. Das UN. Essential fatty acids and osteoporosis [editorial]. Nutrition 2000; 16:386–390.
62. Kruger MC, Horrobin DF. Calcium metabolism, osteoporosis and essential fatty acids: a review. Prog Lipid Res 1997; 36:131–151.
63. Watkins BA, Lippman HE, Le Bouteiller L, Li Y, Seifert MF. Bioactive fatty acids: role in bone biology and bone cell function. Prog Lipid Res 2001; 40:125–148.
64. Das UN. Interaction(s) between essential fatty acids, eicosanoids, cytokines, growth factors and free radicals: relevance to new therapeutic strategies in rheumatoid arthritis and other collagen vascular disease. Prostaglandins Leukot Essent Fatty Acids 1991; 44:201–210.
65. Li Y, Seifert MF, Ney DM, et al. Dietary conjugated linoleic acids alter serum IGF-I and IGF binding protein concentrations and reduce bone formation in rats fed (*n*-6) or (*n*-3) fatty acids. J Bone Miner Res 1999; 14:1153–1162.
66. Teegarden D, Lyle RM, Proulx WR, Johnston CC, Weaver CM. Previous milk consumption is associated with greater bone density in young women. Am J Clin Nutr 1999; 69:1014–1017.
67. Weaver CM, Peacock M, Johnston CC Jr. Adolescent nutrition in the prevention of postmenopausal osteoporosis. J Clin Endocrinol Metab 1999; 84:1839–1843.
68. Kruger MC, Coetzer H, de Winter R, Gericke G, van Papendorp DH. Calcium, gamma-linolenic acid and eicosapentaenoic acid supplementation in senile osteoporosis. Aging (Milano) 1998; 10:385–394.
69. Schlemmer CK, Coetzer H, Claassen N, Kruger MC. Oestrogen and essential fatty acid supplementation corrects bone loss due to ovariectomy in the female Sprague Dawley rat. Prostaglandins Leukot Essent Fatty Acids 1999; 61:381–390.
70. Claassen N, Coetzer H, Steinmann CM, Kruger MC. The effect of different *n*-6/*n*-3 essential fatty acid ratios on calcium balance and bone in rats. Prostaglandins Leukot Essent Fatty Acids 1995; 53:13–19.
71. Wohl GR, Loehrke L, Watkins BA, Zernicke RF. Effects of high-fat diet on mature bone mineral content, structure, and mechanical properties. Calcif Tissue Int 1998; 63:74–79.
72. Sakaguchi K, Morita I, Murota S. Eicosapentaenoic acid inhibits bone loss due to ovariectomy in rats. Prostaglandins Leukot Essent Fatty Acids 1994; 50:81–84.
73. Kruger M, Claassen N, Smuts C, Potgeiter H. Correlation between essential fatty acids and parameters of bone formation and degradation. Asia Pacific J Clin Nutr 1997; 6:235–238.

74. Coetzer H, Claassen N, van Papendorp DH, Kruger MC. Calcium transport by isolated brush border and basolateral membrane vesicles: role of essential fatty acid supplementation. Prostaglandins Leukot Essent Fatty Acids 1994; 50:257–266.
75. Pacifici R, Rifas L, Teitelbaum S, et al. Spontaneous release of interleukin 1 from human blood monocytes reflects bone formation in idiopathic osteoporosis. Proc Natl Acad Sci USA 1987; 84:4616–4620.
76. Lowik CW, van der Pluijm G, Bloys H, et al. Parathyroid hormone (PTH) and PTH-like protein (PLP) stimulate interleukin-6 production by osteogenic cells: a possible role of interleukin-6 in osteoclastogenesis. Biochem Biophys Res Commun 1989; 162:1546–1552.
77. Votta BJ, Bertolini DR. Cytokine suppressive anti-inflammatory compounds inhibit bone resorption in vitro. Bone 1994; 15:533–538.
78. Kumar G, Das U, Kumar V. Effect of *n*-6 and *n*-3 fatty acids on the proliferation and secretion of TNF and IL-2 by human lymphocytes in vitro. Nutr Res 1992; 12:815–823.
79. Kumar GS, Das UN. Effect of prostaglandins and their precursors on the proliferation of human lymphocytes and their secretion of tumor necrosis factor and various interleukins. Prostaglandins Leukot Essent Fatty Acids 1994; 50:331–334.
80. Santoli D, Zurier R. Prostaglandin E precursor fatty acids inhibit human IL-2 production by a prostaglandin E-independent mechanism. J Immunol 1989; 143:1303–1309.
81. Purasiri P, Murray A, Richardson S, Heys SD, Horrobin D, Eremin O. Modulation of cytokine production in vivo by dietary essential fatty acids in patients with colorectal cancer. Clin Sci (Colch) 1994; 87:711–717.
82. Wachman A, Bernstein DS. Diet and osteoporosis. Lancet 1968; 1:958–959.
83. Sellmeyer DE, Stone KL, Sebastian A, Cummings SR. A high ratio of dietary animal to vegetable protein increases the rate of bone loss and the risk of fracture in postmenopausal women. Study of Osteoporotic Fractures Research Group. Am J Clin Nutr 2001; 73:118–122.
84. Margen S, Chu JY, Kaufmann NA, Calloway DH. Studies in calcium metabolism. I. The calciuretic effect of dietary protein. Am J Clin Nutr 1974; 27:584–589.
85. Schuette SA, Zemel MB, Linkswiler HM. Studies on the mechanism of protein-induced hypercalciuria in older men and women. J Nutr 1980; 110:305–315.
86. Heaney RP, Recker RR. Effects of nitrogen, phosphorus, and caffeine on calcium balance in women. J Lab Clin Med 1982; 99:46–55.
87. Hegsted M, Linkswiler HM. Long-term effects of level of protein intake on calcium metabolism in young adult women. J Nutr 1981; 111:244–251.
88. Spencer H, Kramer L, Osis D. Do protein and phosphorus cause calcium loss? J Nutr 1988; 118:657–660.
89. Lutz J, Linkswiler HM. Calcium metabolism in postmenopausal and osteoporotic women consuming two levels of dietary protein. Am J Clin Nutr 1981; 34:2178–2186.
90. Hegsted DM. Calcium and osteoporosis. J Nutr 1986; 116:2316–2319.
91. Abelow BJ, Holford TR, Insogna KL. Cross-cultural association between dietary animal protein and hip fracture: a hypothesis. Calcif Tissue Int 1992; 50:14–18.
92. Schurch MA, Rizzoli R, Slosman D, Vadas L, Vergnaud P, Bonjour JP. Protein supplements increase serum insulin-like growth factor-I levels and attenuate proximal femur bone loss in patients with recent hip fracture. A randomized, double-blind, placebo-controlled trial. Ann Intern Med 1998; 128:801–809.
93. Delmi M, Rapin CH, Bengoa JM, Delmas PD, Vasey H, Bonjour JP. Dietary supplementation in elderly patients with fractured neck of the femur. Lancet 1990; 335:1013–1016.
94. Tkatch L, Rapin CH, Rizzoli R, et al. Benefits of oral protein supplementation in elderly patients with fracture of the proximal femur. J Am Coll Nutr 1992; 11:519–525.
95. Bastow MD, Rawlings J, Allison SP. Benefits of supplementary tube feeding after fractured neck of femur: a randomised controlled trial. Br Med J (Clin Res Ed) 1983; 287:1589–1592.

96. Feskanich D, Willett WC, Stampfer MJ, Colditz GA. Protein consumption and bone fractures in women. Am J Epidemiol 1996; 143:472–479.
97. Munger RG, Cerhan JR, Chiu BC. Prospective study of dietary protein intake and risk of hip fracture in postmenopausal women. Am J Clin Nutr 1999; 69:147–152.
98. Hannan MT, Tucker KL, Dawson-Hughes B, Cupples LA, Felson DT, Kiel DP. Effect of dietary protein on bone loss in elderly men and women: the Framingham Osteoporosis Study. J Bone Miner Res 2000; 15:2504–2512.
99. Promislow JH, Goodman-Gruen D, Slymen DJ, Barrett-Connor E. Protein consumption and bone mineral density in the elderly : the Rancho Bernardo Study. Am J Epidemiol 2002; 155:636–644.
100. Meyer HE, Pedersen JI, Loken EB, Tverdal A. Dietary factors and the incidence of hip fracture in middle-aged Norwegians. A prospective study. Am J Epidemiol 1997; 145:117–123.
101. Marsh AG, Sanchez TV, Michelsen O, Chaffee FL, Fagal SM. Vegetarian lifestyle and bone mineral density. Am J Clin Nutr 1988; 48:837–841.
102. Dawson-Hughes B, Harris SS. Calcium intake influences the association of protein intake with rates of bone loss in elderly men and women. Am J Clin Nutr 2002; 75:773–779.
103. Messina M, Messina V. Soyfoods, soybean isoflavones, and bone health: a brief overview. J Ren Nutr 2000; 10:63–68.
104. Kritz-Silverstein D, Goodman-Gruen DL. Usual dietary isoflavone intake, bone mineral density, and bone metabolism in postmenopausal women. J Womens Health Gend Based Med 2002; 11:69–78.
105. Arjmandi BH. The role of phytoestrogens in the prevention and treatment of osteoporosis in ovarian hormone deficiency. J Am Coll Nutr 2001; 20:398S–402S; discussion 417S–420S.
106. Alekel DL, Germain AS, Peterson CT, Hanson KB, Stewart JW, Toda T. Isoflavone-rich soy protein isolate attenuates bone loss in the lumbar spine of perimenopausal women. Am J Clin Nutr 2000; 72:844–852.
107. Ho SC, Chan SG, Yi Q, Wong E, Leung PC. Soy intake and the maintenance of peak bone mass in Hong Kong Chinese women. J Bone Miner Res 2001; 16:1363–1369.
108. Setchell KD, Brown NM, Lydeking-Olsen E. The clinical importance of the metabolite equol-a clue to the effectiveness of soy and its isoflavones. J Nutr 2002; 132:3577–3584.
109. Setchell KD. Phytoestrogens: the biochemistry, physiology, and implications for human health of soy isoflavones. Am J Clin Nutr 1998; 68:1333S–1346S.
110. Fernandes G, Sun D, Krishnan A, Zaman K, R L. Soy protein and *n*-3 fatty acids prevent receptor activator of NF-kB ligand (RANKL) expression and osteoporosis in ovariectomized mice. FASEB J 2002; 16:A625.
110a. Fernandes G, Lawrence R, Sun D. Protective role of *n*-3 lipids and soy protein in osteoporosis. Prostaglandins Leukot Essent Fatty Acids. 2003; 68:361–372.
111. Heaney RP. Constructive interactions among nutrients and bone-active pharmacologic agents with principal emphasis on calcium, phosphorus, vitamin D and protein. J Am Coll Nutr 2001; 20:403S–409S; discussion 417S–420S.
112. Lacey DL, Timms E, Tan HL, et al. Osteoprotegerin ligand is a cytokine that regulates osteoclast differentiation and activation. Cell 1998; 93:165–176.
113. Kong YY, Yoshida H, Sarosi I, et al. OPGL is a key regulator of osteoclastogenesis, lymphocyte development and lymph-node organogenesis. Nature 1999; 397:315–323.
114. Suda T, Takahashi N, Udagawa N, Jimi E, Gillespie MT, Martin TJ. Modulation of osteoclast differentiation and function by the new members of the tumor necrosis factor receptor and ligand families. Endocr Rev 1999; 20:345–357.
115. Olsen NJ, Kovacs WJ. Gonadal steroids and immunity. Endocr Rev 1996; 17:369–384.
116. Bebo BF Jr, Schuster JC, Vandenbark AA, Offner H. Androgens alter the cytokine profile and reduce encephalitogenicity of myelin-reactive T cells. J Immunol 1999; 162:35–40.

117. Benten WP, Lieberherr M, Giese G, Wunderlich F. Estradiol binding to cell surface raises cytosolic free calcium in T cells. FEBS Lett 1998; 422:349–353.
118. Gilmore W, Weiner LP, Correale J. Effect of estradiol on cytokine secretion by proteolipid protein-specific T cell clones isolated from multiple sclerosis patients and normal control subjects. J Immunol 1997; 158:446–451.
119. Kong YY, Feige U, Sarosi I, et al. Activated T cells regulate bone loss and joint destruction in adjuvant arthritis through osteoprotegerin ligand. Nature 1999; 402:304–309.
120. Horwood NJ, Kartsogiannis V, Quinn JM, Romas E, Martin TJ, Gillespie MT. Activated T lymphocytes support osteoclast formation in vitro. Biochem Biophys Res Commun 1999; 265:144–150.
121. Teng YT, Nguyen H, Gao X, et al. Functional human T-cell immunity and osteoprotegerin ligand control alveolar bone destruction in periodontal infection. J Clin Invest 2000; 106:R59–R67.
122. Grcevic D, Lee SK, Marusic A, Lorenzo JA. Depletion of CD4 and CD8 T lymphocytes in mice in vivo enhances 1,25-dihydroxyvitamin D3-stimulated osteoclast-like cell formation in vitro by a mechanism that is dependent on prostaglandin synthesis. J Immunol 2000; 165:4231–4238.
123. Takayanagi H, Ogasawara K, Hida S, et al. T-cell-mediated regulation of osteoclastogenesis by signalling cross- talk between RANKL and IFN-gamma. Nature 2000; 408:600–605.
124. Muthukumar AR, Jolly CA, Zaman K, Fernandes G. Calorie restriction decreases proinflammatory cytokines and polymeric Ig receptor expression in the submandibular glands of autoimmune prone (NZB × NZW)F1 mice. J Clin Immunol 2000; 20:354–361.
125. Jolly C, Muthukumar A, Reddy Avula C, Troyer D, Fernandes G. Life span is prolonged in food-restricted autoimmune-prone (NZB × NZW)F(1) mice fed a diet enriched with (*n*-3) fatty acids. J Nutr 2001; 131:2753–2760.
126. Terada S, Takizawa M, Yamamoto S, Ezaki O, Itakura H, Akagawa KS. Suppressive mechanisms of EPA on human T cell proliferation. Microbiol Immunol 2001; 45:473–481.
127. Schlemmer CK, Coetzer H, Claassen N, et al. Ectopic calcification of rat aortas and kidneys is reduced with *n*-3 fatty acid supplementation. Prostaglandins Leukot Essent Fatty Acids 1998; 59:221–227.
128. Claassen N, Potgieter HC, Seppa M, et al. Supplemented gamma-linolenic acid and eicosapentaenoic acid influence bone status in young male rats: effects on free urinary collagen crosslinks, total urinary hydroxyproline, and bone calcium content. Bone 1995; 16:385S–392S.
129. Manolagas SC, Bellido T, Jilka RL. New insights into the cellular, biochemical, and molecular basis of postmenopausal and senile osteoporosis: roles of IL-6 and gp 130. Int J Immunopharmacol 1995; 17:109–116.
130. Raisz LG. Physiologic and pathologic roles of prostaglandins and other eicosanoids in bone metabolism. J Nutr 1995; 125:2024S–2027S.
131. Yasuda H, Shima N, Nakagawa N, et al. Osteoclast differentiation factor is a ligand for osteoprotegerin/osteoclastogenesis-inhibitory factor and is identical to TRANCE/RANKL. Proc Natl Acad Sci USA 1998; 95:3597–3602.
132. Sun D, Krishnan A, Zaman K, Lawrence R, Fernandes G. Dietary *n*-3 fatty acids decrease osteoclastogenesis and loss of bone mass in ovariectomized mice. J Bone Miner Res 2003; 18: 1206–1216.
133. Calder PC. Polyunsaturated fatty acids, inflammation, and immunity. Lipids 2001; 36:1007–1024.
134. Priante G, Bordin L, Musacchio E, Clari G, Baggio B. Fatty acids and cytokine mRNA expression in human osteoblastic cells: a specific effect of arachidonic acid. Clin Sci (Lond) 2002; 102:403–409.
135. Hamilton LC, Mitchell JA, Tomlinson AM, Warner TD. Synergy between cyclo-oxygenase-2 induction and arachidonic acid supply in vivo: consequences for nonsteroidal antiinflammatory drug efficacy. FASEB J 1999; 13:245–251.

136. Das UN. Estrogen, statins, and polyunsaturated fatty acids: similarities in their actions and benefits—is there a common link? Nutrition 2002; 18:178–188.
137. Das UN. Antiosteoporotic actions of estrogen, statins, and essential fatty Acids. Exp Biol Med 2002; 227:88–93.
138. Iotsova V, Caamano J, Loy J, Yang Y, Lewin A, Bravo R. Osteopetrosis in mice lacking NF-kappaB1 and NF-kappaB2. Nat Med 1997; 3:1285–1289.
139. Zhang YH, Heulsmann A, Tondravi MM, Mukherjee A, Abu-Amer Y. Tumor necrosis factor-alpha (TNF) stimulates RANKL-induced osteoclastogenesis via coupling of TNF type 1 receptor and RANK signaling pathways. J Biol Chem 2001; 276:563–568.
140. Wong BR, Josien R, Lee SY, Vologodskaia M, Steinman RM, Choi Y. The TRAF family of signal transducers mediates NF-kappaB activation by the TRANCE receptor. J Biol Chem 1998; 273:28355–28359.
141. Darnay BG, Haridas V, Ni J, Moore PA, Aggarwal BB. Characterization of the intracellular domain of receptor activator of NF-kappaB (RANK). Interaction with tumor necrosis factor receptor-associated factors and activation of NF-kappab and c-Jun N-terminal kinase. J Biol Chem 1998; 273:20551–20555.
142. Camandola S, Leonarduzzi G, Musso T, et al. Nuclear factor kB is activated by arachidonic acid but not by eicosapentaenoic acid. Biochem Biophys Res Commun 1996; 229:643–647.
143. Erben RG, Raith S, Eberle J, Stangassinger M. Ovariectomy augments B lymphopoiesis and generation of monocyte-macrophage precursors in rat bone marrow. Am J Physiol 1998; 274:E476–483.
144. Masuzawa T, Miyaura C, Onoe Y, et al. Estrogen deficiency stimulates B lymphopoiesis in mouse bone marrow. J Clin Invest 1994; 94:1090–1097.
145. Morony S, Capparelli C, Kostenuik P, Eli A, Scully W, Wiemann B. Osteoprotegerin prevents osteolytic bone destruction in both athymic and syngeneic models of experimental tumor metastasis to bone. J Bone Miner Res 1999; 14(suppl 1):1124.
146. Smithson G, Couse JF, Lubahn DB, Korach KS, Kincade PW. The role of estrogen receptors and androgen receptors in sex steroid regulation of B lymphopoiesis. J Immunol 1998; 161:27–34.
147. Ishimi Y, Miyaura C, Ohmura M, et al. Selective effects of genistein, a soybean isoflavone, on B-lymphopoiesis and bone loss caused by estrogen deficiency. Endocrinology 1999; 140:1893–1900.
148. Miyaura C, Onoe Y, Inada M, et al. Increased B-lymphopoiesis by interleukin 7 induces bone loss in mice with intact ovarian function: similarity to estrogen deficiency. Proc Natl Acad Sci USA 1997; 94:9360–9365.
149. Onoe Y, Miyaura C, Ito M, Ohta H, Nozawa S, Suda T. Comparative effects of estrogen and raloxifene on B lymphopoiesis and bone loss induced by sex steroid deficiency in mice. J Bone Miner Res 2000; 15:541–549.
150. Manabe N, Kawaguchi H, Chikuda H, et al. Connection between B lymphocyte and osteoclast differentiation pathways. J Immunol 2001; 167:2625–2631.

32 Phytoestrogens

Effects on Osteoblasts, Osteoclasts, Bone Markers, and Bone Mineral Density

Lorraine A. Fitzpatrick

1. INTRODUCTION

Phytoestrogens have estrogen-like properties and have been used to prevent or treat numerous diseases. Epidemiological studies suggest that there is a lower incidence of cardiovascular disease and hormone-dependent breast and prostate cancer in Asia compared to Western populations. It has been suggested that these differences may be due to the phytoestrogen-rich diet in Asian countries.

Many studies have focused on the role of phytoestrogens in bone metabolism, with the hope that a "natural" substance could be found to have beneficial effects. The purpose of this chapter is to review the definition of phytoestrogens, determine their ability to alter bone remodeling, and to determine the risks and benefits of these herbal supplements to prevent or treat bone disease.

2. HERBAL SUPPLEMENTS

Herbal supplements are, in part, regulated by the US Food and Drug Administration, so standards of safety and efficacy vary among products. Dietary supplements are widely available through the Internet, at pharmacies, and at health food stores. Herbal supplements are ingested by approximately 24% of the population, and many of these supplements contain phytoestrogens. All dietary supplements, including herbs, were purchased at the cost of $20 billion in the year 2000. In 1994, the US Dietary Supplement Health and Education Act defined a dietary supplement as a product other than tobacco that contains a vitamin, mineral, amino acid, herb, or other botanical or dietary substance produced to supplement the diet by increasing the total dietary intake. A dietary supplement is further defined as a concentrate, metabolite, constituent, extract, or combination of any of the above ingredients. These products cannot have a direct claim to prevent or cure disease, but as is frequently seen in advertisements, many supplements make these claims.

From: *Nutrition and Bone Health*
Edited by: M. F. Holick and B. Dawson-Hughes © Humana Press Inc., Totowa, NJ

Claims should be related only to physiological issues, however, the burden of safety is the responsibility of the FDA, whereas efficacy has been placed on the consumer who purchases and ingests these products. In addition, few products on the market use good manufacturing or good laboratory practices in their composition, resulting in marked differences from batch to batch of a particular herbal supplement. The addition of low-dose pharmaceutical compounds such as prednisone, valium, or estrogen itself has been documented in supplements to enhance their effects, often resulting in unwanted and unexpected side effects.

In 1992, the US Congress mandated and funded an Office of Alternative Medicine at the National Institutes of Health. The purpose of this office is to fund rigorous, scientific clinical trials in alternative therapies. The office has developed into the National Center for Complementary and Alternative Medicine, with a substantially increased budget. It is hoped that through rigorous basic science and placebo-controlled clinical trials, our understanding of the efficacy of these products will improve along with our understanding of their side effects and drug interactions.

3. DEFINITION OF PHYTOESTROGENS

Phytoestrogens are compounds that mimic the effects of placental or ovarian estrogens or their metabolites. Phytoestrogens may be functionally or structurally related to estrogens, resulting in antagonistic, agonistic, or partially agonistic/antagonistic effects at the estrogen receptor. In an individual subject, tremendous variability regarding the estrogen-like properties of phytoestrogens may occur. The subject's gender, endogenous estrogen levels (e.g., premenopausal vs postmenopausal status), rate of metabolism/catabolism, and even colonic microflora will alter the action of phytoestrogens on target tissues *(1–3)*.

There are many types of phytoestrogens and their metabolites. The three main classes include the isoflavones (genistein, daidzein, and their metabolites), the lignans (enterolactone, enterdiol), and coumestans (coumestrol), and are diphenolic in structure. Phytoestrogens are found in foodstuffs as diverse as soy, black cohosh, fennel, anise, ginseng, dong quai, alfalfa, red clover, flax seed, and licorice.

Soy protein is a widely studied food source of phytoestrogens, partly because of its ability to lower serum cholesterol levels. There has been a great deal of attention paid to soy protein isoflavones as a result of their commonality in certain ethnic diets and the reduction in specific diseases that is noted in epidemiological studies across various ethnic groups. Although soy products contain high amounts of isoflavones, these compounds are also found in such diverse foods as meat, cow's milk, whole grains, alcoholic beverages, fruits, and vegetables. In soy protein, the two most common aglycone isoflavones are genistein and daidzein. Daidzein and genistein bear a marked resemblance, structurally, to 17β-estradiol, and bind to the estrogen receptor. They bind weakly to the estrogen receptor α (ERα), and with more than 80% of estradiol's binding affinity to the estrogen receptor β (ERβ). Genistein binds preferentially to ERβ, similar to the binding preference of 17β-estradiol. However, transcriptional activity of genistein was twofold higher than that of 17β-estradiol for ERα and 1.8-fold higher for ERβ *(4)*. These data suggest that at the concentrations achievable by dietary intake of isoflavones, relevant biological responses are possible. One major difference of the phytoestrogens is that they may exhibit agonist or antagonist effects different from that of estrogen, depending on the target tissue *(1,2)*.

In a manner similar to the differences in herbal preparations from batch to batch, soybeans may contain markedly different amounts of isoflavones depending on the type of soil, weather conditions, the strain of soybean, or method of isoflavone extraction. As a result, even ingesting isoflavones from food can result in a varying response, making epidemiological studies difficult to interpret with accuracy.

The lignans are another class of diphenolic phytoestrogens. Lignans are commonly found in the cell walls of plants and are relatively abundant in such foodstuffs as black or green tea, sunflower seed, flaxseed, broccoli, pumpkinseed, garlic, bran, peanuts, and coffee. The lignans exert their actions by their influence on estrogen metabolism.

4. MECHANISMS OF ACTION

4.1. In Vitro Studies of Phytoestrogens: Do They Inhibit Bone Resorption?

Estrogen plays an important role in the skeleton, and it is not surprising that compounds that bind to the ER also influence skeleton growth and differentation. The phytoestrogens have specific effects on osteoclasts and osteoblasts in vitro. Estrogen alters the generation, lifespan, and functional activity of both osteoclasts and osteoblasts. Osteoclasts are responsible for bone resorption, which aids in skeletal remodeling and repair. Osteoclasts are derived from hematopoietic precursors, which fuse to form multinucleated cells capable of resorbing bone. Estrogen decreases osteoclast lifespan through enhancement of apoptosis, and decreases osteoclast activity and function. The mechanism by which osteoclast precursors fuse and differentiate into mature osteoclasts has recently been explored in vitro.

Early studies recognized that estrogen deficiency resulted in an increase in tumor necrosis factor (TNF)-α, a cytokine that increases bone resorption. The discovery of new members of the TNF family has increased our knowledge of the role and interactions that occur in the presence (or absence) of estrogen. Osteoblasts (and stromal cells of the osteoblast lineage) express the receptor activator of nuclear factor κB ligand (RANKL). RANKL is a paracine effector of osteoclast differentiation. RANKL binds to its receptor, RANK, which is located on the osteoclast membrane and transduces signals responsible for osteoclast differentiation and activity. These same stromal-osteoblast cells secrete osteoprotegerin (OPG), a soluble decoy receptor that neutralizes RANKL. Estrogen enhances production of OPG and indirectly decreases production of RANK.

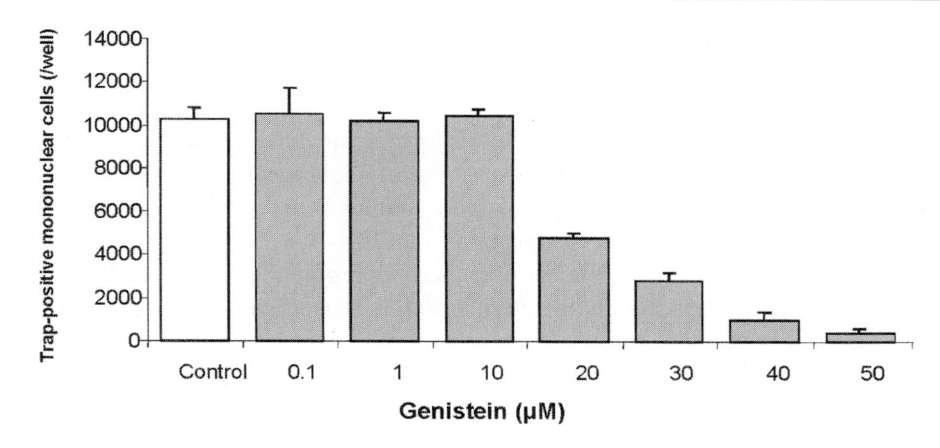

Fig. 1. Inhibition of osteoclast formation by genistein. Mouse osteogenic stromal cells were treated with genistein for 6 h and co-cultured with mouse spleen cells to stimulated differentiation. Cells were stained for tartrate-resistant acid phosphatase, a marker of osteoclasts. Increasing concentrations of genistein inhibited the formation of osteoclast-like cells. (Adapted from ref. *5*.)

Phytoestrogens can also influence the RANK–RANKL–OPG pathway. Osteoclast differentiation was influenced by phytoestrogens, as demonstrated by a study by Yamagishi et al. In a clonal osteogenic cell line designated ST2, osteoclast-like cells are formed when the cells are co-cultured with spleen cells. However, the addition of genistein prevented cell differentiation into osteoclast-like cells (Fig. 1). The mechanism of action is thought to be mediated via the RANKL/RANK/OPG system, since mRNA levels of OPG and RANKL were altered in the presence of genistein. The authors suggested that because genistein inhibited topoisomerase II, a protein kinase inhibitor; this may be the mechanism by which genistein inhibits the formation of osteoclast-like cells *(5)*. Other investigators have demonstrated enhanced apoptosis of osteoclasts by genistein or daidzein that is reversed in the presence of calcium-dependent protein kinases *(6)*.

Daidzein is a soy-derived isoflavone that has different actions from genistein at the cellular level. Although both compounds alter the inward rectifier K^+ current in rat osteoclasts *(7)*, there are other actions of daidzein that either have not been tested or have been tested under less than rigorous circumstances that make interpretation of these data difficult. For example, in contrast to genistein, in isolated osteoclasts, daidzein stimulates protein tyrosine kinase activity *(8)*. There are no clinical studies that evaluate the role of isolated daidzein in bone to date, in spite of the speculation that this phytoestrogen could be applied clinically.

Multinucleated tartrate-resistant acid phosphatase-positive (TRAP+) cells that resorbs bone (osteoclasts) develops in the presence of $1,25(OH)_2D_3$. The number of osteoclasts formed in response to $1,25(OH)_2D_3$ was reduced by $58 \pm 8\%$ by daidzein and by 52 ± 5% by estrogen ($p < 0.01$); these effects were reversed by 10^{-6} M of ICI 182,780, an estrogen receptor antagonist. The area resorbed by mature osteoclasts was reduced by 39 ± 5% by daidzein and by 42 ± 6% by estradiol ($p < 0.01$). Both compounds also inhibited the 1,25$(OH)_2D_3$-induced differentiation of osteoclast progenitors (mononucleated TRAP+ cells), 53 ± 8% by daidzein and 50 ± 7% by estradiol ($p < 0.05$) *(9)*. Moreover, daidzein and estradiol promoted caspase-8 and caspase-3 cleavage and DNA fragmentation of monocytic bone marrow cells. Caspase-3 cleavage was reversed by addition of the estrogen antagonist ICI 182,780. Both compounds upregulated the expression of nuclear estrogen receptors ERα and ERβ. Thus, daidzein, at the same concentration as 17β-estradiol, inhibits osteoclast differentiation and activity *(9)*. This may be caused by, at least in part, by greater apoptosis of osteoclast progenitors mediated via ERs.

4.2. Anabolic Effects of Phytoestrogens on Bone: In Vitro Studies

It has been proposed that phytoestrogens are anabolic to bone, but most of the evidence has been in animal models or in vitro, and clinical trials in humans have demonstrated, at best, a very mild anabolic action to date. At the cellular and molecular levels, the isoflavones may manifest both estrogen receptor-independent, nongenomic/tyrosine kinase pathway and estrogen receptor-dependent actions on osteoblasts. As mentioned above, the OPG–RANKL–RANK system plays an important role in skeletal homeostasis. OPG and the receptor activator NFkB ligand (RANKL) are factors responsible for the modulation of bone remodeling. Receptor activator of nuclear factor-κB ligand (RANKL) is essential for osteoclast formation and activation and enhances bone resorption. In contrast, OPG is a decoy receptor produced by osteoblasts and neutralizes RANKL and prevents bone loss *(10)*. 17β-Estradiol stimulates OPG mRNA levels and protein secretion in human osteoblastic-like cells through activation of ERα. The phytoestrogen genistein stimulates OPG mRNA levels and protein secretion in isolated primary human trabecular osteoblasts *(11)*. This effect was blocked by an antiestrogen, suggesting that the effect is mediated via the ER. RANKL, however, was not consistently expressed by these human osteoblast cells and was not modulated by genistein. The isoflavonoids regulate the expression and production of osteoclastogenesis-regulatory cytokines by osteoblasts such as interleukin (IL)-6 and OPG. Genistein or daidzein treatment of human fetal osteoblast-like cells resulted in significant decreased production of IL-6 and increased OPG mRNA expression at physiological concentrations *(12)*. In fetal osteoblast-like cells transfected with the ERα, the response of IL-6 and OPG production to isoflavones was enhanced, and the addition of an antiestrogen to the cells reduced or eliminated the cytokine production, suggesting that these effects are mediated via the estrogen receptor. The decrease in IL-6 and increase in OPG synthesis suggests that the isoflavones may alter the osteoblast in a manner that indirectly affects osteoclasts *(12,13)*.

To evaluate the effect of unique soy phytochemicals, Burow and colleagues isolated phytoalexins, a chemically heterogeneous group of substances that belong to various subclassifications of flavonoids *(14)*. Phytoalexins are low-molecular-weight antimicrobial compounds synthesized and retained in plants in response to stress *(15)*. The investigators induced the soybean phytoalexinsglyceollins I–III by the fungus *Aspergillus sojae,* a fungus commonly used in their fermentation of soybeans to produce soy sauce and miso. The glyceollins isomers I–III are similar to the phytoestrogen coumestrol and are derived from the precursor daidzein. Using an estrogen-responsive reporter gene assay in MCF-7 human breast carcinoma cells, there were differences in the relative estrogenic activity. The normal soy extract did not contain the phytoalexins and did not exhibit antiestrogenic activity. In contrast, the soy extract induced by the *Aspergillus sojae* decreased estrogen activity in spite of an increase in the presence of coumestrol and glyceollins I–III. Further testing in MCF-7 cells transfected with the ER revealed an antagonistic effect on the glyceollins. The glyceollins had greater affinity for ERα than for ERβ, in contrast to the results seen for other phytoestrogens such as genistein *(14)*.

Using an ethanol extract of soybeans, which contains many compounds in addition to the isoflavones, increased survival, DNA synthesis, alkaline phosphatase activity, and collagen synthesis were noted in MC3T3-E1 osteoblast-like cells. In the same study, osteoblast apoptosis was inhibited, so that these osteoblast-like cells had increased survival and differentiation *(16)*. Other investigators have found an increase in collagen synthesis in human bone cell proliferation. Genistein stimulated thymidine incorporation in a dose-dependent manner in isolated human bone cells and markedly reduced cellular tyrosyl phosphorylation. Collagen synthesis and alkaline phosphatase activity was enhanced in this in vitro model, suggesting an anabolic action on oteoblasts (Fig. 2) *(17)*. Isolated genistein stimulated proliferation of MC3T3-E1 osteoblast-like cells and protected against oxidative cellular damage by free radicals. Thus, there are more cellular actions of the phytoestrogens. A summary of the actions of genistein, a soy isoflavone, is listed in Table 1.

4.3. Animal Studies With Phytoestrogens

Many studies have demonstrated the ability of the soy-based phytoestrogens, genistein and daidzein, to prevent bone loss after ovariectomy in animal models. In an ovariectomized rat model, the effects of the soy isoflavone glycosides daidzin, genistin, and glycitin were compared to replacement with estrone. Daidzin, genistin, and glycitin significantly prevented bone loss in ovariectomized rats, in a manner similar to that seen with estrone. At the same dose, glycitin and daidzin also prevented the ovariectomized-induced uterine atrophy and increases in body weight gain, abdominal fat, serum cholesterol, and triglyercide levels and urinary excretion of pyridinoline and deoxypyridinoline. Thus, these glycosides behaved in a manner similar to estrone. Genistin may have a different mechanism of action, as at a higher dose, it blocked the ovariectomized-induced uterine atrophy but not any of the other estrogen-like effects *(18)*. Other types of phytoestrogens such as coumestrol

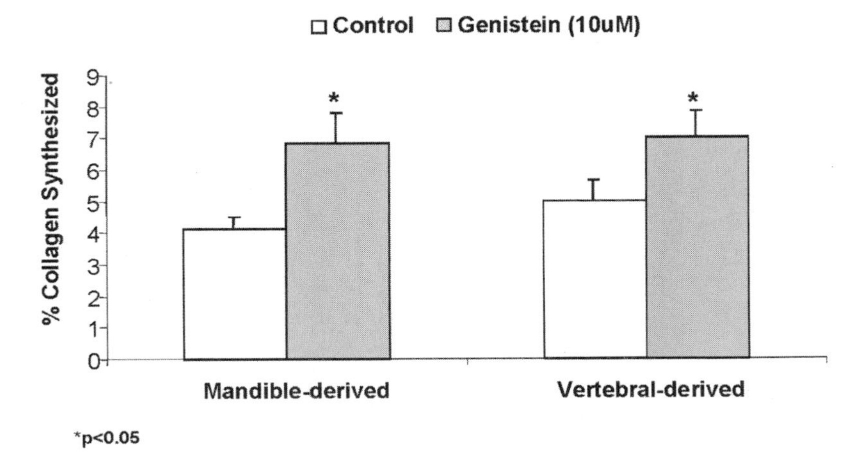

Fig. 2. In vitro anabolic actions of genistein. Normal human mandible- or vertebral-derived bone cells were isolated and cultured. After 48 h exposure to genistein or solvent control, collagen synthesis was increased. (Adapted from ref. *17.*)

Table 1
General Cellular Actions of Genistein

Binds to ER (ERβ > ERα)
Initiates gene transcription (ERα > ER β)
Alters expression of PR, OTR, and AR
Inhibits protein tyrosine kinases
Inhibits DNA topoisomerases I and II
Increases the activity of antioxidant enzymes
Inhibits aromatase and 5α-reductase
Inhibits 17β-hydroxysteroid dehydrogenase
Enhances or inhibits proliferation in estrogen-responsive cells
Contradictory reports about interactions with binding globulins

ER, estrogen receptor; PR, prolactin receptor; OTR, oxytocin receptor; AR, androgen receptor; E, estrogen.

and zearalanol also significantly reduced ovariectomized-induced bone loss in rats as assessed by measurement of bone mineral density (BMD) at the spine, femur, and total body. Pyridinoline (Pyd) and deoxypyridinoline (Dpyd), nonreducible crosslinks derived from lysine and hydroxylysine residues of mature collagen, are markers of bone resorption. The ovariectomy-induced rise in these markers was attenuated by coumestrol and estrogen *(19).*

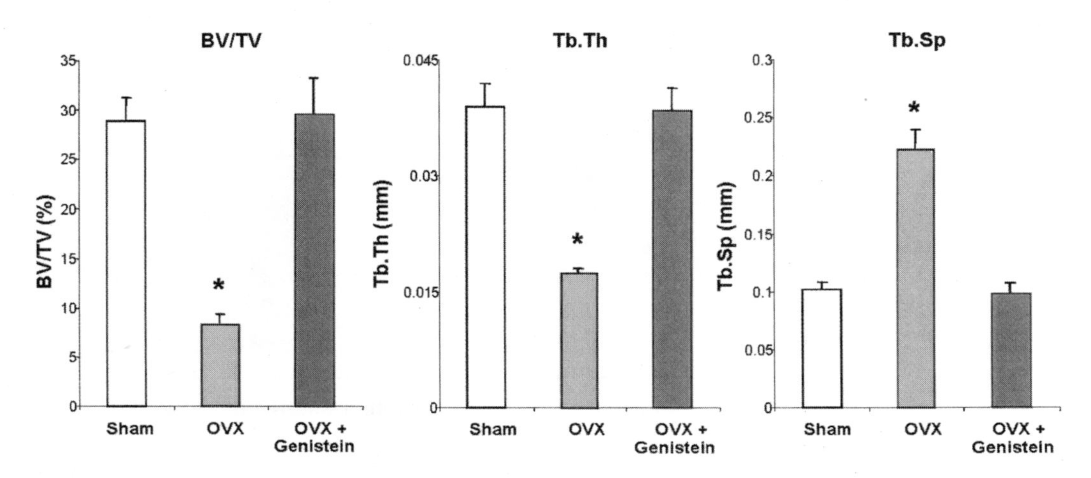

Fig. 3. The effect of genistein on ovariectomy-induced bone loss in mice. Eight-week-old *ddy* mice were sham-operated or ovariectomized and genistein or 17-β-estradiol was administered for 2 wk. Three-dimensional analysis of trabecular architecture by microcomputed tomography was performed of the femoral distal metaphysis. Images were reconstructed to determine morphometric indices of bone volume fraction (BV/TV), trabecular thickness (Tb.Th), and trabecular separation (Tb.Sp). All three parameters were restored in animals treated with genistein. *, significantly different from the sham-operated mice ($p < 0.01$). (From ref. *22.*)

To ascertain the mechanism of action of genistein on the skeleton, Fanti et al. gave ovariectomized rats geinstein and found a higher number of osteoblasts and higher rates of bone formation (per tissue volume). The most profound effect was on cancellous bone. No effect was noted on osteoclast number or Dpyd measurements. Genistein blocked the increased production of TNF-α that occurs after ovariectomy in these animals, and the authors suggest that inhibition of this cytokine may be one mechanism by which genistein reduces ovariectomy-induced bone loss *(20)*. The effect of genistein on bone is dose-dependent and occurs at doses that are 10-fold lower than doses that have been noted to induce uterine hypertrophy *(21)*.

Another proposed mechanism of genistein action is through the regulation of B-lymphopoiesis. B-lymphocytes increase after ovariectomy, and these cells produce bone-resorbing cytokines such as IL-1, IL-6, and TNF-α, which are involved in bone loss associated with estrogen deficiency. Sham-operated or ovariectomized female mice were administered genistein or 17β-estradiol for 2–4 wk. The number of bone marrow cells increased after ovariectomy, and were predominantly pre-B-cells. The increased B-lymphopoiesis after ovariectomy was completely abrogated by either 17β-estradiol or genistein. Genistein also restored the trabecular bone volume and trabecular thickness (Fig. 3) and had no uterine effects *(22)*.

Other sources of phytoestrogens have also proven to prevent ovariectomy-induced bone loss in rats. Safflower seed has been used in Korea to promote bone formation and prevent postmenopausal bone loss. In ovariectomized rats either fed a diet containing defatted safflower seeds or injected with 17β-estradiol, the safflower seed partially prevented the ovariectomy-induced bone loss when compared to sham-operated controls. Safflower seeds also induced weak uterotrophic action, consistent with the effect of an estrogen-like compound. When a polyphenolic extract from safflower seeds was added to osteoblast-like cells in culture, proliferation was stimulated and the effect was as potent as that with 17β-estradiol or the phytoestrogen, genistein *(23)*. Safflower seeds are part of a herbal formulation called Honghwain that is widely used to prevent bone loss. This herbal preparation prevented ovariectomy-induced bone loss in rats, and was thought to be antiresorptive in action. Further investigation suggests that this herbal preparation contains a Src family kinase inhibitor, since Honghwain reduced COX-2 mRNA and PGE2 production induced by IL-1β, TNF-α, and IL-6 in mouse fetal osteoblasts *(24)*.

Another rich source of isoflavones is flaxseed *(Linum usitatissimum)*. Flaxseed oil is a polyunsaturated oil and a rich source of omega-3 fatty acids. Specifically, flaxseed oil contains α-linolenic acid, which is converted to EPA. Flaxseed is also a source of the phytoestrogen secoisolariciresinol-diglucoside (SDG), which is converted in the intestine to two major lignans, enterodiol and enterolactone. Many effects have been attributed to flaxseed, including lipid lowering, anticarcinogenic effects, and laxative effects. Two hypotheses have been formed regarding the potential mechanism of action of flaxseed on bone. Due to its phytoestrogen content, it is thought that the lignans act as estrogen agonists. Another, not necessarily exclusionary, possibility is that the increased levels of omega-3 fatty acids may reduce bone-harming cytokine production and increase calcium absorption, resulting in slowing bone loss. Because of its content of phytoestrogens, a few studies have evaluated the effect of flaxseed on bone metabolism. To determine if flaxseed would compromise bone acquisition or strength, male rats from birth to postnatal d 21 were fed a phytoestrogen-free diet, a diet containing 10% flaxseed, or a diet containing the only the lignan SDG without the flaxseed protein or fat. Bone Mineral Content (BMC), BMD, bone quality, and bone strength were assessed. Ultimate bending stress and Young's modulus, measures of bone strength, were reduced at postnatal d 50 in the rats that received the 10% flaxseed diet. There were no differences in BMC, BMD, or bone size among the groups. However, at postnatal d 132, no differences among the groups were noted in any parameters tested, suggesting that as the pups matured, there were no long-term effects *(25)*. In a slightly different experiment by the same investigator, SDG purified from flaxseed in two doses equivalent to 50 or 100 g flaxseed/kg diet, was provided to female rats during lactation or continued through postnatal d 50 and d 132. Compared with the basal diet, bones from rats exposed to continuous 50- or 100-g flaxseed diets had stronger femurs at postnatal d 50, without changes in BMC. No differences in strength were noted at the later time point (postnatal d 132), but BMC was increased in the normal-diet group compared to rats exposed to SDG during lactation or into adulthood. The authors proposed that the female skeleton was more sensitive to SDG early in life and suggested that this was due to the estrogen-like action of the lignan *(26)*. In an 8-wk study that compared dietary ground flaxseed or defatted flaxseed meal in weanling female Sprague-Dawley rats, plasma alkaline phosphatase was significantly lower in both groups compared to rats fed a control, isocaloric diet *(27)*.

4.4. Dietary Phytoestrogens: Do They Affect Bone Markers and BMD?

Numerous studies have been conducted in which dietary supplementation of phytoestrogens are performed, but there are several problems with these types of studies. Few studies are double-blind in nature, due to the inability to readily disguise the phytoestrogen-containing foodstuff in order to create a comparable placebo. Other issues revolve around the increased caloric intake required to supplement a Western diet with a significant amount of isoflavones, resulting in unwanted weight gain in the study subjects.

Dried plums are rich in phenolic and flavonoid compounds and are high in oxygen radical-absorbance capacity. These compounds could protect bone by scavenging free radicals and preventing oxidative damage to the skeleton. In addition, dried plums contain many trace minerals, including boron and selenium, which may also have specific skeletal benefits. Fifty-eight postmenopausal women, not on hormone therapy (HT), were recruited and randomized to receive either 100 g dried plums or 75 g dried apples (equivalent in terms of calories, carbohydrates, fat, and fiber) daily for 3 mo. Their mean age was 54.3 yr, and 38 completed the study. Consumption of dried plums resulted in a significant increase in serum insulin-like growth factor (IGF)-1 and a nonsignificant increase in insulin-like growth factor binding protein (IGFBP)-3. Both groups of women had a significant increase in alkaline phosphatase activity, but those ingesting dried plums had a significant elevation of bone-specific alkaline phosphatase (BSAP) activity. No changes in serum calcium, magnesium, or phosphorus were noted in either group. There were also no differences in serum TRAP activity, urinary Dpyd, or hormone status as assessed with levels of estradiol and estrone, or vaginal maturation index. Overall, the increased levels of IGF-1 and BSAP in the dried-plum group suggest an anabolic effect on bone, but the effects were not mediated through the estrogen receptor. Longer-term studies are necessary to assess whether these anabolic effects will have substantial effects on bone mineral density *(28)*.

The effect of dietary phytoestrogens has been evaluated in both epidemiological studies and in clinical trials. It has been predicted from animal studies that phytoestrogens have a protective effect against bone loss after ovariectomy. In China, phytoestrogen intake is markedly increased compared to that in the United States. In 650 Chinese women, aged 19 to 86 yr, phytoestrogen intake was assessed by a food frequency questionnaire. One of the disadvantages of this approach is that it is dependent the subjects' recall capabilities. The questionnaire consists of 33 most frequently consumed food items, including 9 soy-containing items. The investigators measured serum estradiol levels and lumbar spine and femoral neck BMD.

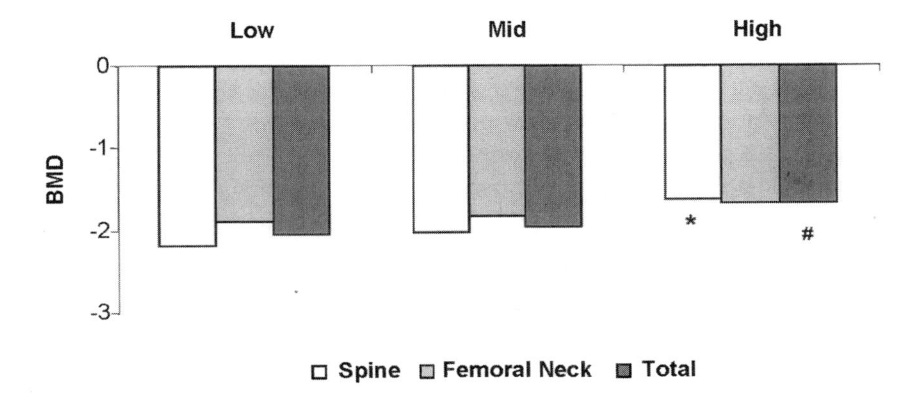

Fig. 4. To test the effect of dietary isoflavone intake on BMD, 650 Chinese women, aged 18 to 86 yr, were enrolled in a study. Data were analyzed by tertile of isoflavone intake that was assessed by dietary questionnaire recall. A significant increase in BMD T-score was noted in women with the highest tertile of isoflavone intake at the lumbar spine and total hip. *, $p < 0.001$; #, $p < 0.02$. (Adapted from ref. *29*.)

Women were then analyzed according to whether they were premenopausal (293 women, average age 37.5 yr) or postmenopausal (average age 63 yr). As is not unusual in China, less than 5% of the cohort were current or past users of HT. When subjects were analyzed according to tertiles of dietary phytoestrogen intake, those taking HT were excluded. Analysis suggested that there was a protective effect of isoflavone intake on BMD as measured at the lumbar spine or hip (Fig. 4). Dietary recall of other phytoestrogens such as coumestrol, lignans, and flavonoids were not correlated with BMD measurements. In addition, differences were found in the pre- vs postmenopausal subjects. Premenopausal subjects had no association between BMD values and phytoestrogen intake, and serum estradiol levels were similar among the tertiles of isoflavone intake. In the postmenopausal women, those with high isoflavone intake were likely to have reduced bone turnover. It was also noted that postmenopausal women with a higher intake of isoflavones had significantly lower serum parathyroid hormone levels. This may be one mechanism by which bone turnover was reduced in the postmenopausal population. The estimated amount of dietary isoflavone necessary to provide a bone protective effect was estimated at 53.3 mg/d (the mean value for the highest tertile of isoflavone intake). The authors suggested that phytoestrogens work as an estrogen agonist in conditions where estrogen is depleted, as in the postmenopausal women, but may be more antagonistic in action in the presence of endogenous estradiol, as in the premenopausal population *(29)*. An alternative hypothesis is that in the estrogen-replete population, the estrogen receptor may be saturated such that additional phytoestrogens in the diet do not exert specific effects on target tissues.

In a similar dietary recall study, Korean postmenopausal subjects assessed their dietary intake of isoflavones and validated the results with a 24-h urine collection for measurement of urinary phytoestrogen excretion. BMD measurements of the lumbar spine and femoral neck were performed. Mean age of the 75 women was 58 ± 1.1 yr. The daily urinary excretion of enterolactone, enterodiol, daidzein, genistein, equol, and apigenin were measured and correlated with BMD results. Patients with a diagnosis of osteoporosis by the World Health Organization criteria had significantly higher urinary apigenin and lower enterolactone levels compared to controls. There were no statistical differences between groups in other measurements. In addition, the total urinary phytoestrogen level did not correlate with BMD at any site. Symptoms related to menopause and lipid profiles also did not correlate with urinary phytoestrogen excretion. There was a skewness in the distribution of urinary phytoestrogen levels, which is not surprising since gastrointestinal flora determine serum and urine phytoestrogen concentrations, with great variability among individuals *(30)*.

It is believed that the soybean products that are ingested by Japanese women, which usually contain more fermented soy products, protect them from osteoporosis. To examine the relationship between soy protein intake and bone metabolism in postmenopausal Japanese women, a cross-sectional study was performed in 85 postmenopausal Japanese women. Approximately 60% of the women were osteopenic or osteoporotic as assessed by measurement of lumbar spine BMD. Bone biochemical markers, BMD, and daily nutritional state was assessed. Dietary intake was assessed over a period of three consecutive days. The kinds and weights of food ingested were recorded, and the protein, calcium, and soy protein intake was calculated. Correlations between each nutritional factor and the lumbar spine BMD revealed significant correlations between soy protein intake and BMD *T*-score and *Z*-score of the lumbar spine. There were also positive correlations between energy intake and protein intake. Surprisingly, calcium intake was not associated with any BMD measurements. For biochemical bone analytes, energy and protein intake were negatively associated with urinary pyridinoline measurements, but not with urinary deoxypyridinoline levels. Soy protein intake was associated only with urinary deoxypyridinoline concentrations, but not with alkaline phosphatase, urinary pyridinoline, or intact osteocalcin. By doing a stepwise multiple regression analysis, soy protein intake was a significant variable as an independent factor that contributed to the increase in BMD. Energy, protein, and calcium intake were not significant in preventing the reduction of BMD. Soy protein was also the major contributor to the suppression of bone resorption *(31)*.

In US women, different results regarding dietary soy intake and bone metabolism have been published. Unsupplemented dietary isoflavone consumption among women eating a Western diet was explored in 208 postmenopausal women living in southern California. Subjects were aged 45 to 74 yr and were enrolled in the Soy Health Effects (SHE) study. The SHE study was a randomized, double-blind, placebo-controlled clinical trial to investigate the association of isoflavone use on multiple outcome variables. Subjects had to be at least 2 yr postmenopausal, not on estrogen therapy for at least 3 mo, and not on other drugs used to treat osteoporosis. The average age of the subjects was 56 ± 6 yr, and they averaged 11 yr postmenopause. The majority of the subjects (78%) were Caucasian. The smaller number of women who were Asian (5.3%) were more likely to report genistein consumption of more than 1000 μg/d. Women with the highest daily intake of dietary genistein had urinary *N*-telopeptide (NTx) concentrations that were 18% lower as compared to women who reported no daily genistein consumption. No other bone markers showed significant differences among women ingesting large amounts of genistein vs lower amounts. Multiple regression analysis examined the association of isoflavone intake with markers of turnover and BMD after adjusting for age, BMI, total calcium intake, ethnicity, years postmenopause, smoking status, exercise status, and alcohol intake. Trends toward associations were observed between NTx levels and genistein, daidzein, and total isoflavone intake, but no statistically significant differences were found. There is also a nonsignficant association of total isoflavone consumption and total spine BMD ($p = 0.07$). In this study, the Block Food-Frequency Questionnaire assessed the major sources of isoflavones. Unfortunately, the Block Food-Frequency Questionnaire assesses diet only within the previous year, so assessment of a longer duration of use may provide different results. There may be a selection bias in that the study recruited volunteers interested in dietary soy supplementation *(32)*.

One of the most notable studies completed that assessed dietary soy intake was the Study of Women's Health Across the Nation (SWAN). This is a US-based, multisite, multiethnic, longitudinal cohort study with more than 3000 participants aged 42–52 yr at baseline. One of the major endpoints of SWAN is to study determinants of bone health. Participant's dietary isoflavone consumption and BMD were measured. At baseline, participants were premenopausal and were comprised of five ethnic groups: African American ($n = 935$), Caucasian ($n = 1550$), Chinese ($n = 250$), Hispanic ($n = 286$), and Japanese ($n = 281$). Women could not be currently using hormones related to ovarian function at the time. Not surprisingly, 31% of Caucasian women and 45% of African-American women recorded no genistein consumption. When considering only Caucasian and African-American women who reported eating genistein- containing foods, the estimated values still remained very low (14 μg for Caucasians and 4 μg for African-Americans). The Japanese and Chinese women reported higher intakes of genistein, but with moderate skewness of the data. Genistein consumption was substantially higher in Japanese women compared to Chinese women. In Chinese women, adjusted mean lumbar spine and femoral neck BMD values were similar at each tertile of genistein intake. No differences among pre- compared to perimenopausal women were found as to the effect of genistein on BMD. For Japanese women, the association between BMD and genistein was dependent on menopausal status. Premenopausal, but not perimenopausal, Japanese women, whose genistein intakes were greater, had higher BMD values at the spine and femoral neck. Pairwise comparisons revealed that the adjusted mean spine BMD of the women in the highest tertile of genistein intake was almost 8% higher than the mean BMD of women in the lowest tertile of genistein intake. A striking dose–response relationship was noted between dietary genistein intake and BMD among premenopausal Japanese women. Thus, a dose–response association between genistein and BMD was observed for premenopausal, but not perimenopausal, Japanese women. No effect of genistein on BMD was evident in Chinese women. Caucasian and African-American women consumed too low levels of genistein for evaluation. After mathematical modeling, the authors suggested that there may be an interaction term between ethnicity and genistein, since two groups of women had different BMD responses to similar amounts of dietary genistein. They propose that Japanese women could have a higher density of osteoblast estrogen receptors compared to Chinese women and, thus, have a greater bone effect for a given dose. The Japanese diet also may contain more readily absorbed forms of isoflavones, since their diet is higher in fermented soy products, which contain aglycones. Japanese women could possibly have higher amounts of β-galactosidase enzyme and are able to metabolize phytoestrogens more efficiently compared to their Chinese counterparts. Last but not least, it is important to understand that measurement of genistein exposure in the study also was dependent on the Block Food Frequency-Questionnaire, which has its own limitations as discussed above. The incompleteness of the database of isoflavone food sources is one of the major issues in all of these dietary recall studies. Nevertheless, the study did find a strong, positive dose–response relation between dietary soy isoflavones and BMD in premenopausal Japanese women *(33)*.

5. INTERPRETATION AND COMPARISON OF CLINICAL TRIALS

One of the major issues in comparing studies with isoflavones or assessing efficacy is the varying amounts of phytoestrogens that may be present. Soybeans, for example, may vary according to the soil, climate, and type of extraction that is utilized. For example, when processing some types of food products made from soy, ethanol extraction is commonly used. This process removes all the phytoestrogens, so some textured food products may not contain these compounds. Low-fat and nonfat soymilk may be relatively deplete of isoflavones due to the extraction process utilized in the production of the final product. Fermenting, which is used to produce miso and tempeh, will also result in different proportions of aglycones and glucosides. Temperatures during the filling of the soybean seed pod will alter isoflavone concentration, with lower amounts present with lower temperatures *(34)*. Even the date of sowing the seeds can be influential on the isoflavone content *(35)*. Thus, a given food product may vary batch to batch. In addition, individual absorption and metabolism varies greatly, so that it becomes difficult to appraise study results critically. Gender, estrogen status, intestinal flora, and metabolism and catabolism rates vary widely among individuals. Thus, the bioavailablity and pharmacokinetics of the isoflavones can vary widely. In spite of the fact that several randomized, placebo-

Table 2
Causes of Variations of Phytoestrogen Content in Clinical Studies

Type of soil
Date of sowing of crop
Climate
Temperature at the time of filling of the seed pod
Method of extraction
Processing, fermentation
Cooking method
Cooking temperature
Gender
Target tissue
Absorption
Colonic microflora
Estrogen status of subject
Bioavailability
Metabolism/catabolism
Pharmacokinetics

controlled clinical trials have evaluated the effect of isolated isoflavones on bone markers and bone mineral density, it is not possible to compare these various trials to each other, due to the fact that the different preparations of isoflavones have varying composition and activity. Thus, it should not be surprising when there is apparent "conflicting" results in the literature, which actually reflect the differences in preparation and metabolism of the product (Table 2).

5.1. Randomized, Placebo-Controlled Trials of Compounded Phytoestrogens

One study that highlights the differences in endogenous estrogen levels and the ingestion of an estrogen-like isoflavone has recently been published. Wangen et al. performed two randomized, crossover studies that utilized isolated soy protein isolates with varying isoflavone content. In a 3-mo trial, three soy protein isolates containing 8, 65, and 130 mg/d (designated control, low-, and high-isoflavone diets, respectively) were provided as a beverage powder supplement to premenopausal and postmenopausal women. In the premenopausal cohort, the diet intervention began on d 2 of the menstrual cycle and continued for three cycles plus 9 d. IGF-1 and IFGBP-3 concentrations were increased in premenopausal women on the low-isoflavone diets. At specific phases of the menstrual cycle, deoxypyridinoline cross-link levels were altered: levels were higher in the earlier follicular phase compared with the mid-follicular phase during the low isoflavone diet ($p = 0.04$), and levels were higher in the early follicular phase compared with the periovulatory phase during the high-isoflavone diet phase ($p = 0.003$). Postmenopausal women were studied for 93 d, and bone-specific alkaline phosphatase levels were reduced in subjects on the isoflavone-containing diets. There were nonstatistical trends toward decreased levels of osteocalcin, IGF-1, and IGFBP-3 with increasing concentrations of isoflavones. Overall, however, it did not appear that the small changes in bone markers noted in this study were clinically relevant. The differences in the pre- and postmenopausal women emphasized the potential differences that may occur in a given hormone milieu *(36)*.

The study described here raises questions regarding the effect of endogenous estrogen on the actions of phytoestrogens, and the results should not be surprising considering the varying effects of selective estrogen receptor modulators on bone metabolism in premenopausal women compared to postmenopausal women. Khalil et al. evaluated the effects of phytoestrogens on bone metabolism in men. It has been documented that estrogen is critical to skeletal growth and development in men, so one could predict that phytoestrogens might also effect skeleton metabolism in men. Healthy men, aged 59.2 ± 17.6 yr, were assigned to consume 40 g of soy protein or milk-based protein daily for 3 mo. The trial was a double-blind, randomized, controlled, parallel trial, but the data were analyzed for men younger than 65 yr of age or older than 65 yr of age due to the reduction in BMD that occurs in men with aging. Serum IGF-1 was increased in men of all ages consuming the soy protein compared to men randomized to the milk protein. No differences were found in bone-specific alkaline phosphatase or urinary deoxypyridinoline excretion. When the data were analyzed by age group, the results remained consistent. The short duration of this study may limit the interpretation, and longer studies with BMD as an endpoint would be useful to understand the benefit, if any, of soy protein on skeletal parameters in men *(37)*.

In a similar study design, these investigators examined middle-aged and older women provided 40 g/d of soy protein or casein in a double-blind study. The mean age of these subjects was 59.2 yr, and the data were analyzed for those over 65 yr of age compared to those less than 65 yr of age. After 3 mo, serum IGF-1 levels were increased in the soy protein-supplemented group, regardless of age. Urinary deoxypyridinoline excretion was also decreased in the soy protein group, regardless of age, suggesting an antiresorptive effect of the soy protein in comparison to casein. No differences were noted in serum levels of bone-specific alkaline phosphatase or tartrate-resistant acid phosphatase *(38)*. In a manner similar to the study by Wengan et al. described above, differences were found when data were analyzed according to serum estradiol levels, suggesting that endogenous levels of estradiol may alter the effect of the soy phytoestrogens on bone. In women with levels of estradiol that were deficient (<17 pg/mL), there was a significant ($p < 0.05$) reduction of 33% in urinary deoxypyridinoline excretion.

In animal studies, it has been proposed that soy isoflavones exert their beneficial effects on bone due to improved intestinal calcium absorption. The enhanced calcium absorption appears to be independent of 1,25-dihydroxyvitamin D concentrations *(39)*. Estrogen enhances intestinal calcium absorption in a manner similar to that caused by calcitriol, but an estrogen antagonist blocks the effect of estrogen. To test whether the isoflavones exert an effect on the skeleton through enhancement of calcium absorption in the intestine, 15 postmenopausal women were randomized to receive, in a crossover design, soy protein enriched with isoflavones, soy protein devoid of isoflavones, or a casein-whey control. Radioisotopic tracer studies were used to evaluate calcium kinetics. The ingestion of soy protein reduced urinary calcium by 33% regardless of isoflavone content. The authors suggested that it was the lower content of sulfur-containing amino acids in the soy protein that led to the reduced calcium excretion *(40)*.

One of the earliest reports of the effects of soy protein and isoflavones on bone density was a 6-mo study published by Potter and co-workers *(41)*. Sixty-six hypercholesterolemic, postmenopausal women were randomized to isolated soy protein containing varying concentrations of isoflavones with 40 g of protein plus either 1.39 mg/g protein or 2.25 mg isoflavones per gram of soy protein, or casein and nonfat dry milk. The soy protein was supplemented with calcium to make it equivalent to the calcium in milk. The test proteins were incorporated into food items such as muffins, breads, soups, and milks. BMC and BMD were measured at the lumbar spine and proximal femur by DXA. At the end of the 24-wk period, lumbar spine BMD and BMC increased significantly, by 2% in the group receiving the highest dose of isoflavones and soy protein. No differences in BMD or BMC at the proximal femur or total body were found. Most investigators have not seen changes at this early time point. In a very small randomized study, postmenopausal women were randomized to 40 g/d of soy protein with negligible isoflavones, 90 mg/d isoflavones, or 160 mg/d of isoflavones for 1 yr. Increases in BMD and BMC of 2% and 2.5% were observed. These changes were not statistically significant, and the authors propose that this was due to the small numbers of subjects, as the women not treated with isoflavones lost 2.9% and 0.6% of their lumbar spine BMC and BMD, respectively *(42)*.

There are fewer randomized trials in premenopausal women, in spite of the fact that soy is widely recommended to attenuate night sweats and hot flashes during the pre- and perimenopausal years. If soy isoflavones increase BMD in premenopausal women, one could propose this intervention to increase peak bone mass. To test this hypothesis, healthy, normally menstruating women aged 21–25 yr were blindly randomized to placebo (n = 13) or soy protein supplement (n = 15). Approximately 90 mg/d of isoflavones were included in the treatment group. DXA measurements of the spine and hip were completed at 6 and 12 mo. No changes in BMD were noted in either group at 12 mo, and there were no changes in the subjects' menstrual patterns *(43)*. This study provides support for the lack of changes in skeletal metabolism in premenopausal subjects in dietary overall or epidemiological-based studies. It is possible that in the estrogen-replete state, the soy isoflavones have no further effect due to saturation of ERα, ERβ, and various transcription factors by estrogen.

In contrast to animal studies, little effect has been noted in biomarkers of bone metabolism in trials using flaxseed in humans. Fifty-eight postmenopausal women were randomly assigned to either 40 g of ground whole flaxseed or a wheat-based comparative control over a 3-mo period. All study participates were provided with calcium (1000 mg) and vitamin D supplements. Of the 58 initially randomized, only 36 completed the study. Some of the dropouts were due to gastrointestinal problems (three from the flaxseed group and six from the wheat group) or lack of palatability of the diet (six from the flaxseed group and three from the wheat group). Although the flaxseed group had a significant decrease in serum total and non-high density lipoprotein cholesterol levels, no differences were noted in IGF-1, IGFBP-3, alkaline phosphatase, bone-specific alkaline phosphatase, tartrate-resistant acid phosphatase, calcium, urinary Dypd, and helical peptide (derived from the helical region of the α1 chain of type 1 collagen) *(44)*.

Since genistein has been shown to be the most efficacious soy phytoestrogen in animal models, Morabito et al. *(44a)* tested the effects of isolated genistein in early-postmenopausal women. This randomized, placebo-controlled, double-blind trial included 90 healthy women between the ages of 47 and 57 yr. Subjects were assigned HT (17β-estradiol plus norethisterone acetate) or genistein, 54 mg/d, or placebo. After 1 yr, there were no differences in serum levels of calcium, PTH, or 25(OH)D. No changes in lipid levels were noted in the placebo or genistein-treated group. Phytoestrogen plasma levels increased in subjects taking genistein, but there were no changes in serum estradiol concentrations in the genistein-treated group. Urinary excretion of Pyr and Dpyr was reduced in the genistein treated groups at 6 mo (–54% and –55%, respectively, $p < 0.001$) and at 12 mo (–42% and –44%, respectively, $p < 0.001$). There were no differences in these markers between the hormone-treated and the genistein-treated groups Marked increases in bone alkaline phosphatase (23%) and bone gla protein (29%) were noted at 6 mo, and these levels remained significantly elevated at 12 mo in the genistein group. In contrast, the postmenopausal women on HT had significant reductions in bone alkaline phosphatase and BGP (–20% and –22%, respectively at 12 mo, $p < 0.001$). These changes were also reflected in the changes in BMD. Both HT and genistein increased BMD at the femoral neck (3.6% for genistein and 2.4% for HT, compared to –0.65% for the placebo group). Similar changes were noted at Ward's triangle and in the lumbar spine (Fig. 5). Using an univariate model, independent predictors of increased BMD were initial BMD, and urinary DYPD levels for the lumbar spine. For the increase in femoral neck BMD, in addition to the two factors listed above, serum bone alkaline phosphatase levels also predicted femoral neck BMD. For safety monitoring, mammograms and transvaginal ultrasound were performed, and no significant changes were noted in the genistein-treated group. As expected, vaginal bleeding occurred in the women randomized to receive HT. This trial demonstrates some differences in bone markers in women treated with HT compared to those treated with genistein. There was a reduction in bone resorption markers and an increase in bone formation markers in the presence of genistein. The reduction

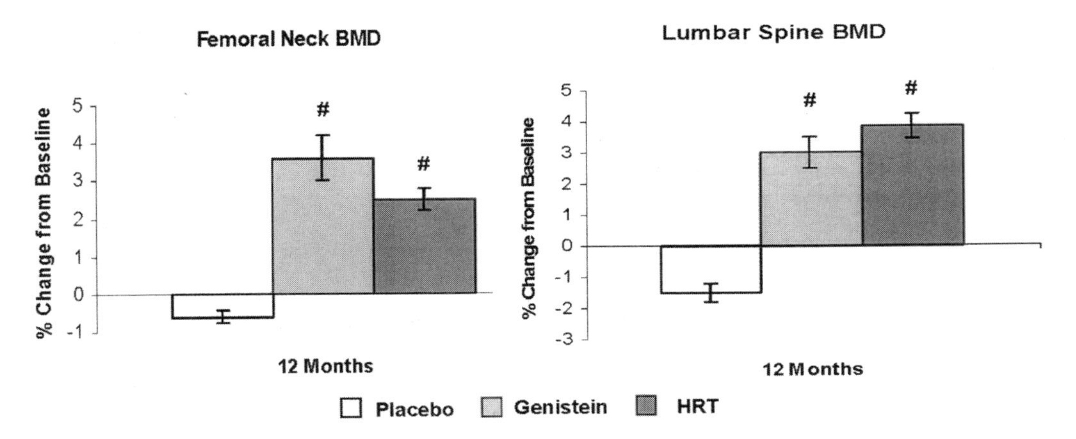

Fig. 5. Changes in BMD in early postmenopausal women with genistein and hormone therapy. In this randomized, double-blind, placebo-controlled study, 90 healthy women between the ages of 47 and 57 yr were assigned to receive continuous hormone therapy, genistein, or placebo for 1 yr. Genistein and hormone therapy significantly increased BMD at the femoral neck and lumber spine. #, $p < 0.01$ vs placebo. (Adapted from ref. *44a.*)

of bone resorption markers was similar to that seen with estrogen treatment, and suggests that genistein may work through the estrogen receptor to dampen bone resorption. Indirect effects could also account for the antiresorptive effects of genistein. In contrast to the effects seen with estrogen, genistein enhanced the markers of bone formation, consistent with the anabolic effect seen in tissue culture and in animals models. Whether this effect is indirect and mediated through protein tyrosine kinases or a direct effect involving ERβ is unknown. In addition, genistein prevented bone loss without adverse effects on the breast or uterus.

6. SAFETY OF ISOFLAVONES

Since phytoestrogens have estrogen-like activity, questions regarding safety are often raised. Some have voiced concerns that these compounds could act as endocrine disrupters. There are some reports of reproductive dysfunction in animals that ingest large quantities of phytoestrogens. Fritz et al. studied rat offspring that nursed from dams fed 250 mg/kg of dietary genistein from 2 wk before pregnancy to 21 d postpartum and found no adverse effects in the pup development or in the estrus of the dam *(45)*. There are no available data on the use of these compounds in women who have breast cancer, for example, yet many patients obtain synthetic or natural phytoestrogens over the Internet or in health food stores. Supplementation of soy formula for infants has been used for decades without known toxic effects. Less information is available in adolescents, but one short-term study provided isoflavones or identical placebo to 16-yr-old boys for 6 wk. There were large increases in urinary daidzein and genistein levels, but no effect noted on bone growth or turnover *(46)*. Except for the actvation of phytoestrogen preparations, most products appear not to have short-term harmful effects. Longer-term studies are essential to be certain of the safety of these compounds that have estrogen-like actions.

7. CONCLUSION

Phytoestrogens have been used to prevent or treat numerous diseases owing to their estrogen-like properties. Many studies have focused on the role of phytoestrogens and bone metabolism, with the hope that a natural substance would be beneficial. There is little regulation of herbal supplements in the United States, so compounds vary widely in their composition, type of phytoestrogen, affinity for the estrogen receptor, efficacy, purity, and side effects. In vitro studies of phytoestrogen on bone resorption suggests that there may be an antiresorptive effect, mediated in part, by the RANK–RANKL–OPG pathway. Other in vitro studies suggest that phytoestrogens are anabolic to bone, but clinical recognition of this finding has not been well documented. In randomized clinical trials, several types of phytoestrogens have proven to have effects on serum and urine measurements of bone resorption and inhibition. Studies vary, similar to the manner in which preparations vary, on the effect on bone mineral density. Long-term safety has yet to be established.

REFERENCES

1. Fitzpatrick LA. Selective estrogen receptor modulators and phytoestrogens: new therapies for the postmenopausal women. Mayo Clin Proc 1999; 74:601–608.
2. Vincent A, Fitzpatrick LA. Soy isoflavones: are they useful in menopause? Mayo Clin Proc 2000; 75:1174–1184.
3. Fitzpatrick LA. Phytoestrogens—mechanism of action and effect on bone markers and bone mineral density. Endocrinol Metab Clin N Am 2003; 32:233–252.
4. Kuiper GG, Lemmen JG, Carlsson B, et al. Interaction of estrogenic chemicals and phytoestrogens with estrogen receptor beta. Endocrinology 1998; 139:4252–4263.
5. Yamagishi T, Otsuka E, Hagiwara H. Reciprocal control of expression of mRNAs for osteoclast differentiation factor and OPG in osteogenic stromal cells by genistein: evidence for the involvment of topoisomerase II in osteoclastogenesis. Endocrinology 2001; 142:3632–3637.
6. Gao YH, Yamaguchi M. Suppressive effect of genistein on rat bone osteoclasts: apoptosis is induced through Ca^{2+} signaling. Biol Pharm Bull 1999; 22:805–809.
7. Okamoto F, Okabe K, Kajiya H. Genistein, a soybean isoflavone, inhibits inward rectifier K^+ channels in rat osteoclasts. Jpn J Physiol 2001; 51:501–509.
8. Gao YH, Yamaguchi M. Suppressive effect of genistein on rat bone osteoclasts: involvement of protein kinase inhibition and protein tyrosine phosphatase activation. Int J Mol Med 2000; 5:261–267.
9. Rassi CM, Lieberherr M, Chaumaz G, Pointillart A, Cournot G. Down-regulation of osteoclast differentiation by daidzein via caspase 3. J Bone Miner Res 2002; 17:630–638.
10. Teitelbaum SL. Bone resorption by osteoclasts. Science 2000; 289:1504–1508.
11. Viereck V, Gründker C, Blaschke S, Siggelkow H, Emons G, Hofbauer LC. Phytoestrogen genistein stimulates the production of osteoprotegerin by human trabecular osteoblasts. J Cell Biochem 2002; 84:725–735.

12. Chen XW, Garner SC, Anderson JJB. Isoflavones regulate interleukin-6 and osteoprotegerin synthesis during osteoblast cell differentiation via an estrogen-receptor- dependent pathway. Biochem Biophys Res Commun 2002; 295:417–422.
13. Liu J, Burdette JE, Xu H, et al. Evaluation of estrogenic activity of plant extracts for the potential treatment of menopausal symptoms. J Agric Food Chem 2001; 49:2472–2479.
14. Burow MD, Boue SM, Collins-Burow BM, et al. Phytochemical glyceollins, isolated from soy, mediate antihormonal effects through estrogen receptor alpha and beta. J Clin Endocrinol Metab 2001; 86:1750–1758.
15. Darvill AG, Albersheim P. Phytoalexins and their elicitors: a defense against microbial infection in plants. Annu Rev Plant Physiol Plant Mol Biol 1984; 35:243–275.
16. Amato P, Christophe S, Mellon PL. Estrogenic activity of herbs commonly used as remedies for menopausal symptoms. Menopause 2002; 9:145–150.
17. Yoon HK, Chen K, Baylink DJ, Lau K-HW. Differential effects of two protein tyrosine kinase inhibitors, tyrphostin and genistein, on human bone cell proliferation as compared with differentiation. Calcif Tissue Int 1998; 63:243–249.
18. Uesugi T, Toda T, Tsuji K, Ishida H. Comparative study on reduction of bone loss and lipid metabolism abnormality in ovariectomized rats by soy isoflavones, daidzin, genistein, and glycitin. Biol Pharm Bull 2001; 24:368–372.
19. Draper CR, Edel MJ, Dick IM, Randall AG, Martin GB, Prince RL. Phytoestrogens reduce bone loss and bone resorption in oophorectomized rats. J Nutr 1997; 127:1795–1799.
20. Fanti P, Monier-Faugere MC, Geng Z, et al. The phytoestrogen genistein reduces bone loss in short-term ovariectomized rats. Osteopor Int 1998; 8:274–281.
21. Ishimi Y, Arai N, Wang X, et al. Difference in effective dosage of genistein on bone and uterus in ovariectomized mice. Biochem Biophys Res Commun 2000; 274:697–701.
22. Ishimi Y, Miyaura C, Ohmura M, et al. Selective effects of genistein, a soybean isoflavone, on B-lymphopoiesis and bone loss caused by estrogen deficiency. Endocrinology 1999; 140:1893–1900.
23. Kim HJ, Bae YC, Park RW, et al. Bone-protecting effect of safflower seeds in ovariectomized rats. Calcif Tissue Int 2002; 71:88–94.
24. Yuk TH, Kang JH, Lee SR, et al. Inhibitory effect of *Carthamus tinctorius* L. seed extracts on bone resorption mediated by tyrosine kinase, COX-2 (cyclooxygenase) and PG (prostaglandin) E2. Am J Chin Med 2002; 30:95–108.
25. Ward WE, Yuan YV, Cheung AM, Thompson LU. Exposure to flaxseed and its purified lignan reduces bone strength in young but not older male rats. J Toxicol Environ Health A 2001; 63:53–65.
26. Ward WE, Yuan YV, Cheung AM, Thompson LU. Exposure to purified lignan from flaxseed *(Linum usitatissimum)* alters bone development in female rats. Br J Nutr 2001; 86:499–505.
27. Babu US, Mitchell GV, Wiesenfeld P, Jenkins MY, Gowda H. Nutritional and hematological impact of dietary flaxseed and defatted flaxseed meal in rats. Int J Food Sci Nutr 2000; 51:109–117.
28. Arjmandi BH, Khalil DA, Lucas EA, et al. Dried plums improve indices of bone formation in postmenopausal women. J Womens Health Gend Based Med 2002; 11:61–68.
29. Mei J, Yeung SSC, Kung AWC. High dietary phytoestrogen intake is associated with higher bone mineral density in postmenopausal but not premenopausal women. J Clin Endocrinol Metab 2001; 86:5217–5221.
30. Kim MK, Chung BC, Yu VY, et al. Relationships of urinary phyto-estrogen excretion to BMD in postmenopausal women. Clin Endocrinol 2002; 56:321–328.
31. Horiuchi T, Onouchi T, Takahashi M, Ito H, Orimo H. Effect of soy protein on bone metabolism in postmenopausal Japanese women. Osteopor Int 2000; 11:721–724.
32. Kritz-Silverstein D, Goodman-Gruen DL. Usual dietary isoflavone intake, bone mineral density, and bone metabolism in postmenopausal women. J Womens Health Gend Based Med 2002; 11:69–78.

33. Greendale GA, Fitzgerald G, Huang M-H, et al. Dietary soy isoflavones and bone mineral density: Results from the Study of Women's Health Across the Nation. Am J Epidemiol 2002; 155:746–754.
34. Tsukamoto C, Shimada S, Igita K, et al. Factors affecting isoflavone content in soybean seeds: changes in isoflavones, saponins, and composition of fatty acids at different temperatures during seed development. J Agric Food Chem 1995; 43:1184–1192.
35. Aussenac T, Lacombe S, Daydé J. Quantification of isoflavones by capillary zone electrophoesis in soybean seeds: effect of variety and environment. Am J Clin Nutr 1998; 68(suppl):1480S–1485S.
36. Wangen KE, Duncan AM, Merz-Demlow BE, et al. Effects of soy isoflavones on markers of bone turnover in premenopausal and postmenopausal women. J Clin Endocrinol Metab 2000; 85:3043–3048.
37. Khalil DA, Lucas EA, Juma S, et al. Soy protein supplementation may exert beneficial effects on bone in men. FASEB J 2001; 15:A727.
38. Arjmandi BH, Khalil DA, Lucas EA, et al. Soy protein with its isoflavones improves bone markers in women particularly those with low estrogen status. J Bone Miner Res Suppl 2001; 16:S533.
39. Arjmandi BH, Khalil DA, Hollis BW. Soy protein: its effects on intestinal calcium transport, serum vitamin D, and insulin-like growth factor-I in ovariectomized rats. Calcif Tissue Int 2002; 70:483–487.
40. Spence LA, Lipscomb ER, Cadogan J, et al. Effects of soy isoflavones on calcium kinetics in postmenopausal women. J Bone Miner Res Suppl 2001; 16:S532.
41. Potter SM, Baum JA, Teng H, Stillman RJ, Shay NF, Erdman JW Jr. Soy protein and isoflavones: their effects on blood lipids and bone density in postmenopausal women. Am J Clin Nutr 1998; 68(suppl):1375S–1379S.
42. Wong WW, Ellis KJ, Jahorr F. Soy isoflavones my stimulate nitric oxide production and reduce lumbar spine bone loss in postmenopausal women. FASEB J 2001; 15:A1088.
43. Anderson JJ, Chen X, Boass A, et al. Soy isoflavones: no effects on bone mineral content and bone mineral density in healthy, menstruating young adult women after one year. J Am Coll Nutr 2002; 21:388–393.
44. Edralin A, Lucas EA, Wild RD, et al. Flaxseed improves lipid profile without altering biomarkers of bone metabolism in postmenopausal women. J Clin Endocrinol Metab 2002; 87:1527–1532.
44a. Morabito N, Crisafulli A, Vergara C, et al. Effects of genistein and hormone-replacement therapy on bone loss in early post-menopausal women: a randomized double-blind placebo-controlled study. J Bone Miner Res 2002; 17:1904–1912.
45. Fritz WA, Coward L, Wang J, Lamartiniere CA. Dietary genistein: perinatal mammary cancer prevention, bioavailability and toxicity testing in the rat. Carcinogenesis 1998; 19:2151–2158.
46. Jones G, Hynes K, Dalais F, Parameswaran V, Greenaway T, Dwyer T. Phytoestrogens, physical activity, vitamin D, bone turnover and short term growth in adolescent boys. J Bone Miner Res Suppl 2001; 16:S560.

VII

Secondary Osteoporosis/Diseases

33 Eating Disorders and Their Effects on Bone Health

Madhusmita Misra and Anne Klibanski

1. EATING DISORDERS: HOW BIG IS THE PROBLEM?

The prevalence of eating disorders has increased in the industrialized world in the last few decades *(1,2)*, and 0.2–1% of all adolescent girls *(3)* and 1–4% *(4)* of all college-aged women in the United States suffer from anorexia nervosa (AN), an eating disorder associated with significant bone loss *(5–17)* and increased fracture risk *(11,18,19)*. Often undiagnosed, the true prevalence is likely higher, and variants of AN are estimated to be twice that diagnosed strictly by DSM-IV criteria *(20,21)*. In addition, AN has been increasingly reported in males, who now comprise 5–15% of this population *(22)*, and also demonstrate significant bone loss *(23)*. In a recent study from Canada, disordered eating attitudes and behaviors were reported in more than 27% of girls aged 12–18 yr *(24)*, with significant increases in Drive for Thinness, Body Dissatisfaction, and Bulimia subscales (Eating Disorders Inventory) with age. Table 1 summarizes the characteristic features of eating disorders *(25)*.

Adolescence is a time when peak bone mass accrues, with maximum bone accrual occurring between 11 and 14 yr in girls and 13 and 16 yr in boys *(26)*. Almost 25% of peak bone mass is formed in the 2 yr surrounding peak height velocity, and 90% of peak bone mass is achieved by the time an individual is 18 yr old *(27)*. Adolescence is thus a critical time of life in optimizing bone health, and insults suffered at this time may cause permanent deficits. Unfortunately, adolescence is also a common time for the onset of eating disorders, which is now the third most common chronic illness experienced by adolescent girls *(1)*.

2. EFFECT ON BONE DENSITY AND BONE TURNOVER

2.1. Anorexia Nervosa

AN consistently decreases bone mass in women *(5–14)*, in adolescent girls *(15–17)*, and boys *(23)*. The degree of bone loss is severe. Low bone mineral density (BMD) has been demonstrated both at the lumbar spine and the hip, suggesting that

From: *Nutrition and Bone Health*
Edited by: M. F. Holick and B. Dawson-Hughes © Humana Press Inc., Totowa, NJ

Table 1
Eating Disorders

	Diagnostic Criteria for Eating Disorders
Anorexia nervosa	Body weight < 85% ideal body weight Body mass index < 17.5 kg/m^2 in older adolescents Intense fear of weight gain Inaccurate perception of own body size, weight, or shape Amenorrhea for at least three consecutive cycles (in postmenarchal girls and women)
Bulimia nervosa	Recurrent binge eating (at least two times/wk for 3 mo) Recurrent purging, excessive exercise, or fasting (at least two times/wk for 3 mo) Excessive concern about body weight or shape Absence of anorexia nervosa
Binge eating disorder (BED)	Recurrent binge eating (at least two times/wk for 6 mo) Marked distress with at least three of the following (eating very rapidly, eating until uncomfortably full, eating when not hungry, eating alone, feeling disgusted or guilty after a binge) No recurrent purging, excessive exercise, or fasting Absence of anorexia nervosa
Eating disorder not otherwise specified (EDNOS)	Clinically important disordered eating, inappropriate weight control, or excessive concern about body weight or shape that does not meet all the criteria for anorexia nervosa, bulimia nervosa, or binge eating disorder

Reprinted with permission from the *Diagnostic and Statistical Manual of Mental Disorders*, Fourth Edition.

bone loss occurs both in trabecular and cortical bone. Osteoporosis may occur in as many as 38% and osteopenia in 92% of women with AN *(28)*. The fracture risk in adult women with AN is seven times that of healthy matched controls *(11)*, and over 50% of all women with AN have a BMD that is below the fracture threshold *(14)*. A population based cohort study in 1999 reported a 57% cumulative incidence of fractures at the hip, spine, and radius in women with AN 40 yr after the diagnosis of this eating disorder *(18)*. Vestergaard et al. *(29)* recently reported increased fracture risk in patients with eating disorders, with the highest incidence rate ratio (IRR) of fractures observed after diagnosis of AN, followed by eating disorders not otherwise specified (EDNOS) and bulimia (1.98, 1.7, and 1.44, respectively). Zipfel et al. *(30)* demonstrated a lower bone density in women with the binge eating/purging subtype compared to the strictly restrictive form of AN.

In adolescents with AN, profound and rapid bone loss occurs in a large proportion of girls suffering from this disorder. Bachrach et al. *(13)* reported that more than half of adolescent girls with AN with osteoporosis had been diagnosed for less than 1 yr. Bone loss is also more severe when AN begins in adolescence than when it begins in adult life, even with a comparable duration of illness *(12)*. The extent of bone loss depends on duration of amenorrhea *(12,13,16,31–34)*, body mass index (BMI; *13,33–36*), lean body mass *(15,16,36)*, and fat mass *(15)*.

Although rarely useful clinically, surrogate markers of bone turnover in this population have provided a clearer understanding of underlying pathophysiological processes leading to decreased BMD. An uncoupling of bone formation and bone resorption has been described in disorders associated with significant weight loss. Low levels of markers of bone formation have been found in women with AN *(14,35)*, while levels of markers of bone resorption are elevated *(14,30,35,37)*. Similarly, in adolescents with AN, low bone formation has been reported *(16,38)*. However, increased bone resorption has been demonstrated less frequently in this younger population, suggesting a generalized reduction in bone turnover. In both adults and adolescents, the net result is decreased accumulation of bone mass and thus low bone density.

2.2. Bulimia Nervosa

Although women with AN and concomitant bulimia have low bone mass, normal-weight bulimic patients may not *(30,39,40)*. Bone density is higher among normal-weight bulimic women who exercise compared to sedentary women *(40)*. Davies et al. *(41)* examined bone mineral content (BMC) in women with AN alone, bulimia alone, and AN with bulimia. They reported lowest BMC in the AN alone group, followed by women with AN with bulimia, and demonstrated that BMC in women with bulimia alone was not different from healthy controls.

2.3. EDNOS

Clinically, there is a large phenotypic section among patients with eating disorders. Of 76 women who qualified for the diagnosis of EDNOS, Marino and Zanarini *(42)* noted a history of restricting without low weight in 20%, binging without purging in 37%, purging without binging in 37%, and low weight without loss of menses in 33%. At least one study reported that in 33 patients with various eating disorder subtypes, bone density was lowest in the EDNOS group, followed by the group having AN, and was highest in the group with bulimia *(43)*.

3. FACTORS CONTRIBUTING TO LOW BONE DENSITY IN EATING DISORDERS ASSOCIATED WITH WEIGHT LOSS

3.1. Hypoestrogenism

Eating disorders in women may be associated with amenorrhea, especially when accompanied by weight loss. Amenorrhea of at least 3 mo duration is, in fact, one of the diagnostic features of AN by DSM-IV criteria. The hypogonadism in AN is attributed to acquired gonadotropin-releasing hormone (GnRH) deficiency, and estrogen levels are typically low in adolescent girls and adult women with

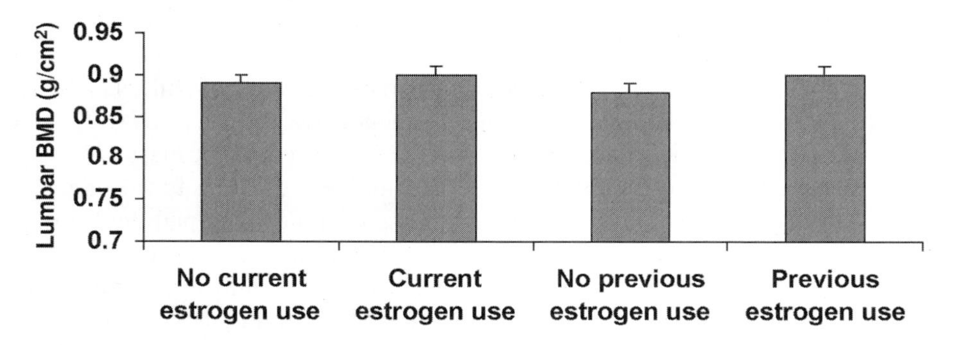

Fig. 1. Lumbar bone mineral density by estrogen use in women with anorexia nervosa. No differences were seen in lumbar bone mineral density (BMD) among women currently using estrogen, currently not using estrogen, women with a history of previous estrogen use, and with no previous history of estrogen use. (Adapted from ref. *28*. Copyright 2000 ACP-ASIM. All rights reserved.)

AN *(16,44)*. Bone density correlates inversely with the duration of amenorrhea *(12,13,16,31–34,45)*, suggesting that hypoestrogenism and its duration contribute to bone loss in this disorder. Audi et al. *(45)* also demonstrated an inverse relationship between estrogen levels and markers of bone resorption in pubertally mature adolescent girls with AN.

Estrogen inhibits bone resorption by an alteration in local concentrations of cytokines involved in osteoclast differentiation, activation, and apoptosis. Estrogen inhibits release of tumor necrosis factor (TNF)-α, interleukin (IL)-1, IL-6, and prostaglandin E2 (PGE2), and stimulates release of transforming growth factor (TGF)-β and osteoprotegerin *(46)*. The net effect is inhibition of differentiation of osteoclast precursors to mature osteoclasts, inhibition of activation of mature osteoclasts, and stimulation of osteoclast apoptosis. It is uncertain if estrogen exerts any effect on osteoblasts, although a role for estrogen in inhibiting glucocorticoid-induced osteoblast apoptosis has been described *(47)*.

Estrogen deficiency also accompanies acquired GnRH deficiency in women with hypothalamic amenorrhea resulting from stress or hyperprolactinemia. The extent of bone loss in AN, however, is far more severe at all sites than in comparable cases of hypothalamic amenorrhea that are not associated with significant weight loss *(36)*. Administration of estrogen to women with hypothalamic amenorrhea without significant weight loss may improve bone density *(48)*. Results have been much less promising in women with AN. In one study, lumbar, but not hip bone density was reported to be higher in women who received oral contraceptives for a 30-mo period compared to women who did not, although lumbar BMD was still significantly lower than in normals *(49)*. In a cross-sectional study of 130 women, we reported no difference in BMD at multiple sites in women with or without estrogen exposure *(28)* (Fig. 1). Consistent with these data, we found

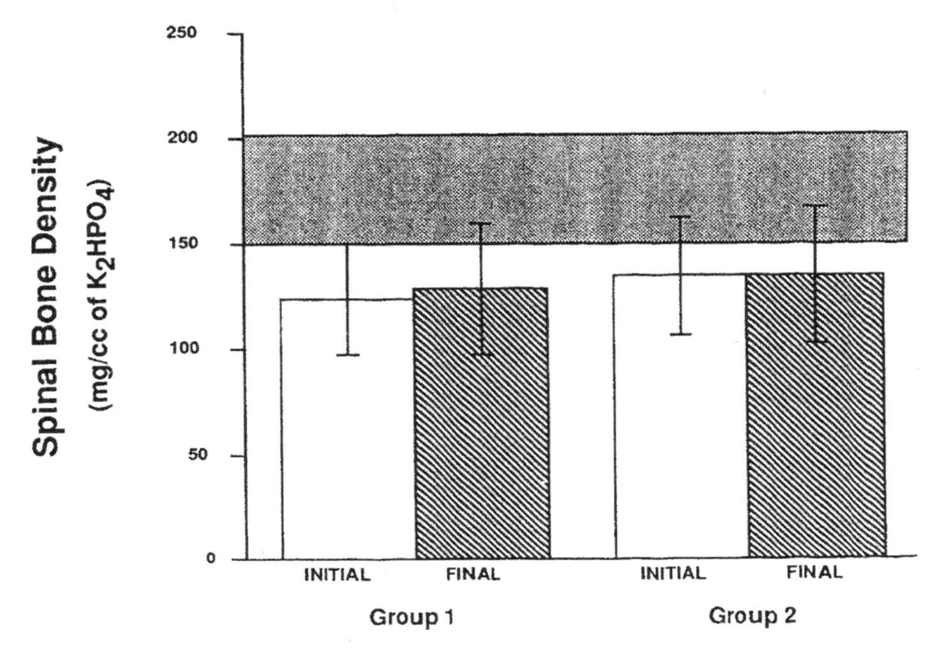

Fig. 2. Effects of estrogen administration (group 1) vs placebo (group 2) on bone density in women with anorexia nervosa. No differences in bone mineral density were observed before and after 18 mo of estrogen–progestin administration in a group of women with anorexia nervosa in a randomized, prospective, placebo-controlled trial. (Reprinted with permission from ref. *44*. Copyright © 1995 The Endocrine Society. All rights reserved.)

no difference in BMD after 18 mo of estrogen–progestin therapy in women with AN in a double-blind, placebo-controlled trial *(44)* (Fig. 2). In women with very low weights, however, a post-hoc analysis showed that estrogen therapy might have some protective effects. Karlsson et al. *(50),* however, in a nonrandomized study, reported some beneficial effects of conjugated equine estrogen administration to women with AN, such that treated women had higher bone densities than the untreated group ($p = 0.05$). Golden et al. *(51)* examined bone density in adolescent girls with AN before and after administration of an oral contraceptive pill (containing 20–35 μg estrogen) for a mean duration of 23 mo, and found no effects. Munoz et al. *(52)* reported no effect of an estrogen–progesterone combination in AN. Neither of these studies, however, was randomized, blinded, or placebo-controlled.

The lack of significant improvement in bone density with estrogen therapy suggests that hypogonadism is not the major cause of low BMD in women and girls with AN. Studies examining bone turnover have uniformly demonstrated decreased bone formation in AN *(14,16,35).* Markers of bone resorption are

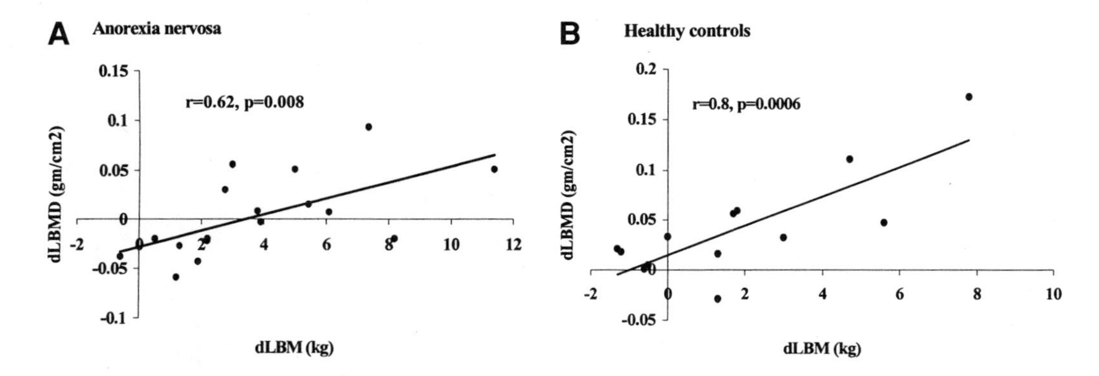

Fig. 3. Correlation of changes in lean body mass (dLBM) with changes in bone density (dLBMD). A strong correlation was observed between change in lean body mass over 1 yr and change in bone mineral density over the same period both in adolescent girls with anorexia nervosa (**A**) and healthy controls (**B**). (Reprinted with permission from ref. *38*. Copyright © 1999 The Endocrine Society. All rights reserved.)

increased in women with AN *(14,30,35,37)*, but are decreased in adolescent girls with this disease *(16)*. Estrogen acts primarily by inhibiting bone resorption; its effects on bone formation are less clear. It is possible that without adequate bone formation, the antiresorptive effects of estrogen are not sufficient to provide enough of a positive balance to result in increased bone mass.

3.2. Undernutrition, Low Insulin-Like Growth Factor-I (IGF-I) Levels, and Growth Hormone (GH) Resistance

Undernutrition plays a key role in development of low bone mass in AN. Strong correlations have been noted between sensitive markers of nutritional status and bone density. A low BMI *(13,33–36,53)*, decreased lean body mass *(15,16,36,53)* and fat mass *(15)*, and a poor caloric intake *(54)* are all associated with bone loss in AN. Of these markers, lean body mass appears to be the major contributor to bone density, both in individuals with AN and in healthy controls *(16,36,38)*. We have reported significant correlations between changes in lean body mass over a 1-yr period and changes in bone density *(38)* (Fig. 3). The effect of lean body mass may relate to biomechanical forces exerted by muscle mass on developing bone. Bachrach et al. *(55)* demonstrated an improvement in bone density with weight recovery in adolescent girls with AN even before resumption of normal gonadal function, underscoring the critical role played by nutritional factors in bone loss in this disorder.

The acute profound effect of undernutrition on bone metabolism in normals was demonstrated in a study by Grinspoon et al. *(56)* when acute fasting for 4 d led to a 50% reduction in markers of bone formation in healthy young women. In adults as well as adolescents with AN, osteocalcin and bone-specific alkaline phosphatase correlate strongly with markers of nutritional status such as BMI and percent body fat *(14,45)*, and with levels of IGF-I *(16,45)*. Markers of bone resorption are elevated in AN in adults *(14,35)*, and correlate negatively with BMI (35). Audi et al. *(45)* demonstrated an inverse correlation between markers of bone resorption and BMI in their study of adolescent girls with AN. Nutritional rehabilitation in adolescent girls with AN is associated with an increase in bone formation markers *(38,57)*. Bone resorption markers, on the other hand, have been noted to decrease in one study *(57)* and increase in another study *(38)*.

A nutritional acquired growth hormone "resistance" has been described in adults *(58–60)* and adolescents *(61,62)* with AN, with low levels of IGF-I *(58–62)*, IGF binding protein-3 (IGFBP-3) *(58,60–62)*, and growth hormone-binding protein (GHBP) *(58,61,62)*, despite high *(58–60,62)* or normal *(61)* values of GH. Weight recovery is associated with a normalization of GH secretion *(61,62)* and an increase in IGF-I levels *(14,58,61,63)*.

Both GH and IGF-I play an important role in bone formation. IGF-I is also nutritionally regulated, and thus serum levels are decreased in states of undernutrition including AN. GH directly stimulates proliferation of osteoblast precursors, and both directly and through IGF-I stimulates the differentiation of osteoblast precursors into active osteoblasts *(64)*. The role of the GH–IGF-I axis on bone resorption is less clear, though a direct inhibitory effect on osteoclasts has been described. IGF-I also stimulates collagen synthesis by stimulation of the Type II IGF-I receptor on osteoblasts *(65)*, and is necessary for longitudinal bone growth *(66)*.

GH deficiency states are associated with low BMD, and BMD increases with physiological GH replacement *(67)*. In addition, adolescent girls with AN have been reported to have decreased adult height *(68,69)* despite high GH values, which supports a state of acquired resistance to GH action. Studies have demonstrated that IGF-I levels predict bone loss and bone turnover in adults and adolescents with AN *(14,38)*.

Grinspoon et al. *(14)* reported a dose-dependent effect of short-term rhIGF-I administration on bone formation markers without any effect on bone resorption markers. Recently, we have shown that administration of rhIGF-I for 9 mo increased bone density in adult women with AN *(70)*. Of importance, an augmentation of this effect was seen in the group receiving birth control pills in addition to rhIGF-I. Like previous studies *(28,44,51,52)*, use of birth control pills alone did not result in an increase in bone density in this population. These data suggest that an effective therapy in this disease would optimally combine both anabolic and antiresorptive therapies.

3.3. Other Hormonal Factors

The contribution of a number of other hormonal factors to the bone loss in AN including cortisol, testosterone, and leptin have been examined. Women with AN may have high cortisol values *(60,71,72)*. Hypercortisolemia can result in osteopenia and osteoporosis. However, only 22% of women with AN and severe bone loss had elevated cortisol values in one study *(14),* suggesting that this is not the major contributor to bone loss in AN. Soyka et al. *(16)* and Audi et al. *(45)* did not find an elevation of urinary free cortisol levels in adolescent girls with AN.

A role for free testosterone and dehydroepiandrosterone sulfate (DHEAS) in maintaining bone mass in hypogonadal women has been reported by Miller et al. *(73),* and free testosterone values have been demonstrated to be significantly lower in adolescent girls with AN as compared to healthy controls *(38).* Like estrogen, the major effect of testosterone is a decrease in bone resorption; in fact, much of the action of testosterone occurs indirectly through its aromatization to estrogen. However, in addition to increasing the lifespan of osteoblasts and osteoclasts by inhibiting apoptosis of these cells (effects similar to estrogen), testosterone also affects osteoblast differentiation and proliferation (unlike estrogen) *(74,75).* We noted a positive correlation between changes in levels of free testosterone and changes in markers of bone formation over a 1-yr follow-up period in adolescent girls with AN in association with weight gain *(38).* Gordon et al. *(76)* administered oral DHEA to young women with AN for 3 mo and reported a decrease in a marker of bone resorption (NTX) and an increase in a marker of bone formation (osteocalcin) following study completion. Therefore, the loss of endogenous androgens seen in AN may contribute to the observed osteoporosis.

Recent reports have suggested a role for leptin in bone metabolism. Serum levels of leptin are very low in individuals with the restrictive form of AN *(16,54),* whereas levels are substantially higher in the subgroup of women with AN who also purge *(77).* A positive correlation between bone mass and leptin levels (independent of body weight and fat mass) was recently reported by Pasco et al. *(78)* in nonobese women. Also, leptin administration decreased trabecular bone loss in osteopenic ovariectomized rats *(79).* There are other reports, however,that suggest that leptin negatively regulates bone density. Sato et al. *(80)* demonstrated an inverse correlation between a marker of bone formation and serum leptin values in adult men. In animal experiments, Ducy et al. *(81)* showed that inactivating mutations of the leptin gene and leptin receptor gene resulted in increases in bone mass, whereas intraventricular leptin infusion caused bone loss. The role of leptin in the bone loss seen in AN is thus uncertain, and further studies are necessary to better understand this.

3.4. Calcium Metabolism

Most reported studies have not found an association between calcium intake and bone density in this disorder *(13,16,28,31).* Abrams et al. *(82)* performed calcium kinetic studies in adolescent girls with AN and demonstrated decreased calcium absorption and increased calcium excretion in these adolescents, suggesting an altered state of calcium metabolism. Castro et al. *(34)* reported that a calcium intake of less than 600 mg/d was a significant predictor of bone loss in AN. In a study of adolescent girls with AN and healthy adolescent girls *(16),* we showed that calcium intake was lower than the recommended daily allowance in 53% girls with AN compared to 67% of healthy controls, and vitamin D intake was lower than the recommended daily allowance in 53% of girls with AN compared with 78% controls. Thus, a higher proportion of healthy adolescent girls than girls with AN had a less than recommended intake of calcium and vitamin D. Similarly, we found no significant difference in calcium or vitamin D intake between adults with AN and controls, although more patients with AN were taking calcium supplements *(83)*. In addition, supplementation of calcium and vitamin D did not improve bone density in adult women with AN *(44)*.

Castro et al. *(34)* demonstrated that less than 3 h of physical activity per week was a risk factor for bone loss in AN, but other studies *(13,16,31)* did not find an association of bone density and physical activity levels in this disorder.

4. NATURAL HISTORY OF BONE LOSS IN AN AND EFFECTS OF WEIGHT RECOVERY

Studies have demonstrated that despite improvement in bone density with weight gain in AN, osteopenia persists in a large proportion of women and adolescents *(11,32,55,84)*. We followed adult women with AN for a period of $1^1/_2$ yr, and demonstrated persistence of low bone density in half of all weight-recovered women at the end of the study *(44)*. In contrast, women with AN who did not recover weight had further decreases in bone density in this period, suggesting that weight gain likely prevented further bone loss. In a group of 19 women with a past history of AN who had been weight recovered for 21 yr or more, Hartman et al. *(85)* reported that femoral bone densities were still significantly lower than in normals. Zipfel et al. *(30)* demonstrated an increase in BMD and T scores at the lumbar spine with weight recovery, with a decrease in prevalence of osteopenia from 35% to 13% and of osteoporosis from 54% to 21%, but a large proportion of weight-recovered subjects continued to have low bone mass. Thus decrease in bone density is significant and often permanent in spite of weight recovery, and underscores the importance of early diagnosis and aggressive treatment of this disorder.

Studies in adolescents with AN likewise have failed to demonstrate a consistent improvement in bone mass with weight recovery. Bachrach et al. *(55)* found that half the girls with AN who were weight recovered still had bone densities more than 2 SDs below the mean for age, although girls with weight recovery did demonstrate an improvement in lumbar and total BMD. Similarly, in a more recent report, Jagielska et al. *(17)* prospectively studied 27 girls with AN, and demonstrated that although an initial decrease in bone density occurred in the first 7 mo, sustained weight recovery in 11 subjects was associated with an improvement in lumbar and total BMD. We studied 19 adolescent girls with AN for a period of 1 yr and found significantly lower bone densities in weight-recovered AN compared to healthy controls at the end of the study period. No increase in BMD occurred with weight recovery (Fig. 4). However, weight-recovered girls had significant increases in

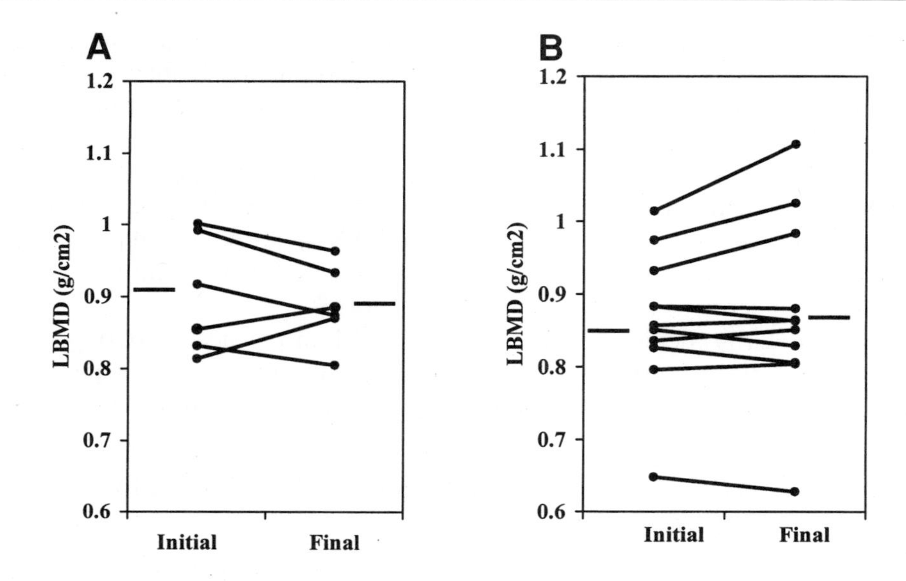

Fig. 4. Initial and final bone mineral density in recovered and nonrecovered adolescent girls with anorexia nervosa. No differences were observed between initial and final bone mineral density in adolescent girls with anorexia nervosa who did not recover weight **(A)** and who recovered weight **(B)** in a year-long prospective study. (Reprinted with permission from ref. *38*.

markers of bone formation and bone resorption over the study period as compared to controls, whereas nonrecovered AN showed no such increase *(38)*. In addition, an increase in a marker of bone formation (bone-specific alkaline phosphatase) in the first 6 mo of the study correlated with an increase in lumbar BMC over the next 6 mo of the study period, suggesting that a longer period of weight recovery may be necessary before significant increases in bone density are measurable.

Thus, it appears that a sustained period of weight recovery is necessary before a definite increase in bone density can occur *(15,17,31)*.

5. MANAGEMENT OF BONE HEALTH IN EATING DISORDERS

A decrease of 1 SD in bone density results in a doubling of fracture risk. With the high prevalence of bone loss in girls and women suffering from AN, bulimia associated with AN, and EDNOS, monitoring bone density is of utmost importance in individuals suffering from eating disorders, especially restrictive eating disorders. The increasing availability of dual-energy X-ray absorptiometry (DXA) has made this an accurate and readily available method of monitoring BMD without radiation exposure issues. Databases are now available to obtain *Z*-scores by age and race in adolescent girls *(86)*. A limitation of DXA readings is that it reports area rather than volumetric BMD (BMC/cross-sectional area of bone in cm^2, rather than BMC/volume of bone in cm^3). However, formulas are available to obtain an estimate of volumetric density or bone mineral apparent density based on the individual's height *(87)*. In children, adjustments based on maturity and bone age readings become necessary when bone age is different from the chronological age.

Although quantitative computed tomography (QCT) provides a direct estimate of volumetric bone density, it is less reproducible and thus less useful than DXA in comparing bone densities over time. In addition, there is an increased exposure to ionizing radiation by this method, and it thus remains largely a research tool.

Calcaneal ultrasound (CUS) is the newest noninvasive method now accepted for assessing bone density in adults. Its small size, low cost, and the absence of ionizing radiation make it an attractive alternative to DXA and QCT. The heel is made of cancellous bone similar to the spine, thus calcaneal bone density readings could act as a surrogate for lumbar spine bone mineral density. In adults, heel ultrasound measurements have been demonstrated to correlate with DXA and predict fracture risk *(88,89)*. Resch et al. *(90)* reported highly significant correlations between BUA and bone density at the spine and the hip in adult women with AN. Lum et al. *(91)* reported moderate correlations between CUS and DXA measurements of the spine, hip, and total body in subjects 9–25 yr old. Databases will need to be established for age before this technique can replace the more conventional methods of bone density assessment currently used for an adolescent population.

Markers of bone formation and bone resorption are important research tools in understanding the dynamics of bone loss in AN *(92)*. However, their clinical utility is limited by the variation in these markers within groups and individuals and also by circadian and day-to-day variations. There might, however, exist a place for bone turnover markers in monitoring therapy for osteoporosis *(93)*. The variations in these markers with age, pubertal stage, growth velocity, hormonal, and nutritional status make them even less useful in a clinical setting in children and adolescents *(94,95)*.

5.1. Treatment

No approved treatment other than weight gain reverses bone loss in individuals with eating disorders. Because the underlying disease is difficult to treat and bone loss may be permanent, effective strategies to address this issue are critically needed. Calcium and vitamin D supplementation did not improve bone density in adults *(44)* or adolescents *(16)* with AN. Nevertheless, it is important to maintain an adequate intake of calcium and vitamin D in this population, and we recommend 1300–1500 mg of elemental calcium and 400 IU of vitamin D for all adolescents and adults with AN. Birth control pills/estrogen–progesterone replacement when given alone have failed to improve bone density in AN in many studies *(44,51,52,70)*. We have recently showed that subcutaneous administration of rhIGF-I (30 μgd/kg b.i.d. for 9 mo) was effective in improving bone density in adults with AN *(70)*, especially when given in conjunction with a birth control pill,

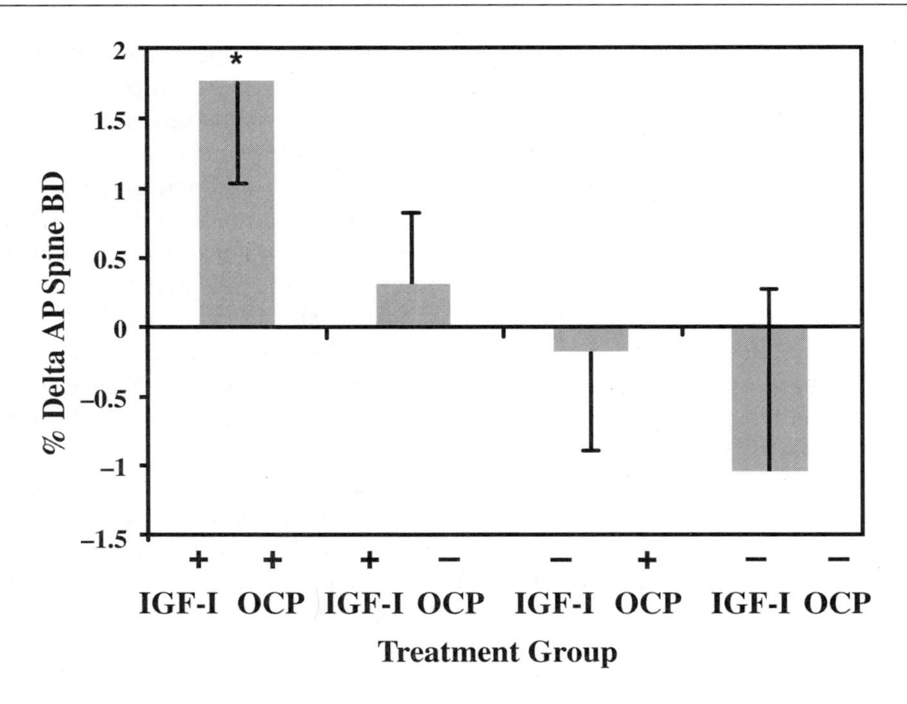

Fig. 5. Effect of rhIGF-I ± OCP on BMD. Administration of recombinant human insulin-like growth factor-I (IGF-I) (30 μg/kg b.i.d. subcutaneously) increased bone density in adult women with anorexia nervosa. The effect was most marked when IGF-I was given with a birth control pill. (Reprinted with permission from ref. *70.* Copyright © 1999 The Endocrine Society. All rights reserved.)

and is a promising new development (Fig. 5). Larger studies will need to be done to determine the therapeutic impact of this approach.

Bisphosphonates inhibit bone loss by an inhibitory effect on osteoclasts and are highly effective in treating postmenopausal osteoporosis. Investigational studies of bisphosphonates in adults with AN are underway *(96).* However, the long-term efficacy and safety of these agents in this disease are currently unknown.

6. CONCLUSION

Bone loss is a common and serious complication of restrictive eating disorders, and it is important to monitor bone mineral density in all individuals suffering from such disorders. The mechanism of bone loss is possibly multifactorial, including hypogonadism, undernutrition, alterations of the GH–IGF-I axis, and in some patients, hypercortisolism. DXA remains the most effective and reproducible method of monitoring bone density. Sustained weight recovery and adequate calcium and vitamin D supplementation are to be emphasized in all individuals suffering from eating disorders. The combination of bone anabolic and antiresorptive therapy is currently the most promising mode of treatment of bone loss in AN and is still under investigation.

REFERENCES

1. Lucas AR, Beard CM, O'Fallon WM, Kurland LT. 50-Year trends in the incidence of anorexia nervosa in Rochester, Minn.: a population-based study. Am J Psychiatry 1991; 148(7):917–922.
2. Pawluck DE, Gorey KM. Secular trends in the incidence of anorexia nervosa: integrated review of population-based studies. Int J Eat Disord 1998; 23:347–352.
3. Crisp AH, Toms DA. Primary anorexia nervosa. Br Med J 1972; 1:334.
4. Pope GH Jr, Huson JI, Yurgelun-Todd D. Prevalence of anorexia nervosa and bulimia in three student populations. Int J Eat Disord 1984; 3:45–51.
5. Ayers JWT, Gidwani GP, Schmidt IM, Gross M. Osteopenia in hypoestrogenemic young women with anorexia nervosa. Fertil Steril 1984; 41:224–228.
6. Rigotti NA, Nussbaum SR, Herzog DB, Neer RM. Osteoporosis in women with anorexia nervosa. N Engl J Med 1984; 311:1601–1606.
7. Brotman AW, Stern TA. Osteoporosis and pathological fractures in anorexia nervosa. Am J Psychiatry 1985; 142:495–496.
8. Szmukler GI, Brown SW, Parsons V, Darby A. Premature loss of bone in chronic anorexia nervosa. Br Med J 1985; 290:26–27.
9. Treasure J, Fogelman I, Russell GFM. Osteopenia of the lumbar spine and femoral neck in anorexia nervosa. Scott Med J 1986; 31:206–207.
10. Savvas M, Treasure J, Studd J, Fogelman I, Monicz C, Brincat M. The effect of anorexia nervosa on skin thickness, skin collagen and bone density. Br J Obstet Gynaecol 1989; 96:1392–1394.
11. Rigotti NA, Neer RM, Skates SJ, Herzog DB, Nussbaum SR. The clinical course of osteoporosis in anorexia nervosa. A longitudinal study of cortical bone mass. JAMA 1991; 265:1133–1138.
12. Biller BMK, Saxe V, Herzog DB, Rosenthal DI, Holzman S, Klibanski A. Mechanisms of osteoporosis in adult and adolescent women with anorexia nervosa. J Clin Endocrinol Metab 1989; 68:548–554.
13. Bachrach LK, Guido D, Katzman D, Litt IF, Marcus R. Decreased bone density in adolescent girls with anorexia nervosa. Pediatrics 1990; 86:440–447.
14. Grinspoon S, Baum H, Lee K, Andersen E, Herzog D, Klibanski A. Effects of short-term rhIGF-I administration on bone turnover in osteopenic women with anorexia nervosa. J Clin Endocrinol Metab 1996; 81:3864–3870.
15. Kooh SW, Noriega E, Leslie K, Muller C, Harrison JE. Bone mass and soft tissue composition in adolescents with anorexia nervosa. Bone 1996; 19:181–188.
16. Soyka LA, Grinspoon S, Levitsky LL, Herzog DB, Klibanski A. The effects of anorexia nervosa on bone metabolism in female adolescents. J Clin Endocrinol Metab 1999; 84:4489–4496.
17. Jagielska G, Wolanczyk T, Komender J, Tomaszewicz-Libudzic C, Przedlacki J, Ostrowski K. Bone mineral content and bone mineral density in adolescent girls with anorexia nervosa—a longitudinal study. Acta Psychiatr Scand 2001; 104:131–137.
18. Lucas AR, Melton LJ 3rd, Crowson CS, O'Fallon WM. Long-term fracture risk among women with anorexia nervosa:a population based cohort study. Mayo Clin Proc 1999; 74:972–977.
19. Espallargues M, Sampietro-Colom L, Estrada MD, et al. Identifying bone mass-related risk factors for fracture to guide bone densitometry measurements: a systematic review of the literature. Osteopor Int 2001; 12:811–822.

20. Hoek HW, Bartelds AI, Bosveld JJ, et al. Impact of urbanization on detection rates of eating disorders. Am J Psychiatry 1995; 152:1272–1278.
21. Shisslak CM, Crago M, Estes LS. The spectrum of eating disorders. Int J Eat Disord 1995; 18:209–219.
22. Andersen AE, Woodward PJ, LaFrance N. Bone mineral density of eating disorder subgroups. Int J Eat Disord 1995; 18:335–342.
23. Castro J, Toro J, Lazaro L, Pons F, Halperin I. Bone mineral density in male adolescents with anorexia nervosa. J Am Acad Child Adolesc Psychiatry 2002; 41:613–618.
24. Jones JM, Bennett S, Olmsted MP, Lawson ML, Rodin G. Disordered eating attitudes and behaviors in teenaged girls:a school-based study. Can Med Assoc J 2001; 165:547–552.
25. Diagnostic and Statistical Manual of Mental Disorders, Fourth Edition. American Psychiatric Association, Washington, DC, 1994.
26. Theintz G, Buchs B, Rizzoli R, et al. Longitudinal monitoring of bone mass accumulation in healthy adolescents:evidence for a marked reduction after 16 years of age at the levels of lumbar spine and femoral neck in female subjects. J Clin Endocrinol Metab 1992; 75:1060–1065.
27. Bachrach LK. Acquisition of optimal bone mass in childhood and adolescence. Trends Endocrinol Metab 2001; 12:22–28.
28. Grinspoon S, Thomas E, Pitts S, et al. Prevalence and predictive factors for regional osteopenia in women with anorexia nervosa. Ann Intern Med 2000; 133:790–794.
29. Vestergaard P, Emborg C, Stoving RK, Hagen C, Mosekilde L, Brixen K. Fractures in patients with anorexia nervosa, bulimia nervosa and other eating disorders—a nationwide register study. Int J Eat Disord 2002; 32:301–308.
30. Zipfel S, Seibel MJ, Lowe B, Beumont PJ, Kasperk C, Herzog W. Osteoporosis in eating disorders:a follow up study of patients with anorexia and bulimia nervosa. J Clin Endocrinol Metab 2001; 86:5227–5233.
31. Hay PJ, Delahunt JW, Hall A, Mitchell AW, Harper G, Salmond C. Predictors of osteopenia in premenopausal women with anorexia nervosa. Calcif Tissue Int 1992; 50:498–501.
32. Iketani T, Kiriike N, Nakanishi S, Nakasuji T. Effects of weight gain and resumption of menses on reduced bone density in patients with anorexia nervosa. Biol Psychiatry 1995; 37:521–527.
33. Baker D, Roberts R, Towell T. Factors predictive of bone mineral density in eating disordered women:a longitudinal study. Int J Eat Disord 2000; 27:29–35.
34. Castro J, Lazaro L, Pons F, Halperin I, Toro J. Predictors of bone mineral density reduction in adolescents with anorexia nervosa. J Am Acad Child Adolesc Psychiatry 2000; 39:1365–1370.
35. Hotta M, Shibasaki T, Sato K, Demura H. The importance of body weight history in the occurrence and recovery of osteopenia inpatients with anorexia nervosa:evaluation by dual X-ray absorptiometry and bone metabolic markers. Eur J Endocrinol 1998; 139:276–283.
36. Grinspoon S, Miller K, Coyle C, et al. Severity of osteopenia in estrogen deficient women with anorexia nervosa and hypothalamic amenorrhea. J Clin Endocrinol Metab 1999; 84:2049–2055.
37. Lennkh C, de Zwaan M, Bailer U, et al. Osteopenia in anorexia nervosa specific mechanisms of bone loss. J Psychiatr Res 1999; 33:349–356.
38. Soyka LA, Misra M, Frenchman A, et al. Abnormal bone mineral accrual in adolescent girls with anorexia nervosa. J Clin Endocrinol Metab 2002; 87:4177–4185.
39. Newton JR, Freeman CP, Hannan WJ, Cowen S. Osteoporosis and normal weight bulimia nervosa—which patients are at risk? J Psychosom Res 1993; 37:239–247.
40. Sundgot-Borgen J, Bahr R, Falch JA, Schneider LS. Normal bone mass in bulimic women. J Clin Endocrinol Metab 1998; 83:3144–3149.
41. Davies KM, Pearson PH, Huseman CA, Greger NG, Kimmel DK, Recker RR. Reduced bone mineral in patients with eating disorders. Bone 1990; 11:143–147.

42. Marino MF, Zanarini MC. Relationship between ENDOS and its subtypes and borderline personality disorder. Int J Eat Disord 2001; 29:349–353.
43. Joyce JM, Warren DL, Humphries LL, Smith AJ, Coon JS. Osteoporosis in women with eating disorders:comparison of physical parameters, exercise, and menstrual status with SPA and DPA evaluation. J Nuclear Med 1990; 31:325–331.
44. Klibanski A, Biller BMK, Schoenfeld DA, Herzog DB, Saxe VC. The effects of estrogen administration on trabecular bone loss in young women with anorexia nervosa. J Clin Endocrinol Metab 1995; 80:898–904.
45. Audi L, Vargas DM, Gussinye M, Yeste D, Marti G, Carrascosa A. Clinical and biochemical determinants of bone metabolism and bone mass in adolescent female patients with anorexia nervosa. Pediatr Res 2002; 51:497–504.
46. Riggs BL. The mechanisms of estrogen regulation of bone resorption. J Clin Invest 2000; 106:1203–1204.
47. Gohel A, McCarthy MB, Gronowicz G. Estrogen prevents glucocorticoid-induced apoptosis in osteoblasts in vivo and in vitro. Endocrinology 1999; 140:5339–5347.
48. Hergenroeder AC, O'Brian Smith E, Shypailo R, Jones LA, Klish WJ, Ellis K. Bone mineral changes in young women with hypothalamic amenorrhea treated with oral contraceptives, medroxyprogesterone, or placebo over 12 months. Am J Obstet Gynecol 1997; 176:1017–1025.
49. Seeman E, Szmukler GI, Formica C, Tsalamandris C, Mestrovic R. Osteoporosis in anorexia nervosa:the influence of peak bone density, bone loss, oral contraceptive use, and exercise. J Bone Miner Res 1992; 7:1467–1474.
50. Karlsson MK, Weigall SJ, Duan Y, Seeman E. Bone size and volumetric density in women with anorexia nervosa receiving estrogen replacement therapy and in women recovered from anorexia nervosa. J Clin Endocrinol Metab 2000; 85:3177–3182.
51. Golden NH, Lanzkowsky L, Schebendach J, Palestro CJ, Jacobson MS, Shenker IR. The effect of estrogen-progestin treatment on bone mineral density in anorexia nervosa. J Pediatr Adolesc Gynecol 2002; 15:135–143.
52. Munoz MT, Morande G, Garcia-Centenara JA, Hervas F, Pozo J, Argente J. The effects of estrogen administration on bone mineral density in adolescents with anorexia nervosa. Eur J Endocrinol 2002; 146:45–50.
53. Turner JM, Bulsara MK, McDermott BM, Byrne GC, Prince RL, Forbes DA. Predictors of low bone density in young adolescent females with anorexia nervosa and other dieting disorders. Int J Eat Disord 2000; 30:245–251.
54. Grinspoon S, Gulick T, Askari H, et al. Serum leptin levels in anorexia nervosa. J Clin Endocrinol Metab 1996; 81:3861–3863.
55. Bachrach LK, Katzman D, Litt IF, Guido D, Marcus R. Recovery from osteopenia in adolescent girls with anorexia nervosa. J Clin Endocrinol Metab 1991; 72:602–606.
56. Grinspoon S, Baum H, Kim V, Coggins C, Klibanski A. Decreased bone formation and increased mineral dissolution during acute fasting in young women. J Clin Endocrinol Metab 1995; 80:3628–3633.
57. Heer M, Mika C, Grzella I, Drummer C, Herpertz-Dahlmann B. Changes in bone turnover in patients with anorexia nervosa during eleven weeks of inpatient dietary treatment. Clin Chem 2002; 48:754–760.
58. Counts DR, Gwirtsman H, Carlsson LMH, Lesem M, Cutler GB. The effect of anorexia nervosa and refeeding on growth hormone binding protein, the insulin like growth factors (IGFs) and the IGF binding proteins. J Clin Endocrinol Metab 1992; 75:762–767.
59. Scacchi M, Pincelli AI, Caumo A, et al. Spontaneous nocturnal growth hormone secretion in anorexia nervosa. J Clin Endocrinol Metab 1997; 82:3225–3229.
60. Stoving RK, Veldhuis JD, Flyvbjerg A, et al. Jointly amplified basal and pulsatile growth hormone (GH) secretion and increased process irregularity in women with anorexia nervosa:indi-

rect evidence for disruption of feedback regulation within the GH-insulin-like growth factor I axis. J Clin Endocrinol Metab 1999; 84:2056–2063.
61. Golden NH, Kreitzer P, Jacobson MS, et al. Disturbances in growth hormone secretion and action in adolescents with anorexia nervosa. J Pediatr 1994; 125:655–660.
62. Argente J, Caballo N, Barrios V, et al. Multiple endocrine abnormalities of the growth hormone and insulin-like growth factor axis in patients with anorexia nervosa:effect of short- and long-term weight recuperation. J Clin Endocrinol Metab 1997; 82:2084–2092.
63. Hotta M, Fukuda I, Sato K, Hizuka N, Shibasaki T, Takano K. The relationship between bone turnover and body weight, serum insulin-like growth factor (IGF) I, and serum IGF-binding protein levels in patients with anorexia nervosa. J Clin Endocrinol Metab 2000; 85:200–206.
64. Ohlsson C, Bengtsson BA, Isaksson OG, Andreassen TT, Slootweg MC. Growth hormone and bone. Endocr Rev 1998; 19(1):55–79.
65. Hock JM, Centrella M, Canalis E. Insulin like growth factor-I has independent effects on bone matrix formation and cell replication. Endocrinology 1988; 122:254–260.
66. Skottner A, Arrhenius-Nyberg V, Kanje M, Fryklund L. Anabolic and tissue repair functions of recombinant insulin-like growth factor I. Acta Paediatr Scand Suppl 1990; 367:63–66.
67. Baum HB, Biller BM, Finkelstein JS, et al. Effects of physiologic growth hormone therapy on bone density and body composition in patients with adult-onset growth hormone deficiency. A randomized, placebo-controlled trial. Ann Intern Med 1996; 125:883–890.
68. Nussbaum M, Baird D, Sonnenblick M, Cowan K, Shenker IR. Short stature in anorexia nervosa patients. J Adolesc Health Care 1985; 6:453–455.
69. Russell GF. Premenarchal anorexia nervosa and its sequelae. J Psychiatr Res 1985; 19:363–369.
70. Grinspoon S, Thomas L, Miller KK, Herzog, DB, Klibanski A. Effects of recombinant human IGF-I and oral contraceptive administration on bone density in anorexia nervosa. J Clin Endocrinol Metab 2002; 87:2883–2891.
71. Salisbury JJ, Mitchell JE. Bone mineral density and anorexia nervosa in women. Am J Psychiatry 1991; 148:768–774.
72. Boyar RM, Hellman LD, Roffwarg H, et al. Cortisol secretion and metabolism in anorexia nervosa. N Engl J Med 1977; 296:190–193.
73. Miller KK, Biller BMK, Hier J, Arena E, Klibanski A. Androgens and bone density in women with hypopituitarism. J Clin Endocrinol Metab 2002; 87:2770–2776.
74. Kasperk CH, Wergedal JE, Farley JR, Linkhart TA, Turner RT, Baylink DJ. Androgens directly stimulate proliferation of bone cells in vitro. Endocrinology 1989; 124:1576–1578.
75. Chen JR, Koousteni S, Bellido T, et al. Gender-independent induction of murine osteoblast apoptosis in vitro by either estrogens or non-aromatizable androgens. J Bone Miner Res 2001; 16 (suppl I):S159 (abstr).
76. Gordon CM, Grace E, Emans SJ, Goodman E, Crawford MH, Leboff MS. Changes in bone turnover markers and menstrual function after short term oral DHEA in young women with anorexia nervosa. J Bone Miner Res 1999; 14:136–145.
77. Mehler PS, Eckel RS, Donahoo WT. Leptin levels in restricting and purging anorectics. Int J Eat Disord 1999; 26:189–194.
78. Pasco JA, Henry MJ, Kotowicz MA, et al. Serum leptin levels are associated with bone mass in nonobese women. J Clin Endocrinol Metab 2001; 86:1884–1887.
79. Burguera B, Hofbauer FC, Thomas J, et al. Leptin reduces ovariectomy induced bone loss in rats. Endocrinology 2001; 142:3546–3553.
80. Sato M, Takeda N, Sarui H, et al. Association between serum leptin concentrations and bone mineral density and biochemical markers of bone turnover in adult men. J Clin Endocrinol Metab 2001; 86:5273–5276.
81. Ducy P, Ambling M, Takeda S, et al. Leptin inhibits bone formation through a hypothalamic relay:a central control of bone mass. Cell 2000; 100:197–207.

82. Abrams SA, Silber TJ, Esteban NV, et al. Mineral balance and bone turnover in adolescents with anorexia nervosa. J Pediatr 1993; 123:326–331.
83. Hadigan CM, Anderson EJ, Miller KK, et al. Assessment of macronutrient and micronutrient intake in women with anorexia nervosa. Int J Eat Disord 2000; 28:284–292.
84. Herzog W, Minne H, Deter C, et al. Outcome of bone mineral density in anorexia nervosa patients 11.7 years after first admission. J Bone Miner Res 1993; 8:597–605.
85. Hartman D, Crisp A, Rooney B, Rackow C, Atkinson R, Patel S. Bone density of women who have recovered from anorexia nervosa. Int J Eat Disord 2000; 28:107–112.
86. Bachrach LK, Hastie T, Wang MC, Narasimhan B, Marcus R. Bone mineral acquisition in healthy Asian, Hispanic, black, and Caucasian youth: a longitudinal study. J Clin Endocrinol Metab 1999; 84:4702–4712.
87. Katzman DK, Bachrach LK, Carter DR, Marcus R. Clinical and anthropometric correlates of bone mineral acquisition in healthy adolescent girls. J Clin Endocrinol Metab 1991; 73:1332–1339.
88. Hans D, Dargent-Molina P, Schott AM, et al. Ultrasonographic heel measurements to predict hip fracture in elderly women: the EPIDOS prospective study. Lancet 1996; 348:511–514.
89. Bauer DC, Gluer C, Cauley JA, et al. Broadband ultrasound attenuation predicts fractures strongly and independently of densitometry in older women. A prospective study. Arch Intern Med 1997; 157:629–634.
90. Resch H, Newrkla S, et al. Ultrasound and X-ray based bone densitometry in patients with anorexia nervosa. Calcif Tissue Int 2000; 66:338–341.
91. Lum CK, Wang MC, Moore E, Wilson DM, Marcus R, Bachrach LK. A comparison of calcaneal ultrasound and dual energy X-ray absorptiometry in healthy North American youths and young adults. J Clin Densitom 1999; 2:403–411.
92. Looker AC, Bauer DC, Chesnut CH 3rd, et al. Clinical use of biochemical markers of bone remodeling: current status and future directions. Osteopor Int 2000; 11:467–480.
93. Miller PD, Baran DT, Bilezikian JP, et al. Practical clinical application of biochemical markers of bone turnover: Consensus of an expert panel. J Clin Densitom 1999; 2:323–342.
94. Mora S, Pitukcheewanont P, Kaufman FR, Nelson JC, Gilsanz V. Biochemical markers of bone turnover and the volume and the density of bone in children at different stages of sexual development. J Bone Miner Res 1999; 14:1664–1671.
95. Szule P, Seeman E, Delmas PD. Biochemical measurements of bone turnover in children and adolescents. Osteopor Int 2000; 11:281–294.
96. Miller KK, Thomas ER, Grinspoon SK, et al. Effects of Risedronate and IGF-I administration on bone metabolism in women with anorexia nervosa. The Endocrine Society 84th Annual Meeting, 2002 (abstr).

34 The Role of Nutrition for Bone Health in Cystic Fibrosis

Kimberly O. O'Brien and Michael F. Holick

Cystic fibrosis (CF) is a common genetic disorder in the United States, affecting approx 1 in 3200 white and 1 in 15,000 black live births *(1)*. This disease results in abnormal sodium and chloride transport resulting from mutations in the CF transmembrane conductance regulator (CFTR) *(2)*. Mutations in the CFTR alter the osmolarity of body secretions, resulting in lung and gastrointestinal (GI) complications. Blocked bronchial airways, lung infections, and GI disturbances involving maldigestion and malabsorption are common consequences of CF.

Improvements in medical care have increased the median age of survival of individuals with CF to over 30 yr of age *(3)*. As the lifespan with individuals with this disease increases, there is a growing awareness of issues related to insufficient bone mass and increased risk of fracture in individuals with this disease *(4–7)*. Net loss of bone mass can result in osteopenia (defined as a bone mineral density [BMD] 1–2.5 standard deviations below the mean expected for a young adult) and osteoporosis (defined as a BMD >2.5 standard deviations below the mean expected for a young adult). The osteopenia observed in CF patients is multifactorial and is influenced by nutritional status, disease severity, glucocorticoid use, hormonal status, inflammation, GI function, mechanical loading, and physical activity patterns.

Nutritional intakes sufficient to acquire and maintain bone mass are essential to promote bone health in individuals with CF. This is particularly important in children with this disease, as many long-term sequelae of CF accelerate bone loss in later life. An average of 26% of total body bone mineral is accrued within the 2-yr period surrounding the period of peak bone mineral acquisition *(8)*. Because of the magnitude of bone growth at this time, special emphasis should be placed on maximizing bone acquisition early in life, and preventative strategies for osteoporosis should be initiated in pediatric populations.

From: *Nutrition and Bone Health*
Edited by: M. F. Holick and B. Dawson-Hughes © Humana Press Inc., Totowa, NJ

1. MAINTENANCE OF ADEQUATE BODY WEIGHT

To consolidate the skeleton and maintain a healthy body weight, optimal nutritional intakes are crucial. Body size is known to be a strong determinant of bone mass in adults, explaining approx 50% of the variance in bone mass among individuals at the population level (after accounting for height) *(9)*. Children and adults with CF are frequently underweight and often do not achieve adequate body mass index (BMI) percentiles *(10)*. According to the 1999 annual data report of the Cystic Fibrosis Foundation, using the Centers for Disease Control growth charts the median body mass, height, and weight percentiles for patients with CF fall at or below the 40th percentile for males and females ages 10–20 yr *(10)*. Promoting optimal weight puts the appropriate loads on bone, maintains muscle mass, and assists the individual in participating in normal activity patterns. In patients with CF, nutritional status has a significant impact on growth, pulmonary function, and survival *(11)*.

In children, low body mass can also delay the onset of puberty and further place this group at risk for decreased BMD. Many studies have reported an approx 2-yr delay in puberty in adolescents with CF *(12–15)*. Optimal nutritional intakes, supplementary feeds, or parenteral intakes should all be utilized as needed to assist in the maintenance of an adequate body weight in patients with CF.

Optimal dietary intakes promote adequate growth and provide the necessary nutrients required for the mineralization and maintenance of bone mass. Specific nutrients required for optimal bone mineralization include calcium, phosphorus, vitamin D, vitamin K, protein, and trace elements such as copper and zinc. The status of many of these nutrients is compromised in children and adults with CF, which may further limit bone mineralization. Several of these nutrients will be summarized here with respect to the data available on status and impact on bone health in CF patients.

2. CALCIUM

Calcium is the principal mineral in bone, comprising more than 30% of bone mineral. The majority (>99%) of the body's calcium content is located in bone. Because calcium physiology is designed to maintain serum concentrations at constant levels, this function will be maintained at the expense of bone. The ability to maintain and consolidate skeletal calcium stores is therefore dependent on optimal calcium intake, absorption of calcium from the GI tract, and sufficient retention of calcium to offset urinary, endogenous fecal, and dermal calcium losses. CF may adversely impact many of these pathways.

2.1. Calcium Absorption

Early studies assumed calcium absorption was reduced in patients with CF based on indirect evidence such as lower-than-expected urinary calcium excretion *(16)*. Few studies to date have directly measured calcium absorption in patients with CF. One early study, published only in abstract form, administered ^{47}Ca to five patients and indicated that calcium absorption was normal in this group *(17)*. Aris et al. *(18)* evaluated calcium homeostasis in adults with CF. They studied the fractional absorption of ^{45}Ca and urinary excretion of calcium in CF patients and normal controls following a high-calcium breakfast. Seven young men and five young women were studied on two separate occasions with and without administration of pancreatic enzymes. Eleven healthy young adults with normal BMD measurements served as controls. Without pancreatic enzymes, subjects with CF showed significantly impaired calcium absorption 5 h after ingestion of the high-calcium ^{45}Ca-labeled diet. Subjects with CF showed a 5-h fractional absorption of 8.9 ± 0.2, compared to 11.8 ± 0.54 the control subjects ($p = 0.02$). The decrease in fractional absorption in the patient with CF paralleled a decrease in calcium excretion after 4 h of 0.16 ± 0.09 mg Ca/mg for subjects with CF, compared to 0.2 ± 0.08 mg Ca/mg creatinine for controls. The addition of pancreatic enzymes did not fully compensate for this deficiency. A recent stable-isotope study of calcium absorption was reported in a group of 23 clinically stable girls with CF (ages 7–18 yr), all of whom were taking pancreatic enzyme replacement as needed and none of whom were receiving oral steroids. Each child received stable isotopes of calcium orally (^{44}Ca) and intravenously (^{42}Ca), and calcium absorption was determined based on the cumulative recovery of these isotopes in urine over a 24-h interval postdosing. In these children, calcium absorption was similar to data reported in healthy girls matched for Tanner stage *(19)*. More data are needed in this group across wider age ranges and under additional clinical and treatment conditions relevant to this patient population.

2.2. Urinary Calcium Excretion

Data on the impact of CF on urinary calcium excretion is often conflicting, and differences in calcium intake between the study and control populations have seldom been taken into account in the interpretation of these data *(20–23)*. Moreover, because the relationship between dietary calcium intake and urinary calcium excretion is not the same between pediatric and adult populations, this may complicate conclusions when both adult and pediatric patients are studied simultaneously. Several studies in patients with CF have suggested that there is an increased risk of Ca oxalate stones (cumulative incidence of 5.7%) and oxalate crystalluria (4.2%) *(24–26)*. This may be associated with enteric hyperoxaluria associated with pancreatic insufficiency and effects of abnormal bile salt metabolism on colonic oxalate absorption.

Other components of the diet influence urinary calcium excretion and may limit the amount of calcium available for bone deposition. This is particularly true for dietary sodium. Individuals with CF are known to be susceptible to heat injury or illness owing to fluid imbalances *(27,28)*. To assist in adequate sodium retention, recommendations have been made to supplement CF patients generously with sodium chloride *(29)*. At present the general recommendation is to give 1/4 teaspoon of salt per day (23 mEq of sodium per day) to infants with CF and 3/4–1 teaspoon of salt per day to older children *(30)*. In adolescents and adults, the recommendation is to salt food liberally and older patients can take non-enteric-coated tablets as needed

(30). In healthy adolescents, dietary sodium is one of the strongest predictors of urinary calcium excretion *(31,32)*. High dietary intakes of sodium may reduce the amount of calcium available for bone deposition *(31)*, a finding that may have more of an adverse impact on bone health in groups already at risk for low bone mass.

To date, studies have not been undertaken in children or adults with CF to examine the degree to which increases in sodium intake may also increase urinary calcium losses. Given the importance of calcium in bone mineralization, this issue needs further study so that recommendations can be made that balance the need for sodium supplementation against its potential impact on bone health.

2.3. Endogenous Fecal Calcium Excretion

Endogenous fecal losses of calcium occur when calcium of nondietary sources is secreted into the GI tract from bile, pancreatic juices, or direct excretion into the GI lumen. In healthy adults and children, endogenous fecal calcium losses typically average 1.5 mg/kg/d, and these losses are only minimally affected by calcium intake *(33–35)*. Alterations in intestinal permeability, intestinal paracellular or transcellular flux, and changes in GI calcium losses from pancreatic or biliary secretions may increase endogenous fecal calcium loses in patients with CF. No published studies to date have reported rates of endogenous fecal calcium excretion in adults or children with CF. However, a recent stable-isotope study reported higher than expected endogenous fecal calcium losses in children with CF *(36)*. Increased endogenous fecal calcium losses have been reported in patients with chronic malabsorption syndromes (severe Crohn's disease and protein-losing enteropathy resulting from intestinal lymphangiectasia) *(37)*. Although few studies have measured these losses directly in individuals with CF, intestinal permeability has been found to be increased in patients with CF *(38)*, and duodenal outputs of calcium in individuals with chronic pancreatitis have been reported to be nearly doubled compared to healthy or diseased controls following saline, cholecystokinin, or secretin infusion *(39–41)*. Excessive fecal bile acid losses have also been demonstrated in patients with CF who have pancreatic insufficiency *(42–44)*.

2.4. Dermal Calcium Losses

Increased dermal losses of sodium are known to occur as a result of CF, making these patients prone to hyponatremia. In CF studies measuring dermal sodium losses to date, calcium losses were not monitored, and the potential for CF to increase dermal calcium losses has not been examined.

2.5. Other Causes of Calcium Loss

Even when optimal calcium retention is achieved, it is necessary to have the appropriate hormonal and cytokine milieu necessary to mineralize bone. Increased levels of osteoclast-stimulatory inflammatory cytokines (interleukin [IL]-6, IL-1, and tumor necrosis factor-α) and increased markers of bone resorption have been reported in CF patients with lung infections *(45)*. These inflammatory responses may further compromise the ability to deposit and retain calcium in bone mineral.

3. VITAMIN D

Vitamin D plays essential roles in calcium homeostasis by increasing the efficiency of intestinal calcium absorption and influencing bone turnover by stimulating osteoblasts to induce the conversion of stem cell monocytes into mature osteoclasts *(46)*.

3.1. Vitamin D Status

Multiple studies have reported vitamin D deficiency among patients with CF *(7,16,47–51)*. This deficiency occurs in spite of the fact that patients with CF are typically supplemented with 400–800 IU of vitamin D per day.

3.2. Vitamin D Absorption

Suboptimal vitamin D status in the face of daily supplementation suggests that the vitamin may be malabsorbed. Fat malabsorption is a frequent finding in patients with CF, and this may limit the intestinal absorption of fat-soluble vitamins such as vitamin D. Patients with fat malabsorption syndromes such as celiac disease and biliary or pancreatic obstruction have been found to malabsorb oral ^{3}H-vitamin D_3 *(52)*. Studies in patients with CF have also reported limited and highly variable absorption of vitamin D following oral vitamin D supplementation *(53,54)*. Lo et al. *(53)*. developed a test procedure for the clinical evaluation of the absorption of vitamin D. They evaluated seven patients with intestinal fat malabsorption syndromes and seven healthy normal subjects. Each subject was given a single oral dose of 50,000 IU (1.25 mg) vitamin D_2. In normal subjects, serum vitamin D concentrations rose from baseline of less than 5 ng/mL to a peak of over 50 ng/mL by 12 h, gradually falling to baseline levels by 3 d. In five of the seven patients with intestinal fat malabsorption, oral administration of 50,000 IU of vitamin D_2 did not raise serum vitamin D concentration above 10 ng/mL. Five of the patients with clinical fat malabsorption secondary to inflammatory bowel disease (IBD), CF, scleroderma, and mucosal villis atrophy essentially were unable to raise their basal serum levels of vitamin D. However, 2 patients with severe IBD and diarrhea, one with ulcerative colitis and one with Crohn's illeal colitis, both of whom required total parenteral nutrition and subsequent illeal colectomy, showed normal absorption of vitamin D. Recently, it was reported that a female adult with severe Crohn's disease, who had only 2 ft of small intestine left and who had severe malabsorption of vitamin D, was able to correct her vitamin D deficiency and osteomalacic symptoms simply by being exposed to a tanning bed three times a week *(54a)*. Thus, an alternative for CF patients is for them to be exposed to either a home unit or a tanning bed in a tanning parlor for the prescribed time recommended by the manufacturer, depending on the individual's skin type. The impairments in vitamin D status may adversely impact bone quality and risk of vertebral fracture *(55)*.

4. VITAMIN K

Vitamin K is essential for bone health, as this vitamin mediates the carboxylation of glutamyl residues on bone proteins, including osteocalcin, the most abundant noncollagenous protein in bone. Because this is also a fat-soluble vitamin, its status may be compromised due to the fat malabsorption that occurs in CF. An increased prevalence of vitamin K deficiency (as determined by prothrombin in vitamin K absence levels [PIVKA-II]) has been found in unsupplemented patients with CF *(56)*. Supplementation studies have found that vitamin K supplementation (with an average of 0.18 mg/d) improved measures of vitamin K status as evidenced by changes in PIVKA-II levels *(57)*. In addition, 4 wk of vitamin K supplementation (5 mg/wk) increased the carboxylation state of osteocalcin in 18 patients with CF *(58)*. Few studies to date have related vitamin K status to measures of bone health in CF populations.

5. ZINC

Zinc is crucial for optimal skeletal maturation, growth, and development *(59,60)*. Zinc also regulates gene expression *(61)*, and is an essential cofactor for bone-related enzymes such as alkaline phosphatase *(59)*. Numerous case-report studies of zinc deficiency have been reported in infants and children with CF. These cases often present as acrodermatitis enteropathica-like rashes *(62–66)*.

Pancreatic-insufficient children with CF (ages 7–17 yr) absorb significantly less dietary zinc in the absence of pancreatic enzymes, presumably because of undigested dietary fat or protein *(67)*. Moreover, pancreatic enzyme replacement therapy (>2 wk) significantly increased plasma zinc levels compared to infants with CF who had not received enzyme replacement therapy *(68)*. Infants who did not receive enzyme replacement therapy had a 29% incidence of deficient plasma zinc concentrations (≤ 9.2 μmol/L) compared to an incidence of 7% in infants receiving pancreatic enzymes *(69)*.

Zinc deficiency in patients with CF may be related in part to alterations in endogenous fecal zinc secretion. Stable isotope studies of endogenous fecal zinc secretion in breast- and formula-fed infants with CF found negative zinc balances in both groups of infants and substantially greater endogenous fecal zinc losses than those typically reported in breast-fed or formula-fed infants. Endogenous fecal zinc losses in CF infants were also correlated to fecal fat excretion *(69)*.

6. COPPER

Copper deficiency can affect bone health, due in part to the essential role copper has in the enzyme involved in lysine and hydoxyproline crosslinking in collagen (lysyl oxidase) *(70,71)*. Copper also stimulates human mesenchymal stem cell differentiation toward the osteogenic lineage *(72)*. Case-control studies have found alterations in copper distribution and significantly lower copper–zinc superoxide dismutase activity in mononuclear and polymorphonuclear cells in CF patients *(73)*. Decreased activity of copper-dependent enzymes, suggestive of abnormal copper homeostasis, has also been reported in pancreatic-insufficient adolescents with CF *(74)*.

7. METABOLIC BONE DISEASE

It is now well recognized that metabolic bone disease is a major affliction of patients with CF *(4–8)*. This in part is the result of the increased life expectancy of patients with CF because of better management of their pulmonary disease. The origin of the metabolic bone disease in patients with CF is not well understood. However, glucocorticoid therapy, potential malabsorption of vitamin D and other nutrients, lack of physical activity, and hypogonadism are all potential contributing factors to this problem.

7.1. Bone Mineral Density in Patients With CF

Flohr et al. *(75)* conducted a cross-sectional study in 75 adult patients with CF (mean age 25.3 yr) to assess the prevalence of low BMD by dual-energy X-ray absorptiometry (DXA). They observed a mean BMD *T*-score ± SEM of –1.4 ± 0.17 at the lumbar spine and –0.54 ± 0.16 at the femoral neck. Based on the definition of osteoporosis as having a *T*-score of greater than –2.5, the authors observed that 27% of the patients had osteoporosis. A multiple regression analysis showed that the forced expiratory volume in 1 s (FEV_1) and the use of oral glucocorticoids were independent predictors of low lumbar BMD, whereas bone mass index (BMI) and the use of oral glucocorticoids were independent predictors of low femoral neck BMD. Elkin et al. *(76)* determined the prevalence of low BMD and vertebral deformities in CF adults with varied disease severity. A total of 107 patients (58 men) aged 18–60 yr underwent a DXA scanning of their hip and spine. Of these, 38% had a *z*-score of <–1, with 13% having *z*-scores of <–2. Of significance was that 17% had evidence of vertebral deformity on radiography, mostly in the thoracic spine. There were reported past fractures in 35%, of which 9% were rib fractures. The amount of daily physical activity and the percent predicted FEV_1 were positively related to BMD.

Henderson and Madsen *(77)* evaluated various measures of growth and body composition in 40 children and young adults (ages 5.7–20.3 yr, mean 11.9 yr). Relative to age-matched normal controls, the authors concluded that CF patients had a deficit in total body bone mineral, averaging 19.1% ± 3.0%. Similarly, Moran et al. *(78)* reported that, of 14 patients with CF and chronic pancreatitis, 10 had osteopenia and 3 had evidence of osteoporosis. In a group of patients with end-stage pulmonary disease who were awaiting lung transplantation, of the 70 patients aged 18–70 yr, Shane et al. *(55)* reported that 30% had osteoporosis at the lumbar spine and 49% at the femoral neck. In addition, 35% had osteopenia of the lumbar spine and 31% at the femoral neck.

7.2. Causes for Osteoporosis and Osteopenia in CF Patients

There are a multitude of potential mechanisms for causing reduced BMD in patients with CF. These include a decrease in physical activity, exposure to glucocorticoids, and malabsorption of nutrients that are critically important for bone health.

7.3. Vitamin D Deficiency in Patients With CF

There is firm evidence that vitamin D deficiency is common among patients who suffer from CF *(7,16,47–50).* Donovan et al. *(7),* for example, reported bone mass and vitamin D deficiency in adults with advanced CF lung disease. Thirty Caucasian men and women with CF were evaluated consecutively following acceptance to a lung transplantation waiting list. A majority of the subjects (80%) were taking 400–800 IU of vitamin D daily in a multivitamin. Twenty of the patients were at the low end of the normal range for 25(OH)D (16 ± 2 ng/mL; normal 10–52 ng/mL). Eight patients (40%) had levels below 10 ng/mL, which is considered to be severe vitamin D deficiency, and seven patients (35%) had levels between 11 and 20 ng/mL, which is also considered to be vitamin D insufficient. The authors found the BMD was lower in patients with low 25(OH)D. Nineteen percent of the patients presented with new vertebral fracture, and 41% of the subjects had historical information that confirmed atraumatic fractures of rib, wrist, and digits. A similar observation was made by Shane et al. *(55).* They followed 70 patients (aged 18–70 yr) with end-stage pulmonary disease, including 11 patients with CF. They observed that 36% of patients with CF and 20% of patients with chronic obstructive pulmonary disease (COPD) suffered from severe vitamin D deficiency as defined as a 25(OH)D of <10 ng/mL. They observed a significant reduction in BMD in the CF patients, and the prevalence rate of vertebral fractures was 25% in CF patients and 29% in COPD patients.

8. NUTRITION, GENETICS, AND DISEASE SEVERITY

Additional information on the impact of CF on metabolism of nutrients required for optimal bone mineralization is needed in order to target interventions to maximize nutrient retention and bone health in this population. It is also important to consider that osteoporosis can occur even in the face of adequate nutrition because of the strong genetic control of peak bone mass. Several genotypes significantly influence bone mass, including those for the vitamin D receptor, calcitonin receptor, parathyroid hormone receptor, and estrogen receptor *(79).* In addition to genetic control of bone acquisition, genotype–phenotype relationships may also influence the severity and clinical implications of diseases such as cystic fibrosis. Better identification of genotypes that may be predictive of low bone mass (both related and unrelated to the CF disease process) will assist in targeting preventative therapies to those at greatest risk for low bone mass or osteoporosis.

9. CONCLUSION

With expected increased life expectancy, patients with CF are at extremely high risk of osteopenia and osteoporosis, which markedly increases their risk of non-traumatic fractures. Although there are many potential causes for metabolic bone disease in patients with CF, individuals with this disease should be counseled on the importance of optimal body weight as it relates to bone health and should receive nutritional counseling if percent ideal body weight falls below the 90th percentile. It is also essential to ensure that patients with CF consume appropriate amounts of calcium and vitamin D. Calcium intakes should, at a minimum, meet the recent adequate intake recommendations for this nutrient. Moreover, it is essential to monitor vitamin D status and correct vitamin D deficiency when present. This can be accomplished by measuring circulating concentrations of 25(OH)D at least once, if not twice a year, in patients with CF. If vitamin D insufficiency is detected, either there should be an increase in the vitamin D intake, either by diet (at least 1000 IU/d) or by pharmaceutical means (usually 50,000 IU vitamin D every 1–2 wk), or the patient should be encouraged to get some sun exposure, or to be exposed to an artificial ultraviolet radiation source that has a ultraviolet B output *(54a)*.

Several studies have suggested that intravenous bisphosphonates—i.e., Pamidronate—may be of some value in patients with CF. However, it is inappropriate to treat patients who suffer from osteomalacia with Pamidronate. This is probably the explanation for why several patients who were treated with intravenous Pamidronate experienced severe bone and joint discomfort. Bisphosphonate therapy should only be considered when the patient has been restored to vitamin D sufficiency.

Finally, it is essential that interventions designed to promote bone health in patients with CF be initiated during childhood. Nutritional counseling and additional interventions as needed in this age group will assist in the attainment of maximal bone accrual during the pubertal growth spurt, thereby promoting bone health and improved quality of life for patients with CF.

REFERENCES

1. Hamosh A, Fitz-Simmons SC, Macek M Jr, Knowles MR, Rosenstein BJ, Cutting GR. Comparison of the clinical manifestations of cystic fibrosis in black and white patients. J Pediatr 1998; 132:255–259.
2. Greger R, Schreiber R, Mall M, et al. Cystic fibrosis and CFTR. Pflugers Arch 2001; 443(suppl 1):S3–S7.
3. Orenstein DM, Rosenstein BJ, Stern RC. Cystic Fibrosis: Medical Care. Lippincott, Williams and Wilkins, Philadelphia, 2000.
4. Conway SP, Morton AM, Oldroyd B, et al. Osteoporosis and osteopenia in adults and adolescents with cystic fibrosis: prevalence and associated factors. Thorax 2000; 55:798–804.
5. Haworth CS, Selby PL, Webb AK, et al. Low bone mineral density in adults with cystic fibrosis. Thorax 1999; 54:961–967.
6. Laursen EM, Molgaard C, Michaelsen KF, Koch C, Muller J. Bone mineral status in 134 patients with cystic fibrosis. Arch Dis Child 1999; 81:235–240.

7. Donovan DSJ, Papadopoulos A, Staron RB, et al. Bone mass and vitamin D deficiency in adults with advanced cystic fibrosis lung disease. Am J Respir Crit Care Med 1998; 157:1892–1899.
8. Bailey DA, McKay HA, Mirwald RL, Crocker PR, Faulkner RA. A six-year longitudinal study of the relationship of physical activity to bone mineral accrual in growing children: the University of Saskatchewan Bone Mineral Accrual Study. J Bone Miner Res 1999; 14:1672–1679.
9. Heaney RP, Abrams S, Dawson-Hughes B, et al. Peak bone mass. Osteopor Int 2000; 11:985–1009.
10. Cystic Fibrosis Foundation. Patient Registry 1999 Annual Report. Cystic Fibrosis Foundation, Bethesda, MD, 2000.
11. Corey M, McLaughlin FJ, Williams M, Levison H. A comparison of survival, growth, and pulmonary function in patients with cystic fibrosis in Boston and Toronto. J Clin Epidemiol 1988; 41:583–591.
12. Landon C, Rosenfeld RG. Short stature and pubertal delay in cystic fibrosis. Pediatrician 1987; 14:253–260.
13. Johannesson M, Landgren BM, Csemiczky G, Hjelte L, Gottlieb C. Female patients with cystic fibrosis suffer from reproductive endocrinological disorders despite good clinical status. Hum Reprod 1998; 13:2092–2097.
14. Stead RJ, Hodson ME, Batten JC, Adams J, Jacobs HS. Amenorrhoea in cystic fibrosis. Clin Endocrinol (Oxf) 1987; 26:187–195.
15. Moshang T, Holsclaw DS Jr. Menarchal determinants in cystic fibrosis. Am J Dis Child 1980; 134:1139–1142.
16. Hahn TJ, Squires AE, Halstead LR, Strominger DB. Reduced serum 25-hydroxyvitamin D concentration and disordered mineral metabolism in patients with cystic fibrosis. J Pediatr 1979; 94:38–42.
17. Simopoulos AP, Taussig LM, Murad E, et al. Parathyroid function in patients with cystic fibrosis (abstr). Pediatr Res 1972; 6:95.
18. Aris RM, Lester GE, Dingman S, Ontjes DA. Altered calcium homeostasis in adults with cystic fibrosis. Osteopor Int 1999; 10:102–108.
19. Schulze KJ, O'Brien KO, Germain-Lee EL, Baer DJ, Leonard A, Rosenstein BJ. Efficiency of calcium absorption is not compromised among clinically stable pre-pubertal and pubertal girls with cystic fibrosis. Am J Clin Nutr 2003; 78:110–116.
20. Hahn TJ, Squires AE, Halstead LR, Strominger DB. Reduced serum 25-hydroxyvitamin D concentration and disordered mineral metabolism in patients with cystic fibrosis. J Pediatr 1979; 94:38–42.
21. Mortensen LA, Chan GM, Alder SC, Marshall BC. Bone mineral status in prepubertal children with cystic fibrosis. J Pediatr 2000; 136:648–652.
22. Salamoni F, Roulet M, Gudinchet F, Pilet M, Thiebaud D, Burckhardt P. Bone mineral content in cystic fibrosis patients: correlation with fat-free mass. Arch Dis Child 1996; 74:314–318.
23. Ionescu AA, Nixon LS, Evans WD, et al. Bone density, body composition, and inflammatory status in cystic fibrosis. Am J Respir Crit Care Med 2000; 162:789–794.
24. Bohles H, Michalk D. Is there a risk for kidney stone formation in cystic fibrosis? Helv Paediatr Acta 1982; 37:267–272.
25. Chidekel AS, Dolan TF. Cystic fibrosis and calcium oxalate nephrolithiasis. Yale J Biol Med 1996; 69:317–321.
26. Turner MA, Goldwater D, David TJ. Oxalate and calcium excretion in cystic fibrosis. Arch Dis Child 2000; 83:244–247.
27. Bar-Or O, Blimkie CJ, Hay JA, MacDougall JD, Ward DS, Wilson WM. Voluntary dehydration and heat intolerance in cystic fibrosis. Lancet 1992; 339:696–699.
28. Orenstein DM, Henke KG, Green CG. Heat acclimation in cystic fibrosis. J Appl Physiol 1984; 57:408–412.
29. Legris GJ, Dearborn D, Stern RC, et al. Sodium space and intravascular volume: dietary sodium effects in cystic fibrosis and healthy adolescent subjects. Pediatrics 1998; 101:48–56.
30. Waring WW. Current management of cystic fibrosis. Adv Pediatr 1976; 23:401–438.

31. O'Brien KO, Abrams SA, Stuff JE, Liang LK, Welch TR. Variables related to urinary calcium excretion in young girls. J Pediatr Gastroenterol Nutr 1996; 23:8–12.
32. Matkovic V, Ilich JZ, Andon MB, et al. Urinary calcium, sodium, and bone mass of young females. Am J Clin Nutr 1995; 417–425.
33. Bronner F, Abrams SA. Urinary and fecal endogenous calcium excretion in the age range of 5–15 y. Am J Clin Nutr 1993; 57:944.
34. Abrams SA, Sidbury JB, Muenzer J, Esteban NV, Vieira NE, Yergey AL. Stable isotopic measurement of endogenous fecal calcium excretion in children. J Pediatr Gastroenterol Nutr 1991; 12:469–473.
35. Heaney RP, Recker RR. Determinants of endogenous fecal calcium in healthy women. J Bone Miner Res 1994; 9:1621–1627.
36. Schulze KJ, O'Brien KO, Germain-Lee EL, Baer DJ, Leonard AL, Rosenstein BJ. Endogenous fecal losses of calcium compromise calcium balance in pancreatic insufficient girls with cystic fibrosis. J Pediatr 2003; 143.
37. Nicolaidou P, Ladefoged K, Hylander E, Thale M, Jarnum S. Endogenous faecal calcium in chronic malabsorption syndromes and in intestinal lymphangiectasia. Scand J Gastroenterol 1980; 15:587–592.
38. Hallberg K, Grzegorczyk A, Larson G, Strandvik B. Intestinal permeability in cystic fibrosis in relation to genotype. J Pediatr Gastroenterol Nutr 1997; 25:290–295.
39. Regan PT, Malagelada JR, DiMagno EP. Duodenal calcium outputs in health and pancreatic disease. Gut 1980; 21:614–618.
40. Nimmo J, Finlayson ND, Smith AF, Shearman DJ. The production of calcium and magnesium during pancreatic function tests in health and disease. Gut 1970; 11:163–166.
41. Hansky J. Calcium content of duodenal juice. Am J Dig Dis 1967; 12:725–733.
42. Weber AM, Roy CC, Chartrand L, et al. Relationship between bile acid malabsorption and pancreatic insufficiency in cystic fibrosis. Gut 1976; 17:295–299.
43. Walters MP, Littlewood JM. Faecal bile acid and dietary residue excretion in cystic fibrosis: age group variations. J Pediatr Gastroenterol Nutr 1998; 27:296–300.
44. O'Brien S, Mulcahy H, Fenlon H, et al. Intestinal bile acid malabsorption in cystic fibrosis. Gut 1993; 34:1137–1141.
45. Aris RM, Stephens AR, Ontjes DA, et al. Adverse alterations in bone metabolism are associated with lung infection in adults with cystic fibrosis. Am J Respir Crit Care Med 2000; 162:1674–1678.
46. Holick MF. Vitamin D: the underappreciated D-lightful hormone that is important for skeletal and cellular health. Curr Opin Endocrinol Diabetes 2002; 8:87–98.
47. Grey V, Lands L, Pall H, Drury D. Monitoring of 25-OH vitamin D levels in children with cystic fibrosis. J Pediatr Gastroenterol Nutr 2000; 30:314–319.
48. Stead RJ, Houlder S, Agnew J, et al. Vitamin D and parathyroid hormone and bone mineralisation in adults with cystic fibrosis [published erratum appears in Thorax 1988 May; 43(5):424]. Thorax 1988; 43:190–194.
49. Hanly JG, McKenna MJ, Quigley C, Freaney R, Muldowney FP, FitzGerald MX. Hypovitaminosis D and response to supplementation in older patients with cystic fibrosis. Q J Med 1985; 56:377–385.
50. Friedman HZ, Langman CB, Favus MJ. Vitamin D metabolism and osteomalacia in cystic fibrosis. Gastroenterology 1985; 88:808–813.
51. Reiter EO, Brugman SM, Pike JW, et al. Vitamin D metabolites in adolescents and young adults with cystic fibrosis: effects of sun and season. J Pediatr 1985; 106:21–26.
52. Thompson GR, Lewis B, Booth CC. Absorption of vitamin D3-3H in control subjects and patients with intestinal malabsorption. J Clin Invest 1966; 45:94–102.
53. Lo CW, Paris PW, Clemens TL, Nolan J, Holick MF. Vitamin D absorption in healthy subjects and in patients with intestinal malabsorption syndromes. Am J Clin Nutr 1985; 42:644–649.
54. Lark RK, Lester GE, Ontjes DA, et al. Diminished and erratic absorption of ergocalciferol in adult cystic fibrosis patients. Am J Clin Nutr 2001; 73:602–606.

54a. Koutkia P, Lu Z, Chen TC, Holick MF. Treatment of vitamin D deficiency due to Crohn's disease with tanning bed ultraviolet B radiation. Gastroenterology 2001; 121:1485–1488.
55. Shane E, Silverberg SJ, Donovan D, et al. Osteoporosis in lung transplantation candidates with end-stage pulmonary disease [see comments]. Am J Med 1996; 101:262–269.
56. Rashid M, Durie P, Andrew M, et al. Prevalence of vitamin K deficiency in cystic fibrosis. Am J Clin Nutr 1999; 70:378–382.
57. Wilson DC, Rashid M, Durie PR, et al. Treatment of vitamin K deficiency in cystic fibrosis: effectiveness of a daily fat-soluble vitamin combination. J Pediatr 2001; 138:851–855.
58. Beker LT, Ahrens RA, Fink RJ, et al. Effect of vitamin K1 supplementation on vitamin K status in cystic fibrosis patients. J Pediatr Gastroenterol Nutr 1997; 24:512–517.
59. Aggett PJ, Comerford JG. Zinc and human health. Nutr Rev 1995; 53:S16–S22.
60. Castillo-Duran C, Cassorla F. Trace minerals in human growth and development. J Pediatr Endocrinol Metab 1999; 12:589–601.
61. Cousins RJ. A role of zinc in the regulation of gene expression. Proc Nutr Soc 1998; 57:307–311.
62. Ghali FE, Steinberg JB, Tunnessen WW. Picture of the month. Acrodermatitis enteropathica-like rash in cystic fibrosis. Arch Pediatr Adolesc Med 1996; 150:99–100.
63. Hansen RC, Lemen R, Revsin B. Cystic fibrosis manifesting with acrodermatitis enteropathica-like eruption. Association with essential fatty acid and zinc deficiencies. Arch Dermatol 1983; 119:51–55.
64. Rosenblum JL, Schweitzer J, Kissane JM, Cooper TW. Failure to thrive presenting with an unusual skin rash. J Pediatr 1985; 107:149–153.
65. Schmidt CP, Tunnessen W. Cystic fibrosis presenting with periorificial dermatitis. J Am Acad Dermatol 1991; 25:896–897.
66. Darmstadt GL, Schmidt CP, Wechsler DS, Tunnessen WW, Rosenstein BJ. Dermatitis as a presenting sign of cystic fibrosis. Arch Dermatol 1992; 128:1358–1364.
67. Easley D, Krebs N, Jefferson M, et al. Effect of pancreatic enzymes on zinc absorption in cystic fibrosis. J Pediatr Gastroenterol Nutr 1998; 26:136–139.
68. Krebs NF, Sontag M, Accurso FJ, Hambidge KM. Low plasma zinc concentrations in young infants with cystic fibrosis. J Pediatr 1998; 133:761–764.
69. Krebs NF, Westcott JE, Arnold TD, et al. Abnormalities in zinc homeostasis in young infants with cystic fibrosis. Pediatr Res 2000; 48:256–261.
70. Rucker RB, Kosonen T, Clegg MS, et al. Copper, lysyl oxidase, and extracellular matrix protein cross-linking. Am J Clin Nutr 1998; 67:996S–1002S.
71. Lowe NM, Lowe NM, Fraser WD, Jackson MJ. Is there a potential therapeutic value of copper and zinc for osteoporosis? Proc Nutr Soc 2002; 61:181–185.
72. Rodriguez JP, Rios S, Gonzalez M. Modulation of the proliferation and differentiation of human mesenchymal stem cells by copper. J Cell Biochem 2002; 85:92–100.
73. Percival SS, Kauwell GP, Bowser E, Wagner M. Altered copper status in adult men with cystic fibrosis. J Am Coll Nutr 1999; 18:614–619.
74. Percival SS, Bowser E, Wagner M. Reduced copper enzyme activities in blood cells of children with cystic fibrosis. Am J Clin Nutr 1995; 62:633–638.
75. Flohr F, Lutz A, App EM, Matthys H, Reincke M. Bone mineral density and quantitative ultrasound in adults with cystic fibrosis. Eur J Endocrinol 2002; 146:531–536.
76. Elkin SL, Fairney A, Burnett S, et al. Vertebral deformities and low bone mineral density in adults with cystic fibrosis: a cross-sectional study. Osteopor Int 2001; 12:366–372.
77. Henderson RC, Madsen CD. Bone mineral content and body composition in children and young adults with cystic fibrosis. Pediatr Pulmonol 1999; 27:80–84.
78. Moran CE, Sosa EG, Martinez SM, et al. Bone mineral density in patients with pancreatic insufficiency and steatorrhea. Am J Gastroenterol 1997; 92:867–871.
79. Rizzoli R, Bonjour JP, Ferrari SL. Osteoporosis, genetics and hormones. J Mol Endocrinol 2001; 26:79–94.

35 Antiepileptic Drugs and Bone Health

Marielle Gascon-Barré

1. INTRODUCTION

Epilepsy is a disorder affecting brain function that leads to the periodic and unpredictable occurrence of seizures. Epilepsy does not, however, refer to a single entity. Indeed, the disease encompasses several forms of seizure disorders that lead to the sudden and disorderly discharge of cerebral neurons. Abnormal movements, convulsions (if the motor cortex is involved), or abnormal perceptions (visual, auditory, or olfactory if the parietal or occipital cortex is involved) accompanied with transient but short impairment of consciousness then occurs *(1)*.

Epilepsy is a common disease, with an estimated 2.5–3.0 million affected individuals in the Unites States. However, the prevalence of epilepsy varies with age and has been estimated to be 0.6–0.9% up to 60 yr of age and then to rise to 1.5% in the eighth decade *(2)*. The abnormalities associated with the disease are, unfortunately, recurrent, and patients usually have to be treated with antiepileptic drugs (AEDs) for the rest of their lives.

The appearance of abnormalities in calcium, vitamin D and bone metabolism in subjects chronically treated with AEDs has been reported by numerous investigators since the original report by Schmid in 1967 *(3)*. The etiology of these abnormalities has not yet been completely understood, but both fundamental and clinical studies have identified several possible contributing factors to explain the observed disturbances in the calcium endocrine system in affected individuals. Most notably, the disturbances observed in AED-treated subjects are often very similar to those of vitamin D insufficiency and even, in some cases, of vitamin D deficiency. Moreover, skeletal disorders have been reported to occur in approx 50% of patients chronically receiving AED therapy, most particularly those living in institutions and those treated with AEDs having an enzyme-inducing property on the hepatic cytochrome P450 (CYP) system *(4–10)*. The skeletal disturbances associated with AED therapy significantly increase the risk of fractures, especially of the hip. Fortunately, the biochemical and radiological abnormalities that are associated with chronic AED therapy are often reversed by vitamin D and calcium administration *(5,11–13)*.

From: *Nutrition and Bone Health*
Edited by: M. F. Holick and B. Dawson-Hughes © Humana Press Inc., Totowa, NJ

Table 1
Antiepileptic Drugs

Enzyme-Inducing AED	*Non-Enzyme-Inducing AED*	*Enzyme-Inhibiting AED*
Carbamazepine (CYP2C9, CYP3A, UGT)	Clonazepan	Oxacarbazepine (CYP2C9)
Gabapentin (UGT)	Ethosuximide	Topiramate (CYP2C19)
Oxacarbazepime (CYP3A4/5, UGT)	Felbamate	Valproic acid (CYP2C9)
Phenobarbital (CYP2C, CYP3A, UGT)	Lamotrigine	
Phenytoin (CYP2C, CYP3A)	Tiagabine	
Primidone (CYP 2C, CYP3A)	Valpoic acid	

Parentheses contain the cytochrome P450 families or the conjugating enzyme UDP-glucuronosylltransferase (UGT) induced or inhibited by the AED. PB, PHT, and primidone were the predominant agents used in the first half of the last century. Advent of CMZ and VPA took place in early 1960 and they are now considered first-line treatments for complexed partial seizures (CMZ) and generalized and partial seizures (VPA). New agents such as felbamate, gabapentin, lamotrigine, and tiagabine are either adjunctive or sole therapy *(28)*.

2. ANTIEPILEPTIC DRUGS

To date, there is no effective prophylaxis or cure for epilepsy, and the most effective mode of treatment for the disease is drug therapy. The clinically available AEDs inhibit seizures or sensory impairments, but therapy is essentially symptomatic and involves the long-term intake of one or more AEDs, particularly if several types of seizures occur. It is reported that complete control of seizures can be achieved in approx 50% of subjects and that an additional 25% can be significantly improved. The degree of success seems to be a function of seizure type, cause, and other factors *(1)*.

Drugs effective against the most common forms of epileptic seizures appear to work either by limiting the sustained repetitive firing of neurons by promoting the inactivity state of voltage-activated Na^+ channels (hydantoins such as phenytoin [PHT] or valproic acid [VPA]), or by increasing pre- or postsynaptic γ-aminobutyric (GABA)-mediated synaptic inhibition (barbiturates, benzodiazepines, gabapendine, tiagabine, tipiramate). Drugs effective against absence seizures limit activation of a voltage-activated Ca^{2+} channel known as T-current *(1)*.

Several AEDs have enzyme-inducing properties, most particularly on the hepatic mixed-function oxydase system of the cytochrome P450 2C and 3A families as well as on several conjugating enzymes (Table 1). This property leads to the induction of these enzymes, and several endogenous and exogenous compounds that are substrates for these cytochromes P450 have an increased clearance rate associated with either an acceleration of their normal metabolism or with the appearance of novel active or inactive metabolites.

AEDs, most particularly the classic AEDs (phenobarbital [PB], PHT, primidone) are often associated with side effects such as mild sedation, which may have an impact on physical activity and time spent outside. Moreover, AEDs are commonly associated with mild laboratory abnormalities that include hepatic enzyme elevation, mostly with hydantoins such as PHT, and mild elevation in ammonia associated with VPA. Serious, but rare, idiosyncratic side effects include aplastic anemia, hepatotoxicity and thrombocytopenia, which have mostly been reported with the classic AEDs. Subtle endocrine abnormalities, including variations in thyroid function test and subclinical effects on peripheral nerve conduction produced by PHT and carbamazepine (CMZ), have also been reported. Allergic reactions to AEDs are common but usually mild. On rare occasions, allergic reactions can progress to severe cutaneous disorders such as Stevens-Johnson syndrome and toxic epidermal necrolysis *(14)*.

As mentioned above, chronic AED therapy has been documented to lead to several side effects on calcium and vitamin D homeostasis, and on bone physiology. Unfortunately, the awareness among neurologists of the potential negative effects of chronic AED therapy on the calcium endocrine system has been reported to be low (41% of pediatric and 28% of adult neurologists routinely evaluate AED-treated patients for bone and mineral diseases), therefore precluding the beneficial effect of a prophylactic intervention to alleviate the negative metabolic effects of AED on calcium and vitamin D metabolism, and on bone *(2,15)*.

3. EFFECTS OF AED ON MINERAL METABOLISM

3.1. Clinical Observations

Several disorders related to calcium metabolism have been reported in AED-treated subjects, and recent evidence indicates that AED therapy is an independent risk factor for reduced bone mineral density (BMD) in epileptic patients *(16)*. The disorders may be mild, ranging from biochemical abnormalities to clinical evidence of bone disease *(4,6,7,10,17,18)*. As illustrated in Fig. 1, the abnormalities include hypocalcemia, hypophosphatemia, elevated serum (se) concentrations of parathyroid hormone (PTH), 1,25-dihydroxyvitamin D [$1,25(OH)_2D$], and alkaline phosphatase (AP), as well as reduced serum concentrations of 25-hydroxyvitamin D [25(OH)D] and 24,25-dihydroxyvitamin D [$24,25(OH)_2D$], decreased intestinal calcium absorption, and reduced urinary calcium excretion. Increased bone turnover is also present, as evidenced by alterations in serum and urinary indices of bone formation (bone-specific AP [BSAP], osteocalcin [OC], C-terminal extension peptide type 1 procollagen [PICP],) and bone resorption (cross-linked carboxy-terminal telopeptide of human type I collagen [ICTP], and total and free deoxypyridinoline). Reduced BMD with cortical bone loss is observed, as well as radiological evidence of rickets and histological evidence of osteoma-

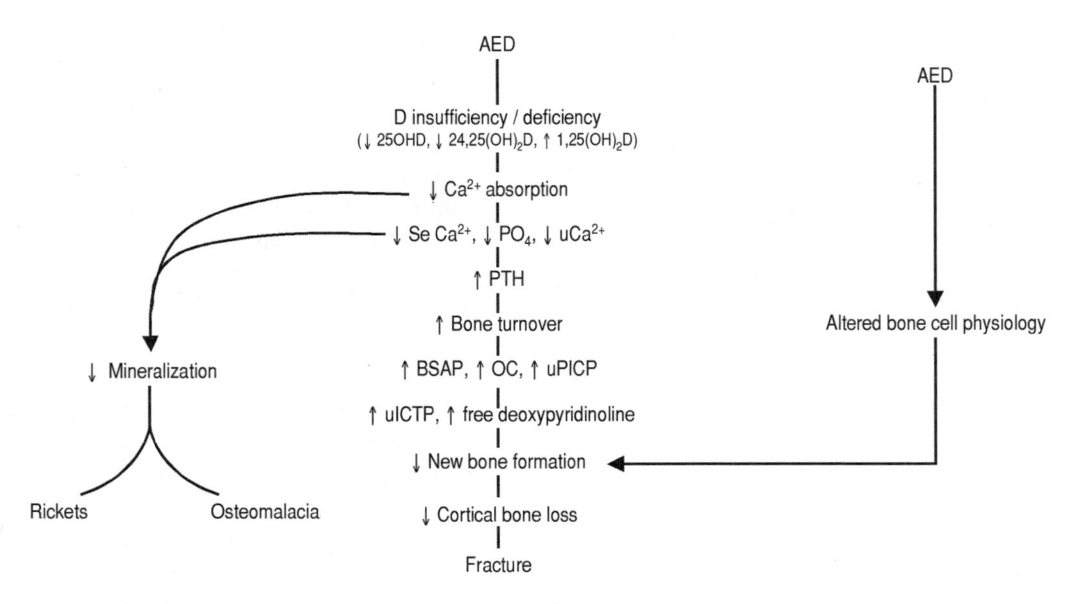

Fig. 1. Response of calcium and vitamin D endocrine status to AED-induced disturbances in vitamin D nutritional status and direct effect of AEDs on bone cell physiology. SE, serum; U, urine.

lacia indicating the presence of a bone mineralization defect *(19)*. The circulating concentrations of the vitamin D-binding protein are, however, unaffected by the chronic exposure to AED. These symptoms are very similar to those observed in vitamin D insufficiency and deficiency. Many etiological factors for these abnormalities have been evoked, such as epilepsy per se, reduced physical activity, inadequate intake of calcium and vitamin D, reduced exposure to ultraviolet light (UV), an effect of AED on vitamin D turnover, as well as a direct effect of AED on bone cell physiology *(20)*. Moreover, in epileptic subjects, but mainly in nonambulatory and institutionalized patients, the combination of lack of sun exposure, poor calcium and vitamin D intake, inadequate physical activity, and AED therapy has been proposed as a co-morbidity factors having a compounding negative effect on calcium, vitamin D and bone metabolism *(21)*.

3.2. Risk Factors

As summarized in Table 2, the most important risk factors for developing disturbances in the calcium endocrine system including reduced BMD and/or bone diseases in epileptic subjects have been identified as age, gender, multiplicity and type of AED (with patients taking enzyme-inducing AEDs tending to have lower BMD than those taking noninducing AEDs, although newer drugs have also been associated with low BMD), length of treatment, type of epilepsy, the severity of symptoms (generalized seizures being associated with more negative effects that

Table 2
Risk Factors for Developing Calcium Homeostasis Imbalances Associated With Chronic AED Therapy in Epileptic Subjects

Risk factors
Risk factors associated with the disease
Type of epilepsy (generalized seizure > focal seizure)
Duration of the disease
Length of treatment
Use of multiple AEDs
Type of AED (enzyme-inducing AED > noninducing AED)
Somatic risk factors
Genetic factors (i.e., mixed-function oxidase induction by AEDs, and AED-induced toxicity)
Gender (women > men)
Age
Physiological state (i.e., pregnancy)
Risk factors associated with nutritional status
Diet not meeting the recommended dietary requirements in several nutrients
Low vitamin D intake
Low calcium intake
Risk factors associated with lifestyle
Inadequate time spent outside, with low exposure to ultraviolet light
Low physical activity
Nonambulatory state
Living in an institution

focal seizures [20]), duration of the disease, decreased physical activity, and low vitamin D intake or exposure to ultraviolet light *(20,22)*. Institutionalized or nonambulatory patients exhibit the greatest risk of increased fractures, especially at the hip *(23–25)*. It is also important to mention the relationship between abnormalities in the calcium endocrine status and epilepsy, as low serum calcium concentrations may aggravate the seizure disorder, while patients are certainly at increased risk of fractures during epileptic seizures.

3.3. Neonates

Several case reports have been published indicating that children born of epileptic mothers on AED therapy exhibit signs indicative of AED toxicity shortly after birth. Indeed, it has been documented that AEDs cross the placenta and that they are found in the neonates' serum. The symptoms reported in these infants include drowsiness, mild jaundice, vomiting, hypotonia as well as hypertonia, tremors, hyperexcitability, and hypocalcemia. It is usually found, however, that following close monitoring, these symptoms disappear and that the neonates usually recover completely *(26)*.

Kuoppala et al. *(27)* also found, in a fœtus, parameters indicating the presence of disturbances in vitamin D and bone metabolism such as increased circulating AP, decreased serum 1,25$(OH)_2$D and 24,25$(OH)_2$D in the presence of normal 25(OH)D concentrations. On the other hand, in a study carried out in 22 women during the third trimester of pregnancy who were on chronic CMZ or PHT therapy and supplemented with 400 IU of vitamin D_3/d, Markestad et al. *(28)* found that the median serum 25(OH)D and 1,25$(OH)_2$D concentrations were lower in patients while that of 24,25$(OH)_2$D was higher, leading to a higher 24,25$(OH)_2$D/25(OH)D ratio in epileptic subjects than in the control group. At delivery, it was found that the serum cord-blood concentrations were similar to those of the mother but that the median serum AP concentrations were higher in cord blood obtained from epileptic subjects than in that obtained from normal healthy women.

Collectively, these observations suggest that AED therapy in pregnant, normally functioning epileptic women, negatively affect the developing fœtus, leading to signs of AED toxicity, disturbances in vitamin D metabolism and calcium homeostasis, which are reflected directly in the newborn child. Unfortunately, no long-term follow-up of the calcium metabolism in these children has been reported. It is expected, however, that in a manner similar to the symptoms of AED toxicity observed in some neonates *(26)*, a normal recovery of vitamin D and calcium metabolism should be expected. To date, the effect of breast feeding by AED-treated mothers on the calcium and vitamin D metabolism of their infants has not yet been evaluated.

3.4. Children and Adolescents

The full impact of AED therapy on the calcium endocrine system and on bone accrual in children and adolescents is presently unknown. Although one study reported no sign of vitamin D metabolism alterations in ambulatory pubertal children on AED therapy for more than 6 yr *(29)*, most studies carried out in epileptic children chronically treated with AED have found some abnormalities in mineral, vitamin D, or bone metabolism, including impairment in bone development.

Indeed, it is generally reported that, in a manner similar to the AED-treated adult population, 50% of children and adolescents have low circulating 25(OH)D concentrations, which do not seem, however, to be correlated with BMD *(20)*. As an example, Erbayat Altay et al. *(30)* found that in CMZ-treated children, serum calcium concentrations were subnormal, those of AP were increased, while BMD remained normal. These findings are in line with the observations of Rieger-Wettengl et al. *(31)* indicating that bone turnover is increased in children treated with AEDs. Urinary calcium has also been found to be lower in patients treated with CMZ than in those receiving VPA *(30)*. In addition, Onoe et al. *(32)* studied 101 outpatient children mainly on multiple AED therapy except for 27 who were on monotherapy. A significant number of these patients were found to have delayed bone development of more than 2 yr when compared to either handicapped or normal control children. Moreover, those with delayed bone development were found to have indices of vitamin D insufficiency such as decreased serum 25(OH)D and increased 1,25$(OH)_2$D concentrations. It was also found that physical activity and duration of AED therapy were negatively correlated with bone development.

3.5. Adults

Signs of vitamin D insufficiency as well as signs of vitamin D deficiency with evidence of osteomalacia, secondary hyperparathyroidism, increased bone turnover, and decreased BMD at the femoral neck, Ward's triangle, calcaneus, and phalanges has often been reported in AED-treated male and female adult subjects *(11–13,19,20,33–39)*. Women, however, seem to be more at risk than men. Indeed, Valimaki et al. *(40)* reported that both male and female patients were found to have low serum 25(OH)D and 1,25$(OH)_2$D concentrations independent of therapy. All patients were also found to have hypocalcemia, but women were also found to have hypocalciuria. Moreover, despite increases in men, intact PTH concentrations and parameters of bone turnover were significantly higher in women than in men.

The influence of associated risk factors or co-morbidity factors on the severity of symptoms associated with AED therapy is also well illustrated by the study of Wark et al. *(40)*, who found that ambulatory mostly male nonepileptic Australian subjects treated with PHT (as an antiarrhythmic agent) for a period of 2 yr had serum 25(OH)D concentrations similar to those of a control group not receiving PHT. Although unchanged by treatment, serum calcium and inorganic phosphate, nevertheless, increased in the post-PHT period, an observation indicative of a subtle effect of the drug on mineral metabolism despite the short-term therapy and the absence of co-morbidity factors.

4. MECHANISMS OF ACTION

Several mechanisms have been proposed to explain the negative skeletal effects of chronic AED therapy in epileptic subjects. A consensus generally exists indicating that long-term AED exposure increases the risk of osteopenia and even of osteomalacia. One of the main etiological factors responsible for these skeletal disturbances is attributed to alterations in vitamin D metabolism, which leads progressively to vitamin D insufficiency and, if not corrected, to vitamin D deficiency in several cases. The alterations in vitamin D metabolism are mainly a consequence of hepatic enzyme induction by drugs such as PB, PHT, and CMZ *(41)*, which lead to an increase in vitamin D turnover including vitamin D catabolism. In addition, some AEDs have now been shown to have a direct effect on bone cell metabolism.

4.1. Vitamin D and Calcium Metabolism

Schaeffer et al. *(42)* have shown that PB does not interfere with vitamin D_3 absorption. In addition, it has been clearly demonstrated that exposure to PB leading to a significant induction of hepatic mixed-function oxidases does not interfere with the hepatic uptake of vitamin D_3 *(43)*. However, most studies indicate that the metabolism of vitamin D is modified by several AEDs and, in experimental animals, the hepatic conversion of vitamin D_3 is increased to give rise to metabolites similar to those found in the serum of AED-treated patients *(41)*.

4.1.1. Hepatic Clearance of Vitamin D

There is no doubt that the hepatic clearance of vitamin D_3 (with production of both active and inactive compounds) is increased by PB. This is well illustrated by the study of Gascon-Barré et al. *(43),* in which short-term PB exposure was found to induce a twofold increase in the hepatic clearance of vitamin D_3. The increased vitamin D_3 clearance was found to be secondary to a three- to fivefold increase in the hepatic $25(OH)D_3$ production rate over that observed during both the pre- or post-PB exposure period. Moreover, it was also found that under PB treatment, the hepatic production of $25(OH)D_3$ was positively correlated to the increase in the hepatic endo/xenobiotic metabolizing enzyme capacity, while no correlation was observed in the absence of PB. These observations suggest that PB induces novel enzyme(s) exhibiting C-25 hydroxylation capacity on vitamin D_3 with, as a consequence, an increased utilization of the substrate vitamin D_3 and hence the presence of increased vitamin D_3 clearance. Moreover, the enzyme induction potential of PB was also found to be heterogeneous and positively correlated with its effect of vitamin D_3 metabolism, indicating that genetic factors are most likely associated with the induction of the mixed-function oxidases, a phenomenon well documented in human subjects *(44)*. These genetic factors may therefore influence the susceptibility of individuals to the effects of AED on vitamin D and bone metabolism.

Silver and co-workers *(45)* also found that PB leads to an increased biliary excretion of mostly water-soluble derivatives of vitamin D_3, including glucuronide congugates, which is also correlated to the choleretic effect induced by PB *(46)*. The increased biliary output of vitamin D_3-derived compounds is therefore an added contributing factor in the observed increased hepatic clearance of the vitamin.

4.1.2. Biphasic Efffects of AED on Vitamin D_3 Metabolism and Induction of a Progressive State of Vitamin D Insufficiency

The increased hepatic clearance of vitamin D_3 associated with enzyme-inducing AEDs may also explain data reported in several studies in which a biphasic effect of these drugs on the conversion of vitamin D_3 into $25(OH)D_3$ was observed. Indeed, an early increase in $25(OH)D_3$, which is then followed by a decrease in its formation, hepatic release, or circulating concentrations, has been reported by several investigators *(47–50)*. This is well illustrated by the study of Hahn et al. *(49),* who reported that during the first 5 d of PB administration in the rat, serum $25(OH)D_3$ and $24,25(OH)_2D_3$ concentrations were significantly increased, while those of $1,25(OH)_2D_3$ remained unchanged. However, after 21 d of treatment, both $25(OH)D_3$ and $24,25(OH)_2D_3$ concentrations and intestinal calcium absorption were significantly decreased, while $1,25(OH)_2D_3$ concentrations were increased by 80%. These observations strongly suggest a progressive PB-mediated state of vitamin D_3 insufficiency despite constant intake of the vitamin. In fact, these authors reported that the effect of PB on vitamin D_3 metabolites was mediated by an accelerated disappearance of vitamin D_3 with values at the 5- and 21-d time points of 50% and 27% of control values, respectively. Others have also found that the tissue retention of vitamin D_3 was decreased by enzyme-inducing AEDs, indicating a increased systemic clearance of the vitamin *(45)*. Similarly, Baran et al. *(50)* found in rat isolated liver perfusion systems that the total hepatic $25(OH)D_3$ output was not perturbed after 4 wk of PB administration but that the efficiency of $25(OH)D_3$ production was decreased after 8 wk of PB exposure in vivo.

Moreover, in a case study in which the serum concentrations of 25(OH)D were evaluated in relation to the circulating concentration of PHT, it was found that at low PHT concentrations 25(OH)D increased, while over time and with increased PHT concentrations, a very significant decrease in serum 25(OH)D levels was observed. Serum $1,25(OH)_2D$ concentrations, on the other hand, were found to be independent of both the dose and serum concentrations of PHT *(51)*. Davie et al. *(52)* also found that the dose of PB and PHT were each inversely related to the 25(OH)D plasma concentrations in institutionalized subjects.

4.1.3. Hepatic Subcellular Localization of Vitamin D_3

The effect of AEDs on the hepatic subcellular localization of vitamin D_3 or its metabolites is still a subject of controversy. Studies have reported that PB mediates an increase in the hepatic radioactivity associated with ^3H-vitamin D_3 in the smooth endoplasmic reticulum, which leads to an increase in the formation of polar metabolites of the vitamin while the radioactivity associated with the mitochondria is decreased *(53)*. These observations confirm that an increased conversion of vitamin D_3 to polar metabolites occurs in the smooth endoplasmic reticulum, the site of the mixed-function oxidases induced by enzyme-inducing AEDs (PB, PHT, and CMZ) in both human and animals *(41)*. On the other hand, Hahn et al. *(49)* have reported that 5 d after the initiation of PB administration in a rat study, hepatic microsomal vitamin D_3 25-hydroxylase activity was not increased while an increased in mitochondrial vitamin D_3 25-hydroxylase activity was observed throughout the 21-d study period. However, evaluation of the response of the rat mitochondrial vitamin D_3 25-hydroxylase transcript *CYP27A* to several cytochrome P450 inducers revealed that the latter is induced by dexamethasone and β-naphatoflavone in liver, and by only dexamethasone in intestine, but that PB was not an inducer of the enzyme *(54)*. It is therefore possible that PB could induce enzyme systems active on the C-25 hydroxylation of vitamin D_3 other than the presently known vitamin D_3 25-hydroxylase CYP27A.

On the other hand, Tomita et al. *(55)* found that PHT and VPA inhibited both the microsomal and the mitochondrial vitamin D_3 25-hydroxylases in rabit livers but not that of the mitochondrial 1α-hydroxylase.

4.1.4. Biotransformation of 25(OH)D

Cinti et al. *(56)* have also found that the microsomal mixed-function oxidase system plays a role in the biotransformation of $25(OH)D_3$ and that $25(OH)D_3$ inhibits the N-demethylation of aminopyrine (an enzyme activity catalyzed by mixed-function oxidases) by tightly binding to cytochrome P450. Indeed, these authors found that $25(OH)D_3$ is a high-affinity substrate for hepatic cytochrome P450, suggesting that the liver microsomal mixed-function oxidases are involved in the biotransformation of $25(OH)D_3$. Indeed, these authors reported that, compared to controls not exposed to PB, PB administration led to a doubling of enzyme activity and to a 40% faster time-dependent drop in in vivo $25(OH)D_3$ concentrations *(57)*.

4.1.5. Increased Vitamin D Turnover and Its Physiological Consequences

The accelerated metabolic turnover of vitamin D_3 compounds is also well illustrated by several studies. Indeed, Norman et al. *(58)* have reported that administration of PB to chicks for a period of 7 d results in reduced ability of single moderate doses of vitamin D_3 or $25(OH)D_3$ to stimulate intestinal calcium absorption and bone resorption. These effects were overcome by increasing the dose of vitamin D_3 or $25(OH)D_3$, but PB did not enhance the response to a physiological dose of $1,25(OH)_2D_3$, indicating that the metabolism of $1,25(OH)_2D_3$ is not directly affected by the drug. In addition, the increased turnover rate of vitamin D_3 associated with a decrease in its endogenous response is well illustrated by the studies of Lukaszkiewicz et al. *(59)* and Gascon-Barré et al. *(48)*, who have found that PB leads to a significant protection against vitamin D_3 toxicity in both humans and animals as well as to an inhibition of the healing of rickets following the administration of a physiological dose of vitamin D_3 in the rat *(60)*.

4.1.6. Summary

Figure 2 summarizes the main hepatic effects of enzyme-inducing AEDs on vitamin D homeostasis. Both endogenously formed vitamin D_3, and vitamin D_2 and vitamin D_3 of exogenous origin will be normally taken up by the liver and transformed into 25(OH)D by the mitochondrial vitamin D 25-hydroxylase CYP27A. However, enzyme-inducing AEDs, by increasing several cytochromes P450 and conjugating enzymes, will promote accelerated vitamin D and 25(OH)D turnover with the formation of active and inactive vitamin D metabolites. They will also promote the biliary excretion of vitamin D metabolites and hence induce accelerated hepatic and systemic clearance of the vitamin. This accelerated clearance will promote all the normal physiological adaptive mechanisms in response to a progressive state of vitamin D insufficiency such as increases in PTH and $1,25(OH)_2D$ concentrations. In turn, the increased $1,25(OH)_2D$ concentrations may lead to an inhibition of the mitochondrial vitamin D_3 25-hydroxylase, CYP27A, as already evidenced in both humans and animals studies *(54,61–63)*, therefore compounding the

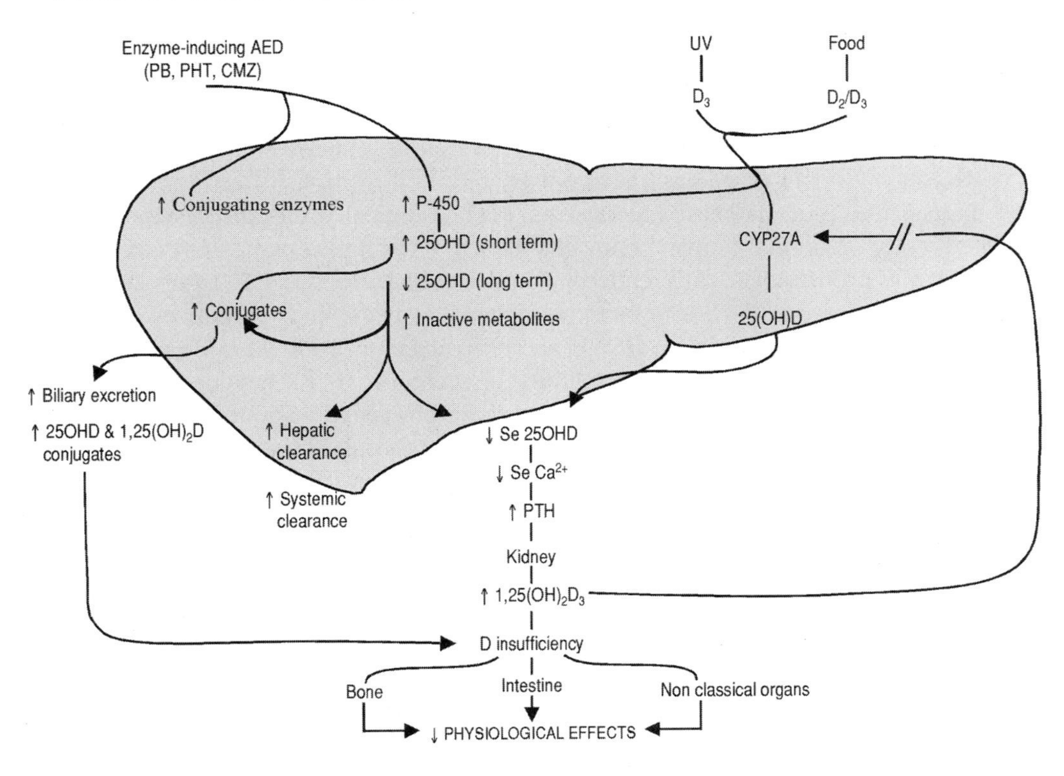

Fig. 2. Metabolic effects of chronic exposure to AEDs on the hepatic handling of vitamin D and of their physiological consequences.

effect of AEDs on the accelerated turnover of vitamin D. Modest pharmacological doses of vitamin D_3 or vitamin D_2 have been shown in many studies to overcome the deleterious effects of enzyme-inducing AEDs on vitamin D metabolism. Vitamin D supplementation, therefore, exerts its beneficial effects by counterbalancing the heightened vitamin D requirement induced by enzyme-inducing AEDs that lead to an increase in the hepatic and systemic clearance of the vitamin.

4.2. Intestinal Calcium Absorption

Several studies suggest that AEDs, by either accelerating vitamin D metabolism or via a direct effect on the intestinal mucosa, directly inhibit intestinal mineral absorption. Koch et al. *(64)* reported that, in the rat, PHT, but not PB, decreases $^{45}Ca^{2+}$ uptake, while Harrison *(65)* found that the response to vitamin D_2 or vitamin D_3 or $25(OH)D_3$ following PB and PHT led to decreased intestinal calcium absorption in everted rat intestine. In this study, serum calcium concentrations were not, however, perturbed by AED exposure.

4.3. Bone Cell Physiology

Despite the fact that the direct effect of AEDs on bone cells still remains controversial, studies are now emerging indicating that AEDs may have a direct effect on bone cell physiology including cell proliferation, indices of bone formation, responses to known physiological agonists, and modifications of calcium fluxes.

Indeed, Lau et al. *(66)* have reported that PHT at clinically relevant concentrations acts directly on normal human bone cells in vitro to stimulate markers of osteogenesis such as proliferation, differentiation, and osteoblastic activity. In addition, these authors reported that PHT also increased serum concentrations of indices of bone formation markers such as OC, BSAP, and procollagen peptide. Vernillo et al. *(67),* on the contrary, observed that PHT inhibits the secretion of OC in a dose-dependent manner in ROS 17.28 cells, while Onodera et al. *(68)* found an increased osteoclast number per area of bone surface in the secondary trabecular bone of rat tibia.

A direct effect of PHT and CMZ on osteoblast proliferation has also been reported by Feldkamp et al. *(69),* but these authors observed an increased proliferation with low dose of PHT while PHT and CMZ led to a decrease in cell growth at concentrations similar to those observed therapeutically. Auszmann et al. *(70),* on the other hand, found that bone cells from 1-d-old mice when exposed to PHT have a reduced PTH-stimulated cAMP response, and Dziak et al. *(71)* reported that, in human osteoblastic osteosarcoma cells SaOS-2 and normal rat osteoblastic cells, PHT decreased basal calcium uptake. In addition, PTH and prostaglandin E_2 increased calcium uptake in SaOS-2 cells but PHT abrogated this increase, indicating, that PHT mediates a decreased response to normal physiological agonists leading to a reduction in osteoblastic calcium influx.

5. CLINICAL EVALUATION OF SUBJECTS ON CHRONIC AED THERAPY

Heller and Sakhaee *(15)* propose that all individuals on chronic AED therapy should be monitored for parameters indicative of disturbances in vitamin D and calcium metabolism. Table 3 presents an adapted summary of their recommendations for baseline evaluation as well as for follow-up evaluation on any abnormal values or symptoms indicative of possible disturbances in the calcium endocrine system including abnormal serum ionized calcium, phosphorus, and AP as well as BMD in adults. The presence of muscle pain, weakness, or tenderness should also be further investigated. Follow-up evaluations should include specific markers of the vitamin D nutritional [25(OH)D] and endocrine [1,25$(OH)_2$D] status, secondary hyperparathyroidism, bone turnover, and a reevaluation of BMD. Referral to a specialist is recommended if abnormal values are found at follow-up.

6. VITAMIN D AND CALCIUM SUPPLEMENTATION

Circulating 25(OH)D concentrations are considered the best marker of the vitamin D nutritional status. Although a debate still exists as to the optimal

Table 3
Guidelines for Evaluation and Treatment

	Baseline Evaluation		*Follow-Up on Abnormal Values, or Patients With Specific Symptoms*[a]		
Group	*Serum*	*Bone*	*Serum*	*Urinary*	*Bone*
Adults	Ca^{2+}	BMD	Ca^{2+}	24-h calcium excretion	BMD
	Phosphorus		Phosphorus	ICTP	
	AP		AP or BSAP	PICP	
			25(OH)D	Free deoxypyridindine	
			$1,25(OH)_2D$		
			iPTH (intact)[b]		
Children	Ca^{2+}		Ca^{2+}	24-h calcium	
	Phosphorus		Phosphorus		
	AP		ALP	PICP	
	PTH (intact)		25(OH)D	ICTP	
			$1,25(OH)_2D$	Free deoxypyridindine	
			iPTH (intact)		

[a] Muscle weakness, pain, tenderness.

[b] Primary hyperparathyroidism should be eliminated.

Source: Adapted from Heller and Sakhaee *(15)*.

serum concentrations of 25(OH)D to support all the endocrine and cellular effects of the vitamin *(72),* recent studies have clearly illustrated that vitamin D insufficiency still exists in the general healthy population of industrial countries *(73–78)*. As mentioned above, chronic AED therapy is an added risk factor for vitamin D insufficiency.

Many studies have shown that reversal of radiographic and biochemical abnormalities and improvement of bone mass occurs in AED-treated subjects, including adolescents and children, given vitamin D and calcium *(5,11–13,38,39,79–83),* although Berry et al. *(84)* reported that despite a normal response to vitamin D supplementation leading to supranormal concentrations of 25(OH)D, hypocalcemia still persisted. These observations led the authors to conclude that low serum calcium in patients on AED therapy is not related to vitamin D deficiency. However, hypocalcemia may persist in the presence of normal or high 25(OH)D concentrations if active mineralization is present and may rather reflect an increased nutritional requirement for calcium during the active bone matrix mineralization period of the vitamin D insufficiency-recovery phase. On the other hand, persistent hypocalcemia in the face of a normal vitamin D status may also be due a defect in bone calcium mobilization induced by the reported direct effect of AEDs on bone cell physiology.

Improvement in the bone response to vitamin D administration is, however, well illustrated by the case study of a 50-year-old patient on a high dose of PB and a small dose of PHT, and a low intake of calcium, vitamin D, little exposure to sunlight, and low physical activity reported by Campbell et al. *(35)*. Indeed, these authors reported the presence of a "brown tumor" of hyperparathyroidism, hence revealing extensive secondary hyperparathyroidism, and osteomalacia. The "brown tumor" healed following vitamin D_3 and calcium supplementation as well as following a reduction in PB and an increase in PHT dosage. Similarly, in an institutionalized patient, limited exposure to UV light or oral administration of vitamin D restored plasma 25(OH)D to normal levels and healed osteomalacia in a subject with tuberous sclerosis *(52)*.

Although some difference in the response to vitamin D_3 and vitamin D_2 have been reported *(85,86)*, the nutritional requirements of vitamin D have been evaluated to range from 10 to 100 μg/d (400 to 4000 IU) in children, adolescents, or adults individuals on chronic AED therapy *(52,79–83)*. These observations indicate that adequate vitamin D intake in AED-treated subjects is essential independent of treatment, age, sex, and activity status to prevent the deleterious effects of these drugs on the vitamin D endocrine system and on bone.

7. CONCLUSION

The reduced BMD observed in a majority of subjects treated chronically with AEDs is most likely the result of multiple factors. The body of work accumulated over the past 35 yr has clearly identified the accelerated hepatic metabolism of vitamin D and 25(OH)D by enzyme-inducing AEDs as a major contributing factor. This accelerated vitamin D and 25(OH)D turnover rate induces an increased nutritional demand for the vitamin, which, if not met, will lead progressively to vitamin D insufficiency and then, over time, to vitamin D deficiency. By contrast, a direct effect of these drugs on $1,25(OH)_2D_3$ metabolism has not been identified, and the elevated concentrations of $1,25(OH)_2D_3$ often observed in chronically treated subjects should be interpreted as resulting directly from the normal adaptation of the vitamin D endocrine system to the AED-induced declining vitamin D nutritional status.

A compounding factor of AEDs on vitamin D and calcium metabolism is the now-emerging realization that some AEDs may also exhibit a direct effect on bone cell physiology. In addition to the normal physiological response of the bone to perturbed vitamin D and calcium homeostasis evidenced by an often-reported state of secondary hyperparathyroidism, the direct effects of AED on bone may negatively amplify bone turnover rate and accelerate bone loss.

Modest vitamin D supplementation or exposure to ultraviolet light has been shown in most studies to greatly improve the vitamin D nutritional status of AED-treated subjects and, in many cases, to improve bone health. By improving their vitamin D nutritional status, these individuals would certainly also benefit from the cellular effects of vitamin D on the immune system and the vitamin D-mediated protection against cancer-inducing agents, an aspect of AED exposure not yet explored in epileptic individuals.

REFERENCES

1. McNamara JO. Drugs effective in the therapy of the epilepsies. In: Hardman JG, Limbird LE, eds. Goodman & Gilman's The Pharmacological Basis of Therapeutics. McGraw-Hill, New York, NY. 2001, pp. 521–547.
2. Valmadrid C, Voorhees C, Litt B, Schneyer CR. Practice patterns of neurologists regarding bone and mineral effects of antiepileptic drug therapy. Arch Neurol 2001; 58:1369–1374.
3. Schmid F. Osteophthien bei antiepileptischer Dauerbehandlung. Fortschr Med 1967; 38:381–382.
4. Lifshitz F, Maclaren NK. Vitamin D-dependent rickets in institutionalized, mentally retarded children receiving long-term anticonvulsant therapy. I. A survey of 288 patients. J Pediatr 1973; 83:612–620.
5. Hahn TJ, Hendin BA, Scharp CR, Boisseau VC, Haddad JG Jr. Serum 25-hydroxycalciferol levels and bone mass in children on chronic anticonvulsant therapy. N Engl J Med 1975; 292:550–554.
6. Tolman KG, Jubiz W, Sannella JJ, et al. Osteomalacia associated with anticonvulsant drug therapy in mentally retarded children. Pediatrics 1975; 56:45–50.
7. Weinstein RS, Bryce GF, Sappington LJ, King DW, Gallagher BB. Decreased serum ionized calcium and normal vitamin D metabolite levels with anticonvulsant drug treatment. J Clin Endocrinol Metab 1984; 58:1003–1009.
8. Sotaniemi EA, Hakkarainen HK, Puranen JA, Lahti RO. Radiologic bone changes and hypocalcemia with anticonvulsant therapy in epilepsy. Ann Intern Med 1972; 77:389–394.
9. Christiansen C, Rødbro P, Lund M. Incidence of anticonvulsant osteomalacia and effect of vitamin D: controlled therapeutic trial. Br Med J 1973; 4:695–701.
10. Sheth RD, Wesolowski CA, Jacob JC, et al. Effect of carbamazepine and valproate on bone mineral density. J Pediatr 1995; 127:256–262.
11. Liakakos D, Papadopoulos Z, Vlachos P, Boviatsi E, Varonos DD. Serum alkaline phosphatase and urinary hydroxyproline values in children receiving phenobarbital with and without vitamin D. J Pediatr 1975; 87:291–296.
12. Sherk HH, Cruz M, Stambaugh J. Vitamin D prophylaxis and the lowered incidence of fractures in anticonvulsant rickets and osteomalacia. Clin Orthoped 1977; 129:251–257.
13. Collins N, Maher J, Cole M, Baker M, Callaghan N. A prospective study to evaluate the dose of vitamin D required to correct low 25-hydroxyvitamin D levels, calcium, and alkaline phosphatase in patients at risk of developing antiepileptic drug-induced osteomalacia. Q J Med 1991; 78:113–122.
14. Harden CL. Therapeutic safety monitoring: what to look for and when to look for it. Epilepsia 2000; 41:S37–S44.
15. Heller JH, Sakhaee K. Anticonvulsant-induced bone disease: a plea for monitoring and treatment. Arch Neurol 2001; 58:1352–1353.
16. Stephen LJ, McLellan AR, Harrison JH, et al. Bone density and antiepileptic drugs: a case-controlled study. Seizure 1999; 8:339–342.
17. Morijiri Y, Sato T. Factors causing rickets in institutionalised handicapped children on anticonvulsant therapy. Arch Dis Child 1981; 56:446–449.
18. Bear MT, Kozlowski BW, Blyler EM, Trahms CM, Taylor ML, Hogan MP. Vitamin D, calcium, and bone status in children with developmental delay in relation to anticonvulsant use and ambulatory status. Am J Clin Nutr 1997; 65:1042–1051.

19. Hoikka V, Alhava EM, Karjalainen P, et al. Carbamazepine and bone mineral metabolism. Acta Neurol Scand 1984; 70:77–80.
20. Farhat G, Yamout B, Mikati MA, Demirjian S, Sawaya R, Fuleihan GEH. Effect of antiepileptic drugs on bone density in ambulatory patients. Neurology 2002; 58:1348–1353.
21. Lamberg-Allardt C, Wilska M, Saraste K-L, Grönlund T. Vitamin D status of ambulatory and nonambulatory mentally retarded children with and without carbamazepine treatment. Ann Nutr Metab 1990; 34:216–220.
22. Gough H, Goggin T, Bissessar A, Baker M, Crowley M, Callaghan N. A comparative study of the relative influence of different anticonvulsant drugs, UV exposure and diet on vitamin D and calcium metabolism in out-patients with epilepsy. Q J Med 1986; 59:569–577.
23. Nilsson OS, Lindholm TS, Elmstedt E, Lindback A, Lindholm TC. Fracture incidence and bone disease in epileptics receiving long-term anticonvulsant drug treatment. Arch Orthoped Trauma Surg 1986; 105:146–149.
24. Pavlakis SG, Chusid RL, Roye DP, Nordli DR. Valproate therapy: predisposition to bone fracture? Pediatr Neurol 1998; 19:143–144.
25. Harrington MG, Hodkinson HM. Anticonvulsant drugs and bone disease in the elderly. J R Soc Med 1987; 80:425–427.
26. Kayemba KS, Beust M, Aboulghit H, Voisin M, Mourtada A. Carbamazepine and vigabatrin in epileptic pregnant woman and side effects in the newborn infant. Arch Pediatr 1997; 4:975–978.
27. Kuoppala T, Tuimala R, Koskinen T. The effects of long-term anticonvulsant drug therapy on vitamin D metabolites and mineral homeostasis in pregnant epileptic women and their newborns. Ann Chir Gynaecol 1985; 197:37–41.
28. Loiseau PJ. Clinical experience with new antiepileptic drugs: antiepileptic drugs in Europe. Epilepsia 1999; 40:S3–S8.
29. Ala-Houhala M, Korpela R, Koivikko M, Koskinen T, Koskinen M, Koivula T. Long-term anticonvulsant therapy and vitamin D metabolism in ambulatory pubertal children. Neuropediatrics 1986; 17:212–216.
30. Erbayat AE, Serdaroglu A, Tumer L, Gucuyener K, Hasanoglu A. Evaluation of bone mineral metabolism in children receiving carbamazepine and valproic acid. J Pediatr Endicrinol Metab 2000; 13:933–939.
31. Rieger-Wettengl G, Tutlewski B, Stabrey A, et al. Analysis of the musculoskeletal system in children and adolescents receiving anticonvulsant monotherapy with valproic acid or carbamazepine. Pediatrics 2001; 108:E107.
32. Onoe S, Mimaki T, Seino Y, Yamaoka K, Yabuuchi H, Yamashita G. Delayed bone development in epileptic children assessed by microdensitometer. Dev Pharmacol Ther 1988; 11:24–31.
33. Pluskiewicz W, Nowakowska J. Bone status after long-term anticonvulsant therapy in epileptic patients: evaluation using quantitative ultrasound of calcaneus and phalanges. Ultrasound Med Biol 1997; 23:553–558.
34. Tjellesen L, Christiansen C. Serum vitamin D metabolites in epileptic patients treated with 2 different anti-convulsants. Acta Neurol Scand 1982; 66:335–341.
35. Campbell JE, Tam CS, Sheppard RH. "Brown tumor" of hyperparathyroidism induced with anticonvulsant medication. J Can Assoc Radiol 1977; 28:73–76.
36. Telci A, Cakatay U, Kurt BB, et al. Changes in bone turnover and deoxypyridinoline levels in epileptic patients. Clin Chem Lab Med 2000; 38:47–50.
37. Okesina AB, Donaldson D, Lascelles PT. Isoenzymes of alkaline phosphatase in epileptic patients receiving carbamazepine monotherapy. J Clin Pathol 1991; 44:480–482.
38. Hunt PA, Wu-Chen ML, Handal NJ, et al. Bone disease induced by anticonvulsant therapy and treatment with calcitriol (1,25-dihydroxyvitamin D_3). Am J Dis Child 1986; 140:715–718.
39. Fischer MH, Adkins WNJr, Liebl BH, VanCalcar SC, Marlett JA. Bone status in nonambulant, epileptic, institutionalized youth. Improvement with vitamin D therapy. Clin Pediatr 1988; 10:499–505.

40. Välimäki MJ, Tiihonen M, Laitinen K, et al. Bone mineral density measured by dual-energy x-ray absorptiometry and novel markers of bone formation and resorption in patients on antiepileptic drugs. J Bone Mineral Res 1994; 9:631–637.
41. Hahn TJ, Birge SJ, Scharp CR, Avioli LV. Phenobarbital-induced alterations in vitamin D metabolism. J Clin Invest 1972; 51:741–748.
42. Schaefer K, Kraft D, von Herrath D, Opitz A. Intestinal absorption of vitamin D_3 in epileptic patients and phenobarbital-treated rats. Epilepsia 1972; 13:509–519.
43. Gascon-Barré M, Vallières S, Huet PM. Influence of phenobarbital on the hepatic handling of [^{3}H]vitamin D_3 in the dog. Am J Physiol 1986; 251:G627–G635.
44. Honkakoski P, Negishi M. Regulation of cytochrome P450 *(CYP)* genes by nuclear receptors. Biochem J 2000; 347:321–337.
45. Silver J, Neale G, Thompson GR. Effect of phenobarbitone treatment on vitamin D metabolism in mammals. Clin Sci Mol Med 1974; 46:433–448.
46. Gascon-Barré M, Glorieux FH. The effect of phenobarbital (PB) treatment on the biliary excretion of cholecalciferol (D_3) and 25-hydroxycholecalciferol (25(OH)D_3) in the D-repleted rat. In: Norman AW, Schaefer K, Coburn JW, et al., eds. Vitamin D, Biochemical, Chemical and Clinical Aspects Related to Calcium Metabolism. de Gruyter, Berlin, New York, 1977, pp. 781–783.
47. von Herrath D, Kraft D, Schaefer K, Koeppe P. Influence of phenobarbital and diphenylhydantoin on vitamin D metabolism and calcium retention in rats. Res Exp Med 1972; 158:194–204.
48. Gascon-Barré M, Côté MG. Effects of phenobarbital and diphenylhydantoin on acute vitamin D_3 toxicity in the rat. Toxicol Appl Pharmacol 1978; 43:125–135.
49. Hahn TJ, Halstead LR. Sequential changes in mineral metabolism and serum vitamin D metabolite concentrations produced by phenobarbital administration in the rat. Calcif Tissue Int 1983; 35:376–382.
50. Baran DT. Effect of phenobarbital treatment on metabolism of vitamin D by rat liver. Am J Physiol 1983; 245:E55–E59.
51. Gascon-Barré M, Villeneuve JP, Lebrun LH. Effect of increasing doses of phenytoin on the plasma 25-hydroxyvitamin D and 1,25-dihydroxyvitamin D concentrations. J Am Coll Nutr 1984; 3:45–50.
52. Davie MW, Emberson CE, Lawson DE, et al. Low plasma 25-hydroxyvitamin D and serum calcium levels in institutionalized epileptic subjects: associated risk factors, consequences and response to treatment with vitamin D. Q J Med 1983; 205:79–91.
53. Hahn TJ, Scharp CR, Avioli LV. Effect of phenobarbital administration on the subcellular distribution of vitamin D_3-^{3}H in rat liver. Endocrinology 1974; 94:1489–1495.
54. Theodoropoulos C, Demers C, Petit JL, Gascon-Barré M. High sensitivity of the rat hepatic vitamin D_3-25 hydroxylase *CYP27A* to 1,25-dihydroxyvitamin D_3 administration. Am J Physiol 2003; 284:E138–E147.
55. Tomita S, Ohnishi JI, Nakano M, Ichikawa Y. The effects of anticonvulsant drugs on vitamin D_3-activating cytochrome P-450-linked monooxygenase systems. J Steroid Biochem Mol Biol 1991; 39:479–485.
56. Cinti DL, Golub EE, Bronner F. 25-hydroxycholecalciferol: high affinity substrate for hepatic cytochrome P-450. Biochem Biophys Res Commun 1976; 72:546–553.
57. Cinti DL. 25-hydroxycholecalciferol: high affinity substrate for hepatic cytochrome P-450. Adv Exp Med Biol 1977; 81:441–453.
58. Norman AW, Bayless JD, Tsai HC. Biologic effects of short-term phenobarbital treatment on the response to vitamin D and its metabolites in the chick. Biochem Pharmacol 1976; 25:163–168.
59. Lukaszkiewicz J, Proszynska K, Lorenc RS, Ludwiczak H. Hepatic microsomal enzyme induction: treatment of vitamin D poisoning in a 7 month old baby. Br Med J 1987; 295:1173.
60. Gascon-Barré M, Côté MG. Influence of phenobarbital and diphenylhydantoin on the healing of rickets in the rat. Calcif Tissue Res 1978; 25:93–97.
61. Baran DT, Milne ML. 1,25 dihydroxyvitamin D-induced inhibition of ^{3}H-25 hydroxyvitamin D production by the rachitic rat liver *in vitro*. Calcif Tissue Int 1983; 35:461–464.

62. Bell NH, Shaw S, Turner RT. Evidence that 1,25-dihydroxyvitamin D_3 inhibits the hepatic production of 25-hydroxyvitamin D in man. J Clin Invest 1984; 74:1540–1544.
63. Theodoropoulos C, Demers C, Mirshahi A, Gascon-Barré M. 1,25-dihydroxyvitamin D_3 down-regulates the rat intestinal vitamin D_3-25-hydroxylase *CYP27A*. Am J Physiol 2001; 281:E315–E325.
64. Koch HU, Kraft D, von Herrath D, Schaefer K. Influence of diphenylhydantoin and phenobarbital on intestinal calcium transport in the rat. Epilepsia 1972; 13:829–834.
65. Harrison HC, Harrison HE. Inhibition of vitamin D-stimulated active transport of calcium of rat intestine by diphenylhydantoin-phenobarbital treatment. Proc Soc Exp Biol Med 1976; 153:220–224.
66. Lau KH, Nakade O, Barr B, Taylor AK, Houchin K, Baylink DJ. Phenytoin increases markers of osteogenesis for the human species *in vitro* and *in vivo*. J Clin Endocrinol Metab 1995; 80:2347–2353.
67. Vernillo AT, Rifkin BR, Hauschka PV. Phenytoin affects osteocalcin secretion from osteobastic rat osteosarcoma 17/2.8 cells in culture. Bone 1990; 11:309–312.
68. Onodera K, Takahashi A, Mayanagi H, Wakabayashi H, Kamei J, Shinoda H. Phenytoin-induced bone loss and its prevention with alfacalcidol or calcitriol in growing rats. Calcif Tissue Int 2001; 69:109–116.
69. Feldkamp J, Becker A, Witte OW, Scharff D, Scherbaum WA. Long-term anticonvulsant therapy leads to low bone mineral density-evidence for direct drug effects of phenytoin and carbamazepine on human osteoblast-like cells. Exp Clin Endocrinol Diabetes 2000; 108:37–43.
70. Auszmann JM, Vernillo AT, Fine AS, Rifkin BR. The effect of phenytoin on parathyroid hormone stimulated cAMP activity in cultured murine osteoblasts. Life Sci 1990; 46:351–357.
71. Dziak R, Vernillo A, Rifkin B. The effects of phenytoin on calcium uptake in osteoblastic cells. J Bone Mineral Res 1988; 3:415–420.
72. Malabanan A, Veronikis IE, Holick MF. Redefining vitamin D insufficiency. Lancet 1998; 351:805–806.
73. Fuleihan GEH, Nabulsi M, Choucair M, et al. Hypovitaminosis D in healthy schoolchildren. Pediatrics 2001; 107:1–7.
74. Vieth R, Cole DE, Hawker GA, Trang HM, Rubin LA. Wintertime vitamin D insufficiency is common in young Canadian women, and their vitamin D intake does not prevent it. Eur J Clin Nutr 2001; 55:1091–1097.
75. Rucker D, Allan JA, Fick GH, Hanley DA. Vitamin D insufficiency in a population of healthy western Canadians. Can Med Assoc J 2002; 166:1517–1524.
76. Thomas MK, Lloyd-Jones DM, Thadhani RI, et al. Hypovitaminosis D in medical inpatients. N Engl J Med 1998; 338:777–783.
77. Harris SS, Soteriades E, Coolidge JA, Mudgal S, Dawson-Hugues B. Vitamin D insufficiency and hyperparathyroidism in a low income, multiracial, elderly population. J Clin Endocrinol Metab 2000; 85:4125–4130.
78. Leboff MS, Kohlmeier L, Hurwitz S, Franklin J, Wright J, Glowacki J. Occult vitamin D deficiency in postmenopausal US women with acute hip fracture. JAMA 1999; 281:1505–1511.
79. Silver J, Davies TJ, Kupersmitt E, Orme M, Petrie A, Vajda F. Prevalence and treatment of vitamin D deficiency in children on anticonvulsant drugs. Arch Dis Child 1974; 49:344–350.
80. Pedrera JD, Canal ML, Carvajal J, et al. Influence of vitamin D administration on bone ultrasound measurements in patients on anticonvulsant therapy. Eur J Clin Invest 2000; 30:895–899.
81. Medlinsky HL. Rickets associated with anticonvulsant medication. Pediatrics 1974; 53:91–95.
82. Hahn TJ, Shires R, Halstead LR. Serum dihydroxyvitamin D metabolite concentrations in patients on chronic anticonvulsant drug therapy: response to pharmacologic doses of vitamin D_2. Metab Bone Dis Rel Res 1983; 5:1–6.
83. Tjellesen L, Gotfredsen A, Christiansen C. Effect of vitamin D2 and D3 on bone-mineral content in carbamazepine-treated epileptic patients. Acta Neurol Scand 1983; 68:424–428.

84. Berry JL, Mawer EB, Walker DA, Carr P, Adams PH. Effect of antiepileptic drug therapy and exposure to sunlight on vitamin D status in institutionalised patients. In: Oxley Jeal, ed. Antiepileptic Therapy: Chronic Toxicity of Antiepileptic Drugs. Raven, New York, 1983, pp. 185–192.
85. Tjellesen L, Gotfredsen A, Christiansen C. Different actions of vitamin D_2 and D_3 on bone metabolism in patients treated with phenobarbitone/phenytoin. Calcif Tissue Int 1985; 37:218–222.
86. Tjellesen L, Hummer L, Christiansen C, Rodbro P. Different metabolism of vitamin D2/D3 in epileptic patients treated with pehnobarbitone/phenytoin. Bone 1986; 7:337–342.

36 Glucocorticoid-Induced Osteoporosis

Barbara P. Lukert

Chronic use of glucocorticoids (GCs) is the most common cause of secondary osteoporosis. We will review the current understanding of the clinical features, pathogenesis, and management of glucocorticoid-induced osteoporosis (GCOP), with emphasis on the nutritional aspects of the disorder.

1. CLINICAL FEATURES

The potent anti-inflammatory and immunosuppressive properties of GCs have prompted their extensive use, but side effects are dramatic *(1)*. Patients receiving pharmacological doses of GCs present a distinctive clinical picture: centripetal obesity with peripheral subcutaneous fat atrophy, thinning of the skin with increased fragility and ecchymoses, proximal muscle weakness, fluid retention, hyperglycemia, and, frequently, vertebral fractures. Fractures resulting from bone loss are among the most incapacitating sequela of steroid therapy. Synthetic derivatives designed in an attempt to diminish the detrimental side effect of cortisol, particularly sodium retention, while augmenting the anti-inflammatory effect of the parent compound, have not succeeded in reducing bone loss caused by this group of compounds. The severity of side effects, other than fluid retention, appears to be proportional to their anti-inflammatory potency. The most frequently prescribed GCs are prednisone, prednisolone, methylprednisolone, betamethasone, dexamethasone, and triamcinolone. Patients prone to adverse side effects of GCs appear to be those with the slowest clearance of the drug *(2)*.

1.1. Bone Density

Bone density has been measured in patients taking steroids for a wide variety of diseases. Older retrospective studies reported bone loss varying from 0 to 17%/yr. Prospective data based on results derived from the placebo groups in clinical trials show that most people taking GCs lose bone. A meta-analysis on the use of bisphosphonates in GCOP included 15 studies *(3)*. The mean change in bone mineral density (BMD) of the spine while taking GCs was –2.97%/yr in the placebo group, most of whom were taking calcium and vitamin D. Only 1 of 15 studies failed to observe

From: *Nutrition and Bone Health*
Edited by: M. F. Holick and B. Dawson-Hughes © Humana Press Inc., Totowa, NJ

bone loss. Ten of these studies measured density of the femoral neck also. All studies showed loss of bone in the placebo group, with a mean decline of –2.71%/yr.

1.2. Fracture Risk

The true incidence of osteoporosis-related fractures in patients taking GCs is unknown. Available data again were derived from the control groups of multicenter trials evaluating the efficacy of bisphosphonates, and from a retrospective cohort study. In the control groups for bisphosphonate studies in which the placebo group received calcium and vitamin D supplements, the fracture incidence in postmenopausal women ranged from 13% to 21% *(4–6)*. No fractures were observed in premenopausal women, and the incidence in men varied from 2.1% to 23.5%. In a retrospective cohort study comparing patients taking oral steroids to those using inhaled steroids, the fracture risk increased within 3 mo after the initiation of oral GCs. This rapid onset on fractures is not surprising when one considers that bone appears to be lost most rapidly during the first few months of GC therapy *(7)*. The risk was dose-dependent, with an increase in risk for vertebral fracture in patients taking as little as 2.5 mg of prednisone daily. The relative risk (RR) for vertebral fracture was 1.55, 2.59, and 5.18 for doses of less than 2.5, 2.5–7.5, and more than 7.5, respectively. The hip fracture risk was 0.99 relative to control in patients taking less than 2.5 mg prednisone daily, rising to 1.77 on daily doses of 2.5–5.0 mg/d, and to 2.27 on doses of 7.5 mg or greater *(8)*.

Although postmenopausal women are at highest risk for fractures, both men and women, old and young, blacks and Caucasians, lose bone, and if they take GCs long enough it is likely that they will be at risk for fractures. Steroid-induced bone loss occurs primarily in trabecular bone *(9)*.

2. HISTOLOGY

Histologically, there is reduction of mean wall thickness of trabecular bone packets, with a consequent decrease in bone volume. Parameters of bone resorption are elevated, with increases in eroded surface, osteoclast-covered surface, and increased osteoclast number *(9)*. The number of apoptotic osteoblasts and osteocytes is significantly increased in bone biopsies from patients taking GCs.

3. PATHOGENESIS

GCOP is multifactorial in origin. Abnormalities in calcium absorption and the renal handling of calcium, gonadal hormone secretion, and direct effect of GCs on bone all contribute to bone loss.

3.1. Parathyroid Hormone (PTH), Vitamin D, and Calcium Absorption

3.1.1. Hyperparathyroidism

Mild hyperparathyroidism has been demonstrated in patients taking long-term GCs in some (Lukert, 1976), *(9b–11)*, not all studies *(12,13)*. Short-term administration of glucocorticoids to normal volunteers caused increased in both PTH *(14)* and 1,25$(OH)_2$D *(15)*. Hyperparathyroidism produced by long-term administration has been attributed to inhibition of gastrointestinal absorption of calcium and hypercalciuria *(10,16)*, resulting in negative calcium balance. However, two recent studies demonstrating elevations in both PTH and 1,25$(OH)_2$D may shed important light on the etiology of secondary hyperparathyroidism induced by GC. High doses of GCs have acute effects on the kidney, causing phosphaturia and increased 1,25$(OH)_2$D levels before an elevation in PTH level is observed *(17)*. Serum 1,25$(OH)_2$D levels rise significantly within 8 h after intravenous methylprednisolone, whereas serum PTH rises slowly, peaking at 2 wk after starting GCs, a time when 1,25$(OH)_2$D levels have fallen. Thus, during short-term GC administration the rise in serum 1,25$(OH)_2$D appears to be independent of PTH. PTH levels are persistently elevated in the presence of high levels of calcium and 1,25$(OH)_2$D in patients taking GCs chronically, suggesting that PTH secretion may not be appropriately inhibited by calcium and 1,25$(OH)_2$D *(11)*. These findings are in agreement with earlier animal studies *(18)*. A recent review of the role of secondary hyperparathyroidism in the pathogenesis of GC-induced bone loss concluded that the effects of hyperparathyroidism are minor *(19)*. However, work many years ago showing that patients taking steroids had higher PTH levels *and* higher levels of urine cAMP suggests that the higher levels of PTH were physiologically significant *(20)*. The importance of hyperparathyroidism in the etiology of GC-induced bone loss is still controversial.

3.1.2. Vitamin D Metabolites

Previous studies reporting normal serum levels of 1,25$(OH)_2$D during long-term GC therapy *(21,22)* did not compare steroid-treated to disease-matched patients. Serum 1,25$(OH)_2$D levels tend to be higher in disease-matched patients taking GCs *(11,20)*. The effect of GCs on serum 25(OH)D levels remains unclear. Both normal *(23,24)* and low *(25,26)* values of 25(OH)D have been reported. Low values seen in some patients are probably the result of inadequate dietary intake combined with lack of exposure to sunlight rather than to GC-induced changes in absorption or metabolism of vitamin D.

3.1.3. Effect of GC on Intestinal Absorption of Calcium

Pharmacological doses of GCs inhibit intestinal absorption of calcium *(11,16)*. In vitro studies show that GCs decrease intestinal mucosal to serosal flux and increase serosal to mucosal flux *(27,27a)*. The effect is at least partially independent of vitamin D effects *(28–30)*, and only 25% of the impaired calcium absorption in corticosteroid-treated patients can be accounted for by slightly reduced serum 1,25$(OH)_2$D levels *(31)*. Administration of 1,25$(OH)_2$D improves calcium transport but does not return absorption to normal *(32)*.

The role of GC–vitamin D interactions in GC-induced changes in calcium absorption remains unclear. Intestinal receptors for 1,25$(OH)_2$D appear to be increased in GC-treated rats and decreased in mice *(33)*. It has been suggested that

GCs may accelerate the breakdown of $1,25(OH)_2D$ at the mucosal receptor site, but most studies show that neither the localization nor further metabolism of $1,25(OH)_2D$ in the intestinal mucosa is altered.

GCs do not affect calcium/magnesium-dependent adenosine triphosphatase (ATPase) activity *(34)* or brush border uptake of calcium *(35)*. Cortisol does not alter $1,25(OH)_2D$-induced, DNA-dependent RNA polymerase activity in intestinal cells *(36)*. In duodenal organ cultures, dexamethasone in low concentrations increases calcium transport and enhances $1,25(OH)_2D$-dependent calbindin-D_{28K} synthesis by positive co-transcriptional regulation of gene expression. High concentrations of dexamethasone in vitro mimic the action of pharmacological doses of glucocorticoids administered in vivo (inhibition of calcium transport) even though the stimulatory effect on calbindin synthesis is maintained *(37)*. These findings suggest that GC-induced inhibition of calcium absorption is caused by alterations in posttranscriptional events, alterations in basolateral membrane transport, or other factors affecting transport across membranes. The defect in transport is made worse by a high sodium intake *(16)* and by a low calcium intake *(27,37)*, and is improved by sodium restriction *(16)* and supplemental calcium *(38)*.

3.1.4. Renal Excretion of Calcium and Phosphorus

Fasting hypercalciuria and elevated serum levels of PTH are present after only 5 d of GC administration *(39)*. Hypercalciuria is due to increased bone resorption and decreased renal tubular reabsorption of calcium, which occurs despite elevated serum levels of PTH *(38)*. Renal loss of calcium is increased by a high-sodium diet and diminished by sodium restriction and thiazide diuretics *(16)*.

The hyperphosphaturia observed in patients taking GCs is due in part to secondary hyperparathyroidism. The other causative factor is a GC-induced change in Na^+–H^+ exchange activity, which causes a decrease in sodium gradient-dependent phosphate uptake in the proximal tubule *(20,40)*.

3.2. Vitamin K and GCs

Vitamin K is essential for the γ-carboxylation of glutamic acid residues, and plays an important role in the formation of osteocalcin, a bone matrix protein *(41)*.

Treatment with GC causes a marked decrease in osteocalcin *(42)*. When menatetrenone, the most potent form of vitamin K, was given to steroid-treated rats, the fall in urine γ-carboxyglutamic acid was prevented and bone density was preserved *(42a)*.

3.3. Gonadal Hormones

GCs blunt pituitary secretion of luteinizing hormone *(43)* and inhibit follicle-stimulating hormone (FSH)-induced estrogen production by cultured rat granulosa cells from the ovary and testosterone production by the testes *(28)*. These effects result in decreased serum concentrations of estradiol, estrone, dehydroepiandrosterone sulfate, and progesterone *(44–46)*. The reduction in serum testosterone levels in men appears to be related to the dose of prednisone *(47)*. Adrenal production of dehydroepiandrosterone and androstenedione is suppressed due to suppression of adrenocorticotropic hormone and resultant adrenal atrophy *(48)*.

3.4. Effect on Growth Hormone Secretion

Prednisone inhibits pituitary secretion of growth hormone (GH) in response to GH-releasing hormone in normal men *(49)*, but serum concentrations of growth hormone and insulin-like growth factor (IGF)-1 are normal in patients receiving glucocorticoids *(50)*. Despite normal levels, IGF-1 activity measured by bioassay is decreased in the serum of patients with GC excess *(51)*. The lack of biological effect of IGF-1 may be due to an IGF-1 inhibitor that has been found in the serum of GC-treated children. This inhibitory effect may be due to alterations in growth hormone-binding proteins, discussed below.

3.5. Interactions With Factors Controlling Bone Metabolism at the Cellular Level

3.5.1. GCs and PTH

GCs enhance the sensitivity of osteoblasts to PTH and potentiate PTH-mediated inhibition of collagen synthesis, alkaline phosphatase activity, and citrate decarboxylation *(52–54)*. The effect of PTH on the renal tubule is also increased by GCs *(7)*.

3.5.2. GCs and 1,25$(OH)_2$D

The effects of GCs on calcitriol receptors in bone vary with species and growth phase of cell cultures. 1,25$(OH)_2$D receptors are downregulated by GCs in mouse osteoblasts, but upregulated in rat osteoblasts *(33,55)*. Synthesis of osteocalcin, the major noncollagen bone protein, is stimulated by 1,25$(OH)_2$D and inhibited by GCs *(56)*. Studies using osteoblast-like cells in tissue culture show that 1,25$(OH)_2$D induces the human osteocalcin promoter through a vitamin D-response element, and GCs repress the vitamin D-induced promoter. Repression appears to involve a negative GC response element that overlaps the TATA box of the osteocalcin promoter. This suggests a stearic interference mechanism for GC repression *(56)*.

There is an interaction between dexamethasone and 1,25$(OH)_2$D in the modulation of the synthesis of IGF-1 in osteoblast-like cells. Dexamethasone causes a dose-dependent inhibition of IGF-1 release into the medium and 1,25$(OH)_2$D partially antagonizes the inhibition *(57)*.

3.5.3. GCs and Prostaglandins

Locally produced prostaglandins transiently inhibit osteoclast function in normal bone, but the major effect in organ culture is to stimulate bone resorption by increasing replication and differentiation of osteoclasts. Prostaglandin E_2 (PGE_2) at concentrations of 10^{-8}–10^{-7} M stimulates collagen and noncollagen protein synthesis. GCs inhibit the production of PGE_2 in tissue culture *(58)*.

3.5.4. GCs and Cytokines

Interleukin-1 (IL-1) and interleukin-6 (IL-6) induce bone resorption and inhibit formation. It is unlikely that cytokines play a major direct role in GCOP. The bone resorbing activity of IL-1 is partially inhibited by cortisol *(59)*, and cortisol reverses the inhibitory effect of IL-1 on collagen synthesis in neonatal mouse calvaria *(60)*. Cytokines also modulate the 11-hydroxysteroid dehydrogenase enzyme system, which regulates the conversion of cortisone to cortisol and thus may potentiate the effect of GC (see below).

3.5.5. GCs and Growth Factors

Somatomedin-C, also known as IGF-1, is a GH-dependent polypeptide that is synthesized by bone cells and stimulates bone cell replication and collagen synthesis *(61)*. PTH and GH stimulate IGF-1 production by bone cells. Physiological concentrations of cortisol increase IGF-1 receptor binding by rat osteoblast-like cells in tissue culture and enhance the anabolic effects of IGF-1 on collagen synthesis and procollagen messenger RNA levels in cultured fetal rat calvaria *(62)*. Pharmacological levels of cortisol inhibit synthesis of IGF-1 by fetal rat calvariae by reducing IGF-1 transcript levels *(63)*.

Glucocorticoids affect IGF-binding proteins (IGFBP) that enhance or inhibit IGF activity. The stimulation of collagen synthesis that occurs during the first 24 h of exposure to cortisol in tissue culture is inhibited by the presence of IGFBP-2, which sequesters secreted IGF *(64)*. When dexamethasone is added to normal human osteoblast-like cells in tissue culture, IGFBP-5 mRNA decreases and IGFBP-4 and IGFBP-3 mRNA increase. Under normal conditions IGFBP-3 and IGFBP-4 inhibit and IGFBP-5 increases the anabolic effects of the IGF system on bone, thus the observed interactions of GCs with the IGF-binding proteins may partially explain the adverse effect of GCs on bone cell proliferation.

Transforming growth factor (TGF)-β enhances replication of osteoblasts and bone matrix protein synthesis. GCs decrease these anabolic effects of TGF-β on bone by redistributing the binding of TGF-β1 toward extracellular matrix storage sites and away from receptors involved in intracellular signal transduction *(65)*. Interactions between GCs and growth factors and their binding proteins may play a major role in the etiology of GC-induced bone loss.

3.6. Effects on Bone

3.6.1. Specific Effect on Bone Formation

Thus far we have discussed the interaction between GCs and other factors that control bone remodeling. We will now consider the final pathways of the effects of GCs on bone metabolism and gene expression.

GCs have a biphasic effect on bone in organ culture systems. Physiological concentrations and brief exposure enhance the function of differentiated osteoblasts, whereas supraphysiological doses for prolonged periods inhibit synthetic processes *(66)*.

GCs affect the synthesis of a number of noncollagen components of bone. The synthesis of osteocalcin, a major bone matrix protein, is inhibited by GCs *(41)*. Serum levels of osteocalcin are low in patients receiving either oral *(42)* or inhaled GCs *(67)*. Collagenase production, which could further decrease collagen content of bone, is increased by cortisol in mouse calvariae in tissue culture. Synthesis of other components of matrix, such as mucopolysaccharide and sulfated glucosaminoglycans, is decreased in bone cultures and mouse dermal fibroblasts exposed to cortisol *(68)*.

3.6.2. Effects on Bone Resorption

Histomorphometric and calcium kinetic studies suggest that bone resorption is enhanced by GCs *(12,69,70)*. GCs may have a biphasic effect on osteoclasts, similar to the biphasic effect on osteoblasts. Physiological concentrations are required for the late stages of differentiation and function, whereas the generation of new osteoclasts involving cell replication is inhibited by high doses and prolonged exposure *(54)*. Studies of the direct effects of GCs on bone in tissue culture have yielded confusing results. On the one hand, GCs appear to inhibit the recruitment and/or differentiation of bone-resorbing cells in rat bone, but stimulate the activity of existing osteoclasts. However, in mouse bone marrow culture systems, dexamethasone enhanced 1,25$(OH)_2$D-induced osteoclast-like cell formation by inhibiting the endogenous production of granulocyte-macrophage colony-stimulating factor (GM-CSF). GM-CSF appears to function as a negative regulator of osteoclast formation, and inhibiting its production by dexamethasone allows an increase in osteoclast formation. The increased bone resorption found in GIOP appears to involve the receptor activator of NF-κB ligand (RANKL) and osteoprotegerin (OP). RANK-L is an osteoblastic signal that binds to an osteoclast receptor and, in association with CSF-1, induces osteoclastogenesis. OP is a faux receptor that binds RANKL, preventing RANKL binding to the osteoclast receptor and hence prevents osteoclastogenesis. Many agents that induce bone resorption (TNF-α, IL-6, and perhaps PTH) act by inducing RANKL expression. GC increase the expression of RANKL and CSF-1 and decrease OP expression in human osteoblasts and stromal cells in culture *(71,72)*. Animal experiments show that enhanced resorption during GC administration can be prevented by parathyroidectomy *(73)*, suggesting that secondary hyperparathyroidism may play a major role in steroid-induced bone resorption in vivo.

3.6.3. Regulation of the Effects of GCs on Bone Metabolism

The magnitude of the effects of GCs on bone may be determined in part by the activity of 11β-hydroxysteroid dehydrogenases (11β-HSD). 11β-HSDs are two isoenzymes present in osteoblasts, which catalyze the interconversion of inactive cortisone to active cortisol. 11β-HSD type 1 (11β-HSD1) converts cortisone to cortisol in osteoblasts. 11β-HSD2 converts cortisol to inactive cortisone. In osteoblasts, GCs induce 11β-HSD1 and thus may enhance the adverse effects of cortisol on bone formation. Proinflammatory cyotkines in osteoblasts may modulate 11β-hydroxysteroid dehydrogenase isozymes, thus enhancing the effect of GCs *(74)*.

3.6.4. Summary of Effects on Bone

The profoundly rapid loss of bone induced by GCs results from the convergence of systemic effects and direct effect on bone. Defects in calcium transport lead to impaired gastrointestinal absorption of calcium, hypercalciuria, and secondary hyperparathyroidism. PTH increases the number of bone remodeling units. In the absence of gonadal hormones, bone is more susceptible to the effects of PTH. Bone resorption is enhanced and formation is inhibited by glucocorticoids at each remodeling site. Bone is being resorbed and the osteoblasts are incapable of replacing it. This combination of events results in rapid bone loss, which appears to be most marked during the first 6–12 mo of GC therapy.

4. OSTEONECROSIS

Osteonecrosis (aseptic necrosis or avascular necrosis), a serious complication of GC therapy, is estimated to occur in 4–25% of patients taking steroids. The hip is most frequently affected, followed by the head of the humerus, and the distal femur. Pain is the usual presenting symptom and may be mild in chronic forms of the disease. The diagnosis is frequently missed, and a high index of suspicion must be maintained in all patients taking GCs.

The etiology of osteonecrosis remains unclear, but recent findings suggest that apoptosis of osteocytes may play an important role. Histological studies of femurs surgically removed from patients suffering from avascular necrosis of the femoral neck due to GCs show a significant increase in the number of apoptotic osteocytes when compared to the number found in the femoral necks taken from patients undergoing hip replacement for degenerative changes as a result of osteoarthritis *(75)*. Older theories regarding etiology include a vascular cause, proposing that ischemia is caused by microscopic fat emboli; a mechanical theory that attributes ischemic collapse of the epiphysis to osteoporosis and the accumulation of unhealed trabecular microcracks, resulting in fatigue fractures through the epiphysis; and the theory that increased intraosseous pressure caused by fat accumulation as part of Cushing's syndrome leads to mechanical impingement on the sinusoidal vascular bed and decreased blood flow. Osteonecrosis may develop in patients who receive steroids in very high doses for a very short period of time, long-term treatment, or intra-articular injections, but in general, the risk increases with both the dose of GCs and the duration of treatment *(76)*.

5. ASSESSMENT OF PATIENTS TREATED WITH GCs

The aim of assessment is to identify those patients in whom calcium homeostasis and osteoblastic function are most severely impaired by GCs. It is prudent to prescribe the lowest effective dose and to use topical preparations whenever possible, because the magnitude of bone loss may be related to dose *(77)*. Although alternate-day therapy preserves normal function of the pituitary–adrenal axis, it does not prevent bone loss *(78)*. Retrospective studies have shown that patients who developed GCOP had lower radiocalcium absorption, higher fasting urinary levels of calcium and hydroxyproline, and lower levels of gonadal hormones *(77)*. Hypercalciuria may be a marker of negative calcium balance and susceptibility to secondary hyperparathyroidism *(10,16)*. Measuring serum levels of 25(OH)D will identify patients with vitamin D deficiency that may augment bone loss. Measurement of serum estradiol and FSH in women and free testosterone in men will identify those who need gonadal hormone replacement. The degree of inhibition of osteoblastic activity can be assessed by measuring serum osteocalcin levels. Low levels predict greater inhibition and can be expected to result in greater bone loss. Patients who lose bone while taking steroids exhibit loss of diurnal variation of osteocalcin levels. The exception to this is that patients taking cyclosporine along with GCs have elevated osteocalcin levels, indicating a high rate of bone remodeling. Those patients with higher osteocalcin levels appear to lose more bone.

5.1. Radiological Evaluation

The radiological picture of GCOP differs from that of postmenopausal osteoporosis in that the vertebrae are uniformly translucent, as opposed to the "corduroy stripe" pattern observed in the latter *(79)*. A hallmark of GCOP is abundant pseudocallus formation observed at the site of stress fractures at the end plates of collapsed vertebrae, ribs, and pelvis. Bone disease is advanced by the time it is detectable by roentgenograms.

Osteonecrosis is frequently missed by routine roentgenograms, but the bone scan will show a photon-deficient area at the site (doughnut sign), which indicates the presence of a dead central area surrounded by a region of increased activity. Within a few weeks, the site shows a uniformly high level of activity that remains throughout the course of the disease. Magnetic resonance imaging is a sensitive and relatively specific method of detecting early osteonecrosis *(80)*.

5.2. Bone Density

Trabecular bone (spine, hip, distal radius, pelvis, and ribs) and the cortical rim of the vertebrae are lost more rapidly than cortical bone from the extremities. Earliest changes can be detected in the spine and femoral neck using dual-energy X-ray absorptiometry (DXA) or quantitative computerized tomography. Evaluation at 6-mo intervals for the first 2 yr will identify patients who are losing bone rapidly.

5.3. Muscle Strength

Patients taking GCs are prone to myopathy, particularly involving proximal muscles. *(81,82)*. There is a striking association between the presence of steroid myopathy and osteoporosis.

6. PREVENTION AND TREATMENT

There are very few results available from well-designed, prospective studies to guide us in recommendations for prevention and treatment. Our recommendations are based on logic and a limited number of facts.

6.1. Planning GC Therapy

It is prudent to prescribe the lowest effective dose and to use topical preparations whenever possible, because the magnitude of bone loss may be related to dose *(83)*. Alternate-day therapy preserves normal function of the pituitary–adrenal axis, but it does not prevent bone loss *(78)*. One should not be complacent about using topical preparations, because inhaled steroids have been reported to lower markers of bone formation, osteocalcin and/or carboxypropeptide of type I procollagen, in patients with asthma *(84,85)*, and bone loss occurs in women using inhaled steroids *(86)*. Recent studies have shown that steroids administered by epidural injection are absorbed into the systemic circulation in quantities sufficient to suppress the pituitary–adrenal axis, and it seems likely that they also affect bone metabolism. Intra-articular injections have been reported to be associated with aseptic necrosis of adjacent bone, but there are no data available on more generalized skeletal effects *(87)*.

6.2. Treatment and Prevention

Interpretation of the results of studies evaluating the efficacy of any form of treatment for GC-induced osteoporosis requires consideration of a number of factors. First, it is important to separate primary prevention trials, i.e., those who have received GC for less than 4 mo, from secondary prevention, those who have received GC for more than 12 mo. The dose of prednisone required by the patients in the study is an important consideration, since bone loss is dose-dependent. When the endpoint is fracture risk, age and menopausal status must be considered, since fracture risk is highest in postmenopausal women.

6.2.1. Calcium and Vitamin D

Evidence suggests that vitamin D plus calcium does not prevent bone loss in patients who started taking GC within 4 mo of the beginning of the study. A 3-yr double-blind, placebo-controlled trial of 62 patients who had started GC 1 mo prior to the start of the study were randomized to receive either placebo or 50,000 IU vitamin D per week plus 1000 mg calcium daily. After 12 mo the change in the BMD of the lumbar spine was –2.6% in the treated group and –4.1% in the placebo group, but at the end of 3 yr the average change in lumbar spine BMD was –2.2% in the treated group and –1.5% in the placebo group. The difference was not significant at either time point *(87a)*.

Vitamin D plus calcium may be effective in preventing bone loss in patients receiving long-term, low-dose treatment (<10 mg/d prednisone). A 2-yr study of 65 patients with rheumatoid arthritis receiving daily doses of prednisone, mean dose 5.6 mg/d, were randomized to a placebo group or a group receiving 1000 mg/d calcium and 500 IU vitamin D_3 daily. The bone density of the placebo-treated group decreased in the lumbar spine and trochanter at rates of –2.0% and –0.9% per year, respectively, while the calcium and vitamin D group gained bone density at rates of 0.72% and 0.85% per year. The densities of the femoral neck did not change. A recent meta-analysis showed that both vitamin D and its metabolites were more effective at preserving bone density than no therapy or calcium alone *(87b)*. Moreover, the efficacy of bisphosphonates was significantly greater when used in combination with vitamin D. Calcitriol and alfacalcidol may have direct antiresorptive and anabolic effects on bone. However, the ability of these metabolites to prevent GC-induced bone loss has been somewhat disappointing. The meta-analysis just mentioned found no statistical difference between estimates of effect for vitamin D and calcitriol. Calcitriol and other potent metabolites of vitamin D should not be used routinely in the management of GC-induced bone loss until matters of dosage and toxicity are better understood.

6.2.2. Vitamin K

The studies of the efficacy of vitamin K to prevent steroid-induced bone loss in humans are somewhat inconsistent, but studies have shown an increase in bone density *(88)* and/or an increase in carboxylated osteocalcin *(89)* when patients taking prednisone were given menatetrenone (*see* chapter on vitamin K).

6.2.3. Antiresorptive Drugs

A drug that limits the depth of the resorption cavity at each remodeling site should minimize the amount of bone loss in spite of continued inhibition of bone formation. Calcitonin and the bisphosphonates alendronate, risedronate, etidronate, and pamidronate are the antiresorptive drugs (other than estrogen) currently available in the United States.

Hormone Replacement and Anabolic Steroids. In the placebo groups for trials focusing on bisphosphonate treatment of GCOP, the most vertebral fractures were seen in postmenopausal women *(5,6,90)*. Bone density increases in the spine when postmenopausal women taking long-term steroids are given hormone replacement therapy (HRT) *(91,92)*. There are no consistent data on the effect on the hip, and there are no trials large enough to assess the effect of HRT on fracture risk in patients taking GCs. Despite evidence for beneficial effects, its is unlikely that HRT will be used routinely for the prevention or treatment of GC-induced bone loss in view of the recent report from the Women's Health Initiative showing an increase in the risk for breast cancer and cardiovascular disease in postmenopausal women taking long-term HRT *(93)*. Premenopausal women whose estradiol levels fall during treatment with GCs should be given HRT.

BMD of the spine increases in hypogonadal men treated with testosterone while receiving GCs, but no hip density or fracture data are available *(47)*.

Calcitonin. A meta-regression analysis comparing the efficacy of drug therapies for the management of GCOP showed that calcitonin given with vitamin D was more effective than no therapy or calcium only, but significantly less effective than bisphosphonates *(94)*.

Bisphosphonates. Etidronate given cyclically and with calcium and vitamin D supplementation (1000 IU/d), improved bone density in preliminary studies *(95)*. No fracture data are available.

Pamidronate given continuously by the oral route was associated with a significant increase in vertebral bone mass in patients taking prednisone at a mean dose of 12.6 mg/d *(96)*. A study comparing pamidronate 90 mg once a year or 90 mg at the beginning of steroid therapy followed by 30 mg intravenously every 3 mo showed that both schedules prevented bone loss *(97)*. A 48-wk, randomized, placebo-controlled study of 10 mg/d alendronate in 477 men and women who were receiving GC therapy showed that bone density increased significantly in the spine (2.9%) and femoral neck (1%), compared to a decrease of –0.4% in the hip, and –1% in the spine in the placebo group. The relative risk for vertebral factures was 0.6 when compared to the placebo group *(5)*. This study was subsequently extended to 2 yr in patients continuing to receive at least 7.5 mg of prednisone daily *(4)*. All patients in both studies also received calcium and vitamin D. The lumbar spine BMD increased by 3.7%, while the placebo group decreased by –0.8%. Femoral neck density was maintained. There were significantly fewer vertebral fractures in the alendronate-treated group compared to placebo (0.7% vs 6.8%, $p = 0.026$) *(4)*.

A multicenter, randomized, double-blind, placebo-controlled, parallel-group study of risedronate was conducted in men and women initiating long-term GC treatment. Patients were treated with 2.5 mg or 5 mg risedronate daily, or placebo. All groups received 500 mg of elemental calcium daily. After 12 mo, the lumbar spine BMD did not change significantly in the treated group compared to baseline, with either the 2.5- or 5-mg dose, while it decreased in the placebo group. A trend toward a decrease in the incidence of vertebral fractures was noted in the 5-mg risedronate group ($p = 0.072$) *(6)*. A study powered to evaluate the fracture prevention efficacy of risedronate showed a 70% reduction in vertebral fracture risk. No fractures were observed in premenopausal women, but fracture incidence was reduced by 66% in men and 73% in postmenopausal women. Nonvertebral fractures were observed with similar incidence among groups *(98)*.

A meta-analysis on the use of bisphosphonates in GCOP included six primary prevention and seven secondary prevention studies, involving 842 participants *(94)*. All of these studies reported data on bone loss from the lumbar spine, whereas eight of the studies reported changes at the femoral neck. All but one study reported significant improvement in lumbar BMD in the treatment group compared to controls. On average, the treatment and placebo groups had a percentage change in bone density that differed by 4%. Four studies reported a significant improvement in femoral neck BMD, whereas the other four reported no significant difference from the control groups. The mean average change in femoral neck BMD differed by 2.1% between the treated and placebo groups.

Four studies reporting the incidence of new vertebral fractures were included in the analysis. Three found a decreased number of fractures, whereas one study found an increased number in the treatment group. In this analysis, the odds ratio of new fracture did not reach statistical significance (0.8). The author concluded that longer follow-up is required to ascertain the efficacy of bisphosphonates in fracture prevention. This analysis did not include large recent trials using risedronate *(98,99)*. A 4% increase in BMD would be expected to decrease fracture risk reported in these studies.

6.2.4. Drugs That Stimulate Bone Formation

The most promising anabolic hormone for treating GC-induced bone loss is PTH. PTH causes exuberant increases in BMD in postmenopausal women who are taking estrogen. A randomized, 12-mo, placebo-controlled trial studied 51 women with chronic inflammatory diseases taking chronic GCs (>5 mg prednisone per day). Women receiving PTH plus estrogen experienced an 11% increase in BMD of the spine measured by DXA, compared to a 1.7% increase in those taking estrogen alone. There were no differences between groups in the density of the hip or forearm. One year after PTH was discontinued, lumbar spine bone density had increased by another 15% *(100,101)*. Although the cellular mechanisms affected by PTH that prevent bone loss in GCOP are unclear, there are several possibilities. PTH stimulates IGF-1 synthesis in bone and prevents GC-induced apoptosis of osteoblasts *(102,103)*.

Sodium fluoride in doses of 50 mg/d stimulates the replication, differentiation, and function of osteoblasts in patients taking steroids, and reverses the steroid-induced decrease in serum osteocalcin levels. This suggests that bone formation is increased by fluoride, and studies have shown a significant increase in bone formation *(104)*.

7. THERAPY OF THE FUTURE

Recombinant human growth hormone prevents protein catabolism and improves protein balance in GC-treated patients *(105)*. However, the effect on bone loss appears less promising. Attempts to stimulate the local production of growth factors in bone that are altered by glucocorticoids, such as IGF-1 or TGF-β, may be useful. The search continues for new agents that stimulate replication, differentiation, and function of osteoblasts. The administration of osteoprotegerin, antibodies to RANKL, or agents that alter the secretion of either may be useful.

There are continuing efforts to develop glucocorticoids that retain anti-inflammatory properties but have less effect on calcium metabolism and bone remodeling. Deflazacort maintains most of the anti-inflammatory effects of prednisone while appearing to have fewer adverse effects on the skeleton *(106)*. In vitro studies show that both deflazacort and cortisol equally inhibit bone DNA and collagen synthesis and degradation and decrease IGF-1 concentrations in bone *(107).* Both prednisone and deflazacort cause a decrease in plasma osteocalcin levels in sheep *(108),* but results in humans have been variable. There is hope for dissociating the anti-inflammatory effects from effects on bone with the discovery of compounds that dissociate transactivation and transexpression *(109).*

8. CONCLUSION

In spite of the fact that GCOP has been recognized for more than 60 yr, it remains a neglected problem. GCs are the mainstay of treatment for a variety of illnesses, and we must be aggressive in our efforts to prevent and treat bone loss, a potentially devastating effect of this very useful class of drugs. GC-induced bone loss is partially reversible *(110).* Attempts to prevent the adverse effects of GCs, based on an understanding of the mechanisms involved, should minimize this devastating complication of GC therapy.

REFERENCES

1. Swartz SL, Dluhy RG. Corticosteroids: clinical pharmacology and therapeutic use. Drugs 1978; 16:238–255.
2. Kozower M, Veatch L, Kaplan MM. Decreased clearance of prednisolone, a factor in the develpment of corticoisteroid side effects. J Clin Endocrinol Metab 1974; 38:407–412.
3. Homik JE, Cranney A, Shea B, et al. A metaanalysis on the use of bisphosphonates in corticosteroid induced osteoporosis. J Rheumatol 1999; 26(5):1148–1157.
4. Adachi J, Saag KG, Delmas PD, et al. Two-year effects of alendronate on bone mineral density and vertebral fracture in patients receiving glucocorticoids. Arthritis Rheum 2001; 44(1):202–211.
5. Saag KG, Emkey R, Schnitzer TJ, Brown JP, Hawkins F. Alendronate for the prevention and treatment of glucocorticoid-induced osteoporosis. N Engl J Med 1998; 339(5):292–299.
6. Cohen S, Levy RM, Keller M, et al. Risedronate therapy prevents corticosteroid-induced bone loss: a twelve-month, multicenter, randomized, double-bind, placebo-controlled, parallel-group study. Arthritis Rheum 1999; 42(11):2309–2318.
7. Gennari C. Glucocorticoids and bone. In: Peck WA, ed. Elsevier, Amsterdam, 1985, pp. 213–232.
8. Van Staa TP, Leufkens HGM, Abenhaim L. Use of oral corticosteroids and risk of fractures. J Bone Miner Res 2000; 15(6):993–999.
9. Dempster DW. Bone histomorphometry in glucocorticoid-induced osteoporosis. J Bone Miner Res 1989; 4(2):137–141.

9b. Luckert BP, Adams JS. Calcium and phosphorus homeostasis in man. Arch Intern Med 1976; 136:1249–1253.

10. Suzuki Y, Ichikawa Y, Saito E, Homma M. Importance of increased urinary calcium excretion in the development of secondary hyperparathyroidism of patients under glucocorticoid therapy. Metabolism 1983; 32:151–156.
11. Bikle DD, Halloran BP, Fong L, Steinbach L, Shellito J. Elevated 1,25-dihydroxyvitamin D levels in patients with chronic obstructive pulmonary disease treated with prednisone. J Clin Endocrinol Metab 1993; 76(2):456–461.
12. Lund B, Storm TL, Lund B. Bone mineral loss, bone histomorphometry and vitamin D metabolism in patients with rheumatoid arthritis on long-term glucocorticoid treatment. Clin Rheumatol 1985; 4:143–149.

13. Slovik DM, Neer RM, Ohman JL, Lowell FC, Potts JT. Parathyroid hormone and 25-hydroxyvitamin D levels in glucocorticoid-treated patients. Clin Endocrinol (Oxf) 1980; 12:243.
14. Fucik RF, Kukreja SC, Hargis GK, Bowser EN, Henderson WJ, Williams GA. Effect of glucocorticoids on function of the parathyroid glands in man. J Clin Endocrinol Metab 1975; 40:152–155.
15. Hodsman AB, Toogood JH, Jennings BH, Fraher LJ, Baskerville JC. Differential effects of inhaled budesonide and oral prednisolone on serum osteocalcin. J Clin Endocrinol Metab 1991; 72:530–540.
16. Adams JS, Wahl TO, Lukert BP. Effects of hydrochlorothiazide and dietary sodium restriction on calcium metabolism in corticosteroid treated patients. Metabolism 1981; 30:217–221.
17. Cosman F, Nieves J, Herbert J, Shen V, Lindsay R. High-dose glucocorticoids in multiple sclerosis patients exert direct effects on the kidney and skeleton. J Bone Miner Res 1994; 9(7):1097–1105.
18. Collins EJ, Garrett ER, Johnston RL. Effect of adrenal steroids on radiocalcium metabolism in dogs. Metabolism 1962; 11:716–726.
19. Rubin MR, Bilezikian JP. The role of parathyroid hormone in the pathogenesis of glucocorticoid-induced osteoporosis—a re-examination of the evidence. J Clin Endocrinol Metab 2002; 87(9):4033–4041.
20. Lukert BP, Adams JS. Calcium and phosphorus homeostasis in man. Arch Intern Med 1976; 136:1249–1253.
21. Seeman E, Kumar R, Hunder GG, Scott M, Heath HI, Riggs BL. Production, degradation, and circulating levels of 1,25-dihydroxyvitamin D in health and in chronic glucocorticoid excess. J Clin Invest 1980; 66:664–669.
22. Reid IR. Pathogenesis and treatment of steroid osteoporosis. Clin Endocrinol (Oxf) 1989; 30:83–103.
23. Hahn TJ, Halstead LR, Teitelbaum SL, Hahn BH. Altered mineral metabolism in glucocorticoid-induced osteopenia. Effect of 25-hydroxyvitamin D administration. J Clin Invest 1979; 64:655–665.
24. Zerwekh JE, Emkey RD, Harris EDJ. Low-dose prednisone therapy in rheumatoid arthritis: effect on vitamin D metabolism. Arthritis Rheum 1984; 27:1050–1052.
25. Chesney RW, Mazess RB. Reduction of serum 1,25-dihydroxyvitamin D3 in children receiving glucocorticoids. Lancet 1978; 2:1123–1125.
26. Klein RG, Arnaud SB, Gallagher JC, DeLuca HF, Riggs BL. Intestinal calcium absorption in exogenous hypercortisonism. Role of 25-hydroxyvitamin D and corticosteroid dose. J Clin Invest 1977; 60:253–259.
27. Favus MJ, Walling MW, Kimberg DV. Effects of 1,25-dihydroxy-cholecalciferol in intestinal calcium transport in cortisone-treated rats. J Clin Invest 1973; 52:1680–1685.
27a. Adams JS, Lukert BP. Effects of sodium restriction on 45-Ca and 22-Na transduodenal flux in cortico-steroid-treated rats. Miner Electrolyte Metab 1980; 4:216–226.
28. Hsueh AJ, Erickson GF. Glucocorticoid inhibition of FSH-induced estrogen production in cultured rat granulosa cells. Steroids 1978; 32:639–648.
29. Kimaru S, Rasmussen H. Adrenal glucocorticoids, adenine nucleotide translocation, and mitochondrial calcium accumulation. J Biol Chem 1977; 252:1217–1225.
30. Shultz TD, Kumar R. Effect of cortisol on [3H] 1,25-dihydroxyvitamin D3 uptake and 1,25-dihydroxyvitamin D3-induced DNA-dependent RNA polymerase activity in chick intestinal cells. Calcif Tissue Int 1987; 40:327–331.
31. Morris HA, Need AG, O'Loughlin PD, Horowitz M, Bridges A, Nordin BEC. Malabsorption of calcium in corticosteroid-induced osteoporosis. Calcif Tissue Int 1990; 46:305–308.
32. Colette C, Monnier L, Pares Herbute N, Blotman F, Mirouze J. Calcium absorption in corticoid treated subjects—effects of a single oral dose of calcitriol. Horm Metab Res 1987; 19:335–338.

33. Chen TL, Cone CM, Morey-Holton E, Feldman D. 1-alpha,25-dihydroxyvitamin D3 receptors in cultured rat osteoblast-like cells. Glucocorticoid treatment increases receptor content. J Biol Chem 1983; 258:4350–4355.
34. Charney AN, Kinsey MD, Myers L. Na-K-activated adenosine triphosphatase and intestinal electrolyte transport. Effect of adrenal steroids. J Clin Invest 1975; 56:653–660.
35. Shultz TD, Bollman S, Kumar R. Decreased intestinal calcium absorption in vivo and normal brush border membrane vesicle calcium uptake in cortisol-treated chickens: evidence for dissociation of calcium absorption from brush border vesicle uptake. Proc Natl Acad Sci USA 1982; 79:3542–3546.
36. Corradino RA, Fullmer CB. Positive cotranscriptional regulation of intestinal calbindin-D28K gene expression by 1,25-dihydroxyvitamin D3 and glucocorticoids. Endocrinology 1991; 128(2):944–950.
37. Aloia JF, Semla HM, Yeh JK. Discordant effects of glucocorticoids on active and passive transport of calcium in the rat duodenum. Calcif Tissue Int 1984; 36:327–331.
38. Reid DM, Kennedy NSJ, Smith MA, Tothill P, Nuki G. Total body calcium in rheumatoid arthritis: effects of disease activity and corticosteroid treatment. Br Med J 1982; 285:330–332.
39. Nielsen HK, Thomsen K, Eriksen EF, Charles P, Storm TL, Mosekilde L. The effect of high-dose glucocorticoid administration on serum bone gamma carboxyglutamic acid-containing protein, serum alkaline phosphatase and vitamin D metabolites in normal subjects. Bone Miner 1988; 4:105–113.
40. Frieberg JM, Kinsella J, Sacktor B. Glucocorticoids increase the Na+–H+ exchange and decrease the Na+ gradient-dependent phosphate-uptake systems in renal brush border membrane vesicles. Proc Natl Acad Sci USA 1982; 79:4932–4936.
41. Price PA, Otsuka JW, Poser JKN, Raman N. Characterization of a gamma-carboxyglutamic acid-containing protein from bone. Proc Natl Acad Sci USA 1976; 73:1447–1451.
42. Lukert BP, Higgins JC, Stoskopf MM. Serum osteocalcin is increased in patients with hyperthyroidism and decreased in patients receiving glucocorticoids. J Clin Endocrinol Metab 1986; 62:1056–1058.
42a. Hara K, Akiyama Y, Okkawa I, Tajima T. Bone 1993; 6:813–818.
43. Sakakura M, Takebe K, Nakagawa S. Inhibition of luteinizing hormone secretion induced by synthetic LRH by long-term treatment with glucocorticoids in human subjects. J Clin Endocrinol Metab 1975; 40:774–779.
44. Doerr P, Pirke KM. Cortisol-induced suppression of plasma testosterone in normal adult males. J Clin Endocrinol Metab 1976; 43:622–629.
45. MacAdams MR, White RH, Chipps BE. Reduction of serum testosterone levels during chronic glucocorticoid therapy. Ann Intern Med 1986; 104:648–651.
46. Crilly RG, Cawood M, Marshall DH, Nordin BEC. Hormonal status in normal, osteoporotic and corticosteroid-treated postmenopausal women. J R Soc Med 1978; 71:733–736.
47. Reid IR, Frances JT, Pybus J, Ibbertson HK. Low plasma testosterone levels in glucocorticoid-treated male asthmatics. Br Med J 1985; 291:574.
48. Montecucco C, Caporali R, Caprotti P, Caprotti M, Notario A. Sex hormones and bone metabolism in postmenopausal rheumatoid arthritis treated with two different glucocorticoids. J Rheumatol 1992; 19(12):1895–1899.
49. Kaufmann S, Jones KL, Wehrenberg WB, Culler FL. Inhibition by prednisone of growth hormone (GH) response to GH-releasing hormone in normal men. J Clin Endocrinol Metab 1988; 67(6):1258–1261.
50. Gourmelen M, Girard F, Binoux M. Serum somatomedin/insulin-like growth factor (IGF) and IGF carrier levels in patients with Cushing's syndrome or receiving glucocorticoid therapy. J Clin Endocrinol Metab 1982; 54:885–892.
51. Unterman TG, Phillips LS. Glucocorticoid effects on somatomedins and somatomedin inhibitors. J Clin Endocrinol Metab 1985; 61:618–626.

52. Chen TL, Feldman D. Glucocorticoid potentiation of the adenosine 3′,5′-monophosphate response to parathyroid hormone in cultured rat bone cells. Endocrinology 1978; 102:589–596.
53. Catherwood BD. 1,25-Dihydrocholecalciferol and glucocorticoid regulation of adenylate cyclase in an osteoblast-like cell line. J Biol Chem 1985; 160:736–743.
54. Wong GL. Basal activities and hormone responsiveness of osteoclast-like and osteoblast-like bone cells are regulated by glucocorticoids. J Biol Chem 1979; 254(14):6337–6340.
55. Chen TL, Cone CM, Morey-Holton E, Feldman D. Glucocorticoid regulation of 1,25(OH)2-vitamin D3 receptors in cultured mouse bone cells. J Biol Chem 1982; 257:13564–13569.
56. Morrison NA, Eisman JA. Role of the negative glucocorticoid regulatory element in glucocorticoid repression of the human osteocalcin promoter. J Bone Miner Res 1993; 8(8):969–975.
57. Chen TL, Mallory JB, Hintz RL. Dexamethasone and 1,25(OH)2 vitamin D3 modulate the synthesis of insulin-like growth factor-I in osteoblast-like cells. Calcif Tissue Int 1991; 48:278–282.
58. Raisz LG, Fall PM. Biphasic effects of prostaglandin E2 on bone formation in cultured fetal rat calvariae: interaction with cortisol. Endocrinology 1990; 126(2):1654–1659.
59. Sato K, Fujii Y, Kasono K, Saji M, Tsushima T, Shizume K. Stimulation of prostaglandin E2 and bone resorption by recombinant human interleukin 1-alpha in fetal mouse bones. Biochem Biophys Res Commun 1986; 138:618–624.
60. Marusic A, Raisz LG. Cortisol modulates the actions of interleukin-1 alpha on bone formation, resorption, and prostaglandin production in cultured mouse parietal bones. Endocrinology 1991; 129(5):2699–2706.
61. Canalis EM, McCarthy TL, Centrella M. Isolation and characterization of insulin-like growth factor I (somatomedin-C) from cultures of fetal rat calvariae. Endocrinology 1988; 122:22.
62. Kream BE, Petersen DN, Raisz LG. Cortisol enhances the anabolic effects of insulin-like growth factor I on collagen synthesis and procollagen messenger ribonucleic acid levels in cultured 21-day fetal rat calvariae. Endocrinology 1990; 126(3):1576–1583.
63. McCarthy TL, Centrella M, Canalis EM. Cortisol inhibits the synthesis of insulin-like growth factor-I in skeletal cells. Endocrinology 1990; 126(3):1569–1575.
64. Kream BE, LaFrancis PM, Fall PM, Feyen JHM, Raisz LG. Insulin-like growth factor binding protein-2 blocks the stimulatory effect of glucocorticoids on bone collagen synthesis. J Bone Miner Res 1992; 7(suppl 1):5100.
65. Centrella M, McCarthy TL, Canalis EM. Glucocorticoid regulation of transforming growth factor B1 activity and binding in osteoblast-enriched cultures from fetal rat bone. Molec Cell Biol 1991; 11(9):4490–4496.
66. Dietrich JW, Canalis EM, Maina DM, Raisz LG. Effect of glucocorticoids on fetal rat bone collagen synthesis in vitro. Endocrinology 1979; 104:715–721.
67. Puolijoki H, Liippo K, Herrala J, Salmi J, Tala E. Inhaled beclomethasone decreases serum osteocalcin in postmenopausal asthmatic women. Bone 1992; 13:285–288.
68. Saarni H, Jalkanen M, Hopsu-Havu VK. Effect of five anti-inflammatory steroids on collagen and glycoaminoglycan synthesis in vitro. Br J Dermatol 1980; 103:167–173.
69. Dempster DW, Arlot MA, Meunier PJ. Mean wall thickness and formation periods of trabecular bone packets in corticosteroid-induced osteoporosis. Calcif Tissue Int 1983; 35:410–417.
70. Meunier PJ, Bressot C. Endocrine influences on bone cells and bone remodeling evaluated by clinical histomorphometry. Endocrinol Calcium Metabol 1982; 445–465.
71. Hofbauer LC, Gori FRBL, Lacey DL, Dunstan CR, Spelsberg TC, Khosla S. Stimulation of osteoprotegerin ligand and inhibition of osteoprotegerin production by glucocorticoids in human osteoblastic lineage cells: potential paracrine mechanisms of glucocorticoid-induced osteoporosis. Endocrinology 1999; 140(10):4382–4389.
72. Ebeling PR, Erbas B, Hopper JL, Wark JD, Rubinfeld AR. Bone mineral density and bone turnover in asthmatics treated with long-term inhaled or oral glucocorticoids. J Bone Miner Res 1998; 13(8):1283–1289.

73. Kukreja SC, Bowser EN, Hargis GK, Henderson WJ, Williams GA. Mechanisms of glucocorticoid-induced osteopenia: role of parathyroid glands. Proc Soc Exp Biol Med 1976; 152:358–361.
74. Cooper MS, Bujalska I, Rabbit EH, et al. Modulation of 11 beta-hydroxysteroid dehydrogenase isozymes by proinflammatory cytokines in osteoblasts: an autocrine switch from glucocorticoid inactivation to activation. J Bone Miner Res 2001; 16(6):1037–1044.
75. Plotkin LI, Weinstein RS, Parfitt AM, Roberson P, Manolagas SC, Bellido T. Prevention of osteocyte and osteoblast apoptosis by bisphosphonates and calcitonin. J Clin Invest 1999; 104(10):1363–1374.
76. Mankin HJ. Nontraumatic necrosis of bone (osteonecrosis). N Engl J Med 1992; 326(22):1473–1478.
77. Reid IR, Heap SW. Determinants of vertebral mineral density in patients receiving long- term glucocorticoid therapy. Arch Intern Med 1990; 150(12):2545–2548.
78. Gluck OS, Murphy WA, Hahn TJ, Hahn BH. Bone loss in adults receiving alternate day glucocorticoid therapy. A comparison with daily therapy. Arthritis Rheum 1985; 24:892–898.
79. Maldague B, Malghem J, DeDeuxchaisnes CN. Radiologic aspects of glucocorticoid-induced bone disease. Adv Exp Med Biol 1984; 171:155–190.
80. Tawn DJ, Watt I. Bone marrow scientography in the diagnosis of post-traumatic avascular necrosis of bone. Br J Radiol 1989; 62:790–795.
81. Askari A, Vignos PJJ, Moskowitz RW. Steroid myopathy in connective tissue disease. Am J Med 1976; 61:485–492.
82. Rebuffe-Scrive M, Krotkiewski M, Elfverson J, Bjorntorp P. Muscle and adipose tissue morphology and metabolism in Cushing's syndrome. J Clin Endocrinol Metab 1988; 67:1122–1128.
83. Reid DM, Kennedy NSJ, Smith MA. Bone loss in rheumatoid arthritis and primary generalised osteoarthrosis: effects of corticosteroids, suppressive antirheumatic drugs and calcium supplements. Br J Rheum 1986; 25:253–259.
84. Jennings BH, Anderson KE, Johansson SA. Assessment of systemic effects of inhaled glucocorticosteroids: comparison of the effects of inhaled budesonide and oral prednisolone on adrenal function and markers of bone turnover. Eur J Clin Pharmacol 1991; 40:77–82.
85. Teelucksingh S, Padfield PL, Tibi L, Gough KJ, Holt PR. Inhaled corticosteroids, bone formation, and osteocalcin. Lancet 1991; 2:60–61.
86. Ip M, Lam K, Yam L, Kung A, Ng M. Decreased bone mineral density in premenopausal asthma patients on long-term inhaled steroids. Chest 1994; 105:1722–1727.
87. Laroche H, Arlet J, Mazieres B. Osteonecrosis of the femoral and humeral heads after intraarticular corticosteroid injections. J Rheumatol 1990; 17:549–551.
87a. Adachi J, Benson W, Bianchi F. Vitamin D and calcium in prevention of corticosteroid-induced osteoporosis: follow-up over three years. J Rheumatol 1996; 23:995–1000.
87b. Amin S, Lavalley MP, Simms RW, Felson DT. The comparative efficacy of drug therapies used for the management of corticosteroid-induced osteoporosis: a meta-regressive. J Bone Miner Res 2002; 17:1512–1526.
88. Yonemura K, Kimura M, Miyaji T, Hishida A. Short-term effect of vitamin K administration on prednisolone-induced loss of bone mineral density in patients with chronic glomerulonehpritis. Calcif Tissue Int 2000; 66(2):123–128.
89. Inoue T, Sugiyama T, Matsubara T, Kawai S, Furukawa S. Inverse correlation between the changes of lumbar bone mineral density and serum undercarboxylated osteocalcin after vitamin K2 (menatetrenone) treatment in children treated with goucocorticoid and alfacalcidol. Endror J 2001; 48(1):11–18.
90. Adachi J, Bensen W, Hodsman AB. Corticosteroid-induced osteoporosis. Semin Arthritis Rheum 1993; 22:375–384.
91. Lukert BP, Johnson BE, Robinson RG. Estrogen and progesterone replacement therapy reduces glucocorticoid-induced bone loss. J Bone Miner Res 1992; 7(9):1063–1069.

92. Hall GM, Daniels M, Doyle DV, Spector TD. Effect of hormone replacement therapy on bone mass in rheumatoid arthritis patients treated with and without steroids. Arthritis Rheum 1994; 37(1499):1505.
93. Writing Group for the Women's Health Initiative Investigators. Risks and benefits of estrogen plus progestin in healthy postmenopausal women. JAMA 2002; 288(3):321–333.
94. Amin S. LM, Simms RW FDT. The comparative efficacy of drug therapies used for the management of corticosteroid-induced osteoporosis; A meta-regression. J Bone Miner Res 2002; 17:1512–1526.
95. Mulder H, Smelder HAA. Effect of cyclical etidronate regimen on prophylaxis of bone loss of glucocorticoid (prednisone) therapy in postmenopausal women. J Bone Miner Res 1992; 7:S331.
96. Reid IR, King AR, Alexander CJ, Ibbertson HK. Prevention of steroid-induced osteoporosis with (3-amino-1-hydroxypropylidene)-1,1-bisphosphonate (APD). Lancet 1988; 1:143–146.
97. Boutsen Y, Jamart J, Esselinckx W, Devogelaer JP. Primary prevention of glucocorticoid-induced osteoporosis with intravenous pamidronate and calcium: A prospective controlled 1-year study comparing a single infusion, and infusion given once every 3 months, and calcium alone. J Bone Miner Res 2001; 16:104–112.
98. Wallach S, Reid DM, Hughes RA, et al. Effects of risedronate treatment on bone density and vertebral fracture in patients on corticosteroid therapy. Calcif Tissue Int 2000; 67:277–285.
99. Reid DM, Hughes RA, Laan R, Sacco-Gibson NA, Wenderoth DH. Efficacy and safety of daily risedronate in the treatment of corticosteroid-induced osteoporosis in men and women: a randomized trial. J Bone Miner Res 2000; 15(6):1006–1012.
100. Lane NE, Sanchez S, Modin GW, Genant HK, Pierini E, Arnaud CD. Parathyroid hormone treatment can reverse corticosteroid-induced osteoporosis. J Clin Invest 1998; 102:1627–1633.
101. Lane NE, Sanchez S, Modin GW, Genant HK, Pierini E, Arnaud CD. Bone mass continues to increase at the hip after parathyroid hormone treatment is discontinued in glucocorticoid-induced osteoporosis: results of a randomized controlled clinical trial. J Bone Miner Res 2000; 27:944–951.
102. Canalis EM, Delaney AM. Mechanisms of glucocorticoid action in bone. Ann N Y Acad Sci 2002; 966:73–81.
103. Bellido T, Plotkin LI, Han L, Manolagas SC, Jilka RL. PTH prevents glucocorticoid-induced apoptosis of osteoblasts and osteoclasts in vitro: direct interference with a private death pathway upstream from caspase-3. Bone 1998; 23:S518.
104. Meunier PJ, Birancon D, Chavassieux P. Treatment with fluoride: bone histomorphometric findings. In: Osteoporosis Christiansen C, Johansen JS, Riis BJ, eds. Osteopress. Copenhagen, 1987, pp. 824–828.
105. Horber FF, Haymond MW. Human growth hormone prevents the protein catabolic side effects of prednisone in humans. J Clin Invest 1990; 86:265–272.
106. Gray RES, Doherty SM, Galloway J, Coulton L, Debroe M, Kanis JA. A double-blind study of deflazacort and prednisone in patients with chronic inflammatory disorders. Arthritis Rheum 1991; 34(3):287–295.
107. Canalis EM, Avioli LV. Effects of deflazacort on aspects of bone formation in cultures of intact calvariae and osteoblast-enriched cells. Bone Miner Res 1992; 7(9):1085–1092.
108. O'Connell SL, Tresham J, Fortune CL. Effects of prednisolone and deflazacort on osteocalcin metabolism in sheep. Calcif Tissue Int 1993; 53:117–121.
109. Vayssiere BM, Dupont S, Choquart A, et al. Synthetic glucocorticoids that dissociate transactivation and AP-1 transrepression exhibit antiinflammatory activity in vivo. Molec Endocrinol 1997; 11(9):1245–1253.
110. Manning PJ, Evans MC, Reid IR. Normal bone mineral density following cure of Cushing's syndrome. Clin Endocrinol 1992; 36:229–234.

Index

A

Acid–base balance, 279–298
  in vitro, 280–281
Acid–base imbalance
  skeleton, 240
Acidosis, 56
  acute, 281–284
  bone ions, 291–294
  chronic, 284–287
    calcium release, 284–285
  gene activity, 288–289
  prostaglandin E2, 289–291
  RANK ligand, 289–291
Acute acidosis, 281–284
Adaptation, 5
Adipocytes, 49, 552–553
Adolescence, 178–191
  calcium, 179f, 181, 182f
    net absorption, 184f
  mineral metabolism
    antiepileptic drugs, 652–653
  physical activity, 519
Adrenal glucocorticoids, 51
Africa, 7, 8, 12, 14, 441
African
  skin pigmentation, 412
African-American children, 28
  lead, 369–370
African-American women, 605
osteoporosis, 69
Age-related bone loss, 217–219
  environmental factors, 219
  heritable factors, 219
  hormonal factors, 218–219
  nutritional factors, 217–218
Aging, 133, 216f
  bone remodeling, 216–217
  ultraviolet radiation, 413
Alcohol, 96–97
  bone density, 500, 501t–503t
  forearm fractures, 506t
  fracture risk, 500–507
  hip fractures, 504t–506t
  skeleton, 499–507
  vertebral fractures, 506t
Alcoholic beverages, 242–243
  vitamin D, 407
Alkaline phosphatase, 604
Alkalosis
  metabolic, 285
Alveolar bone, 129, 130f
  erosion, 131
Alveolar bone loss
  vs systemic bone loss, 131–132
Anabolic agents, 338
Anabolic steroids, 677–678
Androgens, 53
Androstenedione, 243
Angiogenesis, 49
Ankle fractures
  smoking, 498t
Anorexia nervosa, 264, 617–619, 621f, 626f
  bone mineral density, 620f
  diagnostic criteria, 618t
Antiepileptic drugs, 647–661, 648t
  bone cell physiology, 658
  calcium
    intestinal absorption, 657
    metabolism, 653–657
    supplementation, 658–660
  chronic therapy, 658
  mechanism of actions, 653–658
  metabolic effects, 657f
  mineral metabolism, 649–653
    children and adolescents, 652–653
    clinical observations, 649–650
    neonates, 651–652
    risk factors, 650–651, 651t
  vitamin D, 653–657
    biotransformation, 656
    hepatic clearance, 654
    insufficiency, 654–655
    localization, 655
    supplementation, 658–660
    turnover, 656
Apigenin, 604
Ardipithecus ramidus, 6
Areal bone mineral density (aBMD), 20, 262
Arikara, 10
Aromatase, 53
Aseptic necrosis, 674
Asia, 7, 12, 458
Asians
  pregnancy, 151
Aspergillus sojae, 598

Athletes, 264
young
physical activity, 518–519
Atkins Diet, 560
Atrial fibrillation
rheumatic, 472
Australia, 125, 164
Australopithecus, 7
Austria, 14
Autosomal dominant osteopetrosis type I (ADOI), 23
Avascular necrosis, 674

B

Ballet dancers, 264
diagnostic criteria, 618t
Beer
vitamin D, 407
β-estradiol, 601
Beverages, 242–243
alcoholic, 242–243
vitamin D, 407
fluoride, 347–348
Bicarbonate, 294–296, 295f
Binge eating disorder (BED)
diagnostic criteria, 618t
Biological stress, 13
Biomarkers
bone metabolism, 610
Bisphosphonates, 536, 678–679
BMD. *See* Bone mineral density
Body composition, 552–555
Body mass index, 550
Body weight, 549–552
Bone
accretion, 158
carbonate, 284
fat tissue, 554
lean soft tissue, 554
metabolism and repair, 96–97
physiology, 43–57
Bone cells
fluoride, 349
lead, 367–368
Bone densitometry
biomechanical studies, 67–68
rationale, 66–68
Bone density
alcohol, 500, 501t–503t
obesity, 551
smoking, 482–492, 484t–486t, 488t–489t
Bone growth
term-corrected age through adolescence, 166–169
Bone health
adaptation, 5–6
evolutionary aspects, 3–16
Bone ions
acidosis, 291–294
Bone lead
measurement, 371–372
Bone loss
age-related, 217–219
Bone markers
dietary phytoestrogens, 602–606
Bone mass
heritability, 20–23
past populations, 9–10
prehistoric groups, 10
Bone metabolism, 575–576
biomarkers, 610
Bone mineral
dietary factors, 252t
intrauterine accretion, 158
Bone mineral content, 262
formula-fed infants, 160–162
heritability, 21
Bone mineral density
anorexia nervosa, 620f
calcium, 200
clinical use, 68–72
cystic fibrosis, 643–644
dietary phytoestrogens, 602–606
elderly, 109
forearm area, 174f
fractures, 63–79
heritability, 21, 23
historic times, 14
lead, 364f
maximal, 202–203
measuring, 63–64
monitoring therapy, 71–72
nutrition, 87–88
osteoporosis diagnosis, 68–71
patient selection for therapy, 71
research, 78–79
water fluoridation, 354f
Bone modeling, 44
maximal, 187
Bone morphogenetic protein (BMP), 54
Bone remodeling, 44, 45f, 87
activation, 45–46
aging, 216–217
cycle, 44–47
elderly, 212–215
formation, 47
resorption, 46–47

reversal, 47
sodium, 329
Bone strength
heritability, 20–23
terrain, 13
Bone turnover
calcium, 320–322
Bone turnover markers, 22
Boron, 382–383
Breast-fed infants
bone mineral content, 160–162, 161f
Brewed tea
fluoride, 348
Buddhist, 256
Bulimia nervosa, 619
diagnostic criteria, 618t
Burke history
vs 24-h recall, 109t

C

Calcaneal ultrasound (CUS), 627
Calcitonin, 47, 49, 678
Calcitriol
adolescence, 184f
Calcium, 88–89. *See also* Dietary calcium
absorption, 318–319, 558–559
cystic fibrosis, 638–639
adequacy indicators, 199–203
adolescence, 179f, 181, 182f
adult human body content, 328t
balance residuals, 313f
bone mineral density, 200
bone turnover, 320–322
box plot, 181f
chronic acidosis, 284–285
cystic fibrosis, 638–639
dietary factors, 252t
dietary reference intake
term-born infants, 159–160, 160t
endogenous excretion, 319–320
fecal, 311t
adolescence, 191f
fractures, 203–205
influx, 287f
intake association with maximal retention, 199–200
intake recommendations, 206
maternal intake, 159
metabolism, 317f, 318–322, 624–625
antiepileptic drugs, 653–657
net absorption
adolescence, 184f
net flux, 282f, 285f
pharmacotherapy, 205–206
physiology, 197–199
potassium, 331
premature infants, 163–164, 164t
regulation, 56
retention curve, 309f
urinary excretion, 319
Calcium intake
body composition, 555
geographic regions, 12f
growing individuals, 176t
physical activity, 122
VDR gene polymorphisms, 27–30
Calcium isotope tracers, 315t, 321f
Calcium release
acute acidosis, 281–283
hydrogen ion buffering, 295–296
Calcium supplements
antiepileptic drugs, 658–660
bone mass, 189f, 221
fracture prevention, 220–222
lactation, 145
periodontal disease, 131
phosphorus, 336
postpartum, 148f
Cancer
premature mortality, 432f
Candidate genes
osteoporosis, 26t
Carbamazepine, 648t, 649
Carboxyglutamyl, 459
Carotene, 391
Casein, 580f
Catarrhines, 441
Cementum, 129
Cereal grains
calcium, 12
Children
African-American, 28
lead, 369–370
bone accretion, 176
calcium
net absorption, 184f
exercise intervention trials, 519–520
Mexican, 28
mineral metabolism
antiepileptic drugs, 652–653
physical activity, 519
rachitic
Nigeria, 28, 30
renal tubular acidosis, 280
urinary calcium, 185f
vitamin D, 428–429

China, 13, 398, 602, 603
Chinese, 329
Chinese women, 605
Chloride
  adult human body content, 328t
Chronic acidosis, 284–287
  calcium release, 284–285
Clonazepam, 648t
Clothing
  ultraviolet radiation, 413
Computerized National Data Set, 114
Congestive heart failure, 432
Copper, 377–378, 380
  cystic fibrosis, 642–643
  deficiency, 377–378, 378f
    animal studies, 379
    human studies, 379–380
Coronary artery disease, 432
Corticosteroid, 534
Coumestans, 594
Crohn's disease, 430f, 641–642
Cultivated foods
  calcium, 12
Culture, 5
Cushing's syndrome, 674
Cyclooxygenase (COX-2), 55
Cystic fibrosis, 637–645
  body weight, 638
  bone mineral density, 643–644
  calcium, 638–639
    absorption, 638–639
    dermal calcium loss, 640
    endogenous fecal calcium excretion, 640
    urinary calcium excretion, 639–640
  copper, 642–643
  genetics, 644
  metabolic bone disease, 643–644
  osteopenia, 644
  osteoporosis, 644
  vitamin D, 641–642
    absorption, 641–642
    deficiency, 644
  vitamin K, 642
  zinc, 642
Cystic fibrosis transmembrane conductance regulator (CFTR), 637
Cytokines, 54

D

Daidzein, 594, 596–597, 604
  animal studies, 598–602
Dancers
  ballet, 264
Deflazacort, 679
Degenerative disease, 75
Dehydroepiandrosterone sulfate (DHEAS), 624
Dental fluorosis, 345, 351–352
Dental products
  fluoride, 348–349
Deoxypyridinoline, 380, 604
  animal studies, 598–602
Depression
  prevention, 526t
Determined osteoprogenitor cells (DOPC), 48
Dexamethasone, 672
Diary
  dietary, 120
Diet
  bone, 236–237
  calcium content, 8
Dietary Approaches to Stopping Hypertension (DASH), 242
Dietary assessment
  bone studies, 121–125
  meta-analysis, 125
  prospective studies, 123–124
    vs retrospective studies, 124–125
  questionnaire, 125
  retrospective studies, 122–123
Dietary calcium
  decline, 12
  evolutionary perspective, 11–13
  hip fracture, 124
  intake, 108
  WHO recommendation, 186
Dietary diary, 120
Dietary factors
  bone mineralization, 252t
Dietary intake, 120
Dietary phytoestrogens, 602–606
  bone markers, 602–606
  bone mineral density, 602–606
Dietary protein, 578–581
Dietary record, 120
Disodium-monofluorophosphate, 346
Distal forearm fractures
  smoking, 497t
Dual-energy X-ray absorptiometry (DXA), 63, 64–65, 626–627
  artifacts, 75
  clinical interpretation, 72–78
    site selection, 72–78
  clinical reporting, 78
  limitations, 64–65
  monitoring, 78
  osteoporosis

diagnosis, 77–78
positioning mistakes, 74
understanding printouts, 73–77
Dubbo Osteoporosis Study, 108
Duplicate analyzed meal samples, 121
DXA. *See* Dual energy X-ray absorptiometry

E

Early primate evolution, 441
Eating disorder not otherwise specified (EDNOS), 619
diagnostic criteria, 618t
Eating disorders, 617–629
bone health management, 626–628
diagnostic criteria, 618t
low bone density, 619–625
Eating Disorders Inventory, 617
Eicosanoids, 577
Eicosapentaenoic acid (EPA), 578
Elderly, 89, 211–223. *See also* Older men; Older women
bone health, 212–215
bone mineral density, 109
bone remodeling, 212–215
exercise recommendations, 528t
falls, 219–220
fractures, 219–220
prevention, 220–223
nutritional status, 212–215
protein intake
hip fracture, 268
respected positions, 15
vitamin D deficiency, 419–420, 420f
vitamin K, 464
warfarin, 469
Endochondral bone
lead, 365–366
Energy restriction
bone quality, 563
England, 14
Enterodiol, 604
Enterolactone, 604
Environments
expansion, 8–9
Eocene period, 441
Epidermal growth factor (EGF), 49
platelet-derived growth factor, 55
EPIDOS study, 517
Equol, 604
ESHA Food Processor, 120
Eskimo, 10
Essential fatty acids, 576–577
Estradiol, 53, 601
Estrogen, 52–53, 213
bone resorption, 595, 620
Estrogen receptor gene polymorphisms, 30–33
bone mineral density, 32
Etetrinate
adverse effects, 398–399
Ethiopia, 6
Ethosuximide, 648t
Etidronate, 678
Europe, 12, 14, 458
European Vertebral Osteoporosis Study (EVOS), 108
Evolution, 4–6, 4f
mechanisms, 4–5
time line, 6f
Exercise, 264, 515–541
elderly, 528t
estrogen replacement therapy, 535–536
older men, 533–534
older women, 529–533
Exercise programs
practical implementation, 537–541
Extremely low birth-weight (ELBW) infants
bone mineral content, 160
bone mineralization, 165–166

F

Falls, 540t
elderly, 219–220
prevention, 526t
Fan beam scanners, 64
Fatty acids
omega-3, 576–585
Fecal calcium, 311t
adolescence, 191f
Fecal phosphorus, 336
adolescence, 191f
Felbamate, 648t
Femoral neck, 75
Femur
DXA scan, 75f
QTL, 23
Fetus
bone development, 158–159
bone turnover, 140f
calcium needs, 158
intestinal calcium absorption, 140f
renal calcium excretion, 140f
skeleton
nutritional needs, 157–159
Fibroblast growth factor (FGF), 44
Fibromyalgia, 419
Fish
fluoride, 348

Fish oil, 393, 578
Flaxseed (*Linum usitatissimum*), 601
Fluoride, 57
  bone cells, 349
  bone health, 345–358
  brewed tea, 348
  calcified tissues, 349–350
  dental products, 348–349
  fish, 348
  gastrointestinal absorption, 346
  high exposure, 351–352
  with hormone replacement therapy, 356
  insufficiency, 350
  intake, 347–349
  metabolism, 346–347
  molecular mechanism of action, 349
  renal excretion, 346–347
  skeletal accretion, 346
Fluorine, 345
Fluorosis
  dental, 345, 351–352
Food(s)
  acidity, 242
  cultivated
    calcium, 12
  fluoride, 348, 348t
Food composition databases, 114
Food composition tables, 114
Food diary, 110
Food-frequency questionnaire, 106
Food groups, 235–246, 236f
  recommendations, 235–236
Food history, 110
Food intake, 108t
  diet-assessment method, 110t
Food production, 9–13
Food record, 110
Foot fractures
  smoking, 498t
Forearm
  bone mineral area density, 174f
  scan, 72
Forearm fractures
  alcohol, 506t
Formula-fed infants
  bone mineral content, 160–162, 161f
Fracture(s), 75. *See also* Hip fractures; specific area
  bone mineral density, 63–79
  calcium, 203–205
  calcium supplements, 220–222
  elderly, 219–220
  forearm
    alcohol, 506t
  osteoporotic, 11
    rehabilitation after, 534
    risk factor, 526t
  prevention
    elderly, 220–223
  spinal compression, 540t
  vertebral, 75
    alcohol, 506t
  water fluoridation, 355f
Fracture callus, 91
Fracture healing
  nutrition, 88–89
  repair, 86–87
Fracture risk, 15
  alcohol, 500–507
Fracture threshold, 70
Fracture trials, 66–67
Framingham Osteoporosis cohort, 29
Framingham study
  food groups, 237
France, 357
Fruit, 240–242
  bone, 242, 244t

G

Gabapentin, 648t
γ-carboxyglutamyl (Gla), 459
γ-linolenic acid, 578
Gastrointestinal contrast, 77
Gathering, 9–13
Gender
  vitamin K, 464
Gene activity
  acidosis, 288–289
Gene flow, 5
Genetic drift, 5
Genetic markers, 555
Genetics, 19
  cystic fibrosis, 644
Genistein, 594, 595, 596f, 597, 604
  anabolic actions, 599f
  animal studies, 598–602
  bone loss, 600f
  postmenopausal women, 610, 611f
  skeleton, 600
Genome-wide linkage studies, 19
Genome-wide screening, 23
Germany, 353
Ghrelin, 52
Gingiva, 129, 130f
Gingivitis, 129
Gla, 459

Glass
ultraviolet radiation, 413
Glucocorticoid-induced osteoporosis, 667–680
antiresorptive drugs, 677–678
bone, 672–673
bone density, 667–668, 675
bone formation, 672–673
bone metabolism, 673–674
bone resorption, 673
calcium, 676–677
renal excretion, 670
clinical features, 667–668
cytokines, 672
1,25$(OH)_2D$, 671
fracture risk, 668
gonadal hormones, 670–671
growth factors, 672
growth hormone secretion, 671
histology, 668
muscle strength, 675
osteonecrosis, 674
parathyroid hormone, 668–669, 671
pathogenesis, 668–674
phosphorus
renal excretion, 670
prostaglandins, 671
radiological evaluation, 675
vitamin D, 668–669, 676–677
calcium absorption, 668–669
vitamin K, 670, 677
Glucocorticoid response element (GRE), 52
Glucocorticoids
adrenal, 51
Glycogen, 87
GnRH deficiency, 620
Grains
calcium, 12
Growth hormone (GH), 213, 261, 622–623
glucocorticoid-induced osteoporosis, 671
Growth hormone-insulin-like growth factor (GH-IGF) system, 50–51
Growth plate
lead, 365–366, 366f
Growth-regulating hormones, 50–55
Growth retardation
insulin-like growth factor-1, 214

H

Harvard-Willett diet assessment questionnaires, 108
Health diet
guidelines, 236t
Herbal supplements, 593–594
Heritability, 19
age-related bone loss, 219
bone mass, 20–23
bone mineral content, 21
bone mineral density, 21, 23
bone strength, 20–23
defined, 21
High bone mass (HBM), 23
High-protein diet, 8
Hip fractures, 8
alcohol, 504t–506t
dietary calcium, 124
parity, 149f
relative risk, 67f
smoking, 492f, 493t, 494t–495t
water fluoridation, 354f
Hominid clade, 6
Homo, 6
Homo erectus, 7
Homo sapiens, 7
Hong Kong, 329
Hormonal factors
age-related bone loss, 218–219
Hormone replacement therapy (HRT), 30–31, 677–678
bone mineral density, 32
calcium, 205
fluoride, 356
obesity, 551
Human evolution
course, 6–7
Hunting, 9–13
Hydrogen ion buffering, 283
24,25$(OH)_2D_3$, 91
25(OH)D
serum, 426–429
1,25-(OH)D, 49, 50, 91, 197
Hyperparathyroidism, 421, 668–669
Hyperprolactinemia, 620
Hypertension, 431, 432
ultraviolet radiation, 433f
Hypoestrogenism, 619–622
Hypophosphatemia, 337, 421
Hypoprothrombinemia, 466
Hypovitaminosis, 90

I

Iatrogenic fluorosis, 351
Impaired balance
prevention, 526t
Indian, 13
Inducible osteoprogenitor cells (IOPC), 48
Industrial fluorosis, 351

Infectious disease, 13
Insulin-like growth factor-1 (IGF-1), 43–44, 54–55, 96, 179, 212–214, 628f
  adolescence, 191f
  bone mineral density, 270
  calcium phosphate metabolism, 262f
  elderly, 215
  nutrient regulation, 214, 214f
  protein intake, 269
Insulin-like growth factor-2 (IGF-2), 43–44
Insulin-like growth factor binding protein (IGFBP), 602
Interferon-β, 54
Interleukin-1 (IL-1), 54, 620, 672
Interleukin-6 (IL-6), 620, 672
  gene promoter polymorphisms, 33–34
Intestinal malabsorption syndromes, 421, 429–430
Intracellular vitamin D
  trafficking, 452
Intracellular vitamin D-binding proteins, 449–452
Isoflavones, 594, 603f, 604
  safety, 611–612
Isotretinoin
  adverse effects, 398–399

J

Japan, 30, 178, 604
  calcium, 188f
Japanese
  retrospective studies, 122
Japanese women, 605

K

Kenya, 6
Ketogenic diet
  low-carbohydrate, 560
Kidney, 432
Kinetic studies, 316–318
  mathematical modeling, 316–317
Korea, 30, 600–601, 604
K-shell X-ray fluorescence (KXRF), 371–372

L

Lactation, 144–146
  bone mineral, 145–146
  calcium and bone metabolism, 144
  calcium supplements, 145
  osteoporosis
    epidemiological studies, 147–149
  postpartum, 148f
Lacto-ovo-vegetarians (LOV), 249
  defined, 250t
  vs OMNI, 253–256
  vs omnivores, 123
Lamotrigine, 648t
Lateral spine
  scan, 72–73
Lateral vertebral assessment (LVA), 73
Latin America, 12
Lead
  bone cells, 367–368
  bone mineral density, 364f
  clinical studies, 369–370
  drinking water
    bone density, 369f
  DXA, 370–371
  skeletal mass, 368
  skeleton, 364–368
    in vivo, 368–370
Lemurs, 441
Leptin, 52, 553–554, 624
Leukotrienes, 54
Life expectancy, 15
  increase, 14–16
Lignans, 594, 595
Linkage disequilibrium, 32
Linolenic acid, 578
Linum usitatissimum, 601
Lipoprotein receptor related-5 (LRP-5), 55
Liver
  menaquinone, 458
  Local regulatory factors, 54–56
  Longevity
  increase, 14–16
Looser's pseudo-fractures, 421
LOV. *See* Lacto-ovo-vegetarians (LOV)
Low birth-weight (LBW) infants
  copper, 378
Low body weight, 551
Low-carbohydrate ketogenic diet, 560
Lumbar spine
  bone mineral content
    protein intake, 263f
  scan, 73

M

Macrobiotic diet
  defined, 250t
Macrophage colony-stimulating factor (M-CSF), 46f, 581
Madagascar, 441
Magnesium, 237, 339–340
  adult human body content, 328t
  dietary reference intake
    term-born infants, 160t

premature infants, 164t, 165
recommended daily allowance, 339
regulation, 56
Malnutrition
elderly, 87
Maternal calcium intake, 159
neonatal bone, 149–150
Maternal diet
neonate, 149–152
Maternal vitamin D intake
neonatal bone mineral, 150–152
neonatal calcium, 150–152
neonatal vitamin D, 150–152
Matrix Gla protein (MGP), 288, 289, 460
time course, 290f
Maximal bone modeling, 187
Meat
skeleton, 239–240
Medial femoral neck, 75
Medication-exercise interactions, 534–535
Mediterranean Osteoporosis Study (MEDOS), 109, 205
Men
exercise, 533–534
osteoporosis, 69
Menaquinone, 457
Menaquinone-4 (MK-4), 457–458, 461, 465, 469
Menaquinone-6 (MK-6), 469
Menaquinone-7 (MK-7), 464
Menarche, 180
Menkes' kinky hair syndrome, 377
Menopause, 576
Metabolic acidosis
bone, 297f
in vivo effects, 279–280
Metabolic alkalosis, 285
Metabolic balance studies, 308–314
application, 308–309
conducting, 309–312
free-living subjects vs metabolic ward, 313–314
monitoring compliance, 312–313
Metabolic bone disease
cystic fibrosis, 643–644
Metacarpal
cortical area, 180f
cortical bone mass, 183f
Methodists, 254
Mexican children, 28
Microminerals, 377–385
Middle East, 13
Milk
bone mass, 237–238
bone turnover, 140f
fluoride, 347–348
fracture, 239
intestinal calcium absorption, 140f
postmenopausal bone loss, 238–239
renal calcium excretion, 140f
Minerals
adult human body content, 328t
antiepileptic drugs, 649–653
infants, 175f
MK-4, 457–458, 461, 465, 469
MK-6, 469
MK-7, 464
Modern medical care
chronic disease, 15
Modified Block food-frequency method, 110
Multiple sclerosis, 431
Muscle
glucocorticoid-induced osteoporosis, 675
weakness prevention, 526t
Mutations, 5

N

National Center for Complementary and Alternative Medicine, 594
National Health and Nutrition Examination Survey (NHANES), 119, 397
Neanderthals, 7
Neolithic transition, 13
Neonatal bone mineral content, 150f
Neonatal bone mineralization, 165–166
Net endogenous acid production (NEAP), 333
Netherlands, 458
Neuropeptides, 54
New World primates, 442
evolution, 442f
intracellular vitamin D-binding proteins, 451f
Los Angeles zoo
rickets, 444–446
rachitic
biochemical phenotype, 445f
steroid hormone resistance, 443–444
vitamin D, 443–444, 446f
biochemical nature, 446–448
response element-binding protein, 447, 447f, 448f
synthesis, 445f
New Zealand, 178
NHANES, 119, 397
NHANES-II, 107, 108
NHANES-III, 131
Nitric oxide, 54
Nonlactating nonpregnant adolescents, 145

Nurses' Health Study, 239, 500
  Dietary Questionnaire, 106, 107f
Nutrient-fortified formulas, 168
Nutrients, 105–110
  adolescence, 192
Nutrition
  age-related bone loss, 217–218
  bone loss, 10
  fracture healing, 85–97
  oral bone status, 129–134
  orthopedic patients, 87–88
Nutritional assessment, 105–106, 113–126
  analysis of efficacy, 115–117
  dietary assessment, 119–121
  efficacy tests, 118t
  measurements, 116t
  power computations, 117
  randomization, 118–119
  sample size, 117–118
  statistical considerations, 114–119
  study design, 114–115
  study protocol, 115–119
  variables, 113–114
Nutrition Committee for the European Society of Pediatric Gastroenterology, Hepatology and Nutrition, 164
Nutrition Screening Initiative Checklist for Nutrition Risk, 105

O

Obesity, 549–551
  bone, 550–551
  bone density, 551
  bone turnover, 551–552
  DXA, 563f
    measurement error, 562–563
  fracture, 552
  hormone replacement therapy (HRT), 551
  vitamin D deficiency, 421
  weight reduction, 558f
Office of Alternative Medicine, 594
Offspring Cohort of the Framingham Heart Study, 34
Older men
  exercise, 533–534
Older women
  aerobic exercise vs resistance training, 530–532
  exercise, 529–533
  high-impact exercise, 532–533
Omega-3 fatty acids, 576–585
Omnivores, 242
  bone mineral density, 253–257
Omnivorous diet, 249
Oral anticoagulants, 457–474
  bone loss, 470
Oral contraceptives, 535
Orange juice
vitamin D, 409
Orlistat, 560
Orrorin tugenensis, 6
Osteoarthritis
  hip, 75
  pain, 540t
Osteoblast–osteoclast interaction, 46f
Osteoblasts, 48, 49, 212, 552–553, 575–576
  collagen synthesis, 293f
  lead, 367
  synthesis, 214
Osteocalcin, 92–93, 380, 460, 468, 604
Osteoclastogenesis, 584f
Osteoclasts, 86, 576
  lead, 367–368
  lineage, 47–48
Osteocytes, 48
Osteomalacia, 334, 417, 418f
Osteonecrosis
  glucocorticoid-induced osteoporosis, 674
Osteopenia
  cystic fibrosis, 644
  prevention, 526t
Osteophytes, 76f
Osteopontin, 288, 289f
Osteoporosis, 352f, 575–585. *See also* Glucocorticoid-induced osteoporosis
  African-American women, 69
  bone mineral density, 68–71
  calcium therapy, 221t
  cost, 15
  cystic fibrosis, 644
  diagnosis, 77–78
  diet, 240, 576–585
  dual energy X-ray absorptiometry (DXA), 77–78
  elderly, 264
  estrogen, 52–53
  exercise risks, 540t
  family history, 20–22
  fluoride, 355–358
  food-frequency method, 107–109
  historic times, 14
  lactation
    epidemiological studies, 147–149
  peripheral device diagnosis, 69–70
  phosphorus, 337f
  prognosis, 71

protein replenishment, 267–269
skeleton
lead toxicity, 363–372
smoking
estrogen replacement therapy, 487
sodium, 330
undernutrition, 267–269
vitamin D deficiency, 9
vitamin D therapy, 221t
WHO definition, 148
Osteoporosis-pseudoglioma (OPPG), 23, 55
Osteoporotic fractures, 11
rehabilitation after, 534
risk factor, 526t
Osteoprotegerin (OPG), 45, 218, 581, 620, 679
Osterix, 43
Oxacarbazepime, 648t
Oxalates, 88
24,25$(OH)_2D_3$, 91
25(OH)D
serum, 426–429, 427f, 428f
1,25-(OH)D, 49, 50, 91, 197

P

Paleolithic population, 11–12
Pamidronate, 678
Parathyroid hormone (PTH), 49, 89, 213, 280, 679
acidosis, 286f, 287f
glucocorticoid-induced osteoporosis, 668–669, 671
maximal suppression, 201–202
serum, 559f
vitamin D deficiency, 417
Parathyroid hormone-related peptide (PTHRP), 44, 49
Parity
osteoporosis
epidemiological studies, 147–149
PA spine, 71–72
DXA, 74f, 76f
vertebral bodies, 74
Paternal descent
bone mass heritability, 22
Peak bone mass
physical activity, 520–521
Pencil beam scanners, 64
Periodontal disease, 129–130
calcium supplements, 131
Periodontal ligament, 129
Periodontitis, 129, 130–131
Peripheral devices, 66
Peripheral dual energy X-ray absorptiometry (pDXA), 63, 66
Peripheral quantitative computed tomography (pQCT), 63, 66
Phenobarbital, 648t, 649
Phenytoin, 648, 648t
Phosphates, 295f
Phosphorus, 189–190, 237, 334–339
adult human body content, 328t
calcium supplements, 336
dietary reference intake
term-born infants, 159–160, 160t
fecal, 336
adolescence, 191f
glucocorticoid-induced osteoporosis
renal excretion, 670
malnutrition, 92
natural food, 335
premature infants, 163–164, 164t
recommended daily allowance, 336
Phylloquinone
intake, 458, 465–469
liver, 466
serum, 466
Physical activity, 13–14, 516–518
adolescence, 518–521
calcium intake, 122
childhood, 518–521
elderly, 529–534
historic times, 14
osteoporosis, 517
prehistoric times, 13
young women, 525–529
frequency, 527–529
intensity, 527
modality, 525–527
volume, 527
Physical limitations
exercise, 538t–539t
Phytoalexins, 598
Phytoestrogens, 593–612
anabolic effects, 597–598
animal studies, 598–602
bone resorption, 595–597
clinical trials, 606–611, 607t
defined, 594–595
dietary, 602–606
bone markers, 602–606
bone mineral density, 602–606
mechanisms of action, 595–606
randomized trials, 607–611
Plant domestication, 13
Platelet-derived growth factor (PDGF), 55
Platyrrhines, 441, 442
Plums, 602

Polypharmacy
  prevention, 526t
Poor nutrition, 13
Population-based association studies, 25–35
Postmenopausal women
  bone mineral density, 29
  C-terminal crosslinks of Type 1 collagen (CTx), 33
  genistein, 610, 611f
  LOV vs OMNI, 254–255
  sodium, 332f
  vegans
    vs OMNIs and LOVs, 256–257
Postpartum calcium supplements, 148f
Potassium, 331–334
  adult human body content, 328t
  calcium, 294f, 331
  in food, 332
  urine calcium, 333f
Potassium exchange, 284
Potential renal acid load (PRAL), 242, 243t
Prealbumin, 268
Prednisone, 534
Pregnancy
  bone mineral, 142–143
  calcium and bone metabolism, 139–142
Premature breast-fed infants
  whole-body bone mineral content, 168f
Premature infants
  bone growth nutrient needs, 162–165
  calcium, 163–164, 164t
  magnesium, 164t
  skeletal development nutrient needs, 162–169
  whole-body bone mineral content, 167f
Premenopausal women
  bone mineral density
    alcohol, 243
  IL-6 polymorphisms, 34
  LOV vs OMNI, 255–256
  physical activity, 521–529
    experimental trials, 521–525
    meta-analyses, 522t–523t
Preosteoblasts, 212
Previtamin D_3, 412f, 415f
Primary healing, 86
Primates. *See* New World primates
Primidone, 648t, 649
Prostaglandin E2 (PGE2), 54–55, 291f, 620
  acidosis, 289–291
Prostaglandins, 54
  glucocorticoid-induced osteoporosis, 671
Protein, 10, 94–96
  breakdown, 87
  dietary, 578–581
  fracture healing, 95
  skeleton, 239–240
Protein calorie malnutrition, 190
Protein Induced by Vitamin K Absence Factor II (PIVKA-II), 473
Protein intake, 261–272
  bone growth, 261–262
  bone mass, 263–264
  bone mineral density, 265–267
  elderly, 264
  osteoporotic fracture, 265–267
Protein-related bone loss, 271
Protein S, 460–461
Protein supplements, 223
  femur bone loss, 268f
Proton
  net flux, 283f
Provitamin D_3, 412f
Proximal femur, 72
Proximal humerus fractures
  smoking, 497t
PTH. *See* Parathyroid hormone
Pueblo, 10
Pyridinoline, 380
  animal studies, 598–602

Q

Qualitative food frequency, 120
Quantitative clinical nutrition, 307–308
Quantitative computed tomography (QCT), 63, 65–66, 627
  vs DXA, 65
  limitations, 65
Quantitative food frequency, 120–121
Quantitative trait loci (QTL), 19, 24t
  bone mass, 23–25
Quantitative ultrasound (QUS), 63

R

Rachitic bone lesions, 442, 442f
Rachitic children
  Nigeria, 28, 30
Rancho Bernardo Study, 108
Receptor activator of NFKB ligand (RANKL), 45, 50, 212, 417, 582
  acidosis, 289–291
  metabolic acidosis, 292f
Recombinant human growth hormone, 679
Renal failure, 285
Renal tubular acidosis
  children, 280
Resistance training
  intensity, 532

Retinoic acid receptors, 94
Rheumatic atrial fibrillation, 472
Rickets, 176, 415, 416f

S

Sadlermiut Inuit, 10
Safflower seed, 600–601
Salt. *See* Sodium
Sclerosis, 76f
Secondary healing, 86
Second National Health and Nutrition Examination Survey (NHANES-II), 107, 108
Sedentism, 13–14
Selective estrogen receptor modulators (SERM), 53
Semivegetarian diet
  defined, 250t
17β-estradiol, 601
Seventh Day Adventists, 242, 254
Sex hormone-binding globulin (SHBG), 218
Sex hormones, 52–53
Silicon, 384
Simian bone disease, 442–443
Single nucleotide polymorphisms (SNP), 25–26
Skeletal development
  peak bone mass, 174–176
Skeletal fluorosis, 351
Skeletal lead
  systemic toxicity, 371
Skeleton
  acid–base imbalance, 240
  alcohol, 499–507
  development, 43–44
  fetus
    nutritional needs, 157–159
  fluoride insufficiency, 350
  functions, 3
  genistein, 600
  interactions among factors, 213f
  lead, 364–368
    in vivo, 368–370
  lead toxicity, 363–372
  purposes, 212
  smoking, 481–482
    lifespan changes, 482
Skin pigmentation
  vitamin D, 413f
Smoking, 133, 481–499
  ankle fractures, 498t
  bone density, 482–492, 483–485, 484t–486t, 488t–489t, 490
  bone loss, 492t
  distal forearm fractures, 497t
  effect size, 490f
  foot fractures, 498t
  fracture risk, 492–499
  hip fractures, 492f, 493t, 494t–495t
  osteoporosis
    estrogen replacement therapy, 487
  peak bone mass, 482–492
  proximal humerus fractures, 497t
  skeleton, 481–482
    lifespan changes, 482
  vertebral fractures, 496t
Socioeconomic cost
  disease with aging, 15
Socioeconomic status
  elderly, 15
Sodium, 328–329
  adult human body content, 328t
  bone loss, 330
  bone remodeling, 329
  osteoporosis, 330
  postmenopausal women, 332f
  urine calcium, 333f
Sodium exchange, 284
Somatomedin-C, 672
South America, 441
Soy, 580f
Soy Health Effects (SHE) study, 604
Soy protein, 579, 594
Spain, 187
Special diet, 560–561
Spine
  bone density
    premenopausal women, 255
  compression fractures, 540t
  QTL, 23
Steroids
  anabolic, 677–678
Stevens-Johnson syndrome, 649
Stress
  biological, 13
Strontium, 57, 383–384
Study of Osteoporotic Fractures, 517
Study of Women's Health Across the Nation (SWAN), 605
Summer
  zenith angle, 414
Sunlight, 9
Sunscreens, 412–413
Sweden, 14, 398, 482
Sydney Twin Study of Osteoporosis, 21
Syncytium, 48
Synthetic retinoids
  adverse effects, 398–399

Systemic hormones
bone remodeling, 49–50

T

Teeth
fluoride, 349–350
support, 129
Term-born infants
dietary reference intake, 159–160
skeletal development nutrient needs, 159–162
Testosterone, 53, 624
TGF-β, 54, 55, 212, 620
Third National Health and Nutrition Examination Survey (NHANES-III), 131
Thyroid hormone, 51, 534
Thyroxine, 213
Tiagabine, 648t
TNF-α, 54, 86, 271, 582, 595, 620
Tooth loss, 129–130, 131
elderly, 133
Toothpaste
fluoride, 348–349
Total body bone mineral density (TBBMD)
growth, 177f
weight loss, 557f
Total body nitrogen
infants, 175f
Total parenteral nutrition (TPN)
calcium, 331
cholestasis, 379
Toxic epidermal necrolysis, 649
TPN
calcium, 331
cholestasis, 379
Tracer studies, 314–318
application, 314
calcium isotopes, 314–316
kinetic studies, 316–318
Transforming growth factor beta (TGF-β), 54, 55, 212, 620
T-score, 69, 70
Tumor necrosis factor-α (TNF-α), 54, 86, 271, 582, 595, 620
Twenty-four-hour diet recall, 109, 119–120

U

Ultratrace minerals, 384
Ultraviolet B (UVB), 409–411
aging, 413
clothing, 413
nursing home residents, 431f
Undercarboxylated osteocalcin (ucOC), 462, 466
Urinary calcium
adolescence, 186f
children, 185f
sodium, 328
Urinary pyridinoline, 604

V

Valpoic acid, 648, 648t
Vascular endothelial growth factor (VEGF), 55
Vegan diet
defined, 250t
Vegans
vs OMNIs and LOVs, 256–257
Vegetables, 240–242
bone, 242, 244t
Vegetarian diet, 249
acid/alkali balance, 251
calcium, 250–251
defined, 250t
lifestyle factors, 252–253
oxalates, 252
phytates, 252
phytosterols, 252
potassium, 333
protein, 251
vitamin D, 251
Vegetarianism, 241t, 242
bone health, 240–242
bone mineral density, 249–253
women, 249–258
Vegetarians
bone mineral density, 253–257
VEGF, 55
Vertebral fractures, 75
alcohol, 506t
smoking, 496t
Vitamin A, 94, 391–399
animal data, 394–395
high intake, 391–392, 396t
human data, 395–398
follow-up studies, 397–398
interventional studies, 398–399
intake assessment, 393–394
intoxication, 392–393
Vitamin C, 133
Vitamin D, 89–91, 403–435
1,25-dihydroxyvitamin D $1,25(OH)_2D$, 141
adequacy indicators, 201–202
adequate intake vs healthy intake, 425–426
alcoholic beverages, 407
antiepileptic drugs, 653–657
beer, 407, 410f
bone calcium mobilization, 406, 407f
bone mineralization, 406

calcium balance, 133
cancer, 433–434
children, 428–429
deficiency, 9, 415–420
causes, 420–421
cystic fibrosis, 644
diagnosis, 421–424
elderly, 419–420, 420f
epiphyseal plates, 408f
nonskeletal consequences, 430–433
obesity, 421
treatment, 429–430
dietary reference intake
term-born infants, 159–160, 160t
dietary sources, 406–409
elderly, 215
evolution, 403
fractures, 204–205
glucocorticoid-induced osteoporosis, 668–669, 676–677
calcium absorption, 668–669
hormone system, 50
insufficiency, 198f
intake recommendations, 206
intestine, 403–405
intracellular
trafficking, 452
malnourishment, 90
maternal intake
neonatal bone mineral, 150–152
metabolism, 403–405
orange juice, 411f
25-(OH)D, 49, 50, 91, 197
pharmacotherapy, 205–206
physiology, 199
premature infants, 164, 164t
production, 404f
reasonable daily allowance, 425t
serum, 422f, 423f, 424f
serum 25(OH)D, 426–429
subhuman primates, 441–454
sunlight, 409–411
sunscreen, 414f
supplements, 203t, 222
periodontal disease, 131
synthesis, 9
winter, 123
Vitamin D_3, 412f, 415f
cutaneous production, 411–415
$24,25(OH)_2D_3$, 91
Vitamin D-binding proteins
hsp-70 related intracellular, 450f
intracellular, 453f
Vitamin D receptor (VDR) polymorphisms
bone mineral density, 31f
calcium intake, 27–30
Vitamin D response element-binding protein (VDRE-BP), 448–449
Vitamin K, 92–93, 457–474
adequate intake, 458
age-related bone loss, 464
animal models, 461–462
biomarkers, 467t
bone resorption markers, 468
clinical trials, 465
cycle, 459f
cystic fibrosis, 642
elderly, 464
epoxide, 460
fracture healing, 93
glucocorticoid-induced osteoporosis, 670, 677
mechanisms, 459–460
observational studies, 462–465, 463t
optimal dietary intake, 465–469
sources, 457–458
structure, 458f
supplements, 468
urinary calcium excretion, 468–469
Vitamin K_2, 457
Volumetric bone mineral density
heritability, 21

W

Warfarin, 469–473, 471t
bone density, 470
elderly, 469
fracture risk, 470
Water
fluoridation, 347–348
bone mineral density, 354f
fracture prevention, 353–355
fractures, 355f
Weaning
bone mineral, 145–146
bone turnover, 141f
calcium and bone metabolism, 145
intestinal calcium absorption, 141f
renal calcium excretion, 141f
Weighted vest, 537
Weight-lifting, 537
Weight recovery
bone loss, 625–626
Weight reduction, 555–564
calcium, 561–562
fast, 559

involuntary, 555–556
involuntary vs voluntary, 556f
nontraditional methods, 560–561
obesity, 558f
physical activity, 562
vitamin D, 561–562
voluntary, 556–562
Weight regain, 564
Western diet
calcium deficiency, 189
Willett questionnaire, 125
Winter
zenith angle, 414
World Health Organization (WHO) diagnostic criteria
limitations, 70

Y

Young athletes
physical activity, 518–519

Z

Zearalanol, 599
Zenith angle
ultraviolet radiation, 413
Zinc, 237, 381–382
cystic fibrosis, 642
Z-score, 68, 69

About the Series Editor

Dr. Adrianne Bendich is Clinical Director of Calcium Research at GlaxoSmithKline Consumer Healthcare, where she is responsible for leading the innovation and medical programs in support of TUMS and Os-Cal. Dr. Bendich has primary responsibility for the direction of GSK's support for the Women's Health Initiative intervention study. Prior to joining GlaxoSmithKline, Dr. Bendich was at Roche Vitamins Inc., and was involved with the groundbreaking clinical studies proving that folic acid-containing multivitamins significantly reduce major classes of birth defects. Dr. Bendich has co-authored more than 100 major clinical research studies in the area of preventive nutrition. Dr. Bendich is recognized as a leading authority on antioxidants, nutrition, immunity, and pregnancy outcomes, vitamin safety, and the cost-effectiveness of vitamin/mineral supplementation.

In addition to serving as Series Editor for Humana Press and initiating the development of the 15 currently published books in the *Nutrition and Health*™ series, Dr. Bendich is the editor of nine books, including *Preventive Nutrition: The Comprehensive Guide for Health Professionals*. She also serves as Associate Editor for *Nutrition: The International Journal of Applied and Basic Nutritional Sciences*, and Dr. Bendich is on the Editorial Board of the *Journal of Women's Health and Gender-Based Medicine*, as well as a past member of the Board of Directors of the American College of Nutrition.

Dr. Bendich was the recipient of the Roche Research Award, a *Tribute to Women and Industry* Awardee, and a recipient of the Burroughs Wellcome Visiting Professorship in Basic Medical Sciences, 2000–2001. Dr. Bendich holds academic appointments as Adjunct Professor in the Department of Preventive Medicine and Community Health at UMDNJ, Institute of Nutrition, Columbia University P&S, and Adjunct Research Professor, Rutgers University, Newark Campus. She is listed in *Who's Who in American Women*.

About the Editors

Michael F. Holick, PhD, MD is Professor of Medicine, Dermatology, Physiology, and Biophysics, Director of the General Clinical Research Center at the Boston University School of Medicine, and Director of the Bone Health Care Clinic at Boston University Medical Center. He received his PhD, MD degrees from the University of Wisconsin, Madison. Dr. Holick's primary focus is internal medicine/endocrinology. He has also made numerous contributions to the field of the biochemistry, physiology, metabolism, and photobiology of vitamin D for human nutrition. He determined the mechanism for how vitamin D is synthesized in the skin, demonstrated the effect of aging, obesity, latitude, seasonal change, sunscreen use, skin pigmentation, and clothing on this vital cutaneous process, and established global recommendations for exposure to sunlight as a major source of vitamin D. He has helped increase awareness of the pediatric and medical community regarding the extent that vitamin D deficiency exists in the US population, and its role in causing metabolic bone disease, and osteoporosis in adults. He pioneered the use of activated vitamin D compounds for the treatment of psoriasis. His observations provide new insights into the role of sunlight and vitamin D nutrition in prevention of osteoporosis, some common cancers, heart disease, type 1 diabetes, and psoriasis. He has published over 300 peer-reviewed journal articles, book chapters, and reviews. Dr. Holick has received a number of awards including: The American Society for Clinical Nutrition's Mccollum Award for his innovative research in the field of photobiology (1994), the Psoriasis Research Achievement Award from the American Skin Association (2000), the American College of Nutrition's ACN award (2002), and the American Society for Clinical Nutrition's Robert H. Herman Memorial Award in Clinical Nutrition (2003).

Bess Dawson-Hughes, MD is Professor of Medicine at Tufts University and Director of the Bone Metabolism Laboratory at the Jean Mayer United States Department of Agriculture Human Nutrition Research Center on Aging at Tufts University. She is trained in endocrinology and directs the Metabolic Bone Diseases Clinic at the New England Medical Center. Dr. Dawson-Hughes is a member of the advisory council of the National Institute of Arthritis, Musculoskeletal, and Skin Diseases. She is a member of the Board of Trustees and President of the National Osteoporosis Foundation and is the principal investigator of the NIH Osteoporosis and Related Bone Diseases Resource Center in Washington, DC. She has served on the councils of the American Society for Bone and Mineral Research and the American Society of Clinical Nutrition and is currently on the council of the International Bone and Mineral Society. Dr. Dawson-Hughes' research is directed at examining ways in which calcium, vitamin D, and other nutrients influence age-related loss of bone mass and risk of fragility fractures. She has published over 200 peer-reviewed journal articles, book chapters, abstracts, and reviews.